T0327634

Electrochemical Impedance Spectroscopy

THE ELECTROCHEMICAL SOCIETY SERIES

ECS-The Electrochemical Society
65 South Main Street
Pennington, NJ 08534-2839
http://www.electrochem.org

A complete list of the titles in this series appears at the end of this volume.

Electrochemical Impedance Spectroscopy

Second Edition

Mark E. Orazem
University of Florida

Bernard Tribollet
Université Pierre et Marie Curie

Library of Congress Cataloging-in-Publication Data:

Names: Orazem, Mark E. | Tribollet, Bernard.

Title: Electrochemical impedance spectroscopy / Mark E. Orazem, University of
    Florida, Bernard Tribollet, Université Pierre et Marie Curie.
Description: 2nd edition. | Hoboken, New Jersey : John Wiley & Sons, Inc.,
    [2017] | Includes bibliographical references and index.
Identifiers: LCCN 2016039537 (print) | LCCN 2016040277 (ebook) | ISBN
    9781118527399 (cloth) | ISBN 9781119341222 (pdf) | ISBN 9781119340928
    (epub)
Subjects: LCSH: Impedance spectroscopy.
Classification: LCC QD116.I57 O73 2017 (print) | LCC QD116.I57 (ebook) | DDC
    543/.6--dc23
LC record available at https://lccn.loc.gov/2016039537

Dedicated to our families:
Jennifer and Athéna,
Françoise, Julie, and Benjamin

Contents

Preface to the Second Edition xxi

Preface to the First Edition xxiii

Acknowledgments xxvii

The Blind Men and the Elephant xxix

A Brief Introduction to Impedance Spectroscopy xxxiii

History of Impedance Spectroscopy xli

I   Background   1

1   Complex Variables   3

    1.1   Why Imaginary Numbers?   3
    1.2   Terminology   4
          1.2.1   The Imaginary Number   4
          1.2.2   Complex Variables   4
          1.2.3   Conventions for Notation in Impedance Spectroscopy 5
    1.3   Operations Involving Complex Variables   5
          1.3.1   Multiplication and Division of Complex Numbers 6
          1.3.2   Complex Variables in Polar Coordinates 9
          1.3.3   Properties of Complex Variables   15
    1.4   Elementary Functions of Complex Variables 15

     1.4.1   Exponential 15

     1.4.2   Logarithmic 17

     1.4.3   Polynomial 21

     Problems 22

2 Differential Equations 25

  2.1 Linear First-Order Differential Equations 25

  2.2 Homogeneous Linear Second-Order Differential Equations 29

  2.3 Nonhomogeneous Linear Second-Order Differential
      Equations 32

  2.4 Chain Rule for Coordinate Transformations 36

  2.5 Partial Differential Equations by Similarity Transformations 38

  2.6 Differential Equations with Complex Variables 42

     Problems 43

3 Statistics 45

  3.1 Definitions 45

     3.1.1   Expectation and Mean 45

     3.1.2   Variance, Standard Deviation, and Covariance 45

     3.1.3   Normal Distribution 49

     3.1.4   Probability 50

     3.1.5   Central Limit Theorem 52

  3.2 Error Propagation 55

     3.2.1   Linear Systems 55

     3.2.2   Nonlinear Systems 56

  3.3 Hypothesis Tests 59

     3.3.1   Terminology 60

     3.3.2   Student's t-Test for Equality of Mean 61

     3.3.3 F-Test for Equality of Variance 64

     3.3.4   Chi-Squared Test for Goodness of Fit 70

     Problems 71

4 Electrical Circuits 75

  4.1 Passive Electrical Circuits 75

     4.1.1   Circuit Elements 75

          Response to a Sinusoidal Signal 77

          Impedance Response of Passive Circuit Elements 79

     4.1.2   Parallel and Series Combinations 79

4.2   Fundamental Relationships   81

4.3   Nested Circuits   83

4.4   Mathematical Equivalence of Circuits   84

4.5   Graphical Representation of Circuit Response   84

      Problems   87

5   Electrochemistry   89

5.1   Resistors and Electrochemical Cells   89

5.2   Polarization Behavior for Electrochemical Systems   92

      5.2.1   Zero Current   93

              Equilibrium   94

              Nonequilibrium   94

      5.2.2   Kinetic Control   95

      5.2.3   Mixed-Potential Theory   97

      5.2.4   Mass-Transfer Control 104

5.3   Definitions of Potential 107

5.4   Rate Expressions 109

5.5   Transport Processes 113

      5.5.1   Primary Current and Potential Distributions 114

      5.5.2   Secondary Current and Potential Distributions 116

      5.5.3   Tertiary Current and Potential Distributions 116

      5.5.4   Mass-Transfer-Controlled Current Distributions 117

5.6   Potential Contributions 119

      5.6.1   Ohmic Potential Drop 119

      5.6.2   Surface Overpotential 119

      5.6.3   Concentration Overpotential 120

5.7   Capacitance Contributions 122

      5.7.1   Double-Layer Capacitance 122

      5.7.2   Dielectric Capacitance 126

5.8   Further Reading 126

      Problems 127

6   Electrochemical Instrumentation 129

6.1   The Ideal Operational Amplifier 129

6.2   Elements of Electrochemical Instrumentation 131

6.3   Electrochemical Interface 133

      6.3.1   Potentiostat 134

      6.3.2   Galvanostat 135

6.3.3   Potentiostat for EIS Measurement   135

Problems   137

II   Experimental Considerations   139

7   Experimental Methods   141

7.1   Steady-State Polarization Curves   141

7.2   Transient Response to a Potential Step   142

7.3   Analysis in Frequency Domain   143

7.3.1   Lissajous Analysis   144

7.3.2   Phase-Sensitive Detection (Lock-in Amplifier)   151

7.3.3   Single-Frequency Fourier Analysis   152

7.3.4   Multiple-Frequency Fourier Analysis   155

7.4   Comparison of Measurement Techniques   156

7.4.1   Lissajous Analysis   156

7.4.2   Phase-Sensitive Detection (Lock-in Amplifier)   156

7.4.3   Single-Frequency Fourier Analysis   156

7.4.4   Multiple-Frequency Fourier Analysis   156

7.5   Specialized Techniques   157

7.5.1   Transfer-Function Analysis   157

7.5.2   Local Electrochemical Impedance Spectroscopy   158

Global Impedance   160

Local Impedance   160

Local Interfacial Impedance   160

Local Ohmic Impedance   161

Global Interfacial Impedance   161

Global Ohmic Impedance   161

Problems   162

8   Experimental Design   165

8.1   Cell Design   165

8.1.1   Reference Electrodes   165

8.1.2   Flow Configurations   167

Rotating Disk   167

Disk under Submerged Impinging Jet   168

Rotating Cylinders   168

Rotating Hemispherical Electrode   168

8.1.3   Current Distribution   169

8.2   Experimental Considerations   170
      8.2.1   Frequency Range   170
      8.2.2   Linearity   170
      8.2.3   Modulation Technique   182
      8.2.4   Oscilloscope   183
8.3   Instrumentation Parameters   184
      8.3.1   Improve Signal-to-Noise Ratio   184
      8.3.2   Reduce Bias Errors   185
              Nonstationary Effects   186
              Instrument Bias   186
      8.3.3   Improve Information Content   187
      Problems   188

III   Process Models   191

9   Equivalent Circuit Analogs   193

9.1   General Approach   193
9.2   Current Addition   195
      9.2.1   Impedance at the Corrosion Potential   195
      9.2.2   Partially Blocked Electrode   196
9.3   Potential Addition   201
      9.3.1   Electrode Coated with an Inert Porous Layer   201
      9.3.2   Electrode Coated with Two Inert Porous Layers   202
      Problems   205

10 Kinetic Models   207

10.1 General Mathematical Framework   207
10.2 Electrochemical Reactions   209
      10.2.1 Potential Dependent   209
      10.2.2 Potential and Concentration Dependent   213
              Charge-Transfer Resistance   216
              Diffusion Impedance   217
              Cell Impedance   220
10.3 Multiple Independent Electrochemical Reactions   222
10.4 Coupled Electrochemical Reactions   226
      10.4.1 Potential and Surface Coverage Dependent   226
      10.4.2 Potential, Surface Coverage, and Concentration
              Dependent   231

10.5 Electrochemical and Heterogeneous Chemical Reactions   234

       Problems   240

11 Diffusion Impedance   243

11.1 Uniformly Accessible Electrode   244

11.2 Porous Film   245

       11.2.1 Diffusion with Exchange of Electroactive Species   245

       11.2.2 Diffusion without Exchange of Electroactive Species   251

11.3 Rotating Disk   256

       11.3.1 Fluid Flow   256

       11.3.2 Steady-State Mass Transfer   259

       11.3.3 Convective Diffusion Impedance   261

       11.3.4 Analytic and Numerical Solutions   262

              Nernst Hypothesis   262

              Assumption of an Infinite Schmidt Number   263

              Treatment of a Finite Schmidt Number   264

11.4 Submerged Impinging Jet   266

       11.4.1 Fluid Flow   266

       11.4.2 Steady-State Mass Transfer   268

       11.4.3 Convective Diffusion Impedance   268

11.5 Rotating Cylinders   269

11.6 Electrode Coated by a Porous Film   271

       11.6.1 Steady-State Solutions   272

       11.6.2 Coupled Diffusion Impedance   277

11.7 Impedance with Homogeneous Chemical Reactions   278

11.8 Dynamic Surface Films   291

       11.8.1 Mass Transfer in the Salt Layer   292

       11.8.2 Mass Transfer in the Electrolyte   294

       11.8.3 Oscillating Film Thickness   295

       11.8.4 Faradaic Impedance   297

       Problems   300

12 Impedance of Materials   303

12.1 Electrical Properties of Materials   303

12.2 Dielectric Response in Homogeneous Media   304

12.3 Cole–Cole Relaxation   307

12.4 Geometric Capacitance   307

12.5 Dielectric Response of Insulating Nonhomogeneous Media   309

12.6 Mott–Schottky Analysis 311

Problems 317

13 Time-Constant Dispersion 319

13.1 Transmission Line Models 319
  13.1.1 Telegrapher's Equations 321
  13.1.2 Porous Electrodes 322
  13.1.3 Pore-in-Pore Model 328
  13.1.4 Thin-Layer Cell 333
13.2 Geometry-Induced Current and Potential Distributions 337
  13.2.1 Mathematical Development 338
    Blocking Electrode 338
    Blocking Electrode with CPE Behavior 339
    Electrode with Faradaic Reactions 339
    Electrode with Faradaic Reactions Coupled by
        Adsorbed Intermediates 341
  13.2.2 Numerical Method 342
  13.2.3 Complex Ohmic Impedance at High Frequencies 342
  13.2.4 Complex Ohmic Impedance at High and Low
        Frequencies 346
13.3 Electrode Surface Property Distributions 349
  13.3.1 Electrode Roughness 349
    Influence of Roughness on a Disk Electrode 351
    Influence of Surface Roughness on a Recessed
        Electrode 355
  13.3.2 Capacitance 360
    Capacitance Distribution on Recessed Electrodes 361
    Capacitance Distribution on Disk Electrodes 365
  13.3.3 Reactivity 369
13.4 Characteristic Dimension for Frequency Dispersion 369
13.5 Convective Diffusion Impedance at Small Electrodes 370
  13.5.1 Analysis 371
  13.5.2 Local Convective Diffusion Impedance 373
    Low-Frequency Solution 374
    High-Frequency Solution 374
  13.5.3 Global Convective Diffusion Impedance 374
13.6 Coupled Charging and Faradaic Currents 377
  13.6.1 Theoretical Development 378
    Mass Transport in Dilute Solutions 378

         Coupled Faradaic and Charging Currents   379

         Double-Layer Model   380

         Decoupled Faradaic and Charging Currents   382

    13.6.2 Numerical Method   383

         Steady-State Calculations   383

         Double-Layer Properties   383

         Impedance Calculations   384

    13.6.3 Consequence of Coupled Charging and Faradaic

         Currents   385

  13.7 Exponential Resistivity Distributions   389

    Problems   392

14 Constant-Phase Elements   395

  14.1 Mathematical Formulation for a CPE   395

  14.2 When Is a Time-Constant Distribution a CPE?   396

  14.3 Origin of Distributions Resulting in a CPE   399

  14.4 Approaches for Extracting Physical Properties   401

    14.4.1 Simple Substitution   402

    14.4.2 Characteristic Frequency: Normal Distribution   402

    14.4.3 Characteristic Frequency: Surface Distribution   404

    14.4.4 Power-Law Distribution   406

         Bounds for Resistivity   413

         Comparative Analysis   413

  14.5 Limitations to the Use of the CPE   415

    Problems   418

15 Generalized Transfer Functions   421

  15.1 Multi-input/Multi-output Systems   421

    15.1.1 Current or Potential Are the Output Quantity   425

    15.1.2 Current or Potential Are the Input Quantity   427

    15.1.3 Experimental Quantities   429

  15.2 Transfer Functions Involving Exclusively Electrical Quantities   429

    15.2.1 Ring–Disk Impedance Measurements   429

    15.2.2 Multifrequency Measurements for Double-Layer

         Studies   431

  15.3 Transfer Functions Involving Nonelectrical Quantities   434

    15.3.1 Thermoelectrochemical (TEC) Transfer Function   434

    15.3.2 Photoelectrochemical Impedance Measurements   438

15.3.3 Electrogravimetry Impedance Measurements   439

Problems   441

16 Electrohydrodynamic Impedance   443

16.1 Hydrodynamic Transfer Function   445

16.2 Mass-Transport Transfer Function   448

16.2.1 Asymptotic Solution for Large Schmidt Numbers   451

16.2.2 Asymptotic Solution for High Frequencies   451

16.3 Kinetic Transfer Function for Simple Electrochemical

Reactions   453

16.4 Interface with a 2-D or 3-D Insulating Phase   454

16.4.1 Partially Blocked Electrode   455

16.4.2 Rotating Disk Electrode Coated by a Porous Film   457

Steady-State Solutions   458

AC and EHD Impedances   459

Problems   464

IV   Interpretation Strategies   **465**

17 Methods for Representing Impedance   467

17.1 Impedance Format   467

17.1.1 Complex-Impedance-Plane Representation   470

17.1.2 Bode Representation   473

17.1.3 Ohmic-Resistance-Corrected Bode Representation   475

17.1.4 Impedance Representation   476

17.2 Admittance Format   478

17.2.1 Admittance-Plane Representation   480

17.2.2 Admittance Representation   481

17.2.3 Ohmic-Resistance-Corrected Representation   483

17.3 Complex-Capacitance Format   484

17.4 Effective Capacitance   488

Problems   491

18 Graphical Methods   493

18.1 Based on Nyquist Plots   494

18.1.1 Characteristic Frequency   494

18.1.2 Superposition   499

Mass Transfer 499

Evolution of Active Area 501

18.2 Based on Bode Plots 501

    18.2.1 Ohmic-Resistance-Corrected Phase 504

    18.2.2 Ohmic-Resistance-Corrected Magnitude 505

18.3 Based on Imaginary Part of the Impedance 505

    18.3.1 Evaluation of Slopes 505

    18.3.2 Calculation of Derivatives 506

18.4 Based on Dimensionless Frequency 506

    18.4.1 Mass Transport 507

    18.4.2 Geometric Contribution 507

18.5 System-Specific Applications 512

    18.5.1 Effective CPE Coefficient 512

    18.5.2 Asymptotic Behavior for Low-Frequency Mass

          Transport 516

    18.5.3 Arrhenius Superposition 518

    18.5.4 Mott–Schottky Plots 521

    18.5.5 High-Frequency Cole–Cole Plots 522

18.6 Overview 523

Problems 525

19 Complex Nonlinear Regression 527

19.1 Concept 527

19.2 Objective Functions 529

19.3 Formalism of Regression Strategies 530

    19.3.1 Linear Regression 530

    19.3.2 Nonlinear Regression 532

19.4 Regression Strategies for Nonlinear Problems 534

    19.4.1 Gauss–Newton Method 534

    19.4.2 Method of Steepest Descent 534

    19.4.3 Levenberg–Marquardt Method 534

    19.4.4 Downhill Simplex Strategies 535

19.5 Influence of Data Quality on Regression 536

    19.5.1 Presence of Stochastic Errors in Data 537

    19.5.2 Ill-Conditioned Regression Caused by Stochastic

          Noise 537

    19.5.3 Ill-Conditioned Regression Caused by Insufficient

          Range 539

19.6 Initial Estimates for Regression 541

19.7 Regression Statistics 542

    19.7.1 Confidence Intervals for Parameter Estimates 542

    19.7.2 Statistical Measure of the Regression Quality 543

    Problems 543

20 Assessing Regression Quality 545

20.1 Methods to Assess Regression Quality 545

    20.1.1 Quantitative Methods 545

    20.1.2 Qualitative Methods 546

20.2 Application of Regression Concepts 546

    20.2.1 Finite-Diffusion-Length Model 548

        Quantitative Assessment 549

        Visual Inspection 549

    20.2.2 Measurement Model 552

        Quantitative Assessment 553

        Visual Inspection 553

    20.2.3 Convective-Diffusion-Length Model 555

        Quantitative Assessment 557

        Visual Inspection 558

    Problems 563

V   Statistical Analysis   **565**

21 Error Structure of Impedance Measurements 567

21.1 Error Contributions 567

21.2 Stochastic Errors in Impedance Measurements 568

    21.2.1 Stochastic Errors in Time-Domain Signals 569

    21.2.2 Transformation from Time Domain to Frequency
        Domain 571

    21.2.3 Stochastic Errors in Frequency Domain 571

21.3 Bias Errors 575

    21.3.1 Instrument Artifacts 575

    21.3.2 Ancillary Parts of the System under Study 576

    21.3.3 Nonstationary Behavior 576

    21.3.4 Time Scales in Impedance Spectroscopy Measurements 576

21.4 Incorporation of Error Structure 579

21.5 Measurement Models for Error Identification 581

    21.5.1 Stochastic Errors 583

21.5.2 Bias Errors   586

Problems   593

22 The Kramers–Kronig Relations   595

22.1 Methods for Application   595
    22.1.1 Direct Integration of the Kramers–Kronig Relations   597
    22.1.2 Experimental Assessment of Consistency   598
    22.1.3 Regression of Process Models   598
    22.1.4 Regression of Measurement Models   599

22.2 Mathematical Origin   600
    22.2.1 Background   600
    22.2.2 Application of Cauchy's Theorem   604
    22.2.3 Transformation from Real to Imaginary   605
    22.2.4 Transformation from Imaginary to Real   607
    22.2.5 Application of the Kramers–Kronig Relations   609

22.3 The Kramers–Kronig Relations in an Expectation Sense   610
    22.3.1 Transformation from Real to Imaginary   611
    22.3.2 Transformation from Imaginary to Real   612

Problems   614

VI   Overview   **615**

23 An Integrated Approach to Impedance Spectroscopy   617

23.1 Flowcharts for Regression Analysis   617

23.2 Integration of Measurements, Error Analysis, and Model   618
    23.2.1 Impedance Measurements Integrated with Error
        Analysis   619
    23.2.2 Process Models Developed Using Other Observations   620
    23.2.3 Regression Analysis in Context of Error Structure   621

23.3 Application   621

Problems   627

VII   Reference Material   **629**

A Complex Integrals   631

A.1 Definition of Terms   631

A.2 Cauchy–Riemann Conditions   634

A.3 Complex Integration 635
    A.3.1 Cauchy's Theorem 635
    A.3.2 Improper Integrals of Rational Functions 639
    Problems 641

B Tables of Reference Material 643

C List of Examples 645

List of Symbols 651

References 663

Author Index 691

Subject Index 697

Preface to the Second Edition

We are gratified by the response to the first edition of our book. We thank all those who wrote us to alert us to errors and have sought to make appropriate corrections. There are other major changes in the second edition:

- We increased significantly the number of examples and homework problems.

- We adopted a coherent system of notation in which the math italic font is reserved for variables.

- We added a brief introduction in the front matter that provides a somewhat qualitative introduction to electrochemical impedance spectroscopy.

- Within the **Background** part of the book, Chapters 1, 2, 3, and 5 have been greatly expanded. Mathematical development in the rest of the book is now referred to examples in Chapters 1 and 2, and the reader is referred as needed to electrochemical principles in Chapter 5.

- The **Process Models** part of the book has been greatly expanded.

  - The consequences of the circuit models presented in Chapter 9 are described in detail. An example is introduced to show how impedance measurements at potentials slightly above and below the open-circuit potential may be used to determine the role of anodic and cathodic impedances in model development. The chapter demonstrates the use of superposition of impedance data to determine the influence of a partial coverage of the electrode surface by a thin dielectric.

  - The presentation of models based on kinetics has been streamlined in Chapter 10, and the number of cases treated has been increased. This chapter now includes discussion of coupled electrochemical and heterogeneous chemical reactions.

  - The presentation of diffusion impedance in Chapter 11 has been improved. The impedance associated with diffusion through a film now includes diffusion with and without exchange of electroactive species. Other new topics include convective diffusion to an electrode coated by a porous film, impedance with homogeneous chemical reactions, and dynamic surface films in which the film thickness is modulated due to competing dissolution and growth mechanisms.

   – The chapter on semiconducting systems has been replaced by a more general Chapter 12 on the impedance of materials. The new topics include electrical properties of materials, dielectric response in homogeneous media, Cole–Cole relaxation, and geometric capacitance.

   – Chapter 13 on time-constant dispersion has been expanded. More examples are provided for the discussion of transmission line models, and a more thorough discussion is provided for the effects of geometry-induced current and potential distributions. Treatments of electrode surface property distributions and coupled charging and faradaic currents have been introduced.

   – In response to recent developments in the understanding of constant-phase elements, a new Chapter 14 has been introduced. This chapter addresses important topics, such as methods to determine whether a given time-constant distribution is a CPE, the origin of distributions resulting in a CPE, and approaches for extracting physical properties.

• The discussion of graphical methods in the **Interpretation Strategies** part of the book has been reorganized into two chapters: Chapter 17 provides methods for representing impedance data, and Chapter 18 describes graphical methods. The role of complex capacitance representation is expanded in both chapters.

• In the **Statistical Analysis** part of the book, Chapter 22 on the Kramers–Kronig relations has been reorganized.

Mark E. Orazem   Bernard Tribollet
Gainesville, Florida   Paris, France

February 2017

Preface to the First Edition

This book is intended for use both as a professional reference and as a textbook suitable for training new scientists and engineers. As a textbook, this work is suitable for graduate students in a variety of disciplines, including electrochemistry, materials science, physics, electrical engineering, and chemical engineering. As these audiences have very different backgrounds, a portion of the book reviews material that may be known to some students but not to others. There are many short courses offered on impedance spectroscopy, but formal courses on the topic are rarely offered in university settings. Accordingly, this textbook is designed to accommodate both directed and independent learning.

Organization

The textbook has been prepared in seven parts:

Part I Background

This part provides material that may be covered selectively, depending on the background of the students. The subjects covered include complex variables, differential equations, statistics, electrical circuits, electrochemistry, and instrumentation. The coverage of these topics is limited to what is needed to understand the core of the textbook, which is covered in the subsequent parts.

Part II Experimental Considerations

This part introduces methods used to measure impedance and other transfer functions. The chapters in this section are intended to provide an understanding of frequency-domain techniques and the approaches used by impedance instrumentation. This understanding provides a basis for evaluating and improving experimental design. The material covered in this section is integrated with the discussion of experimental errors and noise. The extension of impedance spectroscopy to other transfer-function techniques is developed in Part III.

Part III Process Models

This part demonstrates how deterministic models of impedance response can be developed from physical and kinetic descriptions. When possible, correspondence is drawn between hypothesized models and electrical circuit analogues. The treatment includes electrode kinetics, mass transfer, solid-state systems, time-constant dispersion, models accounting for two- and three-dimensional interfaces, generalized transfer functions, and a more specific example of a transfer-function technique in which the rotation speed of a disk electrode is modulated.

Part IV Interpretation Strategies

This part describes methods for interpretation of impedance data, ranging from graphical methods to complex nonlinear regression. The material covered in this section is integrated with the discussion of experimental errors and noise. Bias errors are shown to limit the frequency range useful for regression analysis, and the variance of stochastic errors is used to guide the weighting strategy used for regression.

Part V Statistical Analysis

This part provides a conceptual understanding of stochastic, bias, and fitting errors in frequency-domain measurements. A major advantage of measurements in the frequency domain is that real and imaginary parts of the response must be internally consistent. The expression of this consistency takes different forms that are known collectively as the Kramers–Kronig relations. The Kramers–Kronig relations and their application to spectroscopy measurements are described. Measurement models, used to assess the error structure, are described and compared with process models used to extract physical properties.

Part VI Overview

The final chapter in this book provides a philosophy for electrochemical impedance spectroscopy that integrates experimental observation, model development, and error analysis. This approach is differentiated from the usual sequential model development for given impedance spectra by its emphasis on obtaining supporting observations to guide model selection, use of error analysis to guide regression strategies and experimental design, and use of models to guide selection of new experiments. These concepts are illustrated with examples taken from the literature. This chapter is intended to illustrate that selection of models, even those based on physical principles, requires both error analysis and additional experimental verification.

Part VII Reference Material

The reference material includes an appendix on complex integration needed to follow the derivation of the Kramers–Kronig relations, a list of tables, a list of examples, a list of symbols, and a list of references.

Pedagogical Approach

The material is presented in a manner that facilitates sequential development of understanding and expertise either in a course or in self-study. Illustrative examples are interspersed throughout the text to show how the principles described are applied to common impedance problems. These examples are in the form of questions, followed by the solution to the question posed. The student can attempt to solve the problem before reading how the problem is solved. Homework problems, suitable either for self-study or for study under direction of an instructor, are developed for each chapter. Important equations and relations are collected in tables, which can be easily accessed. Important concepts are identified and set aside at the bottom of pages as they appear in the text. Readily identifiable icons are used to distinguish examples and important concepts.

As can be found in any field, the notation used in the impedance spectroscopy literature is inconsistent. In treatments of diffusion impedance, for example, the symbol θ is used to denote the dimensionless oscillating concentration variable; whereas, the symbol θ used in kinetic studies denotes the fractional surface coverage by a reaction intermediate. Compromises were necessary to create a consistent notation for this book. For example, the dimensionless oscillating concentration variable was given the symbol θ, and γ was used to denote the fractional surface coverage by a reaction intermediate. As discussed in Section 1.2.3, the book deviates from the IUPAC convention for the notation used to denote the imaginary number and the real and imaginary parts of impedance.

This book is intended to provide a background and training suitable for application of impedance spectroscopy to a broad range of applications, such as corrosion, biomedical devices, semiconductors and solid-state devices, sensors, batteries, fuel cells, electrochemical capacitors, dielectric measurements, coatings, electrochromic materials, analytical chemistry, and imaging. The emphasis is on generally applicable fundamentals rather than on detailed treatment of applications. The reader is referred to other sources for discussion of specific applications of impedance.[1–4]

Remember! 0.1 *The elephant at the left is used to identify important concepts for each chapter. It is intended to remind the student of the parable of the blind men and the elephant.*

The active participation in related short courses demonstrates a rising interest in impedance spectroscopy. As discussed in the preliminary section on the history of the technique, the number of papers published that mention use of electrochemical impedance spectroscopy has increased dramatically over the past 10 years. Nevertheless, the question may be raised: *Why teach a full semester-long course on impedance spectroscopy? It is, after all, just an experimental technique.* In our view, impedance spectroscopy represents the confluence of a significant number of disciplines, and successful training in the use and interpretation of impedance requires a coherent education in the application of each of these disciplines to the subject. In addition to learning about impedance spectroscopy, the student will gain a better understanding of a general philosophy of scientific inquiry.

Mark E. Orazem   Bernard Tribollet
Gainesville, Florida   Paris, France

July 2008

Acknowledgments

The authors met for the first time in 1981 in the research group of John Newman at the University of California, Berkeley. **Mark Orazem** was a graduate student, and **Bernard Tribollet** was a visiting scientist on sabbatical leave from the *Centre National de la Recherche Scientifique* (CNRS) in Paris. We have maintained a fruitful collaboration ever since, and our careers, as well as the content of this book, build on the foundation we received from John. We owe an additional debt of gratitude to many people, including:

- Our families, to whom this book is dedicated and who have embraced our collaboration over the years. We appreciate their immense patience and support.

- *The Electrochemical Society* (ECS), which encouraged Mark Orazem to teach the ECS short course on impedance spectroscopy on an biannual basis, thus allowing us to test the pedagogical approach developed in this book.

- The CNRS, which provided financial support for a sabbatical year spent by Mark Orazem in Paris in 2001/2002.

- The University of Florida, which provided funds to Mark Orazem for a second sabbatical in 2012.

- Graduate students, who provided suggestions and content. Bryan Hirschorn, Vicky Huang, J. Patrick McKinney, Sunil Roy, and Shao-Ling Wu contributed to the first edition. Christopher Alexander, Ya-Chiao Chang, Yu-Min Chen, Christopher Cleveland, Arthur Dizon, Salim Erol, Ming Gao, Morgan Harding, Yuelong Huang, Rui Kong, and Chen You read various iterations of the second edition, helping with examples, solving homework problems, and identifying errors in the text. Research conducted by an undergraduate student, Katherine Davis, contributed to the text.

- Hubert Cachet, Sandro Cattarin, Isabelle Frateur, and Nadine Pébère, who read different chapters of the first edition and provided comments, corrections, and suggestions.

- Michel Keddam, who suggested the presentation of the historical perspective in terms of the categories presented in Table 1.

- Max Yaffe of Gamry Instruments, who provided assistance with the chapter on the Kramers–Kronig relations.

- Christopher Brett, who graciously provided a last technical review before publication of the first edition.

- Mary Yess of the ECS, who has given us unyielding support for this project.

- Dinia Agarwala of the ECS, who designed the cover of both editions.

- Amy Hendrickson, TeXnology Inc., who provided the LaTeX expertise needed to fine tune the visual appearance of the text.

- Laura Carlson, our copy editor, who taught us the proper usage of hyphens, en-dashes, and em-dashes.

- Bob Esposito, our editor from John Wiley & Sons, who helped us through the many steps associated with publishing a book and reassured us that it was good to be considered difficult authors.

- Melissa Yanuzzi, Senior Production Editor, who managed the final stages of publication.

- Our many colleagues and friends who assured us that there would be a demand for this book.

While we have received help and support from many people, the remaining errors and omissions in this text are ours. We will gratefully receive corrections and suggestions from our readers to be implemented in possible future editions of this book.

The Blind Men and the Elephant

Impedance spectroscopy is a complicated area of research that has been subject to significant controversy. As we begin a study of this subject it is well to remember the Buddhist parable of the blind men and the elephant. American poet John Godfrey Saxe (1816-1887) based the following poem on the fable.[5]

The Blind Men and the Elephant

John Godfrey Saxe

It was six men of Indostan
To learning much inclined,
Who went to see the Elephant
(Though all of them were blind),
That each by observation
Might satisfy his mind.

The First approached the Elephant,
And happening to fall
Against his broad and sturdy side,
At once began to bawl:
"God bless me! but the Elephant
Is very like a wall!"

The Second, feeling of the tusk,
Cried, "Ho! what have we here
So very round and smooth and sharp?
To me 'tis mighty clear
This wonder of an Elephant
Is very like a spear!"

The Third approached the animal,
And happening to take
The squirming trunk within his hands,

Thus boldly up and spake:
"I see," quoth he, "the Elephant
Is very like a snake!"

The Fourth reached out an eager hand,
And felt about the knee.
"What most this wondrous beast is like
Is mighty plain," quoth he;
" 'Tis clear enough the Elephant
Is very like a tree!"

The Fifth, who chanced to touch the ear,
Said: "E'en the blindest man
Can tell what this resembles most;
Deny the fact who can,
This marvel of an Elephant
Is very like a fan!?"

The Sixth no sooner had begun
About the beast to grope,
Than, seizing on the swinging tail
That fell within his scope,
"I see," quoth he, "the Elephant
Is very like a rope!"

And so these men of Indostan
Disputed loud and long,
Each in his own opinion
Exceeding stiff and strong,
Though each was partly in the right,
And all were in the wrong!

Moral:
So oft in theologic wars,
The disputants, I ween,
Rail on in utter ignorance
Of what each other mean,
And prate about an Elephant
Not one of them has seen!

The logo for the 2004 International Symposium on Impedance Spectroscopy, shown in Figure 1, was intended to evoke the lessons of the blind men and the elephant. The multiple loops resemble the Nyquist plots obtained in some cases for the impedance of

Figure 1: The logo for the 2004 International Symposium on Impedance Spectroscopy, held in Cocoa Beach, Florida.

corroding systems influenced by formation of surface films. The low-frequency inductive loop was deformed to evoke the image of the elephant's trunk, and the capacitive loops resemble the head and body of the elephant.

Impedance spectroscopy is, of course, not a religion. It is rather an application of a frequency-domain measurement to a complex system that cannot be easily visualized. The quantities measured, e.g., current and potential for electrochemical or electronic systems and stress and strain for mechanical systems, are macroscopic values that represent the spatial average of individual events. These quantities are influenced by the desired physical properties, such as diffusivity, rate constants, and viscosity, but do not provide a direct measure of them.

Application of impedance spectroscopy is very much like feeling an elephant that we cannot see. Measurement of current and potential under a steady state yields some information concerning a given system. By adding frequency dependence to the macroscopic measurements, impedance spectroscopy expands the information that can be extracted from the measurements. Impedance measurements, however, are not sufficient. Additional observations are needed to gain confidence in the model identification.

A Brief Introduction to Impedance Spectroscopy

The objective of this book is to provide a comprehensive introduction to the measurement and analysis of impedance spectra. The objective of this chapter is to provide a brief and somewhat qualitative introduction to the subject.

Consider the system with unknown properties that is shown in Figure 2 and labeled black box. The objective of the exercise is to learn the properties of the box in an effort to understand what it is. A series of measurements may be considered to interrogate the black box by imposing an input and measuring the result. For example, imagine that the box is placed in a dark room and is then subject to light of a specified wavelength. If a response is seen, such as an electrical current, the box may be considered to be photoactive. To explore the kinetics associated with the discharge of current from absorbed photons, the light intensity could be modulated. Such an experimental approach is considered in Section 15.3.2 on page 438. An alternative approach could be to impose an electrical potential and observe the resulting current. Modulation of the input signal would allow exploration of the influence of storage of charge within the box and the kinetics of processes that transform the potential to current.

The relationship between input and output is called a transfer function. Impedance spectroscopy is a special case of a transfer function.

Figure 2: Representation of a black box.

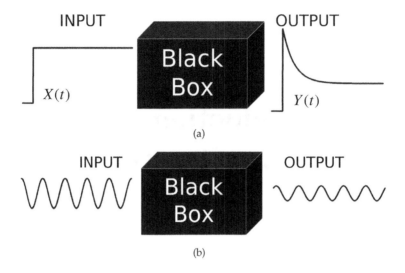

Figure 3: Representation of the system response $Y(t)$ to $X(t)$ defined as: a) step change input; and b) sinusoidal input with frequency ω.

Transfer Function

The transfer function provides a compact description of the input-output relation for a linear time-invariant (LTI) system. Due to the fact that most signals can be decomposed into summation of sinusoids via Fourier series, the response of a system is characterized by the frequency response of the system. A generalized system is illustrated in Figure 3(a). The response to a step input signal $X(t)$ shows different long and short time behaviors that can be represented as the dependence of the transfer function on frequency.

The short-time behavior corresponds to high frequencies, and the long-time behavior corresponds to low frequencies. For an electrochemical system, charging of the electrode-electrolyte interface occurs rapidly and is associated with the high-frequency or short-time response. Diffusion is a slower process with a large time constant and correspondingly a smaller characteristic frequency.

Example 0.1 **Characteristic Frequencies:** *Find the characteristic frequency for double-layer charging, faradaic reactions, and diffusion for a disk electrode with radius $r_0 = 0.25$ cm and a capacity $C_0 = 20$ $\mu F/cm^2$. The disk is immersed in an electrolyte of resistivity $\rho = 10$ Ωcm. Assume that the faradaic reaction has an exchange current density $i_0 = 1$ mA/cm² and that the disk is rotating at $\Omega = 400$ rpm. The kinematic viscosity of the electrolyte is $\nu = 10^{-2}$ cm²/s.*

Solution: *Calculations are presented below for the time constants and characteristic frequencies* *associated with charging, faradaic reaction, and diffusion:*

Double-Layer Charging: *The time constant for charging the electrode surface is given by*

$$\tau_C = C_0 R_e \tag{1}$$

where R_e is the ohmic resistance, given for a disk electrode by equation (5.112) on page 115 in units of Ω as

$$R_e = \frac{1}{4\kappa r_0} = \frac{\rho}{4r_0} \tag{2}$$

or, in units of Ωcm^2,

$$R_e = \frac{\pi r_0 \rho}{4} \tag{3}$$

For the parameters given, the time constant is $\tau_C = 0.04$ ms. The corresponding characteristic angular frequency is given by

$$\omega_C = \frac{1}{\tau_C} \tag{4}$$

and, in units of Hz,

$$f_C = \frac{1}{2\pi\tau_C} \tag{5}$$

Thus, for a time constant $\tau_C = 0.04$ ms, the characteristic frequency is 4.1 kHz.

Faradaic Reaction: *The time constant for a faradaic reaction is given by*

$$\tau_t = C_0 R_t \tag{6}$$

where R_t is the charge-transfer resistance, given for linear kinetics on a disk electrode by equation (5.117) on page 116, or

$$R_t = \frac{RT}{nFi_0} \tag{7}$$

Equations (6) and (7) yield a time constant of 0.51 ms and a characteristic frequency of 310 Hz.

Diffusion: *The time constant for diffusion to a rotating disk electrode is given by*

$$\tau_D = \frac{\delta_N^2}{D_i} \tag{8}$$

where D_i is the diffusion coefficient for the reacting species and δ_N is the diffusion layer thickness given as a function of rotation speed by equation (11.72) on page 262. The time constant for diffusion of a species with a diffusivity of 10^{-5} cm^2/s is equal to 0.41 s. The corresponding characteristic frequency is 0.4 Hz.

Experiments are conducted in the time domain. If the input signal is sinusoidal, as shown in Figure 3(b),

$$X(t) = \overline{X} + |\Delta X| \cos(\omega t) \tag{9}$$

where $\overline{X}$ is the steady-state or time-invariant part of the signal, and $|\Delta X|$ represents the magnitude of the oscillating part of the signal. When $|\Delta X|$ is sufficiently small that

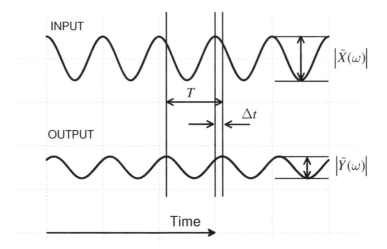

Figure 4: Schematic representation of the calculation of the transfer function for a sinusoidal input at frequency ω. The time lag between the two signals is Δt and the period of the signals is T.

the response is linear, the output will have the form of the input and be at the same frequency, i.e.,

$$Y(t) = \overline{Y} + |\Delta Y| \cos(\omega t + \varphi) \tag{10}$$

where φ is the phase lag between the input and output signals. An alternative representation of the time domain expressions is developed in Example 1.9 on page 20 as

$$X(t) = \overline{X} + \mathrm{Re}\{\widetilde{X}\exp(j\omega t)\} \tag{11}$$

and

$$Y(t) = \overline{Y} + \mathrm{Re}\{\widetilde{Y}\exp(j\omega t)\} \tag{12}$$

respectively, where $\widetilde{X}$ and $\widetilde{Y}$ are complex quantities that are functions of frequency but are independent of time. The transfer function is a function of frequency and is independent of both time and the magnitude of the input signal. While the measurements are made in time domain, the determination of the transfer function is obtained from subsequent analysis.

The calculation of the transfer function at a given frequency ω is presented schematically in Figure 4. The ratio of the amplitudes of the output and input signals yields the magnitude of the transfer function. The phase angle in units of radians can be obtained as

$$\varphi(\omega) = 2\pi \frac{\Delta t}{T} \tag{13}$$

If $\Delta t = 0$, the phase angle is equal to zero. Similarly, the phase angle is equal to zero if $\Delta t = T$. As shown in Figure 4, the output lags the input, and the phase angle has a positive value.

The transfer function is, therefore, characterized by two parameters: the gain

$$|H(\omega)| = \frac{|Y|}{|X|} = \frac{\left|\widetilde{Y}(\omega)\right|}{\left|\widetilde{X}(\omega)\right|} \tag{14}$$

and the phase shift $\varphi(\omega)$. These two parameters can be written in the form of a complex number with a magnitude $|H(\omega)|$ and a phase $\varphi(\omega)$ or with a real part expressed as $|H(\omega)|\cos(\varphi(\omega))$ and an imaginary part $|H(\omega)|\sin(\varphi(\omega))$ (see Chapter 1 on page 3).

Generally the input signal is considered to be a reference for the phase. In this case, the corresponding complex number for the input is real, i.e., $\widetilde{X}(\omega)$ and the output signal is a complex number $\widetilde{Y}(\omega)$ with a magnitude $\left|\widetilde{Y}(\omega)\right|$ and a phase $\varphi(\omega)$. Thus,

$$H(\omega) = \frac{\widetilde{Y}(\omega)}{\widetilde{X}(\omega)} = \frac{\left|\widetilde{Y}(\omega)\right|}{\left|\widetilde{X}(\omega)\right|} (\cos\varphi(\omega) + j\sin\varphi(\omega)) \tag{15}$$

For an electrical or an electrochemical system, the input is usually a potential, the output is a current, and the transfer function is called admittance. In the particular case where the input is a current and the output is a potential, the transfer function is an impedance. The transfer function is, however, a property of the system that is independent of the input signal. As the admittance is the inverse of the impedance,

$$Z(\omega) = \frac{\widetilde{V}(\omega)}{\widetilde{i}(\omega)} \tag{16}$$

Generally only the impedance is considered even if the measurement corresponds to an admittance. The measured impedance can have a strong dependence on the applied frequency. By analyzing the impedance as a function of frequency, a transfer-function model could be defined which takes into account all time constants of the corresponding system.

Example 0.2 Impedance and Ohm's Law: *How can impedance spectroscopy be differentiated from simple application of Ohm's law, i.e., $V = IR$?*

Solution: *The expression $V = IR$ represents a steady-state measurement. For a system consisting of a resistor, shown in Figure 5(a), the measurement of current at an applied potential*

Remember! 0.2 *Impedance is a complex transfer function that relates an electrical output to an electrical input.*

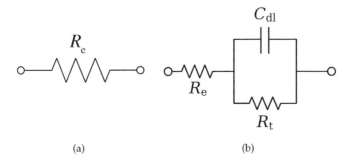

(a)   (b)

Figure 5: Electrical systems: a) a resistor and b) a resistor in series with the parallel combination of a capacitor and a resistor.

yields the value of the resistor, i.e., $\overline{V}/\overline{I} = R_e$. *For potentiostatic impedance measurements, application of an oscillatory potential*

$$V = \overline{V} + |\Delta V|\cos(\omega t) \tag{18}$$

yields a current

$$I = \overline{I} + |\Delta I|\cos(\omega t + \varphi) \tag{19}$$

where φ is the phase lag between the current and potential. As described in Example 1.9 on page 20, equation (18) is mathematically equivalent to

$$V = \overline{V} + \mathrm{Re}\{\tilde{V}\exp(j\omega t)\} \tag{20}$$

and equation (19) is mathematically equivalent to

$$I = \overline{I} + \mathrm{Re}\{\tilde{I}\exp(j\omega t)\} \tag{21}$$

The impedance for the resistor shown in Figure 5(a), expressed as

$$Z = \frac{\tilde{V}}{\tilde{I}} = R_e \tag{22}$$

yields the same information that could be obtained from the steady-state measurement $\overline{V}/\overline{I} = R_e$.

Remember! **0.3** *The measured signals $V(t)$ and $i(t)$ are real quantities; their ratio is also a real quantity which varies with time. This ratio is not the impedance. Thus,*

$$Z(\omega) \neq \frac{V(t)}{i(t)} \tag{17}$$

The circuit presented in Figure 5(b) may be considered to represent a simple potential-dependent electrochemical reaction, as discussed in Section 10.2.1 on page 209. The steady-state measurement of current at an applied potential yields

$$\frac{\overline{V}}{\overline{I}} = R_e + R_t \tag{23}$$

from which the contributions of R_e and R_t cannot be distinguished and the contribution of the parallel capacitance cannot be discerned. In contrast, following the development in Example 4.2 on page 81, the impedance response may be expressed as

$$Z = \frac{\tilde{V}}{\tilde{I}} = R_e + \frac{R_t}{1 + j\omega R_t C_{dl}} \tag{24}$$

from which the parameters R_e, R_t, and C_{dl} may be obtained easily by using the graphical methods outlined in Chapter 18 or by complex nonlinear regression (Chapter 19 on page 527).

Applications of Impedance Spectroscopy

Impedance spectroscopy has been applied to many electrochemical systems. A few examples are given below.

Presence of Adsorbed Intermediates

Some properties can be studied only by measurement of the electrochemical impedance. For example, impedance may be used to demonstrate the existence of an adsorbed reaction intermediate in the form of a fraction of monolayer in the case of iron dissolution in sulphuric acid. The system of interest is the corrosion of a pure iron electrode in an electrolyte containing sulfuric acid. Bockris and coworkers[6] proposed a reaction model in which two consecutive steps are coupled by an adsorbed intermediate. The anodic dissolution of iron can be described in simplified form as

$$Fe \xrightarrow{K_1} Fe^+_{ads} + e^- \tag{25}$$

and

$$Fe^+_{ads} \xrightarrow{K_2} Fe^{2+} + e^- \tag{26}$$

The iron first oxidizes and forms a monovalent intermediate adsorbed on the electrode surface. This reaction is followed by the oxidation of the intermediate. The ferrous ion is soluble and diffuses away from the electrode. Epelboin and Keddam[7] showed that the inductive loops observed in the low-frequency impedance response could be attributed to partial coverage of the iron surface by an adsorbed intermediate. Such a film is too thin to be observed directly. The associated mathematical development is presented in Section 10.4 on page 226.

Identification of Small Rates of Corrosion

The internal corrosion rate of cast iron drinking water pipes, used, for example, for the water distribution network in France, is very small and, in itself, corrosion is generally not a problem. The small rate of corrosion is sufficient to reduce the concentration of free chlorine (the sum of hypochlorous acid HOCl and hypochlorite ions ClO$^-$) introduced in water at the treatment plant in order to maintain microbiological quality. Frateur et al.[8] reported the results of electrochemical impedance measurements that, by use of a model that accounted for anodic metal dissolution, the progressive development of a porous film composed of corrosion product, and oxygen reduction, allowed estimation of corrosion rates. After 28 days of immersion in Evian^{TM} water, the corrosion rate of the cast iron was observed to be about 10 μm/year. This work was supported by surface analysis which showed the presence of corrosion products but could not assess the rate. A corrosion rate this small cannot be assessed by weight-loss measurements. The associated mathematical development is presented in Section 13.1.2 on page 322.

Presence of Viscosity Gradients

Barcia et al.[9] presented electrohydrodynamic impedance measurements showing that, contrary to the usual assumption, electrolyte viscosity is not uniform during electrodissolution of iron in sulfuric acid. The viscosity gradient was later linked to observation of current oscillations in this system.[10] Electrohydrodynamic impedance is a transfer function method in which the current response to modulation of the rotation speed of a disk electrode is measured at constant potential, i.e.,

$$Z_{\text{EHD}}(\omega) = \frac{\tilde{i}(\omega)}{\tilde{\Omega}(\omega)} \tag{27}$$

A detailed discussion of this technique is presented in Chapter 16 on page 443.

Ion and Solvent Insertion into Electroactive Polymers

Gabrielli et al.[11,12] used the coupling of electrogravimetric impedance with the usual electrochemical impedance spectroscopy, described as equation (16), to reveal ion and solvent insertion into polypyrrole electroactive polymers. Electrogravimetric impedance is a transfer function method in which the change of the mass of an electrode is measured in response to modulation of potential. Thus,

$$Z_{\text{EG}}(\omega) = \frac{\tilde{m}(\omega)}{\tilde{V}(\omega)} \tag{28}$$

A description of electrogravimetric impedance is presented in Section 15.3.3 on page 439. The generalized mathematical foundation is presented in Chapter 15 on page 421.

History of Impedance Spectroscopy

Impedance spectroscopy is an electrochemical technique with broad applications that is growing in importance. As seen in Figure 6, the number of papers published in this area has doubled roughly every four or five years. In 2006, 996 journal articles were published that mention the use of electrochemical impedance spectroscopy. In 2015, this number was 3123.

Another measure of the growing importance of electrochemical impedance spectroscopy may be seen in the growth in number of textbooks and monographs covering this subject. These include *Transient Techniques in Electrochemistry* by D. D. Macdonald (1977),[1] *Impedance Spectroscopy: Emphasizing Solid Materials and Systems* by J. R. Macdonald (1987),[3] *Impedance Spectroscopy: Theory, Experiment, and Applications* by Barsoukov and J. R. Macdonald (2005),[4] the first edition of the present book in 2008,[13] *Impedance Spectroscopy: Applications to Electrochemical and Dielectric Phenomena* by Lvovich (2012),[14] *Electrochemical Impedance Spectroscopy and its Applications* by Lasia in 2014,[15] and the third edition of *Bioimpedance and Bioelectricity Basics* by Grimnes and Martinsen in 2015.[16] The first and second editions of *Bioimpedance and Bioelectricity Basics* were published in 2000 and 2008, respectively. Itagaki[17] published the Japanese text *Electrochemical Impedance Method* in 2008, and the second edition was published in 2011.[18]

Timeline

By his application of Laplace transforms to the transient response of electrical circuits, Oliver Heaviside created the foundation for impedance spectroscopy. Heaviside coined the words inductance, capacitance, and impedance and introduced these concepts to the treatment of electrical circuits. His papers on the subject, published in *The Electrician* beginning in 1872, were compiled by Heaviside in book form in 1894.[19,20] From the perspective of the application to physical systems, however, the history of impedance spectroscopy begins in 1894 with the work of Nernst.[21]

Nernst applied the electrical bridge invented by Wheatstone[22,23] to the measurement of the dielectric constants for aqueous electrolytes and different organic fluids.

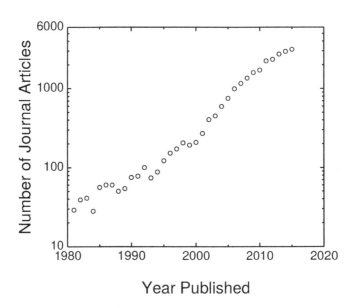

Figure 6: Number of journal articles on electrochemical impedance spectroscopy identified on November 19, 2016, using the Engineering Village search engine. The key words used were ("(impedance or admittance) and (electrochemical)", journal article only).

Nernst's approach was soon employed by others for measurement of dielectric properties[24,25] and the resistance of galvanic cells.[26] Finkelstein[27] applied the technique to the analysis of the dielectric response of oxides. Warburg[28,29] developed expressions for the impedance response associated with the laws of diffusion, developed almost 50 years earlier by Fick,[30] and introduced the electrical circuit analogue for electrolytic systems in which the capacitance and resistance were functions of frequency. The concept of diffusion impedance was applied by Krüger to the capacitive response of mercury electrodes.[31]

In the 1920s, impedance was applied to biological systems, including the resistance and capacitance of cells of vegetables[32] and the dielectric response of blood suspensions.[33–35] Impedance was also applied to muscle fibers, skin tissues, and other biological membranes.[36,37] The capacitance of the cell membranes was found to be a function of frequency,[38] and Fricke observed a relationship between the frequency exponent of the impedance and the observed constant phase angle.[39] In 1941, brothers Cole and Cole showed that the frequency-dependent complex dielectric constant can be represented as a depressed semicircle in a complex admittance plane plot and suggested a formula, consistent with Fricke's law,[39] now known as a constant-phase element.[40]

In 1940, Frumkin[41] explored the relationships among the double-layer structure on mercury electrodes, the capacitance measured by use of a Wheatstone bridge, and the surface tension, following the theoretical underpinnings of the Lippmann equation. Grahame[42,43] expanded this treatment of the mercury electrode, providing a fundamental understanding of the structure of the electrical double layer. Dolin and Ershler[44] applied the concept of an equivalent circuit to electrochemical kinetics for which the circuit elements were independent of frequency. Randles[45] developed an equivalent circuit for an ideally polarized mercury electrode that accounted for the kinetics of adsorption reactions.

In the early 1950s, impedance began to be applied to more complicated reaction systems.[46–48] In subsequent years, Epelboin and Loric[49] addressed the role of reaction intermediates in causing low-frequency inductive loops, de Levie[50] developed transmission line models for the impedance response of porous and rough electrodes, and Newman[51] showed that the nonuniform current and potential distribution of disk electrodes can result in high-frequency time-constant dispersion. Levart and Schuhmann[52] developed a model for the diffusion impedance of a rotating disk that accounted for the influence of homogeneous chemical reactions. Kinetic models accounting for reaction intermediates were addressed in greater detail in publications by Armstrong et al.[53] and Epelboin et al.[54]

Nonlinear complex regression techniques, developed in the early 1970s,[55,56] were applied to impedance data by Macdonald et al.[57,58] and Boukamp.[59] The regression approaches were based on use of equivalent electrical circuits, which became the predominant method for interpretation of impedance data. The experimental investigations turned increasingly toward those associated with technical applications, such as electrodeposition and corrosion.[60–62] Gabrielli et al.[63–65] introduced the concept of a generalized transfer function for impedance spectroscopy. During this time, the Kramers–Kronig relations, developed in the late 1920s,[66,67] were applied for the validation of electrochemical impedance data.[68] Agarwal et al.[69] described an approach that eliminated problems associated with direct integration of the Kramers–Kronig integral equations and accounted explicitly for stochastic errors in the impedance measurement.

Several authors have described methods for generalized deconvolution of impedance data.[70,71] Stoynov and co-workers[72,73] developed a robust method in which calculation of the local derivatives of the impedance with respect to frequency allows visualization of the distribution of time constants for a given spectrum without a priori assumption of a distribution function. Stoynov and Savova-Stoynov[74] described a graphical method of estimating instantaneous impedance projections from consecutive series of impedance diagrams obtained during the time of system evolution.

A conference dedicated to the development of electrochemical impedance spectroscopy techniques was initiated in 1989 in Bombannes, France. The subsequent meetings, held every three years, took place in California (1992), Belgium (1995), Brazil (1998), Italy (2001), Florida (2004), France (2007), Portugal (2010), Japan (2013), and Spain (2016). The special issues associated with these conferences provide unique triennial snapshots of the state of impedance research.[75–83] One driving concern reflected in these volumes is the heterogeneity of electrode surfaces and the correspondence to the use and misuse of constant phase elements. Local impedance spectroscopy, developed by Lillard et al.,[84] may prove to be a useful method for understanding this relationship.

Areas of Investigation

A historical perspective on impedance spectroscopy is presented in Table 1. A brief listing of advances in this field cannot be comprehensive, and many important contributions are not mentioned. The reader may wish to explore other historical perspectives, such as that provided by Macdonald.[85] Chapters written by Sluyters-Rehbach and Sluyters[86] and by Lasia[87] provide excellent overviews of the field. Nevertheless, Table 1 provides a useful guide to the trends in areas related to electrochemical impedance spectroscopy. These areas include the types of systems investigated, the instrumentation used to make the measurements, including changes in the accessible frequency range, the methods used to represent the resulting data, and the methods used to interpret the data in terms of quantitative properties of the system.

Experimental Systems

The early applications of what we now know as impedance spectroscopy were to the dielectric properties of fluids and metal oxides. Impedance measurements performed on mercury electrodes emphasized development of a fundamental understanding of the interface between the electrode and the electrolyte. The mercury drop electrodes were ideal for this purpose because they provided a uniform and easily refreshed interface that could be considered to be ideally polarizable over a broad range of potential. The impedance technique was used to identify an interfacial capacitance that could be compared to the theories for diffuse electrical double layers. In the 1920s, considerable effort was placed on biological systems, including the dielectric properties of blood and the impedance response of cell membranes. In the 1950s, impedance began to be used for studies of anodic dissolution. One may identify a trend from ideal surfaces suitable for fundamental studies to ones associated with technical materials. Impedance became useful for studying processes such as corrosion, deposition of films, and other electrochemical reactions. Heterogeneity of solid electrode surfaces were found to complicate interpretation of impedance spectra in terms of meaningful physical properties. Recently, local impedance spectroscopy has emerged as a means of studying heterogeneous electrode surfaces.

Measurement Techniques

Early experimental techniques relied on use of Wheatstone bridges. The bridge is based on a nulling technique that requires manipulation of an adjustable resistor and capacitor at each frequency to obtain an effective frequency-dependent resistance and capacitance of the cell, from which can be derived an impedance. In time, the mechanical signal generator was replaced by an electronic signal generator, but the frequency range remained limited to acoustic frequencies (kHz to Hz). The ability to record time-domain signals on an oscilloscope enabled measurement to subacoustic frequencies (on the order of mHz). Subsequent development of digital signal analysis allowed automated recording of impedance spectra. These techniques are described in Chapter 7. Development of microelectrodes enabled local measurement of current density and local impedance spectra. These techniques are also described in Chapter 7.

A parallel development has taken place for related transfer-function methods. For electrochemical systems, impedance spectroscopy, which relies on measurement of current and potential, provides the general system response. As described in Chapters 15 and 16, transfer-function methods allow the experimentalist to isolate the portion of the response associated with specific inputs or outputs.

Impedance Representation

The methods used to plot impedance data began with plots of effective resistance and capacitance, reflecting the use of bridges for measurement. These plots gave way to Nyquist and Bode plots, which remain the traditional means of representing impedance data. More recently, authors have promoted the use of Bode plots corrected for ohmic resistance and the use of logarithmic plots of the imaginary impedance as a function of frequency. As described in Chapter 18 (page 493), such plots provide limited yet quantitative interpretation of impedance spectra.

Mathematical Analysis

The impedance response associated with diffusion in an infinite domain and in a solid film was developed in the very early twentieth century. Similarly quantitative models were developed in the 1940s for the capacitive behavior of the double layer. In the middle of the twentieth century, models were being developed that accounted for heterogeneous reactions and adsorbed intermediates. In the 1960s and 1970s, such models were being generalized to account for homogeneous reactions and reactions on porous electrodes. The development of quantitative models is presented in Part III (page 193).

These models provided a quantitative relationship between physicochemical parameters and impedance response, but the application of interpretation strategies did not keep pace with the model development. Interpretation was based on graphical examination of plotted data. In simple cases, plots could be used directly, as described in Chapter 18. For more complicated cases, simulations could be compared graphically to data to reveal qualitative agreement.

Nonlinear regression analysis, described in Chapters 19 and 20, was developed for impedance spectroscopy in the early 1970s. The models were cast in the form of electrical circuits with mathematical formulas added to account for the diffusion impedance associated with simplified geometries.

There were significant difficulties associated with fitting models to impedance data. The electrochemical systems frequently did not conform to the assumptions made in the models, especially those associated with electrode uniformity. Constant-phase elements (CPEs), described in Chapter 14, were introduced as a convenient general circuit element that was said to account for distributions of time constants. The meaning of the CPE for specific systems was often disputed.

In addition, the variance of impedance measurements depends strongly on frequency, and this variation needs to be addressed by the regression strategies employed. An assumed dependence of the variance of the impedance measurement on impedance values was employed in early stages of regression analysis, and this gave rise to some controversy over what assumed error structure was most appropriate. An experimental approach using measurement models, described in Chapter 21, was later developed, which eliminated the need for assumed error structures.

The length of time required to make impedance measurements in the acoustic to subacoustic frequency range may cause the resulting impedance to be influenced by changes in the system properties during the course of the measurement. The Kramers–Kronig relations, described in Chapter 22, were employed to determine whether impedance spectra were corrupted by nonstationary behavior. This approach, too, was controversial due to the need to evaluate the Kramers–Kronig integrals over a frequency range extending from 0 to infinity, requiring extrapolation of the measured data. The use of a measurement model allowed assessment of the degree of consistency with the Kramers–Kronig relations without use of integration procedures.

Access to powerful computers and to commercial software for solving systems of partial-differential equations has facilitated modeling of the impedance response of electrodes exhibiting surface property distributions. Use of these tools, coupled with development of localized impedance measurements, has introduced a renewed emphasis on the study of heterogeneous surfaces. This coupling provides a nice example for the integration of experiment, modeling, and error analysis described in Chapter 23.

Table 1: A time line for the historical development of impedance spectroscopy. Contributions listed for any given range of time add to preceding contributions.

	1894-1920	1920-1952	1952-1960	1960-1972	1972-1990	1990-2016
Experimental system	Dielectric properties	Mercury drop, double layers, biological systems	Anodic dissolution	Electro-crystallization, corrosion, 3-D electrodes	Generators, mixed conductors, redox materials	Heterogeneous surfaces
Measurement techniques	Bridge: mechanical generator	Bridge: electronic generator	Impulse method, oscillograph, Laplace transform	Analogue impedance measurement, potentiostat (AC + DC)	Digital impedance measurement, connection with computer	Local electrochemical impedance spectroscopy (LEIS)
Frequency range	Acoustic, >100Hz	Acoustic, >100Hz	Acoustic and subacoustic, >1mHz	Acoustic and subacoustic, >1mHz	Acoustic and subacoustic, >1mHz	Acoustic and subacoustic, >1mHz
Representation	$R - C$	Electrical equivalent circuits	Nyquist plots	Bode plots		R_e-corrected Bode plots, $\log(|Z_j|)$ vs $\log(f)$
Mathematical analysis	Heaviside theory	Capacitance vs. frequency	$\sqrt{f}$	Fitting	Kramers–Kronig analysis, assumed error structure	Measurement model, measured error structure
Simulation				Mainframe computer	Personal computer	Commercial PDE solvers
Process models	Nernst: dielectrics (1894); Warburg: diffusion (1901); Finkelstein: Solid film (1902)	Randles: double layer and diffusion impedance (1947)	Gerischer: two heterogeneous steps with adsorbed intermediate (1955)	De Levie: porous electrodes (1967)	Schuhmann: homogeneous reactions and diffusion (1964); Gabrielli: generalized impedance (1977)	Isaacs: LEIS (1992)

Part I

Background

Chapter 1

Complex Variables

A working understanding of complex variables is essential for the analysis of experiments conducted in the frequency domain, such as impedance spectroscopy. The objective of this chapter is to introduce the subject of complex variables at a level sufficient to understand the development of interpretation models in the frequency domain. Complex variables represent an exciting and important field in applied mathematics, and textbooks dedicated to complex variables can extend the introduction provided here.[88,89] The overview presented in this chapter is strongly influenced by the compact treatment presented by Fong et al.[90]

1.1 Why Imaginary Numbers?

The terminology used in the study of complex variables, in particular the term imaginary number, is particularly unfortunate because it provides an unnecessary conceptual barrier to the beginning student of the subject. Complex variables are ordered pairs of numbers, where the imaginary part represents the solution to a particular type of equation. As suggested by Cain in his introduction to complex variables,[89] complex numbers can be compared to other ordered pairs of numbers.

Rational numbers, for example, are defined to be ordered pairs of integers. For example, $(3,8)$ is a rational number. The ordered pair (n,m) can be written as $\left(\frac{n}{m}\right)$. Thus the rational number $(3,8)$ can be represented as well by 0.375.

Two rational numbers (n,m) and (p,q) are defined to be equal whenever $nq = pm$. The sum of (n,m) and (p,q) is given by

$$(n,m) + (p,q) = ((nq + pm), mq) \tag{1.1}$$

and the product by

$$(n,m)(p,q) = (np, mq) \tag{1.2}$$

Subtraction and division are defined to be the inverses of the addition and multiplication operations, respectively.

Irrational numbers were introduced because the set of rational numbers could not provide solutions to such equations as $z = \sqrt{2}$. As seen in the subsequent section, the set of real numbers, which encompasses rational and irrational numbers, is not adequate to provide solutions to yet other classes of equations. Thus, complex numbers were introduced, which can be seen in the following sections to be defined as ordered pairs (x, y) of real and imaginary numbers.[89]

1.2   Terminology

The concept of complex variables is used widely in mathematical and engineering analysis. Some definitions and concepts commonly encountered in the field of impedance spectroscopy are presented in this section.

1.2.1   The Imaginary Number

The imaginary number $j = \sqrt{-1}$ is the solution to the algebraic equation

$$z^2 = -1 \tag{1.3}$$

which yields $z = \pm j$. The imaginary number arises as well in the solution to differential equations such as

$$\frac{d^2 y}{dx^2} + by = 0 \tag{1.4}$$

which, as shown in Example 2.3 on page 30, has the characteristic equation

$$m^2 = -b \tag{1.5}$$

with solution

$$m = \pm\sqrt{-b} = \pm j\sqrt{b} \tag{1.6}$$

The homogeneous solution to equation (1.4) is given by

$$y = C_1 \exp\left(j\sqrt{b}x\right) + C_2 \exp\left(-j\sqrt{b}x\right) \tag{1.7}$$

Some useful identities for the complex number j are presented in Table 1.1.

1.2.2   Complex Variables

The solution to the quadratic equation

$$az^2 + bz + c = 0 \tag{1.13}$$

given as

$$z = \frac{-b \pm \sqrt{b^2 - 4ac}}{2a} \tag{1.14}$$

is a complex number if the argument $b^2 - 4ac < 0$. The complex variables can be written as

$$z = z_r + jz_j \qquad (1.15)$$

where z_r and z_j are real numbers that represent the real and imaginary parts of z, respectively. Often the notations $\text{Re}\{z\}$ and $\text{Im}\{z\}$ are used to designate real and imaginary components of the complex number z, respectively.

1.2.3 Conventions for Notation in Impedance Spectroscopy

The IUPAC convention, as described in the overview by Sluyters-Rehbach,[91] is that $\sqrt{-1}$ should be denoted by the symbol i. To avoid confusion with current density, given the symbol i, and the index i used to indicate specific chemical species, we have chosen here to follow the electrical engineering convention in which $\sqrt{-1}$ is given the symbol j.

We also depart from the IUPAC convention in the notation used to denote real and imaginary parts of the impedance. The IUPAC convention is that the real part of the impedance is given by Z' and the imaginary part is given by Z''. We consider that the IUPAC notation can be confused with the use of primes and double primes to denote first and second derivatives, respectively. Thus, we choose to identify the real part of the impedance by Z_r and the imaginary part of the impedance by Z_j.

1.3 Operations Involving Complex Variables

As z is a single value with real and imaginary components, z can be represented as a point on a complex plane, as shown in Figure 1.1. The complex conjugate of a complex number $z = z_r + jz_j$ is defined to be $\bar{z} = z_r - jz_j$. Thus, in Figure 1.1, $\bar{z}$ is seen to be the reflection of z about the real axis.

As is evident in the graphical representation of a complex number in Figure 1.1, two complex numbers are equal if and only if both the real and the imaginary parts are equal. Thus, an equation involving complex variables requires that two equations are

---

Table 1.1: Identities for the imaginary number j.

$$j = \sqrt{-1} \qquad (1.8)$$
$$j^2 = -1 \qquad (1.9)$$
$$j^3 = -j \qquad (1.10)$$
$$j^4 = 1 \qquad (1.11)$$
$$1/j = -j \qquad (1.12)$$

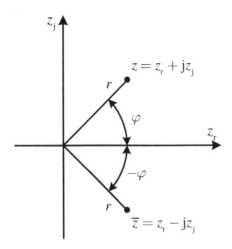

Figure 1.1: Argand diagram showing the position of a complex number and its complex conjugate on a complex plane.

satisfied, one involving the real terms and one involving the imaginary terms. Commutative, associative, and distributive laws hold for complex numbers. Some useful relationships for complex variables are presented in Table 1.2, which demonstrate the commutative, associative, and distributive properties.

The commutative property states that, in addition and multiplication, terms may be arbitrarily interchanged. Thus, equation (1.16) applies for addition and equation (1.17) applies for multiplication of complex numbers z and w. The distributive property is demonstrated by equation (1.18), and the associative property is demonstrated by equation (1.19).

1.3.1   Multiplication and Division of Complex Numbers

Equations (1.20)-(1.24) illustrate the manner in which mathematical operations are carried out in terms of real and imaginary components. These results provide a foundation for the followings series of examples.

Example 1.1 **Multiplication of Complex Numbers:** *Does the imaginary part of the product of two complex numbers equal the product of the imaginary parts?*

Remember! 1.1 *Complex numbers are ordered pairs (x, y) of real and imaginary numbers that represent the solution to a class of problems that cannot be solved using rational and irrational numbers.*

Table 1.2: Relationships for complex variables $z = z_r + jz_j$ and $w = w_r + jw_j$.

$$z + w = w + z \tag{1.16}$$

$$zw = wz \tag{1.17}$$

$$a(z + w) = az + aw \tag{1.18}$$

$$a(zw) = (az)w \tag{1.19}$$

$$z + w = (z_r + w_r) + j(z_j + w_j) \tag{1.20}$$

$$z - w = (z_r - w_r) + j(z_j - w_j) \tag{1.21}$$

$$zw = (z_r w_r - z_j w_j) + j(z_r w_j + z_j w_r) \tag{1.22}$$

$$w\overline{w} = w_r^2 + w_j^2 \tag{1.23}$$

$$\frac{z}{w} = \frac{z\overline{w}}{w\overline{w}}$$
$$= \frac{(z_r w_r + z_j w_j) + j(z_j w_r - z_r w_j)}{w_r^2 + w_j^2} \tag{1.24}$$

Solution: *Consider two numbers $z = z_r + jz_j$ and $w = w_r + jw_j$. The multiplication of z and w follows equation (1.22), i.e.,*

$$zw = (z_r + jz_j)(w_r + jw_j)$$
$$= \left(z_r w_r + j^2 z_j w_j\right) + j\left(z_r w_j + w_r z_j\right) \tag{1.25}$$
$$= (z_r w_r - z_j w_j) + j(z_r w_j + w_r z_j)$$

The imaginary part of zw is $\left(z_r w_j + w_r z_j\right)$, which is not equal to the product of the imaginary parts, i.e.,

$$\left(z_r w_j + w_r z_j\right) \neq z_j w_j \tag{1.26}$$

Thus, the imaginary part of the product of two complex numbers does not equal the product of the imaginary parts.

Example 1.2 Division of Complex Numbers: *In a new experimental technique developed by Antaño-Lopez et al.,[92] an approximate formula for capacitance was used; i.e.,*

$$C = \frac{Y_j(\omega)}{\omega} \tag{1.27}$$

Remember! 1.2 *The notation used in this text provides that $j = \sqrt{-1}$ and that real and imaginary parts of complex numbers are denoted by subscripts r and j, respectively.*

where Y is the complex admittance $Y = Z^{-1}$, *and* ω *is angular frequency. At frequencies sufficiently high to eliminate the contribution of faradaic resistance, the capacitance is shown in Section 17.4 to be obtained correctly from*

$$C = -\frac{1}{\omega Z_j(\omega)} \quad (1.28)$$

where Z_j *is the imaginary part of the complex impedance Z. For capacitive systems,* $Z_j < 0$. *Under what conditions will equation (1.27) be accurate?*

Solution: *Equation (1.27) would agree with equation (1.28) if*

$$Y_j = \text{Im}\{Z^{-1}\} \overset{?}{=} -Z_j^{-1} \quad (1.29)$$

To test the validity of equation (1.29), consider the inverse of the complex number $Z = Z_r + jZ_j$.

$$\frac{1}{Z} = \frac{1}{Z_r + jZ_j} \quad (1.30)$$

Division is possible only after the denominator is converted into a real, rather than complex, number. Both the numerator and the denominator are multiplied by the complex conjugate (see equations (1.23) and (1.24)).

$$
\begin{aligned}
\frac{1}{Z} &= \left\{\frac{1}{Z_r + jZ_j}\right\}\left\{\frac{Z_r - jZ_j}{Z_r - jZ_j}\right\} \\
&= \frac{Z_r - jZ_j}{Z_r^2 + Z_j^2} \\
&= \frac{Z_r}{Z_r^2 + Z_j^2} - j\frac{Z_j}{Z_r^2 + Z_j^2}
\end{aligned} \quad (1.31)
$$

Thus,

$$Y_j = \text{Im}\{Z^{-1}\} = -\frac{Z_j}{Z_r^2 + Z_j^2} \neq -\frac{1}{Z_j} \quad (1.32)$$

Equation (1.29) is satisfied only if $Z_r = 0$. *As discussed in Chapter 10, the real part of the impedance* Z_r *approaches the electrolyte resistance at high frequencies. The capacitance obtained by Antaño-Lopez et al.[92] is correct at high frequencies only if the electrolyte resistance can be neglected, i.e.,* $Z_r^2 \ll Z_j^2$.

Remember! 1.3 *The impedance is a complex number defined to be the ratio of complex potential and complex current.*

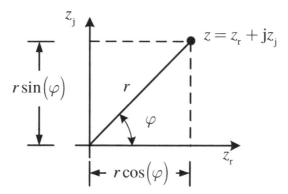

Figure 1.2: Argand diagram showing relationships among complex impedance, magnitude, and phase angle.

Example 1.3 Rectangular Coordinates: *The impedance of a parallel combination of a resistor and a capacitor can be expressed as*

$$Z = \frac{R}{1 + j\omega\tau} \tag{1.33}$$

where τ is the time constant $\tau = RC$. Express equation (1.33) in rectangular coordinates, i.e., find the real and imaginary components of Z.

Solution: *To express equation (1.33) in rectangular coordinates, the denominator must be converted into a real, rather than complex, number. Both the numerator and the denominator are multiplied by the complex conjugate (see equations (1.23) and (1.24)).*

$$\begin{aligned} Z &= \left\{ \frac{R}{1 + j\omega\tau} \right\} \left\{ \frac{1 - j\omega\tau}{1 - j\omega\tau} \right\} \\ &= \frac{R - j\omega R\tau}{1 + \omega^2\tau^2} \\ &= \frac{R}{1 + \omega^2\tau^2} - j\frac{\omega R\tau}{1 + \omega^2\tau^2} \end{aligned} \tag{1.34}$$

Thus,

$$\mathrm{Re}\{Z\} = \frac{R}{1 + \omega^2\tau^2}; \quad \mathrm{Im}\{Z\} = -\frac{\omega R\tau}{1 + \omega^2\tau^2} \tag{1.35}$$

1.3.2 Complex Variables in Polar Coordinates

The transformation from rectangular to polar coordinates is shown schematically in Figure 1.2. The variable r is the modulus or absolute value $|z|$, which always has a positive value. The phase angle is written as $\varphi = \arg(z)$. In the mathematical development (see Section 1.4.1), the phase angle has units of radians; however, it is often presented

Table 1.3: Relationships between polar and rectangular coordinates for the complex variable $z = z_r + jz_j$.

$$z_r = r\cos(\varphi) \tag{1.36}$$
$$z_j = r\sin(\varphi) \tag{1.37}$$
$$r = |z| = \sqrt{z_r^2 + z_j^2} \tag{1.38}$$
$$\varphi = \tan^{-1}\left(\frac{z_j}{z_r}\right) \tag{1.39}$$
$$|z| = \sqrt{z\bar{z}} \tag{1.40}$$
$$z = r\left(\cos(\varphi) + j\sin(\varphi)\right) \tag{1.41}$$
$$z^n = r^n\left(\cos(n\varphi) + j\sin(n\varphi)\right) \tag{1.42}$$
$$z^{1/n} = r^{1/n}\left[\cos\left(\frac{\varphi}{n} + \frac{2\pi k}{n}\right) + j\sin\left(\frac{\varphi}{n} + \frac{2\pi k}{n}\right)\right]; \quad k = 0,1,\ldots,n-1 \tag{1.43}$$

in units of degrees where $90° = \pi/2$. The angle $\arg(z)$ has an infinite number of possible values because any multiple of 2π radians can be added to it without changing the value of z. The value of φ that lies between $-\pi$ and π is called the principal value of $\arg(z)$.

Some useful relationships between polar and rectangular coordinates for complex variables are summarized in Table 1.3. Equation (1.42) is known as De Moivre's theorem. It is valid for all rational values of n.

Example 1.4 **Polar Coordinates:** *The impedance of a parallel combination of a resistor and a capacitor can be expressed as*

$$Z = \frac{R}{1 + j\omega\tau} \tag{1.44}$$

where τ is the time constant $\tau = RC$. Express equation (1.44) in polar coordinates; i.e., find the modulus and phase angle for Z.

Solution: *As shown in Example 1.3, equation (1.44) can be expressed in rectangular coordinates by*

$$\mathrm{Re}\{Z\} = \frac{R}{1 + \omega^2\tau^2}; \quad \mathrm{Im}\{Z\} = -\frac{\omega R\tau}{1 + \omega^2\tau^2} \tag{1.45}$$

Equations (1.38) and (1.39) can be used to convert equation (1.45) into polar coordinates. Thus,

$$r = \sqrt{\left(\frac{R}{1 + \omega^2\tau^2}\right)^2 + \left(\frac{\omega R\tau}{1 + \omega^2\tau^2}\right)^2} \tag{1.46}$$

or

$$r = \sqrt{\frac{R^2}{1 + \omega^2\tau^2}} \tag{1.47}$$

$$\varphi = \tan^{-1}\left[-\frac{\omega R\tau}{1 + \omega^2\tau^2}\frac{1 + \omega^2\tau^2}{R} \right]$$
$$= \tan^{-1}(-\omega\tau) \tag{1.48}$$

The phase angle φ is a function only of frequency ω and the time constant τ. The modulus r depends on the value of R as well as on the frequency ω and the time constant τ.

The solution can be obtained more rapidly by casting the impedance in terms of admittance (see Section 17.2 on page 478). The admittance can be expressed as

$$Y = \frac{1}{Z} = \frac{1 + j\omega\tau}{R} \tag{1.49}$$

The modulus of the admittance is given by

$$|Y| = \sqrt{\frac{1 + \omega^2\tau^2}{R^2}} \tag{1.50}$$

and the argument is given by

$$\varphi_Y = \tan^{-1}(\omega\tau) \tag{1.51}$$

The modulus of the impedance and the corresponding argument can be deduced immediately as

$$|Z| = \frac{1}{|Y|} \tag{1.52}$$

and

$$\varphi = -\varphi_Y \tag{1.53}$$

respectively.

Example 1.5 Square Roots of Complex Variables: *The Warburg impedance associated with diffusion in an infinite medium takes the form $Z = 1/\sqrt{j\omega\tau}$. Find the roots of Z.*

Solution: *The polar form of $z = 1/j\omega\tau = -j/\omega\tau$ is expressed as*

$$z = \frac{1}{\omega\tau}\left(\cos\left(\frac{3\pi}{2}\right) + j\sin\left(\frac{3\pi}{2}\right) \right) \tag{1.54}$$

From equation (1.43)

$$z^{1/2} = \sqrt{\frac{1}{\omega\tau}}\left(\cos\left(\frac{3\pi}{4} + k\pi\right) + j\sin\left(\frac{3\pi}{4} + k\pi\right) \right); \quad k = 0, 1 \tag{1.55}$$

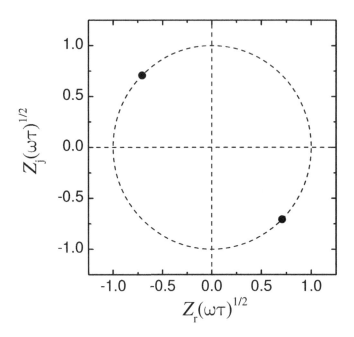

Figure 1.3: Argand diagram showing the two roots of $Z = 1/\sqrt{j\omega\tau}$ as calculated in Example 1.5.

The roots of $1/\sqrt{j\omega\tau}$ are shown in Figure 1.3. The root with $k = 0$ can be rejected using the physical reasoning that the resistance associated with diffusion cannot have a negative sign. Thus, the Warburg impedance can be expressed in rectangular coordinates as

$$Z = \sqrt{\frac{1}{2\omega\tau}} - j\sqrt{\frac{1}{2\omega\tau}} \tag{1.56}$$

The real and the negative imaginary components have the same magnitude and increase according to $\sqrt{1/\omega}$ as frequency tends toward zero.

As shown in the following examples, equation (1.41) is very useful for expressing complex quantities in rectangular coordinates. The imaginary number can be expressed as

$$j = \cos\left(\frac{\pi}{2}\right) + j\sin\left(\frac{\pi}{2}\right) \tag{1.57}$$

where evaluation of equations (1.38) and (1.39) yields $r = 1$ and $\varphi = \pi/2$, respectively

---

Remember! **1.4** *The expression $j = \cos(\pi/2) + j\sin(\pi/2)$ is very useful for finding real and imaginary parts of complex expressions involving the imaginary number raised to a power.*

Example 1.6 Square Roots of Complex Variables 2: *Use equation (1.57) to find real and imaginary parts of the Warburg impedance in the form* $Z = 1/\sqrt{j\omega\tau}$.

Solution: *From equation (1.57), the Warburg impedance may be expressed as*

$$Z = \frac{1}{\sqrt{j\omega\tau}} = \frac{1}{\sqrt{\omega\tau}\left(\cos\left(\frac{\pi}{4}\right) + j\sin\left(\frac{\pi}{4}\right)\right)} \tag{1.58}$$

Multiplication of numerator and denominator by the complex conjugate yields

$$Z = \frac{\cos\left(\frac{\pi}{4}\right) - j\sin\left(\frac{\pi}{4}\right)}{\sqrt{\omega\tau}\left(\cos^2\left(\frac{\pi}{4}\right) + \sin^2\left(\frac{\pi}{4}\right)\right)} \tag{1.59}$$

Introduction of the identity

$$\cos^2(x) + \sin^2(x) = 1 \tag{1.60}$$

yields

$$Z = \frac{\cos\left(\frac{\pi}{4}\right) - j\sin\left(\frac{\pi}{4}\right)}{\sqrt{\omega\tau}} \tag{1.61}$$

As $\cos\left(\frac{\pi}{4}\right) = \sin\left(\frac{\pi}{4}\right) = 1/\sqrt{2}$,

$$Z = \sqrt{\frac{1}{2\omega\tau}} - j\sqrt{\frac{1}{2\omega\tau}} \tag{1.62}$$

This is the same result as obtained in Example 1.5.

Example 1.7 Blocking Constant-Phase Element: *Use equation (1.57) to find real and imaginary parts for a constant-phase element given as*

$$Z_{\text{CPE}} = \frac{1}{(j\omega)^\alpha Q} \tag{1.63}$$

where α is a real number, generally with value $0.5 \leqq \alpha \leqq 1$.

Solution: *Introduction of equation (1.57) yields*

$$Z = \frac{1}{(j\omega)^\alpha Q} = \frac{1}{\omega^\alpha Q\left(\cos\left(\frac{\alpha\pi}{2}\right) + j\sin\left(\frac{\alpha\pi}{2}\right)\right)} \tag{1.64}$$

multiplication by the complex conjugate yields

$$Z = \frac{\cos\left(\frac{\alpha\pi}{2}\right) - j\sin\left(\frac{\alpha\pi}{2}\right)}{\omega^\alpha Q\left(\cos^2\left(\frac{\alpha\pi}{2}\right) + \sin^2\left(\frac{\alpha\pi}{2}\right)\right)} \tag{1.65}$$

From equation (1.60),

$$Z = \frac{\cos\left(\frac{\alpha\pi}{2}\right)}{\omega^\alpha Q} - j\frac{\sin\left(\frac{\alpha\pi}{2}\right)}{\omega^\alpha Q} \tag{1.66}$$

When $\alpha = 1$, the parameter Q can be expressed as a capacitance C, and

$$\lim_{\alpha \to 1} Z_{CPE} = -j\frac{1}{\omega C} \qquad (1.67)$$

which is the impedance for a capacitor.

Example 1.8 **Reactive Constant-Phase Element:** *Use equation* (1.57) *to find real and imaginary parts for a constant-phase element in the expression*

$$Z_{CPE} = \frac{R}{1 + (j\omega)^\alpha QR} \qquad (1.68)$$

where α is a real number, generally with value $0.5 \leq \alpha \leq 1$.

Solution: *Introduction of equation* (1.57) *yields*

$$Z_{CPE} = \frac{R}{1 + (j\omega)^\alpha QR} = \frac{R}{1 + \omega^\alpha QR \left(\cos\left(\frac{\alpha\pi}{2}\right) + j\sin\left(\frac{\alpha\pi}{2}\right)\right)} \qquad (1.69)$$

multiplication by $1 + \omega^\alpha RQ\cos\left(\frac{\alpha\pi}{2}\right) - j\omega^\alpha RQ\sin\left(\frac{\alpha\pi}{2}\right)$ yields

$$Z_{CPE} = \frac{R\left(1 + \omega^\alpha RQ\cos\left(\frac{\alpha\pi}{2}\right)\right) - j\omega^\alpha RQ\sin\left(\frac{\alpha\pi}{2}\right)}{1 + 2\omega^\alpha RQ\cos\left(\frac{\alpha\pi}{2}\right) + \omega^{2\alpha}R^2Q^2} \qquad (1.70)$$

or

$$
\begin{aligned}
Z_{CPE} &= \frac{R\left(1 + \omega^\alpha RQ\cos\left(\frac{\alpha\pi}{2}\right)\right)}{1 + 2\omega^\alpha RQ\cos\left(\frac{\alpha\pi}{2}\right) + \omega^{2\alpha}R^2Q^2} \\
&\quad -j\frac{\omega^\alpha R^2 Q\sin\left(\frac{\alpha\pi}{2}\right)}{1 + 2\omega^\alpha RQ\cos\left(\frac{\alpha\pi}{2}\right) + \omega^{2\alpha}R^2Q^2}
\end{aligned} \qquad (1.71)
$$

When $\alpha = 1$, the parameter Q can be expressed as a capacitance C, and

$$\lim_{\alpha \to 1} Z_{CPE} = \frac{R}{1 + \omega^2 R^2 C^2} - j\frac{\omega R^2 C}{1 + \omega^2 R^2 C^2} \qquad (1.72)$$

Equation (1.72) *is the expression given as equation* (1.34), *where the time constant is expressed as $\tau = RC$.*

The rectangular forms developed in Examples 1.6, 1.7, and 1.8 can be expressed easily in polar form using equations (1.38) and (1.39). For example, the phase angle for the Warburg impedance analyzed in Example 1.6 can be expressed as

$$
\begin{aligned}
\varphi &= \tan^{-1}\left(\frac{Z_j}{Z_r}\right) = \tan^{-1}\left(\frac{-\sqrt{1/2\omega\tau}}{\sqrt{1/2\omega\tau}}\right) \\
&= \tan^{-1}(-1) = -\pi/4 = -45°
\end{aligned} \qquad (1.73)
$$

Table 1.4: Properties for the complex conjugates of $z = z_r + jz_j$ and $w = w_r + jw_j$.

$$\overline{z + w} = \overline{z} + \overline{w} \tag{1.76}$$

$$\overline{zw} = \overline{z}\,\overline{w} \tag{1.77}$$

$$\overline{\left[\frac{1}{z}\right]} = \frac{1}{\overline{z}} \tag{1.78}$$

$$\overline{\overline{z}} = z \tag{1.79}$$

Similarly, the phase angle for the blocking CPE of Example 1.7 can be expressed as

$$
\begin{aligned}
\varphi &= \tan^{-1}\left(\frac{-\sin\left(\frac{\alpha\pi}{2}\right)/\omega^\alpha Q}{\cos\left(\frac{\alpha\pi}{2}\right)/\omega^\alpha Q}\right) \\
&= -\frac{\alpha\pi}{2} = -\alpha 90^\circ
\end{aligned}
\tag{1.74}
$$

The phase angle for the reactive CPE presented in Example 1.8 is given by

$$
\varphi = \tan^{-1}\left(\frac{-\sin\left(\frac{\alpha\pi}{2}\right)}{\frac{1}{\omega^\alpha RQ} + \cos\left(\frac{\alpha\pi}{2}\right)}\right)
\tag{1.75}
$$

As the frequency $\omega \to \infty$, the phase angle approaches $-\alpha 90^\circ$. Each of these types of models can be described as constant-phase elements as the phase angle is either constant or approaches a constant value at high frequency.

1.3.3 Properties of Complex Variables

Some useful properties of the complex conjugates of $z = z_r + jz_j$ and $w = w_r + jw_j$ are presented in Table 1.4, and some relationships for the absolute value of $z = z_r + jz_j$ and $w = w_r + jw_j$ are presented in Table 1.5.

1.4 Elementary Functions of Complex Variables

The definition of many elementary functions can be extended to complex variables. Polynomial, exponential, and logarithmic functions are discussed here.

1.4.1 Exponential

The exponential function e^z is of fundamental importance in impedance spectroscopy. The exponential function is defined such that it retains the properties of the real function e^x that

---

Table 1.5: Properties for the absolute value of $z = z_r + j z_j$ and $w = w_r + j w_j$.

$$|z| = |\bar{z}| \tag{1.80}$$

$$z\bar{z} = |z|^2 \tag{1.81}$$

$$z_r \leq |z| \tag{1.82}$$

$$z_j \leq |z| \tag{1.83}$$

$$|zw| = |z||w| \tag{1.84}$$

$$\left| \frac{1}{z} \right| = \frac{1}{|z|}; \ z \neq 0 \tag{1.85}$$

$$|z + w| \leq |z| + |w| \tag{1.86}$$

---

1. e^z, with argument $z = x + jy$, is single valued and analytic (see Appendix A.1),

2. $de^z/dz = e^z$, and

3. $e^z \to e^x$, when $y \to 0$.

As a consequence of the above requirements, the exponential function with argument $z = x + jy$ can be shown to conform to

$$
\begin{aligned}
e^z &= e^{x+jy} \\
&= e^x \left[\cos(y) + j \sin(y) \right]
\end{aligned} \tag{1.87}
$$

Equation (1.87) can be considered to be the definition of e^z, which can be readily shown to meet the requirements expressed above.

Equation (1.87) is written in the standard polar form, equation (1.41), in which the modulus of e^z is

$$r = |e^z| = e^x \tag{1.88}$$

and the argument, or phase angle, is given by

$$\varphi = \arg\left(e^z\right) = y \tag{1.89}$$

It is evident, then, that the exponential function is periodic, i.e.,

$$e^z = e^{z+j2k\pi} \tag{1.90}$$

for integer values of k.

Any complex number can be written in exponential form. For example, if $x = 0$ and $y = \varphi$, application of equation (1.87) yields

$$\cos(\varphi) + j \sin(\varphi) = e^{j\varphi} \tag{1.91}$$

In addition,

$$e^{-j\varphi} = \cos(-\varphi) + j\sin(-\varphi)$$
$$= \cos(\varphi) - j\sin(\varphi) \tag{1.92}$$

Equations (1.91) and (1.92) yield the Euler formulas

$$\cos(\varphi) = \frac{e^{j\varphi} + e^{-j\varphi}}{2} \tag{1.93}$$

and

$$\sin(\varphi) = \frac{e^{j\varphi} - e^{-j\varphi}}{2j} \tag{1.94}$$

These can be extended for $x \neq 0$ as

$$\cos(z) = \frac{e^{jz} + e^{-jz}}{2} \tag{1.95}$$

and

$$\sin(z) = \frac{e^{jz} - e^{-jz}}{2j} \tag{1.96}$$

Equations (1.95) and (1.96) provide relationships between trigonometric functions and complex variables. These can be manipulated to find relationships with hyperbolic function. Some important definitions and identities are presented in Table 1.6.

1.4.2 Logarithmic

Given

$$z = e^w \tag{1.114}$$

where z and $w = u + jv$ are complex numbers, the natural logarithm can be defined as

$$w = \ln(z) \tag{1.115}$$

As $w = u + jv$, equation (1.114) can be expressed as

$$z = e^u [\cos(v) + j\sin(v)] \tag{1.116}$$

The modulus of equation (1.116) is given as

$$|z| = e^u \tag{1.117}$$

Remember! 1.5 *The exponential representation of a complex number plays an important role in impedance analysis. Remember that* $e^{\pm j\varphi} = \cos(\varphi) \pm j\sin(\varphi)$.

Table 1.6: Trigonometric and hyperbolic relationships for the complex variable $z = z_r + jz_j$.

$$\sin(z) = \left(e^{jz} - e^{-jz} \right) / 2j \tag{1.97}$$

$$\cos(z) = \left(e^{jz} + e^{-jz} \right) / 2 \tag{1.98}$$

$$\frac{d\sin(z)}{dz} = \cos(z) \tag{1.99}$$

$$\frac{d\cos(z)}{dz} = -\sin(z) \tag{1.100}$$

$$\tan(z) = \frac{\sin(z)}{\cos(z)} \tag{1.101}$$

$$\cot(z) = \frac{\cos(z)}{\sin(z)} \tag{1.102}$$

$$\cos^2 z + \sin^2 z = 1 \tag{1.103}$$

$$\cos(z_1 \pm z_2) = \cos(z_1)\cos(z_2) \mp \sin(z_1)\sin(z_2) \tag{1.104}$$

$$\sin(z_1 \pm z_2) = \sin(z_1)\cos(z_2) \pm \cos(z_1)\sin(z_2) \tag{1.105}$$

$$\sinh(z) = \left(e^z - e^{-z} \right) / 2 \tag{1.106}$$

$$\cosh(z) = \left(e^z + e^{-z} \right) / 2 \tag{1.107}$$

$$\tanh(z) = \frac{\sinh(z)}{\cosh(z)} \tag{1.108}$$

$$\coth(z) = \frac{\cosh(z)}{\sinh(z)} \tag{1.109}$$

$$\cos(z) = \cos(x)\cosh(y) - j\sin(x)\sinh(y) \tag{1.110}$$

$$\sin(z) = \sin(x)\cosh(y) + j\cos(x)\sinh(y) \tag{1.111}$$

$$\cosh(z) = \cosh(x)\cos(y) + j\sinh(x)\sin(y) \tag{1.112}$$

$$\sinh(z) = \sinh(x)\cos(y) + j\cosh(x)\sin(y) \tag{1.113}$$

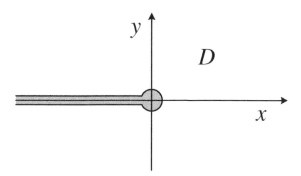

Figure 1.4: Representation of the domain D in which $\text{Ln}(z)$ is analytic.

and the phase angle is given by

$$\arg(z) = v \qquad (1.118)$$

The complex number z can therefore be expressed as

$$z = |z|e^{j\arg(z)} = |z|e^{jv} \qquad (1.119)$$

or

$$\ln(z) = \ln(|z|) + j\arg(z) = \ln(e^u) + jv \qquad (1.120)$$

There are an infinite number of values of $\ln(z)$ because $\arg(z)$ can differ by multiples of 2π. The principal value of $\ln(z)$ is defined for the principal value of $\arg(z)$, both designated by initial capital letters. Thus,

$$\text{Ln}(z) = \ln(|z|) + j\text{Arg}(z) \qquad (1.121)$$

where

$$-\pi < \text{Arg}(z) \leq \pi \qquad (1.122)$$

The function $\text{Ln}(z)$ is not defined at $z = 0$ and is not continuous anywhere on the negative real axis $z = x + 0j$, where $x < 0$. The negative real axis is a line of discontinuity because, on that line, the imaginary part of $\text{Ln}(z)$ has a jump discontinuity of 2π. If a cut is made, as shown in Figure 1.4, to remove the origin and the negative real axis, $\text{Ln}(z)$ is analytic in the resulting domain, and the derivative of $\text{Ln}(z)$ is given by

$$\frac{d\text{Ln}(z)}{dz} = \frac{1}{z} \qquad (1.123)$$

The derivative of $\ln(z)$ is also given by equation (1.123) because $\text{Ln}(z)$ and $\ln(z)$ differ by a constant, $2\pi kj$.

Some functional relationships commonly used in impedance spectroscopy are presented in Table 1.7.

Table 1.7: Functional relationships of complex variables commonly encountered in impedance spectroscopy, where x and y are real numbers, and $z = x + jy$.

$$\exp(jx) = \cos(x) + j\sin(x) \tag{1.124}$$
$$\exp(j(x + y)) = \exp(jx) \cdot \exp(jy) \tag{1.125}$$
$$\cos(\omega t + \varphi) = \mathrm{Re}\left\{\exp(j(\omega t + \varphi))\right\} \tag{1.126}$$
$$= \mathrm{Re}\left\{\exp(j\varphi) \cdot \exp(j\omega t)\right\} \tag{1.127}$$
$$\mathrm{Re}\{\ln(z)\} = \ln|z| \tag{1.128}$$
$$\mathrm{Im}\{\ln(z)\} = \arg(z) \tag{1.129}$$

Example 1.9 Exponential Form: *Show that a time-dependent variable $i(t)$, expressed in terms of the steady-state value $\bar{i}$ and a sinusoidal time-dependent contribution as*

$$i(t) = \bar{i} + |\Delta i|\cos(\omega t + \varphi) \tag{1.130}$$

can be expressed as

$$i(t) = \bar{i} + \mathrm{Re}\left\{\widetilde{i}\exp(j\omega t)\right\} \tag{1.131}$$

where

$$\widetilde{i} = |\Delta i|\exp(j\varphi) \tag{1.132}$$

is a complex number for $\varphi \neq 0$. Equation (1.131) is commonly used in the development of mathematical models for the transfer function, or impedance, response of electrochemical systems.

Solution: *From equation (1.124), the oscillating component of equation (1.130) can be expressed in terms of an exponential as*

$$|\Delta i|\cos(\omega t + \varphi) = |\Delta i|\exp(j(\omega t + \varphi)) - j|\Delta i|\sin(\omega t + \varphi) \tag{1.133}$$

or

$$|\Delta i|\cos(\omega t + \varphi) = \widetilde{i}\exp(j\omega t) - j|\Delta i|\sin(\omega t + \varphi) \tag{1.134}$$

where $\widetilde{i}$ is given by equation (1.132). The quantity on the left-hand side of equation (1.132) must be a real number. Equation (1.134) is formally equivalent to

$$|\Delta i|\cos(\omega t + \varphi) = \mathrm{Re}\left\{|\Delta i|\cos(\omega t + \varphi)\right\} = \tag{1.135}$$
$$\mathrm{Re}\left\{\widetilde{i}\exp(j\omega t) - j\sin(\omega t + \varphi)\right\}$$

The imaginary term $j\sin(\omega t + \varphi)$ does not contribute to the real part of the complex number inside the brackets; thus,

$$|\Delta i|\cos(\omega t + \varphi) = \mathrm{Re}\left\{\widetilde{i}\exp(j\omega t)\right\} \tag{1.136}$$

The above development could be considered to be a verification of equation (1.127) and justifies the treatment of the current and potential response of electrical circuits expressed as equation (4.8).

Example 1.10 Verification of Expression for Impedance: *Show that the impedance may be expressed as* $Z = \widetilde{V}/\widetilde{i}$.

Solution: *From equation (1.132),*

$$\widetilde{i} = |\Delta i| \exp(j\varphi_i) \tag{1.137}$$

where φ_i is the phase associated with the current density signal. Similarly, an expression for $\widetilde{V}$ may be obtained to be

$$\widetilde{V} = |\Delta V| \exp(j\varphi_V) \tag{1.138}$$

The ratio yields

$$\frac{\widetilde{V}}{\widetilde{i}} = \frac{|\Delta V|}{|\Delta i|} \exp(j(\varphi_V - \varphi_i)) \tag{1.139}$$

Equation (1.139) yields a complex number with magnitude $|\Delta V|/|\Delta i|$ and phase $\varphi = \varphi_V - \varphi_i$. This is the impedance.

1.4.3 Polynomial

A polynomial function of degree n is defined to be

$$P_n(z) = a_n z^n + a_{n-1} z^{n-1} + \ldots + a_1 z + a_0 \tag{1.140}$$

where $a_n \neq 0$, a_{n-1}, ..., a_0 are all complex constants. The rational algebraic function

$$w(z) = \frac{P(z)}{Q(z)} \tag{1.141}$$

is defined where $P(z)$ and $Q(z)$ are polynomials.

A continued overview of complex variables is presented in Appendix A in the context of the complex integration used to establish the Kramers–Kronig relations.

Problems

1.1   Calculate the phase angle and modulus for the following:

   (a) $Z = 1/(j\omega C)$

   (b) $Z = R$

1.2   The impedance associated with a single electrochemical reaction on a uniform surface can be expressed as

$$Z(\omega) = R_e + \frac{R_t}{1 + j\omega R_t C_{dl}} \tag{1.142}$$

where R_e is the electrolyte resistance, R_t is the charge-transfer resistance, and C_{dl} is the capacity of the double layer.

   (a)   Find expressions for the real and imaginary parts of the impedance as a function of frequency.

   (b)   Find expressions for the magnitude and phase angle of the impedance.

   (c)   Find expressions for the real and imaginary parts of the admittance as a function of frequency.

1.3   The impedance associated with an ideally polarized (blocking) electrode can be expressed as

$$Z(\omega) = R_e + \frac{1}{j\omega C_{dl}} \tag{1.143}$$

where R_e is the electrolyte resistance and C_{dl} is the capacity of the double layer.

   (a)   Find expressions for the real and imaginary parts of the impedance as a function of frequency.

   (b)   Find expressions for the magnitude and phase angle of the impedance.

   (c)   Find expressions for the real and imaginary parts of the admittance as a function of frequency.

1.4   The impedance associated with a constant phase element can be expressed as

$$Z(\omega) = R_e + \frac{1}{(j\omega)^\alpha Q} \tag{1.144}$$

where α and Q are parameters associated with a constant-phase element (CPE). When $\alpha = 1$, Q has units of a capacitance, i.e., μFcm^{-2}, and represents the capacity of the interface. When $\alpha \neq 1$, the system shows behavior that has been attributed to surface heterogeneity, oxide films, or to continuously distributed time constants for charge-transfer reactions.

   (a)   Find expressions for the real and imaginary parts of the impedance as a function of frequency.

   (b)   Find expressions for the magnitude and phase angle of the impedance.

(c) Find expressions for the real and imaginary parts of the admittance as a function of frequency.

1.5 Consider a situation where the impedance of one layer is given by

$$Z_1(\omega) = \frac{R_1}{1 + j\omega R_1 C_1} \tag{1.145}$$

and the impedance of a second layer is given by

$$Z_2(\omega) = \frac{R_2}{1 + j\omega R_2 C_2} \tag{1.146}$$

Add the two impedances to find the overall impedance of the two layers.

1.6 The pore-in-pore model described in Section 13.1.3 on page 328 yields an impedance

$$Z = \sqrt{R_0 \sqrt{R_1 Z_1}} \tag{1.147}$$

Develop an expression for the real and imaginary parts of the associated impedance and the associated phase angle:

(a) The impedance Z_1 is given by

$$Z_1 = \frac{1}{j\omega C_1} \tag{1.148}$$

(b) The impedance Z_1 is given by

$$Z_1 = \frac{R_1}{1 + j\omega R_1 C_1} \tag{1.149}$$

Chapter 2

Differential Equations

Development of models for impedance requires solution of differential equations. The method of solution requires two steps. In the first, a steady-state solution is obtained, which generally requires solution of ordinary differential equations.

In the second, a solution is obtained for the sinusoidal steady state. Generally, through transformations of the type discussed in Example 1.9, this too requires solutions of ordinary differential equations. While in some cases numerical solution is required, analytic solutions are possible for a large number of problems. Analytic solutions to some typical equations are reviewed in this chapter. For more details, see standard textbooks on engineering mathematics.[90,93]

2.1 Linear First-Order Differential Equations

The general form of a linear first-order differential equation can be expressed as

$$\frac{dy}{dx} + P(x)y = Q(x) \tag{2.1}$$

The homogeneous equation corresponding to equation (2.1), for which $Q(x) = 0$, is given by

$$\frac{dy}{dx} + P(x)y = 0 \tag{2.2}$$

which has the solution

$$\lambda(x) = \exp\left(-\int P(x)dx\right) \tag{2.3}$$

The quantity $\lambda(x)$ is sometimes referred to as being an integrating factor, used to convert equation (2.1) to an exact differential equation.[94]

The solution of the heterogeneous equation (2.1) can be written in the form

$$y = \lambda(x)\Phi(x) \tag{2.4}$$

where $\lambda(x)$ is the solution to the homogeneous equation, i.e., equation (2.3). After simplification, equation (2.1) becomes

$$\lambda(x)\frac{d\Phi}{dx} = Q(x) \tag{2.5}$$

with solution

$$\Phi(x) = \int \left[Q(x)\exp\left[\int P(x)dx\right] \right] dx + C \tag{2.6}$$

The solution to equation (2.1) can be given as

$$y = \frac{\int Q(x)\exp\left[\int P(x)dx\right]dx}{\exp\left[\int Pdx\right]} + \frac{C}{\exp\left[\int Pdx\right]} \tag{2.7}$$

The right-hand side is a function of only x and can be integrated, and C is the constant of integration determined from the boundary condition. In the case that $x \geq 0$, the integration constant C may be expressed as $\Phi(0)$.

The method for solving linear first-order differential equations involves the following steps:

1. Derive the solution to the homogeneous equation $\lambda(x)$.

2. Derive a new unknown function $\Phi(x)$ such that $y = \lambda(x)\Phi(x)$, using numerical integration if a formal analytic solution is not possible.

3. Develop the expression for y.

The method is illustrated in the following examples.

Example 2.1 **Linear First-Order Differential Equation:** *In the development of boundary-layer mass transfer to a planar electrode, a similarity transformation variable (see Section 2.5 on page 38) can be identified through solution of*

$$\beta(x)g^2g' + \frac{1}{2}\beta'(x)g^3 = \ell \tag{2.8}$$

where ℓ is a constant and $g(0) = 0$. Find an expression for $g(x)$.

Solution: *Equation (2.8) can be placed in the form of equation (2.1) by introducing a new variable $f(x) = g(x)^3$ such that*

$$\frac{df}{dx} + \frac{3\beta'(x)}{2\beta(x)}f = \frac{3\ell}{\beta(x)} \tag{2.9}$$

Remember! 2.1 *The general solution to nonhomogeneous linear first-order differential equations can be obtained as the product of a function to be determined and the solution to the homogeneous equation.*

---

where $P(x) = 3\beta'(x)/2\beta(x)$ and $Q(x) = 3\ell/\beta(x)$. The solution to the homogeneous equation can be expressed, following equation (2.3), as

$$\lambda(x) = \exp\left(-\int_0^x \frac{3\beta'(x)}{2\beta(x)}\right) = \exp\left(-\frac{3}{2}\ln\beta\right) = \beta^{-3/2} \tag{2.10}$$

Following equation (2.5), the governing equation for $\Phi(x)$, defined by $f = \lambda(x)\Phi(x)$, is given by

$$\frac{1}{\beta^{3/2}}\frac{d\Phi}{dx} = \frac{3\ell}{\beta(x)} \tag{2.11}$$

or

$$\frac{d\Phi}{dx} = 3\ell\beta(x)^{1/2} \tag{2.12}$$

with solution

$$\Phi(x) = \int_0^x 3\ell\beta^{1/2}dx + \Phi(0) \tag{2.13}$$

As $f(x) = g(x)^3 = \lambda(x)\Phi(x)$,

$$f(x) = \frac{1}{\beta^{3/2}}\int_0^x 3\ell\beta^{1/2}dx + \frac{\Phi(0)}{\beta^{3/2}} \tag{2.14}$$

The constant of integration $\Phi(0)$ is evaluated from the requirement that $g(0) = 0$, and thus, $f(0) = 0$. The constant of integration has a value $\Phi(0) = 0$. Thus,

$$g(x) = \frac{1}{\sqrt{\beta}}\left[\int_0^x 3\ell\beta^{1/2}dx\right]^{1/3} \tag{2.15}$$

This example demonstrates the use of a simple variable transformation to convert a differential equation into a form suitable for application of the methods discussed in this section.

Example 2.2 Convective Diffusion Equation: *The conservation equation near a rotating disk electrode in an electrolyte solution where migration can be neglected can be written as*

$$v_y\frac{d\bar{c}_i}{dy} - D_i\frac{d^2\bar{c}_i}{dy^2} = 0 \tag{2.16}$$

with the boundary conditions

$$\bar{c}_i \to c_i(\infty) \quad \text{for} \quad y \to \infty$$
$$\bar{c}_i = c_i(0) \quad \text{at} \quad y = 0 \tag{2.17}$$

Find an expression for the concentration in the general case and when the normal velocity formula is limited to the first term of its development, i.e.,

$$v_y = -a\frac{\Omega^{3/2}}{\sqrt{\nu}}y^2 = -\alpha y^2 \tag{2.18}$$

where $a = 0.51023$.

Solution: *Equation (2.16) is homogeneous. By introducing the new function λ_i defined as $\lambda_i = d\bar{c}_i/dy$, the second-order differential equation becomes a first-order homogeneous equation written as*

$$v_y \lambda_i - D_i \frac{d\lambda_i}{dy} = 0 \tag{2.19}$$

The solution is

$$\lambda_i = \lambda_i(0) \exp \left(\int_0^y \frac{v_y}{D_i} dy \right) \tag{2.20}$$

where $\lambda_i(0)$ is a constant, representing the value of λ_i for $y = 0$. The solution for concentration $\bar{c}_i$ is

$$\bar{c}_i(y) = \int_0^y \lambda_i(0) \exp \left(-\frac{v_y}{D_i} \right) dy + \bar{c}_i(0) \tag{2.21}$$

The constant $\lambda_i(0)$ is determined according to the boundary conditions to be

$$\lambda_i(0) = \left. \frac{d\bar{c}_i}{dy} \right|_{y=0} = \frac{\bar{c}_i(\infty) - \bar{c}_i(0)}{\int_0^\infty \exp \left(\int_0^y (v_y/D_i)\, dy \right) dy} \tag{2.22}$$

If the normal velocity formula is limited to the first term of its development, i.e., as shown in equation (2.18), equation (2.22) becomes

$$\left. \frac{d\bar{c}_i}{dy} \right|_{y=0} = \frac{\bar{c}_i(\infty) - \bar{c}_i(0)}{\int_0^\infty \exp \left(-\frac{\alpha y^3}{3D_i} \right) dy} \tag{2.23}$$

The variable transformation $u = y^3$ can be used to show that

$$\int_0^\infty e^{-y^3} dy = \Gamma(4/3) = 0.89298 \tag{2.24}$$

where $\Gamma(n)$ represents a tabulated function defined by

$$\Gamma(n) = \int_0^\infty e^{-x} x^{n-1} dx = \frac{1}{n} \Gamma(n+1) \tag{2.25}$$

Thus, equation (2.23) can be expressed as

$$\left. \frac{d\bar{c}_i}{dy} \right|_{y=0} = \frac{\bar{c}_i(\infty) - \bar{c}_i(0)}{\left(\dfrac{3D_i}{\alpha} \right)^{1/3} \Gamma(4/3)} = \frac{\bar{c}_i(\infty) - \bar{c}_i(0)}{\delta_{N,i}} \tag{2.26}$$

where $\delta_{N,i}$ is the Nernst diffusion layer thickness.

Finally, the expression for concentration is given by

$$\bar{c}_i(y) = \frac{\bar{c}_i(\infty) - \bar{c}_i(0)}{\left(\frac{3D_i}{\alpha}\right)^{1/3} \Gamma(4/3)} \int_0^y \exp\left(-\frac{\alpha y^3}{3D_i}\right) dy + \bar{c}_i(0) \tag{2.27}$$

Equation (2.26) represents the starting point for the development of the Levich equation for mass-transfer-limited current to a disk electrode.

2.2 Homogeneous Linear Second-Order Differential Equations

The general homogeneous linear second-order differential equation with constant coefficients can be written as

$$y'' + P(x)y' + Q(x)y = 0 \tag{2.28}$$

Consider here the special case of linear second-order differential equations with constant coefficients, written in general form as

$$ay'' + by' + cy = 0 \tag{2.29}$$

A trial solution for equation (2.29) can be written as

$$y = e^{mx} \tag{2.30}$$

where m is a constant to be determined. Substitution into equation (2.29) yields

$$\left(am^2 + bm + c\right)e^{mx} = 0 \tag{2.31}$$

Equation (2.31) can be satisfied if and only if

$$\left(am^2 + bm + c\right) = 0 \tag{2.32}$$

Equation (2.32) is known as the characteristic or auxiliary equation of equation (2.29). The solution to the quadratic equation (2.32)

$$m = \frac{-b \pm \sqrt{b^2 - 4ac}}{2a} \tag{2.33}$$

provides values for m_1 and m_2. The solution to equation (2.29) is therefore

$$y = C_1 e^{m_1 x} + C_2 e^{m_2 x} \tag{2.34}$$

where C_1 and C_2 are constants to be determined from boundary conditions. If $m_1 = m_2$, equation (2.34) cannot provide a complete solution to equation (2.29). In this case, the solution can be shown to take the form

$$y = C_1 e^{m_1 x} + C_2 x e^{m_1 x} \tag{2.35}$$

In some cases, the roots can be complex. When the roots of the characteristic equation are complex, the general solution takes the form

$$y = C_1 e^{(p+jq)x} + C_2 e^{(p-jq)x} \tag{2.36}$$

where $m_1 = p + jq$ and $m_2 = p - jq$. It is usually convenient to use the Euler relations (1.93) and (1.94) to express the solution as

$$y = e^{px} \left(C_1 \cos(qx) + C_2 \sin(qx) \right) \tag{2.37}$$

where C_1 and C_2 are real constants to be determined from boundary conditions.

Example 2.3 Complex Roots for an ODE: *Find the roots for the differential equation*

$$\frac{d^2 y}{dx^2} + by = 0 \tag{2.38}$$

discussed in Section 1.2.1.

Solution: *Equation (2.38) has the characteristic equation*

$$m^2 = -b \tag{2.39}$$

with solution

$$m = \pm\sqrt{-b} = \pm j\sqrt{b} \tag{2.40}$$

The homogeneous solution to equation (2.38) is given by

$$y = C_1 \exp\left(j\sqrt{b}x \right) + C_2 \exp\left(-j\sqrt{b}x \right) \tag{2.41}$$

The need to find solutions to equations of the form of equation (2.38) is used to motivate the use of imaginary numbers in Section 1.2.1. Equation (2.38) arises in the analysis presented in Section 11.2.1 for diffusion impedance in the absence of convection.

Example 2.4 Diffusion in a Finite Domain: *The steady-state conservation equation for diffusion of species i in a finite medium can be expressed as*

$$D_i \frac{d^2 c_i}{dy^2} = 0 \tag{2.42}$$

with boundary conditions

$$c_i \to c_i(\infty) \quad \text{for} \quad y = \delta$$
$$c_i = c_i(0) \quad \text{at} \quad y = 0 \tag{2.43}$$

We seek an expression for the concentration profile.

Solution: *The solution to equation (2.42) can be obtained by direct integration to be*

$$c_i = \left.\frac{dc_i}{dy}\right|_{y=0} y + c_i(0) \tag{2.44}$$

Application of the boundary condition (2.43) at $y = \delta$ yields

$$c_i = c_i(0) + \frac{y}{\delta}\left(c_i(\infty) - c_i(0)\right) \quad \text{for} \quad 0 \le y \le \delta$$
$$c_i = c_i(\infty) \quad \text{for} \quad y \ge \delta \tag{2.45}$$

The development of the impedance response is presented in Section 11.2.1.

Example 2.5 Transmission Line: *The line voltage $\widetilde{u}(x)$ and the current $\widetilde{i}(x)$ for the transmission line shown in Figure 13.1 on page 320 can be expressed in the frequency domain as*

$$d\widetilde{u}(x) = -Z_1\widetilde{i}(x)dx \tag{2.46}$$

and

$$d\widetilde{i}(x) = -\frac{\widetilde{u}(x)}{Z_2}dx \tag{2.47}$$

respectively, where parameters Z_1 and Z_2 are independent of the position variable x. At $x = 0$, the current is equal to zero. We seek an expression for the impedance of the transmission line.

Solution: *Under the assumption that Z_1 and Z_2 are independent of x, equations (2.46) and (2.47) may be expressed as*

$$\frac{d^2\widetilde{u}(x)}{dx^2} = \frac{Z_1}{Z_2}\widetilde{u}(x) \tag{2.48}$$

The homogeneous solution to equation (2.48) is

$$\widetilde{u}(x) = A\exp\left(x\sqrt{Z_1/Z_2}\right) + B\exp\left(-x\sqrt{Z_1/Z_2}\right) \tag{2.49}$$

and the derivative of $\widetilde{u}(x)$ is expressed as

$$d\widetilde{u}(x) = \sqrt{Z_1/Z_2}\left(A\exp\left(x\sqrt{Z_1/Z_2}\right) - B\exp\left(-x\sqrt{Z_1/Z_2}\right)\right)dx \tag{2.50}$$

Remember! 2.2 *The solution for line voltage and current in a transmission line represents the foundation for modeling the impedance of porous electrodes and thin-layer cells.*

The condition that at $x = 0$, $\widetilde{i}(0) = 0$ means that, from equation (2.46), $\mathrm{d}\widetilde{u}(0) = 0$ and, from equation (2.50), $A = B$.

The overall current produced on a length ℓ is given by

$$\widetilde{i}(\ell) = \int_0^\ell \frac{\widetilde{u}(x)}{Z_2}\mathrm{d}x \tag{2.51}$$

$$= \frac{A}{Z_2}\sqrt{\frac{Z_2}{Z_1}}\left(\exp\left(\ell\sqrt{Z_1/Z_2}\right) + \exp\left(-\ell\sqrt{Z_1/Z_2}\right)\right)$$

The impedance is given by

$$Z = \frac{\widetilde{u}(\ell)}{\widetilde{i}(\ell)} = \frac{A\left(\exp\left(\ell\sqrt{Z_1/Z_2}\right) + \exp\left(-\ell\sqrt{Z_1/Z_2}\right)\right)}{\left(A/\sqrt{Z_1Z_2}\right)\left(\exp\left(\ell\sqrt{Z_1/Z_2}\right) - \exp\left(-\ell\sqrt{Z_1/Z_2}\right)\right)} \tag{2.52}$$

From the hyperbolic relationships for complex variables given in Table 1.6 on page 18,

$$Z = \sqrt{Z_1Z_2}\coth\left(\ell\sqrt{Z_1/Z_2}\right) \tag{2.53}$$

The development presented here supports the treatment of transmission lines presented in Section 13.1 on page 319.

2.3 Nonhomogeneous Linear Second-Order Differential Equations

The nonhomogeneous linear second-order differential equations with constant coefficients can be written in general form as

$$ay'' + by' + cy = f(x) \tag{2.54}$$

If $f(x) = 0$, the equation is called homogeneous. If $f(x) \neq 0$, the equation is called nonhomogeneous. The general solution to equation (2.54) is given as the product

$$y = \lambda(x)\Phi(x) \tag{2.55}$$

where $\lambda(x)$ is the solution to the homogeneous equation.

The homogeneous solution $\lambda(x)$ can be obtained using the methods described in Section 2.2. The unknown function $\Phi(x)$ is defined such that

$$y = \lambda(x)\Phi(x) \tag{2.56}$$

$$y' = \lambda(x)\Phi(x)' + \lambda(x)'\Phi(x) \tag{2.57}$$

$$y'' = \lambda(x)\Phi(x)'' + 2\lambda(x)'\Phi(x)' + \lambda(x)''\Phi(x) \tag{2.58}$$

Remember! 2.3 *The general solution to nonhomogeneous linear second-order differential equations with constant coefficients can be obtained as the product of function to be determined and the solution to the homogeneous equation (see equation (2.55)).*

Equation (2.54) becomes

$$a\left(\lambda\Phi'' + 2\lambda'\Phi' + \lambda''\Phi\right) + b\left(\lambda\Phi' + \lambda'\Phi\right) + c\left(\lambda\Phi\right) = f \tag{2.59}$$

As λ is the solution to the homogeneous equation, equation (2.60) is reduced to

$$a\left(2\lambda'\Phi' + \lambda\Phi''\right) + b\left(\lambda\Phi'\right) = f \tag{2.60}$$

Equation (2.60) can be solved by a reduction of order technique, i.e., by letting $\mu = \Phi'$, such that

$$\mu'a\lambda + \mu\left(2a\lambda' + b\lambda\right) = f \tag{2.61}$$

The resulting first-order equation may be solved numerically if analytic solutions are not available.

The method described here is general and can be applied to higher-order differential equations. The method provides an attractive alternative to the use of particular solutions obtained using trial solutions based on the form of the function $f(x)$ and, in some cases, on the form of the homogeneous solution.[90]

Example 2.6 Accurate Solution for the Convective Diffusion Equation: *In Example 2.2, the velocity expression v_y was limited to the first term of its development (see equation (11.51)). In the present case, the three first terms of the velocity development are taken into account, and then equation (2.16) can be written as*

$$\frac{d^2c_i}{d\zeta^2} + \left(3\zeta^2 - \left(\frac{3}{a^4}\right)^{1/3}\frac{\zeta^3}{Sc_i^{1/3}} - \frac{b}{6}\left(\frac{3}{a}\right)^{5/3}\frac{\zeta^4}{Sc_i^{2/3}}\right)\frac{dc_i}{d\zeta} = 0 \tag{2.62}$$

where $\zeta = y/\delta_i$ and

$$\delta_i = \left(\frac{3}{a}\right)^{1/3}\frac{1}{Sc_i^{1/3}}\sqrt{\frac{v}{\Omega}} \tag{2.63}$$

The boundary conditions are

$$c_i \to c_i(\infty) \quad \text{for} \quad y \to \infty$$
$$c_i = c_i(0) \quad \text{at} \quad y = 0 \tag{2.64}$$

By using a solution in the form

$$c_i = c_{i,0} + \frac{1}{Sc_i^{1/3}}c_{i,1} + \frac{1}{Sc_i^{2/3}}c_{i,2} \tag{2.65}$$

valid for a large value of Sc_i, find an expression for the concentration gradient at the wall.

Solution: *Following Example 2.2, introduction of a new function $\lambda_i = dc_i/d\zeta$ and equation (2.65) yields three differential equations*

$$\frac{d\lambda_{i,0}}{d\zeta} + 3\zeta^2\lambda_{i,0} = 0 \tag{2.66}$$

$$\frac{d\lambda_{i,1}}{d\xi} + 3\xi^2\lambda_{i,1} = \left(\frac{3}{a^4}\right)^{1/3}\xi^3\lambda_{i,0} \tag{2.67}$$

and

$$\frac{d\lambda_{i,2}}{d\xi} + 3\xi^2\lambda_{i,2} = \left(\frac{3}{a^4}\right)^{1/3}\xi^3\lambda_{i,1} + \frac{b}{6}\left(\frac{3}{a}\right)^{5/3}\xi^4\lambda_{i,0} \tag{2.68}$$

where $\lambda_{i,0} = dc_{i,0}/d\xi$, $\lambda_{i,1} = dc_{i,1}/d\xi$, and $\lambda_{i,2} = dc_{i,2}/d\xi$. Equations (2.66), (2.67), and (2.68) represent, respectively, the collected terms of order (1), $(Sc_i^{-1/3})$ and $(Sc_i^{-2/3})$.

Equation (2.66) is a homogeneous equation, and the solution is obtained as shown in Example 2.2 to be

$$\lambda_{i,0} = \frac{dc_{i,0}}{d\xi} = \frac{(c_i(\infty) - c_i(0))}{\Gamma(4/3)}\exp\left(-\xi^3\right) \tag{2.69}$$

where $\lambda_{i,0}$ represents the gradient corresponding to the first term in the concentration expansion. The Γ function is defined by equation (2.25).

The solutions for the homogeneous equations corresponding to equations (2.67) and (2.68) are of the same form, i.e., $\exp(-\xi^3)$. Following the development presented in this section, a new function is introduced of the form $\lambda_{i,1} = \exp(-\xi^3)\Phi_1(\xi)$. Noting that

$$\frac{d\lambda_{i,1}}{d\xi} = -3\xi^2 e^{-\xi^3}\Phi_1 + e^{-\xi^3}\frac{d\Phi_1}{d\xi} \tag{2.70}$$

equation (2.67) becomes

$$\frac{d\Phi_1}{d\xi} = \left(\frac{3}{a^4}\right)^{1/3}\xi^3\frac{c_i(\infty) - c_i(0)}{\Gamma(4/3)} \tag{2.71}$$

yielding

$$\Phi_1(\xi) = \left(\frac{3}{a^4}\right)^{1/3}\frac{c_i(\infty) - c_i(0)}{\Gamma(4/3)}\left(\frac{\xi^4}{4}\right) + \Phi_1(0) \tag{2.72}$$

where $\Phi_1(0)$ is a constant of integration. The concentration term is then given by

$$c_{i,1} = \left(\frac{3}{a^4}\right)^{1/3}\frac{c_i(\infty) - c_i(0)}{4\Gamma(4/3)}\int_0^\xi e^{-\xi^3}\xi^4 d\xi + \int_0^\xi e^{-\xi^3}\Phi_1(0)d\xi \tag{2.73}$$

The constant of integration may be evaluated by observing that $c_{i,1}(\infty) = 0$. Thus,

$$\Phi_1(0) = -\left(\frac{3}{a^4}\right)^{1/3}\frac{c_i(\infty) - c_i(0)}{4\Gamma(4/3)}\frac{\Gamma(5/3)}{3\Gamma(4/3)} \tag{2.74}$$

An evaluation of the concentration derivative at the electrode surface yields

$$\left.\frac{dc_i}{d\xi}\right|_{\xi=0} = \left.\frac{dc_{i,0}}{d\xi}\right|_{\xi=0} + \frac{1}{Sc_i^{1/3}}\left.\frac{dc_{i,1}}{d\xi}\right|_{\xi=0} \frac{c_i(\infty) - c_i(0)}{\Gamma(4/3)}\left(1 - \frac{0.29801}{Sc_i^{1/3}}\right) \tag{2.75}$$

or

$$\left.\frac{dc_i}{d\xi}\right|_{\xi=0} = \frac{c_i(\infty) - c_i(0)}{\Gamma(4/3)}\frac{1}{1 + 0.29801Sc_i^{-1/3}} \tag{2.76}$$

Equation (2.76) represents the correction to the Levich equation to account for two terms in the velocity expansion. For a Schmidt number of 1,000, the second term in the expansion accounts for about 3 percent of the value of the derivative.

The extension to account for the third term in the velocity expansion requires solution of equation (2.68). Introduction of $\lambda_{i,2} = \exp(-\zeta^3)\Phi_2(\zeta)$ into equation (2.68), after cancelation of the $\exp(-\zeta^3)$ term, results in

$$\frac{d\Phi_2}{d\zeta} = \left(\frac{3}{a^4}\right)^{2/3}\frac{\zeta^7}{4}\frac{c_i(\infty) - c_i(0)}{\Gamma(4/3)} + \left(\frac{3}{a^4}\right)^{1/3}\zeta^3\Phi_1(0) \tag{2.77}$$
$$+\frac{b}{6}\left(\frac{3}{a}\right)^{5/3}\zeta^4\frac{c_i(\infty) - c_i(0)}{\Gamma(4/3)}$$

Integration yields

$$\Phi_2(\zeta) = \left(\frac{3}{a^4}\right)^{2/3}\frac{\zeta^8}{32}\frac{c_i(\infty) - c_i(0)}{\Gamma(4/3)} + \left(\frac{3}{a^4}\right)^{1/3}\frac{\zeta^4}{4}\Phi_1(0) \tag{2.78}$$
$$+\frac{b}{6}\left(\frac{3}{a}\right)^{5/3}\frac{\zeta^5}{5}\frac{c_i(\infty) - c_i(0)}{\Gamma(4/3)} + \Phi_2(0)$$

where $\Phi_2(0)$ is the constant of integration. The concentration term is given by

$$c_{i,2} = \int_0^\zeta e^{-\zeta^3}\Phi_2(\zeta)d\zeta \tag{2.79}$$

or

$$c_{i,2} = \frac{c_i(\infty) - c_i(0)}{\Gamma(4/3)}\left(\frac{1}{32}\left(\frac{3}{a^4}\right)^{2/3}\int_0^\zeta \zeta^8 e^{-\zeta^3}d\zeta\right. \tag{2.80}$$
$$\left. -\left(\frac{3}{a^4}\right)^{2/3}\frac{\Gamma(5/3)}{48\Gamma(4/3)}\int_0^\zeta \zeta^4 e^{-\zeta^3}d\zeta + \frac{b}{30}\left(\frac{3}{a}\right)^{5/3}\int_0^\zeta \zeta^5 e^{-\zeta^3}d\zeta\right)$$
$$+\Phi_2(0)\int_0^\zeta e^{-\zeta^3}d\zeta$$

The constant of integration can be evaluated under the assumption that $c_{i,2}(\infty) = 0$ to be

$$\Phi_2(0) = \frac{c_i(\infty) - c_i(0)}{\Gamma(4/3)}\left(\frac{1}{32}\left(\frac{3}{a^4}\right)^{2/3}\frac{\Gamma(3)}{3\Gamma(4/3)}\right. \tag{2.81}$$
$$\left. -\left(\frac{3}{a^4}\right)^{2/3}\frac{\Gamma(5/3)}{48\Gamma(4/3)}\frac{\Gamma(5/3)}{3\Gamma(4/3)} + \frac{b}{30}\left(\frac{3}{a}\right)^{5/3}\frac{\Gamma(2)}{3\Gamma(4/3)}\right)$$

where $\Gamma(3) = 2$ and $\Gamma(2) = 1$. The quantities within the brackets can be evaluated to yield

$$\Phi_2(0) = \frac{c_i(\infty) - c_i(0)}{\Gamma(4/3)}(0.056337) \tag{2.82}$$

An evaluation of the concentration derivative at the electrode surface yields

$$\frac{dc_i}{d\zeta}\bigg|_{\zeta=0} = \frac{c_i(\infty) - c_i(0)}{\Gamma(4/3)}\left(1 - \frac{0.29801}{Sc_i^{1/3}} - \frac{0.05634}{Sc_i^{2/3}}\right) \tag{2.83}$$

or

$$\frac{dc_i}{d\zeta}\bigg|_{\zeta=0} = \frac{1}{\Gamma(4/3)}\left(\frac{c_i(\infty) - c_i(0)}{1 + 0.29801Sc_i^{-1/3} + 0.14515Sc_i^{-2/3} + \mathcal{O}(Sc_i^{-1})}\right) \tag{2.84}$$

Equation (2.84) represents the correction to the Levich equation to account for three terms in the velocity expansion. For a Schmidt number of 1,000, the third term in the expansion accounts for about 0.06 percent of the value of the derivative.

2.4   Chain Rule for Coordinate Transformations

Solution of differential equations is often facilitated by coordinate transformations. Such tranformations are implemented by use of the chain rule summarized in this section. A more complete development can be found in calculus textbooks.[95]

The chain rule provides a means for computing the derivative of the composition of two or more functions. For example, if $y = f(u)$ and $u = g(x)$, the derivative of y with respect to x can be expressed as

$$\frac{dy}{dx} = \frac{dy}{du}\frac{du}{dx} \tag{2.85}$$

where y and u are differentiable functions. This treatment can be expanded for more than two functions.

The chain rule may also be applied to functions of more than one independent variable. For example, if $z = f(x,y)$, the derivative of z with respect to x is obtained from

$$\frac{dz}{dx} = \left(\frac{\partial z}{\partial x}\right)_y \frac{dx}{dx} + \left(\frac{\partial z}{\partial y}\right)_x \frac{dy}{dx} \tag{2.86}$$

or

$$\frac{dz}{dx} = \left(\frac{\partial z}{\partial x}\right)_y + \left(\frac{\partial z}{\partial y}\right)_x \frac{dy}{dx} \tag{2.87}$$

If x and y are orthogonal, then

$$\frac{dy}{dx} = 0 \tag{2.88}$$

and

$$\frac{dz}{dx} = \left(\frac{\partial z}{\partial x}\right)_y \tag{2.89}$$

If, on the other hand, x and y are not orthogonal, then

$$\frac{dy}{dx} \neq 0 \tag{2.90}$$

and all terms in equation (2.87) must be calculated.

Example 2.7 **Diffusion with Changing Film Thickness:** *Diffusion in a film is governed by*

$$\frac{\partial c_i}{\partial t} - D_i \frac{\partial^2 c_i}{\partial y^2} = 0 \tag{2.91}$$

The diffusion impedance response for a film of constant thickness is given in Section 11.2 on page 245. The solution for a film of variable thicknesss $\delta(t)$ can be facilitated by use of a coordinate variable $\chi = y/\delta(t)$. Find the form of equation (2.91) written in terms of y, t, and χ.

Solution: *The transformation of equation (2.91) can be found by sequential application of the chain rule to the terms in the equation. Following equation (2.87),*

$$\left(\frac{\partial c_i}{\partial t} \right)_y = \left(\frac{\partial c_i}{\partial t} \right)_\chi + \left(\frac{\partial c_i}{\partial \chi} \right)_t \frac{d\chi}{dt} \tag{2.92}$$

expansion of the second term of the right-hand side of equation (2.92) yields

$$\left(\frac{\partial c_i}{\partial \chi} \right)_t = \frac{dc_i}{dy} \frac{dy}{d\chi} = \delta(t) \frac{dc_i}{dy} \tag{2.93}$$

where dc_i/dy represents the steady-state concentration gradient within the film, and

$$\frac{d\chi}{dt} = -\frac{y}{\delta(t)^2} \frac{d\delta}{dt} = -\frac{\chi}{\delta(t)} \frac{d\delta}{dt} \tag{2.94}$$

Thus,

$$\left(\frac{\partial c_i}{\partial t} \right)_y = \left(\frac{\partial c_i}{\partial t} \right)_\chi - \chi \frac{dc_i}{dy} \frac{d\delta}{dt} \tag{2.95}$$

Evaluation of the second derivative in equation (2.91) yields

$$\left(\frac{\partial^2 c_i}{\partial y^2} \right)_t = \left(\frac{\partial^2 c_i}{\partial \chi^2} \right)_t \left(\frac{d\chi}{dy} \right)^2 = \frac{1}{\delta(t)^2} \left(\frac{\partial^2 c_i}{\partial \chi^2} \right) \tag{2.96}$$

The transformed equation (2.91) can now be expressed as

$$\frac{\partial c_i}{\partial t} - \chi \frac{dc_i}{dy} \frac{d\delta}{dt} = \frac{D_i}{\delta(t)^2} \left(\frac{\partial^2 c_i}{\partial \chi^2} \right) \tag{2.97}$$

This coordinate transformation was employed in the development of the impedance model presented in Section 11.8 on page 291 for a system influenced by modulation of a salt film thickness.

Example 2.8 **Convective Diffusion with Changing Film Thickness:** *The conservation equation for convective diffusion to a film-covered rotating disk can be expressed in dimensionless coordinates as*

$$\frac{\partial c_i}{\partial \tau_i} - 3 \left(\xi - \frac{\delta(\tau_i)}{\delta_{N,i}} \right)^2 \frac{\partial c_i}{\partial \xi} - \frac{\partial^2 c_i}{\partial \xi^2} = 0 \tag{2.98}$$

where $\tau_i = tD_i/\delta_{N,i}^2$, $\zeta = y/\delta_{N,i}$ and $\delta_{N,i}$ is given by equation (11.72) on page 262. The convective diffusion impedance response associated with modulation of the film thickness $\delta(t)$ can be facilitated by use of a coordinate variable $\eta = \zeta - \delta(\tau_i))/\delta_{N,i}$. Find the form of equation (2.98) written in terms of η and τ_i.

Solution: *The approach is similar to that presented in Example 2.7. Following equation (2.87),*

$$\left(\frac{\partial c_i}{\partial \tau_i}\right)_\zeta = \left(\frac{\partial c_i}{\partial \tau_i}\right)_\eta + \left(\frac{\partial c_i}{\partial \eta}\right)_{\tau_i} \frac{d\eta}{d\tau_i} \tag{2.99}$$

As

$$\frac{d\eta}{d\tau_i} = -\frac{1}{\delta_{N,i}} \frac{d\delta}{d\tau_i} \tag{2.100}$$

equation (2.99) can be expressed as

$$\left(\frac{\partial c_i}{\partial \tau_i}\right)_\zeta = \left(\frac{\partial c_i}{\partial \tau_i}\right)_\eta - \frac{1}{\delta_{N,i}} \frac{d\bar{c}_i}{d\eta} \frac{d\delta}{d\tau_i} \tag{2.101}$$

Evaluation of the first derivative with respect to position in equation (2.98) yields

$$\frac{\partial c_i}{\partial \zeta} = \frac{\partial c_i}{\partial \eta} \tag{2.102}$$

And the second derivative yields

$$\frac{\partial^2 c_i}{\partial \zeta^2} = \frac{\partial^2 c_i}{\partial \eta^2} \tag{2.103}$$

The transformed equation (2.98) can now be expressed as

$$\frac{\partial c_i}{\partial \tau_i} - \frac{1}{\delta_{N,i}} \frac{d\bar{c}_i}{d\eta} \frac{d\delta}{d\tau_i} - 3\eta^2 \frac{\partial c_i}{\partial \eta} - \frac{\partial^2 c_i}{\partial \eta^2} = 0 \tag{2.104}$$

This coordinate transformation was employed in the development of the impedance model presented in Section 11.8 on page 291 for a system influenced by modulation of a salt film thickness.

2.5   Partial Differential Equations by Similarity Transformations

Partial differential equations are generally solved by finding a transformation that allows the partial differential equation to be converted into two ordinary differential equations. A number of techniques are available, including separation of variables, Laplace transforms, and the method of characteristics.

A similarity transformation may be used when the solution to a parabolic partial differential equation, written in terms of two independent variables, can be expressed in terms of a new independent variable that is a combination of the original independent variables. The success of this transformation requires that:

- the governing partial differential equation can be expressed as an ordinary differential equation in terms of the similarity variable,

- the three boundary conditions for the original partial differential equation collapse to form two boundary conditions in terms of the similarity variable, and

- the original independent variables appear in neither the transformed ordinary differential equation nor the transformed and collapsed boundary conditions.

Similarity transformations are often used when the same condition applies when one independent variable is equal to zero and the other independent variable tends toward infinity.

The classic problem of diffusion in an infinite medium can be solved by use of a similarity transformation. The associated impedance response is discussed in Example 11.1 on page 249. A general method for finding the time-dependent concentration profile is presented here in the form of an example.

Example 2.9 **Diffusion in an Infinite Domain:** *The conservation equation for diffusion of species i in an infinite medium can be expressed as*

$$\frac{\partial c_i}{\partial t} - D_i \frac{\partial^2 c_i}{\partial y^2} = 0 \tag{2.105}$$

with boundary conditions

$$\begin{aligned} c_i &\rightarrow c_i(\infty) & \text{for} \quad y &\rightarrow \infty \\ c_i &= c_i(0) & \text{at} \quad y &= 0 \\ c_i &= c_i(\infty) & \text{at} \quad t &= 0 \end{aligned} \tag{2.106}$$

We seek an expression for the time-dependent concentration profile.

Solution: *The method for solving partial differential equations generally involves finding a method to express them as coupled ordinary differential equations. A similarity transformation is possible if c_i can be expressed as a function of only a new variable. This requirement implies that equation (2.105) can be expressed as a function of only the new variable, and that the three conditions (2.106) in time and position can collapse into two conditions in the new variable.*

Remember! 2.4 *Similarity transformations may be used for parabolic partial differential equations when the same condition applies when one independent variable is equal to zero and the other independent variable tends toward infinity.*

The observation that the same condition for c_i applies at $t = 0$ and $y \rightarrow \infty$ suggests that, through a transformation variable of the form $\eta = f(y)/g(t)$, the two boundary conditions at $y = 0$ and $y \rightarrow \infty$ and the initial condition at $t = 0$ could collapse into boundary conditions at $\eta = 0$ (corresponding to $y = 0$) and $\eta \rightarrow \infty$ (corresponding to both $t = 0$ and $y \rightarrow \infty$).

As a trial transformation, let η be given by

$$\eta = \frac{y}{g(t)} \tag{2.107}$$

Application of the chain rule yields

$$\frac{\partial c_i}{\partial t} = \frac{dc_i}{d\eta}\frac{\partial \eta}{\partial t} = \frac{dc_i}{d\eta}\left(-\frac{y}{g^2}\frac{dg}{dt}\right) \tag{2.108}$$

$$\frac{\partial c_i}{\partial y} = \frac{dc_i}{d\eta}\frac{\partial \eta}{\partial y} = \frac{dc_i}{d\eta}\left(\frac{1}{g}\right) \tag{2.109}$$

and

$$\frac{\partial^2 c_i}{\partial y^2} = \frac{d}{d\eta}\left(\frac{dc_i}{d\eta}\right)\frac{1}{g}\frac{\partial \eta}{\partial y} = \frac{d^2 c_i}{d\eta^2}\left(\frac{1}{g^2}\right) \tag{2.110}$$

Introduction of equations (2.107) and (2.110) into equation (2.105) yields

$$\frac{d^2 c_i}{d\eta^2} + \left(\frac{g}{D_i}\frac{dg}{dt}\right)\eta\frac{dc_i}{d\eta} = 0 \tag{2.111}$$

Equation (2.111) will be a function of only η if g satisfies

$$\frac{g}{D_i}\frac{dg}{dt} = \lambda \tag{2.112}$$

where λ is a constant to be determined.

The problem now consists of finding the solution to two ordinary differential equations,

$$\frac{d^2 c_i}{d\eta^2} + \lambda\eta\frac{dc_i}{d\eta} = 0 \tag{2.113}$$

and equation (2.112). The boundary condition for equation (2.112) is that $g = 0$ for $t = 0$. This will allow the conditions at $y \rightarrow \infty$ and $t = 0$ to apply for $\eta = \infty$. The boundary conditions for equation (2.113) are

$$c_i \rightarrow c_i(\infty) \quad \text{for} \quad \eta \rightarrow \infty$$

$$c_i = c_i(0) \quad \text{at} \quad \eta = 0 \tag{2.114}$$

Equation (2.112) is a linear first-order differential equation. It can be solved by use of a variable transformation $h = g^2$ such that

$$\frac{dh}{dt} = \frac{dg^2}{dt} = 2g\frac{dg}{dt} \tag{2.115}$$

Thus

$$\frac{dh}{dt} = 2D_i\lambda \tag{2.116}$$

which can be integrated directly to obtain

$$h = g^2 = 2D_i\lambda t + h(0) \tag{2.117}$$

The requirement that $g = 0$ for $t = 0$ yields $h(0) = 0$; thus,

$$g = \sqrt{2D_i\lambda t} \tag{2.118}$$

and

$$\eta = \frac{y}{\sqrt{2D_i\lambda t}} \tag{2.119}$$

In this case, equation (2.115) could be solved by a straightforward direct integration. Some other cases will require use of the integrating factor discussed in Section 2.1.

Equation (2.113) is a linear second-order differential equation. It can be solved by reduction of order. Let

$$P = \frac{dc_i}{d\eta} \tag{2.120}$$

such that

$$\frac{dP}{d\eta} + \lambda\eta P = 0 \tag{2.121}$$

Integration yields

$$P = P(0)e^{-\lambda\eta^2/2} \tag{2.122}$$

where $P(0)$ is a constant of integration. A second integration performed after substitution of equation (2.120) into equation (2.122) yields

$$c_i = P(0)\int_0^\eta e^{-\lambda\eta^2/2}d\eta + c_i(0) \tag{2.123}$$

where $c_i(0)$ is a constant of integration. The integrals can be placed in standard form by allowing the arbitrary constant to have a value $\lambda = 2$. Evaluation of conditions (2.114) yields

$$\frac{(c_i - c_i(0))}{(c_i(\infty) - c_i(0))} = \frac{\int_0^\eta e^{-\eta^2}d\eta}{\int_0^\infty e^{-\eta^2}d\eta} \tag{2.124}$$

or

$$\frac{(c_i - c_i(0))}{(c_i(\infty) - c_i(0))} = \frac{2}{\sqrt{\pi}}\int_0^\eta e^{-\eta^2}d\eta \tag{2.125}$$

The ratio of integrals appearing here is tabulated and is known as the error function, erf.

$$\frac{(c_i - c_i(0))}{(c_i(\infty) - c_i(0))} = 1 - \text{erf}\left(\frac{y}{\sqrt{4D_i t}}\right) \tag{2.126}$$

This is the solution presented in Example 11.1 on page 249.

2.6   Differential Equations with Complex Variables

In the course of developing models for the impedance response of physical systems, differential equations are commonly encountered that have complex variables. For equations with constant coefficients, solutions may be obtained using the methods described in the previous sections. For equations with variable coefficients, a numerical solution may be required. The method for numerical solution is to separate the equations into real and imaginary parts and to solve them simultaneously.

   An example is provided to illustrate the method.

Example 2.10 Foundation for Warburg Impedance: *The following differential equation appears in the treatment of impedance associated with diffusion:*

$$\frac{d^2 \theta_i}{d \zeta^2} - jK_i\theta_i = 0 \tag{2.127}$$

where θ_i is a dimensionless complex number representing the oscillating contribution to concentration, ζ is a dimensionless position, and K_i is a frequency made dimensionless using the diffusion coefficient for species i. The complete development for this equation can be found in Example 11.1 on page 249.

Solution: *Equation (2.127) is a linear second-order homogeneous differential equation with constant coefficients. It can be solved using the characteristic equation*

$$m^2 - jK_i = 0 \tag{2.128}$$

with solution

$$m = \pm\sqrt{jK_i} \tag{2.129}$$

The corresponding solution takes the form

$$\theta_i = A_i e^{\zeta\sqrt{jK_i}} + B_i e^{-\zeta\sqrt{jK_i}} \tag{2.130}$$

where constants A_i and B_i are to be determined so as to satisfy the boundary conditions. For diffusion in an infinite medium, the boundary conditions are

$$\theta_i = 0 \quad at \quad \zeta \to \infty \tag{2.131}$$

Remember! 2.5 *The method for numerical solution of differential equations with complex variables is to separate the equations into real and imaginary parts and to solve them simultaneously.*

---

and

$$\theta_i = 1 \quad \text{at} \quad \xi = 0 \tag{2.132}$$

Condition (2.131) requires that $A_i = 0$, and condition (2.132) provides that $B_i = 1$. Thus,

$$\theta_i = e^{-\xi \sqrt{jK_i}} \tag{2.133}$$

For diffusion in a finite domain, condition (2.131) is replaced by

$$\theta_i = 0 \quad \text{at} \quad \xi = 1 \tag{2.134}$$

In this case, application of conditions (2.132) and (2.134) provides that

$$0 = A_i e^{\sqrt{jK_i}} + B_i e^{-\sqrt{jK_i}} \tag{2.135}$$

and

$$1 = A_i + B_i \tag{2.136}$$

The resulting solution is

$$\theta_i = \frac{\sinh\left((\xi - 1)\sqrt{jK_i}\right)}{\sinh\left(-\sqrt{jK_i}\right)} \quad \text{for} \quad \xi < 1$$
$$\theta_i = 0 \quad \text{for} \quad \xi \geq 1 \tag{2.137}$$

As an exercise, the reader can verify that equation (2.130) satisfies both real and imaginary parts of equation (2.127). This development represents the starting point for both the Warburg impedance associated with diffusion in a stationary medium of infinite depth and the diffusion impedance associated with a stationary medium of finite depth.

Problems

2.1 Find the general solution for c as a function of y for:

(a) $\frac{dc}{dy} + kc = 0$

(b) $\frac{d^2c}{dy^2} + kc = 0$

(c) $\frac{d^2c}{dy^2} + a\frac{dc}{dy} + kc = 0$

(d) $\frac{d^2c}{dy^2} + a\frac{dc}{dy} + kc = b$

2.2 Show that equation (2.130) satisfies both real and imaginary parts of equation (2.127).

2.3 Verify that equation (2.137) represents the solution for equation (2.127) for diffusion in a finite domain.

2.4 Verify that equation (2.133) represents the solution for equation (2.127) for diffusion in an infinite domain.

Chapter 3

Statistics

A working understanding of statistics is essential for the analysis of experiments conducted in the frequency domain, such as impedance spectroscopy. The objective of this chapter is to provide an overview of concepts and definitions used in statistics at a level sufficient to understand the interpretation of frequency domain data.

3.1 Definitions

Some basic statistical concepts are defined in this section. For a more comprehensive treatment, the reader is directed to standard statistical texts.[96–98]

3.1.1 Expectation and Mean

The sample mean of a quantity x_k, sampled $k = 1 \ldots n_x$ times, is given as

$$\overline{x} = \frac{1}{n_x} \sum_{k=1}^{n_x} x_k \tag{3.1}$$

The quantity x is called a variate and has values x_k that are randomly distributed. As n_x approaches the total population, the sample mean approaches the mean or expectation, i.e., $\mu_x = E\{x\}$. Some useful properties of the expectations of constants and variates are given in Table 3.1.

3.1.2 Variance, Standard Deviation, and Covariance

The sample variance of a quantity x_k, sampled $k = 1 \ldots n_x$ times, is given as

$$s_x^2 = \frac{1}{n_x - 1} \sum_{k=1}^{n_x} (x_k - \overline{x})^2 \tag{3.7}$$

The variance of a discrete population n_t is given in terms of expectations as

$$\sigma_x^2 = E\left\{(x - E\{x\})^2\right\} \tag{3.8}$$

Table 3.1: Properties of the expectation where c is a constant and x and y are random variates.

$$E\{c\} = c \tag{3.2}$$
$$E\{x + c\} = E\{x\} + c \tag{3.3}$$
$$E\{cx\} = cE\{x\} \tag{3.4}$$
$$E\{x + y\} = E\{x\} + E\{y\} \tag{3.5}$$
$$E\{(x + y)^2\} = E\{x\}^2 + E\{y\}^2 + 2E\{x\}E\{y\} \tag{3.6}$$

As n_x approaches the total population, the sample variance approaches the variance, i.e., $s_x^2 \to \sigma_x^2$. For an unbounded population, the sample variance approaches the variance for $n_x \to \infty$.

The sample standard deviation is given by the square root of the sample variance, i.e.,

$$s_x = \sqrt{s_x^2} \tag{3.9}$$

Similarly,

$$\sigma_x = \sqrt{\sigma_x^2} \tag{3.10}$$

Both the standard deviation and variance are positive definite. The standard error

$$SE_x = \frac{s_x}{\sqrt{n_x}} \tag{3.11}$$

is the sample standard deviation scaled by the square root of the sample size n_x. The 95.45% confidence interval for the mean is given by

$$\mu_x = \overline{x} \pm 2\frac{s_x}{\sqrt{n_x}} \tag{3.12}$$

or

$$\overline{x} - 2\frac{s_x}{\sqrt{n_x}} \leq \mu_x \leq \overline{x} + 2\frac{s_x}{\sqrt{n_x}} \tag{3.13}$$

Thus, while the sample standard deviation is governed by the distribution of the population, the size of the confidence interval is governed by both the population distribution and the number of variates sampled.

Remember! 3.1 *The sample standard deviation is governed by the distribution of the population; whereas, the size of the confidence interval is governed by both the population distribution and the number of variates sampled.*

Table 3.2: Properties of the variance where c is a constant and x and y are random variates.

$$\sigma^2(c) = 0 \tag{3.16}$$
$$\sigma^2(x + c) = \sigma_x^2 \tag{3.17}$$
$$\sigma^2(cx) = c^2\sigma_x^2 \tag{3.18}$$
$$\sigma^2(x + y) = \sigma_x^2 + \sigma_y^2 + \sigma_{x,y} \tag{3.19}$$
$$\tag{3.20}$$

The sample covariance is given by

$$s_{x_1 x_2} = \frac{1}{n_x - 1} \sum_{k=1}^{n_x} (x_{1,k} - \overline{x_1})(x_{2,k} - \overline{x_2}) \tag{3.14}$$

and the covariance can be defined to be

$$\sigma_{x_1 x_2} = E\left\{(x_{1,k} - E\{x_1\})(x_{2,k} - E\{x_2\})\right\} \tag{3.15}$$

Unlike the variance, the covariance may have positive or negative values. If x_1 and x_2 are not correlated, $\sigma_{x_1 x_2} = 0$. The techniques discussed in Section 3.2 can be used to demonstrate the properties of the variance given in Table 3.2.

Example 3.1 Mean and Variance for Replicated Measurements: *The data presented in Table 3.3 were obtained at a frequency of 1.2 Hz from replicated impedance measurements for the reduction of ferricyanide on a Pt rotating disk electrode.[99] Calculate the mean and variance for each column of numbers, and calculate the covariance for the real and imaginary parts of the impedance and for the real and imaginary residual errors.*

Solution: *The calculated mean, standard deviation, and standard error are presented in Table 3.4 for the real and imaginary parts of the impedance and for the real and imaginary parts of the residual errors. Following equation (3.12), the 95.45% confidence interval for the mean of the real part of the impedance may be expressed as*

$$\mu_{Z_r} = 48 \pm 1.3 \ \Omega\text{cm}^2 \tag{3.21}$$

Similarly, the 95.45% confidence intervals for the imaginary part of the impedance and the real and imaginary residual errors may be expressed as

$$\mu_{Z_j} = -28.7 \pm 0.13 \ \Omega\text{cm}^2 \tag{3.22}$$

$$\mu_{\epsilon_r} = 0.060 \pm 0.025 \ \Omega\text{cm}^2 \tag{3.23}$$

Table 3.3: Results of replicated impedance measurements for the reduction of ferricyanide on a Pt rotating disk electrode.[99] Data were obtained at a rotation rate of 120 rpm and at a frequency of 1.2 Hz. Residual errors were obtained by regression of a measurement model.[69, 100, 101]

Number	Z_r,Ωcm^2	$-Z_j$,Ωcm^2	ϵ_r,Ωcm^2	ϵ_j,Ωcm^2
2	42.503	28.038	-0.01632	-0.36455
3	43.102	28.488	-0.08698	-0.00219
4	43.963	28.276	0.15081	-0.12454
5	44.715	28.228	0.21895	-0.24488
6	45.200	28.292	0.07457	-0.18878
7	45.833	28.362	0.11958	-0.16687
8	46.325	28.449	0.04344	-0.13545
9	46.823	28.481	0.06956	-0.14148
10	47.278	28.495	0.02965	-0.12911
11	47.751	28.554	0.03330	-0.16709
12	48.178	28.539	0.06175	-0.13928
13	48.499	28.599	0.03468	-0.12994
14	48.884	28.590	-0.02083	-0.20279
15	49.289	28.738	0.06201	-0.13903
16	49.534	28.700	0.04527	-0.12630
17	49.923	28.639	0.09696	-0.09025
18	50.177	28.742	0.05656	-0.04193
19	51.886	28.993	0.06046	-0.04558
20	52.066	29.021	0.06149	-0.05906
21	52.317	29.077	0.08688	-0.02797
22	52.425	29.096	0.02596	-0.04124
23	52.695	29.135	0.12215	-0.02176
24	52.723	29.162	-0.00773	-0.03340
25	53.009	29.209	0.10897	-0.01344

Table 3.4: Statistical quantities corresponding to the data presented in Table 3.3.

Quantity	$Z_r, \Omega cm^2$	$-Z_j, \Omega cm^2$	$\epsilon_r, \Omega cm^2$	$\epsilon_j, \Omega cm^2$
mean, equation (3.1)	48.546	28.663	0.05963	−0.11570
standard deviation, equation (3.9)	3.259	0.330	0.06193	0.08493
standard error, equation (3.11)	0.665	0.0675	0.0126	0.0173
covariance, equation (3.14)	0.984		-0.00033	

and

$$\mu_{\epsilon_j} = -0.116 \pm 0.035 \ \Omega cm^2 \tag{3.24}$$

respectively. The relatively large confidence intervals associated with the real and imaginary parts of the impedance may be attributed to nonstationary behavior of the system over the length of time (greater than 12 hours) needed to make the replicated experiments. Methods to minimize bias errors associated with nonstationary behavior are described in Section 8.3.2 on page 185. Classification of error structure is described in Chapter 21 on page 567, and methods to determine whether bias errors have affected the impedance response are presented in Chapter 22 on page 595.

The covariance presented in Table 3.4 shows that the real and imaginary parts of the impedance are highly correlated; whereas, the real and imaginary parts of the residual errors are not correlated. The lack of correlation of the residual errors supports the use of the standard deviation of residual errors as an estimator for the error structure of impedance data.[102,103]

3.1.3 Normal Distribution

The normal distribution is a function of the mean μ_x, given by equation (3.1), and the standard deviation σ_x, given by equation (3.9), i.e.,

$$f(x) = \frac{1}{\sigma_x \sqrt{2\pi}} \exp\left(-\frac{(x - \mu_x)^2}{2\sigma_x^2}\right) \tag{3.25}$$

The normal distribution function can be written as $N(\mu_x, \sigma_x^2)$. If x has a normal distribution with parameters μ_x and σ_x^2, the variable transformation

$$x_{norm} = \frac{x - \mu_x}{\sigma_x} \tag{3.26}$$

will follow the standard normal distribution[104]

$$f_{norm}(x_{norm}) = \frac{1}{\sqrt{2\pi}} \exp\left(-\frac{x_{norm}^2}{2}\right) \tag{3.27}$$

With respect to equation (3.25), the standard normal distribution function has a mean value $\mu_x = 0$ and a standard deviation $\sigma_x = 1$. The area under a standard normal

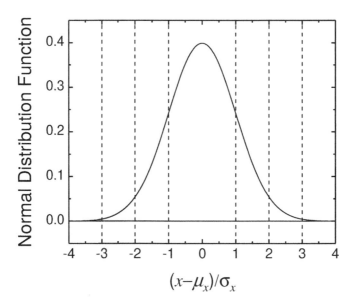

Figure 3.1: The standard normal distribution function, equation (3.25) with mean equal to zero and standard deviation equal to one: 68.27 percent of the distribution lies within $\pm\sigma$; 95.54 percent of the distribution lies within $\pm 2\sigma$; and 99.73 percent of the distribution lies within $\pm 3\sigma$.

distribution is equal to unity. The standard normal distribution function can be written as $N(0,1)$.

The standard normal distribution function, with standard deviation equal to unity and mean equal to zero, is presented in Figure 3.1. There is a 68.27 percent probability (see Section 3.1.4) that a value of x lies within the interval $\mu_x \pm \sigma_x$. There is a 95.45 percent probability that a value of x lies within the interval confidence interval $\mu_x \pm 2\sigma_x$. The central limit theorem described in Section 3.1.5 is often invoked to justify using the normal distribution as a basis for interpreting experimental data.

3.1.4   Probability

The probability is obtained by integrating the distribution function over the appropriate limits. For example, for x being a normal random variable with mean μ_x and standard deviation σ_x, the probability that $x > a$ can be expressed as

$$P(x > a) = \frac{1}{\sigma_x \sqrt{2\pi}} \int_a^{\infty} \exp\left(-\frac{(x-\mu_x)^2}{2\sigma_x^2}\right) dx \qquad (3.28)$$

The probability $P(x > a)$ is shown as the shaded area in Figure 3.2(a). The probability that $x < b$ can be expressed as

$$P(x < b) = \frac{1}{\sigma_x \sqrt{2\pi}} \int_{-\infty}^{b} \exp\left(-\frac{(x-\mu_x)^2}{2\sigma_x^2}\right) dx \qquad (3.29)$$

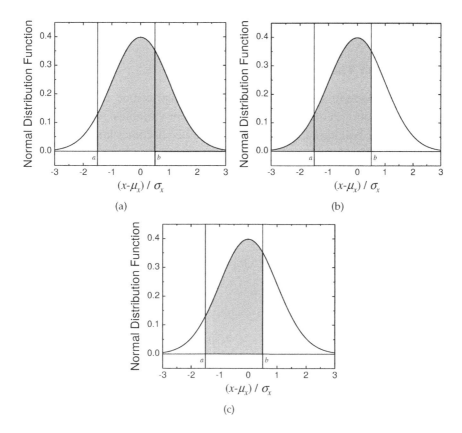

Figure 3.2: Probability distributions: a) $x > a$, equation (3.28); b) $x < b$, equation (3.29); and c) $a < x < b$, equation (3.30).

as shown in Figure 3.2(b), and the probability that $a < x < b$ can be expressed as

$$P(a < x < b) = \frac{1}{\sigma_x \sqrt{2\pi}} \int_a^b \exp\left(-\frac{(x - \mu_x)^2}{2\sigma_x^2}\right) dx \tag{3.30}$$

as shown in Figure 3.2(c). Tables of cumulative standard normal distributions

$$\Phi(x) = \frac{1}{\sqrt{2\pi}} \int_0^x \exp\left(-\frac{x_{norm}^2}{2}\right) dx_{norm} \tag{3.31}$$

are available in reference books and statistics textbooks.[105, 106]

3.1.5 Central Limit Theorem

The central limit theorem states that the sampling distribution of the mean, for any set of independent and identically distributed random variables, will tend toward the normal distribution, equation (3.25), as the sample size becomes large.[104]

Theorem 3.1 (Central Limit Theorem) *Let S represent a sequence of independent discrete random variables. Let $X_1, X_2, \ldots, X_i, \ldots, X_n$ represent n subsets of S, such that each set X_i contains n_x values, has a mean value μ_{X_i}, and has a variance $\sigma_{X_i}^2$. The mean value for the global set S is μ_S and the variance is σ_S^2.*

$$\lim_{n \to \infty} P\left(a < \frac{\mu_{X_i} - \mu_S}{\sigma_{X_i}} < b\right) = \frac{1}{\sqrt{2\pi}} \int_a^b \exp\left(-\frac{x^2}{2}\right) dx \tag{3.32}$$

and μ_{X_i} is distributed according to

$$\lim_{n \to \infty} \mu_{X_i} = N(\mu_S, \frac{\sigma_S^2}{n_x}) \tag{3.33}$$

or

$$\lim_{n \to \infty} \frac{\mu_{X_i} - \mu_S}{\sigma_S / \sqrt{n_x}} = N(0, 1) \tag{3.34}$$

The central limit theorem gives rise to the following important observations:

- The mean of the sampling distribution of means is equal to the mean of the population from which the samples were drawn.

- The variance of the sampling distribution of means is equal to the variance of the population from which the samples were drawn divided by the size of the samples.

- If the original population is distributed normally (i.e., it is bell shaped), the sampling distribution of means will also be normal. Even if the original population is not normally distributed, the sampling distribution of means will increasingly approximate a normal distribution as sample size increases.

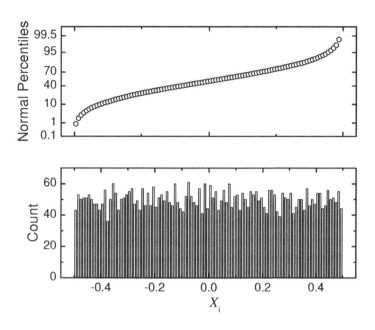

Figure 3.3: The histogram and a cumulative normal distribution plot for 5,000 uniformly distributed random numbers with mean $\mu_S = -8.06 \times 10^{-4}$ and standard deviation $\sigma_S = 0.28736$.

As an illustration of the central limit theorem, consider the distribution of 5,000 random numbers presented in Figure 3.3. The distribution shown in Figure 3.3 represents a global set S that is centered about zero and clearly does not follow a normal distribution. The mean value is $\mu_S = -8.06 \times 10^{-4}$ and the standard deviation is $\sigma_S = 0.28736$. The global set S can be subdivided into $n = 500$ subsets $X_{10,i}$, each containing $n_x = 10$ values. The mean values $\mu_{X_{10,i}}$ shown in Figure 3.4 appear to be normally distributed. The distribution, shown on normal cumulative probability scales, follows a straight line, thus confirming that the distribution of the sample means is normal. The mean of the subset means is equal to the mean of the original global set S, i.e., $\mu_{X_{10}} = \mu_S = -8.06 \times 10^{-4}$.

The standard deviation of the subsets, $\sigma_{X_{10}} = 0.09015$, is smaller than that of the global set S and is a function of the number of data n_x in each subset according to $\sigma_X = \sigma_S / \sqrt{n_x}$. A plot of standard deviation as a function of $1/\sqrt{n_x}$ yields a straight line, shown in Figure 3.5. Thus, the sampling distribution of the mean becomes approximately normal, independent of the distribution of the original variable, and the sampling distribution of the mean is centered at the population mean of the original

Remember! 3.2 *The central limit theorem is often invoked to justify using the normal distribution as a basis for interpreting experimental data.*

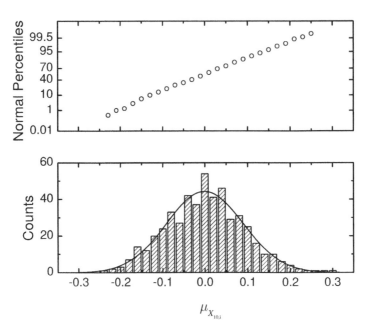

Figure 3.4: The histogram and a cumulative normal distribution plot for the means of subsets of the data shown in Figure 3.3. Each subset contained 10 points. The mean of the subset means is equal to $\mu_{X_{10}} = -8.06 \times 10^{-4}$ and standard deviation $\sigma_{X_{10}} = 0.09015$.

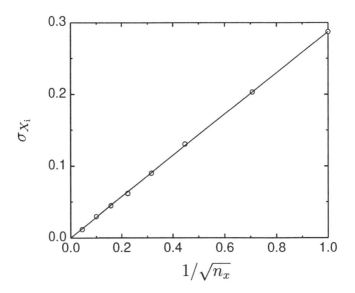

Figure 3.5: Standard deviations σ_{X_i} for the means of subsets of the data shown in Figure 3.3 as a function of $1/\sqrt{n_x}$, where n_x is the number of samples in each subset X_i.

variable. The standard deviation of the sampling distribution of the mean approaches $\sigma_S/\sqrt{n}$.

3.2   Error Propagation

Analytic expressions are available for assessing the propagation of errors through linear systems. Such approaches can be used as well when the variances are sufficiently small that the system can be linearized about its expectation value. Numerical techniques are generally needed to assess the propagation of errors through nonlinear systems.

3.2.1   Linear Systems

Consider a quantity f that is a function of variables $x_1, x_2, \ldots$ Under the assumption that the system is linear, i.e., that derivatives of second and higher order can be neglected, a Taylor series expansion about the mean values of f and $x_1, x_2, \ldots$ yields the kth observation as

$$f_k - \mu_f = \mathrm{E}\left\{\frac{\partial f}{\partial x_1}\right\}(x_{1,k} - \mu_{x_1}) + \mathrm{E}\left\{\frac{\partial f}{\partial x_2}\right\}(x_{2,k} - \mu_{x_2}) + \ldots \tag{3.35}$$

Following equation (3.8), the variance of f may be expressed as

$$\sigma_f^2 = \mathrm{E}\left\{\left(f_k - \mu_f\right)^2\right\} \tag{3.36}$$

Combination of equations (3.35) and (3.36) yields

$$\sigma_f^2 = \left[\mathrm{E}\left\{\frac{\partial f}{\partial x_1}\right\}\mathrm{E}\left\{(x_{1,k} - \mu_{x_1})\right\} + \mathrm{E}\left\{\frac{\partial f}{\partial x_2}\right\}\mathrm{E}\left\{(x_{2,k} - \mu_{x_2})\right\} + \ldots\right]^2 \tag{3.37}$$

Equation (3.37) can be expanded in terms of the variances of respective components as

$$\sigma_f^2 = \mathrm{E}\left\{\frac{\partial f}{\partial x_1}\right\}^2\sigma_{x_1}^2 + \mathrm{E}\left\{\frac{\partial f}{\partial x_2}\right\}^2\sigma_{x_2}^2 + 2\mathrm{E}\left\{\frac{\partial f}{\partial x_1}\right\}\mathrm{E}\left\{\frac{\partial f}{\partial x_2}\right\}\sigma_{x_1 x_2} + \ldots \tag{3.38}$$

Thus, the variance of a general function f can be expressed in terms of the variance of its components as

$$\sigma_f^2 \cong \sigma_{x_1}^2\left(\frac{\partial f}{\partial x_1}\right)^2 + \sigma_{x_2}^2\left(\frac{\partial f}{\partial x_2}\right)^2 + \ldots + \sigma_{x_1,x_2}\left(\frac{\partial f}{\partial x_1}\right)\left(\frac{\partial f}{\partial x_2}\right) + \ldots \tag{3.39}$$

If the errors are not correlated, i.e., if the covariance terms are equal to zero, the variance of f can be expressed as

$$\sigma_f^2 \cong \sigma_{x_1}^2\left(\frac{\partial f}{\partial x_1}\right)^2 + \sigma_{x_2}^2\left(\frac{\partial f}{\partial x_2}\right)^2 + \ldots \tag{3.40}$$

Example 3.2 **Error Propagation:** *As discussed in Chapter 16, regression to electrohydrodynamic impedance data can be used to obtain values for the Schmidt number. If the Schmidt number for reduction of oxygen in 0.5 M NaCl was found to be equal to 510 ± 25, and the kinematic viscosity is equal to $0.89 \times 10^{-2} \pm 0.05 \times 10^{-2}$ cm²/s, calculate the diffusivity and the standard deviation.*[107]

Solution: *The Schmidt number is expressed as* $Sc_{O_2} = v/D_{O_2}$; *thus,* $D_{O_2} = v/Sc_{O_2}$. *Equation (3.40) yields*

$$\sigma_{D_{O_2}}^2 \cong \sigma_v^2 \left(\frac{1}{Sc_{O_2}} \right)^2 + \sigma_{Sc_{O_2}}^2 \left(\frac{v}{Sc_{O_2}^2} \right)^2 \tag{3.41}$$

Upon inserting the numerical values,

$$\sigma_{D_{O_2}}^2 \cong (0.05 \times 10^{-2})^2 \left(\frac{1}{510} \right)^2 + (25)^2 \left(\frac{0.89 \times 10^{-2}}{(510)^2} \right)^2 \tag{3.42}$$

$$= 1.7 \times 10^{-12}$$

The result is $D_{O_2} = 1.75 \pm 0.13 \times 10^{-5}$ *cm²/s.*

The estimate provided in Example 3.2 for the standard deviation may be approximate because the function is not linear with respect to Sc_{O_2}. The influence of nonlinearity is explored in the subsequent section.

3.2.2 Nonlinear Systems

Propagation of errors through nonlinear systems poses three issues. The first is that higher-order derivatives in the Taylor series expansion may make significant contribution to the error analysis. A second issue, demonstrated in Figure 3.6, is that the nonlinear relationships can distort the distributions such that the resulting variable is not normally distributed. The meaning of the variance calculated under these conditions can be questioned. The third is that correlation among parameters may lead to a reduction in the error propagated as compared to that calculated under the assumption that sources of error are not correlated.

Analytic means do not exist to solve the problem of propagation of errors through nonlinear systems. Monte Carlo simulations can be used to assess the magnitude and distribution of propagated errors.

Remember! 3.3 *For linear systems with uncorrelated errors, the variance of a function of independent variables can be estimated from the variances of the independent variables as shown in equation (3.40).*

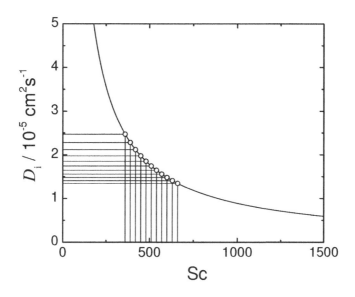

Figure 3.6: Diffusion coefficient as a function of Schmidt number for Example 3.2. The uneven spacing of diffusivity obtained for an even spacing of Schmidt number shows that a normal distribution of Schmidt number values would result in a distorted distribution of diffusivities.

Example 3.3 Continuation of Example 3.2: *Estimate the error in the assessment of the standard deviation for oxygen diffusivity obtained in Example 3.2.*

Solution: *Monte Carlo simulations were performed using two independent sets of 1,000 normally distributed random numbers, one for each of the independent parameters. The values used to introduce noise for the Schmidt number had a standard deviation of 25, and the values used to introduce noise for the kinematic viscosity had a standard deviation of 0.05×10^{-2} cm²/s. A scatter plot for the resulting calculations is shown in Figure 3.7. The resulting histogram for the diffusivity, shown in Figure 3.8, shows that the resulting distribution is normal. The mean value for the diffusivity obtained by use of Monte Carlo simulations to assess propagation of errors was 1.751×10^{-5} cm²/s, and the standard deviation was 1.3×10^{-6} cm²/s. These results are in excellent agreement with those obtained in Example 3.2.*

Example 3.4 Continuation of Example 3.3: *Estimate the error in the assessment of the standard deviation for oxygen diffusivity obtained in Example 3.2 with the exception that the estimated standard deviation for Schmidt number is increased by a factor of 10 such that $Sc_{O_2} = 510 \pm 250$.*

Solution: *Note that the 95.45 percent (2σ) confidence interval for the Schmidt number does not quite include zero, so one may imagine that this value is statistically significant. In practice, the*

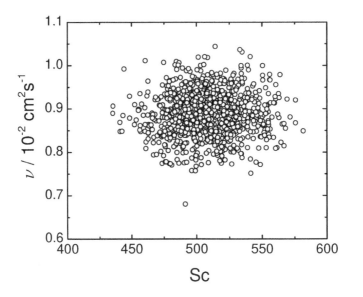

Figure 3.7: Scatter plot showing the use of Monte Carlo simulations for calculation of the diffusion coefficient. Kinematic viscosity is presented as a function of Schmidt number for Example 3.3.

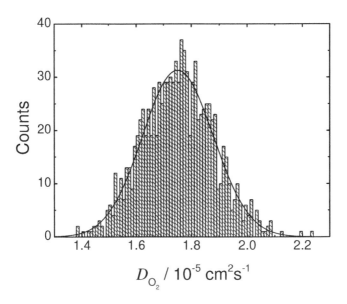

Figure 3.8: Distribution for the oxygen diffusion coefficient obtained by use of Monte Carlo simulations to assess propagation of errors for Example 3.3. The mean value for the diffusivity was 1.751×10^{-5} cm^2/s, and the standard deviation was 1.3×10^{-6} cm^2/s.

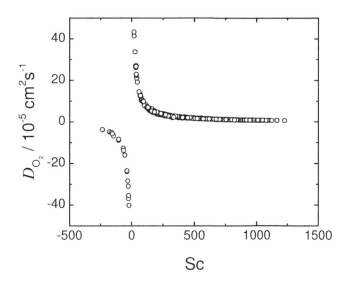

Figure 3.9: Scatter plot showing the use of Monte Carlo simulations for calculation of the diffusion coefficient. Diffusivity is presented as a function of Schmidt number for Example 3.4.

experimentalist should not be satisfied with this level of uncertainty and would want to devise a better experiment or a better model.

Monte Carlo simulations were performed using two independent sets of 1,000 normally distributed random numbers, one for each of the independent parameters. The values used to introduce noise for the Schmidt number had a standard deviation of 250, and the values used to introduce noise for the kinematic viscosity had a standard deviation of $0.05 \times 10^{-2}\ cm^2/s$. A scatter plot for the resulting calculations is shown in Figure 3.9. The calculations clearly trace the nonlinear portion of the curve given in Figure 3.6, and the occasional negative values for Schmidt number yield negative values for the diffusivity.

The resulting histogram for the diffusivity, shown in Figure 3.10, shows that the resulting distribution is not normal. The mean value for the calculated diffusivity is $D_{O_2} = -1.03 \times 10^{-5} \pm 1.1 \times 10^{-3}\ cm^2/s$, but neither the mean nor the standard deviation estimated for this distribution has statistical meaning.

3.3 Hypothesis Tests

A hypothesis test is a procedure for determining whether an assertion about a characteristic of a population is reasonable. The statistical tests described in this section cannot be used to prove a hypothesis, but rather, within a specified level of confidence, to disprove the hypothesis.

Often, the question at hand is whether two distributions have the same mean or the same variance. The question can be posed as: *Are the observed differences in mean or vari-*

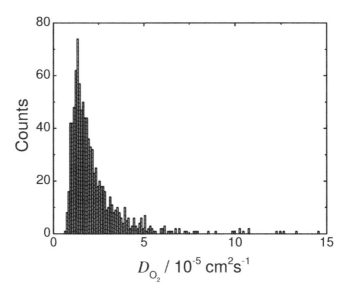

Figure 3.10: Distribution for the oxygen diffusion coefficient obtained by use of Monte Carlo simulations to assess propagation of errors for Example 3.2. The distribution is not normal, and the standard deviation estimated for such a distribution has no statistical meaning as a confidence interval.

ance statistically significant? The Student's *t*-test and the *F*-test can be used, respectively, to resolve these questions.

3.3.1 Terminology

There is a specific terminology used for hypothesis tests. One might ask, for example, whether the rate of corrosion obtained for an aged steel coupon immersed in seawater is equal to that found in city drinking water.

- The null hypothesis is the original assertion. In this case, the null hypothesis is that the rate of corrosion obtained for an aged steel coupon immersed in seawater is equal to that found in city drinking water. The notation used is H_0: $\mu_1 = \mu_2$.

- There are three possibilities for the alternative hypothesis H_1: $\mu_1 > \mu_2$, $\mu_1 < \mu_2$, and $\mu_1 \neq \mu_2$.

- The significance level α is a number between 0 and 1 that represents the probability of incorrectly rejecting the null hypothesis when it is actually true. For a typical significance level of 5 percent, the notation is $\alpha = 0.05$. For this significance level, the probability of incorrectly rejecting the null hypothesis when it is actually true is 5 percent. A smaller value of α provides more protection from this error.

- The p-value is the probability of observing the given sample result under the assumption that the null hypothesis is true. In other words, p is the value of α that will yield the observed statistical test value. Example 3.5 provides an illustration of the relationship among the t-test parameter, α and p. If the p-value is less than α, the null hypothesis can be rejected. The converse is not true. The null hypothesis cannot be accepted on the basis of a p-value that is greater than α.

The statistical tests described in this section cannot be used to prove a given hypothesis. These tests are used instead to identify conditions where the data do not support the hypothesis.

3.3.2 Student's t-Test for Equality of Mean

The Student's t-distribution was developed by William Sealy Gosset, 1876-1937, an English statistician, who published his distribution in 1908 under the pseudonym Student.[108] Student's t-test deals with the problems associated with inference based on small samples. The question is to determine the extent to which the calculated mean and standard deviation based on a small sample may, by chance, differ from the mean and standard deviation that would be obtained from a large distribution. Gosset developed these statistical methods during his employment at the Guinness® brewery in Dublin. The experimental verification of his theories, however, was derived from published measurements of the heights and left-middle-finger lengths of 3,000 criminals.[108]

The Student's probability density function is given by

$$f(t) = \frac{\Gamma\left(\dfrac{\nu+1}{2}\right)}{\sqrt{\nu\pi}\,\Gamma\left(\dfrac{\nu}{2}\right)} \left(1 + \frac{t^2}{\nu}\right)^{-\frac{\nu+1}{2}} \tag{3.43}$$

where ν is the number of degrees of freedom and must be a positive integer. The t-distribution is presented in Figure 3.11 with degree of freedom as a parameter. As $\nu \to \infty$, Student's probability distribution function approaches the normal distribution.

The value of t obtained from equation (3.43) can be compared to the value obtained from the experimental observations. A test of the null hypothesis that the population mean with $\nu = n_x - 1$ degrees of freedom is equal to a specified value μ_0 is obtained in terms of

$$t = \frac{\overline{x} - \mu_0}{s_x / \sqrt{n_x}} \tag{3.44}$$

Remember! 3.4 *Statistical tests cannot be used to prove a given hypothesis. These tests can be used to identify conditions where the data do not support the hypothesis.*

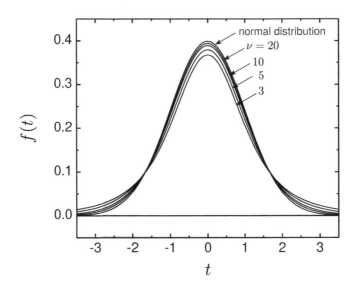

Figure 3.11: Comparison of Student's t-distribution given by equation (3.43) to the normal distribution given by equation (3.25) with degree of freedom as a parameter.

where n_x is the number of experimental data points and $s_x / \sqrt{n_x}$ represents the standard error for the data set. The corresponding expression for comparison of unpaired variates is given as

$$t = \frac{\overline{x}_1 - \overline{x}_2}{\sqrt{\dfrac{s_{x_1}^2}{n_{x_1}} + \dfrac{s_{x_2}^2}{n_{x_2}}}} \tag{3.45}$$

with $\nu = n_{x_1} + n_{x_2} - 2$ degrees of freedom.

When an external factor influences the data sets in a way that is point-by-point identical, the t-test is based on the mean of the differences between the measurements, i.e.,

$$t = \frac{\overline{x_1 - x_2} - \mu_0}{s_{x_1 - x_2} / \sqrt{n_x}} \tag{3.46}$$

where $n_{x_1} = n_{x_2} = n_x$ and $\nu = n_x - 1$. Such a test may be used, for example, for a comparison of two variables obtained as a function of frequency in the same experiment, where frequency may be expected to have the same influence on the two variables.

Table 3.5 is useful for assessing the Student's t-test for equality of mean. Note that, for a 68.27% confidence interval, $t \to 1$ for $\nu \to \infty$ and that $t \to 2$ at $\nu \to \infty$ for a 95.45% confidence interval. Thus, as $\nu \to \infty$, the t-distribution approaches a normal distribution. The 95.45% confidence interval for the mean of a parameter, given by equation (3.12) for a normal distribution is replaced by

$$\mu_x = \overline{x} \pm t_{0.0455} \frac{s_x}{\sqrt{n_x}} \tag{3.47}$$

Table **3.5**: Student's t-test values.

confidence interval	40%	68.27%	90%	95%	95.45%	99%
ν/p	0.6	0.3173	0.10	0.05	0.0455	0.01
3	0.58439	1.1969	2.3534	3.1824	3.3068	5.8409
4	0.56865	1.1417	2.1318	2.7764	2.8693	4.6041
5	0.55943	1.1105	2.0150	2.5706	2.6487	4.0321
10	0.54153	1.0526	1.8125	2.2281	2.2837	3.1693
15	0.53573	1.0345	1.7531	2.1314	2.1812	2.9467
20	0.53286	1.0257	1.7247	2.0860	2.1330	2.8453
25	0.53115	1.0204	1.7081	2.0595	2.1051	2.7874
30	0.53002	1.0170	1.6973	2.0423	2.0868	2.7500
35	0.52921	1.0145	1.6896	2.0301	2.0740	2.7238
40	0.52861	1.0127	1.6839	2.0211	2.0645	2.7045
45	0.52814	1.0113	1.6794	2.0141	2.0571	2.6896
50	0.52776	1.0101	1.6759	2.0086	2.0513	2.6778
55	0.52745	1.0092	1.6730	2.0040	2.0465	2.6682
60	0.52720	1.0084	1.6706	2.0003	2.0425	2.6603
65	0.52698	1.0078	1.6686	1.9971	2.0392	2.6536
70	0.52680	1.0072	1.6669	1.9944	2.0364	2.6479
75	0.52664	1.0067	1.6654	1.9921	2.0339	2.6430
80	0.52650	1.0063	1.6641	1.9901	2.0317	2.6387
85	0.52637	1.0059	1.6630	1.9883	2.0298	2.6349
90	0.52626	1.0056	1.6620	1.9867	2.0282	2.6316
95	0.52616	1.0053	1.6611	1.9853	2.0267	2.6286
100	0.52608	1.0050	1.6602	1.9840	2.0253	2.6259
200	0.52524	1.0025	1.6525	1.9719	2.0126	2.6006
1,000	0.52457	1.0005	1.6464	1.9623	2.0025	2.5808
5,000	0.52443	1.0001	1.6452	1.9604	2.0005	2.5768
10,000	0.52442	1.0001	1.6450	1.9602	2.0003	2.5763
100,000	0.52440	1.0000	1.6449	1.9600	2.0000	2.5759

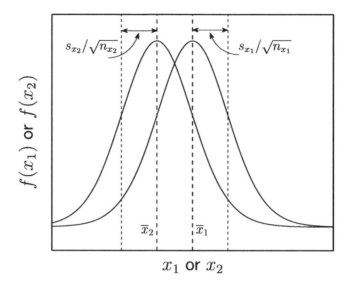

Figure 3.12: Comparison of population distributions for variables x_1 and x_2. The Student's t-test corresponding to equation (3.45) is used to assess whether the null hypothesis, that $\overline{x}_1 = \overline{x}_2$, can be rejected.

for a given value of ν.

A graphical representation of population distributions for variables x_1 and x_2, presented in Figure 3.12, may be used to illustrate the concept of hypothesis tests. The two sample means, $\overline{x}_1$ and $\overline{x}_2$, appear to be different. The question is whether this apparent difference is statistically significant. The Student's t-test corresponding to equation (3.45) is used to assess whether the null hypothesis, that $\overline{x}_1 = \overline{x}_2$, can be rejected.

3.3.3 F-Test for Equality of Variance

The F-test is used to test the null hypothesis that the two population variances corresponding to the two samples are equal.[96] Within each sample, the values are assumed to be independent and to have the identical normal distribution (i.e., the same mean and variance). In addition, the two samples are assumed to be independent of each other.

The value for F is obtained from the experimentally determined variances as

$$F = s_1^2 / s_2^2 \tag{3.48}$$

such that $F > 1$, i.e., $s_1^2 > s_2^2$. The null hypothesis is that the two variances are from the same population, i.e., they are not statistically different.

The significance level at which the hypothesis that the variances are not equal can be rejected is given by

$$Q(F|\nu_1, \nu_2) = \frac{\Gamma\left((\nu_1 + \nu_2)/2\right)}{\Gamma\left(\nu_1/2\right)\Gamma\left(\nu_2/2\right)} \int\limits_0^{\nu_2/(\nu_2+\nu_1 F)} t^{\nu_2/2-1}(1-t)^{\nu_1/2-1}\mathrm{d}t \qquad (3.49)$$

with $\nu_1 = n_{x_1} - 1$ and $\nu_2 = n_{x_2} - 1$ degrees of freedom, respectively. The associated probability density function is given by

$$f(x) = \frac{\Gamma(\frac{\nu_1+\nu_2}{2})(\frac{\nu_1}{\nu_2})^{\frac{\nu_1}{2}} x^{\frac{\nu_1}{2}-1}}{\Gamma(\frac{\nu_1}{2})\Gamma(\frac{\nu_2}{2})(1 + \frac{\nu_1 x}{\nu_2})^{\frac{\nu_1+\nu_2}{2}}} \qquad (3.50)$$

For $\nu_1 = \nu_2 = \nu$, equation (3.50) may be expressed as

$$f(x) = \frac{\Gamma(\nu)x^{\frac{\nu}{2}-1}}{\left(\Gamma(\frac{\nu}{2})\right)^2 (1+x)^{\nu}} \qquad (3.51)$$

Tables are provided for the value of F corresponding to a given number of degrees of freedom ν and for a given confidence level p. If the calculated F is greater than the table value, then the null hypothesis must be rejected. Numerical values useful for assessing the F-test for equality of variance with samples of identical degree of freedom are presented in Table 3.6.

Example 3.5 Evaluation of Impedance Data: *The question of whether the real and imaginary parts of impedance measurements have the same variance or standard deviation has generated significant controversy in the impedance literature. Consider the standard deviations reported by Orazem et al.*[99] *for their impedance data obtained for the reduction of ferricyanide on a Pt rotating disk electrode. The methods of Agarwal et al.*[69, 100] *were used to filter minor lack of replication from 26 repeated impedance experiments. Numerical values are presented in Table 3.7. The results, presented graphically in Figure 3.13, suggest the similarity of values for the standard deviations of the imaginary and real impedance. The question here is whether statistical tests can be used to reject the null hypothesis that standard deviations of the imaginary and real impedance values are equal.*

Solution: *The standard deviations are clearly strong functions of frequency. Thus, the appropriate Student's t-test is that for paired variates, equation (3.46). The calculated t-value is $t_{\exp} = 1.710$. An interpolated table of t-test values is presented as Table 3.8. The significance level at a degree of freedom $\nu = n - 1 = 73$ corresponding to $t_{\exp} = 1.710$ is $p = 0.0915$. The value of t corresponding to the $\alpha = 0.0455$ level is $t_{0.0455} = 2.0348$. As $t_{\exp} < t_{0.0455}$, the null hypothesis that the two means are equal cannot be rejected at the 0.0455 significance level.*

The sample mean of the differences between standard deviations for real and imaginary parts of the impedance is

$$\overline{s_{Z_r} - s_{Z_j}} = 0.00466\ \Omega \qquad (3.52)$$

Table 3.6: F-test values for comparison of variance of samples with equal degrees of freedom, i.e., $\nu_1 = \nu_2 = \nu$.

ν/p	0.3	0.2	0.1	0.05	0.025	0.01
3	1.940	2.936	5.391	9.277	15.439	29.457
4	1.753	2.483	4.107	6.388	9.605	15.977
5	1.641	2.228	3.453	5.050	7.146	10.967
10	1.406	1.732	2.323	2.978	3.717	4.849
15	1.318	1.558	1.972	2.403	2.862	3.522
20	1.268	1.466	1.794	2.124	2.464	2.938
25	1.236	1.406	1.683	1.955	2.230	2.604
30	1.213	1.364	1.606	1.841	2.074	2.386
35	1.196	1.332	1.550	1.757	1.961	2.231
40	1.182	1.308	1.506	1.693	1.875	2.114
45	1.170	1.287	1.470	1.642	1.807	2.023
50	1.161	1.271	1.441	1.599	1.752	1.949
55	1.153	1.256	1.416	1.564	1.706	1.888
60	1.146	1.244	1.395	1.534	1.667	1.836
65	1.140	1.233	1.377	1.508	1.633	1.792
70	1.134	1.224	1.361	1.486	1.604	1.754
75	1.129	1.216	1.346	1.466	1.578	1.720
80	1.125	1.208	1.334	1.448	1.555	1.690
85	1.121	1.201	1.322	1.432	1.534	1.663
90	1.117	1.195	1.312	1.417	1.516	1.639
95	1.114	1.189	1.302	1.404	1.499	1.618
100	1.111	1.184	1.293	1.392	1.483	1.598
200	1.077	1.127	1.199	1.263	1.320	1.391
1,000	1.034	1.055	1.084	1.110	1.132	1.159
5,000	1.015	1.024	1.037	1.048	1.057	1.068
10,000	1.011	1.017	1.026	1.033	1.040	1.048
100,000	1.003	1.005	1.008	1.010	1.012	1.015

Table 3.7: Standard deviations reported by Orazem et al.[99] for their impedance data obtained for the reduction of ferricyanide on a Pt rotating disk electrode. The methods of Agarwal et al.[69, 100] were used to filter minor lack of replication from 26 repeated impedance experiments.

#	f/Hz	s_{Z_r}/Ω	s_{Z_j}/Ω	$s^2_{Z_r}/s^2_{Z_j}$	#	f/Hz	s_{Z_r}/Ω	s_{Z_j}/Ω	$s^2_{Z_r}/s^2_{Z_j}$
1	0.02154	0.35649	0.32284	1.21933	38	25.924	0.02548	0.05162	0.24363
2	0.02592	0.37169	0.28941	1.64949	39	31.408	0.02631	0.04885	0.29019
3	0.03141	0.39259	0.35798	1.20271	40	38.051	0.04043	0.02609	2.40076
4	0.03805	0.2963	0.26128	1.28603	41	55.851	0.03552	0.04002	0.78771
5	0.04610	0.32867	0.31740	1.07226	42	67.67	0.01624	0.04298	0.14275
6	0.05585	0.27794	0.29545	0.88497	43	81.98	0.02826	0.03509	0.64860
7	0.06767	0.25963	0.25954	1.00068	44	120.32	0.03647	0.02799	1.69838
8	0.08198	0.23957	0.20889	1.31536	45	145.78	0.02404	0.03468	0.48056
9	0.09932	0.22973	0.19401	1.40209	46	176.62	0.01793	0.03444	0.27096
10	0.12033	0.16701	0.20924	0.63707	47	213.98	0.02879	0.02481	1.34668
11	0.14578	0.17592	0.18912	0.86524	48	259.24	0.03131	0.01092	8.21824
12	0.17662	0.17040	0.15621	1.18989	49	314.08	0.02308	0.02322	0.98868
13	0.21398	0.13790	0.19223	0.51463	50	380.51	0.00862	0.03234	0.07106
14	0.25924	0.15611	0.09356	2.78387	51	461	0.01852	0.02867	0.41707
15	0.31408	0.15208	0.10010	2.30823	52	558.51	0.02876	0.01917	2.25168
16	0.38051	0.12977	0.07933	2.67563	53	676.7	0.02394	0.02158	1.23128
17	0.461	0.09682	0.10667	0.82375	54	819.8	0.01462	0.02766	0.27936
18	0.55851	0.07737	0.08260	0.87747	55	993.2	0.01562	0.02486	0.39474
19	0.6767	0.06474	0.05871	1.21583	56	1,203.3	0.02378	0.0132	3.24432
20	0.81979	0.08196	0.06187	1.75491	57	1,457.8	0.02292	0.01167	3.85855
21	0.9932	0.05799	0.05996	0.93517	58	1,766.2	0.01015	0.02118	0.22967
22	1.2033	0.06719	0.07872	0.72848	59	2,139.8	0.00779	0.01814	0.18415
23	1.4578	0.10232	0.06678	2.34811	60	2,592.4	0.01460	0.01261	1.33987
24	1.7662	0.11819	0.05019	5.54426	61	3,140.8	0.01076	0.01371	0.61607
25	2.1398	0.06228	0.06882	0.81897	62	3,805.1	0.00618	0.01626	0.14425
26	2.5924	0.06154	0.07755	0.62978	63	4,610	0.00974	0.01076	0.8183
27	3.1408	0.06852	0.05589	1.50346	64	5,585.1	0.01138	0.00427	7.09199
28	3.8051	0.06907	0.04163	2.75322	65	6,766.5	0.00700	0.00569	1.50972
29	4.61	0.03973	0.04051	0.96178	66	8,198	0.00222	0.00743	0.08963
30	5.5851	0.04068	0.04832	0.70865	67	9,932	0.00551	0.00396	1.92994
31	6.767	0.04501	0.04721	0.9089	68	12,033	0.00613	0.00475	1.66243
32	8.198	0.05002	0.0194	6.64847	69	14,578	0.00405	0.00796	0.25897
33	9.932	0.04242	0.02741	2.39524	70	17,662	0.00133	0.0091	0.02151
34	12.033	0.03215	0.04111	0.61162	71	21,397	0.00455	0.00646	0.49491
35	14.578	0.02422	0.03938	0.37827	72	25,924	0.00456	0.00226	4.08208
36	17.662	0.03756	0.02455	2.33978	73	31,407	0.00129	0.0035	0.13629
37	21.398	0.03679	0.02424	2.30361	74	38,051	0.01032	0.00567	3.31491

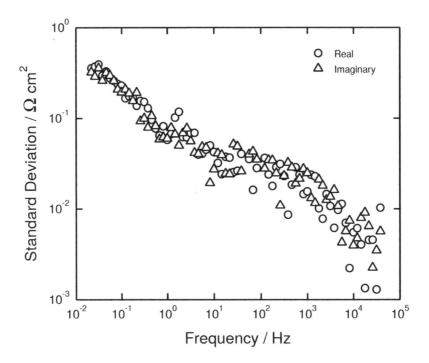

Figure 3.13: Standard deviations reported by Orazem et al.[99] for their impedance data obtained for the reduction of ferricyanide on a Pt rotating disk electrode.

Table 3.8: Student's t-test values for the hypothesis that the standard deviations for real and imaginary parts of the impedance, shown in Table 3.7, are equal. See Example 3.5.

ν/α	0.1	**0.0915**	**0.0455**
70	1.66691	⇑	⇓
73	⇒	**1.710**	**2.0348**
75	1.66543	1.70946	2.0339

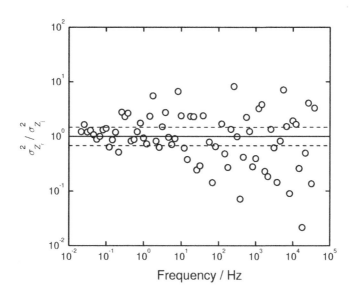

Figure 3.14: Ratio of the variance of the real part of the impedance to the variance of the imaginary part of the impedance for the data for the reduction of ferricyanide on a Pt rotating disk electrode.[99] Dashed lines represent the F-value corresponding to 0.05 significance level and its inverse.

and the sample standard error is

$$SE_{s_{Z_r} - s_{Z_j}} = 0.00273 \ \Omega \tag{3.53}$$

Application of equation (3.47) yields, with $t_{0.0455} = 2.0348$,

$$\mu_{s_{Z_r} - s_{Z_j}} = 0.00466 \pm 0.0056 \ \Omega \tag{3.54}$$

Equation (3.54) shows that the 95.45% confidence interval for the difference between the standard deviations for real and imaginary parts of the impedance includes zero. This result is in agreement with the observation that the null hypothesis that the two means are equal cannot be rejected at the 0.0455 significance level.

A second approach is to consider the F-test for comparison of variance at each frequency. As there were 26 repeated measurements, the degree of freedom for the two calculations of variance is $\nu = 26 - 1 = 25$. At the $\alpha = 0.05$ level, a value of 1.955 is obtained from Table 3.6, and a value of 2.604 is obtained at the $\alpha = 0.01$ level. These critical values can be compared with the ratio presented in Table 3.7. Ordinarily, the F-test parameter is arranged such that the larger variance is divided by the smaller. As is evident in Figure 3.13, the variance of the real part of the impedance is sometimes larger and sometimes smaller than the variance of the imaginary part. The ratios of variances are shown in Figure 3.14 on a logarithmic scale as a function of frequency. While the ratio is scattered about unity, a frequency-by-frequency comparison of the calculated F-test values to the criteria obtained from Table 3.6 suggests that the null hypothesis that the variances are equal can be rejected at the 0.05 significance level on an intermittent basis,

Table 3.9: Student's t-test values for Example 3.5 for a test of $\log_{10}\left(\sigma_{Z_r}^2/\sigma_{Z_j}^2\right)=0$.

ν/α	0.6	0.532	0.0455
70	0.52680	$\Uparrow$	$\Downarrow$
73	$\Rightarrow$	0.628	2.0348
75	0.52664	0.62786	2.0339

with the variance of the real part occasionally larger than that of the imaginary part, and the variance of the imaginary part occasionally larger than that of the real part.

As the uncertainty in the assessment of the variance appears to be large, the Student's t-test may represent the best assessment of the statistics of the error analysis described in Figure 3.13. The ratio of variances is not normally distributed, but the quantity $\log_{10}\left(s_{Z_r}^2/s_{Z_j}^2\right)$ is normally distributed. Evaluation of equation (3.44) for a comparison of the mean of $\log_{10}\left(s_{Z_r}^2/s_{Z_j}^2\right)$ to the expected value of zero yields $|t_{\exp}| = 0.628$. The correspondence of this value to critical values of $t_{0.0455}$ is seen in Table 3.9. The significance level corresponding to $t_{\exp} = 0.628$ at a degree of freedom $\nu = n - 1 = 73$ is $p = 0.532$.

The value of t corresponding to the $\alpha = 0.0455$ level is $t_{0.05} = 2.0348$. As $t_{\exp} < t_{0.0455}$, the null hypothesis that the mean of $\log_{10}\left(s_{Z_r}^2/s_{Z_j}^2\right)$ is equal to zero cannot be rejected at the 0.0455 significance level. The value of $p = 0.532$ represents the the probability of observing the given sample result under the assumption that the null hypothesis is true.

The sample mean of $\log_{10}\left(s_{Z_r}^2/s_{Z_j}^2\right)$ is -0.03519, and the associated sample standard error has a value of 0.056. The 95.45% confidence interval can be expressed as

$$\mu_{\log_{10}\left(s_{Z_r}^2/s_{Z_j}^2\right)} = -0.03519 \pm 0.114 \tag{3.55}$$

As shown in equation (3.55), the 95.45% confidence interval includes zero. This result is in agreement with the observation that the null hypothesis $\log_{10}\left(s_{Z_r}^2/s_{Z_j}^2\right) = 0$ cannot be rejected at the 0.0455 significance level.

The statistical tests described here cannot prove that the variances for real and imaginary parts of the impedance are equal. They show only that, for the given data, the hypothesis that the variances for real and imaginary parts of the impedance are equal cannot be rejected.

3.3.4 Chi-Squared Test for Goodness of Fit

The χ^2 test is used to evaluate the probability that ν random normally distributed variables of unit variance would have a sum of squares greater than χ^2.

$$Q(\chi^2|\nu) = \frac{1}{\Gamma(\frac{\nu}{2})} \int_{\frac{\chi^2}{2}}^{\infty} e^{-t} t^{\nu/2-1} dt \tag{3.56}$$

If the χ^2 statistic originates from comparison of a model to experiment, the degree of freedom is given by

$$\nu = N_{dat} - N_p - 1 \tag{3.57}$$

where N_{dat} represents the number of observations and N_p is the number of adjusted parameters. Tables are provided for the value of χ^2 corresponding to a given number of degrees of freedom and for a given confidence level. If the calculated χ^2 is greater than the table value, then the null hypothesis must be rejected. Numerical values useful for assessing the χ^2-test for equality of mean are presented in Table 3.10.

Example 3.6 Evaluation of Chi-Squared Statistics: *Consider that, for a given measurement, regression of a model to real and imaginary parts of impedance data yielded $\chi^2 = 130$. Measurements were conducted at 70 frequencies. The regressed parameters needed to model the data included the solution resistance and 9 Voigt elements, resulting in use of 19 parameters. Under assumption that the variances used in the evaluation of χ^2 were obtained independently, evaluate the hypothesis that the χ^2 value cannot be reduced by refinement of the model.*

Solution: *As the fit was to both real and imaginary parts of the data, the degree of freedom for this problem is $\nu = n - p = 140 - 19 - 1 = 120$. The χ^2 value corresponding to a 0.05 significance level can be obtained from Table 3.10 to be equal to 146.6, and the probability corresponding to the measured χ^2 value is 0.25. Thus there is a 25 percent probability that the χ^2 statistic could exceed the observed value by chance, even for a correct model.*

Problems

3.1   The steady-state mass-transfer-limited current density for a rotating disk (see Section 11.3 on page 256) can be expressed as

$$i_{lim} = 0.62nFc_i(\infty)D_i^{2/3}\nu^{-1/6}\Omega^{1/2} \tag{3.58}$$

where $c_i(\infty)$ is the concentration of the limiting reacting species i far from the disk, D_i is the diffusivity of the reacting species i, ν is the kinematic viscosity, and Ω is the rotation speed for the disk. Consider that $i_{lim} = 60 \pm 0.5$ mA/cm^2, $c_i(\infty) = 0.1 \pm 0.005$ M, $\nu = 10^{-2} \pm 10^{-4}$ cm^2/s, and $\Omega = 1,000 \pm 1$ rpm. Find the value and standard deviation for the diffusivity of species i in units of cm^2/s.

Remember! 3.5 *The numerical value of the χ^2 statistic for a weighted regression depends on the estimated variance of the data. The numerical value has no meaning if the variance of the data is unknown.*

Table 3.10: χ^2-test values for degree of freedom ν and confidence level p.

ν/p	0.5	0.3	0.2	0.1	0.05	0.01
5	4.35	6.06	7.29	9.24	11.07	15.09
10	9.34	11.78	13.44	15.99	18.31	23.21
15	14.34	17.32	19.31	22.31	25.00	30.58
20	19.34	22.77	25.04	28.41	31.41	37.57
25	24.34	28.17	30.68	34.38	37.65	44.31
30	29.34	33.53	36.25	40.26	43.77	50.89
35	34.34	38.86	41.78	46.06	49.80	57.34
40	39.34	44.16	47.27	51.81	55.76	63.69
45	44.34	49.45	52.73	57.51	61.66	69.96
50	49.33	54.72	58.16	63.17	67.50	76.15
55	54.33	59.98	63.58	68.80	73.31	82.29
60	59.33	65.23	68.97	74.40	79.08	88.38
65	64.33	70.46	74.35	79.97	84.82	94.42
70	69.33	75.69	79.71	85.53	90.53	100.43
75	74.33	80.91	85.07	91.06	96.22	106.39
80	79.33	86.12	90.41	96.58	101.88	112.33
85	84.33	91.32	95.73	102.08	107.52	118.24
90	89.33	96.52	101.05	107.57	113.15	124.12
95	94.33	101.72	106.36	113.04	118.75	129.97
100	99.33	106.91	111.67	118.50	124.34	135.81
120	119.33	127.62	132.81	140.23	146.57	158.95
150	149.33	158.58	164.35	172.58	179.58	193.21
200	199.33	209.99	216.61	226.02	233.99	249.45
500	499.33	516.09	526.40	540.93	553.13	576.49
1,000	999.3	1,023	1,037	1,058	1,075	1,107
10,000	9,999	10,074	10,119	10,182	10,234	10,332
100,000	99,999	100,234	100,376	100,574	100,737	101,043

Table 3.11: Estimated parameter values and standard deviations for a $LiCoO_2|C$ battery at temperatures of 10 and 20 °C at a potential of 4 V.

Parameter	10 °C	20 °C
$C_{a||S}$ / $\mu F\ cm^{-2}$	28±1.8	28.8±0.84
$A_{W,a}$ / $\Omega\ cm^2\ s^{-0.5}$	58.8±0.87	52.5±0.51
$R_{t,a}$ / $\Omega\ cm^2$	0.256±0.0066	0.195±0.0045
$R_{t,S}$ / $\Omega\ cm^2$	1.93±0.013	1.383±0.0073
R_e / $\Omega\ cm^2$	1.357±0.0074	0.845±0.0022
$R_{t,c}$ / $\Omega\ cm^2$	1.19±0.018	0.954±0.0094
α_c	0.912±0.0074	0.938±0.0052
Q_c / $F\ cm^{-2}\ s^{\alpha_c-1}$	0.0662±0.00088	0.0608±0.00065
$\gamma_{d,c}$	0.72±0.015	0.785±0.0087
$A_{d,c}$/ $\Omega\ cm^2\ s^{-\gamma_{d,c}/2}$	0.254±0.0042	0.203±0.0023

3.2 The Schmidt number is expressed as $Sc_i = \nu/D_i$. Find the value and standard deviation for the Schmidt number if the kinematic viscosity is $\nu = 10^{-2} \pm 10^{-4}\ cm^2/s$ and the diffusivity of species i is $D_i = 10^{-5} \pm 10^{-7}\ cm^2/s$.

3.3 Estimate the 95.45 percent confidence interval for the equivalent circuit model given in Figure 4.3(b) over the frequency range of 10 mHz to 10 kHz if the regressed parameter values are expressed as $R_e = 10 \pm 1\ \Omega cm^2$, $R_t = 100 \pm 15\ \Omega cm^2$, and $\tau = R_t C_{dl} = 0.01 \pm 0.001$ s.

3.4 Estimate the 95.45 percent confidence interval for the film thickness extracted by use of the power-law model given as equation (14.33) on page 408, i.e.,

$$Q = \frac{(\varepsilon\varepsilon_0)^\alpha}{g\delta\rho_\delta^{1-\alpha}} \tag{3.59}$$

where

$$g = 1 + 2.88(1-\alpha)^{2.375} \tag{3.60}$$

and $\alpha = 0.91 \pm 0.1$, $Q = 11 \pm 1\ \mu F/s^{(1-\alpha)}cm^2$, and $\rho_\delta = 450\ \Omega cm$. You may assume that $\varepsilon = 12$.

3.5 Erol and Orazem[109] reported regression results that yielded the parameter estimates given in Table 3.11. Calculate the corresponding 95.45% confidence intervals and determine whether zero is included within the confidence interval.

3.6 Provide the equation for a 95% confidence interval that corresponds to equation (3.12) for a 95.45% confidence interval.

3.7 Consider the replicated impedance measurements presented in Table 3.12 for a 5 cm^2 Polymer Electrolyte Membrane (PEM) fuel cell operating at a current of 1 A. The measurements were collected at a frequency of 1 Hz on the same

Table 3.12: Replicated impedance collected at a frequency of 1 Hz on the same Polymer Electrolyte Membrane (PEM) fuel cell using two different sets of instrumentation, the 850C provided by Scribner Associates and the FC350 provided by Gamry Instruments.

	850C		FC350	
Replicate	Z_r, Ω	Z_j, Ω	Z_r, Ω	Z_j, Ω
1	0.17117	−0.041791	0.167993	−0.038176
2	0.18236	−0.043494	0.173985	−0.040603
3	0.18606	−0.044666	0.176629	−0.041786
4	0.18941	−0.045946	0.1827	−0.044806

Table 3.13: Replicated impedance collected at a frequency of 10 Hz on the same PEM fuel cell but using two flow channels.

	Serpentine		Interdigitated	
Replicate	Z_r, Ω	Z_j, Ω	Z_r, Ω	Z_j, Ω
1	0.055822	−0.048797	0.073113	−0.039505
2	0.0583180	−0.053851	0.074073	−0.04012
3	0.05861	−0.05494	0.074244	−0.040369
4	0.058842	−0.055843	0.074362	−0.040548

cell using two different sets of instrumentation, the 850C provided by Scribner Associates and the FC350 provided by Gamry Instruments.

(a) Test the hypothesis that the impedance values obtained by the two instruments are not distinguishable.

(b) Test the hypothesis that the variances obtained by the two instruments are not distinguishable.

3.8   Consider the replicated impedance measurements presented in Table 3.13 for a 5 cm^2 PEM fuel cell operating at a current of 1 A. The measurements were collected at a frequency of 10 Hz on the same cell using a serpentine channel and an interdigitated channel that is believed to be more efficient but also more susceptible to flooding.

(a) Test the hypothesis that the impedance values obtained by the two instruments are not distinguishable.

(b) Test the hypothesis that the variances obtained by the two instruments are not distinguishable.

Chapter 4

Electrical Circuits

Transfer function approaches are general and can be applied to a large variety of electrical, mechanical, and optical systems. For this reason, it is not surprising that the behavior of one system will resemble that of another. Electrochemists take advantage of this similarity by comparing the behavior of electrochemical systems to that of known electrical circuits.

A review of Chapter 1 may be useful. Summaries of relationships among complex impedance, real and imaginary parts of the impedance, and the phase angle and magnitude are found in Tables 1.1, 1.2, and 1.3. A more complete discussion of the use of graphical methods is presented in Chapter 18 on page 493.

4.1 Passive Electrical Circuits

Passive circuit elements are components that do not generate current or potential. A passive electrical circuit is composed only of passive elements. Only an element with two contacts is considered here, which is analyzed by considering the current flowing through and the potential difference between the contacts, shown as open circles in Figure 4.1.

4.1.1 Circuit Elements

Electrical circuits can be constructed from the passive elements shown in Figure 4.1. The fundamental relationship between current and potential for the resistor (Figure 4.1(a)) is

$$V(t) = RI(t) \tag{4.1}$$

where the value of the resistance R represents the fundamental property of the resistor and I is the current. In the notation of this text, I represents the current in units of A and i, such as used in equation (5.15) on page 95, represents the current density in units of A/cm^2. At each point in time, the potential difference between the resistor clamps

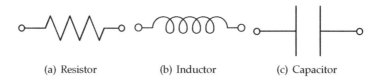

(a) Resistor   (b) Inductor   (c) Capacitor

Figure 4.1: Passive elements that serve as components of an electrical circuit.

is proportional to the current flowing through the resistor. The steady-state current flowing through a resistor is finite and can be obtained from equation (4.1).

The relationship between current and potential difference for the inductor (Figure 4.1(b)) is

$$V(t) = L\frac{dI(t)}{dt} \tag{4.2}$$

Equation (4.2) is the defining equation for the inductor. Under steady-state conditions, $dI(t)/dt = 0$, and, according to equation (4.2), $V(t) = 0$. Thus, the inductor is equivalent to a short circuit under steady-state conditions.

A capacitor (Figure 4.1(c)) is defined by

$$C = \frac{dq(t)}{dV(t)} \tag{4.3}$$

where $q(t)$ is the electric charge. The current flowing through the capacitor is obtained from the derivative of charge with respect to time, i.e.,

$$I(t) = \frac{dq(t)}{dt} \tag{4.4}$$

Then, from equations (4.3) and (4.4):

$$I(t) = C\frac{dV(t)}{dt} \tag{4.5}$$

which provides the relationship between current and potential for the capacitor. Under steady-state condition $dV/dt = 0$, and, according to equation (4.5), $I(t) = 0$. The capacitor is equivalent to an open circuit under steady-state conditions.

The impedance response of electrochemical systems is often normalized to the effective area of the electrode. Such a normalization applies only if the effective area can be well defined and is not used in this chapter on the impedance response of electrical circuits. The capacitance used in this chapter, therefore, has units of F rather than F/cm^2, the resistance has units of Ω rather than Ω cm^2, and the inductance has units of H rather than H cm^2. Some symbols, units, and relationships for quantities used in electrical circuits are presented in Table 4.1.

Table 4.1: Symbols, units, and relationships for quantities used in electrical circuits.

Quantity	Symbol	Units	Abbreviation	Relations
Admittance	Y	Siemen	S or mho	$1\,\text{S} = 1/\Omega$
Angular Frequency	ω	Radian/Second	s^{-1}	$\omega = 2\pi f$
Capacitance	C	Farad	F	$1\,\text{F} = 1\,\text{s}/\Omega$
Charge	q	Coulomb	C	
Current	I	Amp	A	$1\,\text{A} = 1\,\text{C}/\text{s}$
				$1\,\text{A} = 1\,\text{V}/\Omega$
Current Density	i	Amp/Area	A/cm^2	$1\,\text{A}/\text{cm}^2 = 1\,\text{C}/\text{s}\,\text{cm}^2$
Frequency	f	Hertz	Hz	$f = \omega/2\pi$
Impedance	Z	Ohm	Ω	$1\,\Omega = 1\,\text{V}/\text{A}$
				$1\,\Omega = 1\,\text{J}/\text{sA}^2$
Inductance	L	Henry	H	$1\,\text{H} = 1\,\text{s}\Omega$
Potential	Φ	Volt	V	$1\,\text{V} = 1\,\text{J}/\text{C}$
Resistance	R	Ohm	Ω	

Response to a Sinusoidal Signal

The response in current of the passive elements to a pure sinusoidal potential modulation

$$V(t) = |\Delta V| \cos(\omega t) \tag{4.6}$$

can now be considered. The current response is given by

$$I(t) = |\Delta I| \cos(\omega t + \varphi) \tag{4.7}$$

According to the complex number properties described in Table 1.7, e.g., equations (1.124)-(1.127), equation (4.7) can be written as

$$I(t) = \text{Re}\left\{|\Delta I| \exp(\text{j}\varphi)\exp(\text{j}\omega t)\right\}$$
$$= \text{Re}\left\{\tilde{I}\exp(\text{j}\omega t)\right\} \tag{4.8}$$

Remember! 4.1 *Under steady-state conditions, the inductor is equivalent to a short circuit, the capacitor is equivalent to an open circuit, and the resistor allows current in proportion to a potential driving force.*

---

where $\widetilde{I} = |\Delta I| \exp(j\varphi)$. Due to the fact that

$$\frac{d}{dt} \text{Re} \{f(t)\} = \text{Re} \left\{ \frac{df(t)}{dt} \right\} \tag{4.9}$$

where f is a continuous function of t, then

$$\frac{dI(t)}{dt} = \text{Re} \left\{ j\omega \widetilde{I} \exp(j\omega t) \right\} \tag{4.10}$$

In the same way,

$$\frac{dV(t)}{dt} = \text{Re} \left\{ j\omega \widetilde{V} \exp(j\omega t) \right\} \tag{4.11}$$

According to expression (4.2), the response of an inductive element is given by

$$\text{Re} \left\{ \widetilde{V} \exp(j\omega t) \right\} = L\text{Re} \left\{ j\omega \widetilde{I} \exp(j\omega t) \right\} \tag{4.12}$$

As $j = \exp(j\pi/2)$ (see equation (1.87) on page 16),

$$\text{Re} \left\{ \widetilde{V} \exp(j\omega t) \right\} = L\text{Re} \left\{ \omega \widetilde{I} \exp(j(\omega t + \pi/2)) \right\} \tag{4.13}$$

Equation (4.13) shows that the potential difference is out of phase with the current. From expression (4.12),

$$\widetilde{V} = j\omega L \widetilde{I} \tag{4.14}$$

which provides the potential response of an inductor to a sinusoidal signal.

Following equation (4.5), the response of a capacitor is given by

$$\text{Re} \left\{ \widetilde{I} \exp(j\omega t) \right\} = C\text{Re} \left\{ j\omega \widetilde{V} \exp(j\omega t) \right\} \tag{4.15}$$

or

$$\text{Re} \left\{ \widetilde{I} \exp(j\omega t) \right\} = C\text{Re} \left\{ \omega \widetilde{V} \exp(j(\omega t + \pi/2)) \right\} \tag{4.16}$$

The current is out of phase with the potential difference. From expression (4.15), the frequency domain current response to a sinusoidal potential input for a capacitor is given by

$$\widetilde{I} = jC\omega \widetilde{V} \tag{4.17}$$

The corresponding expression for a resistor, according to equation (4.1), can be written as

$$\widetilde{V} = R\widetilde{I} \tag{4.18}$$

The current and potential relationships obtained here are used in the subsequent section to establish the impedance response for each type of circuit element.

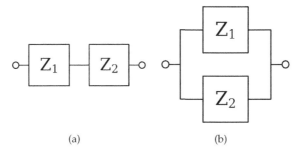

(a)   (b)

Figure 4.2: Combinations of passive elements: a) in series and b) in parallel.

Impedance Response of Passive Circuit Elements

The impedance of a circuit element is defined to be

$$Z = \frac{\widetilde{V}}{\widetilde{I}} \tag{4.19}$$

For a pure resistor, equation (4.19) yields

$$Z_R = R \tag{4.20}$$

For a capacitor

$$Z_C = \frac{1}{j\omega C} \tag{4.21}$$

and for an inductor

$$Z_L = j\omega L \tag{4.22}$$

The impedance responses for resistors, capacitors, and inductors are used to construct the impedance response of circuits.

4.1.2   Parallel and Series Combinations

For two passive elements in series, the same current must flow through the two elements, and the overall potential difference is the sum of the potential difference for each element. Thus, according to the definition of impedance given as equation (4.19), the impedance for the series arrangement shown in Figure 4.2(a) is given by

$$Z = Z_1 + Z_2 \tag{4.23}$$

Remember! 4.2 *All physical time-dependant quantities, e.g., $I(t), V(t), dI(t)/dt,$ and $dV(t)/dt$, are real quantities.*

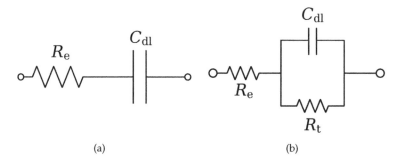

(a)   (b)

Figure 4.3: Electrical circuit consisting of a) a solution resistance in series with a capacitor and b) a solution resistance in series with a Voigt element.

For two passive elements in parallel, the overall current is the sum of the current flowing in each element, and the potential difference is the same for each dipole. Then, according to the impedance definition, the impedance for the parallel arrangement shown in Figure 4.2(b) is given by

$$Z = \left[\frac{1}{Z_1} + \frac{1}{Z_2}\right]^{-1} \tag{4.24}$$

Impedance contributions are additive for elements in series, whereas the inverse of the impedance, or the admittance, is additive for elements in parallel.

Example 4.1 **Impedance in Series:** *Derive an expression for the impedance of the electrical circuit shown in Figure 4.3(a).*

Solution: *The contribution of the resistance term to the impedance is R_e, and the impedance of the capacitor is given by equation (4.21). Following equation (4.23), the impedance of the circuit is therefore given by*

$$Z = R_e + \frac{1}{j\omega C_{dl}} \tag{4.25}$$

which can be rearranged to the form

$$Z = R_e - j\frac{1}{\omega C_{dl}} \tag{4.26}$$

The real part of the impedance is equal to R_e and is independent of frequency. The imaginary part tends toward $-\infty$ as frequency tends toward zero. The dc (zero-frequency) current is equal to zero at any applied potential, and the current at infinite frequency is equal to V / R_e.

Remember! 4.3 *Impedance contributions are additive for elements in series; whereas, the admittance is additive for elements in parallel.*

Example 4.2 Impedance in Parallel: *Derive an expression for the impedance of the electrical circuit shown in Figure 4.3(b).*

Solution: *The contribution of the resistance terms to the impedance are R_e and R_t, respectively. The impedance of the capacitor is given by equation (4.21). Following equations (4.23) and (4.24), the impedance of the circuit is therefore given by*

$$Z = R_e + \frac{1}{1/R_t + j\omega C_{dl}} \tag{4.27}$$

which can be rearranged to the form

$$Z = R_e + \frac{R_t}{1 + j\omega R_t C_{dl}} \tag{4.28}$$

Note that ω is in units of s^{-1}, and that $R_t C_{dl}$ represents the characteristic time constant for the system. The dc (zero-frequency) current is equal to $V/(R_e + R_t)$ at potential V, and the current at infinite frequency is equal to V/R_e.

Examples 4.1 and 4.2 illustrate the manner in which the impedance response of complex arrangements of circuit elements can be derived. In addition to providing an intuitive understanding of the response to a sinusoidal input, these simple circuits often form the basis for a preliminary interpretation of impedance results for electrochemical systems.

4.2 Fundamental Relationships

The impedance response can be described as having real and imaginary components, i.e.,

$$Z = Z_r + jZ_j \tag{4.29}$$

When the input and output are in phase, as shown in equation (4.20), the imaginary part of the impedance has a value of zero, and the impedance has only a real contribution, Z_r. When the input and output are out of phase, as shown in equations (4.21) and (4.22), the real part of the impedance has a value of zero, and the impedance has only imaginary contribution, Z_j.

The relationship between the complex impedance and the phase angle is shown more clearly in the use of phasor diagrams and relationships. The impedance can be expressed as

$$Z = |Z| \exp(j\varphi) \tag{4.30}$$

where $|Z|$ represents the magnitude of the impedance vector and φ represents the phase angle. The relationships among complex impedance, magnitude, and phase angle are shown in Figure 4.4. The magnitude of the impedance can be expressed in terms of real

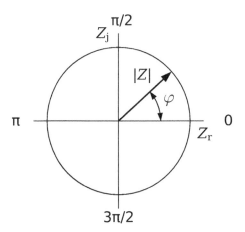

Figure 4.4: Phasor diagram showing relationships among complex impedance, magnitude, and phase angle.

and imaginary components as

$$|Z(\omega)| = \sqrt{Z_r(\omega)^2 + Z_j(\omega)^2} \tag{4.31}$$

The phase angle can be obtained from

$$\varphi(\omega) = \tan^{-1}\left(\frac{Z_j(\omega)}{Z_r(\omega)}\right) \tag{4.32}$$

or through geometric relationships that are evident in Figure 4.4, i.e.,

$$Z_r(\omega) = |Z(\omega)| \cos(\varphi(\omega)) \tag{4.33}$$

and

$$Z_j(\omega) = |Z(\omega)| \sin(\varphi(\omega)) \tag{4.34}$$

The representation of impedance in terms of magnitude and phase angle as functions of frequency on a logarithmic scale are called Bode plots.[110]

Example 4.3 Bode Representation of Elemental Circuits: *Derive an expression for the magnitude and phase angle for the circuit elements shown in Figure 4.1.*

Solution: Resistor. *The impedance for a resistor is given by $Z = R + 0j$; therefore, the magnitude is given by $|Z| = R$ and the phase angle is given by $\varphi = \tan^{-1}(0) = 0$.*

Solution: Capacitor. *The impedance for a capacitor is given by $Z = 0 - j/\omega C$; therefore, the magnitude is given by $|Z| = 1/\omega C$ and the phase angle is given by $\varphi = \tan^{-1}(-\infty) = -\pi/2$. The phase angle for a capacitor is $-90°$.*

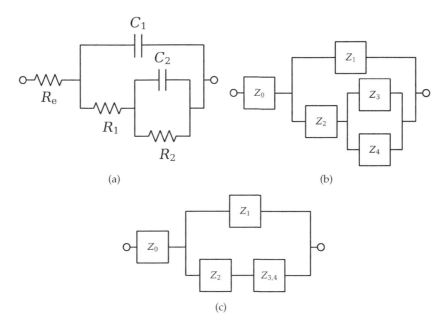

Figure 4.5: Circuits demonstrating the calculation of the impedance of nested circuit elements: a) circuit with resistor and capacitor elements; b) reconstruction of the circuit in terms of the generic impedance Z_k; and c) secondary reconstruction to facilitate calculation.

Solution: Inductor. *The impedance for an inductor is given by $Z = 0 + j\omega L$; therefore, the magnitude is given by $|Z| = \omega L$ and the phase angle is given by $\varphi = \tan^{-1}(\infty) = \pi/2$. The phase angle for an inductor is $+90°$.*

The term constant-phase element (CPE) is applied to a general circuit element that shows a constant phase angle. Thus, the resistor, capacitor, and inductor can all be considered to be constant-phase elements.

4.3 Nested Circuits

The impedance response of more complicated circuits can be readily calculated using combinations of equations (4.23) and (4.24).

Example 4.4 **Impedance Expression for a Nested Circuit:** *Derive an expression for the impedance response of the circuit shown in Figure 4.5(a), and give expressions for the dc current and the current at infinite frequency for an applied potential V.*

Solution: *The circuit can be visualized, in a manner following Figure 4.2, to be a nested series of boxes, as seen in Figure 4.5(b). The parallel arrangement of Z_3 and Z_4 can be combined in* *terms of $Z_{3,4}$ (Figure 4.5(c)). Thus, the impedance of the circuit shown in Figure 4.5(c) can be expressed as*

$$Z = Z_0 + \left[\frac{1}{Z_1} + \frac{1}{Z_2 + Z_{3,4}} \right]^{-1} \tag{4.35}$$

where, following Figure 4.5(b),

$$Z_{3,4} = \left[\frac{1}{Z_3} + \frac{1}{Z_4} \right]^{-1} \tag{4.36}$$

The next step is to introduce the impedance for each individual unit, i.e., $Z_0 = R_e$, $Z_1 = 1/(j\omega C_1)$, $Z_2 = R_1$, $Z_3 = 1/(j\omega C_2)$, and $Z_4 = R_2$.

$$Z_{3,4} = \left[j\omega C_2 + \frac{1}{R_2} \right]^{-1} \tag{4.37}$$

or

$$Z_{3,4} = \frac{R_2}{1 + j\omega R_2 C_2} \tag{4.38}$$

In terms of circuit parameters,

$$Z = R_e + \left[j\omega C_1 + \cfrac{1}{R_1 + \cfrac{R_2}{1 + j\omega R_2 C_2}} \right]^{-1} \tag{4.39}$$

At zero frequency the impedance is equal to $(R_e + R_1 + R_2)$, and at infinite frequency the impedance is equal to R_e.

4.4 Mathematical Equivalence of Circuits

Different electrical circuits possessing the same number of time constants can yield a mathematically equivalent frequency response. For example, the three circuits presented in Figure 4.6 arise from very different physical models and yet can have the same frequency response. Circuit 4.6(a) can describe two resistive layers and has been used as a measurement model, as described in Chapter 21. Circuit 4.6(b) can describe a reaction mechanism comprising two electrochemical steps or a system consisting of a coated electrode. Development of such models is discussed in Chapter 10.

The lack of uniqueness of circuit models creates ambiguity when interpreting impedance response using regression analysis. A good fit does not, in itself, validate the model used. As discussed in Chapter 23, impedance spectroscopy is not a standalone technique. Additional observations are needed to validate a model.

4.5 Graphical Representation of Circuit Response

The impedance response of a resistor in parallel to a capacitor is shown in Figure 4.7(a) as a function of frequency f in units of Hz. When plotted as a function of frequency ω

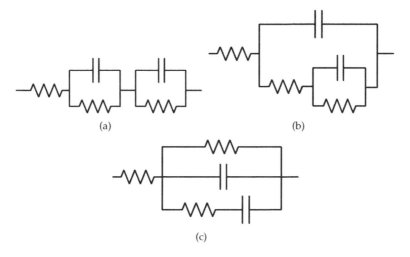

Figure 4.6: Three mathematically equivalent electrical circuits.

in units of s^{-1} (the upper axis), the minimum in the imaginary part of the impedance appears clearly at a characteristic frequency of $\omega_c = 1/\tau_c$. The dashed line corresponds to the characteristic frequency of $1\ s^{-1}$. When plotted against frequency in units of Hertz, the characteristic frequency is shifted by a factor of 2π, i.e., $f_c = 1/2\pi\tau_c$.

The corresponding Bode representation of the impedance response is shown in Figure 4.7(b) as a function of frequency f in units of Hz and frequency ω in units of s^{-1}. When plotted as a function of frequency ω in units of s^{-1}, the phase angle reaches an inflection point $(-45°)$ at a characteristic frequency $\omega_c = 1/\tau_c$. Again, when plotted against frequency in units of Hz, the characteristic frequency is shifted by a factor of 2π.

Graphical methods can be very useful for developing an interpretation of impedance measurements. A more detailed discussion of graphical methods for impedance spectroscopy is presented in Chapter 18 on page 493.

Remember! 4.4 *Electrical circuits are not unique. A good fit to experimental data is not sufficient to validate a model.*

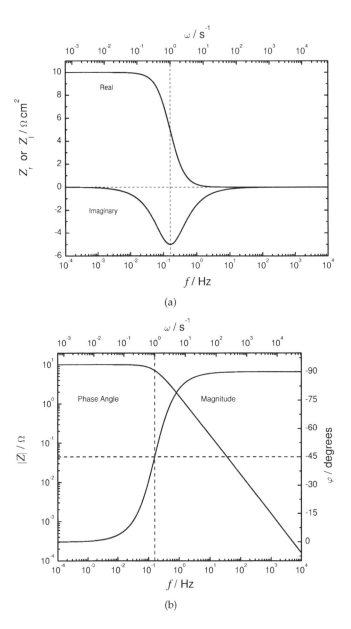

(a)

(b)

Figure 4.7: Graphical representation of the impedance response for a 10 Ω resistor in parallel with a 0.1 F capacitor. The characteristic time constant for the element is 1 s: a) real and imaginary parts of the impedance as functions of frequency and b) Bode representation of magnitude and phase angle as functions of frequency.

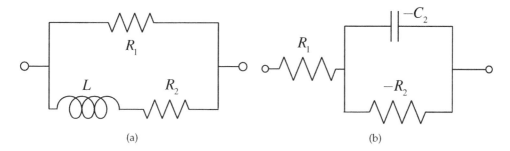

Figure 4.8: Electrical circuits resulting in an inductive loop: a) circuit with inductor and b) circuit with a negative resistance in parallel with a negative capacitance.

Problems

4.1 Show that equation (4.12) for an inductor can lead to equation (4.13). Show that this result demonstrates that current and potential signals are out-of-phase for an inductor.

4.2 Follow the development leading to equation (4.16), and verify that equation (4.16) demonstrates that current and potential signals are out-of-phase for a capacitor.

4.3 With the aid of Table 4.1, show that the units of impedance given in equations (4.21) and (4.22) are ohms.

4.4 Develop an expression for the impedance response of the circuit presented as Figure 4.6(a).

4.5 Develop an expression for the impedance response of the circuit presented as Figure 4.6(b).

4.6 Develop an expression for the impedance response of the circuit presented as Figure 4.6(c).

4.7 The use of inductors to model low-frequency inductive loops in an impedance response is somewhat controversial. Such a circuit is presented in Figure 4.8(a). Demonstrate that the circuit presented as Figure 4.8(b) with negative R and C in the nested element can be mathematically equivalent to a circuit containing an inductor.

4.8 Use a spreadsheet program to plot the impedance response of the circuit presented as Figure 4.3(a) using $R_e = 10\ \Omega$ and $C_{dl} = 20\ \mu F$.

    (a) Plot the real and imaginary parts of the impedance as a function of frequency.

    (b) Plot the results in Bode format (magnitude and phase angle as a function of frequency).

4.9   Use a spreadsheet program to plot the impedance response of the circuit presented as Figure 4.3(b) using $R_e = 10\ \Omega$, $C_{dl} = 20\ \mu F$, and $R_t = 100\ \Omega$.

   (a)   Plot the real and imaginary parts of the impedance as a function of frequency.

   (b)   Plot the results in Bode format (magnitude and phase angle as a function
   of frequency).

Chapter 5

Electrochemistry

A basic understanding of electrochemistry is necessary to develop expertise in application of electrochemical impedance spectroscopy. A brief summary of the field is presented in this chapter. Additional references are cited in Section 5.8.

5.1 Resistors and Electrochemical Cells

As Gileadi observes, the distinguishing feature of an electrochemical reaction is that the associated electrical current is a nonlinear function of electrode potential.[111] This nonlinear current-potential relationship can be contrasted to the behavior of the 1 Ω resistor represented in Figure 5.1(a). If a potential difference of 1 V is applied across the resistor, the resulting current would have a value of 1 A. The relationship between current and potential is linear, such that

$$I = V/R \tag{5.1}$$

as shown in Figure 5.2.

The behavior of the electrochemical cell is strikingly different, as can be seen from an analysis of the electrochemical cell shown schematically in Figure 5.1(b). Imagine that

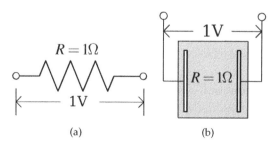

(a) (b)

Figure 5.1: Systems through which current is passed: a) 1 Ω resistor and b) an electrochemical cell with an effective electrolyte resistance of 1 Ω.

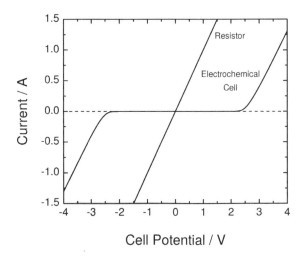

Figure 5.2: Polarization curves for a 1 Ω resistor and for a symmetric electrochemical cell with 1 Ω solution resistance and inert electrodes such as gold or platinum.

the electrodes are made of an inert material such as platinum, and that the electrolyte consists of non-electroactive species such as Na_2SO_4 in distilled water. The current flowing through the cell is influenced, not only by the ohmic resistance of the cell, but also by the potential required to drive charge-transfer reactions. In the cell envisioned here, the only electrochemical reactions that can take place involve decomposition of water into hydrogen and oxygen. To conserve charge, the hydrogen evolution reaction

$$2H_2O + 2e^- \leftrightarrows H_2 + 2OH^- \tag{5.2}$$

must be balanced by oxygen evolution

$$H_2O \leftrightarrows \frac{1}{2}O_2 + 2H^+ + 2e^- \tag{5.3}$$

The resulting current-potential, or polarization, curve is presented in Figure 5.2. At smaller cell potentials, the current is controlled by electrode kinetics. At higher cell potentials, the current increases and is controlled by the ohmic resistance. The current, therefore, appears as a line that is parallel to that of the resistor, but shifted from the origin by a potential of roughly 2 V. Thus, the results shown in Figure 5.2 suggest that a critical cell potential must be exceeded before current can flow.

As will be discussed later in this chapter, within this potential range (or polarization window), current does flow, but, due to unfavorable electrode kinetics, the magnitude of the current may be below detection limits. The rate constants for electrochemical reactions are strongly dependent on the nature of the electrode and the electrolyte in contact with the electrode. Accordingly, the size of the polarization window depends on the electrode material and on the electrolyte. Lovrić[112] reports that the cathodic limits of the mercury electrode in aqueous solutions are between -0.86 V(NHE) at pH=1

Table 5.1: Electrode materials and their potential windows under which faradaic currents cannot be detected. Anodic and cathodic potential limits are referenced to the normal hydrogen electrode. The potential window is independent of reference electrode.

Electrode	Electrolyte	Potential Limits, V(NHE)			Reference
		Anodic	Cathodic	Window	
Au	H_2O	1.53	−0.32	1.85	113
BDD	H_2O	2.30	−1.10	3.40	113, 114
C	1M $HClO_4$	1.65	−0.76	2.40	112
C	1M NaO	1.15	−1.26	2.40	112
Hg	1M $HClO_4$	0.75	−0.86	1.60	112
Hg	1M NaO	0.25	−2.56	2.80	112
Pt	1M $HClO_4$	1.55	−0.01	1.55	112
Pt	1M NaOH	0.99	−0.91	1.90	112

BDD: Boron-doped diamond electrode
C: Glassy carbon electrode

and -2.56 V(NHE) at pH=13. At pH=1, the dominant cathodic reaction is proton reduction, i.e.,

$$2H^+ + 2e^- \leftrightarrows H_2 \tag{5.4}$$

whereas, reaction (5.2) takes place at pH=13.

Evidently, the rate and the identity of the cathodic reaction on mercury depends on electrolyte pH. The anodic limit also depends on the nature of the electrolytic solution. In aqueous nitrate and perchlorate solutions, the anodic limit is 0.45 V(NHE) due to oxidation of the mercury, but the limit is 0.15 V(NHE) in chloride solutions due to the formation and precipitation of calomel.[112] Some reported anodic and cathodic limits for different electrode materials are presented in Table 5.1. The dependence of potential window on electrode material provides evidence that the observed potential window must be attributed to kinetic rather than thermodynamic origins.

The difference, then, between a resistor and an electrochemical cell is that current passes through the resistor through transport of electrons; whereas, in an electrochemical cell, the electrons must be converted to ionic species by electrochemical reactions. As will be shown in this chapter, the rates of electrochemical reactions have an exponential dependence on potential. This effect is shown dramatically for electrodes like those described in Table 5.1 that are considered ideally polarizable over a substantial potential window. If the inert electrodes were replaced by a metal such as iron that dissolves readily, the ideally polarized potential window would not be seen, but the resulting polarization curve would still be nonlinear.

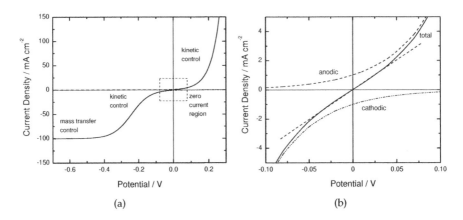

Figure 5.3: Polarization curves for a system with mass-transfer control at cathodic potentials: a) broad range of potential with regions of behavior identified and b) expanded representation of the zero-current region with cathodic and anodic current contributions identified. The potential is referenced to the equilibrium potential.

5.2 Polarization Behavior for Electrochemical Systems

The nonlinear current-potential behavior associated with an electrochemical system is illustrated in Figure 5.3(a). In the case shown here, the anodic (positive) current has an exponential dependence on potential; whereas, the cathodic (negative) current displays an influence of mass-transfer limitations. In Figure 5.3(a), regions are identified for which the current has a value close to zero, the cathodic current is controlled by reaction kinetics, and the cathodic current is controlled by mass transfer.

A generalized expression for a given reaction can be expressed as

$$\sum_i s_i M_i^{z_i} \rightleftarrows ne^- \tag{5.5}$$

where s_i is the stoichiometric coefficient for species i, M_i represents the species i, and z_i is the charge associated with species i. When $s_i = 0$, the species i is not a reactant. For a species that is a reactant for an anodic reaction, $s_i > 0$, and, for species that is a reactant for the cathodic reaction, $s_i < 0$. The relationship between the flux of a given species and the current density associated with reaction (5.5) can be expressed as

$$N_i|_{y=0} = -\frac{s_i}{nF} i_F \tag{5.6}$$

where i_F is the faradaic current density resulting from reaction (5.5).

Remember! 5.1 *Electrochemical reactions, which transfer charge between electrons and soluble species, are required for current to flow in an electrochemical cell.*

Example 5.1 Relationship between Flux and Current Density: *Consider the reaction*

$$H_2O_2 \rightarrow O_2 + 2H^+ + 2e^- \qquad (5.7)$$

Find the direction of the corresponding fluxes and plot concentration gradients for H_2O_2 and O_2.

Solution: *Reaction (5.7) may be written in the form of reaction (5.5) as*

$$H_2O_2 - O_2 - 2H^+ \rightarrow 2e^- \qquad (5.8)$$

where $s_{H_2O_2} = 1$, $s_{O_2} = -1$, and $n = 2$.

As discussed in Section 5.5, the flux may be written as

$$N_i = -D_i \frac{dc_i}{dy} \qquad (5.9)$$

where migration and convection have been neglected. Reaction (5.7) is an anodic reaction, and the associated current density has a positive value. For H_2O_2, the stoichiometric coefficient has a positive value and the associated flux has a negative value, meaning that peroxide is moving in the $-y$ direction. The concentration gradient, obtained from equation (5.9) has a positive value. A representation of the concentration gradient for peroxide within a stagnant layer of thickness δ is presented in Figure 5.4 where the current density was assumed to be equal to $1/2$ of the mass-transfer-limited current density (see Example 5.5).

For O_2, the stoichiometric coefficient has a negative value and the associated flux has a positive value, meaning that oxygen is moving in the $+y$ direction. The concentration gradient, obtained from equation (5.9) has a negative value. A representation of the concentration gradient for oxygen is presented in Figure 5.4.

The properties of electrochemical reactions are explored in the subsequent sections. The generalized rate expression for reaction (5.5) is presented in Section 5.4.

5.2.1 Zero Current

An expanded representation of the zero-current region of Figure 5.3(a) is presented as Figure 5.3(b). The positive current contributed by the anodic reaction is balanced by the negative current contributed by the cathodic reaction. If the anodic and cathodic reactions represent forward and backward rates of the same reaction, a zero current can be obtained under the condition of reaction equilibrium. If the anodic and cathodic reactions represent forward and backward rates of different reactions, a true equilibrium is not reached as neither reaction is equilibrated. Thus, the net current may have a value equal to zero under either equilibrium or nonequilibrium conditions.

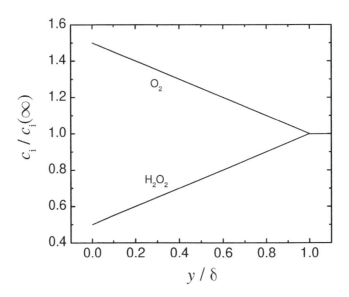

Figure 5.4: Concentration gradients for oxygen and peroxide within a stagnant layer of thickness δ corresponding to reaction (5.7) with a current density equal to $1/2$ of the mass-transfer-limited current density.

Equilibrium

If the current corresponds to a single electrochemical reaction, a zero current is observed if the forward and backward rates for the reaction are equal. For example, if the forward (or anodic) reaction is given by

$$Cu \rightarrow Cu^{2+} + 2e^- \qquad (5.10)$$

the backward (or cathodic) reaction is given by

$$Cu^{2+} + 2e^- \rightarrow Cu \qquad (5.11)$$

If a current i_a is assigned to reaction (5.10), and a current i_c is assigned to reaction (5.11), the net current will be given by

$$i = i_a + i_c \qquad (5.12)$$

where $i_a > 0$ and $i_c < 0$. At equilibrium, $i_a = -i_c$ and $i = 0$. The potential at which the current for a single electrochemical reaction is equal to zero is termed the equilibrium potential. The value for the equilibrium potential can be calculated using thermodynamic arguments.[115,116]

Nonequilibrium

If the zero-current condition arises through a balancing of different reactions, equilibrium is not achieved because the net rate for each reaction is not equal to zero. For example, the corrosion of iron

$$Fe \rightarrow Fe^{2+} + 2e^- \tag{5.13}$$

may be balanced by reduction of oxygen

$$O_2 + 2H_2O + 4e^- \rightarrow 4OH^- \tag{5.14}$$

such that the net current is equal to zero while dissolution of iron continues and oxygen is consumed. This is clearly not an equilibrium condition.

The potential at which the current for multiple electrochemical reactions is equal to zero is termed the mixed potential or, in the case of metal dissolution, the corrosion potential. Concepts of thermodynamics, kinetics, and transport must be applied to calculate values for the mixed or corrosion potential. This condition is described further in Section 5.2.3.

5.2.2 Kinetic Control

The region of kinetic control of electrochemical reactions is characterized by current densities that are exponential functions of potential. For a single reversible reaction, the Butler–Volmer equation

$$i = i_0 \left\{ \exp\left(\frac{(1-\alpha)nF}{RT} \eta_s \right) - \exp\left(-\frac{\alpha nF}{RT} \eta_s \right) \right\} \tag{5.15}$$

is commonly used to describe the influence of potential on the current density. Here, i_0 is the exchange current density, so defined because at $\eta_s = 0$, $i_a = -i_c = i_0$. The surface overpotential η_s represents the departure from an equilibrium potential such that, at $\eta_s = 0$, the total current $i = i_a + i_c$ is equal to zero. The symmetry factor α is the fraction of the surface overpotential that promotes the cathodic reaction. Usually, α is assumed to have a value close to 0.5 and must have a value between 0 and 1.

As shown in Chapter 10 (page 207), electrochemical kinetics plays a major role in the interpretation of impedance spectra. To streamline the discussion of electrochemical kinetics, a more compact notation is used in which

$$b_a = \frac{(1-\alpha)nF}{RT} \tag{5.16}$$

for anodic reactions and

$$b_c = \frac{\alpha nF}{RT} \tag{5.17}$$

for cathodic reactions where b_a and b_c have units of inverse potential. Thus, equation (5.15) can be written

$$i = i_0 \left\{ \exp\left(b_a \eta_s \right) - \exp\left(-b_c \eta_s \right) \right\} \tag{5.18}$$

The parameters b_a and b_c are closely related to the Tafel slope. With the recognition that $\ln(10) = 2.303$, the Tafel slope β_c for a cathodic reaction may be written as

$$\beta_c = \frac{2.303 \times 10^3 RT}{\alpha nF} = \frac{2.303 \times 10^3}{b_c} \tag{5.19}$$

where β_c has units of mV/decade.

The current density presented in Figure 5.3(b) can be described as being a linear function of potential over a narrow range of potential where the magnitude of the current has a value close to zero. Taylor series expansions of the exponential terms in equation (5.15) yield

$$i = i_0 \frac{nF}{RT} \eta_s = i_0 (b_a + b_c) \eta_s \tag{5.20}$$

A similar linear regime can be identified when the zero-current condition arises from the balancing of different anodic and cathodic reactions.

At very positive potentials, the cathodic term is negligible, and the current density can be expressed by

$$i = i_0 \exp (b_a \eta_s) \tag{5.21}$$

At very negative potentials, the anodic term can be neglected, and

$$i = -i_0 \exp (-b_c \eta_s) \tag{5.22}$$

Equations (5.21) and (5.22) are examples of Tafel equations in which the current is an exponential function of potential.

Example 5.2 Butler-Volmer and Tafel Equations: *Show the values of overpotential for which the Butler–Volmer equation may be considered to be approximated by a Tafel equation.*

Solution: *Equation (5.15) may be considered to approach the limiting Tafel behavior expressed in equations (5.21) and (5.22) when the smaller current density represents a small fraction of the current density. A parameter representing this ratio of currents may be expressed as*

$$\zeta_a = 1 - \frac{i_a + i_c}{i_a} = \frac{-i_c}{i_a} \tag{5.23}$$

or

$$\zeta_c = 1 - \frac{i_a + i_c}{i_c} = \frac{-i_a}{i_c} \tag{5.24}$$

For a single electrochemical reaction with $\alpha = 0.5$, equations (5.16) and (5.17) must be equal. Thus, $b = b_a = b_c$, and equation (5.23) yields

$$\zeta_a = \frac{\exp (-b \eta_s)}{\exp (b \eta_s)} = \exp (-2b \eta_s) \tag{5.25}$$

The value of η_s corresponding to a given quantity ζ can be expressed as

$$\eta_{s,\zeta} = -\frac{1}{2b} \ln (\zeta_a) \tag{5.26}$$

Similarly, equation (5.24) yields

$$\eta_{s,\zeta} = \frac{1}{2b} \ln (\zeta_c) \tag{5.27}$$

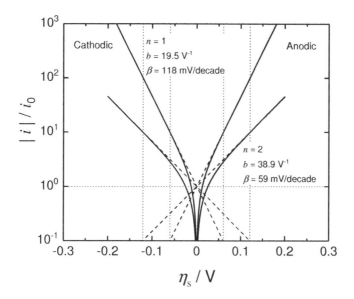

Figure 5.5: Normalized current density as a function of surface overpotential. Dashed lines correspond to equations (5.21) and (5.22), which are, respectively, examples of anodic and cathodic Tafel curves. The dotted vertical lines correspond to $\pm(1/2b)\ln(0.01)$. The intersection of the Tafel lines, shown by the horizontal dotted line, yields the exchange current density i_0.

For $n = 1$ at $T = 298K$, equations (5.16) and (5.17) yield $b = 19.5\ V^{-1}$, corresponding to a Tafel slope of 118 mV/decade. Under the assumption that the smaller current in equation (5.15) is one percent of the current, i.e., $\xi = 0.01$, equation (5.26) yields $\eta_{s,0.01} \approx 0.118\ V$. For $n = 2$ at $T = 298K$, $b = 38.9\ V^{-1}$ ($\beta = 59\ mV/decade$) and $\eta_{s,0.01} \approx 0.59\ V$. The results are presented in Figure 5.5. The intersection of the extrapolated lines for anodic and cathodic currents yields the equilibrium potential and the exchange current density. Tafel behavior is observed at $|\eta_s| > 118\ mV$ for $n = 1$ and at $|\eta_s| > 59\ mV$ for $n = 2$.

Electrochemical experiments are generally performed for systems that are not at equilibrium. Thus, Tafel expressions are often used in the development of kinetic models for impedance spectroscopy.

5.2.3 Mixed-Potential Theory

The Butler–Volmer equation applies for a single electrochemical reaction. A similar mathematical form may be observed for situations in which two or more reactions are taking place, but the kinetic parameters associated with a current equal to zero are not necessarily associated with the equilibrium condition.

Consider, for example, the corrosion of iron in a solution containing 1M $FeSO_4$, i.e.,

$$Fe \rightleftarrows Fe^{2+} + 2e^- \tag{5.28}$$

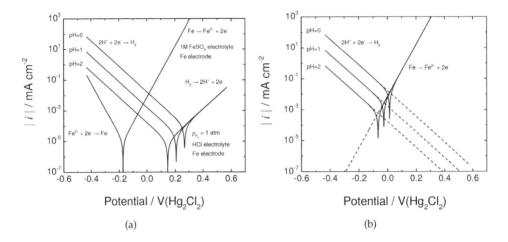

Figure 5.6: Polarization curves for a system with two electrochemical reactions: a) the hydrogen evolution reaction on an iron electrode in HCl at $p_{H_2} = 1$ atm and the corrosion of iron in 1M FeSO$_4$ and b) the hydrogen evolution and corrosion reactions on an iron electrode in HCl electrolytes. Kinetic parameters were taken from McCafferty[117] and Conway.[119]

McCafferty[117] reports an exchange current density for iron in 1M FeSO$_4$ of $i_{0,\mathrm{Fe}} = 1 \times 10^{-8}$ A/cm^2. The equilibrium potential can be obtained from thermodynamic considerations to be $U_{\mathrm{Fe}}^\theta = -0.44$ V(H$_2$) or -0.1724 V(Hg/Hg$_2$Cl$_2$). Jones[118] reports that the Tafel slope is 60 mV/decade. The resulting polarization curve can be expressed in the form of a Butler–Volmer equation and is presented in Figure 5.6(a). Similarly, the hydrogen evolution reaction

$$H_2 \rightleftarrows 2H^+ + 2e^- \tag{5.29}$$

has been reported to have, on an iron electrode in HCl at $p_{H_2} = 1$ atm, an exchange current density of $i_{0,\mathrm{H}_2} = 1 \times 10^{-7}$ A/cm^2 and a Tafel slope of 120 mV/decade.[119] The resulting Butler–Volmer equation is also presented in Figure 5.6(a).

Each reaction is characterized by its own equilibrium potential and exchange current density. The development of the hydrogen reaction is presented in greater detail in Example 5.8. Under the assumption that reaction (5.29) proceeds according to an electrochemical step

$$H_{ads} \underset{k_{-1}}{\overset{k_1}{\rightleftharpoons}} H^+ + e^- \tag{5.30}$$

known as the Volmer reaction and a chemical step

$$2H_{ads} \underset{k_b}{\overset{k_f}{\rightleftharpoons}} H_2 \tag{5.31}$$

known as the Tafel reaction, the corresponding current density can be expressed as a Butler–Volmer equation, shown in equation (5.98). The exchange current density, given by equation (5.99), is dependent on the partial pressure of hydrogen, $P = k_b p_{H_2}/k_f$, and on the concentration of hydrogen ions at the electrode surface. As

$$pH = -\log_{10}(c_{H^+})$$ (5.32)

equation (5.99) can be expressed, for constant hydrogen partial pressure and $\alpha = 0.5$, as

$$i_0 = \frac{F\Gamma}{1+\sqrt{P}}\left(k_1 k_{-1}\sqrt{P}\right)^{1/2} 10^{-pH/2} = K_1 10^{-pH/2}$$ (5.33)

where Γ represents the maximum coverage and P is the scaled partial pressure of hydrogen defined in equation (5.92).

The equilibrium potential, given by equation (5.97), is also dependent on the partial pressure of hydrogen and on the concentration of hydrogen ions at the electrode surface. Equation (5.97) can be expressed in terms of pH as

$$V_{0,H_2} = \frac{RT}{F}\left(-2.303pH + \ln\left(\frac{k_{-1}}{k_1\sqrt{P}}\right)\right)$$ (5.34)

For constant hydrogen partial pressure,

$$V_{0,H_2} = \frac{RT}{F}(K_2 - 2.303pH)$$ (5.35)

The resulting values for current density are given in Figure 5.6(a) with pH as a parameter.

For an iron electrode in a de-aerated aqueous electrolyte containing dilute HCl, the cathodic portion of the iron reaction $Fe^{2+} + 2e^- \rightarrow Fe$ does not occur due to the absence of ferrous ions in the electrolyte. Similarly, the anodic part of the hydrogen reaction does not proceed due to the absence of dissolved hydrogen gas in the electrolyte. Thus, at the condition where the net current is equal to zero, the anodic reaction associated with dissolution of iron is balanced by the cathodic reaction associated with evolution of hydrogen. The corresponding polarization curves are presented in Figure 5.6(b) with pH as a parameter.

The same information is often presented in the form of Evans diagrams, presented in Figure 5.7, from which the corrosion potential V_{corr} and the corrosion current density i_{corr} may be extracted, following the dashed lines meeting at the intersection of the extrapolated Tafel lines.[120] The results presented in Figure 5.7 show clearly that, even at open circuit where the net current is equal to zero, the system is not at equilibrium and the corrosion current density is not equal to zero. As shown in Figure 5.7, the corrosion rate at open circuit increases as the electrolyte becomes more acidic due to the corresponding increase in the hydrogen evolution reaction. Application of a potential that is cathodic to the corrosion potential, as is done for application of cathodic protection, reduces the corrosion current, but, as can be seen in Figure 5.6(b), this current does not become equal to zero. The corrosion potential and corrosion current can also be extracted from Figure 5.6(b), and both representations are found in the literature.

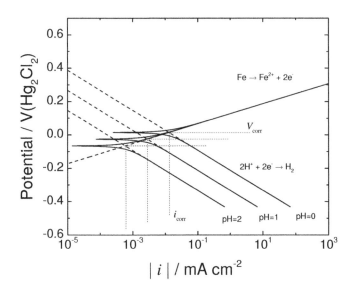

Figure 5.7: Potential as a function of current density for hydrogen evolution and corrosion reactions on an iron electrode in HCl electrolytes. Kinetic parameters were taken from McCafferty[117] and Conway.[119] Dashed lines meeting at the intersection of the extrapolated Tafel lines indicate the corrosion potential E_{corr} and the corrosion current density i_{corr}. Schematic diagrams from which corrosion properties are extracted are called Evans diagrams.[120]

Example 5.3 **Corrosion:** *Consider a system in which the corrosion reaction*

$$\text{Fe} \rightarrow \text{Fe}^{2+} + 2\text{e}^- \tag{5.36}$$

and the hydrogen evolution (or proton reduction) reaction

$$2\text{H}^+ + 2\text{e}^- \rightarrow \text{H}_2 \tag{5.37}$$

take place in an anaerobic environment. Show that the current density for the coupled reactions may be expressed in the form of a Butler-Volmer equation.

Solution: *The current associated with the corrosion and hydrogen evolution reactions may be expressed as*

$$i = K^*_{\text{Fe}} \exp\left(b_{\text{Fe}}(V - V_{0,\text{Fe}})\right) - K^*_{\text{H}_2} \exp\left(-b_{\text{H}_2}(V - V_{0,\text{H}_2})\right) \tag{5.38}$$

where rate constants K^*_{Fe} and $K^*_{\text{H}_2}$ and equilibrium potentials $V_{0,\text{Fe}}$ and V_{0,H_2} for the corrosion and hydrogen evolution reactions, respectively, follow the notation presented in equation (10.14) on page 209. The potential at which the current density is equal to zero is called the corrosion potential. This parameter is defined in Figure 5.7. At $i = 0$,

$$K^*_{\text{Fe}} \exp\left(b_{\text{Fe}}(V_{\text{corr}} - V_{0,\text{Fe}})\right) = K^*_{\text{H}_2} \exp\left(-b_{\text{H}_2}(V_{\text{corr}} - V_{0,\text{H}_2})\right) \tag{5.39}$$

The corrosion potential may be obtained as

$$V_{corr} = \frac{1}{b_{Fe} + b_{H_2}} \left(\ln \left(\frac{K_{H_2}^*}{K_{Fe}^*} \right) + b_{Fe} V_{0,Fe} + b_{H_2} V_{0,H_2} \right) \tag{5.40}$$

The corrosion current i_{corr}, also identified in Figure 5.7, may be obtained from either the anodic or cathodic reaction at the potential V_{corr}, i.e.,

$$
\begin{aligned}
i_{corr} &= K_{Fe}^* \exp\left(b_{Fe}(V_{corr} - V_{0,Fe})\right) \\
&= K_{H_2}^* \exp\left(-b_{H_2}(V_{corr} - V_{0,H_2})\right)
\end{aligned}
\tag{5.41}
$$

Upon introduction of equation (5.40) into the anodic part of equation (5.41),

$$i_{corr} = K_{Fe}^* \left(\frac{K_{H_2}^*}{K_{Fe}^*} \right)^{\frac{b_{Fe}}{b_{Fe} + b_{H_2}}} \exp\left(\frac{b_{Fe} b_{H_2}}{b_{Fe} + b_{H_2}} \left(V_{0,H_2} - V_{0,Fe} \right) \right) \tag{5.42}$$

or

$$i_{corr} = K_{Fe}^{*\frac{b_{H_2}}{b_{Fe} + b_{H_2}}} K_{H_2}^{*\frac{b_{Fe}}{b_{Fe} + b_{H_2}}} \exp\left(\frac{b_{Fe} b_{H_2}}{b_{Fe} + b_{H_2}} \left(V_{0,H_2} - V_{0,Fe} \right) \right) \tag{5.43}$$

The same expression is obtained by starting from the cathodic part of equation (5.41).

The surface overpotential is defined to be

$$\eta_s = V - V_{corr} \tag{5.44}$$

Thus, the expression

$$V = \eta_s + V_{corr} \tag{5.45}$$

is introduced in equation (5.38). After considerable algebraic manipulation, an expression for the current density is found as

$$i = i_{corr} \left\{ \exp(b_{Fe}\eta_s) - \exp(-b_{H_2}\eta_s) \right\} \tag{5.46}$$

This is similar to the Butler–Volmer equation given as equation (5.18), but the exchange current density i_0 is replaced by the corrosion current density i_{corr}, the coefficients b_a and b_c that correspond to the anodic and cathodic parts of a given reaction, respectively, are replaced by coefficients b_{Fe} and b_{H_2} that correspond to different reactions, and the surface overpotential η_s represents a departure from the corrosion potential V_{corr} rather than the equilibrium potential for a given reaction.

It is important to recognize the difference between a system at equilibrium, in which a total current density equal to zero means that all reactions are equilibrated, and a system at open-circuit, in which reactions are not equilibrated but have current contributions that sum to zero. When the corrosion system described in Example 5.3 is at open circuit, the corrosion reaction occurs and the electrons produced are consumed by a cathodic reaction that occurs at the same rate. As the following example shows, similar considerations may apply in the absence of corrosion reactions.

Example 5.4 **Water Oxidation and Reduction:** *Show that current density may be expressed in the form of a Butler–Volmer equation for oxygen and hydrogen evolution reactions on an inert electrode in an electrolyte that contains no dissolved oxygen or hydrogen gasses. The anodic (oxygen evolution) reaction may be expressed as*

$$H_2O \rightarrow \frac{1}{2}O_2 + 2H^+ + 2e^- \tag{5.47}$$

the cathodic (hydrogen evolution) reaction may be expressed as

$$2H_2O + 2e^- \rightarrow H_2 + 2OH^- \tag{5.48}$$

which applies for pH values greater than 7.

Solution: *The current associated with the corrosion and hydrogen evolution reactions may be expressed as*

$$i = K_{O_2}^* \exp\left(b_{O_2}(V - V_{0,O_2})\right) - K_{H_2}^* \exp\left(-b_{H_2}(V - V_{0,H_2})\right) \tag{5.49}$$

The potential for which the net current is equal to zero is obtained by setting $i = 0$, yielding

$$K_{O_2}^* \exp\left(b_{O_2}(V_{\text{mixed}} - V_{0,O_2})\right) = K_{H_2}^* \exp\left(-b_{H_2}(V_{\text{mixed}} - V_{0,H_2})\right) \tag{5.50}$$

The potential at zero current may be obtained as

$$V_{\text{mixed}} = \frac{1}{b_{O_2} + b_{H_2}}\left(\ln\left(\frac{K_{H_2}^*}{K_{O_2}^*}\right) + b_{O_2}V_{0,O_2} + b_{H_2}V_{0,H_2}\right) \tag{5.51}$$

The apparent exchange current density i_{mixed} may be obtained from either the anodic or cathodic reaction at the potential V_{mixed}, i.e.,

$$i_{\text{mixed}} = K_{O_2}^{* \frac{b_{H_2}}{b_{O_2}+b_{H_2}}} K_{H_2}^{* \frac{b_{O_2}}{b_{O_2}+b_{H_2}}} \exp\left(\frac{b_{O_2}b_{H_2}}{b_{O_2} + b_{H_2}}\left(V_{0,H_2} - V_{0,O_2}\right)\right) \tag{5.52}$$

The same expression is obtained by starting from the cathodic part of equation (5.41).
    The surface overpotential is defined to be

$$\eta_s = V - V_{\text{mixed}} \tag{5.53}$$

Thus, after algebraic manipulation, an expression for the current density is found as

$$i = i_{\text{mixed}}\left\{\exp(b_{O_2}\eta_s) - \exp(-b_{H_2}\eta_s)\right\} \tag{5.54}$$

As was seen in Example 5.3, equation (5.54) takes the form of the Butler–Volmer equation, but the exchange current density i_0 is replaced by an apparent exchange current density i_{mixed}, the coefficients b_a and b_c that correspond to the anodic and cathodic parts of a given reaction,

Table 5.2: Kinetic parameters for electrochemical reactions considered in Example 5.4.

Anodic			Cathodic		
$K_{O_2}^* = i_{0,O_2}$	1×10^{-7}	A/cm^2	$K_{H_2}^* = i_{0,H_2}$	1×10^{-10}	A/cm^2
V_{O_2}	1.223	V	V_{H_2}	0	V
β_{O_2}	60	mV/decade	β_{H_2}	120	mV/decade
b_{O_2}	38.38	V^{-1}	b_{H_2}	19.19	V^{-1}

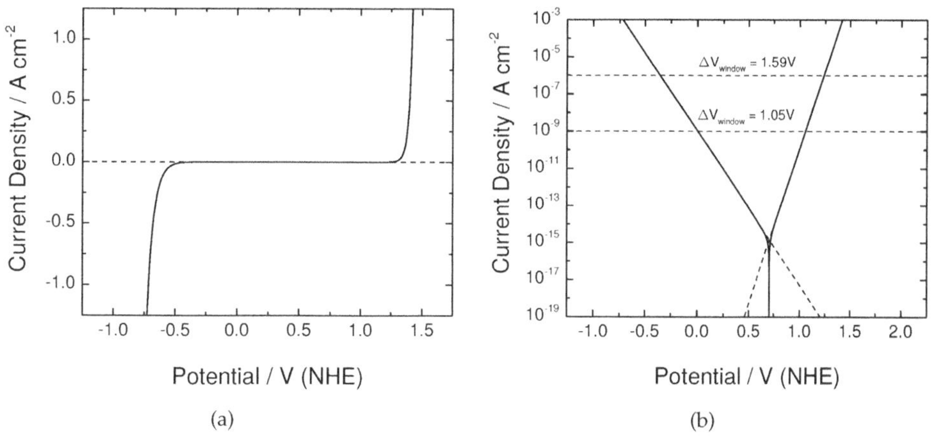

Figure 5.8: Polarization curve for the system considered in Example 5.4 for oxygen and hydrogen evolution reactions on an inert electrode in an electrolyte that contains no dissolved oxygen or hydrogen gasses: a) linear scale for current density and b) logarithmic scale for the absolute vale of current density. Kinetic parameters were taken from McCafferty[117] and Conway.[119]

respectively, are replaced by coefficients b_{O_2} and b_{H_2} that correspond to different reactions, and the surface overpotential η_s represents a departure from the corrosion potential V_{mixed} rather than the equilibrium potential for a given reaction.

Under the assumption that the kinetic parameters are those given in Table 5.2, equation (5.51) yields a value of $V_{mixed} = 0.706\ V(NHE)$, and equation (5.52) yields a value of $i_{mixed} = 1.4 \times 10^{-15}\ A/cm^2$. The resulting polarization curve, presented in linear coordinates in Figure 5.8(a), show a range of potential over which no current appears to flow. The kinetic origin of this potential window is shown in Figure 5.8(b). While a current does not flow at a potential $V = V_{mixed}$ or $\eta_s = 0$, a finite water oxidation reaction is balanced by a finite hydrogen evolution current. The value of this current is very small and much below limits of detection. The size of the potential window can be derived from equation (5.54). The expression

$$\eta_{s,O_2,max} = \frac{1}{b_{O_2}} \ln \left(\frac{i_{detect}}{i_{mixed}} \right) \tag{5.55}$$

provides the maximum value of η_{s,O_2} for which the oxygen evolution current is below a critical *current density* i_{detect}, *and*

$$\eta_{s,H_2,min} = -\frac{1}{b_{H_2}} \ln\left(\frac{i_{detect}}{i_{mixed}}\right) \tag{5.56}$$

provides the minimum value of η_{s,H_2} *for which the magnitude of the hydrogen evolution current is below* i_{detect}. *If* i_{detect} *represents the lower limit for measurement of current density, then*

$$\Delta V_{window} = \left(\eta_{s,O_2,max} - \eta_{s,H_2,min}\right) = \left(\frac{1}{b_{O_2}} + \frac{1}{b_{H_2}}\right) \ln\left(\frac{i_{detect}}{i_{mixed}}\right) \tag{5.57}$$

represents the potential window for which current is not detected. The size of the potential window clearly depends on the magnitude of the detection current density. Under the assumption that the detection current density has a value of 10^{-9} *A/cm^2 (or 1 nA/cm^2), the potential window has a value of 1.06 V; whereas, if the detection current density has a value of* 10^{-6} *A/cm^2 (or 1 μA/cm^2), the potential window has a value of 1.59 V.*

Examples 5.3 and 5.4 show that, while coupled reactions may be placed into the form of Butler–Volmer reactions, the systems are not at equilibrium, even if the net current density is equal to zero. For a single electrochemical reaction, equations (5.16) and (5.17) show that

$$b_a + b_c = \frac{nF}{RT} \tag{5.58}$$

No such relationship exists between the parameters b_{O_2} and b_{H_2} used in Example 5.4 or b_{Fe} and b_{H_2}, used in Example 5.3. It is important to note that two independent Tafel expressions are used to express the reactions involved in a mixed potential situation.

5.2.4 Mass-Transfer Control

The rate of the electrochemical reactions may be limited by the finite rate at which reacting species may be carried to the electrode surface. Such a case is illustrated in Figure 5.3(a) at negative values of potential. For a cathodic reaction that is first order with respect to species i, the current density can be written as

$$i = -k_c nF c_i(0) \exp\left(-b_c \eta_s\right) \tag{5.59}$$

The current density corresponds to the flux density of the reacting species, i.e.,

$$i = -nF D_i \left.\frac{dc_i}{dy}\right|_{y=0} \tag{5.60}$$

Under assumption of a linear concentration gradient in the diffusion layer of thickness δ_i, equation (5.60) becomes

$$i = -nF D_i \frac{c_i(\infty) - c_i(0)}{\delta_i} \tag{5.61}$$

where $c_i(\infty)$ is the concentration of species i in the bulk solution.

A formal treatment for determination of the value of δ_i requires solution of the convective-diffusion equations, as described in Chapter 11 (page 243). By eliminating $c_i(0)$ from equations (5.59) and (5.61), the value of the current density is obtained as

$$i^{-1} = i_{\text{lim}}^{-1} + i_k^{-1} \tag{5.62}$$

where

$$i_{\text{lim}} = -nFD_i c_i(\infty)/\delta_i \tag{5.63}$$

is the mass-transfer-limited current density and

$$i_k = -k_c nF c_i(\infty) \exp(-b_c \eta_s) \tag{5.64}$$

is the kinetic current based on the bulk concentration. Equation (5.62) is referred to as the Koutecky–Levich equation.[121] The numerical value for the mass-transfer-limited current density is influenced by the bulk concentration and diffusivity of the limiting reactant, by the extent of convection, and by the cell geometry.

Equation (5.63) may appear to have been obtained from equation (5.61) under the assumption that $c_i(0) = 0$, which often raises conceptual difficulties because equation (5.59) depends on $c_i(0)$ in such a way that $i = 0$ when $c_i(0) = 0$. In fact, equation (5.63) was obtained under the assumption that $c_i(0) \ll c_i(\infty)$. An expression for $c_i(0)$ for the condition of a mass-transfer-limited current density may be obtained from equation (5.59) as

$$c_i(0) = \frac{i_{\text{lim}}}{-k_c nF \exp(-b_c \eta_s)} = \varepsilon \tag{5.65}$$

where ε is a very small number. Equation (5.65) shows that $c_i(0)$ tends toward zero as $\exp(-b_c \eta_s)$ tends toward infinity. Thus, equation (5.63) is fully consistent with equation (5.59).

Example 5.5 Mass-Transfer-Limited Current Density: *Find the relationship between the concentration of a reacting species at the electrode surface and the corresponding mass-transfer-limited current density.*

Solution: *Equation (5.61) can be expressed in terms of the mass-transfer-limited current density, equation (5.63), as*

$$i = i_{\text{lim}} \left(1 - \frac{c_i(0)}{c_i(\infty)}\right) \tag{5.66}$$

After rearrangement,

$$\frac{c_i(0)}{c_i(\infty)} = 1 - \frac{i}{i_{\text{lim}}} \tag{5.67}$$

As the current density approaches its mass-transfer-limited value, the concentration of the reacting species approaches zero.

Example 5.6 Application of the Koutecky–Levich Equation: *For a rotating disk electrode, the generalized Levich equation indicates that the mass-transfer-limited current density is given by*

$$i_{lim} = 0.62045 \frac{nF}{s_i} D_i^{2/3} \Omega^{1/2} \nu^{-1/6} c_i(\infty) \tag{5.68}$$

where Ω is the disk rotation speed in radian/s. Show how graphical methods may be employed to obtain physically meaningful quantities from current density measured at different potentials and rotation speeds.

Solution: *Two approaches may be convenient, depending on whether measurements are made at fixed potential or at fixed rotation speed.*

a) Fixed rotation speed. *For measurements made at a fixed rotation speed, the current density may be measured as a function of potential. From equations (5.59), (5.64), and (5.67), the current density may be expressed as*

$$i = i_k \frac{c_i(0)}{c_i(\infty)} = i_k \left(1 - \frac{i}{i_{lim}}\right) \tag{5.69}$$

or

$$i = \frac{i_{lim} i_k}{(i_{lim} + i_k)} \tag{5.70}$$

As the potential becomes more negative, $|i_k| \gg |i_{lim}|$ and i approaches i_{lim}. A plot of current as a function of potential will yield the mass-transfer-limited current density. A subsequent plot of i_{lim} as a function of $\sqrt{\Omega}$ will yield a straight line with a slope equal to $0.62045nFD_i^{2/3}\nu^{-1/6}c_i(\infty)$, from which the diffusion coefficient D_i may be obtained.

An example is presented in Figure 5.9 for kinetic data presented by Jovancicevic and Bockris[122] for reduction of oxygen on iron in electrolyte containing 0.15 M H_3BO_3 and 0.0375 M $Na_2B_4O_7$. The kinetic current density was expressed by

$$i_k = i_{0,O_2} \exp\left(b_{c,O_2}(V - V_{0,O_2})\right) \tag{5.71}$$

where $i_{0,O_2} = 1.5 \times 10^{-13}$ A/cm^2, $b_{c,O_2} = 19.2$ V^{-1}, and $V_{0,O_2} = 0.72$ V(NHE). The solubility of oxygen was $c_{O_2}(\infty) = 4 \times 10^{-6}$ mol/cm^3, and the diffusion coefficient was $D_{O_2} = 5 \times 10^{-6}$ cm^2/s. Other relevant parameter values are given in the figure caption. The measurement of current as a function of potential for different rotation speeds yield plots similar to those shown in Figure 5.9(a). The current density measured at the plateau is the mass-transfer-limited current density. The plot of mass-transfer-limited current density as a function of the square root of rotation speed shown in Figure 5.9(b) yields a straight line passing through the origin. As the rotation speed is given in rpm rather than s^{-1}, the slope of the straight line corresponding to equation (5.68) is $0.20078nFD_i^{2/3}\nu^{-1/6}c_i(\infty)$, where $n = 4$.

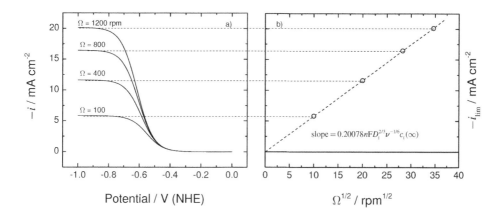

Figure 5.9: Presentation of a Levich plot for reduction of oxygen in borate buffered solutions. The data were taken from Jovancicevic and Bockris,[122] with parameters $i_{0,O_2} = 1.5 \times 10^{-13}$ A/cm^2, $b_{c,O_2} = 19.2$ V^{-1}, $n = 4$, $V_{0,O_2} = 0.72$ V(NHE), $c_{O_2}(\infty) = 4 \times 10^{-6}$ mol/cm^3, and $D_{O_2} = 5 \times 10^{-6}$ cm^2/s: a) Current as a function of potential with rotation speed as a parameter and b) mass-transfer-limited current density as a function of the square root of the rotation speed. The dashed lines connecting the two figures illustrate the manner in which plot (a) is used to provide the mass-transfer-limited current density needed for plot (b).

b) Fixed potential. *For measurements made at a fixed potential, the current density may be measured as a function of rotation speed. As the rotation speed increases, the current becomes increasingly unaffected by mass-transfer limitations. Following equation (5.62), a plot of $1/i$ as a function of $1/\sqrt{\Omega}$ will yield $1/i_k$ in the limit that $1/\sqrt{\Omega}$ approaches zero.*

An example is presented in Figure 5.10 for kinetic data presented presented in Figure 5.9. The kinetic current density i_k and $1/i_k$ are presented in Figure 5.10(a) as a function of potential. Values of $1/i$ obtained at specified potentials are presented in Figure 5.10(b) as functions of the inverse square root of rotation speed. Extrapolation of data in Figure 5.10(b) yields values for the kinetic contribution to the current density shown as $1/i_k$ in Figure 5.10(a).

5.3 Definitions of Potential

The potential of the electrode U is defined to be the difference between the potential of the working electrode Φ_m and the potential of a reference electrode Φ_{ref} located in the bulk of the electrolyte solution, i.e.,

$$U = \Phi_m - \Phi_{ref} \tag{5.72}$$

The cell potential can be expressed in terms of the potential in the electrolyte adjacent to the electrode Φ_0 as

$$U = (\Phi_m - \Phi_0) + (\Phi_0 - \Phi_{ref}) \tag{5.73}$$

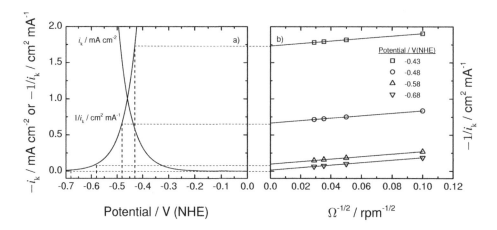

Figure 5.10: Presentation of a Koutecky plot for reduction of oxygen in borate buffered solutions, with parameters based on the data of Jovancicevic and Bockris[122] presented in Figure 5.9: a) the kinetic current density i_k and $1/i_k$ as a function of potential and b) $1/i$ as a function of $1/\sqrt{\Omega}$ with potential as a parameter. The dashed lines connecting the two figures illustrate the manner in which the extrapolation of data in plot (b) yields values for the kinetic contribution to the current density shown as $1/i_k$ in plot (a).

The position at which Φ_0 is evaluated is generally taken to be the inner limit of the electrically neutral diffusion layer (see Figure 5.18 in Section 5.7.1). In this way, the interface is assumed to incorporate the detailed structure of the double layer, including the diffuse region of charge and the inner Helmholtz plane associated with specifically adsorbed charged species.

Equation (5.73) can be written as

$$U = V + iR_e \tag{5.74}$$

where the interfacial potential V is defined by:

$$V = (\Phi_m - \Phi_0) \tag{5.75}$$

and the ohmic potential drop in the electrolyte is given as

$$iR_e = (\Phi_0 - \Phi_{ref}) \tag{5.76}$$

The surface overpotential η_s for a given reaction k is given by

$$\eta_s = V - V_{0,k} \tag{5.77}$$

where $V_{0,k}$ is the equilibrium potential difference that depends on the electrode reaction under consideration. This is the potential driving force used in electrochemical reactions (see equation (10.14) on page 209). The definitions of potentials used in electrochemical systems are summarized in Table 5.3.

Table 5.3: Definitions and notation for potentials used in electrochemical systems.

V	Interfacial potential for the working electrode, $V = \Phi_m - \Phi_0$.
$V_{0,k}$	Interfacial potential at equilibrium for a given reaction k, $V_{0,k} = (\Phi_m - \Phi_0)_{0,k}$.
U	Electrode potential with respect to a reference electrode, $U = \Phi_m - \Phi_{ref}$.
η_s	Surface overpotential for a given reaction k, $\eta_s = V - V_{0,k}$.
η_c	Concentration overpotential defined by equation (5.125).
Φ_m	Electrode potential with respect to an unspecified but common reference potential.
Φ_0	Potential of the electrolyte adjacent to the working electrode with respect to an unspecified but common reference potential.
Φ_{ref}	Potential of a reference electrode with respect to an unspecified but common reference potential.
iR_e	Ohmic potential drop between the solution adjacent to the working electrode and the location of a reference electrode, i.e., $iR_e = \Phi_0 - \Phi_{ref}$.

5.4 Rate Expressions

The rate of the generalized reaction (5.5) can be expressed as

$$r = k_a \exp\left(\frac{(1-\alpha)n\mathrm{F}}{RT}V\right) \prod_i c_i^{p_i}(0) - k_c \exp\left(-\frac{\alpha n\mathrm{F}}{RT}V\right) \prod_i c_i^{q_i}(0) \qquad (5.78)$$

where the exponents on the concentrations are defined in terms of the stoichiometric coefficient s_i. For $s_i = 0$, the species i is not a reactant and $p_i = 0$ and $q_i = 0$. For $s_i > 0$, species i is an anodic reactant for reaction k, and $p_i = s_i$ and $q_i = 0$. For $s_i < 0$, species i is an anodic product (or cathodic reactant) of reaction k and $p_i = 0$ and $q_i = -s_i$.

The rate expression given as equation (5.78) raises questions concerning the location at which concentrations and potentials are measured. The concentration of reacting species, for example, should be that measured at the inner limit of the diffusion layer, as shown in Figure 5.11. This figure also shows a possible double-layer structure in which ihp refers to the inner Helmholtz plane, corresponding to the position associated with physically adsorbed ionic species and ohp refers to the outer Helmholtz plane, corresponding to the plane of closest approach for solvated ionic species. The position at the inner limit of the diffusion layer also can be described as the position just outside

---

Remember! 5.2 *Kinetic expressions of the law of mass action provide the foundation for modeling the charge-transfer resistance commonly encountered in electrochemical impedance spectroscopy.*

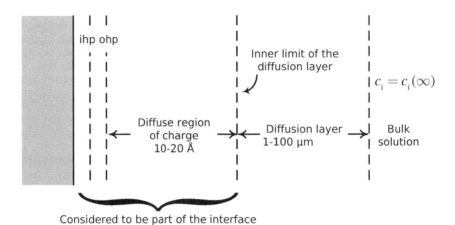

Considered to be part of the interface

Figure 5.11: Schematic representation of an electrode surface showing the presence of a concentration diffusion layer, a diffuse region of charge within the electrolyte and a possible double-layer structure in which ihp refers to the inner Helmholtz plane, corresponding to the position associated with physically adsorbed ionic species and ohp refers to the outer Helmholtz plane, corresponding to the plane of closest approach for solvated ionic species.

the diffuse part of the double layer. The double layer structure, discussed further in Section 5.7, is assumed to be associated with the electrode and, thus, the potential V given in equation (5.78) represents the potential difference between the electrode and the solution at the inner limit of the diffusion layer or just outside the diffuse part of the double layer.

The equilibrium potential can be expressed as

$$V_0 = \frac{RT}{nF}\left[\ln\left(\frac{k_c}{k_a}\right) + \sum_i (q_i - p_i)\ln c_i(\infty)\right] \tag{5.79}$$

where, because equation (5.79) represents an equilibrium condition, there is no concentration gradient and the interfacial concentration $c_i(0)$ is equal to the bulk concentration $c_i(\infty)$. The surface overpotential can be expressed as a departure of V from its equilibrium value, e.g.,

$$\eta_s = V - V_0 \tag{5.80}$$

Equation (5.78) can be expressed in terms of the Butler–Volmer equation (5.15) as

$$i = i_0 \left\{\exp\left(\frac{(1-\alpha)nF}{RT}\eta_s\right) - \exp\left(-\frac{\alpha nF}{RT}\eta_s\right)\right\} \tag{5.81}$$

where

$$i_0 = nFk_c^{1-\alpha}k_a^{\alpha}\prod_i (c_i(0))^{(q_i + \alpha s_i)} \tag{5.82}$$

Here, α, the symmetry factor, is generally assigned a value of 0.5. As the concentration $c_i(0)$ depends on applied potential, the exchange current density is also a function of applied potential.

Example 5.7 Rate Expression for Copper Dissolution: *Develop an expression corresponding to equation (5.82) for the reaction*

$$Cu \rightleftarrows Cu^{2+} + 2e^- \tag{5.83}$$

where copper is assumed to dissolve directly to form the cupric ion.

Solution: *For the reaction (5.83), the stoichiometric coefficients are $s_{Cu} = +1$, $s_{Cu^{2+}} = -1$, and $n = 2$. Thus, reaction (5.83) can be expressed in the form of equation (5.5) as*

$$Cu - Cu^{2+} \rightarrow 2e^- \tag{5.84}$$

The corresponding expression for the reaction rate (5.78) is given as

$$r_{Cu} = \frac{i_{Cu}}{2F} = k_a \exp\left(\frac{F}{RT}V\right) - k_c \exp\left(-\frac{F}{RT}V\right)c_{Cu^{2+}}(0) \tag{5.85}$$

The equilibrium potential can be expressed as

$$V_{0,Cu} = \frac{RT}{2F}\left[\ln\left(\frac{k_c}{k_a}\right) + \ln c_{Cu^{2+}}(\infty)\right] \tag{5.86}$$

and the exchange current density can be written as

$$i_0 = 2Fk_c^{1/2}k_a^{1/2}c_{Cu^{2+}}^{1/2}(0) \tag{5.87}$$

Thus, equation (5.85) can be expressed in the form of a Butler–Volmer equation (5.81).

Example 5.7 represents a simplification of a reaction scheme that would generally involve steps involving reaction of single electrons, e.g., dissolution of copper to form cuprous ions

$$Cu \rightleftarrows Cu^+ + e^- \tag{5.88}$$

and subsequent reaction to form cupric ions

$$Cu^+ \rightleftarrows Cu^{2+} + e^- \tag{5.89}$$

A discussion of the mechanism that results from consideration of reactions (5.88) and (5.89) is presented by Newman and Thomas-Alyea.[116] Electrochemical impedance spectroscopy can often provide evidence for more complicated reaction sequences involving reaction intermediates.

Example 5.8 Rate Expression for the Hydrogen Reaction: *Develop a rate expression in the form of a Butler-Volmer equation for reactions (5.30) and (5.31).*

Solution: *Following Newman and Thomas-Alyea,[116] the rate expression for reaction (5.31) can be expressed as*

$$r = k_f \Gamma^2 \gamma_H^2 - k_b p_{H_2} \Gamma^2 (1 - \gamma_H)^2 \tag{5.90}$$

where Γ represents the maximum coverage, and γ_H represents the fractional coverage of the surface by atomic hydrogen. Under the assumption that reaction (5.31) takes place so readily that it can be considered to be equilibrated,

$$k_f \Gamma^2 \gamma_H^2 = k_b p_{H_2} \Gamma^2 (1 - \gamma_H)^2 \tag{5.91}$$

or

$$\left(\frac{\gamma_H}{1 - \gamma_H}\right) = \left(\frac{k_b p_{H_2}}{k_f}\right)^{1/2} = \sqrt{P} \tag{5.92}$$

Equation (5.92) provides a definition for the scaled partial pressure of hydrogen P.

The current density associated with reaction (5.30) is given by

$$i = F k_1 \Gamma \gamma_H \exp\left(\frac{(1-\alpha)FV}{RT}\right) - F k_{-1} \Gamma (1 - \gamma_H) c_{H^+} \exp\left(\frac{-\alpha FV}{RT}\right) \tag{5.93}$$

From equation (5.92),

$$\gamma_H = \frac{\sqrt{P}}{1 + \sqrt{P}} \tag{5.94}$$

and

$$1 - \gamma_H = \frac{1}{1 + \sqrt{P}} \tag{5.95}$$

Upon introduction of equations (5.94) and (5.95), equation (5.93) yields

$$i = F \frac{\Gamma}{1 + \sqrt{P}} \left\{ k_1 \sqrt{P} \exp\left(\frac{(1-\alpha)FV}{RT}\right) - k_{-1} c_{H^+} \exp\left(\frac{-\alpha FV}{RT}\right) \right\} \tag{5.96}$$

At equilibrium, the net current density is equal to zero. Under the condition that $i = 0$, equation (5.96) can be solved for V, yielding

$$V_{i=0} = V_{0,H_2} = \frac{RT}{F} \ln\left(\frac{k_{-1} c_{H^+}}{k_1 \sqrt{P}}\right) \tag{5.97}$$

Insertion of $V = \eta_s + V_{0,H_2}$ into equation (5.96) yields

$$i = i_0 \left\{ \exp\left(\frac{(1-\alpha)nF}{RT}\eta_s\right) - \exp\left(-\frac{\alpha nF}{RT}\eta_s\right) \right\} \tag{5.98}$$

where

$$i_0 = \frac{F\Gamma}{1 + \sqrt{P}} \left(k_1 \sqrt{P}\right)^\alpha \left(k_{-1} c_{H^+}\right)^{1-\alpha} \tag{5.99}$$

Equation (5.98) is in the form of the Butler-Volmer equation given in equation (5.15). Equations (5.97) and (5.99) show the dependence of the equilibrium potential and the exchange current density, respectively, on the concentration of H^+ ions and the partial pressure of hydrogen.

Table 5.4: Typical values of diffusion coefficients for ions at infinite dilution in water at $T = 25\,°C$.

Cation	z_i	D_i/cm^2s^{-1}	Anion	z_i	D_i/cm^2s^{-1}
H^+	+1	9.312×10^{-5}	OH^-	−1	5.260×10^{-5}
Na^+	+1	1.334×10^{-5}	Cl^-	−1	2.032×10^{-5}
K^+	+1	1.957×10^{-5}	NO_3^-	−1	1.902×10^{-5}
Ag^+	+1	1.648×10^{-5}	SO_4^{2-}	−2	1.065×10^{-5}
Mg^{2+}	+2	0.7063×10^{-5}	$Fe(CN)_6^{3-}$	−3	0.896×10^{-5}
Cu^{2+}	+2	0.72×10^{-5}	$Fe(CN)_6^{4-}$	−4	0.739×10^{-5}

5.5 Transport Processes

Transport processes in dilute electrolytic systems are described here following the treatment of Newman and Thomas-Alyea.[116] Conservation of species provides that

$$\frac{\partial c_i}{\partial t} = -\left(\nabla \cdot \mathbf{N}_i\right) + R_i \tag{5.100}$$

where c_i is the concentration of species i, $\mathbf{N}_i$ is the net flux vector for species i, and R_i is the rate of generation of species i. In a dilute solution, the flux for any species can be written in terms of contributions from convection, diffusion, and migration, i.e.,

$$\mathbf{N}_i = \mathbf{v}c_i - D_i\nabla c_i - z_i u_i c_i F\nabla\Phi \tag{5.101}$$

where $\mathbf{v}$ is the fluid velocity, D_i is the diffusion coefficient, u_i is the mobility, and z_i is the charge of species i. The mobility can be related to the diffusion coefficient by the Nernst–Einstein equation

$$D_i = RTu_i \tag{5.102}$$

Typical values of diffusion coefficients for ions at infinite dilution in water at 25 °C are presented in Table 5.4.

The current density $\mathbf{i}$ is given by the sum of contributions from the flux of each ionic species, i.e.,

$$\mathbf{i} = F\sum_i z_i\mathbf{N}_i \tag{5.103}$$

By imposing conservation of charge,

$$\nabla \cdot \mathbf{i} = 0 \tag{5.104}$$

Remember! 5.3 *The electrolyte resistance commonly encountered in electrochemical impedance spectroscopy arises from transport processes in the bulk electrolyte.*

Table 5.5: Hierarchy of current distribution model assumptions.

Solution	Ohmic Resistance	Kinetic Resistance	Mass-Transfer Resistance
Primary	✓	×	×
Secondary	✓	✓	×
Tertiary	✓	✓	✓
Mass-transfer-limited	×	×	✓

and the condition of electrolyte electroneutrality,

$$\sum_i z_i c_i = 0 \tag{5.105}$$

equations (5.101) and (5.103) yield

$$\nabla \cdot (\kappa \nabla \Phi) + F \sum_i z_i \nabla \cdot (D_i \nabla c_i) = 0 \tag{5.106}$$

In the absence of concentration gradients, equation (5.106) may be simplified to yield Laplace's equation, i.e.,

$$\nabla^2 \Phi = 0 \tag{5.107}$$

In the absence of concentration gradients, equations (5.101) and (5.103) provide

$$\mathbf{i} = -\kappa \nabla \Phi \tag{5.108}$$

which is an expression of Ohm's law. The electrolyte conductivity κ may be expressed, for dilute solutions, as a sum of contributions from each of the ionic species following

$$\kappa = F^2 \sum_i z_i^2 u_i c_i = \frac{F^2}{RT} \sum_i z_i^2 D_i c_i \tag{5.109}$$

Laplace's equation (5.107) does not apply in the presence of concentration gradients because the conductivity κ is not a constant and because the right-hand term in equation (5.106) is not equal to zero. In the absence of concentration gradients and with a uniform value of κ, equation (5.106) reduces to equation (5.107). Thus, a series of approximate solutions may be obtained, as summarized in Table 5.5 and discussed in the subsequent sections. It should be noted that the assumption that there are no concentration gradients is not valid near electrode surfaces. For example, oxygen reduction at an electrode surface causes the pH at the surface of the electrode to reach 10 or 11, even when the pH in the bulk electrolyte is equal to 7.

5.5.1 Primary Current and Potential Distributions

The impedance response can be strongly influenced by the distribution of current and potential at the electrode under study. Some general guidelines can be established to help determine conditions under which a nonuniform distribution can arise.

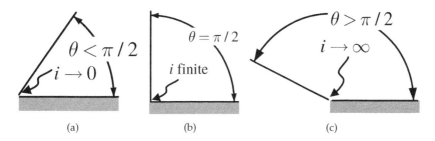

Figure 5.12: Behavior of the primary current distribution following equation (5.110) near the edge of an electrode: a) $i \to 0$ for $\theta < \pi/2$; b) i is finite for $\theta = \pi/2$; and c) $i \to \infty$ for $\theta > \pi/2$.

Under the assumption that the concentrations are uniform within the electrolyte, potential is governed by Laplace's equation (5.107). Under these conditions, the passage of current through the system is controlled by the ohmic resistance to passage of current through the electrolyte and by the resistance associated with reaction kinetics. The primary distribution applies in the limit that the ohmic resistance dominates and kinetic limitations can be neglected. The solution adjacent to the electrode can then be considered to be an equipotential surface with value Φ_0. The boundary condition for insulating surfaces is that the current density is equal to zero.

The primary current distribution generally represents a worst-case scenario for electrode design. At boundaries between an electrode and an insulator, the current density on the electrode surface can be shown, by solving Laplace's equation for potential, to approach

$$i \propto r^{(\pi - 2\theta)/\pi} \tag{5.110}$$

where r is the radial distance from the point of intersection and θ is the angle between the electrode and insulator measured in radians. As shown in Figure 5.12, equation (5.110) shows that the current tends toward zero for angles $\theta < \pi/2$, toward infinity for angles $\theta > \pi/2$, and is uniform only if $\theta = \pi/2$.

For a disk of radius r_0, embedded in an insulating plane and with a counterelectrode infinitely far away, the primary current density was shown by Newman[123] to be given by

$$\frac{i}{\langle i \rangle} = \frac{1}{2\sqrt{1 - \left(\frac{r}{r_0}\right)^2}} \tag{5.111}$$

where $\langle i \rangle$ represents the area-averaged current density. The normalized primary current distribution depends only on the electrode geometry. The current density tends toward infinity at the periphery of the disk. The corresponding primary resistance is given by[124]

$$R_e = \frac{1}{4\kappa r_0} \tag{5.112}$$

The dimensionless primary resistance can be written as $R_e \kappa r_0 = 1/4$.

5.5.2 Secondary Current and Potential Distributions

The secondary distribution applies when kinetic limitations cannot be neglected. The solution adjacent to the electrode can no longer be considered to be an equipotential surface. The condition at the electrode can be replaced by

$$i = -\kappa \left. \frac{\partial \Phi}{\partial y} \right|_{y=0} = f(V) \tag{5.113}$$

where y is the coordinate normal to the electrode surface, V is defined by equation (5.75), and $f(V)$ is a general function that could be given by the Butler–Volmer equation (5.15) or by a Tafel expression, e.g., equations (5.21) or (5.22).

At sufficiently small overpotentials, equation (5.15) can be linearized such that

$$\left. \frac{\partial \Phi}{\partial y} \right|_{y=0} = -JV \tag{5.114}$$

where

$$J = \frac{(b_a + b_c) i_0 \ell}{\kappa} \tag{5.115}$$

(see Problem 5.6) and where ℓ is a characteristic electrode length. For a disk electrode, the electrode radius is the characteristic electrode length, i.e., $\ell = r_0$, and J can be expressed in terms of ohmic and charge-transfer resistances as[125]

$$J = \frac{4}{\pi} \frac{R_e}{R_t} \tag{5.116}$$

where

$$R_t = \frac{1}{(b_a + b_c) i_0} \tag{5.117}$$

The parameter J can be regarded to be a dimensionless exchange current density and is the inverse of the Wagner number.[126] The resulting current distributions, obtained by numerical solution of Laplace's equation, are presented in Figure 5.13. When $J \to \infty$, the ohmic resistance dominates, and the current density follows that of the primary distribution given as equation (5.111).

For Tafel kinetics, valid for $|i| \gg i_0$, the parameter J can be defined to be

$$J = \frac{b_c i_{avg} r_0}{\kappa} \tag{5.118}$$

where J was evaluated for a cathodic reaction. The corresponding current distribution for a disk electrode is similar to that presented for linear kinetics in Figure 5.13.

5.5.3 Tertiary Current and Potential Distributions

Tertiary distributions apply when Laplace's equation is replaced by a series of n equations of the form (5.100) coupled with electroneutrality (5.105) where n represents the number of ionic species in the system. Thus, tertiary distributions relax the assumption that concentrations are uniform. Ohmic, kinetic, and mass-transfer impedances all play a role.

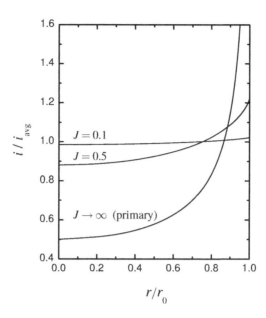

Figure 5.13: Secondary current distribution for linear polarization kinetics on a disk electrode with J, given by equation (5.115), as a parameter. Simulation results taken from Huang et al.[125]

5.5.4 Mass-Transfer-Controlled Current Distributions

Mass-transfer-controlled distributions apply under the assumptions that ohmic and kinetic impedances can be neglected.

Example 5.9 Primary and Secondary Current Distributions: *For the electrode geometries shown in Figure 5.14, indicate whether the primary and secondary current distributions may be expected to be uniform.*

Solution: *A qualitative answer can be obtained by use of equation (5.110) as illustrated in Figure 5.12.*

a) Electrode flush with insulating material. *The angle between the electrode and insulator is π radians or $180°$. This case corresponds to Figure 5.12(c) for which the primary current distribution is not uniform. For linear kinetics, the secondary distribution adds a uniform interfacial charge-transfer resistance, which tends to make the current distribution more uniform, as is shown in Figure 5.13. This electrode configuration corresponds to the rotating disk presented in Section 11.3 and to the submerged impinging jet presented in Section 11.4. These configurations are popular because the electrodes are easily fabricated and polished and because the mass-transfer-limited current can be made to be uniform.*

b) Recessed electrode. *The angle between the electrode and insulator is $\pi/2$ radians or $90°$. This case corresponds to Figure 5.12(b) for which the primary current distribution is*

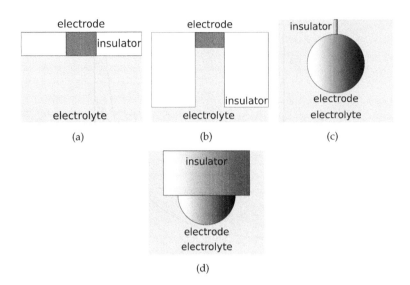

Figure 5.14: Schematic representations of electrode geometries: a) electrode flush with insulating material; b) recessed electrode; c) spherical electrode; and d) hemispherical electrode.

uniform. The degree of uniformity depends on the ratio of the depth of the recess to the electrode diameter. The current distribution may be considered uniform if the depth of the recess is larger than the electrode diameter.[127,128] As the primary distribution is uniform, the secondary distribution is also uniform.

c) Spherical electrode. *If the electrical contact to the sphere can be neglected, Laplace's equation for potential, equation (5.107), is a function only of radial position and is independent of azimuthal and angular directions. Thus, the primary and secondary current distributions will be uniform. The mass-transfer-limited current density for this geometry is not uniform.*

d) Hemispherical electrode. *The result for the spherical electrode can be applied as well for a hemispherical electrode. The primary and secondary current distributions will be uniform. This geometry has been employed as a rotating hemispherical electrode[129–131] and as a hemispherical electrode under a submerged impinging jet.[132,133] For both flow configurations, the mass-transfer-limited current density is not uniform. The hemispherical electrode may be attractive for kinetic studies for which the current density is sufficiently smaller than the average mass-transfer-limited current density. Nisancioglu and Newman[130] suggest that the current density should be smaller than $0.680267 < i_{\lim} >$. Shukla et al.[133] suggest that the corresponding criterion for the submerged impinging jet is that the current density should be smaller than $0.25 < i_{\lim} >$.*

5.6   Potential Contributions

The cell potential can be decomposed into contributions corresponding to different loss terms within the cell. For example, the difference in potential between two electrodes can be expressed as

$$(V_a - V_c) = (V_a - \Phi_{0,a}) + (\Phi_{0,a} - \Phi_{0,c}) + (\Phi_{0,c} - V_c) \qquad (5.119)$$

where $\Phi_{0,a}$ represents the potential at the inner limit of the diffusion layer at the anode and $\Phi_{0,c}$ represents the potential at the inner limit of the diffusion layer at the cathode. The terms collected in the right-hand side of equation (5.119) are developed in terms of ohmic and kinetic contributions in Sections 5.6.1 and 5.6.2, respectively. This book employs a formalism that does not require use of the concentration overpotential. To place this approach into perspective with a commonly used approach, the concentration overpotential is discussed briefly in Section 5.6.3.

5.6.1   Ohmic Potential Drop

The term $(\Phi_{0,a} - \Phi_{0,c})$ represents the potential drop through the electrolyte. As defined here, the numerical evaluation of the potential drop requires accounting for the variation in electrolyte conductivity within the diffusion layer. An alternative approach is to define the ohmic potential drop as being that calculated using Laplace's equation with a uniform solution conductivity. In this case, an additional term is required to account for the influence of the conductivity variation within the diffusion layer on the measured potential. This is incorporated into a concentration overpotential, discussed in Section 5.6.3.

5.6.2   Surface Overpotential

The term $(V_a - \Phi_{0,a})$ can be expressed in terms of the surface overpotential, noting that

$$\eta_{s,a} = (V_a - \Phi_{0,a}) - (V_a - \Phi_{0,a})_0 \qquad (5.120)$$

where $(V_a - \Phi_{0,a})_0$ is the equilibrium difference between the electrode potential and the potential of the electrolyte adjacent to the electrode, generally taken to be at the inner limit of the diffusion layer. If the equilibrium potential is measured at the composition existing at the inner limit of the diffusion layer, $c_i(0)$, then

$$(V_a - V_c) = \eta_{s,a} + (V_a - \Phi_{0,a})_0 + (\Phi_{0,a} - \Phi_{0,c}) - \eta_{s,c} - (V_c - \Phi_{0,c})_0 \qquad (5.121)$$

If, instead, the equilibrium potential is measured at the composition $c_i(\infty)$ of the bulk electrolyte, then an additional term is needed to account for influence of the diffusion layer on the measured potential. This is incorporated into a concentration overpotential, discussed in Section 5.6.3.

5.6.3   Concentration Overpotential

A concentration overpotential is often invoked to account for the influence of concentration distribution on potential. Consider, for example, that the concentration at the inner limit of the diffusion layer is $c_i(0)$. Consistent with equation (5.59), the current for a single anodic reaction can be expressed as

$$i = k_a nF c_i(0) \exp\left(b_a \eta_s\right) \tag{5.122}$$

Equation (5.122) can be written in terms of the bulk concentration as

$$i = k_a nF c_i(\infty) \frac{c_i(0)}{c_i(\infty)} \exp\left(b_a \eta_s\right) \tag{5.123}$$

or

$$i = k_a nF c_i(\infty) \exp\left(b_a \left(\eta_s + \eta_c\right)\right) \tag{5.124}$$

where

$$\eta_c = \frac{1}{b_a} \ln\left(\frac{c_i(0)}{c_i(\infty)}\right) \tag{5.125}$$

Equation (5.125) may be compared to equation (20.17) presented by Newman and Thomas-Alyea.[116] The concentration overpotential is not needed if the concentrations used in the kinetic expressions are those evaluated at the electrode surface.

Example 5.10 Cell Potential Contributions: *Calculate the potential contributions corresponding to Figure 5.2 for two identical inert electrodes in an electrolyte with no dissolved hydrogen or oxygen gasses.*

Solution: *The current associated with the corrosion and hydrogen evolution reactions in an electrolyte with no dissolved hydrogen or oxygen gasses is developed in Example 5.4. For kinetic parameters expressed in Table 5.2, the zero-current potential was found from equation (5.51) to be $V_{mixed} = 0.706$ V(NHE), and the corresponding effective exchange current density was found from equation (5.52) to be $i_{mixed} = 1.4 \times 10^{-15}$ A/cm^2. The cell potential can be expressed for a given current to be*

$$V_{cell} = \eta_{s,anode} + i_{avg} R_e - \eta_{s,cathode} \tag{5.126}$$

where i_{avg} is the current density averaged over the cross-section of the cell.

For the anode, the surface overpotential may be obtained from equation (5.54). At large overpotentials, the Tafel equation applies, and the surface overpotential may be expressed as

$$\eta_{s,anode} = \frac{1}{b_{O_2}} \ln\left(\frac{i_{avg}}{i_{mixed}}\right) \tag{5.127}$$

at smaller overpotentials, the influence of the cathodic reaction may not be neglected, and

$$\eta_{s,anode,n+1} = \frac{1}{b_{O_2}} \ln\left(\frac{i_{avg}}{i_{mixed}} + \exp\left(-b_{H_2} \eta_{s,anode,n}\right)\right) \tag{5.128}$$

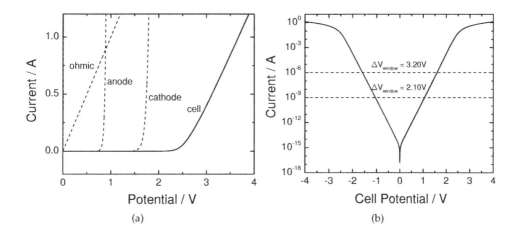

Figure 5.15: Polarization curve for the system considered in Example 5.4 for oxygen and hydrogen evolution reactions on an inert electrode in an electrolyte that contains no dissolved oxygen or hydrogen gasses: a) linear scale for current density and b) logarithmic scale for the absolute value of current density. Kinetic parameters were taken from McCafferty[117] and Conway.[119]

which may be solved iteratively. At still smaller overpotentials, the overpotential may be obtained from

$$\eta_{s,anode,n+1} = \frac{i_{mixed}}{i_{avg}\left(b_{O_2} + b_{H_2}\right)} - \eta_{s,anode,n}^2 \frac{b_{O_2}^2 - b_{H_2}^2}{b_{O_2} + b_{H_2}} \qquad (5.129)$$

The overpotential at the cathode is obtained, keeping in mind that the current at the anode must also flow through the cathode, from

$$\eta_{s,cathode} = -\frac{1}{b_{H_2}} \ln\left(\frac{i_{avg}}{i_{mixed}}\right) \qquad (5.130)$$

Again, an iterative approach is required to obtain accurate solutions when the overpotential is small.

The resulting current is presented as a function of cell potential in Figure 5.2. The contributions corresponding to the ohmic resistance and surface overpotentials are presented in Figure 5.15(a). At small cell potentials, the current is small and contribution of the surface overpotentials dominate the cell potential. At larger cell potentials, the surface overpotentials increase slowly as compared to the ohmic resistance which tends to dominate. At these larger potentials, the slope of the current as a function of potential has a value $dI/dV = 1/R_e$.

The concept that the electrochemical cell does not pass current over a window of cell potentials is explored more fully in Figure 5.15(b). In fact, the system is not at equilibrium and the current that flows at a given potential within the window is constrained by kinetic limitations. As indicated by equation (5.57), the value of the potential window is determined by the level of current that can be detected. For the cell, the size of the potential window is twice the value determined for each individual electrode that comprises the cell. Thus, for a detection current density of 10^{-9} A/cm^2 (or 1 nA/cm^2), the potential window has a value of 2.10 V and for a *detection current density of 10^{-6} A/cm^2 (or 1 µA/cm^2), the potential window has a value of 3.20 V.*

5.7   Capacitance Contributions

The total current through an electrochemical cell consists of a faradaic portion which involves the electrochemical reactions and a time-dependent portion associated with the charging of the interface. Thus,

$$i = i_F + C\frac{dV}{dt} \tag{5.131}$$

where C is the capacitance associated with the electrode. In electrochemical systems, the capacitance may arise due to charge redistribution at interfaces or to dielectric phenomena. The charge redistribution takes place at an electrolytic double layer.

5.7.1   Double-Layer Capacitance

Away from solid surfaces, the electrolyte can be considered to be electrically neutral, and, in the absence of concentration gradients, equation (5.105) is satisfied exactly. If concentration gradients exist for species of differing diffusion coefficients, equation (5.105) is satisfied approximately, but the charge is so small that equation (5.105) can be used. This can be seen from a rearrangement of Poisson's equation as

$$\sum_i z_i c_i = -\frac{\varepsilon \varepsilon_0}{F}\nabla^2 \Phi \tag{5.132}$$

where ε is the dielectric constant of the medium and ε_0 is the permittivity of a vacuum ($\varepsilon_0 = 8.8542 \times 10^{-14}$ F/cm or 8.8542×10^{-14} C/V cm). The constant $\varepsilon \varepsilon_0 / F$ is very small and typically has a value on the order of 10^{-16} equiv/V cm. Thus, for moderate values of $\nabla^2 \Phi$, $\sum_i z_i c_i \approx 0$.

The situation is fundamentally different near an interface due to a significant redistribution of charge. Consider, for example, the potential distribution in an electrochemical cell at open circuit. Consider that a potential can be applied between the two metal electrodes such that no current flows. A situation like this is described in Section 5.1. The electrodes can be considered to be ideally polarized since a potential can be applied without passage of current.

The corresponding potential distribution is given in Figure 5.16. As no current flows

---

Remember! **5.4** *The capacitance commonly encountered in electrochemical impedance spectroscopy may arise from charge redistribution at interfaces or from dielectric phenomena.*

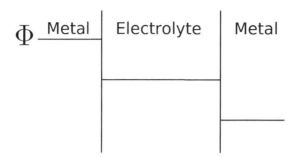

Figure 5.16: Potential distribution for a cell at open circuit consisting of ideally polarized electrodes.

in the cell, the potential must be uniform in the electrolytic solution. The only place where the potential can change in the cell is at the electrode-electrolyte interface. In this region, the second derivative of potential with respect to position $d^2\Phi/dy^2$ must be very large. Equation (5.132) suggests that a substantial redistribution of charge is required to accommodate the abrupt change in potential.

A redistribution of charge is possible because the electrons that accumulate near the metal surface have associated with them a charge. In addition, some ionic species may have a tendency to accumulate preferentially at the electrode-electrolyte interface. Finally, as the interface taken as a whole must be electrically neutral, a diffuse region of charge may be present in the electrolyte adjacent to the electrode.

Experimental measurements on mercury electrodes have contributed significantly to the understanding of the nature of the electrical double layer. The double-layer capacitance obtained by Grahame[42] for a mercury electrode in contact with a NaCl electrolyte at a temperature of 25 °C is presented in Figure 5.17 where potentials are referenced to the electrocapillary maximum or potential of zero charge. The capacitance values range between 14 and 50 $\mu F/cm^2$ and are seen to be strong functions of potential. This range of values and dependence of potential is typical of the results reported by Grahame for mercury electrodes immersed in other electrolytes.[42]

Numerous models of the electrode-electrolyte interface have been developed. The simplest of these is the Helmholtz double-layer model, which posits that the charge associated with a discrete layer of ions balances the charge associated with electrons at the metal surface. The Helmholtz double-layer model predicts incorrectly that the interfacial capacitance is independent of potential. Nevertheless, current models of the charge redistribution at electrode-electrolyte interfaces owe their terminology to the original Helmholtz model.

A schematic representation of an electrical double layer is presented as Figure 5.18. A plane (m) is associated with an excess concentration of electrons near the physical surface of the electrode, represented by a solid line. The inner Helmholtz plane (ihp) is associated with ions that are specifically adsorbed onto the metal surface. The outer Helmholtz plane (ohp) is the plane of closest approach for solvated ions that are free to

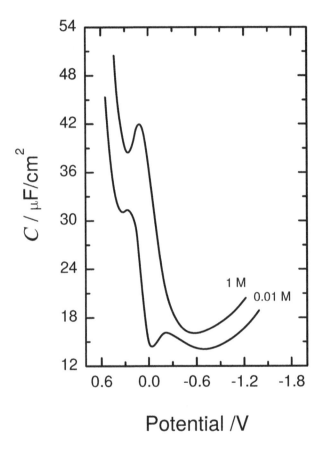

Figure 5.17: Double-layer capacitance obtained by Grahame[42] for a mercury electrode in contact with a NaCl electrolyte at a temperature of 25 °C. Potentials are referenced to the electrocapillary maximum or potential of zero charge.

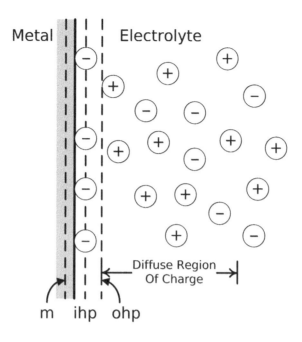

Figure 5.18: The structure of the electrical double layer.

move within the electrolyte. The ions within the electrolyte near the electrode surface contribute to a diffuse region of charge. The diffuse region of charge has a characteristic Debye length,

$$\lambda = \sqrt{\frac{\varepsilon\varepsilon_0 RT}{F^2 \sum_i z_i^2 c_i(\infty)}} \qquad (5.133)$$

which is typically on the order of 1-10 Å for electrolytic solutions.

The charge held in each of the layers must cancel such that

$$q_m + q_{ihp} + q_d = 0 \qquad (5.134)$$

where q_m is the charge of electrons at the electrode surface, q_{ihp} is the charge associated with adsorbed ions, and q_d is the charge held in the diffuse region. There is no plane of charge associated with the outer Helmholtz plane as this represents only an inner limit to the diffuse region of charge.

The potential of the metal referenced to an electrode located outside the diffuse region of charge can be expressed as

$$U - \Phi_{ref} = \left(U - \Phi_{ihp}\right) + \left(\Phi_{ihp} - \Phi_{ref}\right) \qquad (5.135)$$

The capacitance of the interface is defined to be the derivative of charge density with respect to electrode potential at fixed electrochemical potential μ_i and temperature T,

Table 5.6: Typical ranges of values for dielectric constant, film thickness, and capacitance. Dielectric constants for some other materials are presented in Table 12.1 on page 304.

System	Dielectric Constant	Thickness	Capacitance
Double layer on bare metal	–	–	$10 - 50 \ \mu\mathrm{F/cm}^2$
Fe_2O_3 oxide[134]	12	2.7 nm	$4 \ \mu\mathrm{F/cm}^2$
		5.9 nm	$1.8 \ \mu\mathrm{F/cm}^2$
Ni_2O_5 oxide[135]	42	8 nm	$4.6 \ \mu\mathrm{F/cm}^2$
		29 nm	$1.3 \ \mu\mathrm{F/cm}^2$
Human skin[136]	21.2	$15 \ \mu m$	$1.25 \ \mathrm{nF/cm}^2$
Epoxy-polyaminoamide paint[137]	4.9	$21 \ \mu m$	$0.2 \ \mathrm{nF/cm}^2$

i.e.,

$$C = \left(\frac{\partial q}{\partial U} \right)_{\mu_i, T} \tag{5.136}$$

Thus, following equation (5.135), the capacitance of the interface can be expressed in terms of respective contributions as

$$\frac{1}{C} = \frac{1}{C_{\mathrm{m-ihp}}} + \frac{1}{C_{\mathrm{d}}} \tag{5.137}$$

As seen in equation (5.137), the capacitance of an interface is dominated by the part with the smaller capacitance.

5.7.2 Dielectric Capacitance

The capacitance associated with oxide layers and polymeric coatings can be expressed as

$$C = \frac{\varepsilon \varepsilon_0}{\delta} \tag{5.138}$$

where δ is the film thickness, ε is the dielectric constant of the material, and ε_0 is the permittivity of vacuum $\varepsilon_0 = 8.8542 \times 10^{-14}$ F/cm. The capacitance of such oxide and polymeric layers is typically sufficiently small that the contribution of an electrolytic double layer in series can be generally neglected. Some typical values for dielectric constant, film thickness, and capacitance are presented in Table 5.6.

5.8 Further Reading

A number of excellent texts are available. Newman[115] and Newman and Thomas-Alyea[116] provide a comprehensive and mathematically detailed treatment of electrochemical engineering. West[138] provides an approachable introduction to the the treatment presented by Newman and Thomas-Alyea. Prentice[139] provides slightly greater emphasis on applications. Bard and Faulkner[121] emphasize analytical methods, and Bockris and Reddy[140, 141] provide a very approachable introduction to electrochemical processes. Gileadi[111] and Oldham et al.[142] provide excellent treatments of electrode kinetics, and Brett and Brett[143] provide a treatment that includes fundamentals as well as applications, including impedance spectroscopy. Fuller and Harb[144] provide a text suitable for classroom instruction or independent study at the senior undergraduate and beginning graduate student level.

Problems

5.1 Estimate the electrolyte resistance for a 0.25 cm radius disk electrode in 0.1 M NaCl solution at 25 °C.

5.2 Obtain a relationship for the charge-transfer resistance in the Tafel regime.

5.3 Verify equation (5.62).

5.4 Show that the concentration of a reactant at the surface of an electrode can be expressed as a function of current density as

$$\frac{c_i(0)}{c_i(\infty)} = 1 - \frac{i}{i_{lim}} \tag{5.139}$$

5.5 Following equation (5.139), use a spreadsheet program to plot the current density as a function of surface overpotential for $K_c c_i(\infty) = 1 \, \text{mA/cm}^2$, $b_c = 20 \, \text{V}^{-1}$ and i_{lim} equal to -0.1, -1.0, and $-10 \, \text{mA/cm}^2$.

5.6 Use a Taylor's series expansion for the faradaic current about the equilibrium potential to obtain a relationship for the charge-transfer resistance in terms of the parameter $J = n i_0 F / \kappa R T$.

5.7 Estimate the capacitance for an oxide layer on a steel surface that is 50 Å thick.

5.8 Estimate the surface area for an electrode in a 0.1 M NaCl solution if the capacitance is measured to be 120 μF. What might be a reasonable confidence interval for this estimate?

5.9 Use a spreadsheet program to reproduce Figure 5.15, following Example 5.10 on page 120.

Chapter 6

Electrochemical Instrumentation

Operational amplifiers provide the foundation for electrochemical instrumentation. The aim of this chapter is to describe the main properties of an operational amplifier so as to understand the principles of potentiostats and galvanostats and to understand how they can be used for impedance measurements. Operational amplifiers provide a means of setting the potentials at two inputs to be equal with no current passing between the inputs. Current and potential is applied to the output as needed to satisfy the input constraints.

6.1 The Ideal Operational Amplifier

For the purpose of this text, an operational amplifier consists of a series of solid-state components designed to have certain functional characteristics. A schematic representation of an operational amplifier, given in Figure 6.1(a), shows five leads attached to the operational amplifier. The vertical leads, marked V_{S+} and V_{S-}, provide power to the amplifier and are connected to a direct-current power supply. The two leads on the left, termed the noninverting (+) and the inverting (−) input, have potentials V_+ and V_-, respectively. The output potential is V_0.

The amplifier is designed to sense the difference between the voltage signals applied at its two input terminals, to multiply this by a number A_{op}, and to cause the output voltage to be

$$V_0 = A_{op}\,(V_+ - V_-) \tag{6.1}$$

A typical response for an operational amplifier is given in Figure 6.1(b). The output

---

Remember! 6.1 *Operational amplifiers provide the foundation for electrochemical instrumentation.*

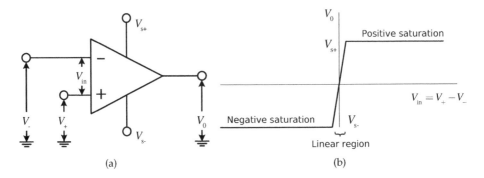

Figure 6.1: The ideal operational amplifier: a) the circuit symbol for an operational amplifier showing the five principal terminals and b) output potential as a function of input potential. The linear range for the output potential is very small.

potential V_0 must have a value between V_{S+} and V_{S-}. For an ideal operational amplifier, the open-loop gain A_{op} is very large (ideally infinite), such that

$$V_+ - V_- = \frac{V_0}{A_{op}} \approx 0 \tag{6.2}$$

The open-loop gain A_{op} for typical operational amplifiers is on the order of 10^4 to 10^6; thus, for supply voltages V_S of 10 to 15 V, the input voltage difference in the linear regime can be on the order of 1 mV and can be as small as a few μV.

The requirement for operation within the linear regime is that

$$|V_+ - V_-| < \left| \frac{V_S}{A_{op}} \right| \tag{6.3}$$

As A_{op} is very large, the linear region of operation is correspondingly very small. The characteristics of an ideal operational amplifier are that:

- The constant A_{op} is very large such that the voltage difference $V_+ - V_- \approx 0$.

- The input impedances are very large such that the currents at the noninverting $(+)$ and the inverting $(-)$ inputs are equal to zero.

- The output potential at saturation is V_{S+} or V_{S-}.

- The output potential in the linear regime is given by equation (6.1).

In use, the objective is to avoid saturation and to operate within the linear regime.

The equations that govern the ideal operational amplifier are expressions of current balances. For the operational amplifier shown in Figure 6.1(a), the current balance is given as

$$i_+ + i_- + i_0 + i_{S+} + i_{S-} = 0 \tag{6.4}$$

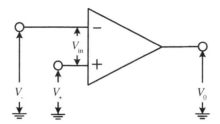

Figure 6.2: The circuit symbol for an operational amplifier showing input and output terminals but omitting the power terminals. The presence of power terminals is nevertheless assumed.

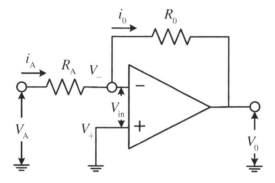

Figure 6.3: Negative feedback: inverting voltage amplifier.

As the input impedance is large, $i_+ = i_- = 0$, and

$$i_0 + i_{S+} + i_{S-} = 0 \qquad (6.5)$$

To reduce clutter in circuit diagrams, it is common to omit the power terminals, for example, as shown in Figure 6.2. The presence of power terminals is nevertheless assumed.

6.2   Elements of Electrochemical Instrumentation

As shown in Figure 6.1(b), the output of the operational amplifier under open-loop conditions tends to be in the saturation region. Operation within the linear region is made possible by inclusion of feedback loops, leading to operational characteristics important for electrochemical instrumentation. This is termed operation under closed-loop conditions.

Example 6.1 Negative Feedback: *Find the electrical characteristics of an ideal operational amplifier with negative feedback, shown schematically in Figure 6.3.*

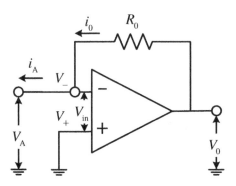

Figure 6.4: The current follower.

Solution: *Due to the fact that the input current is equal to zero, the currents flowing through R_A and R_0 are equal; $i = i_A = i_0$ and*

$$i = \frac{V_- - V_A}{R_A} = \frac{V_0 - V_-}{R_0} \tag{6.6}$$

V_- is equal to zero, due to the fact that the potential of the input + is at the ground. Thus,

$$V_0 = -\frac{R_0}{R_A} V_A \tag{6.7}$$

The output voltage has an opposite sign with respect to the input voltage V_A.

The quantity R_0/R_A is called the closed-loop gain. The requirement for operation within the linear regime for the inverting amplifier is that

$$|V_+ - V_-| < \left| \frac{V_S}{R_0/R_A} \right| \tag{6.8}$$

The effect of the feedback is to reduce the overall gain, to permit correspondingly larger input voltages without saturation, and to replace the open-loop gain with a gain that depends only on passive resistors. The open-loop gain of an operational amplifier depends strongly on temperature and varies from unit to unit. Thus, use of a feedback circuit improves control over the gain of an amplifier. The requirements are that the open-loop gain must be large as compared to the closed loop gain and equation (6.8) should be satisfied, i.e., the system must not be driven to saturation.

The output potential V_0 has the opposite sign as the input potential V_A. A noninverting amplifier is presented in Problem 6.3.

Example 6.2 Current Follower: *Find the operational characteristics of an ideal current follower, shown schematically in Figure 6.4.*

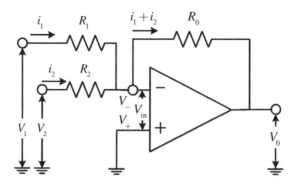

Figure 6.5: The voltage adder.

Solution: *This circuit is very similar to an inverting voltage amplifier without resistance R_A in the input line. The input point A is at a virtual ground potential, and V_o is proportional to the current; i.e., $V_o = R_o i$.*

Example 6.3 Voltage Adder: *Find the operational characteristics of an ideal voltage adder, shown schematically in Figure 6.5.*

Solution: *The example is given with the sum of two voltages, but obviously a larger number of potentials can be added following the same principle. The currents i_1 and i_2 are equal, respectively, to V_1/R_1 and V_2/R_2. Thus, the output voltage is given by*

$$V_o = -R_0(V_1/R_1 + V_2/R_2) \tag{6.9}$$

Different applications can be developed following the relative values of R_0, R_1, and R_2. In particular, for $R_0 = R_1 = R_2$, the output takes the form

$$V_o = -(V_1 + V_2) \tag{6.10}$$

6.3 Electrochemical Interface

Electrochemical interfaces consist of potentiostats and galvanostats. These devices can be described in terms of combinations of operational amplifiers and resistors.

Remember! 6.2 *A basic potentiostat can consist of two operational amplifiers: one to control potential and one to follow current.*

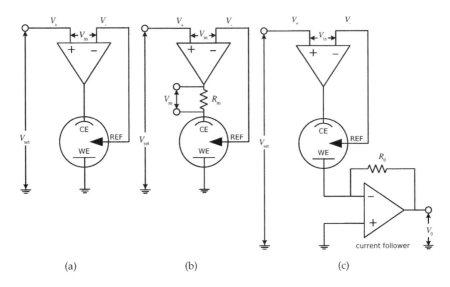

Figure 6.6: The potentiostat: a) simple scheme for controlling the potential of a working electrode with respect to a reference electrode, b) potentiostat with current measurement by potential drop across a measuring resistor, and c) potentiostat with current measurement by use of a current follower.

6.3.1 Potentiostat

The aim of a potentiostat is to maintain a constant potential difference between the working electrode WE and a reference electrode REF. In the simplest scheme the reference electrode is connecting to the inverting input of the operational amplifier as shown in Figure 6.6(a). The potential of the working electrode is at ground potential, and the potential of the reference electrode is held at a potential V_{set}, also referenced to the ground potential. Thus, the potentiostat shown in Figure 6.6(a) controls the potential difference between the working and reference electrodes.

The potentiostat requires as well a means of measuring the current. One approach is to measure the potential difference across a resistance as shown in Figure 6.6(b). The current is given by $I = V_m / R_m$. In the second approach, the current is measured in the working electrode circuit by means of a current follower (Figure 6.6(c)). In this last case, the working electrode is not directly at the ground but at a virtual ground potential.

In each of the different configurations of Figure 6.6, the potential of the working electrode is controlled with respect to the reference electrode. The WE is at the ground, and the potential V between the + entry of the operational amplifier and the ground is the difference of potential between the reference electrode and the WE. There is no potential difference between the entry + and the entry −. The operational amplifier delivers the current through the counterelectrode to have the corresponding difference of potential between the reference electrode and the WE.

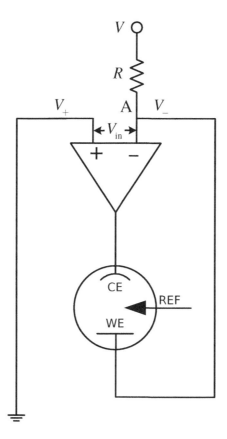

Figure 6.7: A simple scheme for controlling the current through a working electrode.

6.3.2   Galvanostat

A scheme of a galvanostat is given in Figure 6.7. The point A and then the working electrode is at a virtual ground potential. The current I is given by the relation $I = V/R$. The current value could be adjusted by varying either R or V. The potential can be easily measured between the reference electrode and the working electrode.

6.3.3   Potentiostat for EIS Measurement

Impedance measurements may be made under potentiostatic regulation by making the small modifications to the potentiostat shown as Figure 6.6(c). In the example given

Remember! 6.3 *A basic potentiostat for impedance measurements requires a voltage adder to superimpose the sinusoidal signal onto an imposed potential.*

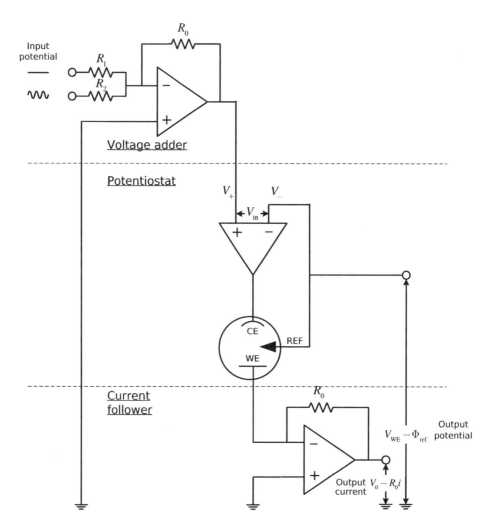

Figure 6.8: Potentiostat for EIS measurement.

as Figure 6.8, a voltage adder is introduced to sum the dc potential corresponding to the polarization point and the ac potential delivered by the generator of the frequency function analyzer. With a judicious choice of R_0, R_1 , and R_2 in equation (6.9), it could be easy to have an ac input with a potential divided by 100 (e.g., $R_0 = R_1 = 100R_2$).

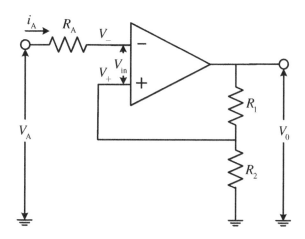

Figure 6.9: Schematic representation of a noninverting amplifier.

Problems

6.1 Devise a circuit for adding a steady baseline potential, a ramped potential, and a sinusoidal perturbation.

6.2 Estimate the maximum error in the term $V - \Phi_{\mathrm{ref}}$ if the operational amplifiers used in Figure 6.8 had an open-loop gain of 10^5 and power source of ± 10 V.

6.3   Show that the output potential for the noninverting amplifier shown in Figure 6.9 can be expressed as

$$V_0 = V_A \left(\frac{R_1 + R_2}{R_2} \right) \tag{6.11}$$

6.4 Devise a system similar to that shown in Figure 6.8 in which impedance measurements are performed under galvanostatic regulation.

Part II

Experimental Considerations

Chapter 7

Experimental Methods

Impedance experiments involve the conversion of time-domain input and output signals into a complex quantity that is a function of frequency. As seen in Figure 6.8, a signal generator is used to drive a potentiostat to induce a perturbed signal. The input signal and the resulting output signal is processed by instrumentation to yield a frequency-dependent transfer function. If the transfer function takes the form of a ratio of potential over current, the transfer function is called an impedance.

The first step in measuring impedance is typically measurement of the steady-state polarization curve. The impedance is then obtained for a specified point on the polarization curve. The instrumentation employed may involve phase-sensitive detection or Fourier analysis to convert time-domain measurements into the frequency domain. In the early days of impedance spectroscopy, Lissajou plots were used to provide a graphical determination of impedance at a given frequency. While no longer used to measure impedance, Lissajous plots provide useful information for the experimentalist. The methods used to obtain impedance from time-domain measurements are presented in the following sections.

7.1 Steady-State Polarization Curves

Steady-state polarization curves, such as that presented in Figure 5.3(a), provide a means of identifying such important electrochemical parameters as exchange current densities, Tafel slopes, and diffusion coefficients. The influence of exchange current density and Tafel slopes on the steady-state current density can be seen in equations (5.18) and (5.19), and the influence of mass transfer and diffusivities on the current density is described in Section 5.2.4. Steady-state measurements, however, cannot provide information on the RC time constants of the electrochemical process. Such properties must be identified by using transient measurements.

Even a slow voltammetric sweep rate may not be sufficient to ensure that the polarization curve represents a steady-state measurement. For these cases, it is necessary

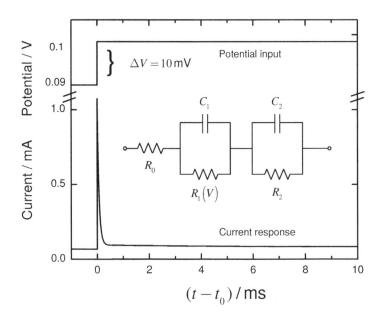

Figure 7.1: The current response of an electrochemical system to a 10 mV step change in applied potential from 0.09 V to 0.1 V for the inserted electrical circuit with parameters $R_0 = 1\ \Omega$, $R_1 = 10^{4-V/0.060}\ \Omega$, $C_1 = 10\ \mu F$, $R_2 = 10^3\ \Omega$, and $C_2 = 20\ \mu F$. The potential dependence of parameter R_1 is consistent with the behavior of the charge-transfer resistance described in Chapter 10.

to measure the current for steps in potential until the current can be seen to stabilize. For some systems, e.g., steel in alkaline electrolytes, the time required to reach a steady state may reach several days.

7.2   Transient Response to a Potential Step

A calculated transient current response to a 10 mV step in potential, introduced at time t_0, is presented in Figure 7.1 for the electrical circuit inserted in the figure. The time constants for the circuit under the conditions of the simulation were $\tau_1 = 0.0021$ s (76 Hz) and $\tau_2 = 0.02$ s (8 Hz). The potential dependence of parameter R_1 is consistent with the behavior of the charge-transfer resistance described in Chapter 10.

The calculated current increases instantaneously and then decreases rapidly. On the linear scale presented in Figure 7.1, it is evident that precise current measurement is needed to observe the features associated with the RC elements in Figure 7.1. Presen-

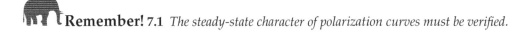

Remember! **7.1** *The steady-state character of polarization curves must be verified.*

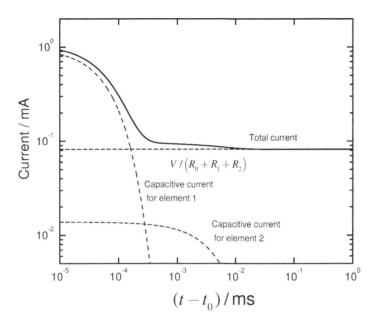

Figure 7.2: The current response in a logarithmic scale of an electrochemical system to a 10 mV step change in applied potential from 0.09 V to 0.1 V applied to the circuit presented in Figure 7.1. The dashed lines show the deconvolution of the current response into components associated with the different circuit elements.

tation in a logarithmic format, as shown in Figure 7.2, allows a clearer representation of the distinct features of the circuit presented in Figure 7.1. The dashed lines show the deconvolution of the current response into the components presented in Figure 7.1. The results presented in Figures 7.1 and 7.2 demonstrate that use of step-transient experiments for identification of the phenomena that govern an electrochemical system requires accurate measurements of current in a very short period of time.

Frequency-domain measurements provide an attractive alternative to use of transient techniques involving steps in potential or current because their to make repeated measurements at a single frequency improves the signal-to-noise ratio and extends the range of characteristic frequencies sampled. These measurements are a type of transient measurement in which the input signal is cyclic.

7.3   Analysis in Frequency Domain

Fourier analysis and phase-sensitive detection are commonly used to convert time-domain signals into the frequency domain. For contextual purposes, the mathematical transformations used by Fourier analysis and phase-sensitive detection instruments are reviewed in the following subsections. Such systems have replaced the Lissajous analysis described in Section 7.3.1. The Lissajous analysis is useful, however, for providing real-time assessment of the quality of impedance measurements and for developing an

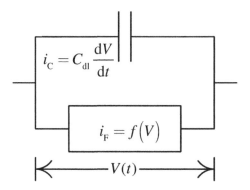

Figure 7.3: Schematic representation of an electrode interface demonstrating the contribution of charging and faradaic current densities.

appreciation for impedance measurements.

7.3.1 Lissajous Analysis

The electronics associated with measurement of the impedance response of a system is presented schematically in Figure 6.8. The input signal can be represented by

$$V = \overline{V} + \Delta V \cos{(2\pi f t)} \tag{7.1}$$

where $\overline{V}$ is an applied bias potential and ΔV is the amplitude of the sinusoidal perturbation. The current response is dependent on the characteristics of the system under study. For example, following the discussion in Section 5.2.2, the faradaic current density can be expressed as

$$i_F = n_a F k_a \exp{(b_a V)} - n_c F k_c \exp{(-b_c V)} \tag{7.2}$$

The current density associated with the capacitive charging of the electrode can be expressed as

$$i_C = C_{dl}\frac{dV}{dt} = -\omega \Delta V C_{dl} \sin{(2\pi f t)} \tag{7.3}$$

As suggested in Figure 7.3, the total current density is the sum of the faradaic and charging contributions.

The simulation results presented in this section were obtained by solution of equations (7.1)-(7.3) with parameters $C_{dl} = 31\ \mu F/cm^2$, $n_a F k_a = 0.5\ mA/cm^2$, $n_c F k_c = 0.5\ mA/cm^2$, $b_a = 19.5\ V^{-1}$, $b_c = 19.5\ V^{-1}$, $\overline{V} = 0\ V$, and $\Delta V = 1\ mV$. The value of the charge-transfer resistance, given by

$$R_t = \frac{1}{\left(b_a n_a F k_a \exp(b_a \overline{V}) + b_c n_c F k_c \exp(-b_c \overline{V})\right)} \tag{7.4}$$

had a value of 51.28 Ωcm^2, yielding a characteristic frequency $f_c = (2\pi R_t C_{dl})^{-1} = 100$ Hz. The current response to the input potential is presented for different applied

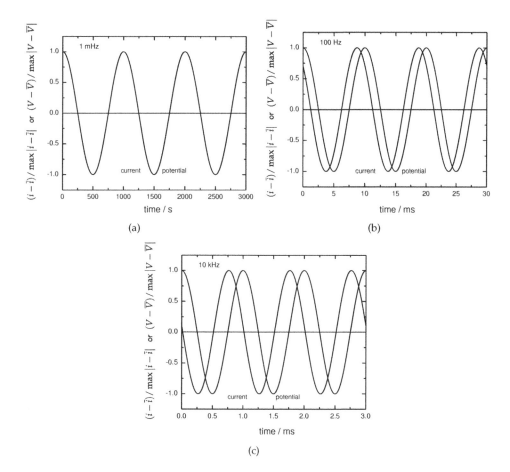

Figure 7.4: The current density response to a sinusoidal potential input for a system with parameters $C_{dl} = 31\ \mu F/cm^2$, $n_a F k_a = n_c F k_c = 0.5\ mA/cm^2$, $b_a = 19.5\ V^{-1}$, $b_c = 19.5\ V^{-1}$, $\overline{V} = 0\ V$, and $\Delta V = 1\ mV$: a) 1 mHz; b) 100 Hz; and c) 10 kHz.

frequencies in Figure 7.4 as functions of time where labels denote the lines representing the potential input and the resulting current density. The signals presented in Figure 7.4 were scaled by the amplitude of the perturbation.

At frequencies much below the characteristic frequency, i.e., 1 mHz, the current density is in phase with the potential perturbation, as shown in Figure 7.4(a). At the characteristic frequency, Figure 7.4(b), the current signal lags the potential input by 45°. At the characteristic frequency, the amplitude of the out-of-phase charging current is equal to the amplitude of the in-phase faradaic current for a linear system. As shown in Figure 7.4(c), the current response lags the potential input by 90° at frequencies much higher than the characteristic frequency.

The phase behavior of the input and output signals is seen more clearly in a Lissajous plot in which the output signal is plotted as a function of the input signal. A Lis-

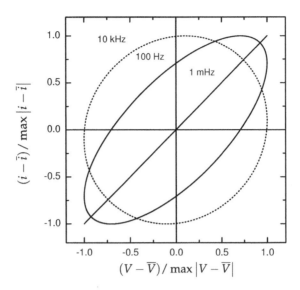

Figure 7.5: Lissajous representation of the signals presented in Figure 7.4 with frequency as a parameter. The signals were normalized by the perturbation amplitude such that the values for current and potential ranged between ±1.

sajous representation of the signals presented in Figure 7.4 is given in Figure 7.5 with frequency as a parameter. The signals were normalized by the perturbation amplitude such that the values for current and potential ranged between ±1. At low frequencies, e.g., 1 mHz, the current density is in phase with the potential perturbation, yielding a straight line in Figure 7.5. At large frequencies, e.g., 10 kHz, the current is dominated by the charging current, which is out of phase with the potential. The resulting Lissajous trace appears as a circle in Figure 7.5 where both the potential and current values were normalized. At the characteristic frequency (100 Hz), the Lissajous plot takes the shape of an ellipse.

Example 7.1 Derivation of Lissajous Ellipse: *Derive the Lissajous ellipse.*

Solution: *Consider input and output signals, respectively, as*

$$V = \Delta V \cos(\omega t) \tag{7.5}$$

and

$$I = \Delta I \cos(\omega t + \varphi) \tag{7.6}$$

Application of the trigonometric identity

$$\cos(\omega t + \varphi) = \cos(\omega t)\cos(\varphi) - \sin(\omega t)\sin(\varphi) \tag{7.7}$$

yields

$$I = \Delta I \left(\cos(\omega t) \cos(\varphi) - \sin(\omega t) \sin(\varphi) \right) \tag{7.8}$$

Another trigonometric identity

$$\cos^2(\omega t) = 1 - \sin^2(\omega t) \tag{7.9}$$

yields

$$\sin(\omega t) = \sqrt{1 - \cos^2(\omega t)} \tag{7.10}$$

Introduction of equation (7.5) yields

$$\sin(\omega t) = \sqrt{1 - \left(\frac{V}{\Delta V} \right)^2} \tag{7.11}$$

Equations (7.5) and (7.11) may be used to eliminate time in equation (7.8), yielding

$$\frac{I}{\Delta I} = \frac{V}{\Delta V} \cos \varphi - \sqrt{1 - \left(\frac{V}{\Delta V} \right)^2} \sin \varphi \tag{7.12}$$

or

$$\frac{I}{\Delta I} - \frac{V}{\Delta V} \cos \varphi = \sqrt{1 - \left(\frac{V}{\Delta V} \right)^2} \sin \varphi \tag{7.13}$$

Both sides of equation (7.13) may be squared, yielding

$$\left(\frac{I}{\Delta I} \right)^2 + \left(\frac{V}{\Delta V} \right)^2 \cos^2 \varphi - 2 \frac{I}{\Delta I} \frac{V}{\Delta V} \cos \varphi = \left[1 - \left(\frac{V}{\Delta V} \right)^2 \right] \sin^2 \varphi \tag{7.14}$$

As

$$\left(\frac{V}{\Delta V} \right)^2 \left(\cos^2 \varphi + \sin^2 \varphi \right) = \left(\frac{V}{\Delta V} \right)^2 \tag{7.15}$$

$$\left(\frac{I}{\Delta I} \right)^2 + \left(\frac{V}{\Delta V} \right)^2 - 2 \frac{I}{\Delta I} \frac{V}{\Delta V} \cos \varphi - \sin^2 \varphi = 0 \tag{7.16}$$

Application of the quadratic formula yields

$$\frac{I}{\Delta I} = \frac{V}{\Delta V} \cos \varphi \pm \sqrt{\left(\frac{V}{\Delta V} \right)^2 \left(\cos^2 \varphi - 1 \right) + \sin^2 \varphi} \tag{7.17}$$

This is the equation for an ellipse such as is shown in Figure 7.5. When $\varphi = 0$, equation (7.17) represents a straight line. When $\varphi = -\pi/2$, equation (7.17) represents a circle.

The magnitude of the time-domain signals and the shape of the ellipse provide information concerning the magnitude of the transfer function and the corresponding phase angle between input and output signals. The method for extracting the impedance response from the Lissajous plot is illustrated using the labeled positions given in Figure 7.6. The frequency is given by the time required to complete a cycle, t_{cycle}, i.e.,

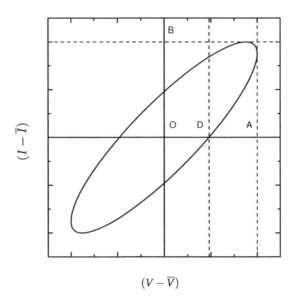

Figure 7.6: The interpretation of a Lissajous plot of time-domain signals in terms of impedance.

$$\frac{1}{t_{\text{cycle}}} = f \qquad (7.18)$$

where f has units of Hz. In frequency domain, the ratio of the potential over the current takes the form of an impedance. The magnitude of the impedance transfer function is therefore obtained by

$$|Z| = \frac{\max|V - \overline{V}|}{\max|I - \overline{I}|} = \frac{OA}{OB} \qquad (7.19)$$

The phase angle is obtained from

$$\sin(\varphi) = -\frac{OD}{OA} \qquad (7.20)$$

or

$$\varphi = \sin^{-1}\left(-\frac{OD}{OA}\right) \qquad (7.21)$$

where the lengths OD, OA, and OB are defined in Figure 7.6.

Example 7.2 Lissajous Analysis: *Use a Lissajous plot to find the impedance at a frequency of 100 Hz for a linear system with capacity $C_{\text{dl}} = 31\ \mu F/cm^2$, charge-transfer resistance $R_t = 51.34\ \Omega cm^2$, and a potential perturbation $\Delta V = 0.01$ V.*

Solution: *The time required for a single cycle is $T = 1/100$ Hz = 0.01 s. The potential over this period of time is given by*

$$V = \Delta V \cos(2\pi f t) \qquad (7.22)$$

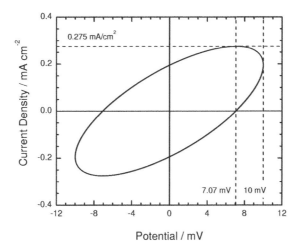

Figure 7.7: The interpretation of a Lissajous plot of time-domain signals in terms of impedance for Example 7.2.

The faradaic current density for a linear system can be expressed as

$$i_F = \frac{V}{R_t} \qquad (7.23)$$

and the current density associated with the capacitive charging of the electrode can be expressed by equation (7.3). The results are presented in Figure 7.7. The magnitude of the impedance can be found as

$$|Z| = \frac{10 \text{ mV}}{0.275 \text{ mA/cm}^2} = 36.4 \text{ } \Omega\text{cm}^2 \qquad (7.24)$$

and the phase angle is given by

$$\varphi = \sin^{-1}\left(-\frac{7.07}{10}\right) = -45° \qquad (7.25)$$

The results presented in Figure 7.7 correspond to the characteristic frequency for the given values of capacitance and charge-transfer resistance of 100 Hz. At this frequency, the phase angle is equal to 45°.

As illustrated in Section 21.2.2, Lissajous plots can be used to explain the mechanism whereby the frequency-domain analysis yields transfer functions with extremely large signal-to-noise ratios, even when the time-domain signals contain a significant level of noise. By measuring the signals over many cycles, the resolution of the ellipse improves by an averaging process, thereby yielding a small stochastic error in the calculated transfer function.

The Lissajous plots presented in Figure 7.5 correspond to the interfacial impedance of a system and do not account for ohmic resistance. This is the reason that the current-potential trace appears as a circle at high frequencies. In fact, the current corresponding to charging the interfacial capacitance is proportional to frequency, as expressed by

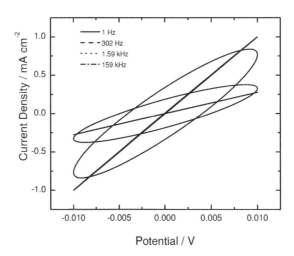

Figure 7.8: Lissajous representation for time-domain signals corresponding to a system with an ohmic resistance $R_e = 10 \ \Omega\text{cm}^2$, exchange current density $nFk_a = nFk_c = 1 \ \text{mA/cm}^2$, $b_a = 19 \ \text{V}^{-1}$, $b_c = 19 \ \text{V}^{-1}$, $\overline{V} = 0 \ \text{V}$, $\Delta V = 10 \ \text{mV}$, and capacitance $C_{dl} = 20 \ \mu\text{F/cm}^2$. The effective charge-transfer resistance was equal to $R_t = 26 \ \Omega\text{cm}^2$, and the characteristic frequency for this system was 302 Hz.

equation (7.3). The influence of the ohmic resistance is to limit the magnitude of the charging current at high frequencies. The resulting Lissajous plots are given in Figure 7.8 for a system with an ohmic resistance $R_e = 10 \ \Omega\text{cm}^2$, effective charge-transfer resistance $R_t = 26 \ \Omega\text{cm}^2$, and capacitance $C_{dl} = 20 \ \mu\text{F/cm}^2$. The input potential perturbation ΔV was 10 mV, yielding a value of $b_a\Delta V = 0.19$ and conforming to a linear response following the criteria developed in Example 8.2. The characteristic frequency for this system was 302 Hz. The Lissajous plot reveals a straight line at low frequency where the current and potential signals are in phase and a straight line again at high frequencies where the current and potential signals are again in phase. The slopes of these lines differ at high and low frequencies because, at low frequencies, the effective resistance is given by $R_e + R_t$, whereas at high frequencies the effective resistance is given by R_e. Comparison of Figures 7.5 and 7.8 illustrates the manner in which the ohmic resistance obscures the behavior of the interfacial processes at high frequencies. This concept motivates the use of the ohmic-resistance-corrected Bode plots discussed in Section 18.2.1.

It is useful to include in the experimental setup an oscilloscope capable of displaying a Lissajous plot. As discussed in Chapter 8, distortions of the elliptical shape can be used to signal nonlinear behavior associated with an input perturbation that is too large. Scatter in the results reveals that the time-domain data may be excessively noisy, reflecting perhaps the need to adjust the instrumental parameters.

7.3.2 Phase-Sensitive Detection (Lock-in Amplifier)

A lock-in amplifier uses phase-sensitive detection, in conjunction with a potentiostat, to measure the complex impedance. The algorithm is fundamentally different from that of the Fourier-based analyzers. These analyzers perform an assessment of the Fourier coefficients of the input and output signals, whereas the lock-in amplifier measures the amplitudes of the two signals and the phase angle of each signal with respect to some reference signal. Thus, the impedance is measured in polar, rather than Cartesian, coordinates.

The input and output signals are treated separately. A reference square wave of unity amplitude is generated at the same frequency as the sinusoidal signal

$$V = \Delta V \sin(\omega t + \varphi_V) \tag{7.26}$$

to be analyzed. The square wave can be represented by the Fourier series

$$S = \frac{4}{\pi} \sum_{n=0}^{\infty} \frac{1}{2n+1} \sin\left[(2n+1)\,\omega t + \varphi_S\right] \tag{7.27}$$

where φ_S is the phase angle of the reference signal. The measured and reference signals are multiplied, resulting in

$$VS = \frac{4\Delta V}{\pi} \sum_{n=0}^{\infty} \frac{1}{2n+1} \sin\left(\omega t + \varphi_V\right) \sin\left[(2n+1)\,\omega t + \varphi_S\right] \tag{7.28}$$

Equation (7.28) can be rewritten, using trigonometric identities, as

$$\begin{aligned} VS &= \sum_{n=0}^{\infty} \frac{2\Delta V}{(2n+1)\,\pi} \left\{ \cos\left[-2n\omega + \varphi_V - (2n+1)\,\varphi_S\right] \right. \\ &\quad \left. - \cos\left[(2n+2)\,\omega t + \varphi_V + (2n+1)\,\varphi_S\right] \right\} \end{aligned} \tag{7.29}$$

The product of the signals can be expanded using the trigonometric identity for the cosine of the sum of two angles and integrated over each cycle. Only the leading term of the series has a nonzero value, thus,

$$\frac{\omega}{2\pi} \int_{0}^{2\pi/\omega} VS dt = \frac{2\Delta V}{\pi} \cos\left(\varphi_V - \varphi_S\right) \tag{7.30}$$

The integral in equation (7.30) has a maximum value when the phase angle of the square wave is equal to the phase angle of the measured signal. In practice, the phase

---

Remember! 7.2 *While use of Lissajous plots for numerical evaluation of impedance can be considered obsolete, it is useful to include an oscilloscope capable of displaying a Lissajous plot in the experimental setup.*

angle of the generated square wave is adjusted such that the integral is maximized. The phase angle of the square wave at the maximum value of the integral yields the phase angle of the measured signal. In addition, the maximum value of the integral can be used to determine the amplitude of the measured signal.

The same procedure is used to analyze the output signal

$$I = \Delta I \sin(\omega t + \varphi_I) \tag{7.31}$$

yielding

$$\frac{\omega}{2\pi} \int_0^{2\pi/\omega} IS\mathrm{d}t = \frac{2\Delta I}{\pi} \cos\left(\varphi_I - \varphi_S\right) \tag{7.32}$$

The phase angle of the generated square wave is adjusted such that the integral is maximized. The phase angle of the square wave at the maximum value of the integral yields the phase angle of the measured signal, and the maximum value of the integral is used to determine the amplitude of the measured signal.

As discussed on page xxxvii, the magnitude of the transfer function is

$$|Y| = \frac{\Delta I}{\Delta V} \tag{7.33}$$

where Y is the admittance. The magnitude of the impedance can be obtained from

$$|Z| = \frac{\Delta V}{\Delta I} \tag{7.34}$$

The corresponding phase angle can be obtained as the difference between the phase angles of the output and input signals, i.e.,

$$\varphi = (\varphi_I - \varphi_S) - (\varphi_V - \varphi_S) \tag{7.35}$$

Carson et al.[145] showed that phase-sensitive detection measurements with a single reference signal biases the error structure of the impedance data due to errors introduced when the square-wave reference signal is in phase with the measured signal. Modern phase-sensitive detection instruments employ more than one reference signal and may thereby avoid this undesired correlation.

7.3.3 Single-Frequency Fourier Analysis

Single-frequency Fourier analyzers make use of the orthogonality of sines and cosines to determine the complex impedance representing the ratio of the response to a single-frequency input signal. A brief outline of the approach is presented in this section.

A periodic function of time can be expressed as a Fourier series, e.g.,[90, 146]

$$f(t) = a_0 + \sum_{n=1}^{\infty} \left(a_n \cos\left(n\omega t\right) + b_n \sin\left(n\omega t\right)\right) \tag{7.36}$$

The trigonometric functions can be expressed in terms of exponentials following equations (1.97) and (1.98) to yield

$$f(t) = \tilde{c}_0 + \sum_{n=1}^{\infty} \left(\tilde{c}_n \exp\left(nj\omega t\right) + \tilde{c}_{-n} \exp\left(-nj\omega t\right) \right) \qquad (7.37)$$

where the coefficients $\tilde{c}_n$ are complex numbers, related to the coefficients a_n and b_n of equation (7.36) by

$$\tilde{c}_n = \frac{a_n - jb_n}{2} \qquad (7.38)$$

$$\tilde{c}_{-n} = \frac{a_n + jb_n}{2} \qquad (7.39)$$

and

$$\tilde{c}_0 = a_0 \qquad (7.40)$$

Equation (7.37) can be written in more compact form as

$$f(t) = \sum_{n=-\infty}^{\infty} \tilde{c}_n \exp\left(nj\omega t\right) \qquad (7.41)$$

where n may take values ranging from $-\infty$ to $+\infty$, and the coefficients $\tilde{c}_n$ can be evaluated from

$$\tilde{c}_n = \frac{1}{T} \int_0^T f(t) \exp\left(-nj\omega t\right) dt \qquad (7.42)$$

where T represents the period of an integer number of cycles at frequency ω. Using equation (1.124), Equation (7.42) can be expressed in terms of trigonometric functions as

$$\tilde{c}_n = \frac{1}{T} \int_0^T f(t) \left(\cos\left(n\omega t\right) - j\sin\left(n\omega t\right)\right) dt \qquad (7.43)$$

Equation (7.43) provides the basis for single-frequency Fourier analysis for impedance measurement.

Linear sinusoidal input and output signals can be expressed in terms of equation (7.36) with $n = 1$. For example, for an input potential

$$V(t) = \Delta V \cos\left(\omega t\right) \qquad (7.44)$$

the output current can be expressed by

$$I(t) = \Delta I \cos\left(\omega t + \varphi_I\right) \qquad (7.45)$$

or

$$I(t) = a_1 \cos\left(\omega t\right) + b_1 \sin\left(\omega t\right) \qquad (7.46)$$

The constant coefficients ΔI and ΔV represent the amplitudes of the respective signals and the parameter φ_I represents the phase lag of the current signal with reference to the input potential signal.

The mapping of the time-domain signals (7.45) and (7.44) to the frequency domain is done via a Fourier complex representation.[2–4] For signals expressed in terms of a cosine, the in-phase or real part of the current signal can be expressed as

$$I_r(\omega) = \frac{1}{T} \int_0^T I(t) \cos(\omega t) dt \tag{7.47}$$

and the imaginary part of the current signal is

$$I_j(\omega) = -\frac{1}{T} \int_0^T I(t) \sin(\omega t) dt \tag{7.48}$$

The real part of the voltage signal is

$$V_r(\omega) = \frac{1}{T} \int_0^T V(t) \cos(\omega t) dt \tag{7.49}$$

and the imaginary part of the voltage signal is

$$V_j(\omega) = -\frac{1}{T} \int_0^T V(t) \sin(\omega t) dt \tag{7.50}$$

Equations (7.47)–(7.50) convert the respective time-domain quantities into the corresponding frequency-domain quantities. Integration is carried out over a period T comprising an integer number of cycles. The use of an integer number of cycles serves to filter errors in the measurement. The complex current $I_r + jI_j$ and potential $V_r + jV_j$ are the coefficients $\tilde{c}_1$ of the Fourier series expressed as equation (7.43).

The impedance is calculated as the complex ratio of the complex representations of the output signal to the input signal. Thus,

$$Z_r(\omega) = \mathrm{Re}\left\{ \frac{V_r + jV_j}{I_r + jI_j} \right\} \tag{7.51}$$

and

$$Z_j(\omega) = \mathrm{Im}\left\{ \frac{V_r + jV_j}{I_r + jI_j} \right\} \tag{7.52}$$

The real and imaginary parts of the impedance are thereby extracted from the same ratio of complex numbers.

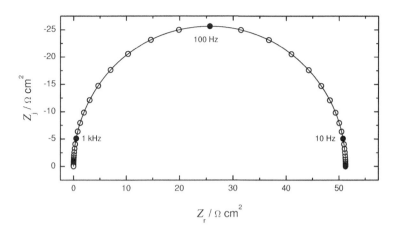

Figure 7.9: The impedance response obtained by Fourier analysis for the calculations presented in Section 7.3.1 (symbols) as compared to the theoretical value (solid line).

Example 7.3 **Fourier Analysis:** *Apply the Fourier analysis to the calculations presented in Section 7.3.1.*

Solution: *For the calculations presented in Section 7.3.1 where potential is the input signal with a phase lag $\varphi_V = 0$, equation (7.49) yields $V_r = \Delta V/2$ and equation (7.50) yields $V_j = 0$. The impedance response obtained by Fourier analysis is given in Figure 7.9. Because the perturbation magnitude was sufficiently small, the results are in good agreement with the expected value for the given kinetic parameters. The zero-frequency asymptotic value for the real impedance is in excellent agreement with the value of 51.28 Ωcm^2 obtained from equation (7.4).*

7.3.4   Multiple-Frequency Fourier Analysis

Impedance transfer functions may be determined through use of an input signal containing more than a single frequency. Such signals may be a tailored multi-sine signal or a signal containing white noise. If the signal amplitude is sufficiently small that the response is linear, the system output response will also be a signal containing the same frequencies as found in the input signal. A fast Fourier transform algorithm may be used to extract the frequency-dependent transfer function. Because all frequencies

Remember! **7.3** *The input and output signals used to generate impedance spectra are functions of time, not frequency. The frequency dependence of the impedance results from the processing of time-domain signals.*

are measured at the same time, the multifrequency approach allows completion of a spectrum in a shorter time than is required for a stepped single-frequency approach.

7.4 Comparison of Measurement Techniques

Each of the frequency-response analysis methods described in the previous section has its place in the experimental arsenal. Their relative merits are summarized in the following sections.

7.4.1 Lissajous Analysis

Lissajous analysis, as an experimental approach for impedance measurement, is obsolete and has been replaced by methods using automated instrumentation. Lissajous plots, however, have great pedagogical value as a means of learning impedance spectroscopy. In addition, as discussed in Section 8.2, use of oscilloscopes is recommended for monitoring the progress of impedance measurements, and oscilloscopes capable of displaying Lissajous plots are particularly useful.

7.4.2 Phase-Sensitive Detection (Lock-in Amplifier)

Phase-sensitive detection is accurate and can be relatively inexpensive. Modern instrumentation uses more than one reference signal and can mitigate the bias in the error structure.

7.4.3 Single-Frequency Fourier Analysis

Sequential measurement of impedance by Fourier analysis provides good accuracy for stationary systems. The sequence of frequencies can be arbitrarily selected, and therefore frequency intervals of $\Delta f / f$, considered to be the most economical use of frequencies, can be employed. Because the measurements at each frequency are independent of each other, frequencies found to be inconsistent with the Kramers–Kronig relations can be deleted.

7.4.4 Multiple-Frequency Fourier Analysis

Multifrequency fast Fourier analysis yields good accuracy for stationary systems and has the advantage that it can be performed more rapidly than an equivalent single-sine measurement. Frequency intervals of Δf and dense sampling at high frequency are required to get good resolution at low frequency. The spectra obtained are always consistent with the Kramers–Kronig relations, so the Kramers–Kronig relations cannot be used to determine whether the measurement was corrupted by instrument artifacts or nonstationary behavior. A correlation coefficient can be calculated and used to determine whether the spectrum is inconsistent with the Kramers–Kronig relations.

Table 7.1: Generalized transfer functions for a rotating disk electrode at fixed temperature.

Fixed Variable	Input	Output	Transfer Function
Rotation speed	Current	Potential	Impedance
Rotation speed	Potential	Current	Admittance
Current	Rotation speed	Potential	Electrohydrodynamic impedance
Potential	Rotation speed	Current	Electrohydrodynamic impedance

7.5 Specialized Techniques

The methods described in this chapter and this book apply to electrochemical impedance spectroscopy. Impedance spectroscopy should be viewed as being a specialized case of a transfer-function analysis. The principles apply to a wide variety of frequency-domain measurements, including non-electrochemical measurements. The application to generalized transfer-function methods is described briefly with an introduction to other sections of the text where these methods are described in greater detail. Local impedance spectroscopy, a relatively new and powerful electrochemical approach, is described in detail.

7.5.1 Transfer-Function Analysis

While the emphasis of this book is on electrochemical impedance spectroscopy, the methods described in Section 7.3 for converting time-domain signals to frequency-domain transfer functions clearly are general and can be applied to any type of input and output. Some generalized transfer-function approaches are described in Chapters 15 and 16.

Four state variables may be defined, for example, for the rotating disk described in Chapter 11. These may include the rotation speed, the temperature, the current, and the potential. At a fixed temperature, three variables remain from which a transfer function may be calculated. As shown in Table 7.1, the generalized transfer functions include impedance, admittance (see Chapter 17 on page 467), and two types of electro-hydrodynamic impedance (see Chapter 16 on page 443).

Remember! 7.4 *The techniques described here for measuring impedance spectra are, in fact, very general and can be used to measure any transfer function.*

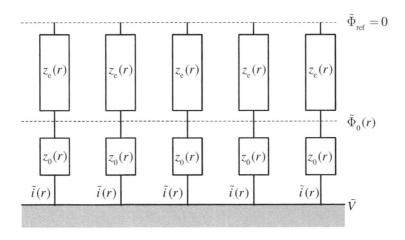

Figure 7.10: The location of current and potential terms that make up definitions of global and local impedance.

7.5.2 Local Electrochemical Impedance Spectroscopy

Local impedance measurements can be considered to be another form of generalized transfer-function analysis. In these experiments, a small probe is placed near the electrode surface. The probe uses either two small electrodes or a vibrating wire to allow measurement of potential at two positions. Under the assumption that the electrolyte conductivity between the two points of potential measurement is uniform, the current density at the probe can be estimated from the measured potential difference ΔV_{probe} by

$$i_{\text{probe}} = \Delta V_{\text{probe}} \frac{\kappa}{d} \qquad (7.53)$$

where d is the distance between the potential sensing probes and κ is the conductivity of the electrolyte.

A schematic representation of the electrode-electrolyte interface is given as Figure 7.10, where the block used to represent the local ohmic impedance reflects the complex character of the ohmic contribution to the local impedance response. The impedance definitions presented in Table 7.2 were proposed by Huang et al.[147] for local impedance variables. These differ in the potential and current used to calculate the impedance. To avoid confusion with local impedance values, the symbol y is used to designate the axial position in cylindrical coordinates.

Remember! 7.5 *Local Electrochemical Impedance Spectroscopy (LEIS) is a relatively new and underutilized technique that is useful for exploring the influence of surface heterogeneities on the impedance response.*

Table 7.2: Definitions and notation for local impedance variables.

Symbol	Meaning	Units
Z	Global impedance (equation (7.54))	Ω or Ωcm^2
Z_r	Real part of global impedance	Ω or Ωcm^2
Z_j	Imaginary part of global impedance	Ω or Ωcm^2
Z_0	Global interfacial impedance (equation (7.61))	Ω or Ωcm^2
$Z_{0,r}$	Real part of global interfacial impedance	Ω or Ωcm^2
$Z_{0,j}$	Imaginary part of global interfacial impedance	Ω or Ωcm^2
Z_e	Global ohmic impedance (equation (7.63))	Ω or Ωcm^2
$Z_{e,r}$	Real part of global ohmic impedance	Ω or Ωcm^2
$Z_{e,j}$	Imaginary part of global ohmic impedance	Ω or Ωcm^2
z	Local impedance (equation (7.56))	Ωcm^2
z_r	Real part of local impedance	Ωcm^2
z_j	Imaginary part of local impedance	Ωcm^2
z_0	Local interfacial impedance (equation (7.58))	Ωcm^2
$z_{0,r}$	Real part of local interfacial impedance	Ωcm^2
$z_{0,j}$	Imaginary part of local interfacial impedance	Ωcm^2
z_e	Local ohmic impedance (equation (7.59))	Ωcm^2
$z_{e,r}$	Real part of local ohmic impedance	Ωcm^2
$z_{e,j}$	Imaginary part of local ohmic impedance	Ωcm^2
$\langle \Phi \rangle$	Spatial average of potential	V
$\bar{\Phi}$	Time average or steady-state value of potential	V
$\langle i \rangle$	Spatial average of current density	A/cm^2
$\bar{i}$	Time average or steady-state value of current density	A/cm^2
y	Axial position variable	cm

Global Impedance

The global impedance is defined to be

$$Z = \frac{\widetilde{V}}{\widetilde{I}} \tag{7.54}$$

where the complex current contribution for a disk electrode is given by

$$\widetilde{I} = \int_0^{r_0} \widetilde{i}(r) 2\pi r \, dr \tag{7.55}$$

The use of an uppercase letter signifies that Z is a global value. The global impedance may have real and imaginary values designated as Z_r and Z_j, respectively. The total current could also be represented by $\widetilde{I} = \pi r_0^2 \langle \widetilde{i}(r) \rangle$ where the brackets signify the area-average of the current density.

Local Impedance

The term local impedance traditionally involves the potential of the electrode measured relative to a reference electrode far from the electrode surface.[84,148] Thus, the local impedance is given by

$$z = \frac{\widetilde{V}}{\widetilde{i}(r)} \tag{7.56}$$

The use of a lowercase letter signifies that z is a local value. The local impedance may have real and imaginary values designated as z_r and z_j, respectively.

The global impedance can be expressed in terms of the local impedance as

$$Z = \left\langle \frac{1}{z} \right\rangle^{-1} \tag{7.57}$$

Equation (7.57) is consistent with the treatment of Brug et al.[149] in which the admittance of the disk electrode was obtained by integration of a local admittance over the area of the disk.

Local Interfacial Impedance

The local interfacial impedance involves the potential of the electrode measured relative to a reference electrode $\Phi_0(r)$ located at the outer limit of the diffuse double layer. Thus, the local interfacial impedance is given by

$$z_0 = \frac{\widetilde{V} - \widetilde{\Phi}_0(r)}{\widetilde{i}(r)} \tag{7.58}$$

The use of a lowercase letter again signifies that z_0 is a local value, and the subscript 0 signifies that z_0 represents a value associated only with the surface. The local interfacial impedance may have real and imaginary values designated as $z_{0,r}$ and $z_{0,j}$, respectively.

Local Ohmic Impedance

The local ohmic impedance involves the potential of a reference electrode $\Phi_0(r)$ located at the outer limit of the diffuse double layer and the potential of a reference electrode located far from the electrode $\widetilde{\Phi}(\infty) = 0$; see Figure 7.10. Thus, the local ohmic impedance is given by

$$z_e = \frac{\widetilde{\Phi}_0(r)}{\widetilde{i}(r)} \tag{7.59}$$

The use of a lowercase letter again signifies that z_e is a local value, and the subscript e signifies that z_e represents a value associated only with the ohmic character of the electrolyte. The local ohmic impedance may have real and imaginary values designated as $z_{e,r}$ and $z_{e,j}$, respectively. The local impedance

$$z = z_0 + z_e \tag{7.60}$$

can be represented by the sum of local interfacial and local ohmic impedances.

The representation of an ohmic impedance as a complex number represents a departure from standard practice. As shown in Section 13.2 on page 337, the local impedance has inductive features that are not seen in the local interfacial impedance. As the calculations assumed an ideally polarized blocking electrode, the result is not influenced by faradaic reactions and can be attributed only to the ohmic contribution of the electrolyte.

Global Interfacial Impedance

The global interfacial impedance is defined for a disk electrode to be

$$Z_0 = 2\pi \left(\int_0^{r_0} \frac{1}{z_0(r)} r \, dr \right)^{-1} \tag{7.61}$$

or, in a more general sense,

$$Z_0 = \left\langle \frac{1}{z_0(r)} \right\rangle^{-1} \tag{7.62}$$

The use of an uppercase letter signifies that Z_0 is a global value. The global interfacial impedance may have real and imaginary values designated as $Z_{0,r}$ and $Z_{0,j}$, respectively.

Global Ohmic Impedance

The global ohmic impedance is defined to be

$$Z_e = Z - Z_0 \tag{7.63}$$

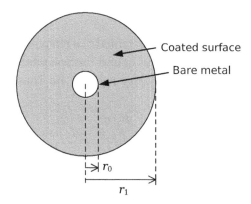

Figure 7.11: Schematic representation of a disk electrode corresponding to Problem 7.5 with bare metal exposed from the origin to a radius r_0 and coated metal surface from radius r_0 to r_1.

The use of an uppercase letter signifies that Z is a global value. As will be shown in subsequent sections, the global ohmic impedance has a complex behavior in a midfrequency range near $K = 1$ (see equation (13.67) on page 339). The global ohmic impedance may have real and imaginary values designated as $Z_{e,r}$ and $Z_{e,j}$, respectively.

Problems

The following problems require use of a spreadsheet program such as Microsoft Excel® or a computational programming environment such as MATLAB®.

7.1   Reproduce the results presented in Figures 7.4 and 7.5.

7.2   Use a Lissajous plot to calculate the phase angle and magnitude of the impedance for the system described in Example 7.2 at frequencies of 1 Hz and 10 kHz. This approach requires calculating the potential perturbation, the charging current, and the faradaic current as functions of time.

7.3   Calculate the ratio of the amplitude of the charging current to the amplitude of the faradaic current for the system described in Example 7.2 at frequencies of 1 Hz, 100 Hz, and 10 kHz.

7.4   Use a Fourier analysis to calculate the impedance as a function of frequency for the system described in Example 7.2. Compare your results to the theoretical value obtained using the methods described in Chapter 4.

7.5   Consider the disk electrode shown in Figure 7.11 with bare metal exposed from the origin to a radius r_0 and coated metal surface from radius r_0 to r_1. Estimate the global interfacial impedance response of this electrode for the following geometries if the coating consists of an organic material of 100 μm thickness and the charge-transfer resistance on the bare metal is 100 Ωcm^2. Hint: Guidelines for estimation of capacitance are given in Table 5.6.

(a) $r_0 = 0.25$ cm and $r_1 = 1.0$ cm.

(b) $r_0 = 0.5$ cm and $r_1 = 1.0$ cm.

(c) $r_0 = 0.75$ cm and $r_1 = 1.0$ cm.

7.6 Develop an expression for the error in the global interfacial impedance response of the bare metal referenced in Problem 7.5 as a function of the relative area of the coated metal r_1^2/r_0^2 and the coating property ε/δ.

7.7 Use a Lissajous analysis to calculate the magnitude and phase angle for the plots given in Figures 7.12.

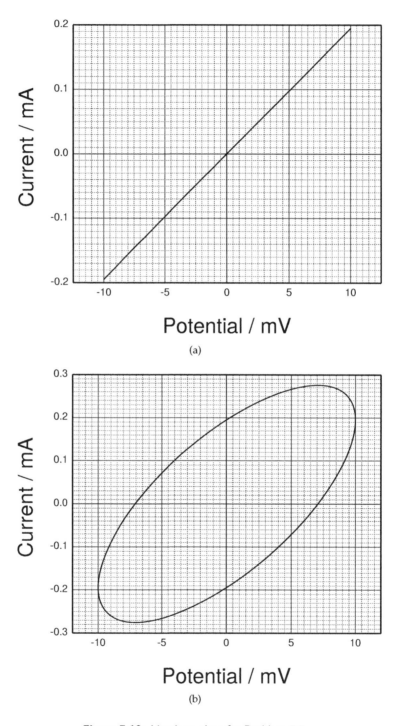

(a)

(b)

Figure 7.12: Lissajous plots for Problem 7.7.

Chapter 8

Experimental Design

Impedance measurements are often used to identify physical phenomena that control an electrochemical reaction and to determine the corresponding physical properties. This chapter provides guidelines for the design of experimental cells, for selection of appropriate impedance parameters, and for selection of appropriate instrument controls.

8.1 Cell Design

Proper cell design is essential to reduce the uncertainty of the interpretation. Reference electrodes can be used to isolate the impedance of cell components, well-defined convective systems can be employed to quantify the role of mass transfer, and electrode configurations can be selected to minimize the role of current and potential distributions across the surface of the electrodes.

8.1.1 Reference Electrodes

As discussed in Section 5.6, the potential drop across an electrochemical cell can be expressed as the sum of contributions

$$V_{WE} - V_{CE} = (V_{WE} - \Phi_{0,WE}) + (\Phi_{0,WE} - \Phi_{0,CE}) - (V_{CE} - \Phi_{0,CE}) \qquad (8.1)$$

Reference electrodes are used to isolate influence of electrodes and membranes. Some typical cell configurations are shown schematically in Figure 8.1. In the two-electrode

---

Remember! 8.1 *Reference electrodes can be used to isolate the impedance of cell components.*

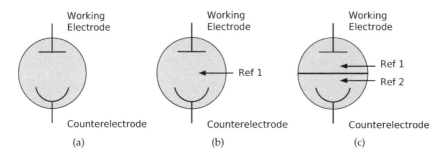

Figure 8.1: Schematic illustration of cell configurations employing reference electrodes to isolate the impedance response of working electrodes and membranes: a) 2-electrode; b) 3-electrode; and c) 4-electrode cell design.

configuration represented in Figure 8.1(a), the impedance

$$Z_{\text{cell}} = \frac{\tilde{V}_{\text{WE}} - \tilde{V}_{\text{CE}}}{\tilde{I}} \tag{8.2}$$

includes the impedance response associated with the working electrode interface, the counterelectrode interface, and the electrolyte between the working and the counter-electrodes. The three-electrode configuration shown in Figure 8.1(b) allows measurement of three impedances; i.e., equation (8.2),

$$Z_{\text{WE}} = \frac{\tilde{V}_{\text{WE}} - \tilde{\Phi}_{\text{ref}}}{\tilde{I}} \tag{8.3}$$

at the working electrode interface and

$$Z_{\text{CE}} = \frac{\tilde{V}_{\text{CE}} - \tilde{\Phi}_{\text{ref}}}{\tilde{I}} \tag{8.4}$$

at the counterelectrode interface. The impedance measurements are related by

$$Z_{\text{cell}} = Z_{\text{WE}} + Z_{\text{CE}} \tag{8.5}$$

Thus, only two of the impedance measurements are independent. For systems containing a membrane separating the working and counterelectrodes, it is useful to use a four-electrode configuration as shown in Figure 8.1(c). The impedance of the membrane can be obtained as

$$Z_{1,2} = \frac{\tilde{\Phi}_{\text{ref},2} - \tilde{\Phi}_{\text{ref},1}}{\tilde{I}} \tag{8.6}$$

The impedance measurements are related by

$$Z_{\text{cell}} = Z_{\text{WE},1} + Z_{\text{CE},2} + Z_{1,2} \tag{8.7}$$

where $Z_{\text{WE},1}$ is the impedance of the working electrode measured with reference Ref1 and $Z_{\text{CE},2}$ is the impedance of the counterelectrode measured with reference Ref2. Only three of the impedance measurements are independent.

Example 8.1 Electrode Connections: *Most potentiostats used for impedance measurements have three or four connections. A potentiostat with three connections has a wire to be connected to the working electrode (WE), a wire to be connected to the reference electrode (Ref), and a wire to be connected to the counterelectrode (CE). A potentiostat with four connections has, in addition to connections to the working and counterelectrode, a wire to be connected to the working sense reference electrode (Ref1) and a wire to be connected to the a wire to be connected to the true reference electrode (Ref2). How should these connections be configured for the three cases descried in Figure 8.1*

Solution: *For a potentiostat with three connections, the wires are connected as follows:*

a) Two electrodes: WE $\cdots\cdots\cdots\cdots$ CE − Ref

b) Three electrodes: WE $\cdots\cdots$ Ref $\cdots\cdots$ CE

c) Four electrodes: *No connection possible.*

For a potentiostat with four connections, the wires are connected as follows:

a) Two electrodes: Ref1 − WE $\cdots\cdots\cdots\cdots$ CE − Ref2

b) Three electrodes: Ref1 − WE $\cdots\cdots$ Ref2 $\cdots\cdots$ CE

c) Four electrodes: WE $\cdots$ Ref1 $\cdots \| \cdots$ Ref2 $\cdots$ CE

8.1.2 Flow Configurations

Use of well-defined convective systems allows the influence of mass transfer to be treated explicitly and quantitatively; thus the interpretation of impedance data in terms of interfacial processes can be emphasized. Several experimental systems are commonly employed.

Rotating Disk

The rotating disk electrode, described in Section 11.3, has the advantage that the fluid flow is well defined and that the system is compact and simple to use. The rotation of the disk imposes a centrifugal flow that in turn causes a radially uniform flow toward the disk. If the reaction on the disk is mass-transfer controlled, the associated current density is uniform, which greatly simplifies the mathematical description. As discussed in sections 5.5.1 and 8.1.3, the current distribution below the mass-transfer-limited current is not uniform. The distribution of current and potential associated with the disk geometry has been demonstrated to cause a frequency dispersion in impedance results. The rotating disk is therefore ideally suited for experiments in which the disk rotation speed is modulated while under the mass-transfer limited condition. Such experiments yield another type of impedance known as the electrohydrodynamic impedance, discussed in Chapter 16.

Disk under Submerged Impinging Jet

The disk subjected to a submerged impinging jet, described in Section 11.4, has, under certain geometric constraints, the same attributes as the rotating disk with the exceptions that the electrode is stationary and the experimental system requires a pump and a flow loop. For electrodes that are within the stagnation region of the impinging flow, the flow is well defined and has been modeled. The flow to the electrode surface is radially uniform; thus, at the mass-transfer-limited condition, the current density is uniform. The geometry induces a current and potential distribution that is similar to that observed for the rotating disk.

Rotating Cylinders

Rotating cylinders, described in Section 11.5, are popular experimental systems because the system setup is relatively simple to use and, at moderate rotation speeds, the flow is turbulent and yields a uniform mass-transfer-controlled current density. Empirical correlations are available that relate the cylinder rotation speed to the mass-transfer coefficient.[150]

Rotating Hemispherical Electrode

The rotating hemispherical electrode, introduced by Chin,[129] has a uniform primary current distribution and would therefore be a suitable configuration for experiments conducted under conditions such that the current distribution is not influenced by the nonuniform accessibility to mass transfer. Nisancioglu and Newman[130] showed that the current distribution in the rotating hemispherical electrode is uniform so long as the average current density has a value smaller than 68 percent of the average mass-transfer-limited value. A refined mathematical model for the convective-diffusion impedance of a rotating hemispherical electrode, developed by Barcia et al.,[131] provided an excellent match to experimental impedance measurements conducted under these conditions.

While a disk electrode under a submerged impinging jet has flow characteristics that resemble those of a rotating disk electrode, the flow experienced by a hemispherical electrode under jet impingement differs greatly from that of a rotating hemisphere. Shukla and Orazem showed through calculations and experiments that boundary-layer separation is observed at an colatitude angle of 54.8°.[132,133] A subsequent analysis of the current distribution below the mass-transfer-limited current indicated that the current distribution on the stationary hemispherical electrode under submerged jet impingement should be uniform so long as the average current density was less than 25 percent of the mass-transfer-limited value.[151] The appearance of boundary-layer separation suggests that the hemispherical electrode under jet impingement is not an appropriate system for electrochemical studies.

Table 8.1: Current distribution characteristics for different electrode designs.

System	Primary Distribution	Mass-Transfer-Controlled Distribution
Rotating disk	Not uniform	Uniform
Disk under submerged impinging jet	Not uniform	Uniform
Partial rotating cylinder	Not uniform	Uniform
Complete rotating cylinder	Uniform	Uniform
Rotating hemisphere	Uniform	Not uniform
Hemisphere under submerged jet	Uniform	Not uniform

8.1.3 Current Distribution

The distribution of impedance along the surface of an electrode greatly complicates the interpretation of the resulting spectra. Such an impedance distribution may result from a variation of surface properties caused, for example, by differences of grain orientation in a polycrystalline material, residual stresses associated with fabrication, or nonuniform distributions of surface films. A distribution of impedance may also be attributed to the current and potential distributions associated with the electrode geometry.

The uncertainty associated with the interpretation of the impedance response can be reduced by using an electrode for which the current and potential distribution is uniform. There are two types of distributions that can be used to guide electrode design. As described in Section 5.5.1, the primary distribution accounts for the influence of ohmic resistance and mass-transfer-limited distributions account for the role of convective diffusion. The secondary distributions account for the role of kinetic resistance which tends to reduce the nonuniformity seen for a primary distribution. Thus, if the primary distribution is uniform, the secondary current distributions will also be uniform. The guidelines illustrated in equation (5.110) and Figure 5.12 can be used to assess the uniformity of current on different electrode geometries. The results are given in Table 8.1. As discussed in Section 13.2, the frequency dispersion associated with a nonuniform primary or secondary current distribution for elementary reactions is not evident at frequencies below a critical value. Thus, the influence of geometry-induced nonuniform current and potential distributions can be avoided by designing experiments for frequencies below the critical value. Figure 13.11 provides a convenient guide for selecting a disk electrode size that will avoid the frequency dispersion effects associated with the geometry-induced current and potential distributions.

Remember! 8.2 *Impedance measurements are sensitive to nonuniform surface reactivity, which may be caused by surface heterogeneities, nonuniform mass transfer, or geometry-induced current and potential distributions.*

8.2 Experimental Considerations

The experimental design parameters described in this section are influenced by the system under investigation, the objective of the investigation, and the capabilities of the instrumentation. The objective is to maximize the information content of the measurement while minimizing bias and stochastic errors.

8.2.1 Frequency Range

The objective of impedance measurements is typically to capture the frequency response of the system under study. To that end, the measured frequency range should include frequencies sufficiently large and frequencies sufficiently small to reach asymptotic limits in which the imaginary impedance tends toward zero. In some cases, for example, blocking electrodes, the low-frequency asymptotic behavior does not exist. In other cases, a true dc limit is not achievable due to nonstationary behavior of the system. Instrument artifacts may limit the performance at high frequency. While experimentalists often routinely base the frequency range on the limits of the instrument, it is important to choose a frequency range that meets the dynamic response of the system under study. The considerations described in Section 13.2 may also constrain the frequency range.

8.2.2 Linearity

As discussed in Section 5.2.2, linearity in electrochemical systems is controlled by potential. The use of a low-amplitude perturbation allows application of a linear model for interpretation of spectra. The correct amplitude represents a compromise between the desire to minimize nonlinear response (by using a small amplitude) and the desire to minimize noise in the impedance response (by using a large amplitude). The amplitude depends on the system under investigation. For systems exhibiting a linear current-voltage curve, a very large amplitude can be used. For systems exhibiting very nonlinear current-voltage curves, a much smaller amplitude is needed.[152–154]

Example 8.2 Guideline for Linearity: *We wish to establish a guideline for the perturbation amplitude needed to maintain linearity under potentiostatic regulation. An electro-*

Remember! 8.3 *Impedance measurements entail a compromise balance between minimizing bias errors, minimizing stochastic errors, and maximizing the information content of the resulting spectrum. The optimal instrument settings and experimental parameters are not universal and must be selected for each system under study.*

---

chemical system that follows a Tafel law is polarized at a potential $\overline{V}$. If a large potential sinusoidal modulation is superimposed, write the current response in the form of a Taylor series and calculate the complete expression of the dc current. By considering only the first three terms of the Taylor series, write the expression of the current under the form of the first three harmonics.

Solution: *For a system that follows Tafel behavior, the current density response to a potential perturbation*

$$V(t) = \overline{V} + \Delta V \cos(\omega t) \tag{8.8}$$

is given by

$$i(t) = K \exp(bV(t)) \tag{8.9}$$

Thus,

$$i(t) = K \exp\left(b(\overline{V} + \Delta V \cos \omega t)\right) \tag{8.10}$$

or

$$i(t) = i_0 \exp(b\Delta V \cos \omega t) \tag{8.11}$$

where

$$i_0 = K \exp(b\overline{V}) \tag{8.12}$$

A Taylor series expansion yields

$$
\begin{aligned}
i(t) \;=\; & i_0 \left(1 + b\Delta V \cos \omega t + \frac{b^2 \Delta V^2 \cos^2 \omega t}{2!} + \frac{b^3 \Delta V^3 \cos^3 \omega t}{3!} + \right. \\
& \left. \cdots + \frac{b^n \Delta V^n \cos^n \omega t}{n!} + \cdots \right)
\end{aligned}
\tag{8.13}
$$

The mean value of the current $i(t)$ is, for T equal to an integer number of cycles,

$$\overline{i}(t) = \frac{1}{T} \int_0^T i(t)\mathrm{d}t \tag{8.14}$$

By taking into account the formula

$$\int \cos^n x \mathrm{d}x = \frac{1}{n} \cos^{n-1} x \sin x + \frac{n-1}{n} \int \cos^{n-2} x \mathrm{d}x \tag{8.15}$$

and observing that $\sin T = 0$,

$$\int_0^T \cos^n x \mathrm{d}x = \frac{n-1}{n} \int_0^T \cos^{n-2} x \mathrm{d}x \tag{8.16}$$

Remember! 8.4 *The optimal perturbation amplitude depends on the polarization curve for the system under study.*

If n is an even number,

$$\int_0^T \cos^n x \, dx = \frac{n-1}{n} \frac{n-3}{n-2} \cdots \frac{1}{2} T \tag{8.17}$$

and if n is an odd number, the value of the integral is equal to zero. Thus, the mean value of $i(t)$ is

$$\bar{i}(t) = i_0 \left(1 + \sum_{n=1}^{\infty} \frac{b^{2n} \Delta V^{2n}}{(2^n n!)^2} \right) \tag{8.18}$$

To have a variation of the dc current lower than 1 percent, ΔV must be lower than $0.2/b$.

Evaluation of the harmonics of the nonlinear current response can be achieved by introduction of the trigonometric expressions

$$\cos 2x = 2 \cos^2 x - 1 \tag{8.19}$$

and

$$\cos 3x = 4 \cos^3 x - 3 \cos x \tag{8.20}$$

By considering only the first three terms of the Taylor series, $i(t)$ becomes

$$i(t) = i_0 \left(\left(1 + \frac{b^2 \Delta V^2}{4} \right) + \left(b \Delta V + \frac{3b^3 \Delta V^3}{24} \right) \cos(\omega t) \right. \tag{8.21}$$
$$\left. + \frac{b^2 \Delta V^2}{4} \cos(2\omega t) + \frac{b^3 \Delta V^3}{24} \cos(3\omega t) \right)$$

Due to the limitation to the first three terms of the Taylor series, only the first term of the series contributes to the mean value (see equation (8.18)). Thus, equation (8.21) shows that the dc current is given by

$$i_0 \left(1 + \frac{b^2 \Delta V^2}{4} \right) \tag{8.22}$$

and the first harmonic or fundamental is given by

$$i_0 \left(b \Delta V + \frac{3b^3 \Delta V^3}{24} \right) \tag{8.23}$$

For ΔV smaller than $0.2/b$, the variation of the dc current is smaller than 1 percent and the variation of the fundamental is smaller than 0.22 percent.

Example 8.2 demonstrates that application of a large-amplitude potential perturbation to a nonlinear system results in harmonics that appear at frequencies corresponding to multiples of the fundamental or applied frequency. A second result of Example 8.2 is the observation that application of a large-amplitude potential perturbation to a nonlinear system changes both the steady-state current density and the fundamental current response. The implication of this result is that the impedance response will also be distorted by application of a large-amplitude potential perturbation.

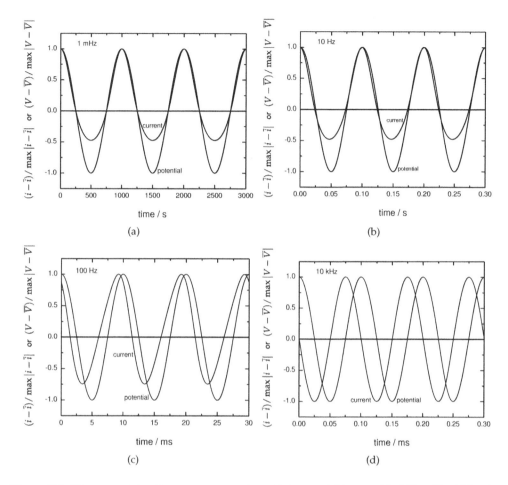

Figure 8.2: The current density response to a sinusoidal potential input with $\Delta V = 40$ mV for the system presented in Section 7.3 with parameters $C_{dl} = 31$ μF/cm^2, $n_a F k_a = n_c F k_c = 0.14$ mA/cm^2, $b_a = 19.5$ V^{-1}, $b_c = 19.5$ V^{-1}, and $\overline{V} = 0.1$ V: a) 1 mHz; b) 10 Hz; c) 100 Hz; and d) 10 kHz. The solid line represents the potential input and the dashed line represents the resulting current density.

The influence of large potential perturbations on the impedance response can be illustrated by an extension of the analysis presented in Section 7.3 for large-amplitude perturbations. The current density response to a 40 mV amplitude ($b_a \Delta V = 0.78$) sinusoidal potential input is presented in Figure 8.2 for the system presented in Section 7.3 with parameters $C_{dl} = 31$ μF/cm^2, $n_a F k_a = n_c F k_c = 0.14$ mA/cm^2, $b_a = 19.5$ V^{-1}, $b_c = 19.5$ V^{-1}, and $\overline{V} = 0.1$ V. Following equation (7.4), these parameters yield a value of charge-transfer resistance $R_t = 51.28$ Ωcm^2 and a characteristic frequency of 100 Hz. The potential and current signals were scaled by the maximum value of the signal.

The current response associated with a 40 mV potential perturbation at a 1 mHz frequency is given in Figure 8.2(a). The indication that the response is nonlinear is given by the observation that the current density is not symmetric about zero, whereas the current response to a 1 mV potential perturbation at a 1 mHz frequency, given in Figure 7.4(a), is symmetric about zero. A similar indication is evident at 10 Hz (Figure 8.2(b)). At the characteristic frequency of 100 Hz, a shift in the phase lag is apparent, as seen in Figure 8.2(c). The shape of the current signal is clearly distorted as compared to the results presented in Figure 7.4(b) for a 1 mV potential perturbation at a 100 Hz frequency. At frequencies much higher than the characteristic frequency, however, the linear charging current dominates over the nonlinear faradaic current, and the resulting current response shows a linear behavior as seen in Figure 8.2(d) for a 40 mV potential perturbation at a 10 kHz frequency. These results resemble closely the results presented in Figure 7.4(c) for a 1 mV potential perturbation at a 10 kHz frequency.

The resulting Lissajous plots are presented in Figure 8.3. The value of $b_a\Delta V$ was 0.0195 for $\Delta V = 1$ mV, 0.39 for $\Delta V = 20$ mV, and 0.78 for $\Delta V = 40$ mV. A perturbation amplitude of 10 mV yields $b_a\Delta V = 0.195$, which is consistent with the guideline developed in Example 8.2. The influence of a large-amplitude potential perturbation is most evident at low frequencies, e.g., Figures 8.3(a) and (b), where the faradaic current is much larger than the charging current. At the characteristic frequency for this system of 100 Hz, the faradaic and charging currents are of the same magnitude. Careful examination of Figure 8.3(c) shows that the shapes of the loops are distorted from being elliptical, but the effect is not as evident as it is in Figures 8.3(a) and (b). At frequencies higher than the characteristic frequency, as shown in Figure 8.3(d), all curves are superposed. The nonlinear behavior arises from the nonlinear behavior of the faradaic current. The charging current, in contrast, is linear. The system behaves as a linear system at high frequencies where the charging current dominates.

The impedance can be calculated from the potential and current time-domain signals using the Fourier analysis presented in Section 7.3.3. The resulting impedance spectra are presented in Nyquist format in Figure 8.4(a). The use of an excessive potential perturbation amplitude causes an error in the impedance response. The error arises from the nonlinear behavior of the faradaic current, which is in phase with the applied potential. This effect is seen most clearly in the real part of the impedance shown in Figure 8.4(b). Nevertheless, as seen in Figure 8.4(c), an error is also seen in the imaginary part of the impedance due to the corresponding shift in the characteristic frequency to larger values.

The percent error in the low-frequency impedance asymptote associated with use of a large-amplitude potential perturbation is given in Figure 8.5 with $b_a\Delta V$ as a parameter. At a value of $b_a\Delta V = 0.2$, the error in the low-frequency impedance asymptote is 0.5 percent. This value can also be calculated using equation (8.21) of Example 8.2.

Example 8.3 Influence of Ohmic Resistance on Linearity: *As an extension of Example 8.2, establish a guideline for the perturbation amplitude needed to maintain linearity under potentiostatic regulation for a system with a nonnegligible ohmic resistance.*

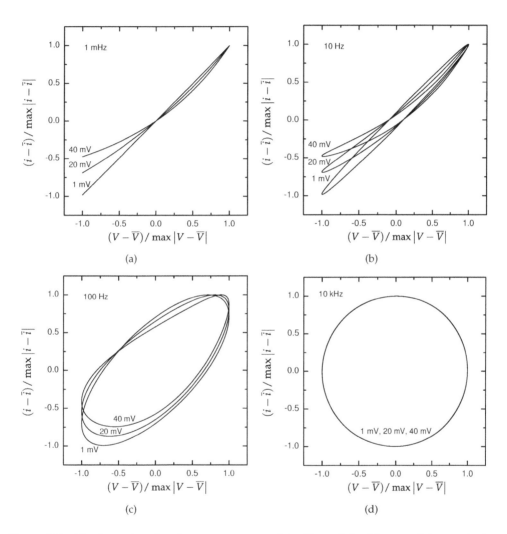

Figure 8.3: Lissajous plots for the system presented in Figure 8.2 with potential perturbation amplitude ΔV as a parameter: a) 1 mHz; b) 10 Hz; c) 100 Hz; and d) 10 kHz. $b_a \Delta V = 0.0195$ for $\Delta V = 1$ mV, $b_a \Delta V = 0.39$ for $\Delta V = 20$ mV, and $b_a \Delta V = 0.78$ for $\Delta V = 40$ mV.

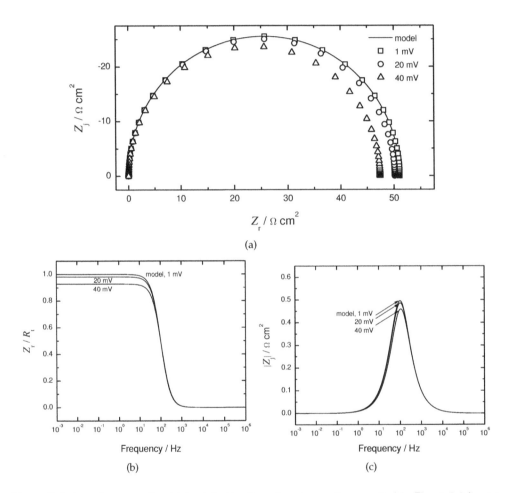

(a)

(b)

(c)

Figure 8.4: Impedance results obtained for the time-domain results presented in Figure 8.3 by use of the Fourier analysis presented in Section 7.3.3 with potential perturbation amplitude ΔV as a parameter: a) Nyquist representation; b) real part of the impedance as a function of frequency; and c) imaginary part of the impedance as a function of frequency.

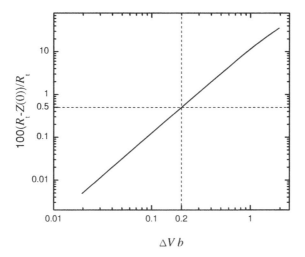

Figure 8.5: The error in the low-frequency impedance asymptote associated with use of a large-amplitude potential perturbation.

Solution: *The presence of an ohmic resistance will reduce the portion of the potential perturbation experienced by the interfacial reactions and double-layer charging. This can be seen in the expression for the impedance of the circuit given in Figure 8.6, i.e.,*

$$Z = R_e + \frac{1}{\dfrac{1}{Z_F} + j\omega C_{dl}} \tag{8.24}$$

The potential drop across the resistor is given by iR_e and the potential drop across the interface η_s is given by $i/(1/Z_F + j\omega C_{dl})$. At large frequencies, the interfacial impedance tends toward zero and the linear ohmic resistance dominates. At low frequency, the capacitive charging has negligible effect and the interfacial impedance is dominated by the faradaic contribution Z_F. This result supports the observation presented in Figure 8.3, which indicates that the nonlinear effects are most significant at low frequency.

The guidelines for the potential perturbation can therefore be determined by the behavior at low frequency. A value $b_a\Delta V \leq 0.2$ can be achieved by setting

$$b_a\Delta U \frac{R_t}{R_e + R_t} \leq 0.2 \tag{8.25}$$

Thus, the influence of ohmic resistance is to increase the allowable magnitude for the potential perturbation.

The Lissajous plot corresponding to the influence of a large (100 mV) potential perturbation is given in Figure 8.7 for the system presented in Figure 7.8. The influence of the nonlinear response is clearly evident as a distortion in the straight lines at low frequency. The response at high frequencies, which are dominated by the ohmic resistance, is linear.

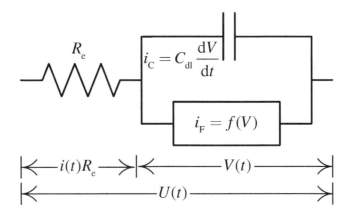

Figure 8.6: Electrical circuit showing the distribution of potential across the ohmic resistance and the interface.

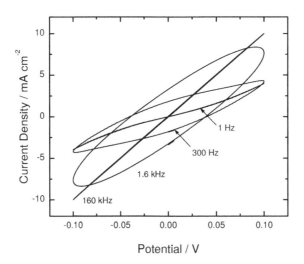

Figure 8.7: Lissajous representation for current and applied potential time-domain signals corresponding to a system presented in Figure 7.8 but with a potential perturbation amplitude $\Delta V = 100$ mV.

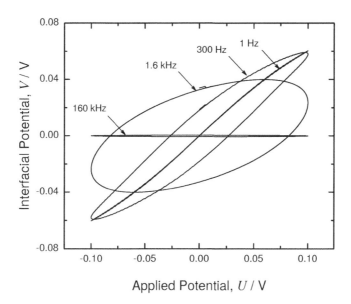

Figure 8.8: Lissajous representation for interfacial potential and applied potential time-domain signals corresponding to a system presented in Figure 8.7.

The reason for the linear response at high frequency can be seen in the Lissajous plot of interfacial potential V as a function of applied potential U, given in Figure 8.8. At low frequencies, the variation in interfacial potential is large and approaches $\Delta U R_t / (R_t + R_e)$; whereas, at high frequencies, the interfacial potential tends toward zero. It is interesting to note that, at low frequencies, the interfacial potential is influenced by the nonlinearity associated with the faradaic reaction.

The frequency dependence of the interfacial potential is shown more clearly in Figure 8.9, where the dimensionless interfacial potential $\Delta V (R_t + R_e) / \Delta U R_t$ is presented as a function of frequency with the magnitude of the applied potential perturbation ΔU as a parameter. The approach to unity at low frequencies for small perturbations shows that the interfacial potential ΔV approaches $\Delta U R_t / (R_t + R_e)$. For larger perturbation amplitudes, the nonlinear response reduces the effective charge-transfer resistance, leading to an asymptotic low-frequency value that is less than unity. At higher frequencies, the 10 mV and 100 mV lines converge, indicating that the high-frequency response is linear for both perturbation amplitudes. The system is controlled by two characteristic frequencies. The characteristic frequency associated with the faradaic reaction, $1/(2\pi R_t C_{dl})$, indicates the frequency at which the charging and faradaic currents are equal, and the characteristic frequency associated with the ohmic resistance, $1/(2\pi R_e C_{dl})$, indicates the frequency at which the capacitive current becomes limited by the ohmic resistance.

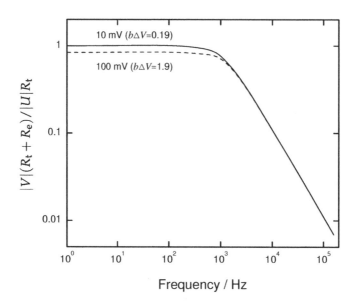

Figure 8.9: The dimensionless interfacial potential $\Delta V(R_t + R_e)/\Delta U R_t$ as a function of frequency with the magnitude of the applied potential perturbation ΔU as a parameter for the system presented in Figure 8.7.

Example 8.4 **Influence of Capacitance on Linearity:** *While the capacitance for a coated surface may be independent of applied potential, as shown in Figure 5.17, the capacitance for a bare electrode may be a function of potential. Explore the influence of a potential dependent capacitance on the linearity of the impedance response.*

Solution: *The potential dependence of capacitance will play the largest role at high frequencies where the current is due to the charging of the interface. To maximize the influence of a potential-dependent capacitance, simulations are performed under assumption that the ohmic resistance may be neglected. The potential-dependent capacitance is assumed to follow*

$$C = C_0(1 + aV) \tag{8.26}$$

where $a = 1.61\ V^{-1}$. This allows a linear change in capacitance of 32 percent for a $\pm100\ mV$ perturbation, which is consistent with the experimental results reported in Figure 5.17 on page 124. All other parameters are the same as developed for Figure 8.2. The resulting Lissajous plots are similar to those presented in Figure 8.2 with the exception that, as seen in Figure 8.10, a distortion is evident at high frequencies. As shown in Figures 8.11(a) and (b), the effect on the impedance response, however, is minimal. As shown in Figure 8.11(a), the real part of the impedance response at low frequencies shows a dependence on potential perturbation amplitude that is consistent with that observed in Figure 8.4(b) for a constant capacitance. The imaginary part of the impedance, shown in Figure 8.11(b), is similar to that shown in Figure 8.4(c) for

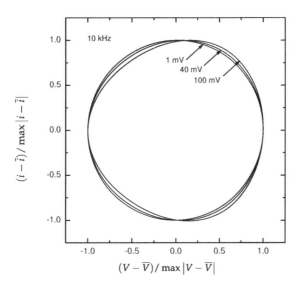

Figure 8.10: Lissajous plot for a capacitance dependent on potential following equation (8.26) at a frequency of 10 kHz.

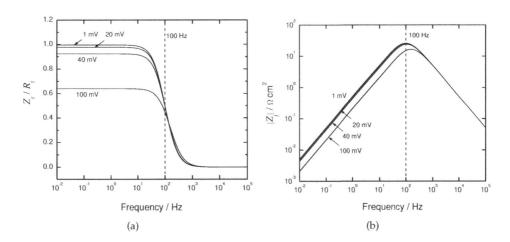

Figure 8.11: Impedance results obtained for the electrochemical system used to generate Figure 8.3 but with an interfacial capacitance that follows equation (8.26) with potential perturbation amplitude ΔV as a parameter: a) normalized real part of the impedance as a function of frequency and b) imaginary part of the impedance as a function of frequency.

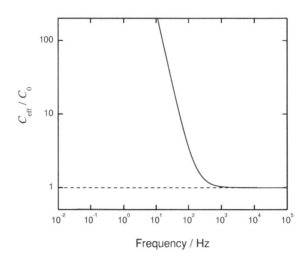

Figure 8.12: The effective capacitance calculated using equation (17.48) and the imaginary impedance given in Figure 8.11(b).

a constant capacitance. As shown in Figure 8.12, the estimation of the capacitance following equation (17.48) (see page 488) yields the correct value for the potential $\overline{V}$ about which the impedance measurement is made.

The optimal perturbation amplitude may be best determined experimentally. Distortions in Lissajous plots at low frequency (see Figure 8.3) may be attributed to a nonlinear response. If the shape is distorted from an ellipse, one should reduce the amplitude. A second approach is to compare the impedance response for several amplitudes as demonstrated in Figure 8.4. If the magnitude of the impedance at low frequency depends on the amplitude of perturbation, the perturbation amplitude is too large.

8.2.3 Modulation Technique

Electrochemical impedance measurements are often performed under potentiostatic regulation. In these measurements the potential is a fixed value with a superimposed (often sinusoidal) perturbation of fixed amplitude. This approach is attractive because, as discussed in Section 8.2.2, linearity in electrochemical systems is controlled by potential.

Galvanostatic regulation is required, however, when the system is to be studied under constant current density. For example, the evaluation of the resistance of skin to iontophoresis is typically done under constant current because the delivery of therapeutic drugs is more directly governed by current density than by potential. Galvanostatic control is also preferred for the use of impedance spectroscopy as a noninvasive tool for periodically observing the condition of a metal coupon held at the corrosion potential for a long period of time. As illustrated by the arrows in Figure 8.13, a drift in the

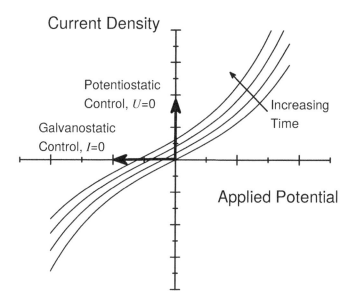

Figure 8.13: Comparison of potentiostatic and galvanostatic modulation for a system in which the open-circuit potential increases with elapsed time.

open-circuit potential during the course of the measurement of a impedance spectrum will, in the case of potentiostatic regulation, result in application of a potential that is anodic (or, depending on the direction of the drift, cathodic) to the true open-circuit potential, thus perturbing the long-term measurement at the zero-current condition. Under galvanostatic control, the desired zero-current condition is maintained throughout the impedance measurement.

The difficulty with fixed-amplitude galvanostatic measurements is that such measurements can result in severe swings in potential, especially at low frequencies where the impedance is large. The amplitude of the potential variation associated with a perturbation of current is given by

$$\Delta V = \Delta I \, |Z(\omega)| \qquad\qquad (8.27)$$

A current perturbation as small as 10 μA can result in potential swings of 1 V for systems with a polarization resistance of $10^5 \Omega$, a value typical of many membranes and some slowly corroding systems. Nonlinear behavior of skin has, for example, been reported for impedance measurements conducted under galvanostatic regulation. An algorithm for a variable amplitude galvanostatic modulation has been described by Wocjik et al.[155,156]

8.2.4 Oscilloscope

It is strongly advised to use an oscilloscope while making impedance measurements. It is useful to monitor the time-domain signals that are processed in the impedance instrumentation. It is particulary useful to monitor the signals in the form of a Lissajous plot as discussed in Section 7.3.1.

8.3 Instrumentation Parameters

The contributions to the error structure of impedance measurements are described in Section 21.1. Impedance measurements entail a compromise between minimizing bias errors, minimizing stochastic errors, and maximizing the information content of the resulting spectrum. The parameter settings described in this section may not apply to all impedance instrumentation.

8.3.1 Improve Signal-to-Noise Ratio

The following steps may be taken to reduce the role of stochastic errors (see Section 21.2) in impedance measurements.

- *Use the optimal current measuring range:* Current is converted to a potential signal through circuitry in the potentiostat. Potentiostats may employ a version of a current follower, for example, as described in Example 6.2. A mismatch between the measured current and the range set in the potentiostat can cause either excessive stochastic noise (if the set range is too high) or bias errors caused by current overloads (if the set range is too low). Some instrumentation vendors require that the user guess the correct current measuring range. Estimating the current range is not difficult for experiments conducted under a dc current. Under open-circuit conditions, the desired current range will vary considerably with frequency. Automatic selection of current measuring range is discouraged if the approach imposes a current on the system that may induce a change in system properties. An algorithm for a noninvasive approach for estimating the current at a given frequency has been described by Wocjik et al.[155,156]

- *Increase the integration time/cycles:* As described in Section 7.3, impedance measurements involve the conversion of time-domain signals to a complex value for each frequency. Stochastic errors in this measurement can be reduced by increasing the time allowed for integration at each frequency. As shown in Figure 21.6(a), the number of cycles required to achieve a set error level at each frequency depends on the frequency of the measurement. Many cycles are required at high frequencies, but only three or four cycles are needed at low frequency. Some instruments allow setting autointegration modes in which the system determines the number of cycles needed to converge to a given criterion. The noise levels can be reduced by selecting the tighter convergence criterion. This choice is termed *long/short integration* on some instrumentation.

- *Increase the amplitude of modulated signal:* As described in Section 8.2.2, the polarization curve for a given system dictates the size of the modulation amplitude that may be used while retaining a linear system response. Many high-impedance systems are characterized by a relatively large linear range of potential. In such cases, the stochastic errors can be reduced significantly by using a large modulation amplitude.

- *Introduce a delay time:* Impedance measurements are taken at the sinusoidal steady state, meaning that the sinusoidal response to the sinusoidal input is unchanging with respect to time. A transient is observed as the system responds to a change from one frequency to another, and this transient is incorporated into the integrated value of the impedance. Pollard and Compte[157] have shown that this transient can introduce as much as a 4 percent error in the impedance response measured by integration over the first cycle. To avoid this undesired error caused by the transient, it is better to introduce a delay of one or two cycles between the change of frequency and impedance measurement.

- *Ignore the first frequency measured:* The impedance measured at the first frequency of measurement is often corrupted by a startup transient. The best option is to ignore the first measured frequency when regressing models to the data.

- *Avoid the line frequency and harmonics:* Modern impedance instruments provide very effective filters for stochastic noise, but these filters are generally inadequate for measurements conducted at the line frequency. The resulting measurements generally appear as outliers in an impedance spectrum, and such outliers have a profound impact on nonlinear regression used to extract parameters from the data. Measurement of impedance should be avoided at line frequency and its first harmonic, i.e., 60 ± 5 Hz and 120 ± 5 Hz in the United States and 50 ± 5 Hz and 100 ± 5 Hz in Europe.

- *Avoid external electric fields:* External devices such as electric motors, pumps, and fluorescent lighting emanate electric fields that can contribute significantly to the apparent noise in a system. This influence, seen most easily in high-impedance systems, can be mitigated by use of a Faraday cage, described in the following section.

8.3.2 Reduce Bias Errors

The steps described in this section may be taken to reduce the role of systematic bias errors (see Section 21.3) in impedance measurements. Bias errors associated with non-stationary effects have greatest impact at low frequencies where each measurement requires a significant amount of time.

Nonstationary Effects

A systematic change in system properties will have a significant influence on measurements made at low frequencies. The time required for measurements at each frequency is discussed in Figures 21.6 and 21.7.

- *Reduce time for measurement:* The total time required to measure an impedance spectrum can be reduced by reducing the time allowed for integration at each frequency, thereby increasing the magnitude of stochastic errors in the measurement. In effect, this approach requires accepting more stochastic noise to achieve a smaller bias error. A second approach is to reduce the number of measured frequencies by reducing the frequency range or the number of frequencies measured per decade. As the greater number of measured frequencies yields better parameter estimates, this approach requires accepting a lesser ability for model discrimination to achieve a smaller bias error.

- *Introduce a delay time:* As discussed above, the transient seen as the system adjusts to a changed modulation frequency yields a bias error in the measured impedance. This undesired error can be avoided by introducing a delay of one or two cycles between the change of frequency and impedance measurement.

- *Avoid the line frequency and harmonics:* As discussed above, measurements made at the line frequency or its first harmonic typically have a significant error that will appear as an outlier when compared to the rest of the spectrum. Measurement within ± 5 Hz of the line frequency and its first harmonic should be avoided.

- *Select an appropriate modulation technique:* Proper selection of modulation technique, discussed in Section 8.2.3, can have a significant impact on presence of bias errors. Use of potentiostatic modulation for a system in which the potential changes with time can increase measurement time on autointegration. The user should consider what should be held constant (e.g., current or potential).

Instrument Bias

Instrument bias errors are often seen at high frequencies, especially for systems exhibiting a small impedance.

- *Use a faster potentiostat:* The influence of high-frequency bias errors can be mitigated by proper selection of potentiostat. The capability of potentiostats to perform measurements at high frequency differs from brand to brand.

- *Use short shielded leads:* High-frequency bias errors can be seen when the cell impedance is of the same order as the internal impedance of the instrumentation. Under these circumstances, it is essential to minimize the impact of ancillary pieces such as wires. Use of short shielded cables is highly recommended.

- *Use a Faraday cage:* A Faraday cage consists of a metallic conductor that surrounds the cell under study and is intended to shield the cell from the influence of external electric fields. The conductor may be in the form of a fine wire mesh or metal sheets. Typically the cage is grounded. It is important to avoid placing electrical components such as motors inside the Faraday cage because such devices can induce the electric fields that the cage is intended to shield. Faraday cages are essential for high-impedance systems that are characterized by a small electrical current. The wires act as an antenna, collecting stray electric fields, which induce a supplementary current. This current may be a significant portion of the signal if the cell current is small.

- *Check the results:* The presence of instrument bias errors can be difficult to discern. The Kramers–Kronig relations may provide a suitable guide, but as discussed in Chapter 22, some instrument-imposed bias errors are Kramers–Kronig transformable. If possible, high-frequency asymptotic values should be compared to independently obtained parameters. A third and highly recommended approach is to measure the impedance response of an electrical circuit exhibiting the same impedance magnitude and characteristic frequencies as seen in the measured impedance response. Systematic instrument errors should be evident in the measured response.

8.3.3   Improve Information Content

The information content of the impedance spectrum can be enhanced by increasing the frequency range, increasing the number of measured frequencies, reducing the magnitude of the bias and stochastic errors, and optimizing the measured frequencies.

- *Broaden the frequency range:* As described in Section 19.5.3, an insufficient frequency range will reduce the ability to identify system characteristics by regression. Typically, an increase in frequency range is constrained at high frequencies by instrument limitations and at low frequencies by nonstationary behavior.

- *Include more frequencies per decade:* The quality of a regression is generally enhanced by increasing the number of measured frequencies, thereby increasing the degree of freedom for the regression. An increased number of measured frequencies requires an increase in the time required for the impedance measurement, thus increasing the potential for nonstationary behavior.

- *Reduce bias and stochastic errors:* The efforts described in Sections 8.3.1 and 8.3.2 to reduce bias and stochastic errors will also improve the information content of the data.

- *Optimize measured frequencies:* The information content of a regression can be enhanced by ensuring the measurements are made at frequencies at which the measurements are sensitive to model parameters. For example, model discrimination will be poor if almost all the impedance data are collected at high frequencies where the impedance approaches an asymptotic value and few measurements are made at lower frequencies that are sensitive to kinetic and transport parameters. There is general agreement that a logarithmic spacing of frequencies maximizes the ability to discriminate between models and to extract model parameters.

Problems

8.1   Following the discussion presented in Example 8.3, estimate the effect an ohmic resistance has on the maximum potential perturbation amplitude that can be applied to an electrochemical system while satisfying the guidelines presented in Example 8.2.

8.2   Estimate the maximum amplitude one should use for a potential perturbation for a system under Tafel kinetics with:

   (a)   A Tafel slope of 60 mV/decade and negligible ohmic resistance. Keep in mind the relationsip between Tafel slope β and Tafel constant b given in equation (5.19).

   (b)   A Tafel slope of 120 mV/decade and negligible ohmic resistance.

   (c)   A Tafel slope of 60 mV/decade, an exchange current density equal to 1 mA/cm^2, an applied potential of 100 mV, and an ohmic resistance of 10 Ωcm^2.

   (d)   A Tafel slope of 60 mV/decade, an exchange current density equal to 1 mA/cm^2, an applied potential of 200 mV, and an ohmic resistance of 10 Ωcm^2.

8.3   Use the methods described in Section 3.2 to estimate the error in the real part of the impedance associated with a 0.22 percent error in the fundamental current response.

8.4   Should the same potential perturbation amplitude be applied for all parts of the polarization curve? Give examples to demonstrate your answer.

8.5   Will a spatial distribution of capacitance lead to a nonlinear current response to a sinusoidal potential input?

8.6   Reproduce the results presented in Figures 8.3 and 8.4. This problem requires use of a spreadsheet program such as Microsoft Excel® or a computational programming environment such as MATLAB®.

8.7   Researchers have reported that, for impedance measurements on human skin under fixed-amplitude galvanostatic modulation, the measurement caused significant changes in skin properties. The magnitude of the skin impedance varied from about 10 Ωcm^2 at high frequency to 100 $k\Omega cm^2$ at low frequency. The perturbation amplitude was 0.1 mA on an exposed skin sample of 1 cm^2 area. Explain the reasons for their observation and suggest an improved experimental protocol.

Part III

Process Models

Chapter 9

Equivalent Circuit Analogs

As described in the subsequent chapters in Part III, models for the impedance response can be developed from proposed hypotheses involving reaction sequences (e.g., Chapter 10 on page 207), mass transfer (e.g., Chapters 11 and 16, pages 243 and 443, respectively), and physical phenomena (e.g., Chapters 13 and 15, pages 319 and 421, respectively). These models can often be expressed in the mathematical formalism of electrical circuits. Electrical circuits can also be used to construct a framework for accounting for the phenomena that influence the impedance response of electrochemical systems. A method for using electrical circuits is presented in this chapter.

9.1 General Approach

The first step in developing an equivalent electrical circuit for an electrochemical system is to analyze the nature of the overall current and potential. For example, in the simple case of the uniformly accessible electrode shown in Figure 9.1(a), the overall potential is the sum of the interfacial potential V plus the ohmic drop $R_e i$. Accordingly, the overall impedance is the sum of the interfacial impedance Z_0 plus the electrolyte resistance R_e,

$$Z = R_e + Z_0 \tag{9.1}$$

At the interface itself, shown in Figure 9.1(b), the overall current is the sum of the faradaic current i_F plus the charging current i_C through the double-layer capacitor C_{dl}. Thus, the interfacial impedance results from the double-layer capacity in parallel with the faradaic impedance Z_F. The corresponding impedance is given by

$$Z = R_e + \frac{Z_F}{1 + j\omega C_{dl} Z_F} \tag{9.2}$$

where the term Z_F, like the boxes in Figure 9.1, are used to designate impedances that cannot be generally described in terms of passive elements such as resistors and capacitors. In the case of a single reaction on a uniform electrode, as is shown in Section 10.2.1

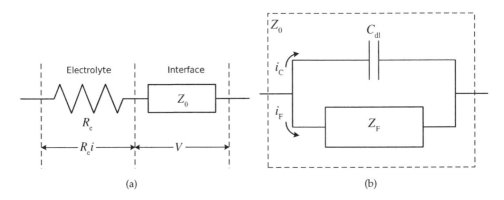

Figure 9.1: Electrical circuit corresponding to a single reaction on a uniformly accessible electrode: a) series combination of the electrolyte resistance and the interfacial impedance and b) parallel combination of the faradaic impedance and the double-layer capacitance, which comprise the interfacial impedance.

(page 209), the faradaic impedance Z_F shown in Figure 9.1(b) can be represented as a charge-transfer resistance. The representation is, however, more complicated for the interfacial response of coupled reactions, reactions involving mass transfer, reactions involving adsorbed species, and reactions on nonuniform surfaces.

Nevertheless, Figure 9.1 illustrates the procedure to be used in more complicated situations. When the current flowing through circuit elements is the same, but the potential drop is different, the respective impedances must be added in series. This case is illustrated in Figure 9.1(a). When the current flowing through circuit elements is different, but the potential drop is the same, the respective impedances must be added in parallel. This case is illustrated in Figure 9.1(b). See Section 4.1.2 for a review of methods for parallel and series addition of circuit components.

The physical understanding of the current paths and potential drops in the system serves to guide the structure of the corresponding electrical circuit. The mathematical expression for the interfacial impedance can be obtained following the development presented in the subsequent chapters. Several examples are given in the following sections to illustrate the procedure.

Remember! 9.1 *In an equivalent electrical circuit, boxes should be used to designate impedances that cannot be generally described in terms of passive elements.*

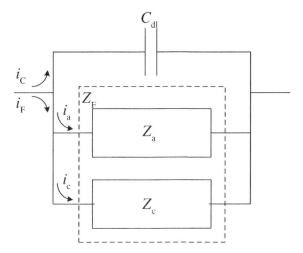

Figure 9.2: Equivalent electrical circuit of the interfacial impedance at the corrosion potential, where i_C represents the charging current, i_a represents the anodic current, and i_c represents the cathodic current.

9.2 Current Addition

Figure 9.1(b) provides an example of the case where the circuit development is based on the addition of current contributions. The examples provided in this section illustrate the application of the principle to more complex situations.

9.2.1 Impedance at the Corrosion Potential

The electrical circuit corresponding to a freely corroding electrode can be developed in two steps. As shown in Figure 9.1(a), the electrolyte resistance will be in series with an interfacial impedance. The interfacial impedance can be developed considering the diagram shown in Figure 9.2. As is the case described in Figure 9.1(b), the total current consists of the sum of charging and faradaic currents. At the corrosion potential, the sum of the anodic and cathodic faradaic currents is equal to zero, i.e., $\bar{i}_a + \bar{i}_c = 0$. The faradaic impedance must, therefore, be a parallel combination of Z_a and Z_c. The contribution of the double-layer capacitance is added in parallel. The corresponding

---

Remember! 9.2 *Electrical circuit components must be added in parallel when the total current is the sum of individual current contributions. Electrical circuit components must be added in series when the total potential drop is the sum of individual contributions.*

Table 9.1: Results of experiments performed to determine if the simplifications to equation (9.3) expressed as equations (9.4) and (9.5) are justified. Experiments are performed at V_{corr}, $V_{corr} + \Delta V$, and $V_{corr} - \Delta V$. Typical values of ΔV are between 10 and 50 mV. Results are presented in terms of the expected change in the impedance as compared to the value measured at the open-circuit potential.

Assumption	$V_{corr} + \Delta V$	$V_{corr} - \Delta V$
$Z_c \gg Z_a$	decrease ⇓	increase ⇑
$Z_c \ll Z_a$	increase ⇑	decrease ⇓
$Z_c \approx Z_a$	decrease ⇓	decrease ⇓
$Z_a \neq f(V)$ & $Z_a \gg Z_c$	increase ⇑	decrease ⇓
$Z_a \neq f(V)$ & $Z_a \ll Z_c$	unchanged ⇔	unchanged ⇔

expression for the impedance may be given as

$$Z = R_e + \cfrac{1}{\cfrac{1}{Z_a} + \cfrac{1}{Z_c} + j\omega C_{dl}} \tag{9.3}$$

Expressions for the impedances Z_a and Z_c must be developed separately according to proposed reaction mechanisms (see Chapter 10 on page 207).

 Example 9.1 **Method to Identify Corrosion Model:** *If $Z_a \gg Z_c$, equation (9.3) can be expressed as*

$$Z = R_e + \frac{Z_c}{1 + j\omega C_{dl} Z_c} \tag{9.4}$$

Conversely, if $Z_c \gg Z_a$, equation (9.3) can be expressed as

$$Z = R_e + \frac{Z_a}{1 + j\omega C_{dl} Z_a} \tag{9.5}$$

How can we determine when this simplification is possible?

Solution: *The recommended approach is to measure the impedance at the open-circuit potential and then at a potential slightly anodic to the open-circuit potential and slightly cathodic to the open-circuit potential, taking care to ensure that a steady-state condition is achieved before making impedance measurements. If both reactions are kinetically controlled, an increase in potential will increase the cathodic impedance Z_c and decrease the anodic impedance Z_a, and a decrease in potential will decrease Z_c and increase the anodic impedance Z_a. The results are presented in Table 9.1 in terms of the expected change in the impedance as compared to the value measured at the open-circuit potential.*

9.2.2 Partially Blocked Electrode

Partial coverage of an electrode by a surface film, e.g., an oxide layer, may block passage of faradaic current. In some cases, the fractional coverage of the surface is influ-

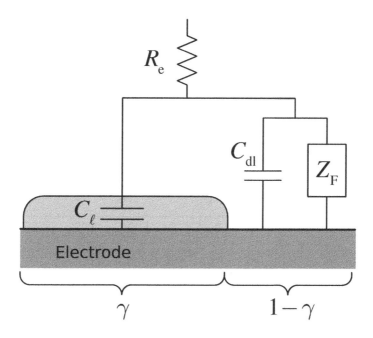

Figure 9.3: Equivalent electrical circuit for a partially blocked surface.

enced by the modulation of potential and must be treated by the methods presented in sections 10.4.1 and 10.4.2.

The case considered here is one where the blocked surface is independent of the potential. As shown in Figure 9.3, the blocked site is assumed to be a perfect insulator with fractional coverage γ, and the fractional area of active surface is $(1 - \gamma)$.

For the system shown in Figure 9.3, the impedance may be expressed as

$$Z = R_{\mathrm{e}} + \frac{Z_{\mathrm{F}}}{1 - \gamma + j\omega\left(\gamma C_{\ell} + (1 - \gamma)C_{\mathrm{dl}}\right)Z_{\mathrm{F}}} \qquad (9.6)$$

As the fractional coverage γ approaches zero, the impedance approaches that of the uncoated surface. As γ approaches unity, the impedance becomes larger, and, for $\gamma = 1$, the impedance is that of a blocking electrode with capacity C_{ℓ}, i.e.,

$$\lim_{\gamma \to 1} Z = R_{\mathrm{e}} + \frac{1}{j\omega C_{\ell}} = R_{\mathrm{e}} - j\frac{1}{\omega C_{\ell}} \qquad (9.7)$$

If the system is not mass-transport limited, the effect of the partial coverage is simply to reduce the active area.

Example 9.2 **Coated Electrode:** *How many independent variables may be obtained from equation (9.6)?*

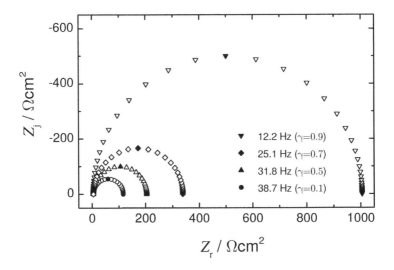

Figure 9.4: Impedance response of the system shown in Figure 9.3 with fractional coverage of a blocking coating as a parameter. The model parameters are $R_e = 5 \ \Omega\text{cm}^2$, $C_\ell = 10 \ \mu\text{F/cm}^2$, $C_{\text{dl}} = 40 \ \mu\text{F/cm}^2$, and $Z_F = R_t = 100 \ \Omega\text{cm}^2$.

Solution: *Equation (9.6) may be written as*

$$Z = R_e + \frac{Z_F/(1-\gamma)}{1 + j\omega \left(\gamma C_\ell + (1-\gamma)C_{\text{dl}}\right) Z_F/(1-\gamma)} \tag{9.8}$$

If the following new parameters are considered:

$$Z_{F,\gamma} = \frac{Z_F}{1-\gamma} \tag{9.9}$$

$$C_{\ell,\gamma} = \gamma C_\ell \tag{9.10}$$

and

$$C_{\text{dl},\gamma} = (1-\gamma)C_{\text{dl}} \tag{9.11}$$

equation (9.8) becomes

$$Z = R_e + \frac{Z_{F,\gamma}}{1 + j\omega \left(C_{\ell,\gamma} + C_{\text{dl},\gamma}\right) Z_{F,\gamma}} \tag{9.12}$$

Thus, from equation (9.6), only three independent variables may be obtained by regression; i.e., R_e, $Z_{F,\gamma}$, and $(C_{\ell,\gamma} + C_{\text{dl},\gamma})$.

To illustrate the role of surface coverage, consider an electrode covered by a 1 nm thick film with a dielectric constant $\epsilon = 11$. Following equation (5.138) on page 126, the corresponding film capacitance has a value near 10 μF/cm^2. If, following the development in Section 10.2.1 on page 209, the faradaic impedance $Z_F = R_t$, then the impedance can be expressed as shown in Figure 9.4 where the double layer capacitance

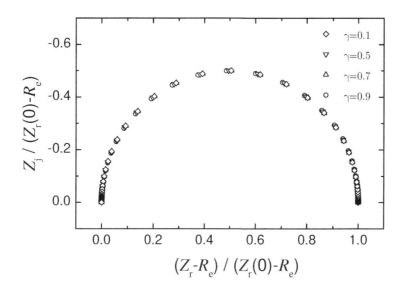

Figure 9.5: The normalized impedance response of the system shown in Figure 9.4 in Nyquist format.

was assumed to have a value of 40 $\mu F/cm^2$ and the charge transfer resitance was assumed to have a value $R_t = 100\ \Omega cm^2$. The measured faradaic impedance is inversely proportional to the active area; thus, if the capacity of the covered surface can be neglected with respect to the double-layer capacity of the active surface, all impedance values are inversely proportional to $(1 - \gamma)$. The normalized impedance, presented in Figure 9.5, is therefore independent of surface coverage γ when presented in Nyquist format, even when, as shown in Figures 9.6(a) and 9.6(b), the characteristic frequency depends on γ. If the capacity of the covered surface cannot be neglected, the overall capacity will be the area-weighted sum of capacitances for coated and uncoated areas, i.e.,

$$C = \gamma C_\ell + (1 - \gamma)C_{dl} \qquad (9.13)$$

For the example given in Figure 9.4, the characteristic frequency depends on fractional coverage according to

$$f_c = \frac{1 - \gamma}{2\pi R_t \left(\gamma C_\ell + (1 - \gamma)C_{dl}\right)} \qquad (9.14)$$

as shown in Figure 9.7.

Superposition of Nyquist plots was used by Baril et al.[158] to demonstrate that

Remember! 9.3 *Superposition of Nyquist plots provides a powerful tool to demonstrate the influence of active surface area.*

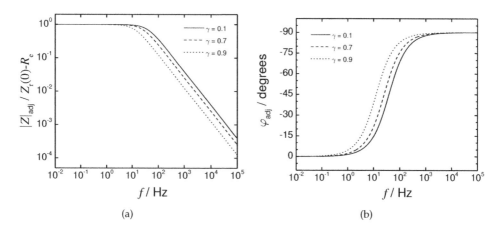

Figure 9.6: The impedance response in Bode format of the system shown in Figure 9.4: a) normalized ohmic-resistance-corrected magnitude (equation (18.12) on page 505) and b) ohmic-resistance-corrected phase angle (equation (18.11) on page 504).

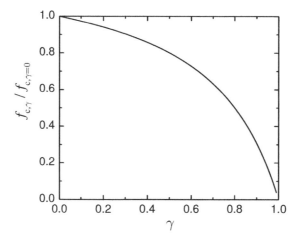

Figure 9.7: The characteristic frequency given by equation (9.14) as a function of γ for the system shown in Figure 9.4.

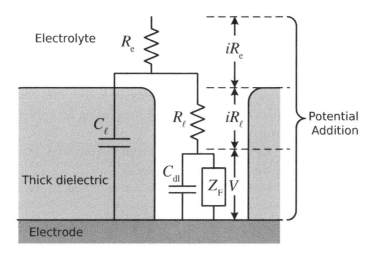

Figure 9.8: Equivalent electrical circuit of the impedance for an electrode coated by a thick dielectric layer with pores exposing the electrode to the electrolyte.

changes in impedance response could be attributed to an increased coverage of an electrode by an oxide layer. If the system is limited by mass-transport,[159,160] the effect of the partial coverage is more complex, and no analytic solution exists for the general case. Electrohydrodynamic (EHD) impedance, discussed in Chapter 16, provides an appropriate technique to analyze the combined effects of partial surface coverage and mass transfer.

9.3 Potential Addition

Figure 9.1(a) provides an example of the case where the circuit development is based on the addition of potential contributions. The examples provided in this section illustrate the application of the principle to more complex situations.

9.3.1 Electrode Coated with an Inert Porous Layer

Surface films commonly form in electrochemical studies, and these films can influence the impedance response. The electrode coated with an inert porous layer may be considered to be an extension of the case described in Section 9.2.2 in which the film is thicker and the fractional surface coverage approaches unity.

The layer shown in Figure 9.8 is considered to be porous, with electrochemical reactions occurring only on the exposed electrode surface at the end of the pore. The faradaic impedance is the same as discussed in Section 9.2.2. However, in the kinetic model, it could be necessary to take into account the fact that in the pore, the concentration of the different species involved in the reaction could differ from the bulk concentration.

The equivalent circuit corresponding to the scheme of the coated electrode is presented in Figure 9.8.[161] At the interface localized at the end of the pore, the corresponding impedance is the parallel combination of Z_F and C_{dl}. Within the pore length, the electrolyte resistance is R_ℓ, and the insulating part of the coating can be considered to be a capacitor C_ℓ, which is in parallel with the impedance in the pore. The capacitance can be related to the permittivity and thickness of the coating according to equation (5.138) with typical values presented in Table 5.6.

The electrolyte resistance R_e is added in series with the previous impedance. The corresponding expression for the impedance is given as

$$Z = R_e + \frac{\left(R_\ell + \dfrac{Z_F}{1 + j\omega C_{dl} Z_F}\right)}{1 - \gamma + j\omega\gamma C_\ell \left(R_\ell + \dfrac{Z_F}{1 + j\omega C_{dl} Z_F}\right)} \tag{9.15}$$

which may be written in terms of variables defined in equations (9.9)-(9.11) as

$$Z = R_e + \frac{\left(R_{\ell,\gamma} + \dfrac{Z_{F,\gamma}}{1 + j\omega C_{dl,\gamma} Z_{F,\gamma}}\right)}{1 + j\omega C_{\ell,\gamma} \left(R_{\ell,\gamma} + \dfrac{Z_{F,\gamma}}{1 + j\omega C_{dl,\gamma} Z_{F,\gamma}}\right)} \tag{9.16}$$

where

$$R_{\ell,\gamma} = \frac{R_\ell}{1 - \gamma} \tag{9.17}$$

The parameters R_e, $R_{\ell,\gamma}$, $Z_{F,\gamma}$, ($C_{\ell,\gamma}$ and $C_{dl,\gamma}$) may be obtained by regression. The only way to obtain a value for the fractional coverage γ is to assume a value for C_{dl}. From C_ℓ, it is possible to deduce the thickness of the dielectric film δ, as shown in Section 5.7.2 (page 126). From R_ℓ, it is possible to evaluate the electrolyte resistivity in the pore from

$$\rho_\ell = \frac{R_\ell}{\delta} \tag{9.18}$$

The assumption of the capacitance of the bare metal is used because the value can be expected to be between 10 and 50 $\mu F/cm^2$ (see Section 5.7 on page 122).

As γ approaches unity, equation (9.15) approaches equation (9.7). If the electrochemical reaction is mass-transport limited, the previous equivalent circuit is still valid, but the faradaic impedance includes a diffusion impedance Z_D as described in Chapter 11.

9.3.2 Electrode Coated with Two Inert Porous Layers

In corrosion systems, a salt film may cover an electrode that is itself covered by a porous oxide layer. If two different layers are superimposed, the geometrical analysis shows that the equivalent circuit corresponds to that described in Section 9.3.1 with an additional series $R_{\ell 2} C_{\ell 2}$ circuit to take into account the effect of the second porous layer.

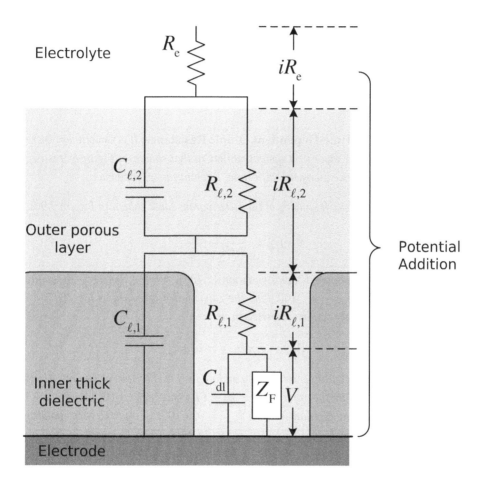

Figure 9.9: Equivalent electrical circuit of the impedance for a electrode coated by a porous layer covering a thick dielectric layer with pores exposing the electrode to the electrolyte.

The circuit shown in Figure 9.9 is approximate because it assumes that the boundary between the inner and outer layers can be considered to be an equipotential plane. This plane will, however, be influenced by the presence of pores. The circuit shown in Figure 9.9 will provide a good representation for systems with an outer layer that is much thicker than the inner layer and with an inner layer that has relatively few pores. The impedance response is given by

$$Z = R_e + \frac{R_{\ell,2}}{1 + j\omega R_{\ell,2}C_{\ell,2}} + \frac{\left(R_{\ell,1,\gamma} + \dfrac{Z_{F,\gamma}}{1 + j\omega C_{dl,\gamma}Z_{F,\gamma}}\right)}{1 + j\omega C_{\ell,1,\gamma}\left(R_{\ell,1,\gamma} + \dfrac{Z_{F,\gamma}}{1 + j\omega C_{dl,\gamma}Z_{F,\gamma}}\right)} \qquad (9.19)$$

Example 9.3 **Time-Dependent Ohmic Resistance:** *It is sometimes observed that the impedance of the outer layer in a system similar to that shown in Figure 9.9 is negligible, but that the ohmic resistance increased with time. Explain this phenomenon.*

Solution: *The characteristic frequency of the outer porous layer shown in Figure 9.9 is*

$$f_{\ell,2} = \frac{1}{2\pi R_{\ell,2}C_{\ell,2}} \qquad (9.20)$$

The capacitance can be estimated using equation (5.138). With a dielectric constant $\epsilon_{\ell,2} = 10$ and coating thickness $\delta_{\ell,2} = 100$ μm, the capacitance takes the value of $C_{\ell,2} = 9 \times 10^{-11}$ F/cm^2. The resistance can be estimated from

$$R_{\ell,2} = \delta_{\ell,2}/\kappa_{\ell,2} = \rho_{\ell,2}\delta_{\ell,2} \qquad (9.21)$$

where $\kappa_{\ell,2}$ is the effective coating conductivity and $\rho_{\ell,2}$ is the effective coating resistivity. For a salt film, the value will be on the order of $\kappa_\ell = 10^{-4}$ $\Omega^{-1}cm^{-1}$ ($\rho_\ell = 10^4$ Ωcm).

    The corresponding time constant will be $\tau_{\ell,2} = 9 \times 10^{-9}$ s, and the corresponding characteristic frequency will be $f_{\ell,2} = 1.8 \times 10^7$ Hz or 18 MHz. This frequency is well above the capabilities of most electrochemical impedance instrumentation. Thus, the capacitive loop corresponding to the outer layer will not be observed experimentally. The resistance of the layer influences measurements at all frequencies; thus, the presence of a growing layer thickness will be manifested as an apparent increase of the ohmic resistance. For the situation described in this example, the circuit shown in Figure 9.9 should be amended as shown in Figure 9.10.[162] The

Remember! 9.4 *While all resistance contributions to an electrical circuit can be observed at low frequencies, the inability to measure at sufficiently high frequencies may make it impossible to obtain all capacitance values.*

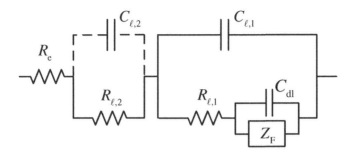

Figure 9.10: Equivalent electrical circuit for the system presented in Figure 9.9 with the capacitance $C_{\ell,2}$ indicated by dashed lines to denote its experimental inaccessibility.

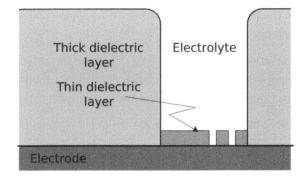

Figure 9.11: Schematic representation for the electrode coated by a thick dielectric layer with pores exposing a thin dielectric layer and bare electrode as discussed in Problem 9.1.

ability to measure the capacitive loop associated with the outer porous layer does not depend on layer thickness, but it is sensitive to the effective conductivity of the layer. The effective conductivity of paints and polymer films is much lower, resulting in a smaller characteristic frequency that is closer to the experimentally accessible range of frequencies.

Problems

9.1   Consider the schematic representation of a coated electrode presented in Figure 9.11. Develop the corresponding equivalent electrical circuit.

9.2   Is the circuit developed for Problem 9.1 mathematically equivalent to that developed for Section 9.3.2?

9.3   Estimate the characteristic frequency for the impedance response of the outer porous layer of the system shown in Figure 9.9 for the following materials:

(a)   Salt film with effective conductivity $\kappa = 10^{-4}\ \Omega^{-1}\mathrm{cm}^{-1}$

(b)   Polymer coating with effective conductivity $\kappa = 10^{-8}\ \Omega^{-1}\mathrm{cm}^{-1}$

(c)   Epoxy coating with effective conductivity $\kappa = 10^{-9}\ \Omega^{-1}\mathrm{cm}^{-1}$

9.4   Show that the characteristic frequency for the impedance of the outer porous layer of the system shown in Figure 9.9 is independent of film thickness. Show the influence of layer thickness on the resistance of the layer.

9.5   Devise a criterion for deciding whether a dielectric film partially covering an electrode should be considered to be thin, as described in Section 9.2.2, or thick, as described in Section 9.3.1.

9.6   Consider that a thin 10 nm thick dielectric film of $Fe_2O_3 - \alpha$ (hematite) covers a metal electrode and that the fractional coverage of the electrode is $\gamma = 0.7$. Estimate the capacitance of the system.

9.7   How is the circuit of Figure 9.9 affected by a nonuniform thickness of the outer salt layer? Can the nonuniformity of the salt-film layer thickness be identified in the impedance response?

9.8   Under the assumption that Figure 9.3 properly describes a partially blocked electrode, will the associated equivalent circuit account for the influence of nonuniform current and potential distributions associated with this geometry?

9.9   Verify the results presented in Table 9.1.

Chapter 10

Kinetic Models

The electrical circuits developed in Chapter 9 made use of boxes and undefined transfer functions Z_F to account for the impedance associated with interfacial reactions. In some cases, the interfacial impedance may be described in terms of such circuit elements as resistors and capacitors, but the nature of the impedance response depends on the proposed reaction mechanism. The objective of this chapter is to explore the relationship between proposed reaction mechanisms and the interfacial impedance response.

10.1 General Mathematical Framework

As discussed in Section 5.2 on page 92, a generalized heterogeneous reaction mechanism can be expressed in symbolic form as

$$\sum_i s_i M_i^{z_i} \leftrightarrows ne^-$$ (10.1)

where the stoichiometric coefficient s_i has a positive value for a reactant, has a negative value for a product, and is equal to zero for a species that does not participate in the reaction.

The current density corresponding to this faradaic reaction can be expressed as a function of an interfacial potential V, as presented in equation (5.75), the surface concentration of bulk species $c_i(0)$, and the surface coverage of adsorbed species γ_k as

$$i_F = f\left(V, c_i(0), \gamma_k\right)$$ (10.2)

where the interfacial potential can be considered to be the difference between the potential of the electrode Φ_m and the potential in the electrolyte adjacent to the electrode Φ_0, measured with respect to the same reference electrode as used to measure the cell potential U (see equation (10.33)).

All oscillating quantities, such as concentration, current, or potential, can be written in the form

$$X = \overline{X} + \text{Re}\left\{\widetilde{X}e^{j\omega t}\right\}$$ (10.3)

where the overbar represents the steady value, j is the imaginary number ($j = \sqrt{-1}$), ω is the angular frequency, and the tilde denotes a complex variable that is a function of frequency. In particular, the current density can be expressed in terms of a steady, time-independent value and an oscillating value (see equation (1.136)) as

$$i_F = \bar{i}_F + \mathrm{Re}\left\{\tilde{i}_F\, e^{j\omega t}\right\} \tag{10.4}$$

A Taylor series expansion about the steady value $\bar{i}_F$ can be written as

$$\tilde{i}_F = \left(\frac{\partial f}{\partial V}\right)_{c_i(0),\gamma_k} \tilde{V} + \sum_i \left(\frac{\partial f}{\partial c_i(0)}\right)_{V,c_{\ell,\ell\neq i}(0),\gamma_k} \tilde{c}_{i,0} + \sum_k \left(\frac{\partial f}{\partial \gamma_k}\right)_{V,c_i(0),\gamma_{\ell,\ell\neq k}} \tilde{\gamma}_k \tag{10.5}$$

where $\tilde{V}$, $\tilde{c}_i(0)$, and $\tilde{\gamma}_k$ are assumed to have a small magnitude such that the higher-order terms can be neglected. Equation (10.5) represents a general result that can be applied to any electrochemical reaction.

In the absence of migration, the flux of species is related to the current density by

$$D_i \left.\frac{\partial c_i}{\partial y}\right|_{y=0} = \frac{s_i i_F}{nF} \tag{10.6}$$

and the oscillating faradaic current density is obtained from equation (10.6) to be

$$\tilde{i}_F = \frac{nFD_i}{s_i} \left.\frac{d\tilde{c}_i}{dy}\right|_{y=0} \tag{10.7}$$

Equation (10.5) can, therefore, be written as

$$1 = \left(\frac{\partial f}{\partial V}\right)_{c_i(0),\gamma_k} Z_F + \sum_i \left(\frac{\partial f}{\partial c_i(0)}\right)_{V,c_{\ell,\ell\neq i}(0),\gamma_k} \left(\frac{s_i \tilde{c}_i(0)}{nFD_i \left.\dfrac{d\tilde{c}_i}{dy}\right|_{y=0}}\right)$$
$$+ \sum_k \left(\frac{\partial f}{\partial \gamma_k}\right)_{V,c_i(0),\gamma_{\ell,\ell\neq k}} \left(\frac{\tilde{\gamma}_k}{\tilde{V}} Z_F\right) \tag{10.8}$$

where Z_F is the faradaic impedance defined by

$$Z_F = \frac{\tilde{V}}{\tilde{i}_F} \tag{10.9}$$

Equation (10.9) may be expressed as

$$Z_F = \frac{R_t + Z_D}{1 + R_t \sum_k \left(\dfrac{\partial f}{\partial \gamma_k}\right)_{V,c_i(0),\gamma_{\ell,\ell\neq k}} \dfrac{\tilde{\gamma}_k}{\tilde{V}}} \tag{10.10}$$

where the charge transfer resistance is defined to be

$$R_t = \left[\left(\frac{\partial f}{\partial V} \right)_{c_i(0),\gamma_k} \right]^{-1}$$

(10.11)

and the diffusion impedance is defined to be

$$Z_D = R_t \sum_i \frac{s_i}{nFD_i} \left(\frac{\partial f}{\partial c_i(0)} \right)_{V,c_{\ell,\ell \neq i}(0),\gamma_k} \left(\frac{-\widetilde{c}_i(0)}{\left. \frac{d\widetilde{c}_i}{dy} \right|_{y=0}} \right)$$

(10.12)

The transfer function $\widetilde{\gamma}_k / \widetilde{V}$ must be derived for each adsorbed species k, as is shown in this chapter.

The general expression (10.5) guides development of impedance models from proposed reaction sequences. The reaction mechanisms considered here include reactions dependent only on potential, reactions dependent on both potential and mass transfer, coupled reactions dependent on both potential and surface coverage, and coupled reactions dependent on potential, surface coverage, and mass transfer. The proposed reaction sequence has a major influence on the frequency dependence of the interfacial faradaic impedance described in Chapter 9.

10.2 Electrochemical Reactions

The electrochemical reactions considered in this section are single-step reactions that may be driven by potential or by both potential and mass transfer considerations. The influence of surface coverage is described in Section 10.4.

10.2.1 Potential Dependent

Consider the dissolution of metal in an aqueous medium

$$M \rightarrow M^{n+} + ne^-$$

(10.13)

represented schematically in Figure 10.1. The steady faradaic current associated with this reaction can be expressed in terms of Tafel kinetics as (5.21)

$$i_M = K_M^* \exp \left(b_M \left(V - V_{0,M} \right) \right)$$

(10.14)

where $K_M^* = nFk_M$ with units of current density, and b_M is $\alpha_M nF/RT$ where n is the number of electrons transferred per mole of reacting species in reaction (10.13), F is Faraday's constant (96,487 Coulombs/equivalent), k_M is the rate constant for the reaction, α_M is the symmetry factor, R is the universal gas constant (8.3147 J/mol K), T is temperature in absolute units, V is the interfacial potential, and $V_{0,M}$ represents the interfacial equilibrium potential as discussed in Section 5.3.

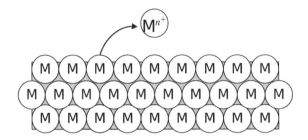

Figure 10.1: Schematic representation of a metal dissolution reaction.

For systems involving single reactions, it is convenient to incorporate the interfacial equilibrium potential $V_{0,M}$ into the effective rate constant as

$$i_M = K_M^* \exp\left(-b_M V_{0,M}\right) \exp\left(b_M V\right) \tag{10.15}$$

or

$$i_M = K_M \exp\left(b_M V\right) \tag{10.16}$$

where $K_M = K_M^* \exp\left(-b_M V_{0,M}\right)$. Under the assumption that the reaction is not influenced by the presence of adsorbed intermediates or layers of corrosion products, the concentration of the reactant can be considered to be a constant embedded within the effective rate constant k_M for the reaction. The current density given in equation (10.14) is a function only of potential. The steady current density increases exponentially as the potential V becomes more positive.

The faradaic current response to a potential perturbation can be expressed as:

$$\widetilde{i}_M = K_M \exp\left(b_M \overline{V}\right) b_M \widetilde{V} = \overline{i}_M b_M \widetilde{V} \tag{10.17}$$

where $\widetilde{V}$ represents the perturbation of potential. A charge-transfer resistance for this reaction can be identified such that

$$\widetilde{i}_M = \frac{\widetilde{V}}{R_{t,M}} \tag{10.18}$$

where

$$R_{t,M} = \frac{1}{K_M \exp\left(b_M \overline{V}\right) b_M} = \frac{1}{\overline{i}_M b_M} \tag{10.19}$$

Equation (10.18) corresponds to equation (10.5) with only the first term of the expression.

The faradaic impedance is obtained from

$$Z_{F,M} = \frac{\widetilde{V}}{\widetilde{i}_M} = R_{t,M} \tag{10.20}$$

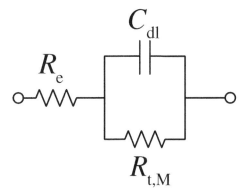

Figure 10.2: Electrical circuit providing the equivalent to the impedance response for a single electrochemical reaction.

From equation (9.1) on page 193, the impedance for a reaction on a planar electrode may be expressed as

$$Z_M(\omega) = R_e + \frac{R_{t,M}}{1 + j\omega R_{t,M} C_{dl}} \tag{10.21}$$

An electrical circuit that yields the impedance response equivalent to equation (10.21) for a single faradaic reaction is presented in Figure 10.2. Equation (10.20) may be used in the other circuits shown in Chapter 9 (page 193) corresponding to other situations. As is shown in Example 10.1, equation (10.21) may also be derived without using the electric circuit framework.

Example 10.1 Impedance Derivation without Electrical Circuits: *Derive the expression given as equation (10.21) for the cell impedance associated with a single potential-dependent reaction without using the electrical circuit framework presented in Chapter 9 (page 193).*

Solution: *The development makes use of some useful relationships presented in Table 10.1. An expression for the total current density can be obtained as the summation of faradaic and charging current densities, i.e.,*

$$i = i_M + C_{dl} \frac{dV}{dt} \tag{10.25}$$

Remember! 10.1 *While some faradaic impedances can be expressed in terms of passive elements, the development of such models from proposed reaction sequences provides insight and physical meaning to the circuit parameters.*

Table 10.1: Some useful relationships for the development of the impedance response associated with faradaic reactions.

$$\tilde{i} = \tilde{i}_F + j\omega C_{dl}\tilde{V} \tag{10.22}$$

$$\tilde{U} = \tilde{i}R_e + \tilde{V} \tag{10.23}$$

$$Z(\omega) = \frac{\tilde{U}}{\tilde{i}} \tag{10.24}$$

where C_{dl} represents the double-layer capacitance. The addition of charging and faradaic currents is illustrated in Figure 9.1(b) on page 194. The potential and current can be expressed as the sum of steady state and oscillating contributions, i.e.,

$$V = \overline{V} + \text{Re}\left\{\tilde{V}e^{j\omega t}\right\} \tag{10.26}$$

Thus, equation (10.25) can be expressed as

$$\overline{i} + \text{Re}\left\{\tilde{i}e^{j\omega t}\right\} = \overline{i}_M + \text{Re}\left\{\tilde{i}_M e^{j\omega t}\right\} + C_{dl}\frac{d}{dt}\text{Re}\left\{\tilde{V}e^{j\omega t}\right\} \tag{10.27}$$

Under steady-state conditions,

$$\overline{i} = \overline{i}_M \tag{10.28}$$

Thus, equation (10.27) can be expressed as

$$\text{Re}\left\{e^{j\omega t}\left(\tilde{i} - \tilde{i}_M - j\omega C_{dl}\tilde{V}\right)\right\} = 0 \tag{10.29}$$

Equation (10.29) can be satisfied only if

$$\tilde{i} = \tilde{i}_M + j\omega C_{dl}\tilde{V} \tag{10.30}$$

corresponding to equation (10.22) in Table 10.1. Insertion of equation (10.18) into equation (10.30) yields

$$\tilde{i} = \tilde{V}\left(\frac{1}{R_{t,M}} + j\omega C_{dl}\right) \tag{10.31}$$

or

$$\frac{\tilde{V}}{\tilde{i}} = \frac{R_{t,M}}{1 + j\omega R_{t,M}C_{dl}} \tag{10.32}$$

which represents the interfacial impedance as described on page 160.

The relationship between the potential evaluated at the electrode surface and the potential measured with respect to a reference electrode some distance away from the electrode is given by

$$U = iR_e + V \tag{10.33}$$

which leads to equation (10.23). Following equation (10.24), the cell impedance corresponding to reaction (10.13) is given by

$$Z_M(\omega) = \frac{\tilde{U}}{\tilde{i}} = R_e + \frac{\tilde{V}}{\tilde{i}} \tag{10.34}$$

Insertion of equation (10.32) yields the cell impedance

$$Z_M(\omega) = R_e + \frac{R_{t,M}}{1 + j\omega R_{t,M} C_{dl}} \tag{10.35}$$

This result is the same as equation (10.21), obtained by use of an electrical circuit. The use of electric circuit frameworks for model development can speed the development of impedance models, especially for the cases that involve partially blocked electrodes and electrodes covered by films.

The advantage of impedance models derived from kinetic arguments over development of electrical circuits by analogy is that the parameters obtained have a physical meaning. For example, the charge-transfer resistance given in equation (10.19) is a function of the steady-state potential $\overline{V}$. By application of equation (5.19) on page 95, equation (10.14) can be expressed in terms of the Tafel slope as

$$\bar{i}_M = K_M \exp\left(\frac{2.303 \times 10^3\ \overline{V}}{\beta_M}\right) \tag{10.36}$$

By introduction of equation (10.36) into equation (10.19), the potential dependence of the charge-transfer resistance can be expressed in terms of the Tafel slope and the steady-state current density as

$$R_{t,M} = \frac{\beta_M}{2.303 \times 10^3 \bar{i}_M} \tag{10.37}$$

Equation (10.37) represents a very important result because the charge-transfer resistance obtained from impedance measurements is related to two well-defined steady-state variables: the steady-state current density and the Tafel slope. Measurement of the impedance at different points along a polarization curve may be used to test whether this model based on a single faradaic reaction provides an adequate description of the system. For example, equation (10.37) allows the value of the steady-state current and the corresponding value of $R_{t,M}$ to be used to calculate the Tafel slope β_M. If β_M changes with potential, then the model based on a single faradaic reaction provides an inadequate description of the system, and another model should be considered.

10.2.2 Potential and Concentration Dependent

Many electrochemical reactions are influenced by the rate of transport of reactants to the electrode surface. Formal treatment of the impedance response for such a system

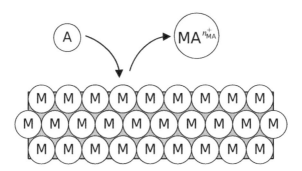

Figure 10.3: Schematic representation of metal dissolution by reaction with an electrolytic species.

requires both the kinetic analysis presented in this chapter and consideration of mass transfer as presented in Chapter 11.

Consider, as an example, the corrosion of a metal in an aqueous medium where the metal reacts with a species A as shown in Figure 10.3. The reaction mechanism is given as

$$M + A \rightarrow MA^{n+} + ne^-$$ (10.38)

The steady-state current density is given by

$$\bar{i}_{MA} = K_{MA}\bar{c}_A(0) \exp\left(b_{MA}\bar{V}\right)$$ (10.39)

where $K_{MA} = nFk_{MA} \exp\left(-b_{MA}V_{0,MA}\right)$ and $K_{MA}\bar{c}_A(0)$ has the units of a current density. The current density is a function of both concentration of species A at the electrode surface and the potential difference between the electrode and the solution adjacent to the electrode. Oxidation of a species A on an inert electrode such as platinum or gold gives the same equation for the current.

In agreement with equation (5.6) on page 92, the current density may be expressed in terms of the flux density of the reacting species, i.e.,

$$i_{MA} = nFD_A \left.\frac{dc_A}{dy}\right|_{y=0}$$ (10.40)

Following equation (10.3), the current density may be expressed as the sum of a steady-state term and a term associated with a sinusoidal variation as

$$i_{MA} = \bar{i}_{MA} + \text{Re}\{\tilde{i}_{MA} \exp(j\omega t)\}$$ (10.41)

Remember! 10.2 *Transfer functions such as impedance provide the relationship between two oscillating variables, e.g., $\tilde{i}$ and $\tilde{V}$. When the expression for $\tilde{i}$ is given in terms of two or more oscillating variables, additional relationships must be found.*

---

where $\tilde{i}_{MA}$ is a complex number. Similarly,

$$c_A = \bar{c}_A + \mathrm{Re}\{\tilde{c}_A \exp(j\omega t)\} \tag{10.42}$$

where $\tilde{c}_A$ is a complex number.

Under assumption of a linear concentration gradient in the diffusion layer of thickness δ_A, the steady-state current density becomes

$$\bar{i}_{MA} = nFD_A \frac{c_A(\infty) - \bar{c}_A(0)}{\delta_A} \tag{10.43}$$

where $c_A(\infty)$ is the concentration of species A in the bulk solution. By eliminating $\bar{c}_A(0)$ from equations (10.39) and (10.43), the value of the current density is obtained as

$$\frac{1}{\bar{i}_{MA}} = \frac{1}{\bar{i}_{\mathrm{lim,MA}}} + \frac{1}{\bar{i}_{\mathrm{k,MA}}} \tag{10.44}$$

where

$$\bar{i}_{\mathrm{lim,MA}} = \frac{nFD_A c_A(\infty)}{\delta_A} \tag{10.45}$$

is the mass-transfer-limited current density and

$$\bar{i}_{\mathrm{k,MA}} = K_{MA} c_A(\infty) \exp(b_{MA}\overline{V}) \tag{10.46}$$

is the kinetic current based on the bulk concentration.

The above analysis follows the development of the Koutecky–Levich equation in Section 5.2.4 on page 104.

An expression for the concentration $\bar{c}_A(0)$ may be obtained from equations (10.39) and (10.43) to be[163]

$$\bar{c}_A(0) = \frac{c_A(\infty)}{\dfrac{K_{MA}\delta_A}{nFD_A} \exp\left(b_{MA}\overline{V}\right) + 1} \tag{10.47}$$

As $\exp\left(b_{MA}\overline{V}\right)$ tends toward infinity, equation (10.47) tends toward

$$\bar{c}_A(0) = \frac{i_{\mathrm{lim,MA}}}{K_{MA} \exp\left(b_{MA}\overline{V}\right)} = \varepsilon \tag{10.48}$$

where ε is a small number. A similar development yielded equation (5.65) on page 105.

The oscillating component of the current density is obtained from equation (10.39) to be

$$\tilde{i}_{MA} = K_{MA} b_{MA} \bar{c}_A(0) \exp\left(b_{MA}\overline{V}\right) \tilde{V} + K_{MA} \exp\left(b_{MA}\overline{V}\right) \tilde{c}_A(0) \tag{10.49}$$

where $\tilde{c}_A(0)$ represents the oscillating component of the concentration of species A evaluated at the electrode surface. Equation (10.49) takes the form of equation (10.5) with $\tilde{\gamma}_k = 0$, or

$$\tilde{i}_{MA} = \frac{1}{R_{\mathrm{t,MA}}} \tilde{V} + K_{MA} \exp(b_{MA}\overline{V}) \tilde{c}_A(0) \tag{10.50}$$

where

$$R_{t,MA} = \frac{1}{K_{MA}\bar{c}_A(0)b_{MA}\exp(b_{MA}\overline{V})} \tag{10.51}$$

Equation (10.51) is used to develop expressions for the charge-transfer resistance and the diffusion impedance.

Charge-Transfer Resistance

Introduction of equation (10.47) to equation (10.51) yields

$$R_{t,MA} = \frac{\dfrac{K_{MA}\delta_A}{nFD_A}\exp\left(b_{MA}\overline{V}\right) + 1}{K_{MA}\bar{c}_A(\infty)b_{MA}\exp(b_{MA}\overline{V})} \tag{10.52}$$

Equation (10.52) may be expressed as

$$R_{t,MA} = R_{t_k,MA} + R_{t_{lim},MA} \tag{10.53}$$

where

$$R_{t_k,MA} = \frac{1}{K_{MA}\bar{c}_A(\infty)b_{MA}\exp(b_{MA}\overline{V})} = \frac{1}{b_{MA}\bar{i}_{k,MA}} \tag{10.54}$$

and

$$R_{t_{lim},MA} = \frac{\delta_A}{b_{MA}nFD_A\bar{c}_A(\infty)} = \frac{1}{b_{MA}\bar{i}_{lim,MA}} \tag{10.55}$$

Thus,

$$R_{t,MA} = R_{t_k,MA} + R_{t_{lim},MA} = \frac{1}{b_{MA}\bar{i}_{MA}} = \frac{1}{b_{MA}}\left(\frac{1}{\bar{i}_{k,MA}} + \frac{1}{\bar{i}_{lim,MA}}\right) \tag{10.56}$$

The charge-transfer resistance may be expressed in terms of the kinetic and limiting currents defined by Koutecky–Levich representation as

$$R_{t,MA} = R_{t_{lim},MA}\left(1 + \frac{\bar{i}_{lim,MA}}{\bar{i}_{k,MA}}\right) \tag{10.57}$$

At positive potentials, $\bar{i}_{k,MA}$ becomes large, and $R_{t,MA}/R_{t_{lim},MA}$ approaches unity. This behavior is in stark contrast with the behavior of the charge-transfer resistance for a reaction that does not depend on concentration, e.g., reaction (10.13). For such a system,

$$R_{t,M} = \frac{1}{K_M b_M \exp(b_M\overline{V})} \tag{10.58}$$

approaches zero at very positive potentials.

Diffusion Impedance

Equation (10.50) may be expressed as

$$1 = \frac{1}{R_{t,MA}} Z_{F,MA} + \frac{K_{MA} \exp(b_{MA}\overline{V})}{nFD_A} \frac{\tilde{c}_A(0)}{\left.\dfrac{d\tilde{c}_A}{dy}\right|_{y=0}} \tag{10.59}$$

where $Z_{F,MA} = \tilde{V}/\tilde{i}_{MA}$ or

$$Z_{F,MA} = R_{t,MA} + R_{t,MA} \frac{K_{MA}\delta_A \exp(b_{MA}\overline{V})}{nFD_A} \left(-\frac{1}{\theta_A'(0)}\right) \tag{10.60}$$

where $\theta_A = \tilde{c}_A/\tilde{c}_A(0)$. Equation (10.60) may be expressed in terms of diffusion resistance $R_{D,MA}$ as

$$Z_{F,MA} = R_{t,MA} + R_{D,MA} \left(-\frac{1}{\theta_A'(0)}\right) \tag{10.61}$$

where

$$R_{D,MA} = R_{t,MA} \frac{K_{MA}\delta_A \exp(b_{MA}\overline{V})}{nFD_A} \tag{10.62}$$

and the diffusion impedance may be represented as

$$Z_{D,MA} = R_{D,MA} \left(-\frac{1}{\theta_A'(0)}\right) \tag{10.63}$$

Equation (10.63) provides a useful notation in the development of the complete impedance representation shown in equation (10.77).

Upon insertion of equation (10.52)

$$R_{D,MA} = R_{t_{lim},MA} \left(\frac{K_{MA}\delta_A}{nFD_A} \exp(b_{MA}\overline{V}) + 1\right) \tag{10.64}$$

The term $R_{D,MA}$ may be expressed as

$$R_{D,MA} = R_{D_1} + R_{t,lim} \tag{10.65}$$

where $R_{t,lim}$ is given by equation (10.55) and

$$R_{D_1} = R_{t,lim} \frac{K_{MA}\delta_A}{nFD_A} \exp(b_{MA}\overline{V}) \tag{10.66}$$

Remember! 10.3 *For an electrochemical reaction influenced by mass-transfer, the charge-transfer resistance tends toward the limiting charge-transfer resistance, defined by equation (10.55), as the corresponding rate constant tends toward infinity.*

The term $R_{\mathrm{D,MA}}$ may be expressed in terms of the kinetic and limiting currents defined by Koutecky–Levich representation as

$$R_{\mathrm{D,MA}} = R_{t_{\lim},\mathrm{MA}}\left(1 + \frac{\bar{i}_{\mathrm{k,MA}}}{\bar{i}_{\lim,\mathrm{MA}}}\right) \tag{10.67}$$

At very negative potentials, $\bar{i}_{\mathrm{k,MA}}$ becomes very small, and $R_{\mathrm{D,MA}}/R_{t_{\lim},\mathrm{MA}}$ approaches unity.

From equations (10.45) and (10.46),

$$\frac{\bar{i}_{\lim,\mathrm{MA}}}{\bar{i}_{\mathrm{k,MA}}} = \frac{nFD_A}{K_{\mathrm{MA}}\delta_A}\exp(b_{\mathrm{MA}}\overline{V}) \tag{10.68}$$

which may be expressed as

$$\ln\left(\frac{\bar{i}_{\lim,\mathrm{MA}}}{\bar{i}_{\mathrm{k,MA}}}\right) = b_{\mathrm{MA}}\overline{V} + \ln\left(\frac{nFD_A}{K_{\mathrm{MA}}\delta_A}\right) \tag{10.69}$$

The total current density, obtained from equation (10.44), can be expressed as

$$\frac{\bar{i}_{\mathrm{MA}}}{\bar{i}_{\lim,\mathrm{MA}}} = \frac{1}{1 + \dfrac{\bar{i}_{\lim,\mathrm{MA}}}{\bar{i}_{\mathrm{k,MA}}}} \tag{10.70}$$

The terms $R_{t,\mathrm{MA}}/R_{t_{\lim},\mathrm{MA}}$ from equation (10.57) and $R_{\mathrm{D,MA}}/R_{t_{\lim},\mathrm{MA}}$ from equation (10.67) are presented in Figure 10.4 as functions of $b_{\mathrm{MA}}\overline{V} + \ln(K_{\mathrm{MA}}\delta_A/nFD_A)$. The dimensionless current density from equation (10.70) is also presented.

For a rotating disk electrode, the Levich equation, presented as equation (5.68) on page 106, can be expressed as

$$\bar{i}_{\lim,\mathrm{MA}} = k_{\Omega,\mathrm{MA}}\Omega^{1/2} \tag{10.71}$$

Thus,

$$R_{t,\mathrm{MA}} = R_{t_{\lim},\mathrm{MA}}\left(1 + \frac{k_{\Omega,\mathrm{MA}}\Omega^{1/2}}{\bar{i}_{\mathrm{k,MA}}}\right) \tag{10.72}$$

and

$$R_{\mathrm{D,MA}} = R_{t_{\lim},\mathrm{MA}}\left(1 + \frac{\bar{i}_{\mathrm{k,MA}}}{k_{\Omega,\mathrm{MA}}\Omega^{1/2}}\right) \tag{10.73}$$

As rotation speed tends toward infinity, $R_{t,\mathrm{MA}} \to \infty$ and $R_{\mathrm{D,MA}}$ tends toward $R_{t_{\lim},\mathrm{MA}}$. As $\Omega \to 0$, $R_{t,\mathrm{MA}} \to R_{t_{\lim},\mathrm{MA}}$ and $R_{\mathrm{D,MA}} \to \infty$.

Example 10.2 **Evaluation of Kinetic Current:** *Show that the assumption that a reaction follows equation (10.39) may be tested by experimental measurement of the mass-transfer-limited current density and evaluation of R_t and R_D at different potentials.*

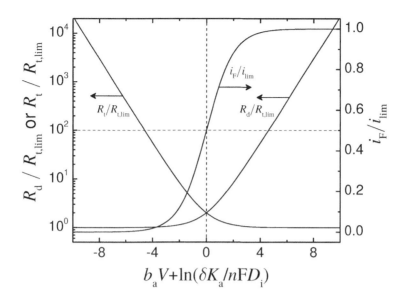

Figure 10.4: The normalized charge-transfer resistance (equation (10.57)), normalized diffusion resistance (equation (10.67)), and normalized current density (equation (10.70)) as function of scaled potential.

Solution: *The values of R_t and R_D obtained by regression provide a method to extract values for the kinetic current, i_k. Equations (10.57) and (10.67) yield*

$$\frac{R_D}{R_t} = \frac{1 + i_k/i_{lim}}{1 + i_{lim}/i_k} \tag{10.74}$$

which, as the mass-transfer-limited current density is a constant at a given rotation rate, can be solved for the kinetic current density as

$$i_k = \frac{-i_{lim}}{2}\left(1 - \frac{R_D}{R_t} + \sqrt{\left(1 - \frac{R_D}{R_t}\right)^2 + 4\frac{R_D}{R_t}}\right) \tag{10.75}$$

If the reaction follows equation (10.39), the kinetic current will be a simple exponential function of potential, as would be expected from equation (10.46). The evolution of a complex mechanism as a function of electrode potential may be followed by use of equation (10.75), yielding more information than can be obtained by use of the Tafel plot. An experimental application is presented by Tran et al.[163]

Remember! **10.4** *For an electrochemical reaction influenced by mass-transfer, the corresponding diffusion resistance (equation (10.62)) tends toward the limiting charge-transfer resistance, defined by equation (10.55), as the current tends toward zero.*

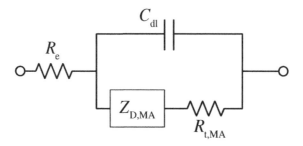

Figure 10.5: Electrical circuit providing the equivalent to the impedance response for a single electrochemical reaction coupled with a mass transfer impedance. This circuit is known as the Randles circuit.[45]

Cell Impedance

The faradaic impedance, given by

$$Z_{F,MA} = \frac{\widetilde{V}}{\widetilde{i}_{MA}} = R_{t,MA} + Z_{D,MA} \tag{10.76}$$

may be introduced in equation (9.2) on page 193. Alternatively, the relationships presented in Table 10.1 (equations (10.22)-(10.24)) can be introduced to obtain a relationship for impedance in terms of capacitance and electrolyte resistance as

$$Z_{MA}(\omega) = R_e + \frac{R_{t,MA} + Z_{D,MA}(\omega)}{1 + j\omega C_{dl}(R_{t,MA} + Z_{D,MA}(\omega))} \tag{10.77}$$

The electrical circuit presented in Figure 10.5 yields the impedance response equivalent to equation (10.77) for a single faradaic reaction coupled with mass transfer. This circuit is known as the Randles circuit.[45] Such a circuit may provide a building block for development of circuit models as shown in Chapter 9 for the impedance response of a more complicated system involving, for example, coupled reactions or more complicated two- or three-dimensional geometries.

The circuit shown in Figure 10.5 illustrates an important feature of equation (10.77). While the diffusion impedance is obtained from a solution of convective diffusion equations, the diffusion impedance itself is a property of the electrode-electrolyte interface.

Example 10.3 **Diffusion with First-Order Reaction:** *Develop an expression for the impedance response for the oxygen reduction at potentials sufficiently cathodic to allow all anodic reactions to be ignored. Under these conditions, the reaction*

$$O_2 + H_2O + 4e^- \rightarrow 4OH^- \tag{10.78}$$

provides an example of a first-order reaction involving only one mass-transfer-limited species.

Solution: *For a simple first-order reaction where only one species is involved, the faradaic current can be written as*

$$i_{O_2} = -K_{O_2} c_{O_2}(0) \exp\left(-b_{O_2} V\right) \tag{10.79}$$

Following equation (5.6) on page 92 and under assumption of a linear concentration gradient in the diffusion layer of thickness δ_{O_2}, the steady-state current density can be written as

$$\bar{i}_{O_2} = -4FD_{O_2} \frac{c_{O_2}(\infty) - \bar{c}_{O_2}(0)}{\delta_{O_2}} \tag{10.80}$$

where $c_{O_2}(\infty)$ is the concentration of oxygen in the bulk solution. An expression for the steady-state oxygen concentration at the electrode surface may be obtained from equations (10.79) and (10.80) as

$$\bar{c}_{O_2}(0) = \frac{c_{O_2}(\infty)}{\dfrac{K_{O_2}\delta_{O_2}}{4FD_{O_2}} \exp\left(-b_{O_2}\bar{V}\right) + 1} \tag{10.81}$$

From equation (10.79), the oscillating current can be expressed as

$$\tilde{i}_{O_2} = K_{O_2} \bar{c}_{O_2}(0) b_{O_2} \exp\left(-b_{O_2}\bar{V}\right) \tilde{V} - K_{O_2}(0) \exp\left(-b_{O_2}\bar{V}\right) \tilde{c}_{O_2}(0) \tag{10.82}$$

As was shown for equation (10.49), equation (10.82) is expressed in terms of three oscillating variables; whereas, the impedance, like any transfer function, is an expression relating two oscillating variables. The additional relationship is expressed as equation (10.40). For reaction (10.78), $s_{O_2} = -1$. Thus,

$$\tilde{i}_{O_2} = -4FD_{O_2} \frac{\tilde{c}_{O_2}(0)}{\delta_{O_2}} \theta'_{O_2}(0) \tag{10.83}$$

where $\theta_{O_2} = \tilde{c}_{O_2}/\tilde{c}_{O_2}(0)$ is the dimensionless oscillating concentration, θ'_{O_2} is the derivative of θ_{O_2} with respect to dimensionless position, $\zeta = y/\delta_{O_2}$, and δ_{O_2} is the diffusion layer thickness. The concentration $\tilde{c}_{O_2}(0)$ can be eliminated in equations (10.82) and (10.83) to yield

$$Z_{F,O_2} = \frac{\tilde{V}}{\tilde{i}_{O_2}} = R_{t,O_2} + Z_{D,O_2} \tag{10.84}$$

where

$$R_{t,O_2} = \frac{1}{K_{O_2} b_{O_2} \bar{c}_{O_2}(0) \exp\left(-b_{O_2}\bar{V}\right)} \tag{10.85}$$

and

$$Z_{D,O_2}(\omega) = \frac{\delta_{O_2}}{4FD_{O_2}\bar{c}_{O_2}(0)b_{O_2}} \left(-\frac{1}{\theta'_{O_2}(0)}\right) \tag{10.86}$$

Equation (10.86) may be expressed as

$$Z_{D,O_2}(\omega) = R_{D,O_2} \left(-\frac{1}{\theta'_{O_2}(0)}\right) \tag{10.87}$$

Expressions for the term $(-1/\theta'_{O_2}(0))$ *are developed in Chapter 11 (see page 243).*

By using equation (10.81), R_{t,O_2} and R_{D,O_2} can be expressed in function of the parameters of the system. The charge transfer resistance is given by

$$R_{t,O_2} = \frac{\frac{K_{O_2}\delta_{O_2}}{4FD_{O_2}}\exp\left(-b_{O_2}\overline{V}\right)+1}{K_{O_2}c_{O_2}(\infty)b_{O_2}\exp\left(-b_{O_2}\overline{V}\right)} \tag{10.88}$$

The charge transfer resistance can be expressed as the sum of two terms

$$R_{t,O_2} = R_{t_{lim},O_2} + R_{t_k,O_2} \tag{10.89}$$

where

$$R_{t_{lim},O_2} = \frac{\delta_{O_2}}{4FD_{O_2}c_{O_2}(\infty)}\frac{1}{b_{O_2}} = -\frac{1}{b_{O_2}\bar{i}_{lim,O_2}} \tag{10.90}$$

and

$$R_{t_k,O_2} = \frac{1}{K_{O_2}c_{O_2}(\infty)b_{O_2}\exp\left(-b_{O_2}\overline{V}\right)} \tag{10.91}$$

The diffusion resistance may be written as

$$R_{D,O_2} = R_{t_{lim},O_2}\left(\frac{K_{O_2}\delta_{O_2}\exp\left(-b_{O_2}V\right)}{4FD_{O_2}}+1\right) \tag{10.92}$$

Following the development in Section 10.2.2, equation (10.89) may be expressed as

$$R_{t,O_2} = R_{t_{lim},O_2}\left(1 + \frac{\bar{i}_{lim,O_2}}{\bar{i}_{k,O_2}}\right) \tag{10.93}$$

and equation (10.92) may be expressed as

$$R_{D,O_2} = R_{t_{lim},O_2}\left(1 + \frac{\bar{i}_{k,O_2}}{\bar{i}_{lim,O_2}}\right) \tag{10.94}$$

where R_{t_{lim},O_2} *is defined by equation (10.90),* $\bar{i}_{lim,O_2}$ *is defined by equation (10.45), and* $\bar{i}_{k,O_2}$ *is defined by equation (10.46).*

10.3 Multiple Independent Electrochemical Reactions

The impedance associated with multiple uncoupled reactions is developed on the principle that currents are additive. As shown in Section 9.2.1 on page 195, the associated faradaic impedances are added in parallel.

 Example 10.4 Iron in Anaerobic Solutions: *Consider the corrosion of iron*

$$Fe \rightarrow Fe^{2+} + 2e^{-} \tag{10.95}$$

at the corrosion potential in an anaerobic aqueous medium with water electrolysis

$$H_2O + e^- \rightarrow \frac{1}{2}H_2 + OH^- \tag{10.96}$$

as the balancing cathodic reaction. Find an expression for the faradaic impedance response.

Solution: *The steady-state anodic current density is given by*

$$\bar{i}_{Fe} = K_{Fe} \exp\left(b_{Fe}\bar{V}\right) \tag{10.97}$$

and the steady-state cathodic current is given by

$$\bar{i}_{H_2} = -K_{H_2} \exp\left(-b_{H_2}\bar{V}\right) \tag{10.98}$$

where the effective rate constants K_{Fe} and K_{H_2} include the respective equilibrium potentials for the corrosion and hydrogen evolution reactions.

The total faradaic current is given by

$$i_F = i_{Fe} + i_{H_2} \tag{10.99}$$

At the corrosion potential, $i_F = 0$. The oscillating component of the current density is given by

$$\tilde{i}_F = \tilde{i}_{Fe} + \tilde{i}_{H_2} \tag{10.100}$$

The steady potential and the oscillating potential are the same for both reactions; thus,

$$\frac{\tilde{i}_F}{\tilde{V}} = \frac{\tilde{i}_{Fe}}{\tilde{V}} + \frac{\tilde{i}_{H_2}}{\tilde{V}} \tag{10.101}$$

or, with the impedance notation,

$$Z_F^{-1} = Z_{F,H_2}^{-1} + Z_{F,Fe}^{-1} \tag{10.102}$$

The techniques developed in Section 10.2.1 can be used to show that

$$Z_{F,Fe} = R_{t,Fe} = \left[K_{Fe}b_{Fe}\exp\left(b_{Fe}\bar{V}\right)\right]^{-1} \tag{10.103}$$

and

$$Z_{F,H_2} = R_{t,H_2} = \left[K_{H_2}b_{H_2}\exp\left(-b_{H_2}\bar{V}\right)\right]^{-1} \tag{10.104}$$

Introduction of equation (9.2) on page 193 yields the overall impedance as

$$Z = R_e + \frac{Z_F}{1 + j\omega Z_F C_{dl}} \tag{10.105}$$

or, following equation (10.102),

$$Z = R_e + \frac{1}{\dfrac{1}{R_{t,Fe}} + \dfrac{1}{R_{t,H_2}} + j\omega C_{dl}} \tag{10.106}$$

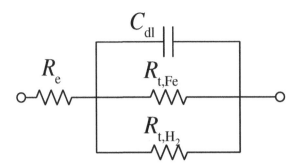

Figure 10.6: Electrical circuit providing the impedance response equivalent corresponding to Example 10.4. The impedance response is equivalent to that for a single electrochemical reaction.

An equivalent circuit providing the impedance response at corrosion potential is given in Figure 10.6. This circuit shows the two parallel impedances corresponding to the anodic process and to the cathodic process, in agreement with the general case described by Figure 9.2 (page 195), but, on an experimental point of view, only the overall charge-transfer resistance R_t can be obtained.

Each charge-transfer resistance in equation (10.106) can be written by using the Tafel slope (see equation (10.37)) as

$$R_{t,Fe} = \frac{\beta_{Fe}}{2.303\bar{i}_{Fe}} \tag{10.107}$$

and

$$R_{t,H_2} = \frac{\beta_{H_2}}{2.303\bar{i}_{H_2}} \tag{10.108}$$

where both β_{H_2} and $\bar{i}_{H_2}$ are negative. At the corrosion potential, $\bar{i}_{corr} = \bar{i}_{Fe} = -\bar{i}_{H_2}$. The overall charge-transfer resistance is given by:

$$R_t = \frac{-\beta_{H_2}\beta_{Fe}}{2.303 \times 10^3 \bar{i}_{corr}(\beta_{Fe} - \beta_{H_2})} \tag{10.109}$$

From this last equation, if the Tafel slopes and the overall charge-transfer resistance are measured, $\bar{i}_{corr}$ can be obtained from

$$\bar{i}_{corr} = \frac{-\beta_{H_2}\beta_{Fe}}{2.303 \times 10^3 R_t(\beta_{Fe} - \beta_{H_2})} \tag{10.110}$$

This relation was first developed by Stern and Geary[164] and is applied by some industrial instrumentation to determine the corrosion rate.

Example 10.5 Iron in Aerobic Solutions: *Consider a system at open circuit in an aerobic aqueous medium in which the anodic reaction corresponds to the corrosion of iron, i.e.,*

$$Fe \rightarrow Fe^{2+} + 2e^- \tag{10.111}$$

and the cathodic reaction is the reduction of oxygen

$$O_2 + 2H_2O + 4e^- \rightarrow 4OH^- \tag{10.112}$$

Find an expression for the impedance response.

Solution: *The steady-state anodic current density is given by equation (10.97). According to equation (10.39), the steady-state cathodic current is given by*

$$\bar{i}_{O_2} = -K_{O_2}\bar{c}_{O_2}(0)\exp\left(-b_{O_2}\overline{V}\right) \tag{10.113}$$

The impedance for the iron dissolution is given by equation (10.103), and the impedance for the oxygen reduction reaction is given by

$$Z_{F,O_2} = R_{t,O_2} + Z_{D,O_2} \tag{10.114}$$

with

$$R_{t,O_2} = \left[K_{O_2}\bar{c}_{O_2}(0)b_{O_2}\exp\left(-b_{O_2}\overline{V}\right)\right]^{-1} \tag{10.115}$$

and

$$Z_{D,O_2} = \frac{\delta_{O_2}}{4FD_{O_2}\bar{c}_{O_2}(0)}\frac{1}{b_{O_2}}\left(-\frac{1}{\theta'_{O_2}(0)}\right) \tag{10.116}$$

Following equation (9.3) on page 196,

$$Z = R_e + \cfrac{1}{\cfrac{1}{R_{t,Fe}} + \cfrac{1}{R_{t,O_2} + Z_{D,O_2}} + j\omega C_{dl}} \tag{10.117}$$

An equivalent circuit providing the impedance response at corrosion potential is given in Figure 10.7. This circuit shows the two parallel impedances corresponding to the anodic and the cathodic processes, in agreement with the general case described by Figure 9.2 (page 195).

Examples 10.4 and 10.5 show two approaches for the development of impedance models for coupled reactions. In Example 10.4, a faradaic impedance is developed for the combined effects of the two reactions, and then an expression from an equivalent circuit based on a single reaction is used to include the effects of the ohmic resistance and the capacitance. In Example 10.5, expressions for the faradaic impedances for the two reactions are inserted into an expression from an equivalent circuit that accounts for multiple reactions.

Remember! 10.5 *The Stern–Geary relationship given as equation (10.110) applies only when both the anodic and the cathodic reactions follow a Tafel relationship. When this restriction is not met, the rate of corrosion must be obtained from the charge-transfer resistance of the corrosion reaction obtained from impedance measurements.*

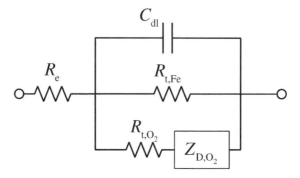

Figure 10.7: Electrical circuit providing the equivalent to the impedance response for iron dissolution and oxygen reduction.

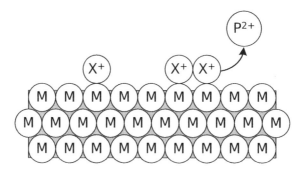

Figure 10.8: Schematic representation of metal dissolution through an adsorbed intermediate.

10.4   Coupled Electrochemical Reactions

In contrast to the independent electrochemical reactions considered in Section 10.3, the electrochemical reactions considered in this section are coupled by reaction intermediates. These reactions may be influenced by potential and surface coverage or by potential, surface coverage, and mass transfer.

10.4.1   Potential and Surface Coverage Dependent

Consider a hypothetical reaction sequence shown in Figure 10.8 in which a metal M dissolves through an adsorbed intermediate X^+ following

$$M \rightarrow X^+ + e^- \tag{10.118}$$

which reacts in a second electrochemical step

$$X^+ \rightarrow P^{2+} + e^- \tag{10.119}$$

to form the final product P. Adsorption of reaction intermediate X^+ obeys a Langmuir isotherm and is characterized by a surface coverage γ_{X^+}. The steady-state current density associated with reaction (10.118) is given by

$$\bar{i}_M = K_M \left(1 - \bar{\gamma}_X\right) \exp\left(b_M \overline{V}\right) \tag{10.120}$$

where γ_X represents the fractional surface coverage by the intermediate X^+. Similar mechanisms were proposed by Epelboin and Keddam[7] for calculating the impedance of iron dissolution through two steps involving an adsorbed FeOH intermediate and by Peter et al.[165] for the impedance model of the dissolution of aluminum in three consecutive steps with two adsorbed intermediates.

The steady-state current density associated with reaction (10.119) is given by

$$\bar{i}_X = K_X \bar{\gamma}_X \exp\left(b_X \overline{V}\right) \tag{10.121}$$

where the rate constant K_X includes the maximum surface concentration of the intermediate X, defined in equation (10.122) as Γ.

The variation of the surface coverage by the intermediate X^+ is given by the expression

$$\Gamma \frac{d\gamma_X}{dt} = \frac{i_M}{F} - \frac{i_X}{F} \tag{10.122}$$

Under a steady-state condition, $d\gamma_X/dt = 0$ and $\bar{i}_M = \bar{i}_X$. By using the corresponding equations (10.120) and (10.121), an expression for the steady-state surface coverage $\bar{\gamma}_X$ is obtained as

$$\bar{\gamma}_X = \frac{K_M \exp\left(b_M \overline{V}\right)}{K_M \exp\left(b_M \overline{V}\right) + K_X \exp\left(b_X \overline{V}\right)} \tag{10.123}$$

Examination of equation (10.123) reveals that, if $K_M \exp\left(b_M \overline{V}\right) \gg K_X \exp\left(b_X \overline{V}\right)$, then $\bar{\gamma}_X \to 1$. If $K_M \exp\left(b_M \overline{V}\right) \ll K_X \exp\left(b_X \overline{V}\right)$, then $\bar{\gamma}_X \to 0$. An expression for the total steady-state current density, given by

$$\bar{i}_F = \bar{i}_M + \bar{i}_X = \frac{2K_M \exp(b_M \overline{V}) K_X \exp(b_X \overline{V})}{K_M \exp(b_M \overline{V}) + K_X \exp(b_X \overline{V})} \tag{10.124}$$

can be found from equations (10.120), (10.119), and (10.123).

The oscillating component of the current density for each reaction is given, respectively, by

$$\tilde{i}_M = \frac{1}{R_{t,M}} \tilde{V} - K_M \exp\left(b_M \overline{V}\right) \tilde{\gamma}_X \tag{10.125}$$

and

$$\tilde{i}_X = \frac{1}{R_{t,X}} \tilde{V} + K_X \exp\left(b_X \overline{V}\right) \tilde{\gamma}_X \tag{10.126}$$

where the charge-transfer resistances are defined by

$$R_{t,M} = \frac{1}{K_M \left(1 - \bar{\gamma}_X\right) b_M \exp\left(b_M \overline{V}\right)} \tag{10.127}$$

and

$$R_{t,X} = \frac{1}{K_X \overline{\gamma}_X b_X \exp\left(b_X \overline{V}\right)} \tag{10.128}$$

According to equation (10.122), the oscillating component of the surface coverage can be expressed as

$$\Gamma F j\omega \tilde{\gamma}_X = \left(\frac{1}{R_{t,M}} - \frac{1}{R_{t,X}}\right) \tilde{V} - \left(K_X \exp\left(b_X \overline{V}\right) + K_M \exp\left(b_M \overline{V}\right)\right) \tilde{\gamma}_X \tag{10.129}$$

yielding

$$\tilde{\gamma}_X = \left(\frac{\dfrac{1}{R_{t,M}} - \dfrac{1}{R_{t,X}}}{\Gamma F j\omega + \left(K_X \exp\left(b_X \overline{V}\right) + K_M \exp\left(b_M \overline{V}\right)\right)}\right) \tilde{V} \tag{10.130}$$

The net faradaic current density, given by the sum of contributions from reactions (10.118) and (10.119), is a function of γ_X and V, and, following equation (10.5),

$$\tilde{i}_F = \tilde{i}_X + \tilde{i}_M \tag{10.131}$$

$$= \left(\frac{1}{R_{t,M}} + \frac{1}{R_{t,X}}\right) \tilde{V} + \left(K_X \exp\left(b_X \overline{V}\right) - K_M \exp\left(b_M \overline{V}\right)\right) \tilde{\gamma}_X$$

The expression of the impedance is

$$\frac{1}{Z_F} = \frac{\tilde{i}_F}{\tilde{V}} \tag{10.132}$$

or

$$\frac{1}{Z_F} = \frac{1}{R_t} + \frac{\left(K_X \exp\left(b_X \overline{V}\right) - K_M \exp\left(b_M \overline{V}\right)\right)\left(R_{t,M}^{-1} - R_{t,X}^{-1}\right)}{\Gamma F j\omega + \left(K_X \exp\left(b_X \overline{V}\right) + K_M \exp\left(b_M \overline{V}\right)\right)} \tag{10.133}$$

where

$$\frac{1}{R_t} = \frac{1}{R_{t,M}} + \frac{1}{R_{t,X}} \tag{10.134}$$

The impedance given in equation (10.133) can be expressed in the form

$$\frac{1}{Z_F} = \frac{1}{R_t} + \frac{A}{j\omega + B} \tag{10.135}$$

where A can have a positive or negative sign according to the constant parameter values and the potential.

The term A may also be written

$$A = \frac{\partial \tilde{i}_F}{\partial \gamma} \frac{\partial \dot{\gamma}}{\partial V} \tag{10.136}$$

where

$$\dot{\gamma} = \frac{\partial \gamma}{\partial t} = \frac{i_M - i_X}{\Gamma F} \tag{10.137}$$

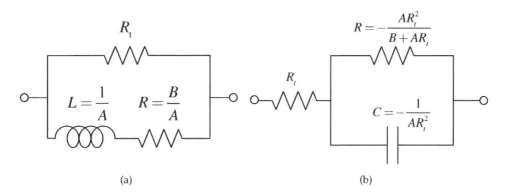

(a) (b)

Figure 10.9: Electrical circuit providing the equivalent to the impedance response for two coupled reactions with surface coverage: a) case of an inductive impedance where $A > 0$ and b) case of a capacitive impedance where $A < 0$.

Thus,

$$A = \left(\frac{1}{\Gamma F} \frac{\partial(\bar{i}_M + \bar{i}_X)}{\partial \bar{\gamma}} \right) \frac{\partial(i_M - i_X)}{\partial \bar{V}} \tag{10.138}$$

If A is positive, the electrical circuit providing the impedance response equivalent to equation (10.135) is a charge-transfer resistance R_t in parallel with an inductance in series with a resistance (Figure 10.9(a)). This inductance has a value of $1/A$, and the resistance has a value of B/A. If A is negative, the impedance can be written under the following expression:

$$Z_F = R_t + \frac{\dfrac{-AR_t^2}{B + AR_t}}{\dfrac{j\omega}{B + AR_t} + 1} \tag{10.139}$$

The electrical circuit providing the impedance response equivalent to the same equation (10.135) is a charge-transfer resistance in series with a Voigt element composed of a capacitance in parallel with a resistance (Figure 10.9(b)). The capacitance has a value of $-1/AR_t^2$ and the resistance has a value of $-AR_t^2/(B + AR_t)$.

The term $(B + AR_t)$ always has a positive value. The easiest way to determine whether the low-frequency loop is inductive or capacitive is to calculate the value of $(R_t - Z_F)$ at zero frequency. If the value is positive, an inductive loop is present; if the value is negative, a capacitive loop appears. Thus the same impedance expression

---

Remember! 10.6 *The easiest way to determine whether the low-frequency loop is inductive or capacitive is to calculate $(R_t - Z_F)$ at zero frequency. If the value is positive an inductive loop is present, and, if the value is negative, a capacitive loop appears.*

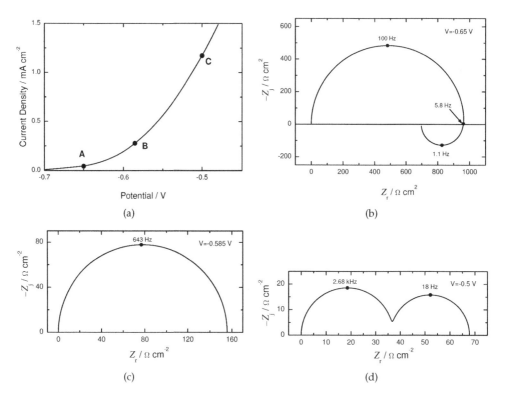

Figure 10.10: Calculated steady-state and impedance response for coupled reactions dependent on potential and surface coverage. a) Simulated current-potential curve following equation (10.124) with the kinetic parameters $K_M = 4F$ A/cm^2, $b_M = 36$ V^{-1}, $K_X = 10^{-6}F$ A/cm^2, $b_x = 10$ V^{-1}, $\Gamma = 2 \times 10^{-9}$ mol/cm^2, and $C_{dl} = 20\mu$F/cm^2. The points A, B, and C correspond to the simulated impedance; b) impedance diagram simulated at the point A ($V = -0.65$ V); c) impedance diagram simulated at the point B ($V = -0.585$ V); and d) impedance diagram simulated at the point C ($V = -0.50$ V).

(10.133) can yield two completely different equivalent circuits according to the potential and the constant parameter values.

To illustrate the previous calculation, some typical simulation results are provided in Figure 10.10. Points A, B, and C labeled on the current-potential curve given in Figure 10.10(a) correspond to potentials at which impedance simulations were performed. A high-frequency capacitive loop coupled with a low-frequency inductive loop is evident at a potential of -0.65 V (Figure 10.10(b)). At a slightly higher potential of -0.585 V, the inductive and capacitive loops merge into a single loop as shown in Figure 10.10(c). Two capacitive loops are observed in Figure 10.10(d) for a potential $V = -0.5$ V.

Kinetic models such as that presented in this section are superior to the use of electrical circuit analogues because the same model can account for the broad range of behavior shown in Figure 10.10. A second advantage is that the examination of the

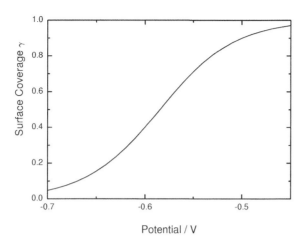

Figure 10.11: Calculated surface coverage γ corresponding to Figure 10.10.

variables such as the surface coverage shown in Figure 10.11 can give insight into the reaction mechanism. In this case, the low-frequency loops are evident when the surface coverage is small.

10.4.2 Potential, Surface Coverage, and Concentration Dependent

The approach developed in the previous sections can be applied to situations such as that shown in Figure 10.12 in which an ionic product species diffuses from the surface and reacts through a backward reaction with the intermediate adsorbed species. This situation resembles the one presented in Section 10.4.1 with the exception that the mass transfer of product species affects the current.

Example 10.6 **Corrosion of Magnesium:** *Consider that the corrosion of magnesium proceeds according to the two-step reaction sequence in which*

$$Mg \xrightarrow{k_1} Mg_{ads}^+ + e^- \tag{10.140}$$

Remember! 10.7 *Low-frequency inductive loops in the impedance response can be attributed to faradaic reactions that involve adsorbed intermediate species. Such systems can be described in terms of electrical circuits that involve inductances.*

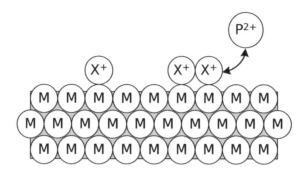

Figure 10.12: Schematic representation of mass-transfer controlled reaction by an ionic species formed through an adsorbed intermediate.

involves production of an adsorbed reaction intermediate (Mg^+_{ads}), which reacts further to form the divalent Mg^{2+} ion, i.e.,

$$Mg^+_{ads} \underset{k_{22}}{\overset{k_2}{\rightleftharpoons}} Mg^{2+} + e^- \tag{10.141}$$

The product Mg^{2+} diffuses through a porous layer of $Mg(OH)_2$ that has a thickness of δ. Find the impedance response for this reaction sequence.

Solution: *On the basis of the reaction model (equations (10.140) and (10.141)), the impedance can be derived under the assumption that the adsorbate Mg^+_{ads} obeys a Langmuir isotherm and that the rate constants of electrochemical reactions are exponentially dependent on potential (e.g., following Tafel's law). Each reaction with index i has a normalized rate constant K_i corresponding to its rate constant k_i by*

$$K_i = k_i F \exp(b_i V) \tag{10.142}$$

Under the assumption that the maximum number of sites per surface unit that can be occupied by the adsorbate Mg^+_{ads} is Γ, the mass and charge balances are expressed in function of the fraction of the surface coverage by the adsorbed species γ as

$$F\Gamma\frac{d\gamma}{dt} = K_1(1-\gamma)\exp(b_1 V) - K_2\gamma\exp(b_2 V) \tag{10.143}$$
$$+ K_{22}c_{Mg^{2+}}(0)(1-\gamma)\exp(-b_{22}V)$$

where the normalized rate constants K_1, K_2, and K_{22} include the maximum coverage Γ. A material balance on Mg^{2+} under the steady-state condition yields

$$F\frac{D_{Mg^{2+}}\bar{c}_{Mg^{2+}}(0)}{\delta_{Mg^{2+}}} = K_2\bar{\gamma}\exp(b_2\bar{V}) - K_{22}\bar{c}_{Mg^{2+}}(0)(1-\bar{\gamma})\exp(-b_{22}\bar{V}) \tag{10.144}$$

where $\delta_{Mg^{2+}}$ is the thickness of the Nernst diffusion layer and $\bar{c}_{Mg^{2+}}(0)$ is the concentration of the Mg^{2+} ion at the electrode interface. The total faradaic current density $i_{F,Mg}$ can be expressed *as*

$$i_{F,Mg} = \left[K_1 (1 - \overline{\gamma}) + K_2 \overline{\gamma} - K_{22} \overline{c}_{Mg^{2+}}(0) \right] \tag{10.145}$$

At steady state, $\overline{\gamma}$ and $\overline{c}_{Mg^{2+}}(0)$ are thus given by

$$\overline{\gamma} = \frac{K_1 \left(\dfrac{D_{Mg^{2+}}}{\delta_{Mg^{2+}}} + K_{22} \right)}{K_1 \left(\dfrac{D_{Mg^{2+}}}{\delta_{Mg^{2+}}} + K_{22} \right) + K_2 \dfrac{D_{Mg^{2+}}}{\delta_{Mg^{2+}}}} \tag{10.146}$$

and

$$\overline{c}_{Mg^{2+}}(0) = \frac{\overline{\gamma} K_2}{\dfrac{D_{Mg^{2+}}}{\delta_{Mg^{2+}}} + K_{22}} \tag{10.147}$$

respectively. The faradic impedance Z_F, is thus calculated by linearizing the mathematical expressions (10.143), (10.144), and (10.145) for small sine wave perturbations to obtain

$$(F\Gamma j\omega + K_1 + K_2) \widetilde{\gamma} = \tag{10.148}$$
$$\left[(1 - \gamma) K_1 b_1 - \gamma K_2 b_2 - c_{Mg^{2+}}(0) K_{22} b_{22} \right] \widetilde{V} + K_{22} \widetilde{c}_{Mg^{2+}}(0)$$

and

$$\widetilde{i}_{F,Mg} = (K_2 - K_1) \widetilde{\gamma} \tag{10.149}$$
$$+ \left[(1 - \gamma) K_1 b_1 + \gamma K_2 b_2 + c_{Mg^{2+}}(0) K_{22} b_{22} \right] \widetilde{V} - K_{22} \widetilde{c}_{Mg^{2+}}(0)$$

Equation (10.149) has the form of the general expression (10.5).

Since the specie Mg^{2+} diffuses toward the electrode surface, the resulting concentration perturbation $\widetilde{c}_{Mg^{2+}}(0)$ is obtained from the finite-length diffusion impedance, represented by $-1/\theta'_{Mg^{2+}}(0)$ and given in equation (11.20). The resulting impedance response is given by

$$Z_{F,Mg} = \frac{\widetilde{V}}{\widetilde{i}_{F,Mg}} = \frac{1 + K_{22} \left(-1/\theta'_{Mg^{2+}}(0) \right) \left(1 - \dfrac{K_2 - K_1}{F\Gamma j\omega + K_1 + K_2} \right)}{A \left(\dfrac{(r_1 - r_2)(K_2 - K_1)}{F\Gamma j\omega + K_1 + K_2} + (r_1 + r_2) \right)} \tag{10.150}$$

with $r_1 = (1 - \gamma) K_1 b_1$ and $r_2 = K_2 b_2 \gamma + K_{22} b_{22} c_{Mg^{2+}}(0)$. An example of a simulated diagram is given in Figure 10.13. The impedance response is characterized by three loops: a charge-transfer resistance loop at high frequency, a diffusion impedance loop proportional to $-1/\theta'_{Mg^{2+}}(0)$, and an inductive loop at low frequency.

For this system, the impedance of the cathodic reaction is large and, following the arguments described in Section 9.2.1 on page 195, does not contribute to the observed impedance.

The reaction sequence described in Example 10.6 represents a simplification of a model developed by Baril et al.[158] who included an additional reaction

$$2Mg^+ + 2H_2O \rightarrow 2Mg^{2+} + 2OH^- + H_2 \tag{10.151}$$

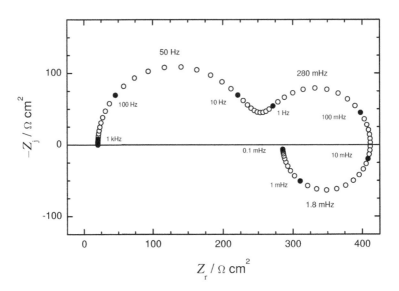

Figure 10.13: Simulated impedance diagram following the expression given in equation (10.150) for the faradaic impedance. This faradaic impedance is simulated in parallel with a high-frequency capacitance. (Taken from Baril et al.[158])

Reaction (10.151) is a chemical reaction that results in hydrogen evolution at potentials associated with Mg dissolution. Due to the anomalous production of hydrogen at anodic rather than cathodic potentials, this reaction was termed the Negative Difference Effect (NDE). The result presented in Example 10.6 can be obtained from that presented by Baril et al.[158] by setting their k_3 from reaction (10.151) equal to zero.

10.5 Electrochemical and Heterogeneous Chemical Reactions

The coupling of faradaic and heterogeneous chemical reactions is presented in this section. The coupling of faradaic and homogeneous chemical reactions requires more detailed development of mass-transfer concepts and, accordingly, is presented in Section 11.7 on page 278.

Example 10.7 **Diffusion of Two Species:** *At sufficiently high overpotentials, the deposition of zinc from zincate solutions can be assumed to follow a simplified reaction scheme as a chemical step*

$$Zn(OH)_4^{2-} \rightleftarrows Zn(OH)_3^- + OH^- \qquad (10.152)$$

followed by an electrochemical step

$$Zn(OH)_3^- + 2e^- \rightarrow Zn + 3OH^- \qquad (10.153)$$

Under the assumption that reaction (10.152) is fast, an empirical relation has been established for zinc deposition from basic media at the equlibrium potential as[166]

$$i_{Zn} = K_{Zn} \frac{c_{Zn(OH)_3^-}}{c_{OH^-}} \tag{10.154}$$

Develop an expression for the impedance response, taking into account diffusion of the reactant $Zn(OH)_3^-$ *toward the electrode and of the product* OH^- *away from the electrode.*

Solution: *The formal dependence of equation (10.154) on potential can be obtained under the assumption of a Tafel behavior, e.g.,*

$$i_{Zn} = K_{Zn} \frac{c_{Zn(OH)_3^-}}{c_{OH^-}} \exp\left(b_{Zn} V\right) \tag{10.155}$$

Following equation (10.5),

$$\tilde{i}_{Zn} = \frac{1}{R_{t,Zn}} \tilde{V} + \frac{K_{Zn}}{\bar{c}_{OH^-}} \exp\left(b_{Zn}\overline{V}\right) \tilde{c}_{Zn(OH)_3^-} \tag{10.156}$$

$$- K_{Zn} \frac{\bar{c}_{Zn(OH)_3^-}}{\left(\bar{c}_{OH^-}\right)^2} \exp\left(b_{Zn}\overline{V}\right) \tilde{c}_{OH^-}$$

where the charge-transfer resistance is given by

$$R_{t,Zn} = \frac{1}{K_{Zn} b_{Zn} \dfrac{\bar{c}_{Zn(OH)_3^-}(0)}{\bar{c}_{OH^-}(0)} \exp\left(b_{Zn}\overline{V}\right)} \tag{10.157}$$

or

$$R_{t,Zn} = \frac{1}{i_{Zn} b_{Zn}} \tag{10.158}$$

From the stoichiometry of reaction (10.153), the current density is related to the diffusive flux of $Zn(OH)_3^-$ *by*

$$\tilde{i}_{Zn} = -2 F D_{Zn(OH)_3^-} \left.\frac{d\tilde{c}_{Zn(OH)_3^-}}{dy}\right|_{y=0} \tag{10.159}$$

Similarly, the current density is related to the diffusive flux of OH^- *by*

$$\tilde{i}_{Zn} = +\frac{2}{3} F D_{OH^-} \left.\frac{d\tilde{c}_{OH^-}}{dy}\right|_{y=0} \tag{10.160}$$

The convective-diffusion impedance can be found to be

$$Z_D = Z_{D,Zn(OH)_3^-} + Z_{D,(OH)^-} \tag{10.161}$$

where

$$Z_{D,Zn(OH)_3^-} = R_{t,Zn} \frac{K_{Zn} \exp\left(b_{Zn}\overline{V}\right)}{2 F \bar{c}_{OH^-}(0) D_{Zn(OH)_3^-}} \frac{-\tilde{c}_{Zn(OH)_3^-}(0)}{\left.\dfrac{d\tilde{c}_{Zn(OH)_3^-}}{dy}\right|_{y=0}} \tag{10.162}$$

and

$$Z_{\text{D,OH}^-} = R_{\text{t,Zn}} \frac{3 K_{\text{Zn}} \exp\left(b_{\text{Zn}} \overline{V}\right) \overline{c}_{\text{Zn(OH)}_3^-}(0)}{2F \left(\overline{c}_{\text{OH}^-}(0)\right)^2 D_{\text{OH}^-}} \frac{-\widetilde{c}_{\text{OH}^-}(0)}{\left.\dfrac{\text{d}\widetilde{c}_{\text{OH}^-}}{\text{d}y}\right|_{y=0}} \tag{10.163}$$

Thus, the impedance response is given by

$$Z = R_{\text{e}} + \frac{R_{\text{t,Zn}} + Z_{\text{D,Zn(OH)}_3^-} + Z_{\text{D,(OH)}^-}}{1 + j\omega C_{\text{dl}} \left(R_{\text{t,Zn}} + Z_{\text{D,Zn(OH)}_3^-} + Z_{\text{D,(OH)}^-}\right)} \tag{10.164}$$

The corresponding electrical circuit has the appearance of the Randles circuit shown in Figure 10.5, but has two diffusion elements in series with the charge-transfer resistance.

Example 10.8 **Corrosion of Copper in Chloride Solutions:** *The kinetics and mechanisms of the anodic dissolution of copper in acid and neutral solutions containing chloride ions have been described by many authors.[167–173] A kinetic model that agrees with all published experimental data is given by*

$$\text{Cu} + \text{Cl}^- \underset{K_{-1}}{\overset{K_1}{\rightleftharpoons}} \text{CuCl} + \text{e}^- \tag{10.165}$$

and

$$\text{CuCl} + \text{Cl}^- \underset{k_{\text{b}}}{\overset{k_{\text{f}}}{\rightleftharpoons}} \text{CuCl}_2^- \tag{10.166}$$

where CuCl *is an insoluble adsorbed species. Find the impedance response for this reaction sequence.*

Solution: *The faradaic current may be obtained from*

$$\frac{i_{\text{F,Cu}}}{F} = K_1 c_{\text{Cl}^-}(0)\,(1 - \gamma)\exp(bV) - K_{-1}\Gamma\gamma\exp(-bV) \tag{10.167}$$

where γ *is the fractional surface coverage of* CuCl, Γ *is the maximum surface coverage by* CuCl *in units of moles per unit area, and* $b = 0.5\text{F}/RT$. *Application of a conservation equation for the adsorbed intermediate yields*

$$\begin{aligned}
\Gamma \frac{\text{d}\gamma}{\text{d}t} = \;& K_1 c_{\text{Cl}^-}(0)\,(1 - \gamma)\exp(bV) - K_{-1}\Gamma\gamma\exp(-bV) \\
& - k_{\text{f}}\Gamma\gamma c_{\text{Cl}^-}(0) + k_{\text{b}}\,(1 - \gamma)\,c_{\text{CuCl}_2^-}(0)
\end{aligned} \tag{10.168}$$

where $c_{\text{Cl}^-}(0)$ *and* $c_{\text{CuCl}_2^-}(0)$ *represent concentrations at the electrode surface. Barcia et al.[174] showed by electrohydrodynamical impedance (see chapter 16 on page 443) and by using the graphical method described by Tribollet et al.[175] (see Section 18.5.2 on page 516) that the mass transport corresponds to a species with a Schmidt number of 2,000. As the diffusion of* Cl^-

would lead to a Schmidt number of 500, the mass transfer limitation can be attributed to $CuCl_2^-$. Thus, $c_{Cl^-}(0) = c_{Cl^-}(\infty)$ in equations (10.167) and (10.168).

The steady flux of $CuCl_2^-$ is related to the current density by

$$\frac{\bar{i}_{F,Cu}}{F} = D_{CuCl_2^-}\frac{\bar{c}_{CuCl_2^-}(0)}{\delta_{CuCl_2^-}} \tag{10.169}$$

where $\delta_{CuCl_2^-}$ is the diffusion layer thickness of $CuCl_2^-$. At steady state, $d\gamma/dt = 0$; thus, combination of equations (10.168) and (10.169) yields

$$\frac{\bar{i}_{F,Cu}}{F} = k_f\Gamma\bar{\gamma}\,c_{Cl^-}(\infty) - k_b(1-\gamma)\frac{\bar{i}_F\delta_{CuCl_2^-}}{FD_{CuCl_2^-}} \tag{10.170}$$

In reference 174, the assumption was made that $\gamma \ll 1$. With this assumption, equation (10.167) can be expressed as

$$\frac{\bar{i}_{F,Cu}}{F} = K_1 c_{Cl^-}(\infty)\exp(bV) - K_{-1}\Gamma\bar{\gamma}\exp(-bV) \tag{10.171}$$

and equation (10.170) can be expressed as

$$\frac{\bar{i}_{F,Cu}}{F} = k_f\Gamma\bar{\gamma}\,c_{Cl^-}(\infty) - k_b\frac{\bar{i}_F\delta_{CuCl_2^-}}{FD_{CuCl_2^-}} \tag{10.172}$$

From equation (10.172), the steady-state surface coverage can be expressed as

$$\bar{\gamma} = \frac{\dfrac{\bar{i}_{F,Cu}}{F}\left(1 + k_b\dfrac{\delta_{CuCl_2^-}}{D_{CuCl_2^-}}\right)}{k_f\Gamma c_{Cl^-}(\infty)} \tag{10.173}$$

Insertion of equation (10.173) in equation (10.167) yields

$$\frac{F}{\bar{i}_{F,Cu}} = \frac{1}{K_1\bar{c}_{Cl^-}(\infty)\exp(b\overline{V})} + \frac{K_{-1}\exp(-2b\overline{V})}{K_1k_f\left(\bar{c}_{Cl^-}(\infty)\right)^2\exp(b\overline{V})}\left(1 + \frac{k_b\delta_{CuCl_2^-}}{D_{CuCl_2^-}}\right) \tag{10.174}$$

By elimination of the faradaic current in equation (10.173), the steady-state surface concentration may be expressed as

$$\Gamma\bar{\gamma} = \frac{K_1\bar{c}_{Cl^-}(\infty)\exp(b\overline{V})\left(1 + \dfrac{k_b\delta_{CuCl_2^-}}{D_{CuCl_2^-}}\right)}{k_f\bar{c}_{Cl^-}(\infty) + K_{-1}\exp(-b\overline{V})\left(1 + \dfrac{k_b\delta_{CuCl_2^-}}{D_{CuCl_2^-}}\right)} \tag{10.175}$$

Equations (10.174) and (10.175) represent the steady-state behavior of the system.

Under a sinusoidal perturbation and under the assumptions used to generate equations (10.174) and (10.175), equations (10.167) and (10.170) may be expressed as

$$\tilde{i}_{F,Cu} = \frac{\tilde{V}}{R_t} - FK_{-1}\Gamma\exp(-b\overline{V})\tilde{\gamma} \tag{10.176}$$

where

$$\frac{1}{R_t} = F\left(K_1\bar{c}_{Cl^-}(\infty)b\exp\left(b\overline{V}\right) + K_{-1}\Gamma\bar{\gamma}b\exp\left(-b\overline{V}\right)\right) \tag{10.177}$$

Equation (10.168) can be expressed as

$$j\omega\Gamma\tilde{\gamma} = \frac{\tilde{V}}{FR_t} - \left(K_{-1}\exp(-b\overline{V}) + k_f c_{Cl^-}(\infty)\right)\Gamma\tilde{\gamma} + k_b\tilde{c}_{CuCl_2^-}(0) \tag{10.178}$$

Equations (10.176) and (10.178) are written in terms of four oscillating variables: $\tilde{V}$, $\tilde{i}$, $\tilde{\gamma}$, and $\tilde{c}_{CuCl_2^-}(0)$. An additional equation is needed to allow calculation of transfer functions. Equation (10.169) may be expressed as

$$D_{CuCl_2^-}\frac{d\tilde{c}_{CuCl_2^-}}{dy}\Bigg|_{y=0} = k_f c_{Cl^-}(\infty)\Gamma\tilde{\gamma} - k_b\tilde{c}_{CuCl_2^-}(0) \tag{10.179}$$

Equation (10.179) is written in terms of reaction (10.166) because, under non-steady-state conditions, the rates of reactions (10.165) and (10.166) are not equal.

Following the development in Chapter 11 (page 243),

$$\frac{\tilde{c}_{CuCl_2^-}(0)}{\dfrac{d\tilde{c}_{CuCl_2^-}}{dy}\Bigg|_{y=0}} = \delta_{CuCl_2^-}\left(\frac{-1}{\theta'_{CuCl_2^-}(0)}\right) \tag{10.180}$$

Equations (10.176), (10.178), (10.179), and (10.180) constitute a set of four equations with four unknowns, e.g., $\tilde{i}/\tilde{V}$, $\tilde{\gamma}/\tilde{V}$, $\tilde{c}_{CuCl_2^-}(0)/\tilde{V}$, and $\left(d\tilde{c}_{CuCl_2^-}/dy\right)\Big|_{y=0}/\tilde{V}$.

The solution for the faradaic impedance is given by

$$Z_{F,Cu} = R_t + \cfrac{R_t K_{-1}\exp(-b\overline{V})}{j\omega + k_f c_{Cl^-}(\infty)\left(\cfrac{\dfrac{D_{CuCl_2^-}}{\delta_{CuCl_2^-}}\left(\dfrac{-1}{\theta'_{CuCl_2^-}(0)}\right)}{k_b + \dfrac{D_{CuCl_2^-}}{\delta_{CuCl_2^-}}\left(\dfrac{-1}{\theta'_{CuCl_2^-}(0)}\right)}\right)} \tag{10.181}$$

For a rotating disk electrode, the diffusion layer thickness can be expressed in the form of equation (11.72) on page 262 to be

$$\delta_{CuCl_2^-} = \left(\frac{3}{a}\right)^{1/3}\frac{1}{Sc_{CuCl_2^-}^{1/3}}\sqrt{\frac{\nu}{\Omega}} \tag{10.182}$$

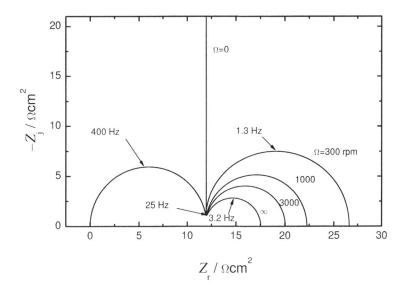

Figure 10.14: Simulation of the impedance represented by equation (10.181) with disk rotation rate Ω in units of rpm as a parameter. The parameters were $R_t = 12 \ \Omega cm^2$, $C_{dl} = 30 \ \mu F/cm^2$, $k_f c_{Cl^-}(\infty) = 1 \ s^{-1}$, $k_b = 0.007 \ cm/s$, $D/\delta = 0.0006\Omega^{1/2} \ cm/s$ with Ω in units of s^{-1}, and $R_t K_{-1} \exp(-b\overline{V}) = 5 \ \Omega/s$.

where Ω is the rotation speed. For $\Omega \to \infty$, $\delta_{CuCl_2^-} \to 0$ and

$$\lim_{\Omega \to \infty} Z_{F,Cu} = R_t + \frac{R_t K_{-1} \exp(-bV)}{j\omega + k_f c_{Cl^-}(\infty)} \tag{10.183}$$

For $\Omega \to 0$, $\delta_{CuCl_2^-} \to \infty$ and

$$\lim_{\Omega \to 0} Z_{F,Cu} = R_t + \frac{R_t K_{-1} \exp(-bV)}{j\omega} \tag{10.184}$$

Simulations corresponding to equation (10.181) are given in Figure 10.14 where Z_F was placed in parallel with a double-layer capacitance $C_{dl} = 30 \ \mu F/cm^2$. Other parameters are presented in the caption for Figure 10.14. The impedance diagram is composed of two half-circles, the first semicircle corresponds to the charge-transfer resistance in parallel to the double-layer capacitance and the second semicircle depends on mass transport. While this impedance diagram can be described by an equivalent circuit in terms of resistors and capacitors, these parameters cannot have a physical meaning due to fact that the diffusion impedance appears in a complicated way in the formula. This simulation is in good agreement with the experimental data presented in Figure 3 of Barcia et al.[174]

The examples presented in this chapter demonstrate the importance of developing models based on hypothesized reaction mechanisms. As shown in Section 10.4, the same mechanism may yield low-frequency inductive or capacitive loops, depending on the relative values of parameters. As shown in Example 10.7, data described by a simple Randles circuit may conceal the effects of two diffusion impedances. As shown in Example 10.8, an apparent capacitive loop may depend on mass transfer, as shown by its dependence of the rotation rate of a disk electrode. While models may be described usefully in terms of an equivalent electrical circuit, the origin of the model should lie in proposed kinetic and physical phenomena.

Problems

10.1   Develop an expression for the faradaic impedance for the reaction

$$Ag \rightarrow Ag^+ + e^- \tag{10.185}$$

10.2   Develop an expression for the faradaic impedance for the reaction

$$Ag^+ + e^- \rightarrow Ag \tag{10.186}$$

in which the concentration of Ag^+ is influenced by mass transfer. Use a spreadsheet program to plot the current density as a function of potential. Use the same estimated parameters to plot the impedance response at a potential corresponding to $1/4$, $1/2$, and $3/4$ of the mass-transfer-limited current density.

10.3   Should the Stern–Geary relation expressed as equation (10.110) be used to assess the corrosion rate of magnesium described in Example 10.6? If not, how should the corrosion rate be estimated?

10.4   Consider the reaction sequence

$$M + A \rightarrow MA^+_{ads} + e^- \tag{10.187}$$

where MA^+_{ads} is an adsorbed intermediate that reacts according to

$$MA^+_{ads} + A \rightarrow MA^{2+}_2 + e^- \tag{10.188}$$

Derive the faradaic impedance taking into account the mass-transfer limitation to the species A.

10.5   Derive the faradaic impedance for the anodic dissolution of copper at low overpotential in a chloride medium where the reactions proceed according to

$$Cu + Cl^- \rightleftarrows CuCl_{ads} + e^- \tag{10.189}$$

where $CuCl_{ads}$ is an adsorbed intermediate that reacts with chloride ions to form $CuCl_2^-$, i.e.,

$$CuCl_{ads} + Cl^- \rightleftarrows CuCl_2^- \tag{10.190}$$

The mass-transfer limitation is due only to $CuCl_2^-$.

10.6 Re-derive the faradaic impedance developed in Example 10.6 taking into account the NDE reaction (10.151).

10.7 Perform the calculations needed to generate Figure 10.10.

10.8 Explain why K_M^*, defined in equation (10.14), has a value that is independent of the reference electrode.

10.9 If $V_{ref1} - V_{ref2} = 0.4$ V, give the expression of $K_{M,ref2}$ with respect to $K_{M,ref1}$ where $K_{M,refi}$ is the value corresponding to an experiment performed with reference electrode ref$i$. Show that the value of the charge-transfer resistance R_t is independent of the reference electrode used.

10.10 The oxidation of peroxide on a platinum disk electrode may be expressed as

$$H_2O_2 \rightarrow O_2 + 2H^+ + 2e^- \tag{10.191}$$

(a) Develop an expression for the oscillating faradaic current $\tilde{i}_F$ in terms of variables $\tilde{V}$ and $\tilde{c}_{H_2O_2}(0)$.

(b) Develop an expression for the faradaic impedance in terms of the charge-transfer resistance and the diffusion impedance, providing clear relationships for charge-transfer resistance and diffusion impedance in terms of kinetic parameters.

(c) Provide a relationship between charge-transfer resistance, Tafel slope, and peroxide oxidation current.

10.11 Consider the cathodic reduction of oxygen on a Pt catalyst, e.g.,

$$O_2 + 2H^+ + 4e^- \rightarrow 2H_2O \tag{10.192}$$

Consider also that the Pt catalyst can be oxidized by the reversible reaction

$$Pt + H_2O \rightleftharpoons PtO + 2H_2 + 2e^- \tag{10.193}$$

(a) Can reaction (10.192) yield a low-frequency inductive loop in the impedance response? You need not derive the associated faradaic impedance, but you should explain your reasoning.

(b) Find the steady-state fractional coverage by PtO, $\overline{\gamma}_{PtO}$.

(c) Develop an expression for the faradaic impedance response associated with reactions (10.192) and (10.193). You may assume that reaction (10.192) proceeds rapidly on the Pt catalyst but does not proceed on the PtO, Thus, the effective rate constant for reaction (10.192) can be expressed as $K_{eff} = K_{O_2}(1 - \gamma_{PtO})$. In addition, mass-transfer limitations to hydrogen ions and water may be neglected, but the influence of mass transfer on the oxygen concentration may not be ignored.

Chapter 11

Diffusion Impedance

The development of kinetic models presented in Sections 10.1 on page 207, 10.2.2 on page 213, and 10.4.2 on page 231 require expressions for the concentrations of reacting species at the electrode surface. The development was expressed in terms of an inverse dimensionless concentration gradient given as $-1/\theta_i'(0)$. The objective of this chapter is to explore conditions and systems for which expressions for $-1/\theta_i'(0)$ can be developed.

Experimental systems used for electrochemical measurements should be selected to take maximum advantage of well-understood phenomena such as mass transfer so as to focus attention on the less-understood phenomena such as electrode kinetics. For example, the study of electrochemical reactions in stagnant environments should be avoided because concentration and temperature gradients give rise to natural convection, which has an effect on mass transfer that is difficult to characterize. It is better to engage in such experimental investigations in systems for which mass transfer is well defined. To simplify interpretation of the impedance data, the electrode should be uniformly accessible to mass transfer.

Some issues pertaining to mass transfer to electrodes are described in Section 5.5 on page 113, and the associated issues for cell design are considered further in Section 8.1.2 on page 167. In many cases, a uniformly accessible electrode cannot be used. The time-constant dispersion that can arise as a result of nonuniform mass transfer is discussed in Section 13.5 on page 370.

Remember! 11.1 *Experimental systems used for electrochemical measurements should be selected to take maximum advantage of well-understood phenomena such as mass transfer so as to focus attention on the less-understood phenomena such as electrode kinetics.*

11.1 Uniformly Accessible Electrode

A uniformly accessible electrode is an electrode where, at the interface, the flux and the concentration of a species produced or consumed on the electrode are independent of the coordinates that define the electrode surface. The mass flux at the interface is obtained by solving the material balance equation. If migration can be neglected, the material balance equation for dilute electrolytic solutions is reduced to the convective-diffusion equation. For an axisymmetric electrode, the concentration derivatives with respect to the angular coordinate are equal to zero, and the convective-diffusion equation can be expressed in cylindrical coordinates as

$$\frac{\partial c_i}{\partial t} + v_r\frac{\partial c_i}{\partial r} + v_y\frac{\partial c_i}{\partial y} = D_i\left\{\frac{1}{r}\frac{\partial}{\partial r}\left(r\frac{\partial c_i}{\partial r}\right) + \frac{\partial^2 c_i}{\partial y^2}\right\} \tag{11.1}$$

with the boundary conditions:

$$c_i \to c_i(\infty) \quad \text{for} \quad y \to \infty \tag{11.2}$$

$$f\left(c_i(0), \frac{\partial c_i}{\partial y}\bigg|_{y=0}\right) = 0 \quad \text{for} \quad y = 0 \tag{11.3}$$

where c_i is the concentration of the species i and D_i is the corresponding diffusion coefficient. The condition expressed as equation (11.3) is imposed, not only by the electrochemical reaction taking place at the electrode surface, but also by the type of regulation imposed, e.g., galvanostatic or potentiostatic. This condition corresponds generally to a fixed concentration or a fixed concentration gradient at the electrode surface.

If a flow exists with axial velocity v_y independent of the radial coordinate and if the boundary condition at $y = 0$ is also independent of the radial coordinate, then the concentration is only a function of y and the convective-diffusion equation is reduced to

$$\frac{\partial c_i}{\partial t} + v_y\frac{\partial c_i}{\partial y} - D_i\frac{\partial^2 c_i}{\partial y^2} = 0 \tag{11.4}$$

Equation (11.4) represents a uniformly accessible electrode because the concentration is a function only of time t and the axial position variable y. The boundary condition for $y = 0$ (see equation (11.2)) must be also uniform on the electrode and independent of the radial coordinate.

The simplest uniformly accessible electrode is a planar electrode under conditions where the convection can be neglected. When the convection cannot be neglected, it

---

Remember! 11.2 *The diffusion impedance is a part of the interfacial impedance, but its value depends on the concentration profiles throughout the domain.*

is necessary to impose a flow that yields, with respect to the electrode surface, a uniform normal velocity component. Since the work of Levich,[176] the rotating disk system has been well known to provide a uniformly accessible electrode. Electrodes placed in some other flow geometries, such as within the stagnation region of a submerged impinging jet cell or as a rotating cylinder, can also be uniformly accessible. Electrodes placed in the above flow geometries can have more complex electrode/solution interfaces and yet can still be considered to be uniformly accessible. Examples include an electrode coated by a porous layer or an electrode in the presence of a viscosity gradient. For a uniformly accessible electrode the problem is reduced to one dimension: the distance to the interface.

11.2 Porous Film

The diffusion impedance of a coated (or film-covered) electrode under the condition where the diffusion process in the bulk electrolyte is negligible is developed here for diffusion through a solid film. In first case considered, the electroactive species is exchanged between the film and the bulk electrolyte. In the second case, and diffusion inside the film occurs without exchange of electroactive species with the bulk solution.

11.2.1 Diffusion with Exchange of Electroactive Species

In a porous layer, mass transfer occurs only by diffusion and migration, and convection may be neglected. With an excess of supporting electrolyte, the migration can be neglected and the mass conservation equation is reduced to

$$\frac{\partial c_i}{\partial t} - D_i \frac{\partial^2 c_i}{\partial y^2} = 0 \tag{11.5}$$

which is known as Fick's second law. For steady-state diffusion through a solid film, the boundary conditions may be given as

$$\bar{c}_i = \bar{c}_i(0) \quad \text{at} \quad y = 0 \tag{11.6}$$

$$\bar{c}_i = \bar{c}_i(\infty) \quad \text{at} \quad y = \delta_f \tag{11.7}$$

A schematic representation of diffusion through a film with exchange of electroactive species between the film and the electrolyte is shown in Figure 11.1.

Under steady-state conditions, $\partial c_i / \partial t = 0$, yielding

$$D_i \frac{d^2 \bar{c}_i}{dy^2} = 0 \tag{11.8}$$

Successive integrations, using boundary conditions (11.6) and (11.7), yields the concentration gradient as

$$\frac{d\bar{c}_i}{dy} = \frac{\bar{c}_i(\infty) - \bar{c}_i(0)}{\delta_f} \tag{11.9}$$

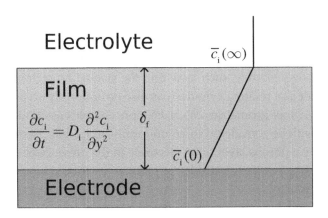

Figure 11.1: Schematic representation of diffusion through a film with exchange of electroactive species between the film and the electrolyte. Under steady-state conditions, the concentration follows equation 11.10.

and the concentration as

$$\bar{c}_i = \bar{c}_i(0) + (\bar{c}_i(\infty) - \bar{c}_i(0)) \frac{y}{\delta_f} \tag{11.10}$$

Equation (11.10) provides the steady-state concentration distribution expected for diffusion with exchange of electroactive species through a film of thickness δ_f. The boundary condition given as equation (11.40) applies when there is no exchange of the electroactive species with the surrounding electrolyte.

Under a sinusoidal perturbation, the concentration can be expressed in terms of a steady or time-independent value and an oscillating value (see Example 1.9 on page 20 or equation (10.3) on page 207) as

$$c_i = \bar{c}_i + \mathrm{Re}\left\{ \tilde{c}_i e^{j\omega t} \right\} \tag{11.11}$$

Substitution of equation (11.11) into equation (11.5) yields

$$\mathrm{Re}\left\{ j\omega\tilde{c}_i e^{j\omega t} \right\} - D_i \frac{\mathrm{d}^2\bar{c}_i}{\mathrm{d}y^2} - D_i\mathrm{Re}\left\{ \frac{\mathrm{d}^2\tilde{c}_i}{\mathrm{d}y^2} e^{j\omega t} \right\} = 0 \tag{11.12}$$

As the steady-state solution is represented by equation (11.8),

$$\mathrm{Re}\left\{ j\omega\tilde{c}_i e^{j\omega t} - D_i \frac{\mathrm{d}^2\tilde{c}_i}{\mathrm{d}y^2} e^{j\omega t} \right\} = 0 \tag{11.13}$$

The exponential term $e^{j\omega t}$ in equation (11.13) can be canceled, thus eliminating the explicit dependence on time. Equation (11.13) can be satisfied only if

$$j\omega\tilde{c}_i - D_i \frac{\mathrm{d}^2\tilde{c}_i}{\mathrm{d}y^2} = 0 \tag{11.14}$$

Equation (11.14) can be written in terms of the dimensionless oscillating concentration $\theta_i(y) = \tilde{c}_i/\tilde{c}_i(0)$ and dimensionless position $\xi = y/\delta_f$ as

$$\frac{d^2\theta_i}{d\xi^2} - jK_i\theta_i = 0 \tag{11.15}$$

where

$$K_i = \frac{\omega\delta_f^2}{D_i} \tag{11.16}$$

represents a dimensionless frequency for species i.

The general solution to equation (11.15) is given in Example 2.3 on page 30 to be

$$\theta_i = C_1 \exp\left(\xi\sqrt{jK_i}\right) - C_2 \exp\left(-\xi\sqrt{jK_i}\right) \tag{11.17}$$

The appropriate transmissive boundary conditions for a diffusion layer of finite thickness are

$$\theta_i = 0 \quad \text{at} \quad \xi = 1$$
$$\theta_i = 1 \quad \text{at} \quad \xi = 0 \tag{11.18}$$

Because θ_i is a complex number, the boundary conditions expressed as equation (11.18) are that $\text{Re}\{\theta_i\} = 0$ and $\text{Im}\{\theta_i\} = 0$ at $\xi = 1$ and that $\text{Re}\{\theta_i\} = 1$ and $\text{Im}\{\theta_i\} = 0$ at $\xi = 0$. Evaluation of the integration constants C_1 and C_2 yields

$$\theta_i = \frac{\sinh\left((\xi - 1)\sqrt{jK_i}\right)}{\sinh\left(-\sqrt{jK_i}\right)} \quad \text{at} \quad \xi < 1$$
$$\theta_i = 0 \quad \text{at} \quad \xi \geq 1 \tag{11.19}$$

The dimensionless diffusion impedance identified in Section 10.2.2 on page 213 is given by

$$\frac{-1}{\theta_i'(0)} = \frac{\tanh\left(\sqrt{jK_i}\right)}{\sqrt{jK_i}} \tag{11.20}$$

Equation (11.20) is the diffusion impedance for a film of finite thickness δ_f.

The diffusion impedance is presented in Figure 11.2 in Nyquist format with dimensionless frequency $K_i = \omega\delta_f^2/D_i$ as a parameter. The solid line is the diffusion impedance for a film of finite thickness δ_f. The peak in the negative imaginary part of the diffusion impedance is seen at a dimensionless frequency $K_i = 2.5$. When the frequency tends toward zero, $\tanh\left(\sqrt{jK_i}\right)$ tends toward $\sqrt{jK_i}$ and $-1/\theta_i'(0)$ tends toward unity. When the frequency tends toward infinity, $\tanh\left(\sqrt{jK_i}\right)$ tends toward unity and $-1/\theta_i'(0)$ tends toward $1/\sqrt{jK_i}$, which is the expression of the Warburg impedance, developed in Example 11.1.

The tendency of the diffusion impedance to approach the Warburg impedance is seen more clearly in Figure 11.3, where the magnitudes of the real and imaginary parts of the diffusion impedance are presented as a function of dimensionless frequency. The

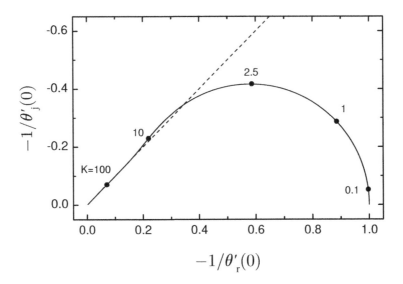

Figure 11.2: Diffusion impedance in Nyquist format with dimensionless frequency $K_i = \omega \delta_f^2 / D_i$ as a parameter. The solid line is the diffusion impedance for a film of finite thickness δ_f, equation (11.20) and the dashed line is the Warburg impedance for diffusion in an infinite domain, given as equation (11.28).

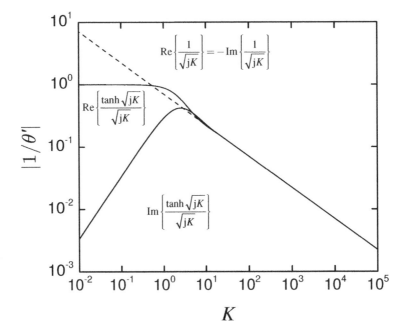

Figure 11.3: Real and imaginary parts of the diffusion impedance as a function of dimensionless frequency $K_i = \omega \delta_f^2 / D_i$. The solid line is the diffusion impedance for a film of finite thickness δ_f, equation (11.20) and the dashed line is the Warburg impedance for diffusion in an infinite domain, equation (11.28).

magnitudes of the real and imaginary parts of the diffusion impedance approach the corresponding values for the Warburg impedance for frequencies $K_i > 10$.

The diffusion impedance response for an electrochemical system influenced by mass transfer can be obtained by inserting the dimensionless quantity given as equation (11.20) into equation (10.61), shown on page 217. The general approach presented here has the benefit that models for mass transfer impedance may be employed without affecting the kinetic development presented in Chapter 10.

Example 11.1 **Warburg Impedance for Stagnant Diffusion Layer:** *Find the diffusion impedance in the absence of imposed convection and under conditions such that natural convection may be ignored.*

Solution: *In stagnant environments, the boundary conditions for equation (11.5) may be given as*

$$
\begin{aligned}
c_i &\to c_i(\infty) \quad \text{for} \quad y \to \infty \\
c_i &= c_i(0) \quad \text{at} \quad y = 0 \\
c_i &= c_i(\infty) \quad \text{at} \quad t = 0
\end{aligned}
\tag{11.21}
$$

No steady-state solution is possible. As shown in Example 2.9 on page 39, the concentration is given in terms of a similarity variable as[177]

$$
\Theta_i = \frac{c_i(y) - c_i(0)}{c_i(\infty) - c_i(0)} = 1 - \text{erf}\left(\frac{y}{\sqrt{4D_i t}}\right)
\tag{11.22}
$$

In principle, impedance measurements are possible only for stationary systems, i.e., those systems for which a steady solution is possible. However, after a sufficient time, the concentration profile near the electrode can be considered to be stationary with respect to the time required for the impedance measurement.

For diffusion in an infinite domain, there is no characteristic length such as was identified for a film of thickness δ_f. Equation (11.14) can be expressed in terms of the dimensionless oscillating concentration $\theta_i(y) = \tilde{c}_i/\tilde{c}_i(0)$ as

$$
j\frac{\omega}{D_i}\theta_i - \frac{d^2\theta_i}{dy^2} = 0
\tag{11.23}
$$

Following Example 2.3 (page 30), the general solution to equation (11.23) is given by

$$
\theta_i = C_1 \exp\left(y\sqrt{j\omega/D_i}\right) - C_2 \exp\left(-y\sqrt{j\omega/D_i}\right)
\tag{11.24}
$$

which can be solved subject to the boundary conditions

$$
\begin{aligned}
\theta_i &\to 0 \quad \text{as} \quad y \to \infty \\
\theta_i &= 1 \quad \text{at} \quad y = 0
\end{aligned}
\tag{11.25}
$$

Thus, $C_1 = 0$ and

$$\theta_i = e^{-y\sqrt{j\omega/D_i}} \tag{11.26}$$

The derivative with respect to position y is given by

$$\left.\frac{d\theta_i}{dy}\right|_{y=0} = -\sqrt{j\omega/D_i} \tag{11.27}$$

and, following equation (10.12) on page 209 and the definition of θ_i,

$$\frac{-\tilde{c}_i(0)}{\left.\frac{d\tilde{c}_i}{dy}\right|_{y=0}} = \frac{1}{\sqrt{j\omega/D_i}} \tag{11.28}$$

Equation (11.28) is known as the Warburg impedance.[28]

Example 11.2 Insensitivity of Warburg Impedance to Film Thickness: *At high frequency, the film impedance given in equation (11.20) approaches the Warburg impedance, i.e.,*

$$\lim_{\omega\to\infty} \frac{-1}{\theta_i'(0)} = \frac{1}{\sqrt{jK_i}} = \frac{1}{\sqrt{j\omega\delta_f^2/D_i}} \tag{11.29}$$

which appears to be influenced by film thickness δ_f; whereas, the development in Example 11.1 shows that the Warburg impedance is independent of film thickness. Reconcile this apparent discrepancy.

Solution: *Equation (11.29) can be expressed as*

$$\frac{-1}{\theta_i'(0)} = \left.\frac{d\xi}{d\theta_i}\right|_{\xi=0} = \left.\frac{dy}{d\theta_i}\right|_{y=0}\frac{1}{\delta_f} \tag{11.30}$$

Thus,

$$\left.\frac{dy}{d\theta_i}\right|_{y=0}\frac{1}{\delta_f} = \frac{1}{\sqrt{j\omega\delta_f^2/D_i}} \tag{11.31}$$

or

$$\left.\frac{dy}{d\theta_i}\right|_{y=0} = \frac{1}{\sqrt{j\omega/D_i}} \tag{11.32}$$

Equation (11.32) is the same as equation (11.27). Another way to show the lack of influence of film thickness on the Warburg impedance that appears as the high-frequency limit of equation (11.20) is to insert equation (11.29) into the definition of diffusion impedance given as equation (10.60) on page 217 and to note the cancelation of the δ terms.

Example 11.3 Propagation of Concentration Fluctuations into Film: *Exploration of time-dependent concentration profiles as a function of frequency provides another way*

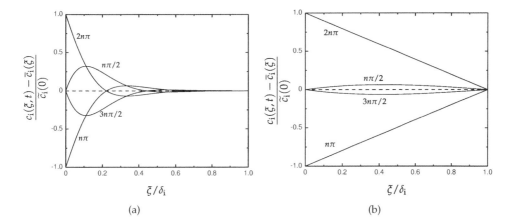

Figure 11.4: Oscillating concentration as a function of position with time as a parameter for a finite stagnant diffusion layer: a) $K_i = 100$ and b) $K_i = 1$.

to appreciate the independence of the Warburg impedance on film thickness and the observation that the diffusion impedance is a complex number that contains both magnitude and phase information. Calculate the concentration profiles corresponding to high and low-frequency behavior.

Solution: *The concentration is given by equation (11.11). The oscillating contribution to the concentration, given by* $\text{Re}\{\widetilde{c}_i e^{j\omega t}\}$, *is presented in Figures 11.4(a) and (b) for dimensionless frequencies of 100 and 1, respectively, with dimensionless time as a parameter. At the higher frequency, the concentration perturbation does not extend to the limits of the diffusion layer. At the lower frequency, the perturbation extends to the limit of the diffusion layer, and an abrupt change is seen at $\xi = 1$, the limit of the stagnant film.*

The concentration profile expected for a system at one-half of the mass-transfer-limited current and for a concentration perturbation of 20 percent at the interface is presented in Figure 11.5 with dimensionless time as a parameter. At the higher frequency, the propagation of the disturbance away from the electrode surface lags behind the perturbation at the surface. At lower frequencies, the concentration far from the surface responds with almost no phase lag. In the limit that $K_i \to 0$, the phase lag tends toward zero and, correspondingly, the imaginary part of the impedance tends toward zero.

The concentration perturbation for $K_i = 100$ is given in Figure 11.6 as a function of time with dimensionless position as a parameter. The extent to which the concentration at any position lags behind the concentration at the surface is a function of position. The variation of phase lag with position is consistent with the propagation of a wave through a dissipating medium.

11.2.2 Diffusion without Exchange of Electroactive Species

Diffusion in films without exchange of electroactive species with the external electrolyte can appear in a variety of electrochemical systems, including some batteries.

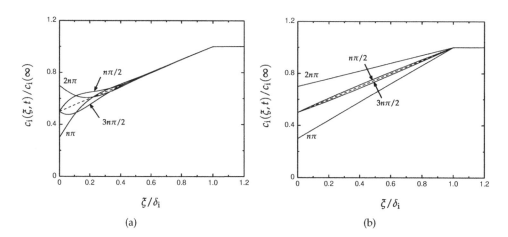

Figure 11.5: Concentration as a function of position with time as a parameter for a finite stagnant diffusion layer: a) a) $K_i = 100$ and b) b) $K_i = 1$.

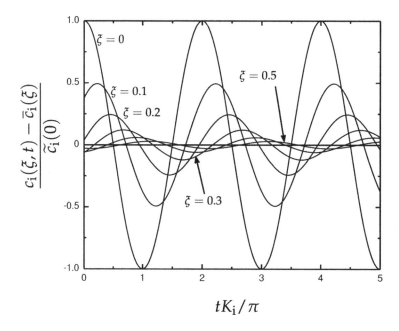

Figure 11.6: Oscillating concentration for $K_i = 100$ as a function of time with position as a parameter for a finite stagnant diffusion layer.

Electrolyte

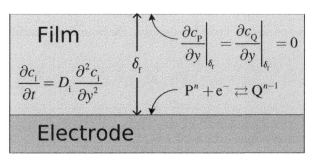

Figure 11.7: Schematic representation of diffusion through a film without exchange of electroactive species between the film and the electrolyte.

This mass transport problem was solved by Ho et al.[178] for lithium batteries and by Gabrielli et al.[179] for an electrode coated by a redox polymer film. Following Gabrielli et al.,[179] a redox reaction is considered at the electrode, e.g.,

$$P + e^- \rightleftarrows Q \tag{11.33}$$

In the absence of convection and migration, the concentrations of species P and Q are governed by equation (11.5). The redox species are linked to the film and must stay inside the film; therefore,

$$\left.\frac{\partial c_P}{\partial y}\right|_{y=\delta_f} = \left.\frac{\partial c_Q}{\partial y}\right|_{y=\delta_f} = 0 \tag{11.34}$$

This situation is presented schematically in Figure 11.7.

Under steady-state conditions, equation (11.5) becomes equation (11.8); thus, concentrations $\bar{c}_P$ and $\bar{c}_Q$ are uniform and do not depend on the position variable y. Reaction (11.33) leads to an expression for the faradaic current as

$$i_F = k_f c_P - k_b c_Q \tag{11.35}$$

where the rate constants k_f and k_b are functions of potential as shown in Chapter 10 on page 207.

The total concentration of the redox couple is

$$c_P + c_Q = c^* \tag{11.36}$$

Since the steady-state concentrations $\bar{c}_P$ and $\bar{c}_Q$ are constant, the steady-state faradaic current density given as equation (11.35) is equal to zero. Equations (11.35) and (11.36) yield

$$\bar{c}_P = \frac{k_b c^*}{k_f + k_b} \tag{11.37}$$

and

$$\bar{c}_Q = \frac{k_f c^*}{k_f + k_b} \tag{11.38}$$

As the electroactive species are linked to the film, the boundary conditions for equation (11.5) become

$$c_i = c_i(0) \quad \text{at} \quad y = 0 \tag{11.39}$$

$$\frac{dc_i}{dy} = 0 \quad \text{at} \quad y = \delta_f \tag{11.40}$$

where the index i corresponds to either P or Q.

The general solution to equation (11.15) is still given by equation (11.17), but the reflective boundary conditions are given by

$$\frac{d\theta_i}{d\xi} = 0 \quad \text{at} \quad \xi = 1$$

$$\theta_i = 1 \quad \text{at} \quad \xi = 0 \tag{11.41}$$

The resulting dimensionless diffusion impedance, given by

$$\frac{-1}{\theta_i'(0)} = \frac{\coth\left(\sqrt{jK_i}\right)}{\sqrt{jK_i}} \tag{11.42}$$

is compared in Figure 11.8 to the Warburg impedance and to the diffusion impedance for a film in which exchange of electroactive species with the electrolyte occurs.

When the frequency tends toward infinity, $\coth\left(\sqrt{jK_i}\right)$ tends toward unity and, as shown in Figure 11.8, $-1/\theta_i'(0)$ tends toward $1/\sqrt{jK_i}$, which is the expression of the Warburg impedance (see Example 11.1). When the frequency tends toward zero, $\coth\left(\sqrt{jK_i}\right)$ tends toward $(1/\sqrt{jK_i} + \sqrt{jK_i}/3)$. The term $-1/\theta_i'(0)$ tends toward $1/3 + 1/jK_i$, which is the behavior of a capacitor in series with a resistor of value $1/3$. These limiting behaviors are shown in Figure 11.8.

At dimensionless frequencies $K > 10$, all three impedance models overlap. Thus, if the film thickness is very large, equations (11.20) and (11.42) may correspond in the entire measurable experimental frequency range to the Warburg impedance. In such a case, the nature of the problem under investigation should determine which expression is introduced into equation (10.61), given on page 217, to obtain a complete solution for the impedance.

Example 11.4 Diffusion or Capacitance: *What value of capacitance would correspond to the low-frequency behavior shown for diffusion through a film in the absence of exchange of electroactive species?*

Solution: *The low-frequency response shown in Figure 11.8 yields*

$$-\frac{1}{\theta'(0)} = \frac{1}{3} - j\frac{1}{K_i} \tag{11.43}$$

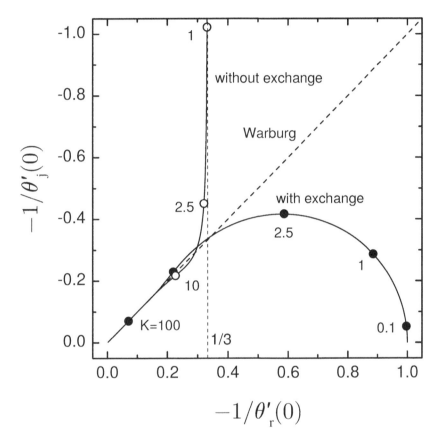

Figure 11.8: Diffusion impedance in Nyquist format with dimensionless frequency $K_i = \omega \delta_f^2 / D_i$ as a parameter. The solid line is the diffusion impedance for a film of finite thickness δ_f, either equation (11.20) for exchange of electroactive species with the electrolyte or equation (11.42) in the absence of exchange of electroactive species with the electrolyte. The dashed line is the Warburg impedance for diffusion in an infinite domain given as equation (11.28).

or

$$-3\frac{1}{\theta'(0)} = 1 - j\frac{3D_i}{\omega\delta_i^2} \tag{11.44}$$

This could be compared to a corresponding dimensionless expression for the impedance of a resistor of value R in series with a capacitor of value C, i.e.,

$$\frac{Z}{R} = 1 - j\frac{1}{\omega RC} \tag{11.45}$$

A relationship for capacitance can be obtained upon equating equations (11.44) and (11.45), i.e.,

$$C = \frac{\delta_i^2}{3D_i R} \tag{11.46}$$

where R represents the high-frequency asymptote for the low-frequency capacitive portion of the impedance diagram.

11.3   Rotating Disk

A schematic illustration of the flow field generated by a rotating disk is presented in Figure 11.9. The rotation of the disk causes a spiral movement of the fluid, seen in Figure 11.9(a), which results in a net velocity toward the disk and in the radial direction. A projection of the trajectories onto a plane at a fixed axial position, shown in Figure 11.9(b), shows that the dominant velocity component is in the θ-direction. The corresponding radial and axial velocities are much smaller. The rotating disk, therefore, can be regarded to be an inefficient pump that draws fluid toward the disk surface and ejects it in the radial direction. The popularity of the rotating disk electrode in the use of frequency-domain techniques has motivated development of sophisticated models for interpretation of experimental data.

11.3.1   Fluid Flow

The steady flow created by an infinite disk rotating at a constant angular velocity in a fluid with constant physical properties was first studied by von Kármán.[180] The solution was sought by using a separation of variables using a dimensionless distance

$$\zeta = y\sqrt{\Omega/\nu} \tag{11.47}$$

Remember! **11.3** *The Warburg impedance, equation (11.28), applies for diffusion in an infinite stagnant domain. This expression applies as a high-frequency limit for diffusion in a finite domain.*

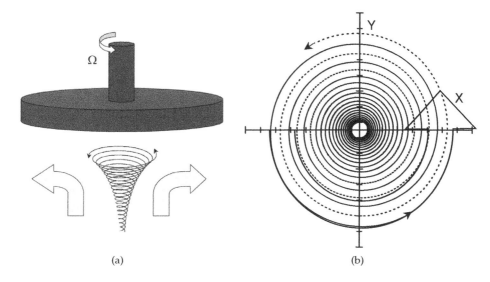

Figure 11.9: Flow patterns associated with a rotating disk: a) a three-dimensional representation of flow trajectories that lead to a net flow toward the disk and in the radial direction with an expanded axial scale to allow visualization of the flow trajectories and b) a projection of the flow trajectories onto a plane at a fixed axial position.

and dimensionless radial velocity

$$v_r = r\Omega F(\zeta) \tag{11.48}$$

angular velocity

$$v_\theta = r\Omega G(\zeta) \tag{11.49}$$

and axial velocity

$$v_y = \sqrt{\nu\Omega} H(\zeta) \tag{11.50}$$

where ν is the kinematic viscosity and Ω the rotation speed.

Upon introduction of equations (11.48), (11.49), and (11.50), the equation of continuity and the Navier-Stokes equations can be solved numerically.[115,180] As shown by Cochran,[181] the variables F, G, and H can be written as two sets of series expansions for small and large values of ζ, respectively. The series solutions for small values of ζ are especially relevant to the mass-transfer problem. In particular, the derivatives at $\zeta = 0$ are essential in order to determine the first coefficient of the series expansions. The other coefficients are deduced from the first one by using the equation of continuity and the Navier-Stokes equations,

$$H = -a\zeta^2 + \frac{\zeta^3}{3} + \frac{b}{6}\zeta^4 + \ldots \tag{11.51}$$

$$F = a\zeta - \frac{\zeta^2}{2} - \frac{b}{3}\zeta^3 + \ldots \tag{11.52}$$

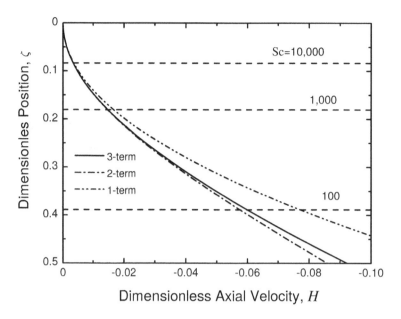

Figure 11.10: Distribution of axial velocity for a region near a rotating disk. The characteristic length for mass transfer is indicated by dashed lines for Schmidt numbers of 10,000, 1,000, and 100, respectively.

and

$$G = 1 + b\zeta + \frac{a}{3}\zeta^3 + \dots \tag{11.53}$$

where a $=$ 0.510232618867 and b $=$ $-$0.615922014399.[182] The axial velocity can be expressed in terms of equations (11.50) and (11.51) as

$$v_y = -\sqrt{\nu\Omega}\left(a\zeta^2 - \frac{\zeta^3}{3} - \frac{b}{6}\zeta^4 + \dots\right) \tag{11.54}$$

As shown in Figure 11.10. the contributions of the second and third terms in the velocity expansion become more significant farther from the disk electrode.

The relevance of the error in calculating the velocity using only the first term in the velocity expansion is illustrated in Figure 11.10 through inclusion of the characteristic distances for mass transfer, i.e., $\zeta = (aSc_i/3)^{-1/3}\xi$, where the characteristic length for mass transfer is given by equation (2.63). The ratio ζ/ξ has a value of 0.18 for $Sc_i = 1,000$. This means that the distance over which the concentration varies is roughly 1/5 of the distance over which the velocity varies. The error in velocity caused by using only the first term in the velocity expansion is -0.2 percent of the free-stream velocity (or -12 percent of the value at $\zeta = 0.18$).

11.3.2   Steady-State Mass Transfer

The governing equation for mass transfer to a rotating disk in the absence of migration can be expressed as equation (11.4). Under the assumption of a steady state,

$$v_y \frac{d\bar{c}_i}{dy} - D_i \frac{d^2\bar{c}_i}{dy^2} = 0 \tag{11.55}$$

which can be solved with the boundary conditions

$$\bar{c}_i \to c_i(\infty) \quad \text{for} \quad y \to \infty$$
$$\bar{c}_i = c_i(0) \quad \text{at} \quad y = 0 \tag{11.56}$$

The solution to equation (11.55) is presented in Example 2.2 on page 27 for the case that the normal velocity is limited to the first term of its development, i.e.,

$$v_y = -a \frac{\Omega^{3/2}}{\sqrt{\nu}} y^2 = -\alpha y^2 \tag{11.57}$$

The resulting concentration is given as a function of position as

$$\bar{c}_i(y) = \frac{\bar{c}_i(\infty) - \bar{c}_i(0)}{\left(\dfrac{3D_i}{\alpha}\right)^{1/3} \Gamma(4/3)} \int_0^y \exp\left(-\frac{\alpha y^3}{3D_i}\right) dy + \bar{c}_i(0) \tag{11.58}$$

Equation (11.58) can be expressed in dimensionless form as

$$\Theta_i = \frac{c_i - c_i(0)}{c_i(\infty) - c_i(0)} = \frac{1}{\Gamma(4/3)} \int_0^\zeta \exp(-\zeta^3) d\zeta \tag{11.59}$$

where $\zeta = y/\delta_i$ and the characteristic length for mass transfer of species i is

$$\delta_i = \left(\frac{3D_i}{\alpha}\right)^{1/3} = \left(\frac{3}{a}\right)^{1/3} \frac{1}{Sc_i^{1/3}} \sqrt{\frac{\nu}{\Omega}} \tag{11.60}$$

The dimensionless concentration gradient at the electrode surface is

$$\left.\frac{d\Theta_i}{d\zeta}\right|_0 = \frac{1}{\Gamma(4/3)} = \frac{\delta_i}{\delta_{N,i}} \tag{11.61}$$

The corresponding concentration profile is presented in Figure 11.11.

Equation (11.55) can be expressed in terms of dimensionless position ζ as

$$\frac{d^2\bar{c}_i}{d\zeta^2} + \left(3\zeta^2 - \left(\frac{3}{a^4}\right)^{1/3} \frac{\zeta^3}{Sc_i^{1/3}} - \frac{b}{6}\left(\frac{3}{a}\right)^{5/3} \frac{\zeta^4}{Sc_i^{2/3}}\right) \frac{d\bar{c}_i}{d\zeta} = 0 \tag{11.62}$$

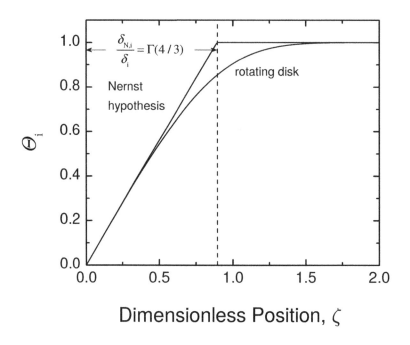

Figure 11.11: Concentration distribution near a rotating disk electrode.

where all three terms are used in the velocity expansion in equation (11.54). The boundary conditions in terms of ζ are

$$\bar{c}_i \to c_i(\infty) \quad \text{for} \quad \zeta \to \infty$$
$$\bar{c}_i = c_i(0) \quad \text{at} \quad \zeta = 0 \tag{11.63}$$

The solution is given in Example 2.6 on page 33.

The dimensionless mass transfer rate can be expressed in terms of the gradient of dimensionless concentration Θ_i at the electrode surface as[183]

$$\frac{1}{Sc_i}\Theta'(0) = \frac{0.62045 Sc_i^{-2/3}}{1 + 0.29801 Sc_i^{-1/3} + 0.14515 Sc_i^{-2/3} + O(Sc_i^{-1})} \tag{11.64}$$

Equation (11.64) may be regarded as being a small extension of equation (2.84), developed in Example 2.6 on page 33. The corrections to the current density were obtained by accounting for additional terms in the velocity expansion given as equation (11.54). Most electrolytic systems have a Schmidt number on the order of $1,000$, thus, the error in equation (11.64) caused by neglecting the second and higher terms in the velocity expansion is generally less than 3 percent. The errors have been shown to be significantly larger for frequency-domain calculations.

11.3.3 Convective Diffusion Impedance

The mathematical models for the impedance associated with convective diffusion to a disk electrode are developed here in the context of a generalized framework in which a normalized expression accounts for the influence of mass transfer.

Substitution of the definition for concentration (equation (11.11)) into the expression for conservation of species i (equation (11.4)) in one dimension yields

$$j\omega\widetilde{c}_i e^{j\omega t} + v_y \frac{d\overline{c}_i}{dy} + v_y \frac{d\widetilde{c}_i}{dy} e^{j\omega t} - D_i \frac{d^2\overline{c}_i}{dy^2} - D_i \frac{d^2\widetilde{c}_i}{dy^2} e^{j\omega t} = 0 \qquad (11.65)$$

Upon cancellation of the steady-state terms and division by the term $e^{j\omega t}$, equation (11.65) can be expressed as

$$j\omega\widetilde{c}_i + v_y \frac{d\widetilde{c}_i}{dy} - D_i \frac{d^2\widetilde{c}_i}{dy^2} = 0 \qquad (11.66)$$

Following Tribollet and Newman,[184] the dimensionless form of the equation governing the contribution of mass transfer to the impedance response of the disk electrode is developed here in terms of dimensionless position $\zeta = y/\delta_i$, where δ_i is defined in equation (11.60). The dimensionless frequency is given by

$$K_i = \frac{\omega}{\Omega}\left(\frac{9\nu}{a^2 D_i}\right)^{1/3} = \frac{\omega}{\Omega}\left(\frac{9}{a^2}\right)^{1/3} Sc_i^{1/3} = \frac{\omega\delta_i^2}{D_i} \qquad (11.67)$$

By introducing the dimensionless concentration $\theta_i(\zeta) = \widetilde{c}_i/\widetilde{c}_i(0)$, equation (11.66) becomes

$$\frac{d^2\theta_i}{d\zeta^2} + \left(3\zeta^2 - \left(\frac{3}{a^4}\right)^{1/3}\frac{\zeta^3}{Sc_i^{1/3}} - \frac{b}{6}\left(\frac{3}{a}\right)^{5/3}\frac{\zeta^4}{Sc_i^{2/3}}\right)\frac{d\theta_i}{d\zeta} - jK_i\theta_i = 0 \qquad (11.68)$$

where the three terms given in equation (11.51) were included in the expansion for axial velocity. A solution to equation (11.68) can be found that satisfies the boundary conditions

$$\theta_i \to 0 \quad \text{as} \quad \zeta \to \infty$$
$$\theta_i = 1 \quad \text{at} \quad \zeta = 0 \qquad (11.69)$$

which show that the concentration perturbation imposed by the impedance measurement is diminished far from the electrode.

Analytic solutions are possible for only simplifications of equation (11.68). Numerical solutions are otherwise required, but the formulation of the problem in dimensionless terms allows application of the result to different conditions. The approaches made to obtain a solution of equation (11.68) differ in the types of assumptions made concerning the velocity expansions.

11.3.4   Analytic and Numerical Solutions

Significant effort has been expended to develop analytic solutions for equation (11.68). These solutions differ primarily in the manner in which the convective contribution is approximated. The different models and their influence on the impedance response are described in this section.

Nernst Hypothesis

The concentration profile presented in Figure 11.11 may be used to illustrate the Nernst hypothesis. The concentration gradient at the electrode surface

$$\frac{d\Theta_i}{d\zeta}\bigg|_{\zeta=0} = \frac{1}{\Gamma(4/3)} \tag{11.70}$$

is extrapolated through the entire diffusion layer. The thickness of the Nernst diffusion layer can be obtained from

$$\frac{dc_i}{dy}\bigg|_{y=0} = \frac{c_i(\infty) - c_i(0)}{\Gamma(4/3)\left(\dfrac{3}{a}\right)^{1/3}\dfrac{1}{Sc_i^{1/3}}\sqrt{\dfrac{\nu}{\Omega}}} = \frac{c_i(\infty) - c_i(0)}{\delta_{N,i}} \tag{11.71}$$

where $\delta_{N,i}$ is the Nernst diffusion layer thickness given by

$$\delta_{N,i} = \Gamma(4/3)\left(\frac{3}{a}\right)^{1/3}\frac{1}{Sc_i^{1/3}}\sqrt{\frac{\nu}{\Omega}} = 1.61\frac{1}{Sc_i^{1/3}}\sqrt{\frac{\nu}{\Omega}} \tag{11.72}$$

A linear concentration profile corresponds to the situation of pure diffusion in a porous layer of thickness $\delta_{N,i}$. The impedance response is given in equation (11.20) where the film thickness δ_f in equation (11.16) is replaced by $\delta_{N,i}$, i.e.,

$$\frac{-1}{\theta_i'(0)} = \frac{\tanh\left(\sqrt{j\omega\dfrac{\delta_{N,i}^2}{D_i}}\right)}{\sqrt{j\omega\dfrac{\delta_{N,i}^2}{D_i}}} \tag{11.73}$$

As shown in Figure 11.11, the concentration profile corresponding to the Nernst hypothesis differs from the correct one at distances close to $y = \delta_{N,i}$ and the two profiles superimpose close to the electrode.

A comparison is presented in Figure 11.12 of the dimensionless diffusion impedance for a Nernst stagnant diffusion layer and the convective diffusion impedance obtained under assumption of an infinite Schmidt number, discussed in Section 11.3.4. Due to the relationship between propagation distance of a perturbation and frequency (see Example 11.3), the Nernst hypothesis yields incorrect results at low frequencies, although the asymptotic behavior at large frequencies is in agreement with that obtained by a correct solution for the convective-diffusion impedance.

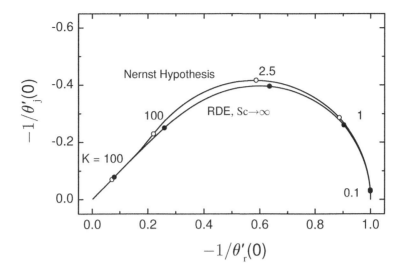

Figure 11.12: Dimensionless diffusion impedance obtained for a rotating disk under the Nernst hypothesis and under assumption of an infinite Schmidt number.

Assumption of an Infinite Schmidt Number

Under the assumption that the Schmidt number is infinitely large, the axial velocity can be approximated by the first term of the expansion given as equation (11.51). Under steady-state conditions, neglect of higher-order terms in the expansion causes an error on the order of about 3 percent in the value of the mass-transfer-limited current density. As discussed later, the errors caused by neglecting higher-order terms in the expansion can be significantly larger in the frequency domain.

The convective-diffusion equation (11.68) is reduced to

$$\frac{d^2\theta_i}{d\zeta^2} + 3\zeta^2\frac{d\theta_i}{d\zeta} - jK_i\theta_i = 0 \qquad (11.74)$$

A number of analytic solutions have been proposed. Deslouis et al.[185] developed a method that, after an approximation, reduces the problem to the canonical equation for Airy functions. Tribollet and Newman[186] gave a solution under the form of two series: one for $K < 10$ and one for $K > 10$. The two series overlapped well.

🐘 **Remember! 11.4** *The formula for impedance obtained under the Nernst hypothesis, as given by equation (11.20), provides a poor model for convective-diffusion impedance.*

Treatment of a Finite Schmidt Number

For a Schmidt number of $1,000$, use of an infinite Schmidt number approximation in the evaluation of the Schmidt number from impedance data resulted in 24.4 percent error in the estimation of Sc_i.[175] The consequence of neglecting higher-order terms in the velocity expansion is therefore much more significant than is seen for the steady-state case. The complete solution for the convective-diffusion impedance requires, in general, a numerical solution. The discussion here follows that presented by Tribollet and Newman.[184]

Several authors have addressed the influence of a finite value of the Schmidt number on expressions for the convective-diffusion impedance. Levart and Schuhmann[187] showed that the concentration term could be expressed as a series expansion in $Sc_i^{1/3}$, i.e.,

$$\theta_i\left(\zeta, Sc_i, K\right) = \theta_{i,0}\left(\zeta, K\right) + \frac{\theta_{i,1}\left(\zeta, K\right)}{Sc_i^{1/3}} + \frac{\theta_{i,2}\left(\zeta, K\right)}{Sc_i^{2/3}} + \dots \tag{11.75}$$

where $\theta_{i,0}$, $\theta_{i,1}$, and $\theta_{i,2}$ are the solutions of the corresponding coupled differential equations

$$\frac{d^2\theta_{i,0}}{d\zeta^2} + 3\zeta^2\frac{d\theta_{i,0}}{d\zeta} - jK_i\theta_{i,0} = 0 \tag{11.76}$$

$$\frac{d^2\theta_{i,1}}{d\zeta^2} + 3\zeta^2\frac{d\theta_{i,1}}{d\zeta} - jK_i\theta_{i,1} = \left(\frac{3}{a^4}\right)^{1/3}\zeta^3\frac{d\theta_{i,0}}{d\zeta} \tag{11.77}$$

and

$$\frac{d^2\theta_{i,2}}{d\zeta^2} + 3\zeta^2\frac{d\theta_{i,2}}{d\zeta} - jK_i\theta_{i,2} = \frac{b}{6}\left(\frac{3}{a}\right)^{5/3}\zeta^4\frac{d\theta_{i,0}}{d\zeta} + \left(\frac{3}{a^4}\right)^{1/3}\zeta^3\frac{d\theta_{i,1}}{d\zeta} \tag{11.78}$$

subject to boundary conditions

$$\theta_{i,0}(\infty) \to 0; \quad \theta_{i,1}(\infty) \to 0; \quad \text{and} \quad \theta_{i,2}(\infty) \to 0 \tag{11.79}$$

and

$$\theta_{i,0}(0) = 1; \quad \theta_{i,1}(0) = 0; \quad \text{and} \quad \theta_{i,2}(0) = 0 \tag{11.80}$$

The diffusion impedance for a finite Schmidt number is obtained from

$$-\frac{1}{\theta_i'(0)} = \frac{-1}{\theta_{i,0}'(0) + \theta_{i,1}'(0)Sc_i^{-1/3} + \theta_{i,2}'(0)Sc_i^{-2/3}} \tag{11.81}$$

which can be expressed as

$$-\frac{1}{\theta_i'(0)} = -\frac{1}{\theta_{i,0}'(0)} + \frac{\theta_{i,1}'(0)}{\left(\theta_{i,0}'(0)\right)^2}\frac{1}{Sc_i^{1/3}} \tag{11.82}$$

$$-\frac{1}{\theta_{i,0}'(0)}\left[\left(\frac{\theta_{i,1}'(0)}{\theta_{i,0}'(0)}\right)^2 - \frac{\theta_{i,2}'(0)}{\theta_{i,0}'(0)}\right]\frac{1}{Sc_i^{2/3}}$$

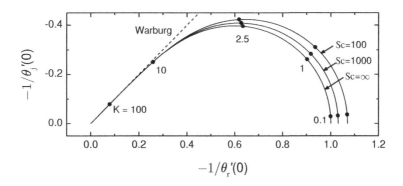

Figure 11.13: Dimensionless diffusion impedance obtained for a rotating disk under assumption of infinite Schmidt number and for a finite Schmidt number equal to 100 and 1000. At high frequencies, the finite and infinite Schmidt numbers converge and are in agreement with the Warburg impedance given in equation (11.28) and represented by a dashed line.

such that

$$Z_{(0)} = -\frac{1}{\theta'_{i,0}(0)} \tag{11.83}$$

$$Z_{(1)} = \frac{\theta'_{i,1}(0)}{\left(\theta'_{i,0}(0)\right)^2} \tag{11.84}$$

and

$$Z_{(2)} = -\frac{1}{\theta'_{i,0}(0)}\left[\left(\frac{\theta'_{i,1}(0)}{\theta'_{i,0}(0)}\right)^2 - \frac{\theta'_{i,2}(0)}{\theta'_{i,0}(0)}\right] \tag{11.85}$$

The convective-diffusion impedance is obtained directly as a function of the Schmidt number from

$$\frac{-1}{\theta'_i(0)} = Z_{(0)} + \frac{Z_{(1)}}{Sc_i^{1/3}} + \frac{Z_{(2)}}{Sc_i^{2/3}} + \dots \tag{11.86}$$

A comparison between the dimensionless convective diffusion obtained under assumption of infinite and finite Schmidt numbers is presented in Figure 11.13 for Schmidt numbers equal to 100, corresponding to diffusion of hydrogen ions, and 1000, typical of most ionic species. At high frequencies, the finite and infinite Schmidt numbers converge and are in agreement with the Warburg impedance given in equation (11.28) and represented in Figure 11.13 by a dashed line.

Tabulated values of $Z_{(0)}$, $Z_{(1)}$, and $Z_{(2)}$ have been presented by Tribollet and Newman as a function of $pSc_i^{1/3}$ where p is the frequency made dimensionless by the steady rotation rate of the disk electrode, $p = \omega/\Omega$.[184] In terms of p, $K_i = 3.258pSc_i^{1/3}$. The relative contributions of the terms can be seen in Figure 11.14, presented here as functions of K_i.

A similar development was provided by Tribollet and Newman[184] for electrohydrodynamic impedance. The use of look-up tables facilitates regression of models to

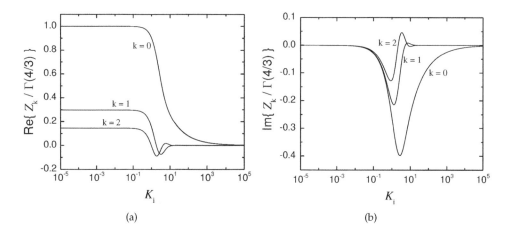

Figure 11.14: Values for dimensionless contributions $Z_{(0)}$, $Z_{(1)}$, and $Z_{(2)}$ to the diffusion impedance (see equation (11.86)) as a function of dimensionless frequency K_i: a) real part and b) imaginary part.[184]

experimental data that take full account of the influence of a finite Schmidt number on the convective-diffusion impedance. Use of only the first term in equation (11.86) yields a numerical solution for an infinite Schmidt number. Tribollet and Newman report use of the first two terms in equation (11.86).[184] The low level of stochastic noise in experimental data justifies use of the three-term expansion reported here.

11.4   Submerged Impinging Jet

The axisymmetric impinging jet, shown in Figure 11.15(a), is a very attractive, if somewhat underemployed, system for electrochemical investigations. Within the stagnation region, shown in Figure 11.15(b), the axial velocity is independent of radial position and convective diffusion to the disk is uniform, much as is seen for the rotating disk electrode.[188–190] Because the mass-transfer rate is uniform for an electrode that lies entirely within the stagnation region, differential mass-transfer cells are not established. Thus, the impinging jet has the attractive uniform accessibility to mass transfer seen for the rotating disk. In contrast to the rotating disk, the electrode is stationary and is therefore suitable for in situ observation.[190–193]

11.4.1   Fluid Flow

The fluid flow within the stagnation region of the electrode in an impinging jet cell is well-defined.[188,194–197] The submerged impinging jet geometry can be made to give uniform mass-transfer rates across a disk electrode within the stagnation region. The stagnation region is defined to be the region surrounding the stagnation point in which

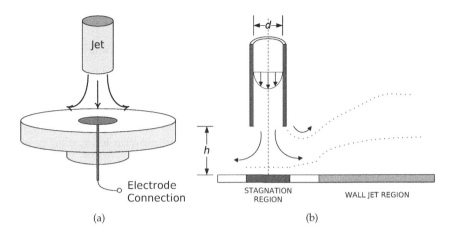

Figure 11.15: A disk electrode subjected to a submerged impinging jet: a) schematic illustration; b) identification of flow regimes.

the axial velocity, given by

$$v_y = -\sqrt{a_{JJ}\nu}\phi\left(\eta\right) \tag{11.87}$$

is independent of radial position, and the radial velocity is given by

$$v_r = \frac{a_{JJ}r}{2}\frac{d\phi\left(\eta\right)}{d\eta} \tag{11.88}$$

where a_{JJ} is the hydrodynamic constant that is a function only of geometry and fluid velocity, r and y are the radial and axial positions, respectively, ν is the kinematic viscosity, and ϕ is the stream function that is given in terms of dimensionless axial position $\eta = y\sqrt{a_{JJ}/\nu}$ as[188]

$$\phi\left(\eta\right) = 1.352\eta^2 - \frac{1}{3}\eta^3 + 7.2888 \times 10^{-3}\eta^6 + \ldots \tag{11.89}$$

Esteban et al.[189] used ring electrodes to find that the stagnation region extends to a radial distance roughly equal to the inside radius of the nozzle. A refined analysis by Baleras et al.[198] showed that the stagnation region becomes smaller at large values of the ratio of nozzle height h to nozzle diameter D.

Within the stagnation region, the surface shear stress τ_{ry} is given by

$$\tau_{ry} = -1.312r(\mu\rho)^{\frac{1}{2}}a_{JJ}^{\frac{3}{2}} \tag{11.90}$$

where μ and ρ are the viscosity and density of the fluid, respectively. The hydrodynamic constant a_{JJ} can be determined experimentally using ring or disk electrodes at the mass-transfer-limited condition and is proportional to the jet velocity.

11.4.2   Steady-State Mass Transfer

The equation that governs steady-state mass transfer to the impinging jet electrode is equation (11.55), the same as that for the rotating disk. The boundary conditions are given as equation (11.21), and the solution depends on the number of terms used in the velocity expansion. The only difference between the solution presented for the rotating disk and here for the impinging jet is that the axial velocity for the disk is expressed as equation (11.51), whereas the expansion for the impinging jet system is given as equation (11.89).

11.4.3   Convective Diffusion Impedance

Under the assumption that the electrode is uniformly accessible, the equation governing mass transfer in the frequency domain is given by

$$\frac{d^2\theta_i}{d\zeta^2} + \left(3\zeta^2 - \left(\frac{3}{1.352^4}\right)^{1/3}\frac{\zeta^3}{Sc_i^{1/3}} + \cdots\right)\frac{d\theta_i}{d\zeta} - jK_i\theta_i = 0 \qquad (11.91)$$

where

$$K_i = \frac{\omega}{a_{JJ}}\left(\frac{9}{(1.352)^2}\right)^{1/3}Sc_i^{1/3} = 1.70123\frac{\omega Sc_i^{1/3}}{a_{JJ}} \qquad (11.92)$$

represents a dimensionless frequency, and $\zeta = y/\delta_i$ is a dimensionless position where

$$\delta_i = \left(\frac{3}{1.352}\right)^{1/3}\frac{1}{Sc_i^{1/3}}\sqrt{\frac{\nu}{a_{JJ}}} = 1.180\frac{1}{Sc_i^{1/3}}\sqrt{\frac{\nu}{a_{JJ}}} \qquad (11.93)$$

is a characteristic length for transport of species i. Terms in ζ of order greater or equal to 6 were neglected in equation (11.91).

Following Tribollet and Newman,[184] the concentration term can be expressed as a series expansion in $Sc_i^{-1/3}$ as

$$\theta_i = \theta_{i,0} + \frac{\theta_{i,1}}{Sc_i^{1/3}} + \frac{\theta_{i,2}}{Sc_i^{2/3}} + \cdots \qquad (11.94)$$

where $\theta_{i,k}$ represents the solution to

$$\frac{d^2\theta_{i,0}}{d\zeta^2} + 3\zeta^2\frac{d\theta_{i,0}}{d\zeta} - jK_i\theta_{i,0} = 0 \qquad (11.95)$$

$$\frac{d^2\theta_{i,1}}{d\zeta^2} + 3\zeta^2\frac{d\theta_{i,1}}{d\zeta} - jK_i\theta_{i,1} = \left(\frac{3}{(1.352)^4}\right)^{1/3}\zeta^3\frac{d\theta_{i,0}}{d\zeta} \qquad (11.96)$$

and

$$\frac{d^2\theta_{i,2}}{d\zeta^2} + 3\zeta^2\frac{d\theta_{i,2}}{d\zeta} - jK_i\theta_{i,2} = \left(\frac{3}{(1.352)^4}\right)^{1/3}\zeta^3\frac{d\theta_{i,1}}{d\zeta} \qquad (11.97)$$

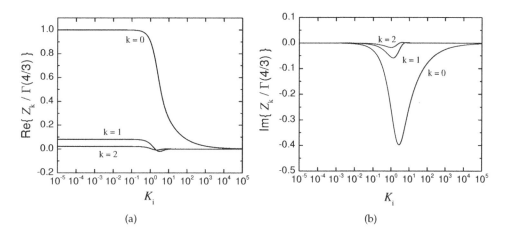

Figure 11.16: Values for dimensionless contributions $Z_{(0)}$, $Z_{(1)}$, and $Z_{(2)}$ to the diffusion impedance (see equation (11.98)) as functions of K_i: a) real part and b) imaginary part.

In this development, terms of order Sc_i^{-1} and higher were neglected.

The convective-diffusion impedance can be tabulated directly as a function of the Schmidt number as

$$-\frac{1}{\theta_i'(0)} = Z_{(0)} + \frac{Z_{(1)}}{Sc_i^{1/3}} + \frac{Z_{(2)}}{Sc_i^{2/3}} + \dots \tag{11.98}$$

Tabulated values of $Z_{(0)}$, $Z_{(1)}$, and $Z_{(2)}$ were found as a function of K_i. The relative contributions of the terms can be seen in Figure 11.16.

11.5 Rotating Cylinders

The rotating cylinder is a popular tool for electrochemical research because it is convenient to use and both the primary and mass-transfer-limited current distributions are uniform.[199] A schematic representation of the rotating cylinder is presented in Figure 11.17. At very low rotation speeds, the fluid flows in concentric circles around the rotating cylinder, satisfying a no-slip condition at the rotating inner cylinder and at the stationary outer cylinder. Since there is no velocity component in the radial direction, there is no convective enhancement to mass transfer. This simple flow pattern becomes unstable at higher rotation speeds, and a cellular flow pattern (termed Taylor vortices) is observed. Taylor vortices provide an irregular enhancement to mass transfer. At still higher rotation speeds, the flow becomes fully turbulent. Mass-transfer studies with rotating cylinders are conducted in the turbulent flow regime because the turbulence provides a uniform enhancement to mass transfer.

Two configurations have been described in the literature. When the electrode encompasses the entire length of the inner cylinder, as shown in Figure 11.17(a), the mass-transfer-controlled and primary current distributions are both uniform at any applied

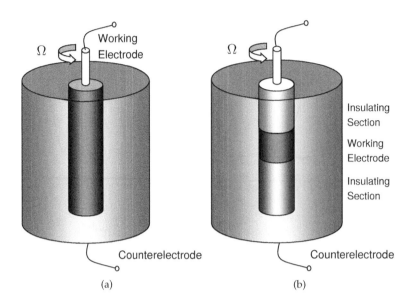

Figure 11.17: Schematic representation of a rotating cylinder electrode: a) entire cylinder used as working electrode and b) band-shape cylindrical coupon used as a working electrode. The geometry (a) provides a uniform current and potential distribution at and below the mass-transfer-limited current, and geometry (b) is useful for studies conducted at the open-circuit condition.

potential. Often, such a geometry is not practical, and a coupon electrode (see Figure 11.17(b)) is used that has insulating cylinders above and below the active surface. The mass-transfer-controlled current distribution for the geometry of Figure 11.17(b) is still uniform, but the primary distribution is not uniform. The nonuniform primary distribution poses no significant problem for corrosion experiments conducted at the open-circuit condition because the ohmic potential drop is insignificant when the net current is equal to zero.

Analytic expressions cannot be derived for the turbulent flow regime, but empirical correlations are available that relate the cylinder rotation speed to the mass-transfer coefficient for a given geometry. The correlation of Eisenberg et al.[150] is given by

$$\mathrm{Sh}_i = 0.0791 \left(\mathrm{Re} \frac{d_R}{d_L} \right)^{0.7} \mathrm{Sc}_i{}^{0.356} \tag{11.99}$$

where d_R is the diameter of the inner, rotating cylinder, d_L is the diameter of the cylinder at which the current is limited by mass transfer (usually the inner cylinder), Sh_i is the Sherwood number, related to the mass-transfer coefficient k_M by $\mathrm{Sh}_i = k_M d_R / D_i$, Re is the Reynolds number for the cylinder geometry given by $\mathrm{Re} = \Omega d_R^2 / 2\nu$, and Sc_i is the Schmidt number given by $\mathrm{Sc}_i = \nu / D_i$.

In principle, development of the convective-diffusion impedance for this system

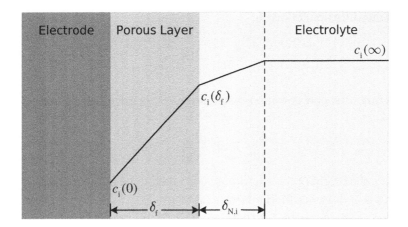

Figure 11.18: Schematic representation of an electrode covered by a porous layer: δ_f is the porous layer thickness, $\delta_{N,i}$ is the Nernst diffusion layer, and c_i is the concentration of the electroactive species which reacts at the interface and diffuses through the porous layer and through the electrolyte.

requires solution in the frequency domain to

$$\frac{\partial c_i}{\partial t} + \frac{1}{r}\left[\frac{\partial}{\partial r}\left(r\left(D_i + D_i^t(r)\right)\frac{\partial c_i}{\partial r}\right)\right] = 0 \qquad (11.100)$$

where $D_i^t(r)$ represents the eddy diffusivity that arises from the enhancement to mass transfer caused by turbulent eddies. The dependence of $D_i^t(r)$ on position can be estimated from comparison to the universal velocity profile established for turbulent flow in pipes. As a result of the approximate character of the treatment of mass transfer, the rotating cylinder is used primarily for cases where qualitative comparisons are satisfactory, for example, in the ranking of the resistance to corrosion for different environments or of the resistance to corrosion of different metals in the same environment.

11.6 Electrode Coated by a Porous Film

Porous nonreacting layers covering reacting metallic interfaces may slow down the rate of transfer for diffusing species. This decrease includes the effect of the diffusivity $D_{f,i}$ as well as that of the layer thickness δ_f.

A schematic representation of the system under investigation is presented in Figure 11.18. The concentration gradient is distributed between the fluid and the porous layer. In addition, the metal-layer interface is assumed to be uniformly reactive. Two material balance equations can be written as follows:

1. In the porous layer, the concentration distribution $c_i^{(1)}$ is determined only by

molecular diffusion following

$$\frac{\partial c_i^{(1)}}{\partial t} = D_{f,i}\frac{\partial^2 c_i^{(1)}}{\partial y^2} \tag{11.101}$$

2. In the fluid, the concentration distribution $c_i^{(2)}$ is governed by convective diffusion, i.e.,

$$\frac{\partial c_i^{(2)}}{\partial t} = D_i\frac{\partial^2 c_i^{(2)}}{\partial y_e^2} - v_y\frac{\partial c_i^{(2)}}{\partial y_e} \tag{11.102}$$

For simplicity, the origin of the coordinate y is taken to be at the metal-layer interface and that of y_e is at the layer-fluid interface ($y_e = y - \delta_f$).

Associated with equations (11.101) and (11.102) are boundary conditions that express the continuity of the concentration fields and of the fluxes for the steady state as well as for the time-dependent quantities. For $y = \delta_f$ or $y_e = 0$,

$$c_i^{(1)}(\delta_f) = c_i^{(2)}(0) \tag{11.103}$$

and

$$D_{f,i}\frac{\partial c_i^{(1)}}{\partial y} = D_i\frac{\partial c_i^{(2)}}{\partial y_e} \tag{11.104}$$

At $y_e \to \infty$, the concentration tends towards the value far from the electrode, i.e., $c_i^{(2)} \to c_i(\infty)$.

11.6.1 Steady-State Solutions

Under the steady-state condition, equation (11.101) is reduced to the simple form

$$\frac{\partial^2 \bar{c}_i^{(1)}}{\partial y^2} = 0 \tag{11.105}$$

which leads to

$$\bar{J}_i = D_{f,i}\frac{\bar{c}_i^{(1)}(\delta_f) - \bar{c}_i^{(1)}(0)}{\delta_f} = D_{f,i}\frac{\bar{c}_i^{(1)}(\delta_f)}{\delta_f} \tag{11.106}$$

in which the assumption that $\bar{c}_i^{(1)}(0) = 0$ is considered to apply for a mass-transfer-limited reaction.

The flux in the electrolyte is given by

$$\bar{J}_i = D_i\frac{c_i(\infty) - \bar{c}_i^{(2)}(0)}{\delta_{N,i}} \tag{11.107}$$

Elimination of $\bar{c}_i^{(2)}(0)$ in equations (11.106) and (11.107) yields

$$\bar{J}_i = \frac{c_i(\infty)}{\dfrac{\delta_{N,i}}{D_i} + \dfrac{\delta_f}{D_{f,i}}} \tag{11.108}$$

Equation (11.108) can be written as

$$\frac{1}{\bar{J}_i} = \frac{1}{\bar{J}_{L,i}} + \frac{1}{\bar{J}_{\Omega\to\infty,i}} \tag{11.109}$$

in which

$$\bar{J}_{L,i} = D_i \frac{c_i(\infty)}{\delta_{N,i}} \tag{11.110}$$

is the limiting diffusion flux on a metallic surface, free from porous layer, and

$$\bar{J}_{\Omega\to\infty,i} = D_{f,i} \frac{c_i(\infty)}{\delta_f} \tag{11.111}$$

is the limiting flux when the entire concentration gradient is located within the porous layer, i.e., when $\Omega \to \infty$.

The interest of using reciprocal values is that the experimental plot of $1/\bar{J}_i$ as a function of $\Omega^{-1/2}$ must be a straight line parallel to the line corresponding to Levich's result for the mass-transfer-limited case, which passes through the origin. The ordinate value of the intercept of this straight line at $\Omega^{-1/2} = 0$ is $1/J_{\Omega\to\infty,i}$. Similar approaches are developed in Example 5.6 on page 106 to extract the kinetic current from a current that is influenced by convective diffusion to a rotating disk electrode (see Figure 5.10 on page 108).

Example 11.5 Electrode Covered by Porous Film: *Show that a plot of $1/\bar{J}_i$ as a function of $\Omega^{-1/2}$ is a straight line with an intercept of $1/J_{\Omega\to\infty,i}$.*

Solution: *From equations (11.109)-(11.111),*

$$\frac{1}{\bar{J}_i} = \frac{\delta_{N,i}}{D_i c_i(\infty)} + \frac{\delta_f}{D_{f,i} c_i(\infty)} \tag{11.112}$$

Following equation (2.26) on page 28 (see also equation (11.72)), the Nernst diffusion layer thickness is given by

$$\delta_{N,i} = 1.61 \frac{1}{Sc_i^{1/3}} \sqrt{\frac{\nu}{\Omega}} \tag{11.113}$$

Thus

$$\frac{1}{\bar{J}_i} = \frac{1.61\sqrt{\nu}}{Sc_{ii} D_i c_i(\infty)} \frac{1}{\sqrt{\Omega}} + \frac{\delta_f}{D_{f,i} c_i(\infty)} \tag{11.114}$$

$1/\bar{J}_i$ is presented in Figure 11.19 as a function of $1/\sqrt{\Omega}$. The intercept is $1/J_{\Omega\to\infty,i} = \delta_f/D_{f,i} c_i(\infty)$, and the slope is $1.61\sqrt{\nu}/Sc_i^{1/3} D_i c_i(\infty)$.

The flux through the stationary film and the bulk region is controlled by two parameters, i.e.,

$$\psi = \frac{D_{f,i} \delta_{N,i}}{D_i \delta_f} \tag{11.115}$$

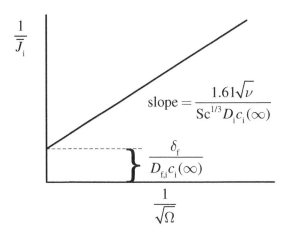

Figure 11.19: Schematic representation showing that a plot of $1/\bar{J}_i$ as a function of $\Omega^{-1/2}$ yields a straight line with an intercept of $1/\bar{J}_{\Omega\to\infty,i} = \delta_f/D_{f,i}c_i(\infty)$.

and the ratio of diffusion lengths, $\delta_{N,i}/\delta_f$. Under the assumption that the diffusivity within the film of porosity ϵ is related to the bulk diffusivity by[115,200]

$$D_{f,i} = D_i\epsilon^{1.5} \tag{11.116}$$

equation (11.115) can be expressed as

$$\psi = \frac{\delta_{N,i}}{\delta_f}\epsilon^{1.5} \tag{11.117}$$

Thus, concentration profiles within the coating and the bulk region can be seen to be controlled by the coating porosity ϵ and by the ratio of diffusion lengths $\delta_{N,i}/\delta_f$.

The concentration distributions obtained for a film-covered rotating disk electrode are presented in Figure 11.20(a) for $\delta_{N,i}/\delta_f = 2$ and with coating porosity as a parameter. As the coating porosity tends toward zero, the resistance of the coating to diffusion dominates, and the concentration at the interface between the coating and the bulk solution is equal to the bulk concentration $c_i(\infty)$. In this case, the diffusion impedance for the electrochemical system is given by that developed in Section 11.2.1 for diffusion through a solid film. At larger values of coating porosity, both convective diffusion and diffusion through the film are important and should be treated.

The coating on the electrode can be deposited (e.g., a polymer coating) or it can be produced by the electrochemical reactions. For many electrochemical systems, the growth of a porous film causes it to become the dominant resistance to mass transfer, as is shown in Figure 11.20(b) for a porosity $\epsilon = 0.3$ and with relative film thickness $\delta_{N,i}/\delta_f$ as a parameter. The position variable is scaled to $\delta_{N,i}$ to emphasize the changing dimension of the coating layer.

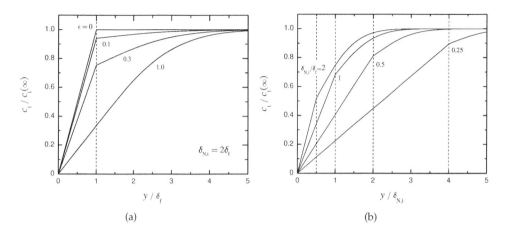

Figure 11.20: Concentration distribution for a coated rotating disk electrode: a) results obtained for $\delta_{\mathrm{N,i}}/\delta_{\mathrm{f}} = 2$ with coating porosity as a parameter; b) results obtained for a porosity $\epsilon = 0.3$ and relative coating thickness $\delta_{\mathrm{N,i}}/\delta_{\mathrm{f}}$ as a parameter.

Example 11.6 Diffusion through a Film with Porosity = 1: *Does the lack of an abrupt change in slope for a coating porosity equal to 1.0 in Figure 11.20(a) mean that the role of a coating can be ignored when calculating the flux and diffusion impedance? Note that some coatings, for example, a biofilm formed by microorganisms, have a porosity approaching 1.0.*[201]

Solution: *While the concentration profile shown for $\epsilon = 1.0$ in Figure 11.20(a) resembles that for an uncoated rotating disk, the flux is affected by the fact that, within the coating, the velocity is equal to zero. The flux to the surface can be given in terms of the concentration at the coating/bulk solution interface as*

$$\mathbf{N}_{\mathrm{i}} = D_{\mathrm{i}}\epsilon^{1.5}\frac{c_{\mathrm{i}}(\delta_{\mathrm{f}})}{\delta_{\mathrm{f}}} \tag{11.118}$$

or, for $\epsilon = 1$

$$\mathbf{N}_{\mathrm{i}} = D_{\mathrm{i}}\frac{c_{\mathrm{i}}(\delta_{\mathrm{f}})}{\delta_{\mathrm{f}}} \tag{11.119}$$

For a given film thickness δ_{f}, the important parameter is the concentration at the film surface. If this concentration tends toward zero, the film has a negligible influence on mass transfer. Otherwise, the influence of the film cannot be neglected.

The concentration at the interface between the film and the bulk electrolyte indicates whether the diffusion impedance of the system can be assumed to be that of the stationary film. The interface concentration, given by

$$\frac{c_{\mathrm{i}}(\delta_{\mathrm{f}})}{c_{\mathrm{i}}(\infty)} = \frac{1}{1+\psi} \tag{11.120}$$

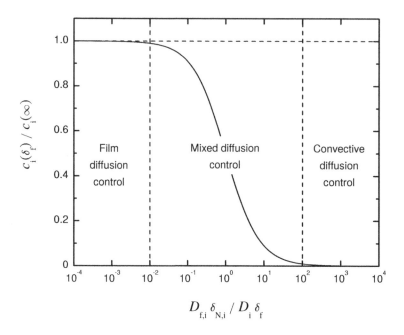

Figure 11.21: Concentration at the interface between the coating and the bulk solution as a function of $\psi = D_{f,i}\delta_{N,i}/D\delta_f$.

is a function only of $\psi = D_{f,i}\delta_{N,i}/D_i\delta_f$. Regions of mass-transfer control can be identified, as shown in Figure 11.21. When the dimensionless interface concentration approaches 1, the film diffusion impedance discussed in Section 11.2.1 dominates. When the dimensionless interface concentration approaches 0, the film diffusion impedance can be neglected, and the convective-diffusion response discussed in Section 11.3 dominates. In the intermediate region, both convective diffusion and diffusion through the film are important, as discussed in Section 11.6.2.

Example 11.7 Continuation of Example 11.6: *Can the diffusion impedance of a film with $\epsilon = 1.0$ be ignored?*

Solution: *As seen from equation (11.119), for a given film thickness δ_f, the important parameter is the concentration at the film surface. If this concentration tends toward zero, the film has a negligible influence on mass transfer. Otherwise, the influence of the film cannot be neglected. As*

$$\lim_{\epsilon \to 1} \frac{D_{f,i}\delta_{N,i}}{D_i\delta_f} = \frac{\delta_{N,i}}{\delta_f} \tag{11.121}$$

the role of a film of fixed thickness of $\epsilon = 1.0$ depends solely on the rotation speed of the disk. As the disk speed increases, $\delta_{N,i}$ decreases, and the film diffusion becomes more important.

11.6.2 Coupled Diffusion Impedance

In the porous layer, the fluctuating part of equation (11.101) may be written as

$$\frac{\partial^2 \tilde{c}_i^{(1)}}{\partial y^2} - \frac{j\omega}{D_{f,i}} \tilde{c}_i^{(1)} = 0 \tag{11.122}$$

The solution is:

$$\tilde{c}_i^{(1)} = C_1 \exp\left(\sqrt{\frac{j\omega}{D_{f,i}}} y^2\right) + C_2 \exp\left(-\sqrt{\frac{j\omega}{D_{f,i}}} y^2\right) \tag{11.123}$$

where C_1 and C_2 are integration constants obtained from the boundary conditions.

At the fluid-porous-layer interface ($y_e = 0$), the concentration gradient and the concentration are linked by the relationship

$$\left.\frac{\partial \tilde{c}_i^{(2)}}{\partial y_e}\right|_{y_e=0} = \frac{\tilde{c}_i^{(2)}(0)}{\delta_{N,i}} \theta'(0) \tag{11.124}$$

Then, from the boundary conditions at $y = \delta_f$, the constants C_1 and C_2 may be eliminated and the general expression is obtained as

$$\left.\frac{\partial \tilde{c}_i^{(1)}(0)}{\partial y}\right|_0 = \frac{-\tilde{c}_i^{(1)}}{\delta_f} \frac{\left(\dfrac{j\omega\delta_f^2}{D_{f,i}}\right) Z_D Z_{D,f} + \dfrac{D_i}{D_{f,i}} \dfrac{\delta_f}{\delta_{N,i}}}{Z_D} \tag{11.125}$$

where $Z_D = -1/\theta'(0)$ is the dimensionless convective diffusion in the solution (see equation (11.20)) and

$$Z_{D,f} = \frac{\tanh \sqrt{j\omega\delta_f^2/D_{f,i}}}{\sqrt{j\omega\delta_f^2/D_{f,i}}} \tag{11.126}$$

is the dimensionless diffusion impedance for a finite stagnant diffusion layer (see equation (11.20)).

It may be easily verified that when the layer effect is gradually decreased (i.e., $\delta_f \to 0$ and $D_{f,i} \to D_i$) one finds again the relation (11.73). At the opposite extreme, when $\Omega \to \infty$, then $\delta \to 0$, the relation becomes

$$\left.\frac{\partial \tilde{c}_i^{(1)}}{\partial y}\right|_0 = -\frac{\tilde{c}_i^{(1)}(0)}{\delta_f} \frac{1}{Z_{D,f}} \tag{11.127}$$

As shown in Figure 11.21, for a broad range of parameters for film-covered electrodes, the diffusion impedance must account for both convective diffusion associated with the external imposed flow and diffusion through a stagnant layer. Following Deslouis et al.,[202] the net diffusion impedance can be expressed as being composed of contributions from film and convective-diffusion terms

$$Z_{D,net} = \frac{Z_D + \frac{D_i}{D_{f,i}} \frac{\delta_f}{\delta_{N,i}} Z_{D,f}}{Z_{D,f} Z_D \left(j\omega \frac{\delta_f^2}{D_{f,i}} \right) + \frac{D_i}{D_{f,i}} \frac{\delta_f}{\delta_{N,i}}} \tag{11.128}$$

where the inner term $Z_{D,f}$ is the finite-length diffusion impedance corresponding to a porous layer of thickness δ_f, and the outer term Z_D may correspond to either a stagnant or a convective region of finite thickness $\delta_{N,i}$. In this case, the effective diffusion impedance $Z_{D,net}$ is a function of the time constant $\tau_{i,f} = \delta_f^2 / D_{f,i}$, the ratio $D\delta_f / D_{f,i}\delta_{N,i}$, and the Schmidt number Sc_i. The coupling of film and convective-diffusion impedances is also developed in Section 16.4.2 on page 457 for electrohydrodynamic impedance measurements.

Example 11.8 Diffusion Impedances in Series: *Following equation (4.23), the impedance corresponding to two resistors in series is equal to the sum of the resistances. Why is it incorrect to treat diffusion through two layers by adding two diffusion impedances?*

Solution: *The issue is that the condition at the interface between the two diffusion layers is not treated correctly when diffusion impedances, such as presented in equation (11.20), are added. The solutions to the two sets of equations are coupled by the continuity of concentration at the interface, as seen in Figure 11.20.*

11.7 Impedance with Homogeneous Chemical Reactions

The impedance response at an electrode surface may also be influenced by homogeneous reactions. Remita et al.[203] have shown that, for a deaerated aqueous electrolyte containing dissolved carbon dioxide, the electrochemical hydrogen evolution reaction is enhanced by the homogeneous dissociation of CO_2 following

$$H_2O + CO_{2,aq} \rightleftharpoons HCO_3^- + H^+ \tag{11.129}$$

and

$$HCO_3^- \rightleftharpoons CO_3^{2-} + H^+ \tag{11.130}$$

Remember! 11.5 *The impedance associated with diffusion through a series of layers cannot be modeled by adding the diffusion impedances for each individual layer. Such a treatment does not account correctly for the conditions at the interface between the layers.*

These reactions explain the high corrosivity of solutions containing dissolved CO_2. In related work, Tran et al.[204] demonstrated that homogeneous dissociation of acetic acid (CH_3COOH) has a similar contribution, with the reaction

$$HAc_{aq} \rightleftarrows Ac^- + H^+ \tag{11.131}$$

contributing H^+ ions that participate in cathodic reactions.

Coupled electrochemical and homogeneous reactions are also involved in sensors used to monitor glucose concentrations for management of diabetes.[205,206] In the presence of glucose oxidase enzyme, glucose $(C_6H_{12}O_6)$ and oxygen are converted into peroxide and gluconolactone $(C_6H_{10}O_6)$, i.e.,

$$C_6H_{12}O_6 + GOX - FAD \quad \rightleftarrows \quad C_6H_{10}O_6 - GOX - FADH_2 \tag{11.132}$$
$$C_6H_{10}O_6 - GOX - FADH_2 \quad \rightarrow \quad C_6H_{10}O_6 + GOX - FADH_2 \tag{11.133}$$

where $GOX - FAD$ and $GOX - FADH_2$ are, respectively, the oxidized and reduced form of glucose oxidase and $C_6H_{10}O_6 - GOX - FADH_2$ represents an intermediate complex. The oxidized form of glucose oxidase is regenerated by

$$GOX - FADH_2 + O_2 \quad \rightleftarrows \quad GOX - FAD - H_2O_2 \tag{11.134}$$
$$GOX - FAD - H_2O_2 \quad \rightarrow \quad GOX - FAD + H_2O_2 \tag{11.135}$$

where $GOX - FAD - H_2O_2$ represents an intermediate complex. The peroxide that is formed reacts at the electrode surface following

$$H_2O_2 \rightarrow 2H^+ + O_2 + 2e^- \tag{11.136}$$

The measured current is proportional to the glucose concentration.

The evaluation of the impedance response associated with the coupling of homogeneous and heterogeneous electrochemical reactions is obtained through numerical calculations. A general homogeneous reaction may be expressed as[207]

$$AB \underset{k_b}{\overset{k_f}{\rightleftarrows}} A^- + B^+ \tag{11.137}$$

Consider that one of the ionic species such as B^+ is electroactive and that this species is consumed at the electrode. The reaction may be expressed as

$$B^+ + e^- \rightarrow B \tag{11.138}$$

The corresponding current may be expressed as

$$i_{B^+} = -K_{B^+} c_{B^+}(0) \exp(-b_{B^+} V) \tag{11.139}$$

As the electroactive species is consumed, a concentration gradient of B^+ must exist near the electrode.

Far from the electrode, the species AB, A$^-$, and B$^+$ are equilibrated, thus

$$K_{eq} = \frac{k_f}{k_b} = \frac{c_{A^-}(\infty)c_{B^+}(\infty)}{c_{AB}(\infty)} \tag{11.140}$$

The concentrations may be obtained in terms of a constant c^* which is given by

$$c^* = c_{AB}(\infty) + c_{B^+}(\infty) \tag{11.141}$$

As $c_{A^-}(\infty) = c_{B^+}(\infty)$, the different concentrations can be determined from

$$(c_{A^-}(\infty))^2 + K_{eq}c_{A^-}(\infty) - K_{eq}c^* = 0 \tag{11.142}$$

with the solution

$$c_{A^-}(\infty) = c_{B^+}(\infty) = \frac{-K_{eq} + \sqrt{K_{eq}^2 + 4K_{eq}c^*}}{2} \tag{11.143}$$

and

$$c_{AB}(\infty) = c^* - c_{A^-}(\infty) \tag{11.144}$$

The equilibrium relationships developed above provide useful dependencies among the species participating in the homogeneous reaction, but these apply far from the electrode surface and are not sufficient to establish values for the concentration gradient of B$^+$ near the electrode.

To solve the problem it is necessary to integrate the conservation equation for each species, e.g.,

$$\frac{\partial c_i}{\partial t} = D_i \frac{\partial^2 c_i}{\partial y^2} - v_y \frac{\partial c_i}{\partial y} - R_i \tag{11.145}$$

where R_i represents the rate of production of species i by homogeneous reactions. In the present case,

$$\begin{aligned} R_{A^-} = R_{B^+} = -R_{AB} &= k_f c_{AB}(y) - k_b c_{A^-}(y)c_{B^+}(y) \\ &= k_b \left(K_{eq}c_{AB}(y) - c_{A^-}(y)c_{B^+}(y) \right) \end{aligned} \tag{11.146}$$

where reaction (11.137) is not assumed to be equilibrated in a region near the electrode surface. The form of equation (11.146) makes the problem nonlinear. The boundary conditions at the interface are

$$\frac{\partial c_{AB}}{\partial y}\bigg|_{y=0} = \frac{\partial c_{A^-}}{\partial y}\bigg|_{y=0} = 0 \tag{11.147}$$

for the nonreacting species, and, following equation (5.6) on page 92,

$$-FD_{B^+} \frac{\partial c_{B^+}}{\partial y}\bigg|_{y=0} = i_{B^+} \tag{11.148}$$

for the reacting species B^+. The concentration field for each species and the current-potential curve can be obtained by solving numerically the set of the three steady-state nonlinear differential equations with the corresponding boundary conditions.

From this steady-state solution, the corresponding impedance can be derived for each specified frequency. Following the development in earlier parts of this chapter, the governing equation in the bulk phase is given by

$$jw\tilde{c}_i = D_i \frac{d^2\tilde{c}_i}{dy^2} - v_y \frac{d\tilde{c}_i}{dy} - \tilde{R}_i \tag{11.149}$$

where

$$\tilde{R}_{A^-} = \tilde{R}_{B^+} = -\tilde{R}_{AB} \tag{11.150}$$
$$= k_f \tilde{c}_{AB}(y) - k_b \bar{c}_{A^-}(y)\tilde{c}_{B^+}(y) - k_b \tilde{c}_{A^-}(y)\bar{c}_{B^+}(y)$$

The corresponding set of equations is linear, but, as shown in equation (11.150), it involves steady-state concentrations. Thus, solution of (11.149) requires solution of the steady-state equation. At $y \to \infty$, the boundary conditions are

$$\tilde{c}_{A^-} = \tilde{c}_{B^+} = \tilde{c}_{AB} = 0 \tag{11.151}$$

At $y = 0$,

$$\frac{d\tilde{c}_{AB}}{dy}\bigg|_{y=0} = \frac{d\tilde{c}_{A^-}}{dy}\bigg|_{y=0} = 0 \tag{11.152}$$

for the nonreacting species, and

$$\tilde{c}_{B^+}(0) = 1 \tag{11.153}$$

for the reacting species B^+. The value of $\tilde{c}_{B^+}(0)$ may be chosen arbitrarily because the governing equations for the impedance response are linear, even when the steady-state problem is nonlinear.

The oscillating current associated with B^+ may be expressed as

$$\tilde{i}_{B^+} = \left(\frac{\partial i_{B^+}}{\partial V}\right)_{c_{B^+}(0)} \tilde{V} + \left(\frac{\partial i_{B^+}}{\partial c_{B^+}(0)}\right)_V \tilde{c}_{B^+}(0) \tag{11.154}$$

Following equation (5.6) on page 92, the flux expression for B^+,

$$i_{B^+} = \frac{nFD_{B^+}}{s_{B^+}} \frac{dc_{B^+}}{dy}\bigg|_{y=0} = -FD_{B^+} \frac{dc_{B^+}}{dy}\bigg|_{y=0} \tag{11.155}$$

yields a second equation for the oscillating current density as

$$\tilde{i}_{B^+} = -FD_{B^+} \frac{d\tilde{c}_{B^+}}{dy}\bigg|_{y=0} \tag{11.156}$$

Table 11.1: Species and associated parameter values for the system described in Example 11.9 in which a homogeneous reaction (11.137) contributes to a heterogeneous reaction (11.138).

Species	$c_i(\infty)$, mol/cm^3	z_i	D_i, cm^2/s
AB	0.01	0	1.684×10^{-5}
A$^-$	0.0001	-1	1.957×10^{-5}
B$^+$	0.0001	1	1.902×10^{-5}

Equation (11.154) can be divided by equation (11.156), yielding

$$1 = \left(\frac{\partial i_{B^+}}{\partial V}\right)_{c_{B^+}(0)} \frac{\widetilde{V}}{\widetilde{i}_{B^+}} + \left(\frac{\partial i_{B^+}}{\partial c_{B^+}(0)}\right)_V \frac{-\widetilde{c}_{B^+}(0)}{FD_{B^+} \left.\frac{d\widetilde{c}_{B^+}}{dy}\right|_{y=0}} \tag{11.157}$$

Thus, the impedance may be expressed as

$$Z_{F,B^+} = R_{t,B^+} + Z_{D,B^+} \tag{11.158}$$

where, from the respective derivatives of equation (11.139),

$$R_{t,B^+} = \frac{1}{K_{B^+} b_{B^+} \bar{c}_{B^+}(0) \exp\left(-b_{B^+}\overline{V}\right)} \tag{11.159}$$

and

$$Z_{D,B^+} = \frac{R_{t,B^+} K_{B^+} \exp\left(-b_{B^+}\overline{V}\right)}{FD_{B^+}} \left(-\frac{\widetilde{c}_{B^+}(0)}{\left.\frac{d\widetilde{c}_{B^+}}{dy}\right|_{y=0}}\right) \tag{11.160}$$

The concentration distributions required to assess equation (11.160) can be obtained for each frequency from the numerical solution of equation (11.149).

Example 11.9 Impedance with Homogeneous Reactions: *Calculate the impedance response for a system in which a homogeneous reaction (11.137) contributes to a heterogeneous reaction (11.138). The system parameters are presented in Tables 11.1 and 11.2.*

Solution: *As described in Section 11.7, the calculation of impedance requires solution of the steady-state problem. The system of equations was solved numerically using the FORTRAN programming language and Newman's BAND algorithm.[208] The polarization curve corresponding to the parameters given in Tables 11.1 and 11.2 is presented in Figure 11.22. The heterogeneous reaction rate increases as the potential becomes more negative, reaching a mass-transfer-limited plateau for potentials smaller than -2 V. The large magnitude of the mass-transfer-limited current density may be attributed to the production of B$^+$ by the homogeneous reaction.*

Table 11.2: System and kinetic parameter values for the system described in Example 11.9.

Parameter	Value	Units
Disk rotation rate, Ω	2,000	rpm
Kinematic viscosity, ν	0.01	cm^2/s
Homogeneous equilibrium constant, K_{eq}	10^{-6}	mol/cm^3
Homogeneous rate constant, k_b	10^7	$cm^3/mol\ s$
Heterogeneous rate constant, K_{B+}	2×10^{-12}	A/cm^2
Heterogeneous constant, b_{B+}	19.9	V^{-1}

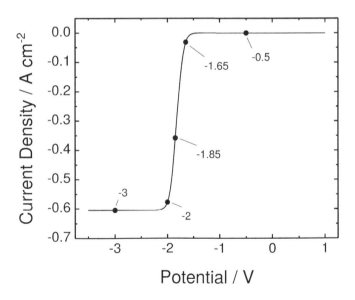

Figure 11.22: Polarization curve calculated for system parameters presented in Tables 11.1 and 11.2. Labeled potential values correspond to steady-state concentration profiles presented in Figure 11.23.

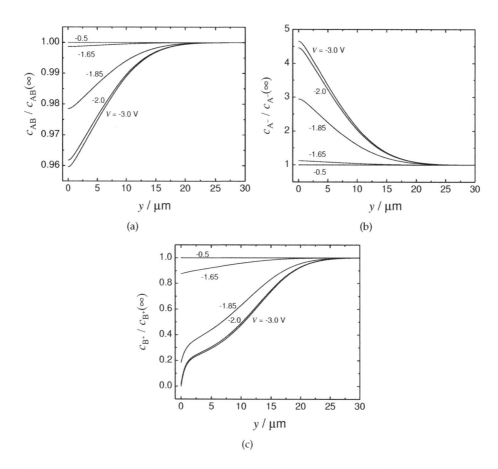

Figure 11.23: Calculated steady-state concentration distributions corresponding to system parameters presented in Tables 11.1 and 11.2: a) AB; b) A^-; and c) B^+.

Labeled potential values in Figure 11.22 correspond to steady-state concentration profiles presented in Figure 11.23. Concentrations were scaled by the corresponding values at $y \to \infty$ to emphasize the relative changes in values. The concentration of AB, shown in Figure 11.23(a), decreases to a value that is 96 percent of the bulk value as potential approaches -3 V. In contrast, the concentration of A^- shown in Figure 11.23(b) reaches a value that is almost 5 times its bulk value at the mass-transfer-limited current density. The normalized concentration distribution of B^+ is presented in Figure 11.23(c). The concentration of B^+ at the electrode surface approaches a value of zero as the mass-transfer-limited current density is approached. The concentration distribution shown in Figure 11.23(c) should be compared to that presented in Figure 11.11 for a system that is not influenced by homogeneous reactions. The sharp profile that appears close to the electrode surface in Figure 11.23(c) is consistent with a large current density.

The influence of the homogeneous production of B^+ on the polarization curve can be seen in Figure 11.24(a) where a homogeneous rate constant of $k_b = 10^7$ cm^3/mol s yields a mass-transfer-limited current density that is 4.7 times larger than that in the absence of homogeneous

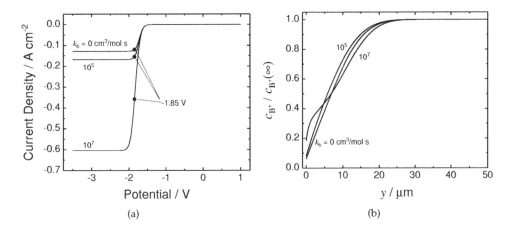

Figure 11.24: Calculations corresponding to system parameters presented in Tables 11.1 and 11.2 with homogeneous rate constant k_b as a parameter: a) polarization curve; and b) concentration distribution for B^+ at a potential of $V = -1.85$ V.

reactions. The concentration of B^+ is presented in Figure 11.24(b) as a function of position. In the absence of homogeneous reaction, the concentration profile resembles that shown in Figure 11.11. The slope at the electrode-electrolyte interface becomes larger as the homogeneous rate constant increases. For $k_b = 10^7$ cm³/mol s, the steeper slope at the electrode corresponds to a shallower slope at intermediate values of y.

The convective diffusion impedance corresponding to the steady-state results presented in Figures 11.22 and 11.23 is presented in Figure 11.25 with applied potential as a parameter. As compared to the dimensionless diffusion impedance presented in Figure 11.13 for finite Schmidt numbers and in the absence of homogeneous reactions, the diffusion impedance presented in Figure 11.25 is smaller in magnitude and decreases as the rate of the heterogeneous consumption of B^+ increases. Also, two asymmetric capacitive loops are seen in Figure 11.25 as compared to a single loop in Figure 11.13. The low-frequency loop has a characteristic frequency $K = 2.5$, which is in agreement with the characteristic frequency associated with diffusion in the absence of homogeneous reactions. The high-frequency loop can then be associated with the homogeneous reaction. The characteristic frequency for the high-frequency loop is on the order of $K = 1000$, suggesting that the characteristic dimension for the reaction is much smaller than the Nernst diffusion-layer thickness.

The characteristic frequencies may be seen more clearly in a plot of the magnitude of the imaginary impedance as a function of dimensionless frequency shown in Figure 11.26. As shown in Figure 11.27, the characteristic dimension for the region in which the homogeneous production of B^+ takes place is smaller than 1 µm. For the parameters given in Tables 11.1 and 11.2, the convective diffusion-layer thickness has a value of 15.5 µm. Comparison of the characteristic frequencies for the two loops yields a value of 0.77 µm, which is in agreement with the results presented in Figure 11.27.

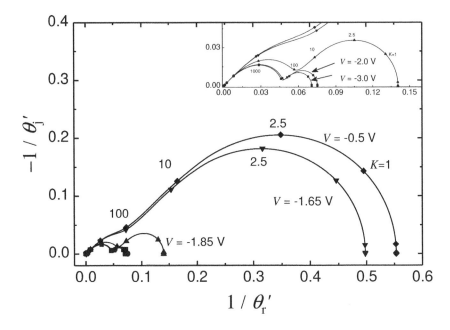

Figure 11.25: Dimensionless convective diffusion impedance for the system presented in Figure 11.22 with applied potential as a parameter.

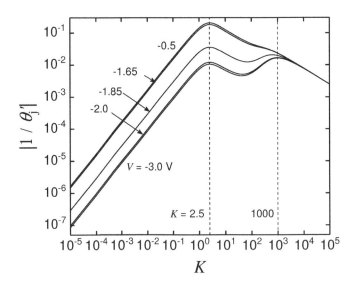

Figure 11.26: The imaginary part of the dimensionless convective diffusion impedance for the system presented in Figure 11.22 as a function of dimensionless frequency with applied potential as a parameter.

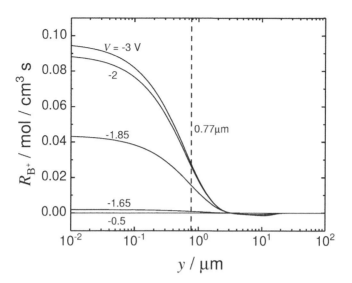

Figure 11.27: The rate of homogeneous production of B^+ for the system presented in Figure 11.22 as a function of position with applied potential as a parameter.

In the limit of a very fast homogeneous reaction rate, reactions (11.137) and (11.138) may be expressed as

$$AB + e^- \rightarrow A^- + B \qquad (11.161)$$

In this case, only one loop is visible in the convective diffusion impedance plots corresponding to the diffusion of AB to the electrode surface.

Example 11.10 Gerischer Impedance: *In the presence of supporting electrolyte, the concentration of the anion* A^- *may sufficiently large to be considered constant, making reaction (11.137) first order with respect to species AB and* B^+. *Under the assumptions that the diffusion coefficients for AB and* B^+ *are equal and that convection may be ignored, develop an analytic expression for the diffusion impedance associated with a heterogeneous reaction influenced by a homogeneous reaction. This derivation makes use of the Nernst stagnant diffusion layer hypothesis developed in Section 11.3.4.*

Solution: *When the concentration of* A^- *may be considered invariant, the rate of production of species AB and* B^+ *by reaction (11.137) may be expressed as*

$$R_{B^+} = -R_{AB} = k_f c_{AB}(y) - k_b c_{B^+}(y) \qquad (11.162)$$

The conservation equations for species AB and B^+ *may be expressed as*

$$\frac{\partial c_{AB}}{\partial t} = D \frac{\partial^2 c_{AB}}{\partial y^2} - k_f c_{AB} + k_b c_{B^+} \qquad (11.163)$$

and

$$\frac{\partial c_{B^+}}{\partial t} = D\frac{\partial^2 c_{B^+}}{\partial y^2} + k_f c_{AB} - k_b c_{B^+} \tag{11.164}$$

respectively, where

$$D = D_{AB} = D_{B^+} \tag{11.165}$$

The sum of equations (11.163) and (11.164) yields

$$\frac{\partial}{\partial t}\left(c_{AB} + c_{B^+}\right) = D\frac{\partial^2}{\partial y^2}\left(c_{AB} + c_{B^+}\right) \tag{11.166}$$

After algebraic manipulation, the difference between equations (11.163) and (11.164) may be expressed as

$$\frac{\partial}{\partial t}\left(c_{AB} - \frac{c_{B^+}}{K_{eq}}\right) = D\frac{\partial^2}{\partial y^2}\left(c_{AB} - \frac{c_{B^+}}{K_{eq}}\right) - k\left(c_{AB} - \frac{c_{B^+}}{K_{eq}}\right) \tag{11.167}$$

where $k = k_f + k_b$ and $K_{eq} = k_f/k_b$. As equations (11.166) and (11.167) are linear, the solution for the corresponding diffusion impedance does not require a solution for the steady state.

Equations (11.166) and (11.167) may be expressed in frequency domain as

$$\frac{j\omega}{D}\left(\tilde{c}_{AB} + \tilde{c}_{B^+}\right) = \frac{\partial^2}{\partial y^2}\left(\tilde{c}_{AB} + \tilde{c}_{B^+}\right) \tag{11.168}$$

and

$$\frac{j\omega}{D}\left(\tilde{c}_{AB} - \frac{\tilde{c}_{B^+}}{K_{eq}}\right) = \frac{\partial^2}{\partial y^2}\left(\tilde{c}_{AB} - \frac{\tilde{c}_{B^+}}{K_{eq}}\right) - \frac{k}{D}\left(\tilde{c}_{AB} - \frac{\tilde{c}_{B^+}}{K_{eq}}\right) \tag{11.169}$$

respectively. The general solution to equations (11.168) and (11.169) may be expressed, following Example 2.3 on page 30, as

$$\left(\tilde{c}_{AB} + \tilde{c}_{B^+}\right) = C_1\exp(-s_1 y) + C_2\exp(s_1 y) \tag{11.170}$$

and

$$\left(\tilde{c}_{AB} - \frac{\tilde{c}_{B^+}}{K_{eq}}\right) = C_3\exp(-s_2 y) + C_4\exp(s_2 y) \tag{11.171}$$

respectively, where

$$s_1^2 = \frac{j\omega}{D} \tag{11.172}$$

and

$$s_2^2 = \frac{j\omega + k}{D} \tag{11.173}$$

The boundary conditions for diffusion through a film of finite thickness δ are:

$$\tilde{c}_{AB}(\delta) = 0 \tag{11.174}$$

$$\tilde{c}_{B^+}(\delta) = 0 \tag{11.175}$$

$$\tilde{c}_{B^+}(0) = 1 \tag{11.176}$$

and

$$\left.\frac{d\tilde{c}_{AB}}{dy}\right|_{y=0} = 0 \tag{11.177}$$

The corresponding expressions are

$$0 = C_1 \exp(-s_1 y) + C_2 \exp(s_1 y) + K_{eq}C_3 \exp(-s_2 y) + K_{eq}C_4 \exp(s_2 y) \tag{11.178}$$

$$0 = C_1 \exp(-s_1 y) + C_2 \exp(s_1 y) - C_3 \exp(-s_2 y) - C_4 \exp(s_2 y) \tag{11.179}$$

$$1 + \frac{1}{K_{eq}} = C_1 + C_2 - C_3 - C_4 \tag{11.180}$$

and

$$0 = -C_1 s_1 + C_2 s_1 - K_{eq}C_3 s_2 + K_{eq}C_4 s_2 \tag{11.181}$$

respectively.

The dimensionless diffusion impedance may be expressed as

$$-\frac{1}{\theta'_{B^+}} = -\frac{1}{\delta}\frac{\tilde{c}_{B^+}(0)}{\dfrac{d\tilde{c}_{B^+}}{dy}\bigg|_{y=0}} \tag{11.182}$$

$$= \frac{1}{K_{eq}+1}\frac{\tanh\sqrt{(j\omega+k)\dfrac{\delta^2}{D}}}{\sqrt{(j\omega+k)\dfrac{\delta^2}{D}}} + \frac{K_{eq}}{K_{eq}+1}\frac{\tanh\sqrt{j\omega\dfrac{\delta^2}{D}}}{\sqrt{j\omega\dfrac{\delta^2}{D}}} \tag{11.183}$$

or

$$-\frac{1}{\theta'_{B^+}} = \frac{1}{K_{eq}+1}\frac{\tanh\sqrt{jK+k_{dim}}}{\sqrt{jK+k_{dim}}} + \frac{K_{eq}}{K_{eq}+1}\frac{\tanh\sqrt{jK}}{\sqrt{jK}} \tag{11.184}$$

where $K = \omega\delta^2/D$ and $k_{dim} = k\delta^2/D$. The results are presented in Figure 11.28. The first term on the right-hand-side of equation (11.184) corresponds to a modified Gerischer impedance (see, for example Boukamp and Bouwmeester[209]). The second term on the right-hand-side of equation (11.184) corresponds to the diffusion impedance. The results shown in Figure 11.28 are similar to those presented in Figure 11.25 for a rotating disk electrode.

At high frequencies,

$$-\frac{1}{\theta'_{B^+}} = \frac{1}{K_{eq}+1}\frac{1}{\sqrt{jK+k_{dim}}} + \frac{K_{eq}}{K_{eq}+1}\frac{1}{\sqrt{jK}} \tag{11.185}$$

$$= \frac{1}{K_{eq}+1}Z_G + \frac{K_{eq}}{K_{eq}+1}Z_W \tag{11.186}$$

where Z_G is the Gerischer impedance[210] and Z_W is the Warburg impedance. While the form of equation (11.186) applies for diffusion in an infinite domain such as discussed in Example 11.1, the Gerischer impedance shows a finite low-frequency asymptote

$$\lim_{K\to 0} Z_G = \frac{1}{\sqrt{k_{dim}}} \tag{11.187}$$

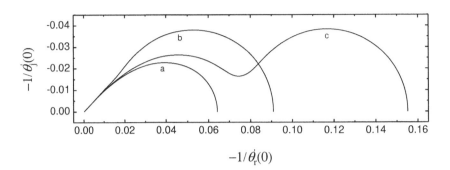

Figure 11.28: The dimensionless diffusion impedance given by equation (11.184) in Nyquist format for $K_e q = 0.1$ and $k\delta^2/D = 200$: a) the first term on the right-hand-side of equation (11.184) corresponding to a modified Gerischer impedance; b) the second term on the right-hand-side of equation (11.184) corresponding to the diffusion impedance; and c) the sum of impedances (a) and (b).

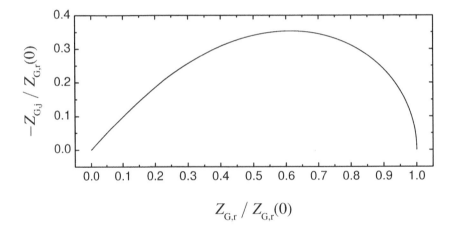

Figure 11.29: The normalized Gerischer impedance in Nyquist format.

due to the contribution of local homogeneous generation of reacting species. The shape of the normalized Gerischer impedance, given in Figure 11.29, is independent of the value of k_{dim}, but the characteristic frequency depends on k_{dim} following

$$K_c = 1.7783 k_{\text{dim}} \tag{11.188}$$

or, as frequency and $k = k_f + k_b$ are made dimensionless using the same terms,

$$\omega_c = 1.7783 k \tag{11.189}$$

Lasia provides a different derivation of the Gerischer impedance in Section 4.10 of reference 15.

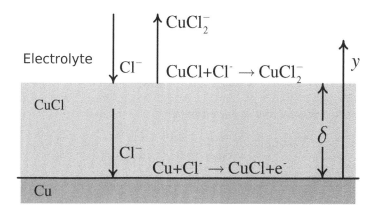

Figure 11.30: Schematic representation of a copper electrode covered by a CuCl layer of thickness δ.

11.8 Dynamic Surface Films

The treatment presented in Section 11.2 of the diffusion impedance associated with a porous film is predicated on the assumption that the film thickness is independent of time. Surface films, however, are often the result of dynamic processes where the film is formed on the electrode surface and dissolves at the film-electrolyte interface. If the formation of the film is the result of electrochemical reactions, the film thickness may be time dependent. Thus, in response to a modulated potential, the film thickness is modulated, and this modulation will influence the diffusion impedance through the film. An example, based on the work of Barcia et al.[174] is presented below.

Example 11.11 Dissolution of Copper through a CuCl Salt Layer: *Develop a model for the impedance response for dissolution of copper on the mass-transfer-limited plateau in 1 M HCl electrolyte where the CuCl film layer thickness is modulated in response to the potential perturbation. The system is represented by the schematic illustration shown in Figure 11.30.*

Solution: *At the Cu-CuCl interface, designated to be $y = 0$, the chloride ion reacts to form CuCl following*

$$Cu + Cl^- \rightarrow CuCl + e^- \qquad (11.190)$$

This reaction is assumed to be limited by the transport of chloride ions through the film, thus, the current density at $y = 0$ is given by

$$\frac{i_{Cu}}{F} = +D_{Cl^-,f}\left.\frac{dc_{Cl^-}}{dy}\right|_{y=0} = D_{Cl^-,f}\frac{c_{Cl^-}(\infty) - c_{Cl^-}(0)}{\delta} \qquad (11.191)$$

where δ is the film thickness, $D_{\mathrm{Cl}^-,\mathrm{f}}$ is the diffusion coefficient for Cl^- within the film, and the concentration of Cl^- at $y = \delta$ was assumed to be equal to the concentration at $y = \infty$.

At the CuCl layer-electrolyte interface, $y = \delta$, the layer dissolves according to

$$\mathrm{CuCl} + \mathrm{Cl}^- \rightarrow \mathrm{CuCl}_2^- \tag{11.192}$$

The current density evaluated at $y = \delta$ is given by

$$\frac{i_{\mathrm{CuCl}_2^-}}{\mathrm{F}} = k_2 c_{\mathrm{Cl}^-}(\infty) - k_{-2} c_{\mathrm{CuCl}_2^-}(\delta) \tag{11.193}$$

Equation (11.192) represents a chemical dissolution of the film. The current density can also expressed in terms of diffusion of CuCl_2^- from the film surface as

$$\frac{i_{\mathrm{CuCl}_2^-}}{\mathrm{F}} = -D_{\mathrm{CuCl}_2^-} \left. \frac{d c_{\mathrm{CuCl}_2^-}}{dy} \right|_{y=\delta} = D_{\mathrm{CuCl}_2^-} \frac{c_{\mathrm{CuCl}_2^-}(\delta)}{\delta_{\mathrm{N},\mathrm{CuCl}_2^-}} \tag{11.194}$$

where $\delta_{\mathrm{N},\mathrm{CuCl}_2^-}$ is the diffusion layer thickness for CuCl_2^- within the electrolyte. Combination of equations (11.193) and (11.194) yields

$$\frac{i_{\mathrm{CuCl}_2^-}}{\mathrm{F}} = \frac{k_2 c_{\mathrm{Cl}^-}(\infty)}{1 + k_{-2}\delta_{\mathrm{N},\mathrm{CuCl}_2^-} / D_{\mathrm{CuCl}_2^-}} \tag{11.195}$$

If $k_{-2}\delta_{\mathrm{N},\mathrm{CuCl}_2^-} / D_{\mathrm{CuCl}_2^-} \gg 1$, $i_{\mathrm{CuCl}_2^-}$ varies according to the square root of rotation speed, in accordance with the Levich equation given as (5.68) on page 106. Under the assumption that reaction (11.190) is mass transfer limited, equation (11.195) is independent of potential.

The general expression for the thickness of the CuCl layer is given by

$$\rho(1-\varepsilon) \frac{\partial \delta}{\partial t} = D_{\mathrm{Cl}^-,\mathrm{f}} \left. \frac{\partial c_{\mathrm{Cl}^-}}{\partial y} \right|_{y=0} - D_{\mathrm{CuCl}_2^-} \left. \frac{\partial c_{\mathrm{CuCl}_2^-}}{\partial y} \right|_{y=\delta} \tag{11.196}$$

where ε is the layer porosity, and ρ is the layer molar density. At steady state, the layer thickness is given by

$$\delta = \frac{D_{\mathrm{Cl}^-,\mathrm{f}}}{D_{\mathrm{CuCl}_2^-}} \frac{k_{-2}}{k_2} \delta_{\mathrm{N},\mathrm{CuCl}_2^-} \tag{11.197}$$

Equation (11.197) shows that the thickness of the CuCl salt layer varies with the diffusion layer thickness in the electrolyte. Mass transfer characteristics must be established for both the electrolyte and the salt layer.

11.8.1   Mass Transfer in the Salt Layer

The conservation equation for Cl^- within the salt film can be expressed as

$$\frac{\partial c_{\mathrm{Cl}^-}}{\partial t} = D_{\mathrm{Cl}^-,\mathrm{f}} \frac{\partial^2 c_{\mathrm{Cl}^-}}{\partial y^2} \tag{11.198}$$

Equation (11.198) can be expressed in terms of the time-dependent film thickness by introducing a normalized position variable

$$\chi = \frac{y}{\delta(t)} \tag{11.199}$$

The development presented in Example 2.7 on page 37 can be used to express equation (11.198) as

$$\frac{D_{Cl^-,f}}{\delta^2} \frac{\partial^2 c_{Cl^-}}{\partial \chi^2} = \left(\frac{\partial c_{Cl^-}}{\partial t} \right)_\chi - \chi \frac{d\bar{c}_{Cl^-}}{dy} \frac{\partial \delta}{\partial t} \tag{11.200}$$

Following the form presented as equation (10.3) on page 207, equation (11.200) can be expressed in frequency domain as

$$\frac{\partial^2 \tilde{c}_{Cl^-}}{\partial \chi^2} = j\mu \tilde{c}_{Cl^-} - \chi \frac{d\bar{c}_{Cl^-}}{dy} j\mu \tilde{\delta} \tag{11.201}$$

where $\mu = \omega \delta^2 / D_{Cl^-,f}$.

   The general solution for equation (11.201) is

$$\tilde{c}_{Cl^-} = \frac{d\bar{c}_{Cl^-}}{dy} \tilde{\delta} + A \exp \left(\sqrt{j\mu} \chi \right) + B \exp \left(-\sqrt{j\mu} \chi \right) \tag{11.202}$$

where A and B are constants obtained from consideration of boundary conditions. At the interface between the salt layer and the electrolyte, $y = \delta$ or $\chi = 1$, the concentration of Cl^- is constant and equal to its value at $y = \infty$. Thus,

$$\tilde{c}_{Cl^-}(\delta) = 0 = \frac{d\bar{c}_{Cl^-}}{dy} \chi \tilde{\delta} + A \exp \left(\sqrt{j\mu} \right) + B \exp \left(-\sqrt{j\mu} \right) \tag{11.203}$$

The expression for $y = \chi = 0$ can be simplified to obtain

$$\tilde{c}_{Cl^-}(0) = A + B \tag{11.204}$$

and

$$\left. \frac{\partial \tilde{c}_{Cl^-}}{\partial \chi} \right|_{y=0} = \frac{d\bar{c}_{Cl^-}}{dy} \tilde{\delta} + \sqrt{j\mu} (A - B) \tag{11.205}$$

From the definition of χ expressed in equation (11.199),

$$\frac{\partial \tilde{c}_{Cl^-}}{\partial \chi} = \frac{\partial \bar{c}_{Cl^-}}{\partial y} \tilde{\delta} + \bar{\delta} \frac{\partial \tilde{c}_{Cl^-}}{\partial y} \tag{11.206}$$

Thus, equation (11.205) can be expressed as

$$\left. \frac{\partial \tilde{c}_{Cl^-}}{\partial y} \right|_{y=0} = \frac{\sqrt{j\mu}}{\bar{\delta}} (A - B) \tag{11.207}$$

Equations (11.204) and (11.207) are not sufficient to obtain values for A and B. It is necessary to solve balance equations for $CuCl_2^-$ in the electrolyte and for film thickness.

11.8.2   Mass Transfer in the Electrolyte

The solution here follows the development presented in Section 11.3.3 with the exception that the position viable is referenced to the time-dependent film thickness, i.e.,

$$\eta = \frac{y - \delta(t)}{\delta_{N,i}} \tag{11.208}$$

where, to create a more compact presentation, the subscript i refers to $CuCl_2^-$. Following the approach presented in Example 2.8 on page 37, the governing equation may be expressed as

$$\frac{\partial c_i}{\partial \tau_i} - \frac{1}{\delta_{N,i}} \frac{d\bar{c}_i}{d\eta} \frac{d\delta}{d\tau_i} - 3\eta^2 \frac{\partial c_i}{\partial \eta} - \frac{\partial^2 c_i}{\partial \eta^2} = 0 \tag{11.209}$$

where $\tau_i = t D_i / \delta_{N,i}^2$, and $\delta_{N,i}$ is given by equation (11.72). Equation (11.209) may be expressed in frequency domain as

$$\frac{d^2 \tilde{c}_i}{d\eta^2} + 3\eta^2 \frac{d\tilde{c}_i}{d\eta} - jK\tilde{c}_i = -jK \frac{1}{\delta_{N,i}} \frac{d\bar{c}_i}{d\eta} \tilde{\delta} \tag{11.210}$$

where $K = \omega \delta_{N,i}^2 / D_i$. For $c_i = \lambda_i \theta_{i,0}$ where $\theta_{i,0}$ is the solution to the homogeneous part, equation (11.210) may be written as

$$\frac{d^2 \lambda_i}{d\eta^2} + \left(\frac{2}{\theta_{i,0}} \frac{d\theta_{i,0}}{d\eta} + 3\eta^2 \right) \frac{d\lambda_i}{d\eta} = -\frac{jK}{\theta_{i,0} \delta_{N,i}} \frac{d\bar{c}_i}{d\eta} \tilde{\delta} \tag{11.211}$$

The solution to the homogeneous part of equation (11.211) is

$$\frac{d\lambda_{i,0}}{d\eta} = \frac{\exp\left(-\eta^3\right)}{\theta_{i,0}^2} \tag{11.212}$$

The solution to equation (11.211) is of the form

$$\frac{d\lambda_i}{d\eta} = \gamma_i \frac{\exp\left(-\eta^3\right)}{\theta_{i,0}^2} \tag{11.213}$$

where

$$\frac{d\gamma_i}{d\eta} = -jK\theta_{i,0} \exp\left(\eta^3\right) \frac{d\bar{c}_i}{d\eta} \frac{\tilde{\delta}}{\delta_{N,i}} \tag{11.214}$$

As shown in Example 2.2 on page 27, the steady-state concentration gradient may be expressed as

$$\frac{d\bar{c}_i}{d\eta} = \frac{\bar{c}_i(\infty) - \bar{c}_i(0)}{\Gamma(4/3)} \exp\left(-\eta^3\right) = \left.\frac{d\bar{c}_i}{d\eta}\right|_0 \exp\left(-\eta^3\right) \tag{11.215}$$

Thus,

$$\frac{d\gamma_i}{d\eta} = -jK\theta_{i,0} \left.\frac{d\bar{c}_i}{d\eta}\right|_0 \frac{\tilde{\delta}}{\delta_{N,i}} \tag{11.216}$$

and

$$\gamma_i = -jK \left.\frac{d\bar{c}_i}{d\eta}\right|_0 \frac{\tilde{\delta}}{\delta_{N,i}} \left(\int_0^\eta \theta_{i,0} d\eta + C_1 \right) \tag{11.217}$$

where C_1 is a constant of integration.

The solution to equation (11.211) may be expressed as

$$\gamma_i = -jK \left.\frac{d\bar{c}_i}{d\eta}\right|_0 \frac{\tilde{\delta}}{\delta_{N,i}} \left(\int_0^{\eta_1} \left(\int_0^\eta \theta_{i,0} d\eta + C_1 \right) \frac{\exp\left(-\eta^3\right)}{\theta_{i,0}} d\eta_1 \right) + C_2 \tag{11.218}$$

where C_2 is a constant of integration. The solution to equation (11.210) is

$$\tilde{c}_i = C_2 \theta_{i,0} - jK \left.\frac{d\bar{c}_i}{d\eta}\right|_0 \frac{\tilde{\delta}}{\delta_{N,i}} \theta_{i,0} \left(\int_0^{\eta_1} \left(\int_0^\eta \theta_{i,0} d\eta + C_1 \right) \frac{\exp\left(-\eta^3\right)}{\theta_{i,0}} d\eta_1 \right) \tag{11.219}$$

A similar development is presented for electrohydrodynamic impedance in Section 16.2 on page 448.

At $\eta = 0$, $\theta_{i,0} = 1$, and $C_2 = \theta_i(0)$. For $\eta \to \infty$, $\tilde{c}_i$ tends toward zero, yielding $C_1 = -\int_0^\infty \theta_{i,0} d\eta$. The derivative at the interface is given by

$$\left.\frac{d\tilde{c}_i}{dy}\right|_0 = \tilde{c}_i(0) \frac{\theta'_{i,0}}{\delta_{N,i}} + jK \left.\frac{d\bar{c}_i}{d\eta}\right|_0 \frac{\tilde{\delta}}{\delta_{N,i}} W_i \tag{11.220}$$

where $W_i = \int_0^\infty \theta_{i,0} d\eta$ must be obtained from numerical calculations. An expression for W_i is presented in equation (16.30) on page 450. The second term on the right-hand side of equation (11.220) accounts for the influence of the modulated film thickness on the convective diffusion impedance. At the interface between the film and the electrolyte,

$$D_i \left.\frac{d\tilde{c}_i}{dy}\right|_0 = k_{-2} \tilde{c}_i(0) \tag{11.221}$$

Equation (11.220) becomes

$$\left.\frac{d\tilde{c}_i}{dy}\right|_0 = \frac{jK \left.\dfrac{d\bar{c}_i}{dy}\right|_0 \dfrac{\tilde{\delta}}{\delta_{N,i}} W_i}{1 - \dfrac{\theta'_{i,0}}{\delta_{N,i}} \dfrac{D_i}{k_{-2}}} \tag{11.222}$$

The next step in the development is to develop an expression for the oscillating film thickness, $\tilde{\delta}$.

11.8.3 Oscillating Film Thickness

To maintain the compact presentation, the subscript i refers to $CuCl_2^-$. After inclusion of equation (11.207), equation (11.196) may be written in frequency domain as

$$\rho\left(1-\varepsilon\right) j\omega\tilde{\delta} = D_{Cl^-,f} \frac{\sqrt{j\mu}}{\tilde{\delta}} (A - B) - D_i \left.\frac{d\tilde{c}_i}{dy}\right|_0 \tag{11.223}$$

After inclusion of equation (11.222),

$$\tilde{\delta} \left(\rho \left(1 - \varepsilon\right) j\omega + D_i \frac{jK \left.\frac{d\bar{c}_i}{dy}\right|_0 \frac{W_i}{\delta_{N,i}}}{1 - \frac{\theta'_{i,0}}{\delta_{N,i}} \frac{D_i}{k_{-2}}} \right) = D_{Cl^-,f} \frac{\sqrt{j\mu}}{\tilde{\delta}} (A - B) \tag{11.224}$$

Equation (11.202) may be written as

$$\tilde{\delta} = - \frac{A \exp\left(\sqrt{j\mu}\right) + B \exp\left(-\sqrt{j\mu}\right)}{\dfrac{d\bar{c}_{Cl^-}}{dy}} \tag{11.225}$$

The quantity $F_1(\omega)$ may be defined such that

$$F_1(\omega) = \rho \left(1 - \varepsilon\right) j\omega + D_i \frac{jK \left.\frac{d\bar{c}_i}{dy}\right|_0 \frac{W_i}{\delta_{N,i}}}{1 - \frac{\theta'_{i,0}}{\delta_{N,i}} \frac{D_i}{k_{-2}}} \tag{11.226}$$

represents the left-hand side of equation (11.224) divided by $\tilde{\delta}$. Introduction of dimensionless frequency $K = \omega \delta_{N,i}^2 / D_i$ yields

$$F_1(\omega) = j\omega \left(\rho \left(1 - \varepsilon\right) + D_i \frac{\delta_{N,i} \left.\frac{d\bar{c}_i}{dy}\right|_0 \frac{W_i}{}}{1 - \frac{\theta'_{i,0}}{\delta_{N,i}} \frac{D_i}{k_{-2}}} \right) \tag{11.227}$$

Introduction of equation (11.225) into equation (11.224) yields

$$-\left(\frac{A \exp\left(\sqrt{j\mu}\right) + B \exp\left(-\sqrt{j\mu}\right)}{\dfrac{d\bar{c}_{Cl^-}}{dy}} \right) F_1(\omega) = D_{Cl^-,f} \frac{\sqrt{j\mu}}{\tilde{\delta}} (A - B) \tag{11.228}$$

Upon separation of terms involving constants A and B,

$$A \left(D_{Cl^-,f} \frac{\sqrt{j\mu}}{\tilde{\delta}} + \frac{\exp\left(\sqrt{j\mu}\right)}{\dfrac{d\bar{c}_{Cl^-}}{dy}} F_1(\omega) \right) = \tag{11.229}$$

$$B \left(D_{Cl^-,f} \frac{\sqrt{j\mu}}{\tilde{\delta}} - \frac{\exp\left(-\sqrt{j\mu}\right)}{\dfrac{d\bar{c}_{Cl^-}}{dy}} F_1(\omega) \right)$$

The ratio A/B is expressed as

$$\frac{A}{B} = \frac{\left(D_{\text{Cl}^-,\text{f}} \dfrac{\sqrt{j\mu}}{\bar{\delta}} - \dfrac{\exp\left(-\sqrt{j\mu}\right)}{\dfrac{d\bar{c}_{\text{Cl}^-}}{dy}} F_1(\omega) \right)}{\left(D_{\text{Cl}^-,\text{f}} \dfrac{\sqrt{j\mu}}{\bar{\delta}} + \dfrac{\exp\left(\sqrt{j\mu}\right)}{\dfrac{d\bar{c}_{\text{Cl}^-}}{dy}} F_1(\omega) \right)} \tag{11.230}$$

By adding unity to both sides of equation (11.230),

$$\frac{A}{B} + 1 = \frac{\left(2D_{\text{Cl}^-,\text{f}} \dfrac{\sqrt{j\mu}}{\bar{\delta}} + \dfrac{F_1(\omega)}{\dfrac{d\bar{c}_{\text{Cl}^-}}{dy}} \left(\exp\left(\sqrt{j\mu}\right) - \exp\left(-\sqrt{j\mu}\right) \right) \right)}{\left(D_{\text{Cl}^-,\text{f}} \dfrac{\sqrt{j\mu}}{\bar{\delta}} + \dfrac{\exp\left(\sqrt{j\mu}\right)}{\dfrac{d\bar{c}_{\text{Cl}^-}}{dy}} F_1(\omega) \right)} \tag{11.231}$$

Similarly,

$$\frac{A}{B} - 1 = \frac{\left(-\dfrac{F_1(\omega)}{\dfrac{d\bar{c}_{\text{Cl}^-}}{dy}} \left(\exp\left(\sqrt{j\mu}\right) + \exp\left(-\sqrt{j\mu}\right) \right) \right)}{\left(D_{\text{Cl}^-,\text{f}} \dfrac{\sqrt{j\mu}}{\bar{\delta}} + \dfrac{\exp\left(\sqrt{j\mu}\right)}{\dfrac{d\bar{c}_{\text{Cl}^-}}{dy}} F_1(\omega) \right)} \tag{11.232}$$

The terms $A/B + 1$ and $A/B - 1$ are used in the development of the expression for faradaic impedance.

11.8.4 Faradaic Impedance

Following the development in Section 10.2.2 on page 213, the oscillating current is related to potential and concentration according to

$$\tilde{\imath}_{\text{Cu}} = K_1 \bar{c}_{\text{Cl}^-}(0) b_{\text{Cu}} \exp\left(b_{\text{Cu}} \overline{V}\right) \tilde{V} + K_1 \exp\left(b_{\text{Cu}} \overline{V}\right) \tilde{c}_{\text{Cl}^-}(0) \tag{11.233}$$

and

$$\tilde{\imath}_{\text{Cu}} = F D_{\text{Cl}^-,\text{f}} \left. \frac{d\tilde{c}_{\text{Cl}^-}}{dy} \right|_0 \tag{11.234}$$

The faradaic impedance may be given as

$$Z_{\text{F}} = R_{\text{t}} - R_{\text{t}} \frac{K_1 \exp\left(b_{\text{Cu}} \overline{V}\right) \tilde{c}_{\text{Cl}^-}(0)}{F D_{\text{Cl}^-,\text{f}} \left. \dfrac{d\tilde{c}_{\text{Cl}^-}}{dy} \right|_0} \tag{11.235}$$

where

$$R_t = \frac{1}{K_1 \bar{c}_{Cl^-}(0) \exp\left(b_{Cu}\overline{V}\right)} \qquad (11.236)$$

Introduction of equations (11.204) and (11.207) yields

$$Z_F = R_t - R_t \frac{K_1 \exp\left(b_{Cu}\overline{V}\right)}{FD_{Cl^-,f}} \frac{\left(\dfrac{A}{B}+1\right)}{\dfrac{\sqrt{j\mu}}{\bar{\delta}}\left(\dfrac{A}{B}-1\right)} \qquad (11.237)$$

Introduction of equations (11.231) and (11.232) yields

$$Z_F = R_t - R_t \frac{K_1 \exp\left(b_{Cu}\overline{V}\right)}{FD_{Cl^-,f}} \frac{\left(2D_{Cl^-,f}\dfrac{\sqrt{j\mu}}{\bar{\delta}} + \dfrac{F_1(\omega)}{\dfrac{d\bar{c}_{Cl^-}}{dy}}\sinh\left(\sqrt{j\mu}\right)\right)}{\dfrac{\sqrt{j\mu}}{\bar{\delta}}\left(-\dfrac{F_1(\omega)}{\dfrac{d\bar{c}_{Cl^-}}{dy}}\cosh\left(\sqrt{j\mu}\right)\right)} \qquad (11.238)$$

Equation (11.238) can be simplified to

$$Z_F = R_t - R_t \frac{K_1 \exp\left(b_{Cu}\overline{V}\right)}{F}\left(\frac{\dfrac{dc_{Cl^-}}{dy}}{F_1(\omega)\cosh\sqrt{j\mu}} + \frac{\bar{\delta}\tanh\sqrt{j\mu}}{D_{Cl^-,f}\sqrt{j\mu}}\right) \qquad (11.239)$$

where $F_1(\omega)$ is given by equation (11.227).

The resulting impedance diagrams take the form presented in Figure 11.31(a). The faradaic impedance is characterized by a capacitive loop followed by an inductive loop and a nearly vertical line. This simulation may be compared to experimental results shown in Figure 11.31(b) that were obtained by Barcia et al.[174] for the dissolution of copper in 1 M HCl solution at the anodic plateau of the polarization curve.

The systems described in Chapter 10 (page 207) are influenced by modulation of potential, current, concentrations at the electrode-electrolyte interface, and fractional coverage of the surface. Most of the development presented in this chapter emphasize methods to extract expressions for the diffusion impedance. Section 11.7 describes the manner in which homogeneous reactions may influence the diffusion impedance. Example 11.11 illustrates the coupling between mass transfer and the dynamic growth and dissolution of a film that results in a modulation of film thickness.

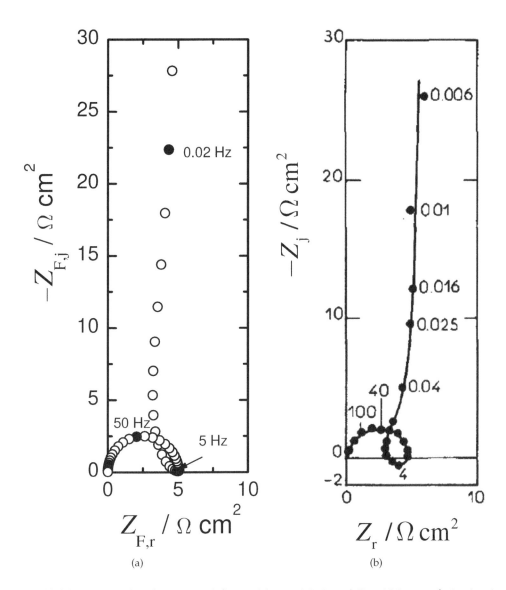

Figure 11.31: Nyquist plots for systems influenced by modulation of film thickness: a) simulated faradaic impedance corresponding to equation (11.239); and b) experimental data taken from Barcia et al.[174] for the dissolution of copper in 1 M HCl solution at the anodic plateau of the polarization curve.

Problems

11.1 Starting with the material balance equation, verify the development of equation (11.20).

11.2 Starting with the material balance equation, verify the development of equation (11.42).

11.3 Show that, for disk rotating in an incompressible fluid, $-\nabla \cdot N_i = 0$ gives rise to equation (11.55) and $-\nabla \cdot N_i \neq -\dfrac{dN_i}{dy}$.

11.4 Plot, on an impedance plane format, the impedance obtained for a Nernst stagnant diffusion layer and the impedance obtained for a rotating disk electrode under assumption of an infinite Schmidt number. Show that, while the behaviors of the two models at high and low frequencies are in agreement, the two models do not agree at intermediate frequencies.[211] Explain.

11.5 It can be useful to have a general analytic expression for velocity that applies for regions both close to and far from the disk electrode. Velocity expansions appropriate for positions close to the disk are given as equations (11.51) and (11.52) for axial and radial velocities, respectively. For positions far from the disk electrode, the respective velocity expansions can be expressed as[176]

$$
\begin{aligned}
H &= -\alpha + \frac{2A}{\alpha}\exp(-\alpha\zeta) - \frac{A^2 + B^2}{2\alpha^3}\exp(-2\alpha\zeta) \\
&\quad + \frac{A(A^2 + B^2)}{6\alpha^5}\exp(-3\alpha\zeta) + \dots
\end{aligned}
\tag{11.240}
$$

and

$$
\begin{aligned}
F &= A\exp(-\alpha\zeta) - \frac{A^2 + B^2}{2\alpha^2}\exp(-2\alpha\zeta) \\
&\quad + \frac{A(A^2 + B^2)}{4\alpha^4}\exp(-3\alpha\zeta) + \dots
\end{aligned}
\tag{11.241}
$$

where $\alpha = 0.88447411$, $A = 0.92486353$, and $B = 1.20221175$.[212] Other constants and definitions are presented in Section 11.3.1. Find an interpolation formula that applies for all values of ζ.

11.6 Starting with equation (11.81), derive equations (11.83), (11.84), and (11.85) as used in equation (11.86) for the rotating disk. Do these relationships apply as well for equation (11.98) that was developed for the impinging jet?

11.7 Show that the line represented in Figure 11.19 is parallel to the line corresponding to the Levich equation for the mass-transfer-limited current density to a rotating disk.

11.8 The equation governing mass-transfer for an impinging jet can be expressed as

$$\frac{\partial c_i}{\partial t} + v_y \frac{\partial c_i}{\partial y} - D_i \frac{\partial^2 c_i}{\partial y^2} = 0 \tag{11.242}$$

(a) Starting with $c_i = \bar{c}_i + \mathrm{Re}\left\{\tilde{c}_i e^{j\omega t}\right\}$, develop the equation we need to solve for $\tilde{c}_i$. You need not solve the equation.

(b) Identify the boundary conditions needed to solve the equation you develop.

(c) Explain the relationship between $\tilde{c}_i$ and the diffusion impedance. In other words, how would you use the solution to the equation you develop to obtain the diffusion impedance?

11.9 Develop a solution for the steady-state distribution of concentration and reaction rate for the system described in Example 11.10.

11.10 Following Example 11.10, derive an expression for the dimensionless diffusion impedance for an electrochemical reaction influenced by a chemical reaction in an infinite domain.

11.11 Use equation (11.189) to estimate the thickness of the reaction zone for an electrochemical reaction influenced by a chemical reaction in an infinite domain.

11.12 Specify the boundary conditions used for the solution to equation (11.122) for the electrode covered by a porous film. In other words, below equation (11.124), the boundary conditions at $y = \delta_f$ are invoked to find expressions for the constants C_1 and C_2. What are these boundary conditions?

Chapter 12

Impedance of Materials

In many cases, the impedance measurement provides information concerning the properties of materials. For example, the Young model, developed in Section 13.7 (page 389), and the power-law model, developed in Section 14.4.4 (page 406) demonstrate the use of impedance spectroscopy to explore the distribution of resistivity in a material. At high frequencies, the impedance measurement may be used to explore dielectric relaxation. This chapter provides an overview of the impedance of materials, including dielectric relaxation, geometric capacitance, and Mott–Schottky analysis.

12.1 Electrical Properties of Materials

Materials may be classified by considering their electrical conduction. Materials with large values of conductivity κ (small values of resistivity ρ) are termed conductors; metals and metal alloys are conductors. Ceramics, oxides, glasses, and polymers with small values of conductivity (large values of resistivity) are termed insulators. Semiconductors have intermediate values of conductivity. This classification is very general, and some oxides or some polymers could be conductors.

The resistance of a material depends on its geometry. For a sample of length L and area A, the resistance is

$$R = \frac{L}{\kappa A} = \frac{L\rho}{A} \tag{12.1}$$

Electrical resitivities and dielectric constants are presented in Table 12.1 for some typical materials. Silver, gold, copper, aluminum, iron, lead, and manganese are termed conductors, but the resistivity varies significantly from one metal to another. The variation in resistivity given for silicon and germanium arise from different concentrations of dopants. The resistivity of commercial semiconductors depends on doping levels and ranges between 10^{-2} and 10^6 Ωcm.[215] Quartz, alumina, and fusion-bonded epoxy are given in Table 12.1 as examples of insulators.

An electrical insulator is a dielectric material which can be polarized by an applied

Table 12.1: Values for resistivity and dielectric constant at 298 K for solids. Unless otherwise indicated, values are taken from reference 213. Typical ranges of values for dielectric constant, film thickness, and capacitance are presented in Table 5.6 on page 126.

Resistivity, Ωcm		Category	Dielectric Constant	Material
1.617×10^{-6}		conductor		Silver, Ag
1.712×10^{-6}		conductor		Copper, Cu
2.255×10^{-6}		conductor		Gold, Au
2.709×10^{-6}		conductor		Aluminum, Al
9.87×10^{-6}		conductor		Iron, Fe
2.11×10^{-5}		conductor		Lead, Pb
1.44×10^{-4}		conductor		Manganese, Mn
5.2×10^{-3}		conductor	20	Magnetite, Fe_3O_4
		semiconductor	12	Hematite, Fe_2O_3-α
0.04	[214]	semiconductor	11.9	Silicon, Si (high purity)
200	[214]	semiconductor	16	Germanium, Ge (high purity)
1–5×10^3		semiconductor	7.6	Cuprite, Cu_2O
6×10^5		semiconductor	18.1	Melaconite, CuO
1×10^8	[215]	insulator	4.1	Quartz, SiO_2
10^{15}–10^{17}	[214]	insulator	2.3	polyethylene

electric field E. A charge density is produced and is directly proportional to the electric field. The proportionality constant is ε_0 for the case of vacuum with $\varepsilon_0 = 8.854 \times 10^{-14}$ C/Vcm (see Section 5.7 on page 122). For a general material, the proportionality constant is $\varepsilon\varepsilon_0$ where ε is a dimensionless material constant called relative permittivity or the dielectric constant.

If ferroelectric and piezoelectric properties are not considered, a material is characterized on an electrical point of view by two parameters, the resistivity (or conductivity) and the permittivity. For a conductor, if the area is not too small and if the length is not too large, the impedance of the conductor may be considered to be negligible with respect to the accuracy of the usual measurement.

12.2 Dielectric Response in Homogeneous Media

At high frequencies, the electrolyte may show a dielectric response. For example, the dielectric relaxation of water based on data presented by Kaatze[216] is shown in Figure 12.1(a) with temperature as a parameter. The complex relative permittivity follows a Debye relaxation function, i.e.,

$$\epsilon(\omega) = \frac{\epsilon(0) - \epsilon(\infty)}{1 + j\omega\tau} \qquad (12.2)$$

where $\epsilon(0)$ represents the limiting value at low frequency, $\epsilon(\infty)$ is the value at high frequency, and τ is the relaxation time constant. The time constant for dielectric relaxation of water and the corresponding characteristic frequency are provided in Figure

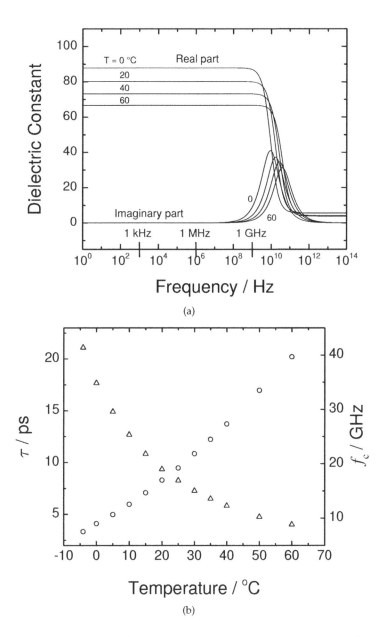

Figure 12.1: Dielectric relaxation of water with Debye model parameters taken from Kaatze:[216] a) relative permittivity as a function of frequency with temperature as a parameter and b) relative time constant and corresponding frequency as a function of temperature.

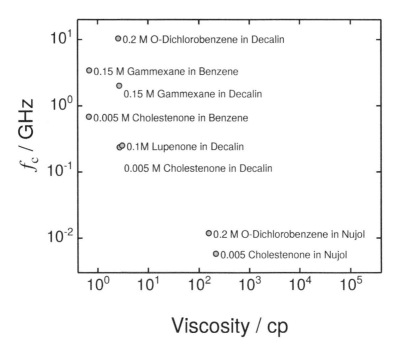

Figure 12.2: The characteristic frequency for dielectric relaxation as a function of viscosity for large dipolar molecules dissolved in nonpolar solvents. Data taken from Daniel.[217]

12.1(b) as functions of temperature. The dielectric relaxation takes place at frequencies in the GHz range, much above the usual frequency range for electrochemical impedance measurements. Water is an associated liquid in which forces between molecules are strong and directional. The relaxation process involves fluctuation of polarization and dissociation of water into ions.[217]

The dielectric relaxation associated with orientation of dipoles has a strong dependence on viscosity. For example, the characteristic frequency for dielectric relaxation of large dipolar molecules dissolved in nonpolar solvents is presented in Figure 12.2 as a function of viscosity.

Reorientation of dipoles may take place as well in crystalline solids. The dielectric relaxation time constant for D-Camphor ($C_{10}H_{16}O$) is 11 ps (14 GHz) at 80 °C.[217] Shang and Umana[218] report that the characteristic dielectric relaxation time constant for asphalt is on the order of 100 to 400 ps (1.6-0.4 GHz). The characteristic dielectric relaxation time constant for pure single-crystal ice at a temperature of -10.8 °C de-

Remember! 12.1 *Dielectric relaxation is typically associated with frequencies larger than the usual frequency range associated with electrochemical impedance spectroscopy.*

pends on orientation and has a value of 5.8×10^{-5} s in the c-direction and 6.1×10^{-5} s in the direction normal to the c-direction. These values correspond to characteristic frequencies of 2.7 kHz and 2.6 kHz, respectively.

12.3   Cole–Cole Relaxation

The Cole–Cole model,[40,219] i.e.,

$$\epsilon = \epsilon_\infty + \frac{\epsilon_s - \epsilon_\infty}{1 + (j\omega\tau_c)^\alpha} \tag{12.3}$$

is often applied for systems with distributed characteristic relaxation time constants. In equation (12.3), α is defined such that ideal Debye relaxation is observed when $\alpha = 1$. The dielectric relaxation of water has an almost ideal Debye relaxation with a Cole–Cole parameter $\alpha = 0.98$.[217] While the dielectric relaxation of D-Camphor at 80 °C follows equation (12.2), the relaxation follows equation (12.3) at -20 °C with $\alpha = 0.07$.[217] Other distribution functions used to describe dielectric relaxation are presented in the literature.[3,4,217]

12.4   Geometric Capacitance

An electrochemical cell is composed of three electrodes. The electrolyte resistance, which is obtained by solving Laplace's equation (see Section 13.2, page 337), corresponds to the resistance between the working electrode and the counter electrode. The geometrical capacitance C_g formed by the working electrode and the counterelectrode is parallel to the electrolyte resistance. For the usual system, the influence of the geometric capacitance is out of range of the frequency domain explored for an electrochemical impedance measurement. In some particular cases, for example when the resistivity of the electrolyte is very large, the characteristic frequency

$$f_c = 1/(2\pi R_e C_g) \tag{12.4}$$

is within the accessible frequency domain.[220]

In the general case, the characteristic frequency given by equation (12.4) has a weak dependence on cell geometry. For the system described in Example 12.1, the characteristic frequency is independent of the distance between electrodes and the area. Thus, the impedance normalized by the ohmic resistance can be superposed in a Nyquist plot, shown in Figure 12.3. As the dielectric constant does not change very much from one liquid to another, the characteristic frequency depends most strongly with electrolyte resistivity. For a resistivity $\rho = 2 \times 10^{10}$ Ωcm, corresponding to a synthetic lubricating oil, the characteristic frequency is on the order of 22 Hz. For pure water ($\rho = 18.2$ $M\Omega$cm and $\varepsilon = 78$ at 25 °C), the characteristic frequency is on the order of 1.2 kHz. For typical dionized water with $\rho = 1$ $M\Omega$cm, the characteristic frequency is

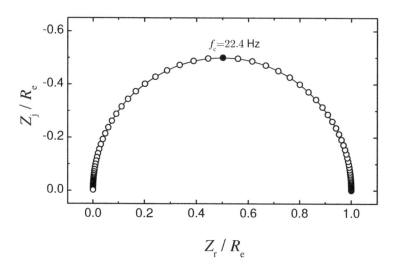

Figure 12.3: Geometrical impedance response normalized by the ohmic resistance R_e for the system described in Example 12.1.

23 kHz, and for seawater ($\rho = 25$ Ωcm), the characteristic frequency is on the order of 1 GHz. Thus, for normal electrochemical measurements with frequencies ranging up to 10 or 100 kHz, the geometric capacitance cannot be discerned; thus, the influence of the electrolyte is seen as only an ohmic resistance. This effect is similar to that described in Example 9.3 (page 204) where an ohmic resistance may be observed to increase with time due to growth of a porous layer.

Example 12.1 Geometric Capacitance: *An electrochemical cell consists of two electrodes which fill the cross-section of a cylinder, as shown in Figure 12.4, with a length $L = 1$ cm and a cross-sectional area $A = 1$ cm^2. The cylindrical design of the electrochemical cell was chosen to avoid the influence of nonuniform current and potential distributions described in Section 13.2 (page 337). The electrolyte has a relative permittivity (or dielectric constant) $\varepsilon = 4$ and a resistivity $\rho = 2 \times 10^{10}$ Ωcm. Find the geometric impedance response.*

Solution: *The electrolyte resistance is given by*

$$R_e = \frac{\rho L}{A} \tag{12.5}$$

Remember! 12.2 *For impedance measurements performed in a medium with large resistivity, a capacitive loop that extends at high frequency to the origin in a Nyquist plot may be associated with a geometric capacitance.*

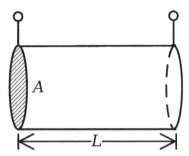

Figure 12.4: Cylindrical cell used to obtain the geometrical impedance as described in Example 12.1.

and the capacitance is

$$C_g = \frac{\varepsilon \varepsilon_0 A}{L} \tag{12.6}$$

For this system, the ohmic resistance has a value $R_e = 2 \times 10^{10}$ Ω, and the capacitance has a value $C_g = 3.53 \times 10^{-13}$ F. The characteristic frequency, given by equation (12.4), may be expressed as

$$f_c = \frac{1}{2\pi\rho\varepsilon\varepsilon_0} \tag{12.7}$$

yields a value of 22.4 Hz. The corresponding impedance response, presented in Figure 12.3, has a high-frequency limit at the origin in a Nyquist plot.

Example 12.1 demonstrates that even a very small capacitance may be measurable if the parallel resistance is sufficiently large.

12.5 Dielectric Response of Insulating Nonhomogeneous Media

In nonhomogeneous media, the permittivity and the resistivity of the media vary according to position. Under the assumption that the variation of the electrical parameters is in the direction normal to the electrode, a local value of resistivity may be designated as $\rho(y)$, and a local value of dielectric constant may be given as $\varepsilon(y)$. The resistance associated with a thin differential element dy may be expressed as $\rho(y)dy$, and the capacitance may be expressed as $\varepsilon\varepsilon_0/dy$. The impedance of this thin differential element is

$$dZ = \frac{\rho(y)dy}{1 + j\omega\rho(y)\varepsilon(y)\varepsilon_0} \tag{12.8}$$

The corresponding equivalent circuit representation is given in Figure 12.5. The impedance of the dielectric layer is obtained by integrating over the thickness of the layer, e.g.,

$$Z = \int_0^\delta \frac{\rho(y)}{1 + j\omega\rho(y)\varepsilon(y)\varepsilon_0} dy \tag{12.9}$$

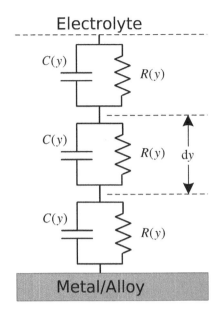

Figure 12.5: Equivalent circuit corresponding to an oxide layer with an axial distribution of dielectric and resistive properties.

The variation of electrical properties in the film causes a frequency dispersion that is frequently observed and is analysed in Chapters 13 (page 319) and 14 (page 395).

When the frequency tends towards zero, in each thin differential layer, the resistance $\rho(y)\mathrm{d}y$ is much smaller than the impedance of the corresponding capacitance $-j\varepsilon(y)\varepsilon_0/\omega\mathrm{d}y$; thus, the impedance of the film is a resistance

$$R_{\text{film}} = \int\limits_0^\delta \rho(y)\mathrm{d}y \tag{12.10}$$

In contrast, when the frequency tends towards infinity, the resistance $\rho(y)\mathrm{d}y$ can be neglected with respect to the impedance of the corresponding capacitance, expressed as $-j\mathrm{d}y/\omega\varepsilon(y)\varepsilon_0$, and the impedance is the impedance of a capacitor, i.e.,

$$\frac{1}{C_{\text{film}}} = \int\limits_0^\delta \frac{1}{\varepsilon(y)\varepsilon_0}\mathrm{d}y \tag{12.11}$$

Exponential and power-law distributions of resistivity have been invoked to develop models for time-constant dispersion that are developed in Sections 13.7 (page 389) and 14.4.4 (page 406), respectively.

12.6 Mott–Schottky Analysis

A distribution of charge is observed at the interface between phases.[221,222] For example, at an electrode-electrolyte interface, as described in Section 5.7.1 (page 122), a diffuse region of charge is seen in the electrolyte that is balanced by charge held at the electrode-electrolyte interface such that the interface, taken as a whole, is electrically neutral. Poisson's equation

$$\frac{d^2\Phi}{dy^2} = -\frac{F}{\varepsilon\varepsilon_0}\sum_i z_i c_i \qquad (12.12)$$

provides a relationship between the concentration of charged species and potential within the diffuse part of the double layer in an electrolyte. As the diffuse part of the double layer is very thin in typical electrolytes, impedance models simply assign a capacitance to the electrode-electrolyte interface. A diffuse region of charge is not expected in metallic electrodes due to the large conductivity of the metal. The diffuse region of charge plays an important role, however, in the semiconducting class of materials. For a discussion of semiconductor physics, the reader is directed to textbooks on the subject.[223–225]

Mott–Schottky analysis is a graphical technique that relates a single-frequency measurement to semiconductor properties. The frequency of the measurement is selected to exclude contributions of confounding phenomena. For example, impedance measurements on a semiconductor diode at a sufficiently high frequency exclude the influence of leakage currents and of electronic transitions between deep-level and band-edge states. As discussed in Section 17.4, the capacitance can be extracted from the imaginary part of the impedance as

$$C = \frac{1}{\omega Z_j} \qquad (12.13)$$

The problem is reduced to one of identifying the relationship between semiconductor properties and the capacitance as a function of applied potential.

The capacitance of the space-charge region C_{sc} is determined through

$$C_{sc} = \left.\frac{\partial q_{sc}}{\partial \Phi}\right|_{y=0} \qquad (12.14)$$

where the charge density of the space-charge region q_{sc} is related to potential through equation (12.12) where the concentration c_i accounts for the contributions of electrons, holes, and fixed charge.

Remember! 12.3 *Mott-Shottky theory provides a relationship between the experimentally measured capacitance, the doping level, and the flat-band potential.*

Solution of equation (12.12) is facilitated by expressing the concentration of electrons and holes in terms of potential as

$$n = N \exp\left(F\Phi/RT\right) \tag{12.15}$$

and

$$n = P \exp\left(-F\Phi/RT\right) \tag{12.16}$$

for electrons and holes, respectively. Thus, Poisson's equation can be expressed as

$$\frac{d^2\Phi}{dy^2} = -\frac{F}{\varepsilon\varepsilon_0}\left[Pe^{-F\Phi/RT} - Ne^{F\Phi/RT} + (N_d - N_a)\right] \tag{12.17}$$

where Φ is the electrostatic potential, $(N_d - N_a)$ is the doping level, and P and N are the hole and electron concentrations, respectively, at the flat-band potential. Deep-level states were not included in the expression for charge density. The occupancy of deep-level states is a function of potential, and a similar development can be made to take such states into account.[226]

The charge density used in equation (12.17), i.e.,

$$\rho_{sc}(\Phi) = Pe^{-F\Phi/RT} - Ne^{F\Phi/RT} + (N_d - N_a) \tag{12.18}$$

is now an explicit function of potential rather than position. Integration of Poisson's equation is facilitated by posing equation (12.17) in terms of the electric field

$$E = -\frac{d\Phi}{dy} \tag{12.19}$$

The second derivation of potential with respect to position can be expressed in terms of a derivative with respect to potential as

$$\frac{d^2\Phi}{dy^2} = -\frac{dE}{dy} = -\frac{dE}{d\Phi}\frac{d\Phi}{dy} = E\frac{dE}{d\Phi} = \frac{1}{2}\frac{dE^2}{d\Phi} \tag{12.20}$$

Thus,

$$\frac{dE^2}{d\Phi} = -\frac{2}{\varepsilon\varepsilon_0}\rho_{sc}(\Phi) \tag{12.21}$$

Integration yields

$$E^2 = -\frac{2}{\varepsilon\varepsilon_0}\int_{\Phi_{fb}}^{\Phi}\rho_{sc}(\Phi)d\Phi \tag{12.22}$$

where the flat-band potential Φ_{fb} is the potential in the electrically neutral region far from the interface.

The charge held within the space-charge region is given by

$$q_{sc} = \int_0^\infty \rho_{sc}dy \tag{12.23}$$

Under the assumption that there are no surface states or specific adsorption of charged species, the space charge q_{sc} in a semiconductor in contact with an electrolyte is balanced by the charge in the diffuse part of the double layer q_d; thus, $q_{sc} = q_d$. Gauss's law can therefore be used to provide a boundary condition for the electric field at the surface of the semiconductor as

$$E(\Phi(0)) = -\left.\frac{d\Phi}{dy}\right|_{y=0} = \frac{q_{sc}}{\varepsilon\varepsilon_0} \tag{12.24}$$

Thus

$$q_{sc} = \varepsilon\varepsilon_0 E(\Phi(0)) \tag{12.25}$$

The space-charge capacitance is given by

$$C_{sc} = -\frac{dq_{sc}}{d\Phi(0)} = \varepsilon\varepsilon_0 \frac{dE(\Phi(0))}{d\Phi(0)} \tag{12.26}$$

or

$$C_{sc} = -\frac{\rho_{sc}(\Phi(0))}{E(\Phi(0))} \tag{12.27}$$

Equation (12.27) provides a relationship between the capacitance of the space-charge region of the semiconductor, the electric field at the interface, and the charge density at the interface.

Under the convention that the potential is referenced to the flat-band potential Φ_{fb}, i.e., the potential is equal to zero far from the interface where the electric field is also equal to zero, integration of equation (12.22) yields

$$E^2 = \frac{2RT}{\varepsilon\varepsilon_0}\left[P(e^{-F\Phi/RT} - 1) + N(e^{F\Phi/RT} - 1) - \frac{F\Phi}{RT}(N_d - N_a)\right] \tag{12.28}$$

A general expression for capacity can be found to be

$$\frac{1}{C_{sc}^2} = \frac{2RT}{F^2\varepsilon\varepsilon_0}\frac{P(e^{-F\Phi(0)/RT} - 1) + N(e^{F\Phi(0)/RT} - 1) - \frac{F\Phi(0)}{RT}(N_d - N_a)}{\left[-Pe^{-F\Phi(0)/RT} + Ne^{F\Phi(0)/RT} - (N_d - N_a)\right]^2} \tag{12.29}$$

The values for N and P can be evaluated at the flat-band potential under the assumptions of electroneutrality (i.e., $\rho_{sc} = 0$) and equilibrium, i.e.,

$$np = n_i^2 \tag{12.30}$$

where n_i is the intrinsic concentration. Thus,

$$N = \frac{1}{2}\left[(N_d - N_a) + \sqrt{(N_d - N_a)^2 + 4n_i^2}\right] \tag{12.31}$$

and

$$P = \frac{1}{2}\left[-(N_d - N_a) + \sqrt{(N_d - N_a)^2 + 4n_i^2}\right] \tag{12.32}$$

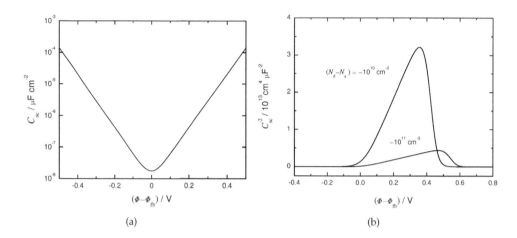

Figure 12.6: Calculated capacitance for a GaAs Schottky diode (see Table 12.2 for relevant parameters for GaAs semiconductors): a) capacitance as a function of potential referenced to the flat-band potential for an intrinsic semiconductor and b) $1/C_{sc}^2$ as a function of potential referenced to the flat-band potential for a lightly doped p-type semiconductor.

Table 12.2: Physical properties for GaAs at 300 K

Energy gap	E_g	1.424	eV
Intrinsic carrier concentration	n_i	2.1×10^6	cm^{-3}
Effective conduction band density of states	N_c	4.7×10^{17}	cm^{-3}
Effective valence band density of states	N_v	9.0×10^{18}	cm^{-3}
Dielectric constant (static)	ϵ	12.9	
Dielectric constant (high frequency)	ϵ	10.89	
Mobility of electrons	μ_n	8500	$cm^2V^{-1}s^{-1}$
Mobility of holes	μ_p	400	$cm^2V^{-1}s^{-1}$
Diffusion coefficient of electrons	D_n	200	cm^2s^{-1}
Diffusion coefficient of holes	D_p	10	cm^2s^{-1}

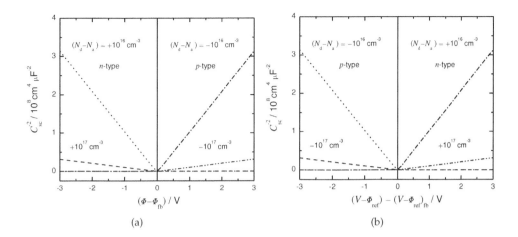

Figure 12.7: Mott–Schottky plot of $1/C_{sc}^2$ as a function of potential referenced to the flat-band potential for a GaAs Schottky diode: a) potential referenced to the ohmic contact; and b) potential referenced to a reference electrode located in the electrolyte according to the IUPAC convention for semiconductor electrodes.[227,228]

Equations (12.31) and (12.32) can be inserted into equation (12.29) to provide an expression for capacitance as a function of potential, with doping level as a parameter.

The capacitance is presented in Figure 12.6(a) for an intrinsic GaAs semiconductor diode. Some relevant physical properties of GaAs are presented in Table 12.2. The capacitance is symmetric with respect to potential referenced to the flat-band potential. At modest doping levels, as shown in Figure 12.6(b) for p-type semiconductors, the plot of $1/C_{sc}^2$ has a significant linear portion with respect to potential. At more positive potentials, the plot deviates from a straight line due to the contribution of minority carriers.

Mott-Shottky plots of $1/C_{sc}^2$ as a function of potential are particularly useful at larger doping levels. Calculated values for $1/C_{sc}^2$ are presented in Figure 12.7(a) for a GaAs diode with potential referenced to the ohmic contact. The corresponding values are presented in Figure 12.7(b) for potentials referenced to a reference electrode located in the electrolyte in accordance with the IUPAC convention for semiconductor electrodes.[227,228] The principal distinction between the two plots is that the positive slopes correspond to a p-type semiconductor in Figure 12.7(a) and to an n-type semiconductor in Figure 12.7(b). All concentration terms were included in the analysis.

As seen in Figure 12.7, $1/C_{sc}^2$ is linear over a broad range of potential. For an n-type semiconductor, the linear portion for Figure 12.7(a) is given by

$$\frac{1}{C_{sc}^2} = -\frac{2(\Phi(0) - \Phi_{fb} + RT/F)}{\varepsilon\varepsilon_0 F(N_d - N_a)} \qquad (12.33)$$

The linear portion for Figure 12.7(a) is given by

$$\frac{1}{C_{sc}^2} = -\frac{2(\Phi(0) - \Phi_{fb} - RT/F)}{\varepsilon\varepsilon_0 F(N_d - N_a)} \tag{12.34}$$

for a *p*-type semiconductor.

Example 12.2 Mott-Schottky Plots: *Derive equation (12.33) for an n-type semiconductor, beginning with equation (12.29).*

Solution: *For an n-type semiconductor, $(N_d - N_a)$ is positive and much larger than the intrinsic concentration n_i; thus, $N \to (N_d - N_a)$ and $P \to 0$. In the denominator, the concentration of holes $P \exp(-F\Phi(0)/RT)$ can be neglected. In the numerator, $P(\exp(-F\Phi(0)/RT) - 1)$ can also be neglected. The linear portion of the curve lies at negative potentials, referenced to the flat-band potential; thus, $\exp(F\Phi(0)/RT) \to 0$ and*

$$N(\exp(-F\Phi(0)/RT) - 1) \approx -(N_d - N_a) \tag{12.35}$$

Equation (12.29) can be expressed as

$$\frac{1}{C_{sc}^2} = \frac{2RT}{F^2\varepsilon\varepsilon_0} \frac{-(N_d - N_a)\left(\frac{F\Phi(0)}{RT} + 1\right)}{(N_d - N_a)^2} \tag{12.36}$$

or

$$\frac{1}{C_{sc}^2} = -\frac{2(\Phi(0) - \Phi_{fb} + RT/F)}{\varepsilon\varepsilon_0 F(N_d - N_a)} \tag{12.37}$$

A similar development at positive potentials will yield equation (12.34).

The assumptions implicit in the Mott–Schottky theory are:

- The potential is restricted to the range where both the majority (in this case, electrons) and the minority carriers (holes) are negligible as compared to the doping level. These constraints are violated at small and large magnitudes of potential (referenced to the flat-band potential), respectively. This potential range becomes increasingly restrictive as the doping level decreases, and the technique is unusable for semi-insulating materials.

- The semiconductor electrode must be ideally polarizable over the potential range of interest. This means that there is no leakage current or faradaic reaction to allow charge transfer across the semiconductor-electrolyte interface. This restriction is not too important if measurements are taken at sufficiently high frequency that the effects of faradaic reactions are suppressed.

- Electron and hole concentrations follow a Boltzmann distribution, i.e., activity coefficient corrections can be neglected.

Equations (12.33) and (12.34) form the basis of a method, described in Section 18.5.4, used to extract doping levels and flat-band potentials for semiconducting materials.

It should be noted that Mott–Schottky theory applies specifically for cases in which the measured capacitance is independent of frequency. For single-crystal semiconductors influenced by deep-level electronic states, the space-charge capacitance should be extracted from a model accounting for the influence of the deep-level states. An example is provided as Figure 17.10 on page 483. The results obtained by direct application of Mott–Schottky analysis to nonhomogeneous materials must be considered to be qualitative.

Problems

12.1 Calculate the impedance response for the system described in Example 12.1 for $L = 2$ cm and for $L = 10$ cm. Present the results as a Nyquist plot. What is the corresponding ohmic resistance, geometric capacitance, and characteristic frequency?

12.2 Estimate the frequency at which a geometric capacitance is apparent for:

(a) an electrolyte containing 0.1 M NaCl at a temperature of 25 °C.

(b) an electrolyte containing 1.0 M NaCl at a temperature of 25 °C.

12.3 Calculate the impedance of a film with a resistivity distribution given as equation (13.171).

12.4 Develop the relationship needed to convert the mobility given in Table 12.2, e.g., μ_n, to diffusivity.

12.5 Calculate the Debye length in units of μm expected for an n-type GaAs semiconductor with a dopant concentration of 10^{16}. Compare the value you obtain to the Debye length obtained for an electrolytic system with a NaCl concentration of 0.1 M.

12.6 The capacity of the space-charge region can be related to the dopant concentration (or fixed charge) in a semiconductor. The space-charge region is essentially equivalent to the diffuse double layer treated in electrolytes with the exception that ionized impurities are present that, at room temperatures, are immobile. For this case, Poisson's equation becomes

$$\nabla^2 \Phi = -\frac{F}{\varepsilon \varepsilon_0} \{ n - p - (N_d - N_a) \} \tag{12.38}$$

Show that the capacity can be related to doping level $(N_d - N_a)$ and potential by the Mott–Schottky relationship

$$\frac{1}{C^2} = -\frac{2}{F \varepsilon \varepsilon_0} \frac{(V + RT/F)}{(N_d - N_a)} \tag{12.39}$$

12.7 The capacity of the electrolytic diffuse double layer is often ignored when Mott–Schottky plots are used to characterize semiconductor-electrolyte interfaces. Under what conditions is this assumption justified?

12.8 Mott–Schottky plots are often generated by using measurements at a single frequency, often 1 kHz. Explain the limitations of this approach.

Chapter 13

Time-Constant Dispersion

The impedance models developed in Chapters 9-12 are based on the assumption that the electrode behaves as a uniformly active surface where each physical phenomenon or reaction has a single-valued time constant. The assumption of a uniformly active electrode is generally not valid. Time-constant dispersion can be observed due to variation along the electrode surface of reactivity or of current and potential. Such a variation is described in Chapter 14 as resulting in a surface distribution of time constants. Time-constant dispersion can also be caused by a distribution of time constants that reflect a local property of the electrode, resulting in a distribution in a direction normal to the electrode.

This chapter describes the impedance response described by transmission-line models, including that associated with porous electrodes, hierarchical porous electrodes comprising macro-, meso-, and micro-pores, and thin-layer cells. Other systems require solution of appropriate governing equations such as Laplace's equation for geometry-induced current and potential distributions, convective diffusion equations for nonuniform mass transfer, and multicomponent transport for coupled charging and faradaic currents. Other distributed-time-constant behavior can result from exponential distributions of resistivity in a film. Constant-phase elements are described in Chapter 14 on page 395.

13.1 Transmission Line Models

Transmission line models owe their origin and name to the development of mathematical theory for the performance of submarine telegraph lines. The speed at which signals could be transmitted through above-ground telegraph lines was limited by the capacitive coupling of the wires to the ground over which the wires were suspended. As this capacitance was relatively small, the reduction in speed was considered acceptable. William Thomson (later Lord Kelvin) considered that, because the capacitive coupling would be to the seawater adjacent to the cable, the advent of submarine cables required

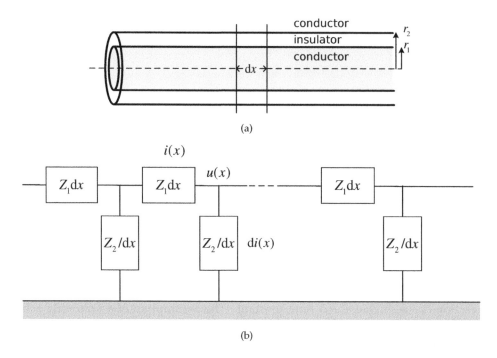

(a)

(b)

Figure 13.1: Schematic representation of the telegraph cable considered by Thomson[230] and Heaviside:[231–233] a) representation showing a differential element with length dx and b) elementary components of a transmission line. For the wire shown in a, $Z_1 = R_0$ and $Z_2 = 1/j\omega C_{\text{wire}}$.

a more careful analysis of the problem.[229] He modeled a submerged telegraph line as a conducting wire of radius r_1 separated from a conductive ocean by an insulating concentric cylinder of radius r_2. He expressed the capacitance per unit length of the coated wire by[230]

$$C_{\text{wire}} = \frac{2\pi\varepsilon\varepsilon_0}{\ln(r_2/r_1)} \tag{13.1}$$

where ε is the dielectric constant of the coating.

The wire was represented by an inner cylinder, shown in Figure 13.1(a), having a resistance per unit length R_0. The potential drop over a distance dx is given by

$$\frac{du}{dx} = R_0 i(x) \tag{13.2}$$

The effect of capacitance on a distance dx is given by

$$\frac{di}{dx} = C_{\text{wire}} \frac{du}{dt} \tag{13.3}$$

The resulting conservation equation may be written as

$$\frac{\partial^2 u}{\partial x^2} = R_0 C_{\text{wire}} \frac{\partial u}{\partial t} \tag{13.4}$$

where u is the potential.

An arbitrary input at one end of the cable can be expressed in terms of frequency by a Fourier series (see equation (7.36) on page 152). Thomson showed that each of these frequency components would diffuse through the wire at different speeds, resulting in frequency dispersion. Thus, an impulse would broaden by the time it reached the other end of the cable, requiring a significant reduction in transmission speed to resolve the pulses. Indeed, in accordance with Thomson's predictions, the first message of the 1858 trans-Atlantic cable, containing 99 words consisting of 509 letters, required 17 hours and 40 minutes for the transmission.[234] Heaviside extended the theory for telegraphy in a series of three papers,[231–233] expressing the problem in frequency rather than time domain and including the effect of inductance per unit length of the cable. Heaviside later generalized this analysis in terms of transmission lines.[235]

13.1.1 Telegrapher's Equations

The telegrapher's equations are a pair of linear differential equations which describe the voltage and current on an electrical transmission line with distance and time. An equivalent circuit of a transmission line is given in Figure 13.1 where Z_1 is the impedance per unit length with units Ω/cm and Z_2 is the impedance for a unit length with units Ωcm.

The line voltage $u(x)$ and the current $i(x)$ can be expressed in the frequency domain as

$$d\widetilde{u}(x) = -Z_1 \widetilde{i}(x) dx \tag{13.5}$$

and

$$d\widetilde{i}(x) = -\frac{\widetilde{u}(x)}{Z_2} dx \tag{13.6}$$

respectively. When the impedances Z_1 and Z_2 are independent of the distance x, the second order steady-state telegrapher's equations may be found to be

$$\frac{d^2\widetilde{u}(x)}{dx^2} = \frac{Z_1}{Z_2}\widetilde{u}(x) \tag{13.7}$$

and

$$\frac{d^2\widetilde{i}(x)}{dx^2} = \frac{Z_1}{Z_2}\widetilde{i}(x) \tag{13.8}$$

Equations (13.7) and (13.8) are fundamental to transmission line theory. The solutions have the form (see Example 2.5 on page 31)

$$\widetilde{u}(x) = A \exp\left(x\sqrt{Z_1/Z_2}\right) + B \exp\left(-x\sqrt{Z_1/Z_2}\right) \tag{13.9}$$

and

$$\widetilde{i}(x) = C \exp\left(x\sqrt{Z_1/Z_2}\right) + D \exp\left(-x\sqrt{Z_1/Z_2}\right) \tag{13.10}$$

where A, B, C, and D are constants obtained from application of appropriate boundary conditions.

Example 13.1 Submarine Telegraph Line Model: *Find expressions for the Tele-grapher's equation parameters Z_1 and Z_2 that correspond to Thomson's model for a submarine telegraph line expressed as equation (13.4).*

Solution: *Equation (13.4) must be converted from time-domain to frequency domain. Thus, by inserting*

$$u = \bar{u} + \mathrm{Re}\{\tilde{u}\exp(\mathrm{j}\omega t)\} \tag{13.11}$$

and cancelling the steady-state portion, one obtains

$$\frac{\mathrm{d}^2\tilde{u}}{\mathrm{d}x^2} = R_0 C_{\mathrm{wire}}\left(\mathrm{j}\omega\tilde{u}\right) \tag{13.12}$$

By inspection,

$$Z_1 = R_0 \tag{13.13}$$

and

$$Z_2 = \frac{1}{\mathrm{j}\omega C_{\mathrm{wire}}} = -\mathrm{j}\frac{1}{\omega C_{\mathrm{wire}}} \tag{13.14}$$

The term Z_1 represents the resistance of the wire and the term Z_2 represents the capacitive coupling through the insulation to the seawater.

13.1.2 Porous Electrodes

Porous electrodes are used in numerous industrial applications because they have the advantage of an increased effective active area. A porous electrode can be obtained by such different techniques as pressing metal powder or dissolution.[236] This type of porous electrode structure is also observed on some corroded electrodes.[8] It is impor-tant to recognize that a porous electrode is not the same as a porous layer. The structure may be the same, but, while the pore walls are electroactive for a porous electrode, the pore walls are inert for a porous layer.

The random structure of the porous electrode, illustrated in Figure 13.2(a), leads to a distribution of pore diameters and lengths. Nevertheless, the porous electrode is usually represented by the simplified single-pore model shown in Figure 13.2(b) in which pores are assumed to have a cylindrical shape with a length ℓ and a radius r. The impedance of the pore can be represented by the transmission line presented in

---

Remember! 13.1 *A porous electrode is not the same as a porous layer. The structure may be the same, but, while the pore walls are electroactive for a porous electrode, the pore walls are inert for a porous layer.*

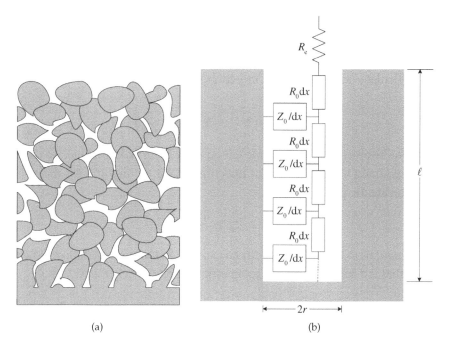

Figure 13.2: Schematic representations of a porous electrode: a) porous electrode with irregular channels between particles of electrode material and b) transmission line inside a cylindrical pore.

Figure 13.2(b) where R_0 is the electrolyte resistance for one-unit length pore, with units of Ωcm^{-1}, Z_0 is the interfacial impedance for a unit length pore, with units of Ωcm, r is the pore radius in cm, and ℓ is the pore length in cm. The specific impedances R_0 and Z_0 can be expressed in function of the pore radius as

$$R_0 = \frac{\rho}{\pi r^2} \tag{13.15}$$

and

$$Z_0 = \frac{Z_{\text{eq}}}{2\pi r} \tag{13.16}$$

respectively, where Z_{eq} is the interfacial impedance per surface unit, with units of Ωcm^2, and ρ is the electrolyte resistivity, with units of Ωcm. The terms R_0 and Z_0 correspond, respectively, to Z_1 and Z_2 in the general transmission line model (see example 2.5 on page 31).

In the general case, Z_0 and R_0 are functions of the distance x. This dependence is due to the potential distribution or/and to the concentration distribution in the pore. The general solution can be obtained only by a numerical calculation of the corresponding transmission line. For example, the impedance of a porous electrode in the presence of a concentration gradient was numerically studied by Keddam et al.[237] but only a totally irreversible charge-transfer reaction was considered and the ohmic drop in the pore was neglected. A complete numerical calculation in the presence of a concentration gradient and a potential drop in the pores was developed later by Lasia.[238]

With the restrictive assumption that Z_0 and R_0 are independent of the distance x, de Levie[50] calculated analytically the impedance of one pore to be

$$Z_{pore} = (R_0 Z_0)^{1/2} \coth \left(\ell \sqrt{\frac{R_0}{Z_0}} \right) \tag{13.17}$$

The derivation is presented in Example 2.5 on page 31. The impedance of the overall electrode is obtained by accounting for the ensemble of n pores and for the electrolyte resistance outside the pore, i.e.,

$$Z = R_e + \frac{Z_{pore}}{n} \tag{13.18}$$

The set of equations (13.15)-(13.18) yields an expression for the impedance of the porous electrode Z that is a function of three geometrical parameters ℓ, r, and n as

$$Z = R_e + \frac{(\rho Z_{eq})^{1/2}}{\sqrt{2}\pi n r^{3/2}} \coth \left(\ell \sqrt{\frac{2\rho}{r Z_{eq}}} \right) \tag{13.19}$$

Several limiting behaviors can be seen in equation (13.19). For example, when the argument to the coth function, $\ell\sqrt{2\rho/rZ_{eq}}$ is sufficiently large, $\coth(\ell\sqrt{2\rho/rZ_{eq}})$ tends toward unity and $(Z - R_e)$ tends toward $(\rho Z_{eq})^{1/2}/\sqrt{2}\pi n r^{3/2}$ In this particular case, the pores behave as though they are semi-infinitely deep. Other limiting behaviors of equation (13.19) are explored in Problems 13.8 and 13.9. The shape of the pores influences the value of the impedance,[239] but, in the high-frequency range, this geometrical influence disappears and the impedance is proportional to $(Z_{eq})^{1/2}$.

If the interfacial impedance Z_{eq} is a simple charge-transfer resistance in parallel with a double layer capacitance, then the phase shift of this impedance tends towards $-90°$ when the frequency tends towards infinity. If the impedance is proportional to $Z_{eq}^{1/2}$ then, according to Table 1.3 on page 10, the phase shift tends towards $-45°$ when the frequency tends towards infinity. In term of CPE behavior, this corresponds to $\alpha = 0.5$ (see Chapter 14 on page 395).

For an interfacial impedance expressed as a charge-transfer resistance in parallel with a double layer capacitance, i.e.,

$$Z_{eq} = \frac{R_t}{1 + j\omega C R_t} \tag{13.20}$$

equation (13.19) can be written as

$$Z = R_e + \frac{A}{\sqrt{1 + j\omega C R_t}} \coth \left(B\sqrt{1 + j\omega C R_t} \right) \tag{13.21}$$

where

$$A = \frac{\sqrt{\rho R_t}}{\sqrt{2}\pi n r^{3/2}} \tag{13.22}$$

and

$$B = \ell\sqrt{\frac{2\rho}{rR_{\text{t}}}} \tag{13.23}$$

The impedance depends on only four parameters, R_{e}, A, B, and CR_{t}. The individual parameters such as R_{t}, C, n, r, ρ, and ℓ cannot be determined separately by regression analysis.

Example 13.2 Corrosion of Cast Iron in Drinking Water: *The internal corrosion rate of drinking water pipes is very small and, in itself, corrosion is generally not a problem. But free chlorine (FCl) (the sum of hypochlorous acid HOCl and hypochlorite ions ClO^-) introduced in water at the treatment plant in order to maintain microbiological quality, gradually disappears throughout the distribution system, which necessitates rechlorination. In order to optimize rechlorination procedures, the different sources of chlorine consumption must be identified. The most often invoked and investigated causes of chlorine decay are the chemical bulk oxidation of organic compounds dissolved in water and the reactions with biofilms on the pipe surface. Furthermore, chlorine reacts with the pipe materials themselves in the corrosion process of cast iron pipes. The corrosion has been invoked as an important source of chlorine decay and thus the corrosion rate must be evaluated.[8] Derive a model for the impedance response of an iron electrode, taking into account the material presented in Chapters 9, 10, and 11 and treating the iron as a porous electrode with pores filled with corrosion product as shown in Section 9.3.2.*

Solution: *Free chlorine can be directly involved in the corrosion process and reduced at the metal-water interface, following the electrochemical reaction, given for acidic pH by*

$$HOCl + H^+ + 2e^- \rightleftarrows Cl^- + H_2O \tag{13.24}$$

coupled with the anodic dissolution of ferrous material

$$Fe \rightleftarrows Fe^{2+} + 2e^- \tag{13.25}$$

On the other hand, chlorine can be chemically reduced by ferrous ions produced by reaction (13.25), according to the homogeneous reaction (in acidic media)

$$2Fe^{2+} + HOCl + H^+ \rightleftarrows 2Fe^{3+} + Cl^- + H_2O \tag{13.26}$$

In aerated and chlorinated waters, the rate of reaction (13.24) can be considered to be negligibly small, and the single cathodic process coupled with the dissolution of iron is the reduction of dissolved oxygen, written in acidic media as

$$\frac{1}{2}O_2 + 2H^+ + 2e^- \rightleftarrows H_2O \tag{13.27}$$

The analysis of the corrosion products suggests the scheme presented in Figure 13.3 for the cast iron–drinking water interface:

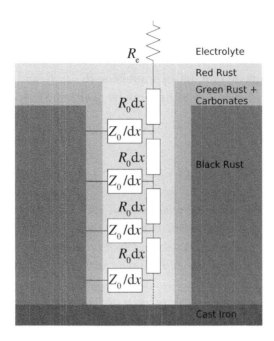

Figure 13.3: Schematic representation of a cast iron–water interface. (Taken from Frateur et al.[8])

- *On top, the red rust layer explains the absence of hydrodynamic effects after two days of immersion. This layer, which is an electronic insulator but an ionic conductor, does not play any role on the kinetics.*

- *Below the reddish layer, an electronically conductive layer of black rust, pictured as an arrangement of macropores, covers the metal except at the end of the pores. The flattened aspect of the diagrams reflects the presence of this macroporous layer.*

- *The black rust is covered by a very compact microporous layer, made up of green rust and calcium carbonates. This film influences the high-frequency loop of the impedance diagrams.*

From this physical model, an electrical model of the interface can be given. Free corrosion is the association of an anodic process (iron dissolution) and a cathodic process (electrolyte reduction). Therefore, as discussed in Section 9.2.1, the total impedance of the system near the corrosion potential is equivalent to an anodic impedance Z_a in parallel with a cathodic impedance Z_c with a solution resistance R_e added in series as shown in Figure 13.4(a). The anodic impedance Z_a is simply depicted by a double-layer capacitance in parallel with a charge-transfer resistance (Figure 13.4(b)). The cathodic branch is described, following the method of de Levie,[50] by a distributed impedance in space as a transmission line in the conducting macropore (Figure 13.3). The interfacial impedance of the microporous layer Z_0 is given in Figure 13.4(c). The term R_f represents the ohmic resistance of the electrolyte through the film, and the term C_f represents the film capacitance. The ohmic resistance R_f is in series with the parallel arrangement of the

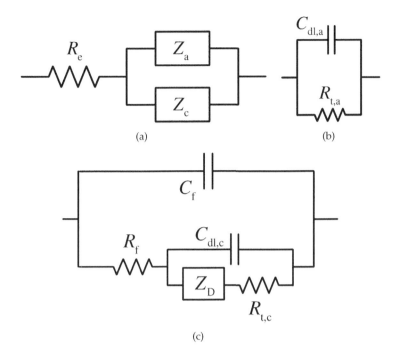

Figure 13.4: Equivalent circuit for: a) the total impedance of the cast iron–water interface; b) the anodic impedance; and c) the interfacial impedance of the microporous layer.

cathodic double-layer capacitance $C_{dl,c}$ and the faradaic branch consisting of a cathodic charge-transfer resistance $R_{t,c}$ in series with a diffusion impedance Z_D (see also Figure 9.8). The term Z_D, which describes the radial diffusion in the macropores, i.e., through the red rust, is given by equation (11.20).

Thus, the anodic surface corresponds to the end of the macropores, whereas the cathodic reaction occurs at the end of the micropores, which are located at the walls of the macropores. It should be noted that this physical-electrochemical model describes the behavior of cast iron at any time of immersion.

The calculation gives, for the cathodic impedance, the general form shown in equation (13.17), i.e.,

$$Z_c = \sqrt{R_0 Z_0}\coth\left(\frac{\ell}{\lambda}\right)$$ (13.28)

with

$$\lambda = \sqrt{\frac{Z_0}{R_0}}$$ (13.29)

and where ℓ is the mean length of the macropores and λ is the penetration depth of the electrical signal. When ℓ is small with respect to λ, the macropores respond like a flat electrode and the cathodic impedance tends to Z_0/ℓ. In this case, the angle made by the diffusion impedance is equal to 45°. When ℓ/λ becomes large, the macropores behave as though they were semi-

infinitely deep. Thus, $\coth(\ell/\lambda)$ *tends to unity, and* Z_c *equals* $\sqrt{R_0 Z_0}$*, which gives an angle of about* $22.5°$ *in the so-called Warburg domain.*

With the model illustrated in Figures 13.3 and 13.4, the impedance diagrams were analyzed by using a nonlinear least-squares regression procedure to extract physically meaningful parameters. For each time of immersion, the electrolyte resistance was measured separately and fixed in order to decrease the number of unknown parameters. The error structures identified by means of the measurement model described in Chapter 21 were used to weight the data during regression of the physical model.

The results of the fitting for water containing 2 *mg* l^{-1} *of FCl are presented in Nyquist format in Figures 13.5(a), (b), and (c), respectively, for different times of immersion. The model fits the experimental data very well, even under conditions where the diffusion loop is badly defined (i.e., at long times of immersion). Therefore, despite the large number of parameters imposed by the physical model, each parameter could be determined with a narrow confidence interval. The calculated diagrams show that the HF loop is, in fact, composed of two capacitive loops: one related to the microporous film and the other to the cathodic charge transfer. Due to the similar values of the time constants* $R_{t,c}C_{dl,c}$ *and* $R_f C_f$*, the two corresponding half-circles were nearly indistinguishable. The low-frequency loop characterizes the diffusion of the solute as well as the anodic charge transfer. After 28 days of immersion, the theoretical diagram makes an angle of* $45°$ *in the very-high-frequency range and the angle of the diffusion impedance is* $22.5°$*, which means that the pores behave as though they are semi-infinitely deep at these frequencies.*

The anodic charge-transfer resistance could be extracted from the fitting procedure. Thus the method provided a reliable value of the corrosion current and rate. This corrosion rate is about 10 *micrometers per year, which is not negligible if the chlorine consumption is considered.*

13.1.3   Pore-in-Pore Model

Gourbeyre et al.[240] and Itagaki et al.[241] have proposed that pores can appear inside a macro-pore structure as illustrated in Figure 13.6. Gourbeyre et al.[240] considered the response associated with a hierarchy with two pore sizes. This work was expanded by Itagaki et al.[241] who considered a hierarchy with up to three pore sizes.

The pores were assumed to be semi-infinite, therefore the impedance of the first pore, called macropore by Itagaki, is $\sqrt{R_0 Z_0}$, in agreement with the de Levie theory, where Z_0 is the impedance on the wall of the first pore. If the pore intersects with a large number of smaller mesopores, the impedance Z_0 becomes the impedance of the second pore. Under the assumption that this pore is semi-infinite, $Z_0 = \sqrt{R_1 Z_1}$, where R_1 is the electrolyte resistance per unit pore length in the mesopore and Z_1 is the interfacial impedance for a unit length pore in the mesopore. If a hierarchy of two pore sizes is considered, the impedance of one macropore is given by

$$Z = \sqrt{R_0 \sqrt{R_1 Z_1}} \qquad (13.30)$$

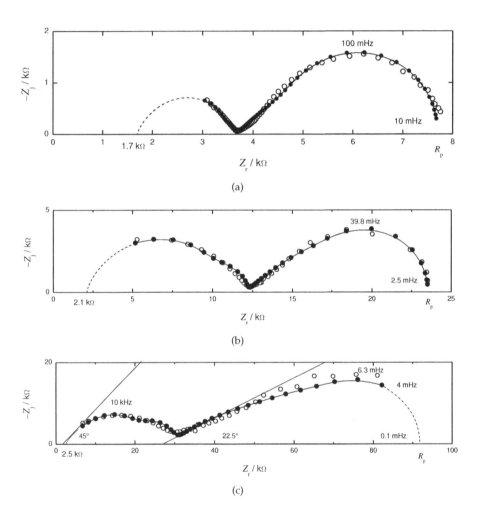

Figure 13.5: Regression results for the impedance diagrams of the cast iron rotating disk electrode after (a) 3, (b) 7, and (c) 28 days of immersion in Evian^{TM} water with 2 mg l^{-1} of FCl. (•) Experimental data and (○) fitted values using the equivalent circuits given in Figure 13.4. (Taken from Frateur et al.[8])

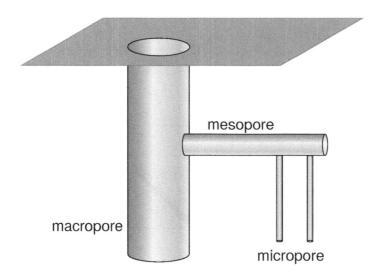

Figure 13.6: Schematic representation of a pore-within-a-pore hierarchy as suggested by Itagaki et al.[241]

If Z_1 corresponds to a blocking electrode or to a charge transfer resistance in parallel to a double layer capacitance, the phase shift of Z_1 tends towards 90° when the frequency tends towards infinity. As Z is proportional to $\sqrt{\sqrt{Z_1}}$, the phase shift of equation (13.30) tends towards 22.5° when the frequency tends towards infinity. In term of CPE behavior, this means that $\alpha = 0.25$ (see Chapter 14 on page 395).

If a triple-pore structure is under consideration, the previous impedance Z_1 is the impedance of the micropore shown in Figure 13.6. The impedance of one macro-pore becomes

$$Z = \sqrt{R_0 \sqrt{R_1 \sqrt{R_2 Z_2}}} \tag{13.31}$$

Z is proportional to $Z_2^{1/8}$, and the phase shift tends towards 11.25° when the frequency tends towards infinity. In term of CPE behavior, $\alpha = 0.125$ (see Chapter 14 on page 395).

Example 13.3 Impedance of Duplex Coating: *Gourbeyre et al.[240] studied duplex coatings which combine an "active" protection provided by a metal deposit, playing the role of sacrificial anode to cathodically protect the substrate, with "passive" protection of an organic layer, playing the role of a diffusion barrier to hinder transport of ionic species. The corrosion behavior of these duplex coating was studied in a chlorinated solution for long immersion times. Impedance spectra are presented for different immersion times in NaCl solution (6g/L) in Figures 13.7(a) and 13.7(b). The impedance diagram obtained after an immersion time shorter than 21 days could be described by only one flattened capacitive loop (Figure 13.7(a)). The value of the polarization resistance, obtained by extrapolation to zero frequency, decreased*

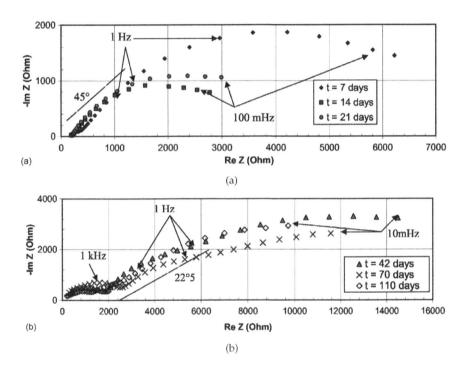

Figure 13.7: Nyquist representation of the evolution of the impedance of the duplex coating, ZnAl, and epoxy paint with time of immersion in NaCl electrolyte (6 g/L) as a parameter: a) elapsed time from 7 to 21 days and b) elapsed time from 42 to 110 days (Taken from Gourbeyre et al.[240]).

during the first 2 weeks of immersion. From an immersion time longer than 42 days, a new flattened loop appeared in high frequencies (Figure 13.7(b)). Usually this high-frequency loop could be attributed to the presence of a porous layer (see Section 13.1.2). In the present case, it could be the result of the formation of corrosion products because this loop did not exist at short immersion times. The polarization resistance at 42 days of immersion is about three times larger than the corresponding resistance at 21 days, which suggests a strong decrease of the corrosion due to the blocking effect of the corrosion products. Find an interpretation of theses results based on the pore-in-pore hierarchy.

Solution: *The high-frequency resistance is of the order of magnitude of a few hundred ohms for all immersion times. This value is very large with respect to the electrolyte resistance, which is of the order of a few ohms. As described in Section 9.3.2 on page 202, the high-frequency resistance could be the sum of the electrolyte resistance and the resistance of the pore inside the organic layer. The loop corresponding to the resistance of the pore and to the capacitance of the film is out of range of the present measurement due to the limited frequency range.*

At short immersion times (Figure 13.7(a)), the Nyquist diagrams present a slope of 45°. Thus it is possible to assume that the sample behaves like a porous electrode with pores of semi-infinite length. The corresponding impedance is given by

$$Z = \sqrt{R_0 Z_0} \tag{13.32}$$

where R_0 has the usual meaning and Z_0 corresponds to the impedance of the charge-transfer resistance R_t, placed in parallel with the double layer capacitance C_{dl}, i.e.,

$$Z_0 = \frac{R_t}{1 + j\omega R_t C_{dl}} \tag{13.33}$$

If the Zn-Al electrode corresponds to n pores, the overall impedance is

$$Z = R_{HF} + \frac{1}{n}\sqrt{\frac{R_0 R_t}{1 + j\omega R_t C_{dl}}} \tag{13.34}$$

where R_{HF} is the value of the extrapolation of the impedance at infinite frequency. Equation (13.34) contains three unknown lumped parameters: R_{HF}, $R_0 R_t / n^2$, and $C_{dl} R_t$. The individual parameters cannot be determined independently by regression.

For longer immersion times, the loop in low frequencies is more flattened than the corresponding loop for short immersion time (see Figure 13.7(b)). A slope of 22.5° appears on the Nyquist plot. For these longer immersion times another loop is observed at higher frequencies. Corrosion products, which could be at the origin of the high-frequency loop, appear inside the pore with a capacity C_p and a resistance R_{cp}. In addition it was possible to assume that for long immersion times the corrosion occurs in smaller pores called secondary pores or mesopores. According to the de Levie theory, the expression of the impedance becomes

$$Z_{sp} = \sqrt{\frac{R_1 R_t}{1 + j\omega C_{dl} R_t}} \tag{13.35}$$

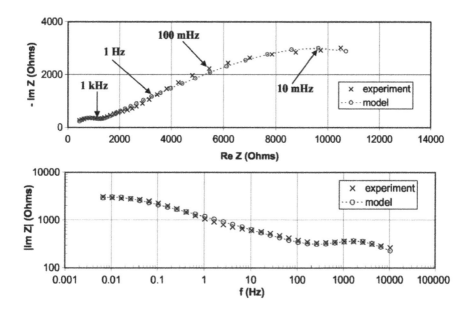

Figure 13.8: Correlation between the impedance measured after 110 days of immersion and the calculated impedance according to the model described in Example 13.3.

where Z_{sp} is the impedance of the secondary pore and R_1 is the electrolyte resistance per unit length inside the secondary pore. The overall impedance is given by the expression

$$Z = R_{HF} + \frac{1}{n}\sqrt{\frac{R_0\left(R_{cp} + Z_{sp}/n_1\right)}{1 + j\omega C_{cp}\left(R_{cp} + Z_{sp}/n_1\right)}} \tag{13.36}$$

Six unknown parameters are identified in equation (13.36). These can be expressed as: R_{HF}, $R_0 R_{cp}/n^2$, $R_0^2 R_1 R_t/n^4 n_1^2$, $C_{dl}R_t$, $C_p R_{cp}$, and $C_p^2 R_1 R_t$.

*   The experimental results obtained after 110 days of immersion are compared in Figure 13.8 to the regression results for the physical model represented by equations (13.35) and (13.36). The agreement is satisfactory in the entire frequency range for both the Nyquist diagram and for the plot of the imaginary part versus the frequency in logarithmic coordinates. This agreement was obtained with a physical model which involved only six adjusted parameters.*

13.1.4   Thin-Layer Cell

Thin-layer cells are experimental devices used to simulate some specific situations, such as crevice corrosion. They are usually achieved by confining a thin electrolyte layer (the thickness of which is less than a few hundred of micrometers) between the surface of a working electrode and a parallel insulating wall, mechanically set to reach the desired thickness. The most classical thin-layer cells have a cylindrical geometry and involve large disk electrode. The geometrical accuracy is generally limited by the difficulty of achieving a strict parallelism between the electrode surface and the confining wall. Remita et al.[242] used a positioning control procedure based on the measurement of the electrolyte resistance to quantify and minimize parallelism errors in these cells. Another possibility consists of using an ultramicroelectrode (UME) as a working electrode to reduce the dimension of the thin-layer cell. Such a thin-layer-cell configuration can be obtained with a scanning electrochemical microscope (SECM) used in feedback mode.[243]

The existence of a radial potential distribution should be taken into account for modelling suitably the impedance response on a large disk electrode located in a cylindrical thin-layer cell. A transmission line model can be adapted to the geometry of the cylindrical thin-layer cell.[244] In this development, the electrolyte resistivity is assumed to be homogeneous in the entire liquid film confined in the vicinity of the working electrode surface and the interfacial impedance of the electrode Z_{int} is assumed to be homogeneous over the electrode surface (i.e., no reactivity or capacitance distributions exist on the electrode surface).

This model sketched in Figure 13.9, is very similar to that of de Levie[50] for describing impedance response of electrodes located in a cylindrical thin-layer cell. The cylindrical geometry of the thin-layer cell is taken into account by introducing the dependency on the radial coordinate of the ring shape electrode surface elements $dS_{elec}(r)$ and on the lateral surface of the cylindrical electrolyte film confined within the cell $dS_{lat}(r)$, i.e.,

$$dS_{elec}(r) = 2\pi r dr \tag{13.37}$$

and

$$dS_{lat}(r) = 2\pi r \delta \tag{13.38}$$

where r is the radial coordinate and δ is the film thickness.

The resistance and admittance can be expressed as

$$dR = R_e dr \tag{13.39}$$

and

$$dY = \frac{dr}{Z_0} \tag{13.40}$$

respectively, where dR is the resitance of a length element of the liquid layer, R_e is the electrolyte resistance, dY is the admittance of a ring-shape surface element $dS_{elec}(r)$, and Z_0 is the impedance of the interface for a radial unit length of electrode. Equations (13.39) and (13.40) can be written as

$$dR = \frac{\rho}{2\pi r \delta} dr \tag{13.41}$$

and

$$dY = \frac{2\pi r dr}{Z_{int}} \tag{13.42}$$

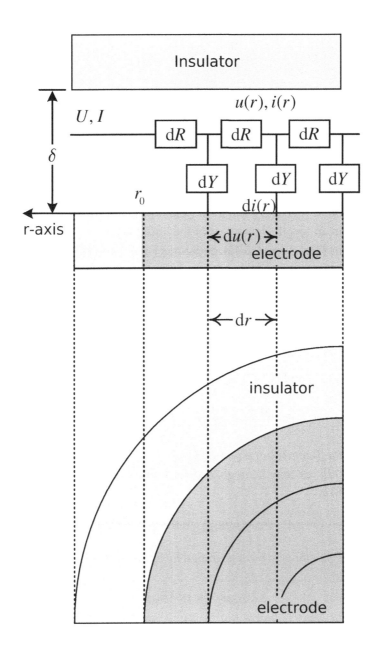

Figure 13.9: Schematic representation of a cylindrical thin-layer cell.[244]

where ρ is the electrolyte resistivity.

The local currents i and di flowing in the radial and normal directions of the electrode surface through the electrolyte layer, expressed in frequency domain, are linked by charge conservation, i.e.,

$$d\tilde{i}(r) = \tilde{i}(r + dr) - \tilde{i}(r) \tag{13.43}$$

These currents can also be expressed using Ohms law according to

$$\tilde{i} = -\frac{d\tilde{u}}{dR} \tag{13.44}$$

and

$$d\tilde{i} = -\tilde{u}dY \tag{13.45}$$

where $\tilde{u}$ and d$\tilde{u}$ are, respectively, the local potential within the thin-layer cell and the local ohmic drop in an length element dr of the liquid layer.

Using equations (13.41) and (13.42), equations (13.44) and (13.45) may be written as

$$\frac{d\tilde{u}}{dr} = -\frac{\rho}{2\pi r \delta}\tilde{i} \tag{13.46}$$

and

$$\frac{d\tilde{i}}{dr} = -\frac{2\pi r}{Z_{\text{int}}}\tilde{u} \tag{13.47}$$

Using equations (13.44) and (13.47), equation (13.46) becomes

$$\frac{d^2\tilde{u}}{dr^2} + \frac{1}{r}\frac{d\tilde{u}}{dr} - \frac{\rho}{\delta Z_{\text{int}}}\tilde{u} = 0 \tag{13.48}$$

which resembles the Telegrapher equation (13.7).

The term $(1/r)d\tilde{u}/dr$ may be neglected for a thin-layer cell. The general solution for equation (13.48) is

$$\tilde{u} = A\exp\left(r\alpha_{\text{cell}}\right) + B\exp\left(-r\alpha_{\text{cell}}\right) \tag{13.49}$$

where A and B are constants to be determined by application of the appropriate boundary conditions, and

$$\alpha_{\text{cell}} = \sqrt{\rho/\delta Z_{\text{int}}} \tag{13.50}$$

From consideration of the cylindrical symmetry of the cell,

$$\left.\frac{d\tilde{u}}{dr}\right|_{r=0} = 0 \tag{13.51}$$

Thus, $A = B$, and

$$\tilde{u} = 2A\cosh\left(r\alpha_{\text{cell}}\right) \tag{13.52}$$

The global impedance of the transmission line may be expressed in terms of the potential at the periphery of the electrode, $\widetilde{U} = \widetilde{u}(r_0)$ and the global current flowing through the interface $\widetilde{I} = \widetilde{i}(r_0)$ as

$$Z = \frac{\widetilde{U}}{\widetilde{I}} \pi r_0^2 \tag{13.53}$$

The global current may be written as

$$\widetilde{I} = \int_0^{r_0} \mathrm{d}\widetilde{i} = -\int_0^{r_0} \frac{2\pi r}{Z_{\text{int}}} \widetilde{u} \, \mathrm{d}r \tag{13.54}$$

Combination of equations (13.52) and (13.54) yields

$$\widetilde{I} = \frac{4\pi A}{Z_{\text{int}}} \left(\frac{r_0}{\alpha_{\text{cell}}} \sinh(r_0 \alpha_{\text{cell}}) + \frac{1}{\alpha_{\text{cell}}^2} (1 - \cosh(r_0 \alpha_{\text{cell}})) \right) \tag{13.55}$$

The impedance of the electrode is given by

$$Z = \frac{r_0^2 Z_{\text{int}}}{2 \left(\frac{r_0}{\alpha_{\text{cell}}} \tanh(r_0 \alpha_{\text{cell}}) + \frac{1}{\alpha_{\text{cell}}^2} \left(\frac{1}{\cosh(r_0 \alpha_{\text{cell}})} - 1 \right) \right)} \tag{13.56}$$

When $\alpha_{\text{cell}} \to 0$, i.e., when $\delta \to \infty$ or $\rho \to 0$, $Z \to Z_{\text{int}}$. When $\alpha_{\text{cell}} \to \infty$, i.e., when $\delta \to 0$ or $\rho \to \infty$, the cell may be treated as a one-dimensional system and Z is proportional to $\sqrt{Z_{\text{int}}}$, a result that is similar to de Levie's impedance for a one-dimensional porous electrode, as shown in equation (13.17).

13.2 Geometry-Induced Current and Potential Distributions

The geometry of an electrode frequently constrains the distribution of current density and potential in the electrolyte adjacent to the electrode in such a way that both cannot simultaneously be uniform. The disk electrode, for example, consisting of a circular electrode embedded in an insulating plane, is commonly used in electrochemical measurements. The disk geometry may be associated with rotating or stationary electrodes. A discussion of the convective diffusion impedance associated with a rotating disk electrode is presented in Section 11.3 on page 256, and the impinging jet electrode is discussed in Section 11.4 on page 266.

Newman showed, in 1966, that the electrode-insulator interface gives rise to nonuniform current and potentials distributions for currents that are below the mass-transfer-limited current.[124, 123] The geometry-induced nonuniform current and potential distributions influences the transient response of the disk electrode. Nisancioglu and Newman[245, 246] developed a solution for the transient response of a faradaic reaction on a nonpolarizable disk electrode to step changes in current. The solution to Laplace's equation was performed using a transformation to rotational elliptic coordinates and a series expansion in terms of Lengendre polynomials. Antohi and Scherson have recently expanded the solution to the transient problem by expanding the number of terms used in the series expansion.[247]

Newman also provided a treatment for the impedance response of a disk electrode which showed that the capacity and ohmic resistance were functions of frequency above a critical value of frequency.[51] More recently, comprehensive numerical and experimental studies of the impedance response of the disk electrode have been presented that accounted for the response of ideally polarized electrodes,[147] electrodes showing local CPE behavior,[248] electrodes with faradaic reactions,[125] and electrodes with reactions involving adsorbed intermediates.[249,250]

13.2.1 Mathematical Development

Under the assumption that mass transfer may be neglected, the electrical potential in the electrolyte is governed by Laplace's equation

$$\nabla^2 \Phi = 0 \tag{13.57}$$

The system may be assumed to have cylindrical symmetry such that the potential in solution is dependent only on the radial position r along the electrode surface and the normal position y. Following Example 1.9 on page 20, the potential can be separated into steady and time-dependent parts as

$$\Phi = \overline{\Phi} + \mathrm{Re}\{\widetilde{\Phi} \exp{(\mathrm{j}\omega t)}\} \tag{13.58}$$

where $\overline{\Phi}$ is the steady-state solution for potential and $\widetilde{\Phi}$ is the complex oscillating component which is a function of position only. The form of Laplace's equation to be solved, subject to appropriate boundary conditions, is

$$\frac{1}{r}\frac{\partial}{\partial r}\left(r\frac{\partial\widetilde{\Phi}}{\partial r}\right) + \frac{\partial^2\widetilde{\Phi}}{\partial y^2} = 0 \tag{13.59}$$

Equation (13.59) can be solved using numerical methods.

The boundary condition for potential at $r \to \infty$ is that $\widetilde{\Phi} \to 0$, and the boundary condition on the insulating surface is that $\partial\widetilde{\Phi}/\partial y = 0$. The distinction between different conditions simulated is seen in the boundary condition for the electrode surface.

Blocking Electrode

For a blocking electrode,[147] the boundary condition at the electrode surface may be given as

$$\mathrm{j}\omega C_0(\widetilde{V} - \widetilde{\Phi}_0) = -\kappa\frac{\partial\widetilde{\Phi}}{\partial y}\bigg|_{y=0} \tag{13.60}$$

where Φ_0 is the potential in the solution adjacent to the electrode, κ is the conductivity of the electrolyte, and C_0 is the capacitance of the disk. In dimensionless form,

$$jK(\tilde{V} - \tilde{\Phi}_0) = -r_0 \left.\frac{\partial \tilde{\Phi}}{\partial \xi}\right|_{\xi=0} \tag{13.61}$$

where K is the dimensionless frequency

$$K = \frac{\omega C_0 r_0}{\kappa} \tag{13.62}$$

and $\xi = y/r_0$. The current is only that required to charge or discharge the electrode. The characteristic dimension of the disk electrode, expressed as

$$\ell_{c,\text{disk}} = r_0 \tag{13.63}$$

may be compared to characteristic dimensions for other geometries in Section 13.4.

Blocking Electrode with CPE Behavior

For a blocking electrode with CPE behavior (see Chapter 14 on page 395),[248] the interfacial impedance is given as

$$Z_{\text{CPE}} = \frac{\tilde{V} - \tilde{\Phi}_0}{\tilde{i}} = \frac{1}{(j\omega)^\alpha Q} \tag{13.64}$$

On application of the Euler identity, equation (1.57) on page 12,

$$Z_{\text{CPE}} = \frac{1}{\omega^\alpha Q}\left(\cos\left(\frac{\alpha\pi}{2}\right) - j\sin\left(\frac{\alpha\pi}{2}\right)\right) \tag{13.65}$$

Thus,

$$K(\tilde{V} - \tilde{\Phi}_0)\left(\cos\left(\frac{\alpha\pi}{2}\right) + j\sin\left(\frac{\alpha\pi}{2}\right)\right) = -r_0 \left.\frac{\partial \tilde{\Phi}}{\partial \xi}\right|_{\xi=0} \tag{13.66}$$

where K is the dimensionless frequency defined for a system with CPE behavior to be

$$K = \frac{Q\omega^\alpha r_0}{\kappa} \tag{13.67}$$

It is important to note that Q has units of $s^\alpha/\Omega\text{cm}^2$ or $F/s^{(1-\alpha)}\text{cm}^2$.

Electrode with Faradaic Reactions

For a faradaic reaction taking place at the electrode (see Section 10.2.1 on page 209), the boundary condition at the electrode is expressed in frequency domain as[125]

$$jK(\tilde{V} - \tilde{\Phi}_0) + J(\tilde{V} - \tilde{\Phi}_0) = -r_0 \left.\frac{\partial \tilde{\Phi}}{\partial \xi}\right|_{\xi=0} \tag{13.68}$$

where $\widetilde{V}$ represents the imposed perturbation in electrode potential referenced to an electrode at infinity and K is defined by equation (13.62) under the assumption of pure capacitive behavior for the double layer. Under the assumption of linear kinetics, valid for $\bar{i} \ll i_0$, the parameter J is defined to be

$$J = \frac{(\alpha_a + \alpha_c)\,\mathrm{F}i_0 r_0}{RT\kappa} \tag{13.69}$$

For Tafel kinetics, valid for $\bar{i} \gg i_0$, the parameter J is defined to be

$$J(\overline{V} - \overline{\Phi}_0) = \frac{\alpha_c \mathrm{F}\,|\bar{i}(\overline{V} - \overline{\Phi}_0)|\,r_0}{RT\kappa} \tag{13.70}$$

As $\overline{\Phi}_0$ depends on radial position, the value of J expressed in equation (5.118) is a function of radial position.

The charge-transfer resistance for linear kinetics can be expressed in terms of parameters used in equation (5.115) as

$$R_t = \frac{RT}{i_0 \mathrm{F}(\alpha_a + \alpha_c)} \tag{13.71}$$

and, in terms of parameters used in equation (5.118),

$$R_t = \frac{RT}{|\bar{i}(\overline{V} - \overline{\Phi}_0)|\,\alpha_c \mathrm{F}} \tag{13.72}$$

For linear kinetics, R_t is independent of radial position, but, under Tafel kinetics, as shown in equation (13.72), R_t depends on radial position. From a mathematical perspective, the principal difference between the linear and Tafel cases is that J and R_t are held constant for the linear polarization; whereas, for the Tafel kinetics, J and R_t are functions of radial position determined by solution of the nonlinear steady-state problem.

The ohmic resistance of a blocking disk electrode embedded within an insulating plane, in units of $\Omega\mathrm{cm}^2$ is given by[124]

$$R_e = \frac{1}{4\kappa r_0}\pi r_0^2 = \frac{\pi r_0}{4\kappa} \tag{13.73}$$

Huang et al.[125] provided a relationship between the parameter J and the charge-transfer and ohmic resistances as

$$J = \frac{4}{\pi}\frac{R_e}{R_t} \tag{13.74}$$

Large values of J are seen when the ohmic resistance is much larger than the charge-transfer resistance, and small values of J are seen when the charge-transfer resistance dominates. The definition of parameter J in equation (13.74) is the reciprocal of the Wagner number,[126] which is a dimensionless quantity that measures the uniformity of the current distribution in an electrolytic cell.

Electrode with Faradaic Reactions Coupled by Adsorbed Intermediates

As the reaction sequences become more complicated, it becomes difficult to express the model results in a general manner. Wu et al.[249] analyzed the impedance response associated with two successive charge-transfer steps involving an intermediate species adsorbed on the electrode surface, i.e.,

$$M \rightarrow X_{ads}{}^{+} + e^{-} \tag{13.75}$$

and

$$X_{ads}{}^{+} \rightarrow P^{2+} + e^{-} \tag{13.76}$$

The reactant could be a metal M which dissolves to form an adsorbed intermediate $X_{ads}{}^{+}$, which then reacts to form the final product P^{2+}. The reactions were assumed to be irreversible, and diffusion processes were considered negligible. The model development presented in Section 10.4.1 on page 226 leads to an expression for the faradaic impedance Z_F at a given potential given by

$$\frac{1}{Z_F} = \frac{1}{R_t} + \frac{A}{j\omega + B} \tag{13.77}$$

where the charge-transfer resistance R_t is defined by

$$\frac{1}{R_t} = \frac{1}{R_{t,M}} + \frac{1}{R_{t,X}} = b_M|\bar{i}_M| + b_X|\bar{i}_X| \tag{13.78}$$

and parameters A and B are potential dependent variables given by

$$A = \frac{\partial \bar{i}_F}{\partial \bar{\gamma}}\frac{\partial \bar{\gamma}}{\partial \bar{V}} = \frac{(R_{t,M}^{-1} - R_{t,X}^{-1})[K_X \exp(b_X(\overline{V} - \overline{\Phi}_0)) - K_M \exp(b_M(\overline{V} - \overline{\Phi}_0))]}{\Gamma F} \tag{13.79}$$

and

$$B = -\frac{\partial \dot{\gamma}}{\partial \bar{\gamma}} = \frac{K_X \exp(b_X(\overline{V} - \overline{\Phi}_0)) + K_M \exp(b_M(\overline{V} - \overline{\Phi}_0))}{\Gamma F} \tag{13.80}$$

The sign of B is always positive, and the sign of A may be positive or negative.

The features of the impedance change according to the sign of A. When A is positive, the impedance plot in the Nyquist plane of Z_F in parallel with the double-layer capacitance (C_0) shows a high-frequency capacitive loop corresponding to R_t in parallel with C_0, i.e., $R_t||C_0$, with a low-frequency inductive loop corresponding to the second term of equation (13.77). When A is negative, the impedance plot shows a high-frequency capacitive loop $(R_t||C_0)$ with a low-frequency capacitive loop, also corresponding to the second term of equation (13.77). When A is equal to zero, the two terms of the numerator in equation (13.79) cancel, i.e., $\partial \bar{i}_F/\partial \bar{\gamma} = 0$. In this case, the reaction current density is not dependent on the surface coverage, and, therefore, the impedance plot shows a single capacitive loop corresponding to $R_t||C_0$ with no low-frequency loop.

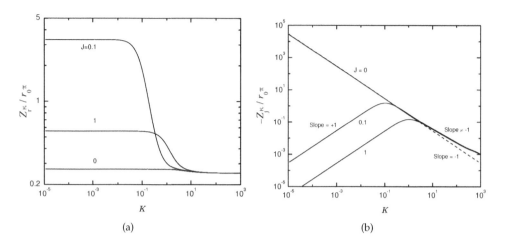

Figure 13.10: Calculated representation of the impedance response for a disk electrode under assumption of Tafel kinetics with J as a parameter: a) real part and b) imaginary part. The value $J = 0$ corresponds to an ideally capacitive blocking electrode.

13.2.2 Numerical Method

Numerical solutions are required for the equations posed in this section. Two methods were used to solve the sets of equations. In one method, following Newman,[51] Laplace's equation was transformed to rotational elliptic coordinates. The equations were solved under assumption of a uniform capacitance C_0 using the collocation package PDE2D® developed by Sewell.[251] The calculations were performed for different domain sizes, and the results were obtained by extrapolation to an infinite domain size.

The equations were also solved in cylindrical coordinates using a finite-elements package COMSOL®.® The domain size was 2,000 times larger than the disk electrode dimension in order to meet the assumption that the counterelectrode was located infinitely far from the electrode surface. The results obtained by the two packages were in excellent agreement for dimensionless frequencies $K < 100$, providing a frequency domain covering the entire electrochemical response of the system.

13.2.3 Complex Ohmic Impedance at High Frequencies

One result of the simulations based on use of Laplace's equation to account for potential in the electrolyte was that the ohmic contribution has a complex character. This may be seen at high frequencies for systems that do not involve an adsorbed reaction intermediate. For systems that involve an adsorbed reaction intermediate, frequency dispersion may be seen at both high and low frequencies.

The global impedance response, calculated under assumption of Tafel kinetics, is presented in Figure 13.10 as a function of dimensionless frequency with J as a param-

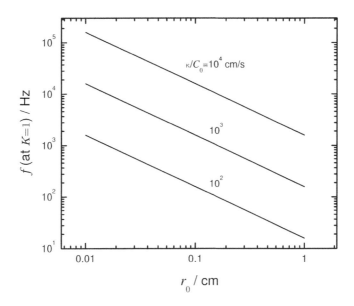

Figure 13.11: The frequency $K = 1$ at which the current distribution influences the impedance response with κ/C_0 as a parameter. (Taken from Huang et al.[147])

eter. The value $J = 0$ corresponds to an ideally capacitive blocking electrode. At low frequencies, values for the real part of the impedance differ for impedance calculated under the assumptions of linear and Tafel kinetics, whereas, the values of the imaginary impedance calculated under the assumptions of linear and Tafel kinetics are superposed for all frequencies. The slopes of the lines presented in Figure 13.10(b) are equal to $+1$ at low frequencies but differ from -1 at high frequencies. The slopes of these lines in the high-frequency range can be related to the exponent α used in models for CPE behavior (see Chapter 14 on page 395).[252]

Two characteristic frequencies are evident in Figure 13.10. The characteristic frequency $K = 1$ is associated with the influence of current and potential distributions. The characteristic frequency $K/J = 1$ is associated with the R_tC_0-time constant for the faradaic reaction.

The frequency $K = 1$ at which the current and potential distributions begin to influence the impedance response can be expressed as

$$f_{c,\text{disk}} = \frac{1}{2\pi} \frac{\kappa}{C_0 r_0} \tag{13.81}$$

The frequency $f_{c,\text{disk}}$ at which the current distribution influences the impedance response is shown in Figure 13.11 with κ/C_0 as a parameter. As demonstrated in Example 13.4, the influence of high-frequency geometry-induced time-constant dispersion can be avoided for reactions that do not involve adsorbed intermediates by conducting experiments below the characteristic frequency given in equation (13.81). The characteristic frequency can be well within the range of experimental measurements. The value $\kappa/C_0 = 10^3$ cm/s, for example, can be obtained for a capacitance $C_0 = 10 \ \mu\text{F}/\text{cm}^2$ (corresponding to the value expected for the double layer on a metal electrode) and conductivity $\kappa = 0.01$ S/cm (corresponding roughly to a 0.1 M NaCl solution). Equation (13.81) suggests that time-constant dispersion should be expected above a frequency of 630 Hz on a disk with radius $r_0 = 0.25$ cm. The results presented in Figure 13.11 suggest that, by using an electrode that is sufficiently small, the experimentalist may be able to avoid the frequency range that is influenced by current and potential distributions.

Example 13.4 Characteristic Frequency for Disk: *Consider an experimental system involving a disk electrode covered by an oxide film in 1.0 M KCl solution at room temperature for which impedance measurements are desired to a maximum frequency of 10 kHz. Estimate the maximum radius for a disk electrode that will avoid the influence of high-frequency geometry-induced time-constant dispersion. The oxide film may be assumed to have a thickness of 3 nm and a dielectric constant $\varepsilon = 12$.*

Solution: *The characteristic frequency given in equation (13.81) depends on the ratio κ/C_0. The conductivity can be taken from reference books[213] or estimated using equation (5.109) on page 114 and values of diffusivity taken from Table 5.4 on page 113. From reference 213, the conductivity is $\kappa = 0.108$ S/cm. The double-layer capacitance for the oxide-covered electrode may be estimated from equation (5.138) (page 126) to be $C_0 = 3.5 \ \mu\text{F}/\text{cm}^2$. Thus, $\kappa/C_0 = 3.05 \times 10^4$ cm/s, and, following*

$$r_0 = \frac{1}{2\pi} \frac{\kappa}{C_0 f_{\text{c,disk}}} \tag{13.82}$$

the maximum disk radius is 0.49 cm. This result can also be obtained from Figure 13.11.

For linear kinetics, the global interfacial impedance, defined by equation (7.61) on page 161, is independent of radial position and is given by

$$Z_0 = \frac{R_t}{1 + j\omega C_0 R_t} \tag{13.83}$$

The global ohmic impedance Z_e is obtained from the global impedance Z by the expression

$$Z_e = Z - Z_0 \tag{13.84}$$

The real and imaginary parts of Z_e, obtained for linear kinetics, are given in Figures 13.12(a) and 13.12(b), respectively, as functions of dimensionless frequency K with J as a parameter. In the low-frequency range, $Z_e\kappa/r_0\pi$ is a pure resistance with a numerical value that depends weakly on J. All curves converge in the high-frequency range such that $Z_e\kappa/r_0\pi$ tends toward $1/4$. The imaginary part of the global ohmic impedance shows a non-zero value in the frequency range that is influenced by the current and potential distributions. The global ohmic impedance is presented in Nyquist format in Figure 13.13. For $K < 0.1$, the ohmic impedance is a real number, as shown in Figure

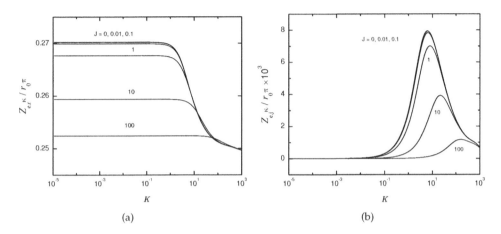

Figure 13.12: Calculated global ohmic impedance as a function of dimensionless frequency with J as a parameter: a) real part and b) imaginary part.

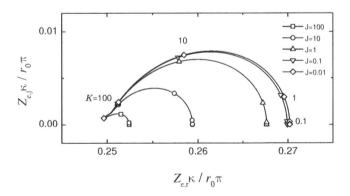

Figure 13.13: Calculated global ohmic impedance taken from Figure 13.12 in Nyquist format.

13.12(a). At frequencies $K > 0.1$, the ohmic contribution is represented by a complex number. The influence of high-frequency geometry-induced time-constant dispersion can be avoided for reactions that do not involve adsorbed intermediates by conducting experiments below the characteristic frequency given in equation (13.81).

It is useful to compare the concept of a complex ohmic impedance to Newman's treatment of frequency dispersion.[51] A blocking interface with a uniform frequency-independent capacitance C_0 has an interfacial impedance $Z_0 = 1/j\omega C_0$. This interfacial impedance is independent of the electrode geometry. The measured overall impedance, which includes the ohmic contribution, is $Z = Z_e + 1/j\omega C_0$, where the capacitance C_0 is independent of frequency and Z_e is termed the ohmic impedance. Newman, in contrast, represented the overall impedance as the sum of a frequency-dependent resistance R_{eff} in series with a frequency-dependent capacitance C_{eff}. The two descriptions of the same phenomena give

$$Z_e + \frac{1}{j\omega C_0} = R_{eff} + \frac{1}{j\omega C_{eff}} \tag{13.85}$$

or

$$Z_e = R_{eff} - \frac{j}{\omega}\left(\frac{C_0 - C_{eff}}{C_0 C_{eff}}\right) \tag{13.86}$$

The observation that Z_e is frequency dependent is in perfect agreement with Newman's result.[51] When the frequency tends towards infinity, the current distribution corresponds to the primary current distribution, and

$$\lim_{\omega \to \infty} Z_e = \frac{1}{4\kappa r_0} \tag{13.87}$$

in agreement with Newman's formula. The complex character of the ohmic impedance is a property of the electrode geometry, interfacial impedance, and electrolyte with conductivity κ. A complex ohmic impedance, for example, is not seen for recessed electrodes for which both the current density and interfacial potential are uniform.

13.2.4   Complex Ohmic Impedance at High and Low Frequencies

The results presented here are based on simulations presented by Wu et al.[249] A steady-state polarization curve is presented in Figure 13.14(a) where dashed boxes are used to represent the range of steady-state current density and potential corresponding to the three cases shown in Figure 13.15. The current distribution becomes more nonuniform at larger values of potential. For $\langle A \rangle = -0.83 \ \Omega^{-1}cm^{-2}s^{-1}$, the parameter J has the largest value, meaning, as shown in equation (13.74), that the ohmic resistance is much larger than the charge-transfer resistance. The corresponding adsorption isotherm given in Figure 13.14(b) shows an inflection at $\langle A \rangle = 0$, representing the stronger dependence of the surface coverage on the interfacial potential; whereas, at $\langle A \rangle = -0.83 \ \Omega^{-1}cm^{-2}s^{-1}$, the isotherm crosses the largest potential interval, indicating the most nonuniform potential distribution on the electrode surface.

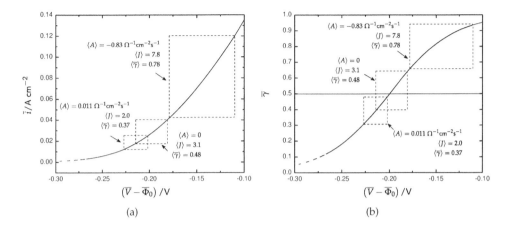

Figure 13.14: The variations of a) steady-state current density and b) steady-state surface coverage density with the interfacial potential. Dashed squares are used to identify the range of current and surface coverage corresponding to the simulations performed at $\overline{V} = -0.15\text{V}$ ($\langle A \rangle = 0.011$ $\Omega^{-1}\text{cm}^{-2}\text{s}^{-1}$), $\overline{V} = -0.1\text{V}$ ($\langle A \rangle = 0$ $\Omega^{-1}\text{cm}^{-2}\text{s}^{-1}$), and $\overline{V} = 0.1\text{V}$ ($\langle A \rangle = -0.83$ $\Omega^{-1}\text{cm}^{-2}\text{s}^{-1}$). The position $r = 0$ corresponds to the lower-left corner of each box. Taken from Wu et al.[249]

The global impedance represents an averaged response of the electrode. The calculated results of global impedance for the cases with $\langle A \rangle > 0$, $\langle A \rangle = 0$, and $\langle A \rangle < 0$ are presented in Nyquist format in Figures 13.15(a), 13.15(b), and 13.15(c),

respectively. The solid lines in Figure 13.15 represent the simulation results obtained by solving Laplace's equation subject to the boundary conditions that account for the time-constant dispersion associated with the electrode geometry. The dashed curves represent the global impedances calculated as

$$Z = R_e + \frac{1}{1/Z_F + j\omega C_0} \tag{13.88}$$

The faradaic impedance Z_F was calculated from equation (13.77) in terms of the surface-averaged parameters given from equation (13.78) to (13.80) and therefore did not account for the influence of electrode geometry. The geometry of the disk electrode is shown to distort the global impedance response. The geometry-induced distortion of impedance response and corresponding depressions of semicircles at high and low frequencies are more obvious in Figure 13.15(c) where $\langle A \rangle < 0$.

The interfacial impedance corresponding to the calculations of Figure 13.15 are presented in Figure 13.16. No frequency dispersion is evident. The ohmic impedance, obtained by subtraction, is given in Figure 13.17. In contrast to the behavior seen in Figure 13.13, the effects of frequency dispersion are seen at both high and low frequencies. The overlapping of circles on the real axis represents the frequency range where the ohmic contribution may be expressed as a real number.

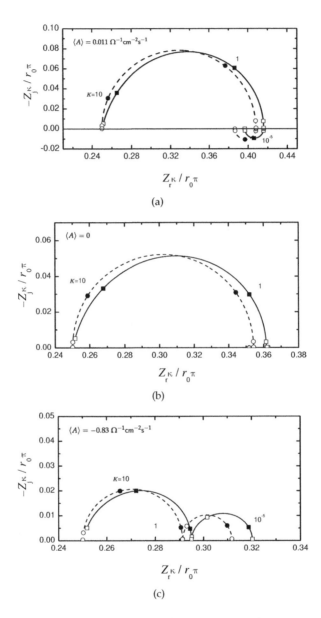

Figure 13.15: Calculated Nyquist representation of the global impedance response for a disk electrode considering the influence of electrode geometry (solid lines) and in the absence of geometry effect (dashed lines): a) $\langle A \rangle > 0$; b) $\langle A \rangle = 0$; and (c) $\langle A \rangle < 0$. Taken from Wu et al.[249]

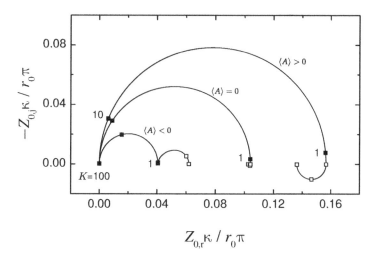

Figure 13.16: The global interfacial impedance for $\langle A \rangle > 0$, $\langle A \rangle = 0$, and $\langle A \rangle < 0$ for the results presented in Figure 13.15. Taken from Wu et al.[253]

13.3 Electrode Surface Property Distributions

The frequency dispersion described in Section 13.2 is attributed to the geometry of the electrode. Frequency dispersion may also be associated with a spatial distribution of properties along the electrode surface. This section provides a discussion of the influence of surface heterogeneity on impedance response. The surface heterogeneity considered includes roughness, surface distribution of capacitance, and surface distribution of rate constants for electrochemical reactions.

13.3.1 Electrode Roughness

In early experiments on solid electrodes, micro-scale surface roughness was believed to contribute to the non-ideality of electrochemical measurements.[50] Borisova and Ershler[254] were the first to observe that roughness influenced electrochemical measurements. They found that the extent of frequency dispersion was reduced by melting a metal electrode and letting it cool to form into a droplet, suggesting that the smoother surface led to a more ideal response.

Following the development of fractal theory,[255] there was an attempt to correlate the fractal dimensions of the surface to the CPE exponent α.[256,257] Fractal geometry was shown to cause frequency dispersion, however a correlation between the fractal dimension and variance from ideality could not be found. Pajkossy[258–260] showed experimentally that annealing can reduce the degree of frequency dispersion, even though the roughness of the surface remained the same, and thereby concluded that the frequency dispersion cannot be due to the geometric effect solely but may also have a contribution of atomic-scale heterogeneities, distinguishing between the effects of a dis-

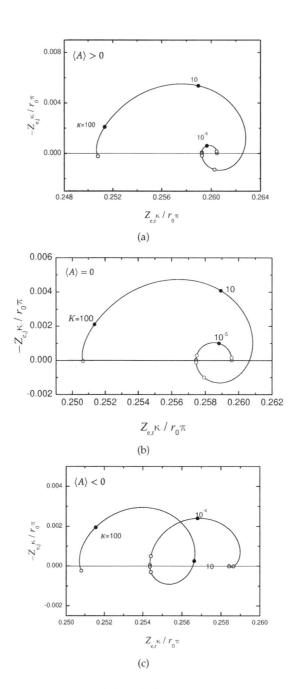

Figure 13.17: Nyquist representation of the calculated global ohmic impedance response for the results presented in Figure 13.15: a) $\langle A \rangle > 0$; b)$\langle A \rangle = 0$; and (c)$\langle A \rangle < 0$. Taken from Wu et al.[253]

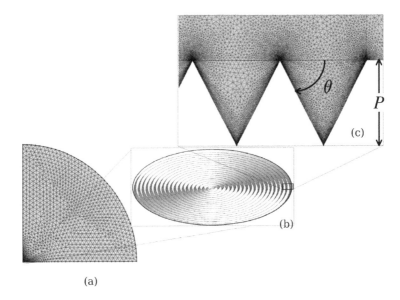

Figure 13.18: Schematic representation showing the finite-element mesh used for the disk electrode simulations: a) entire domain; b) grooved surface of the electrode; and c) detail of the grooved electrode. Taken from Alexander et al.[262]

tribution of solution resistance and a distribution of surface properties. Emmanuel[261] used an analytic-continuation method to calculate the impedance of a two-dimensional Hull cell and simulated the effect of a linear distribution of solution resistance assuming a uniform capacitance. The results of his work showed that frequency dispersion occurred at high frequencies while the calculations yielded ideal behavior at low frequencies. Similar results were reported by Alexander et al.[262] for a rough disk electrode.

The mathematical development follows that described in Section 13.2.1 with boundary conditions corresponding to a blocking electrode (Section 13.2.1). The roughness of the electrode was quantified in terms of the roughness factor or rugosity, f_r, obtained by dividing the true polarizable area by the geometric area.[263] A smooth electrode has a roughness factor of unity; whereas, roughness factors much greater than unity may be attributed to porous electrodes. For a uniformly rough surface with concentric V-shaped grooves, the roughness factor may be expressed as $f_r = 1/\cos\theta$, where θ represents the contact angle between the rough surface and a smooth plane.

Influence of Roughness on a Disk Electrode

To explore the role of surface roughness, Alexander et al.[262] simulated a smooth disk electrode, a rough disk electrode, and a recessed rough disk electrode. The calculations were performed in axisymmetric cylindrical coordinates for a quarter of a circle domain, shown in Figure 13.18(a), which represents the electrolyte. The counterelectrode

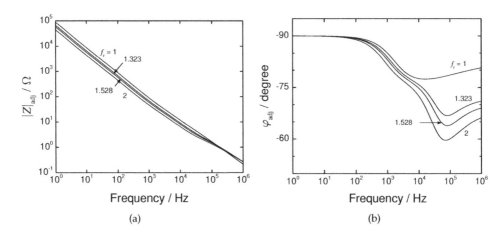

Figure 13.19: Calculated global impedance as a function of frequency for a rough disk electrode with roughness factor as a parameter: a) ohmic-resistance-corrected magnitude and b) ohmic-resistance-corrected phase angle. Taken from Alexander et al.[262]

was located at a position $\sqrt{r^2 + y^2} = 500$ cm, making the domain size more than 1,000 times larger than the 0.24 cm radius of the disk. A free triangular mesh was used with a greater density of elements near the working electrode. The grooved surface of the disk is shown in Figure 13.18(b), and an expanded view of the grooves is shown in Figure 13.18(c). The angle θ used for the calculation of roughness factor is also shown in Figure 13.18(c).

Calculated ohmic-resistance-corrected Bode plots (see Section 18.2 on page 501) are presented in Figure 13.19 as functions of frequency with the roughness factor as a parameter. The ohmic-resistance-corrected phase angle shown in Figure 13.19(b) emphasizes the phase response of the electrode. The phase angle is equal to $-90°$ at the low-frequency limit, corresponding to the behavior of a capacitive electrode. At higher frequencies, the ohmic-resistance-corrected phase angle reveals two deviations from an ideal capacitive response.

A clearer separation of features can be obtained by use of a phase angle that is based on only the imaginary part of the impedance, i.e.,

$$\varphi_{dZj} = \frac{d \log |Z_j|}{d \log f} \times 90° \qquad (13.89)$$

The derivative of the logarithm of the imaginary impedance with respect to the logarithm of the frequency, used previously to estimate the exponent for a constant-phase element,[252] is expressed here as a phase angle. The imaginary-impedance-derived phase angle is presented in Figure 13.20(a) as a function of frequency with the roughness factor as a parameter. The results for the smooth electrode, i.e., $f_r = 1$, showed nonideal behavior at frequencies greater than 800 Hz due to the nonuniform current distribution caused by the electrode configuration.[147] For the rough electrodes, two

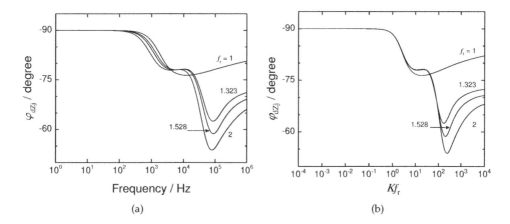

Figure 13.20: Imaginary-impedance-derived phase angle calculated from the impedance presented in Figure 13.19 for a rough disk electrode: a) as a function of frequency and b) as a function of dimensionless frequency Kf_r. Taken from Alexander et al.[262]

distinct deviations from ideal behavior were observed. The lower-frequency deviation for a 0.24 cm radius electrode with a roughness factor of 2 occurred at frequencies equal to and greater than 300 Hz, which is 500 Hz less than that observed for the smooth electrode. The magnitude of the difference in frequency between the lower-frequency deviation of the rough electrode compared to the smooth electrode is proportional to the roughness factor. The higher-frequency deviation occurred at roughly 20 kHz.

The characteristic frequency at which the current and potential distributions on a smooth disk electrode within an insulating plane begin to influence the impedance response may be expressed as

$$f_c = \frac{\kappa}{2\pi C_0 r_0} \tag{13.90}$$

For a rough electrode, equation (13.90) must be modified to be

$$f_c = \frac{\kappa}{2\pi C_0 f_r r_0} \tag{13.91}$$

where the characteristic dimension for the rough disk is given as

$$\ell_{c,\text{disk}} = f_r r_0 \tag{13.92}$$

Equation (13.91) yields a characteristic frequency that is slightly smaller than the characteristic frequency associated with the influence of the nonuniform current distributions on a smooth disk electrode.

The imaginary-impedance-derived phase angle is presented in Figure 13.20(b) as a function of dimensionless frequency in which the characteristic length is given by equation (13.92). The superposition of curves indicates that the low-frequency decrease in phase angle seen at $Kf_r = 1$ can be attributed to the coupled effect of disk geometry

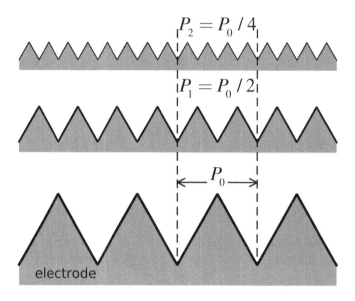

Figure 13.21: Schematic representation showing the manner in which the roughness period was varied for a fixed roughness factor equal to 2. The three configurations have the same surface area. Taken from Alexander et al.[262]

and surface roughness. The magnitude of the higher-frequency deviation increased as the roughness factor increased, but the characteristic frequency was not altered. This provides indication that the higher-frequency deviation may also be dependent on the roughness factor.

A series of simulations were performed in which the depth and period of the roughness was increased while the roughness factor was fixed at $f_r = 2$, as illustrated in Figure 13.21. The corresponding imaginary-impedance-derived phase angle is presented in Figure 13.22(a) as a function of Kf_r. The results show that as the roughness depth increased, the characteristic frequency of the roughness decreased, but the magnitude of the deviation from ideality remained consistent. Also, the portion of the plot that is superposed is the low-frequency deviation representative of the coupled effect of the nonuniform current distribution due to the disk geometry and the surface roughness.

The dimensionless frequency corresponding to the roughness was found to be[262]

$$Kf_r^2 P/r_0 = \frac{\omega C_0 f_r^2 P}{\kappa} \tag{13.93}$$

The imaginary-impedance-derived phase angle is shown in Figure 13.22(b) as a function of the dimensionless frequency given as equation (13.93). Similar superposition was observed for simulations of grooves separated by smooth surfaces and for rectangular indentations. The results suggest that the characteristic dimension associated with the roughness is on the order of

$$\ell_{c,\text{roughness}} = f_r^2 P \tag{13.94}$$

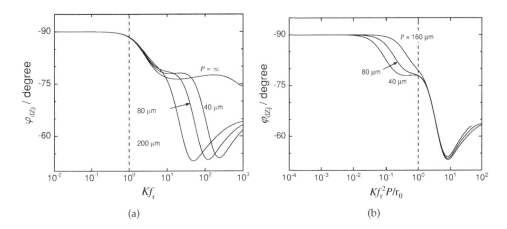

Figure 13.22: The imaginary-impedance-derived phase angle calculated from impedance simulations of a rough disk electrode within an insulating plane with the roughness period as a parameter and the roughness factor held constant: a) Phase angle as a function of frequency modified by the roughness factor and b) Phase angle as a function of dimensionless frequency derived from the approximated ohmic resistance. Taken from Alexander et al.[262]

This result is seen more clearly for recessed disk electrodes.

Influence of Surface Roughness on a Recessed Electrode

Simulations were performed on a rough recessed electrode to isolate the effect of the surface roughness from that of the disk geometry. The recessed electrode was configured such that the insulating plane was perpendicular to the electrode surface as shown in Figure 13.23. The depth of the recess was three times the disk diameter, ensuring that the primary current distribution was uniform across the electrode surface.[127] A rough surface was simulated by adding concentric V-shaped grooves to the electrode surface, as is illustrated in magnified view in Figure 13.23(c).

The imaginary-impedance-derived phase angle, presented in Figure 13.24(a) as a function of dimensionless frequency with the roughness period as a parameter, shows that the frequencies at which deviation from an ideal response occurred decreased as the depth of the roughness increased. The length scale used to make the frequency dimensionless was the product of the roughness factor and the radius of the disk.

The imaginary-impedance-derived phase angle is presented in Figure 13.24(b) as a function of the dimensionless frequency that is based on the characteristic dimension $f_r^2 P$, shown in equation (13.94). The results for three different roughness periods, seen in Figure 13.24(a), superpose when plotted as a function of $K f_r^2 P / r_0$. The product of the roughness factor and the geometric disk radius is the characteristic dimension for a rough disk. The period of the roughness multiplied by the square of the roughness factor is the appropriate characteristic length associated with the roughness itself.

The deviation from an ideal response of a rough recessed electrode at high fre-

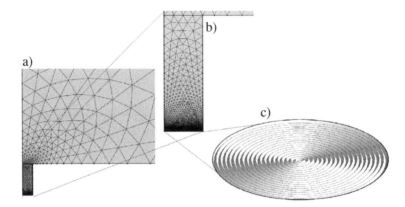

Figure 13.23: Schematic representation showing the finite-element mesh used for the recessed disk electrode simulations: a) portion of the domain emphasizing the recessed electrode; b) the recessed electrode; and c) detail of the grooved electrode. Taken from Alexander et al.[262]

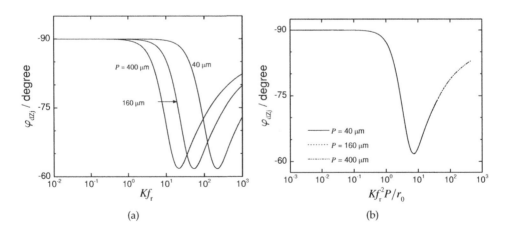

Figure 13.24: Imaginary-impedance-derived phase angle calculated from the simulated impedance of a rough recessed electrode with the roughness period as a parameter: a) as a function of dimensionless frequency Kf_r and b) as a function of dimensionless frequency Kf_r^2P/r_0. Taken from Alexander et al.[262]

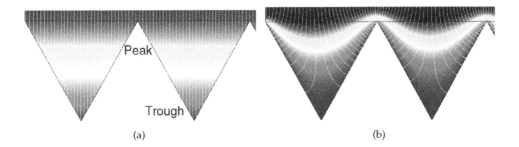

Figure 13.25: The current paths obtained as $\text{Re}\{\widetilde{i}\exp(jwt)\}$ at a fixed time, t: a) at $K = 10^{-5}$ and b) at $K = 10^3$. The potential distribution within the electrolyte adjacent to the rough surface is shown by the grey-scale representation. Taken from Alexander et al.[262]

quency may be explained by the nonuniform current and potential distribution along the electrode surface presented in Figure 13.25. The results for $K = 10^{-5}$ are presented in Figure 13.25(a), and the results at $K = 10^3$ are presented in Figure 13.25(b). The streamlines represent the current distribution and the potential distribution is shown as a grey-scale gradient. At low frequencies, the potential varied uniformly with the depth of the roughness such that the current lines were parallel to the y-axis. The potential distribution at high frequencies followed the surface profile, and the current did not reach the deepest parts of the rough surface. The change in current distribution with frequency caused the observed frequency dispersion.

The effective capacitance of the electrode/electrolyte interface may be determined from the imaginary part of the impedance as

$$C_{\text{eff}} = \frac{-1}{\omega Z_j} \tag{13.95}$$

The ratio of the effective capacitance and the input capacitance is provided in Figure 13.26 for the recessed electrode as a function of dimensionless frequency Kf_r with the roughness factor as a parameter. The current reached all parts of the rough electrode at low frequencies, making the capacitance ratio equal to the roughness factor. At high frequencies, the penetration depth of the current was smaller than the depth of the roughness, and the capacitance ratio decreased due to the nonuniform current distribution associated with the combined effects of the disk geometry and the surface roughness. The effective capacitance of the smooth disk, $f_r = 1$, was equal to the input capacitance, C_0, for all of the frequencies simulated.

An inherent difficulty in determining the characteristic frequency associated with roughness is that the shape of the roughness greatly influences the ohmic resistance.[258] Nevertheless, the characteristic frequency at which roughness begins to influence the impedance may be roughly estimated as

$$f_c = \frac{\kappa}{2\pi C_0 f_r^2 P} \tag{13.96}$$

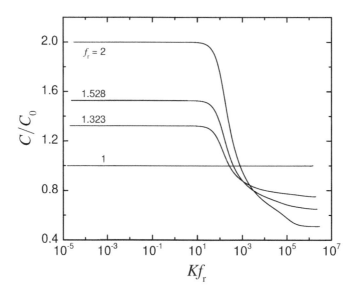

Figure 13.26: The ratio of the calculated effective capacitance and the input capacitance as a function of dimensionless frequency Kf_r with the roughness factor as a parameter for recessed disk electrodes. Taken from Alexander et al.[262]

which may be compared to the characteristic frequency developed for a rough disk geometry in equation (13.91). The frequency at which time-constant dispersion was observed is presented in Figure 13.27 as a function of the characteristic dimension for roughness $f_r^2 P$ and the characteristic dimension for a rough disk $f_r r_0$.

Example 13.5 Characteristic Frequency for Roughness: *Consider the experimental system described in Example 13.4 for which impedance measurements are desired to a maximum frequency of 10 kHz. Estimate the maximum radius for a disk electrode that will avoid the influence of high-frequency geometry-induced time-constant dispersion for an electrode with a roughness factor of 5 and an average width or period of roughness equal to 4 μm (approximately half of the average particle size of No. 1000 grit paper). Estimate the frequency at which the roughness will cause frequency dispersion.*

Solution: *The disk radius corresponding to the characteristic frequency for a rough disk electrode is given by*

$$r_0 = \frac{1}{2\pi} \frac{\kappa}{C_0 f_r f_{c,\text{disk}}} \tag{13.97}$$

The maximum disk radius is 0.09 cm, substantially smaller than the radius of 0.49 cm predicted in Example 13.4 for a smooth electrode. This result can also be obtained from Figure 13.27. The frequency at which roughness causes frequency dispersion may be estimated from equation (13.96) to be 490 kHz. This value is much larger than the frequency of 10 kHz specified in the problem.

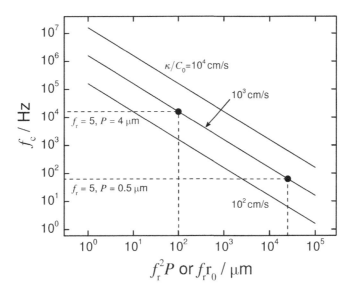

Figure 13.27: The characteristic frequencies, f_c, associated with dimensionless frequencies $K(f_r^2 P/r_0) = 1$ or $K(f_r) = 1$ at which the surface roughness influences the impedance as a function of either the roughness depth or the product of roughness factor and electrode radius with κ/C_0 as a parameter. Taken from Alexander et al.[262]

The results presented here demonstrate that surface roughness on solid electrodes influences the impedance response when coupled with the effect of nonuniform current distributions due to disk electrode geometry. For small roughness factors, roughness by itself causes frequency dispersion only at frequencies greater than that due to the geometry of disk electrodes.

However, since the roughness factor is a parameter contained within the characteristic dimension of the disk geometry and the surface roughness, there will exist cases in which the roughness of the disk will cause frequency dispersion at frequencies lower than the effect of the disk geometry. Porous electrodes provide an example in which the roughness factor can be on the order of 1,000 and the effect of disk geometry would not be seen. Specifically, when $f_r P$ is greater than r_0, frequency dispersion due to the surface roughness will control.

The imaginary-impedance-derived phase angle is presented in Figure 13.28 for a smooth disk electrode, a rough disk electrode within an insulating plane with $f_r = 2$ and $P = 40$ μm, and a rough recessed disk electrode with $f_r = 2$ and $P = 40$ μm. The geometry of the smooth disk electrode within an insulated plane caused deviation from ideal behavior at $K f_r = 1$ due to the nonuniform current distribution. The line representing the rough, recessed electrode showed frequency dispersion at $K f_r = 20$, which is numerically equal to $K f_r^2 P/r_0 = 1$. The rough electrode within the insulating plane showed the effect of both the disk geometry and the roughness. The roughness effect occurred at $K f_r = 20$, and the disk geometry effect was observed at $K f_r = K = 1$.

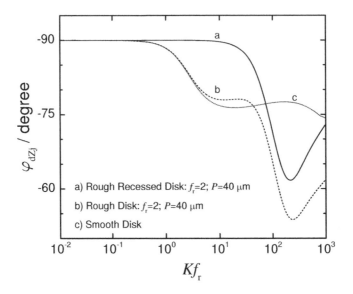

Figure 13.28: The imaginary-impedance-derived phase angle calculated from impedance simulations of a rough recessed electrode ($f_r = 2$ and $P = 40$ μm), a rough disk electrode within an insulating plane ($f_r = 2$ and $P = 40$ μm), and a smooth disk electrode within an insulating plane. Taken from Alexander et al.[262]

Alexander et al.[262] showed that the spacing between grooves can be accommodated by replacing f_r^2 with $f_r f_p$ where f_p is the surface area of the groove divided by the area of the groove mouth. Therefore, the characteristic frequency at which the roughness will begin to influence the impedance is

$$f_c = \kappa / 2\pi C_0 f_r f_p P \tag{13.98}$$

The development is general and can be applied to different shapes of roughness.

13.3.2 Capacitance

Alexander et al.[264] simulated a disk electrode (see Figure 13.18) and a recessed disk electrode (see Figure 13.23). A capacitance distribution was simulated by a Fourier series that represented a square wave distribution, as shown in Figure 13.29 as a function of radial position. The Fourier series[90] was expressed as

$$C_0(r) = \langle C \rangle + \sum_{n=1}^{\infty} (C_{max} - C_{min}) \cos(2n\pi r / P) \tag{13.99}$$

where the constants C_{max} and C_{min} represented the maximum and minimum capacitance. The average capacitance of the electrode surface $\langle C \rangle$ was calculated as

$$\langle C_0 \rangle = \frac{2}{r_0^2} \int_0^{r_0} C_0 r dr \tag{13.100}$$

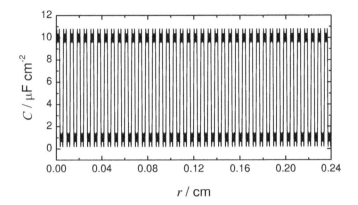

Figure 13.29: Capacitance distribution as a function of radial position based on a square wave represented by a Fourier series with a period of 60 μm. Taken from Alexander et al.[264]

where r_0 is the radius of the disk and C_0 is given by equation (13.99). The capacitance as a function of the radial position was integrated over the surface of the electrode and then divided by the area of the electrode. The period P of the square wave may be representative of an elemental grain size in which the capacitance is assumed to be relatively uniform across a grain and then jump to another value at an adjacent grain. Simulations were performed with 5, 10, 20, and 40 terms of the Fourier series. Numerical artifacts were observed at high frequencies when the series was truncated at 5 and 10 terms. As more terms were added to the series the artifacts shifted to higher frequencies, and with 20 terms in the series this behavior was eliminated from the simulated frequency range. The Fourier series was therefore truncated after 20 terms. The mathematical development follows that described in Section 13.2.1 with boundary conditions corresponding to a blocking electrode (Section 13.2.1).

Capacitance Distribution on Recessed Electrodes

The simulated impedance response of a recessed disk electrode with the square wave capacitance distribution shown in Figure 13.29 is presented in Figure 13.30 as a function of frequency with the period of distribution as a parameter. The minimum and maximum capacitance values expressed in equation (13.99) were set to 1 μF/cm^2 and 10 μF/cm^2, respectively, to represent a surface with an oxide film and a bare metal surface. The conductivity of the solution was 10^{-5} S/cm. The impedance was representative of an ideal capacitor at frequencies below 100 kHz, shown by the vertical lines that are perpendicular to the real axis. The figure inset shows a magnified view of the impedance at high frequencies. As the period of the capacitance distribution increases, the frequency at which dispersion begins decreases.

The deviation from the expected impedance response may be explained by the current and potential distribution along the electrode surface, which is presented in Figure

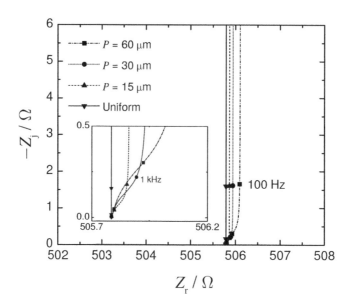

Figure 13.30: The impedance in Nyquist format of a recessed disk electrode with the square wave capacitance distribution shown in Figure 13.29 and the period of distribution as a parameter. Taken from Alexander et al.[264]

13.31. In the case of blocking electrodes, the solution resistance controls the current distribution at high frequencies while the interfacial impedance controls at low frequencies. The potential distribution at low frequencies is represented by the color gradient in Figure 13.31(a). The streamlines corresponded to the path of the modulus of the oscillating current expressed, for a given frequency, as

$$|\tilde{i}| = \sqrt{\tilde{i}_r^2 + \tilde{i}_j^2} \qquad (13.101)$$

in which $\tilde{i}_r$ and $\tilde{i}_j$ represent the real and imaginary parts of the oscillating current. The grey-scale representation of potential distribution was adjusted to emphasize the variations near the electrode surface. The regions of the electrode surface with higher capacitance have a lower impedance and the current flows more easily through these points. Despite the nonuniform current distribution, the impedance response at low frequencies was indicative of a pure capacitor with a value of the averaged surface capacitance. The potential distribution along the electrode surface at 100 kHz, presented in Figure 13.31(b), was more uniform.

The modulus of the oscillating current density along the electrode surface is presented in Figure 13.32 as a function of radial position. The low-frequency current response, presented in Figure 13.32(a), showed a variation of current proportional to the variation in surface capacitance and small values of current. The high frequency response, presented in Figure 13.32(b), showed much higher values of current; however, the distribution was much more uniform. The small variations at locations where the

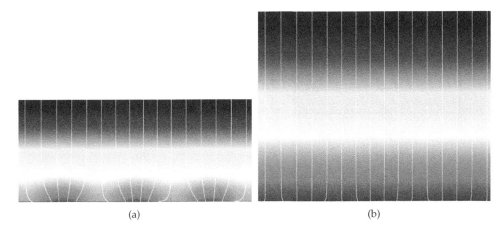

(a)   (b)

Figure 13.31: The current paths near the surface of a recessed electrode exhibiting a square-wave distribution of capacitance obtained as $\sqrt{\tilde{i}_r^2 + \tilde{i}_j^2}$: a) at 10 mHz and b) at 100 kHz. The potential distribution within the electrolyte adjacent to the rough surface is shown by the grey-scale representation. Taken from Alexander et al.[264]

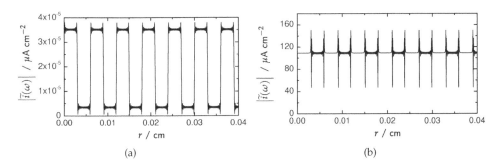

(a)   (b)

Figure 13.32: Normal current distribution at the electrode surface due to a nonuniform capacitance distribution with a period of 60μm of a recessed electrode as a function of radial position: a) current distribution at 10mHz and b) current distribution at the high frequency limit of the simulations $f = 100$ kHz. Taken from Alexander et al.[264]

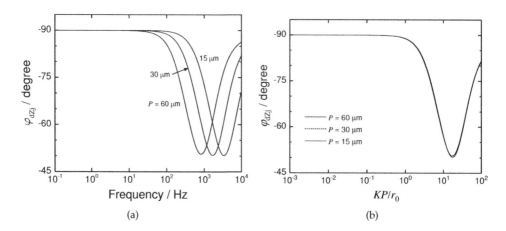

Figure 13.33: Imaginary-impedance-derived phase angle calculated from the impedance data in Figure 13.30: a) as a function of frequency and b) as a function of dimensionless frequency based on the averaged capacitance and the period of the distribution. Taken from Alexander et al.[264]

capacitance changes from one value to the other can be attributed to the finite number of terms in the Fourier series.

The imaginary-impedance-derived phase angle defined in equation (13.89) is presented in Figure 13.33(a) as a function of frequency with the period of distribution as a parameter. The distribution of capacitance along the surface of the electrode did not influence the imaginary impedance at low frequencies as indicated by the phase angle, $\varphi_{dZj} = -90°$. However, frequency dispersion did occur in all cases at high frequencies with a phase angle reaching approximately $-50°$. The deviation from the expected capacitive behavior occurred at lower frequencies as the period of the distribution increased. If the period of the distribution is taken as the grain size within a polycrystalline surface, grain sizes on the order of 1 μm and less should not lead to frequency dispersion at frequencies less than 1 kHz.

The dimensionless frequency corresponding to a distribution of capacitance with period P may be expressed as

$$KP/r_0 = \frac{\omega \langle C_0 \rangle P}{\kappa} \qquad (13.102)$$

The imaginary-impedance-derived phase angle is presented in Figure 13.33(b) as a function of dimensionless frequency. The results were superposed, confirming that the period is the appropriate characteristic length to use to describe surface heterogeneity of capacitance. The characteristic frequency at which a nonuniform capacitance began to influence the impedance was determined to be $KP/r_0 = 1$.

Since impedance is often used to measure interfacial capacitance, it is important to ensure that a distribution of capacitance along the surface does not complicate the use of this technique. The effective capacitance of the electrode/electrolyte interface may be determined from the imaginary part of the impedance as shown in equation

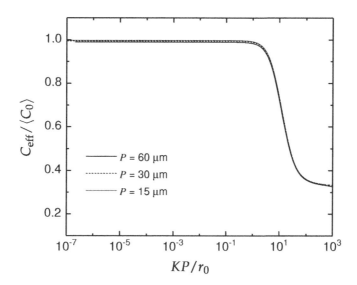

Figure 13.34: The ratio of the calculated effective capacitance and the surface-averaged input capacitance as a function of dimensionless frequency $K4P/\pi r_0$ with the period of the distribution as a parameter for recessed disk electrodes. Taken from Alexander et al.[264]

(13.95). The ratio of the calculated capacitance and the surface-averaged capacitance is shown in Figure 13.34 as a function of dimensionless frequency with the period of the distribution as a parameter. In all cases, the simulated capacitance closely matched the surface-averaged capacitance of the electrode at low frequencies.

Capacitance Distribution on Disk Electrodes

The geometry of a disk electrode embedded within an insulating plane causes frequency dispersion at high frequencies due to the nonuniform current distribution at high frequency. As the frequency increases to infinity, the magnitude of the oscillating component of the modulated current at the periphery of the disk approaches infinity while the current at the center of the disk remains finite. The simulated current distribution at an electrode surface containing the capacitance distribution in Figure 13.29 is provided as a function of radial position for 10 mHz in Figure 13.35(a) and for 100 kHz in Figure 13.35(b). The current distribution at low frequencies resembled the square-wave distribution features. At high frequencies, the effect of the disk geometry overshadowed the effect of the capacitance distribution since the current at the periphery approached infinity.

The imaginary-impedance-derived phase angle is provided in Figure 13.36(a) as a function of frequency. At low frequencies the phase angle was constant with a value equal to $-90°$. Frequency dispersion became apparent at approximately 1 Hz for all periods of capacitance distribution due to the geometry of the disk. At higher frequencies, dispersion due to the nonuniform capacitance distribution occurred such that increases

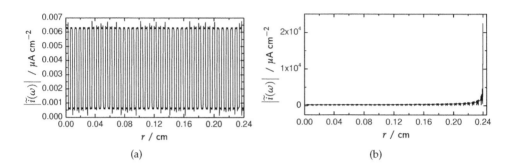

Figure 13.35: Normal current distribution at a disk electrode surface as a function of radial position: a) current distribution at 10 mHz and b) current distribution at 100 kHz. Taken from Alexander et al.[264]

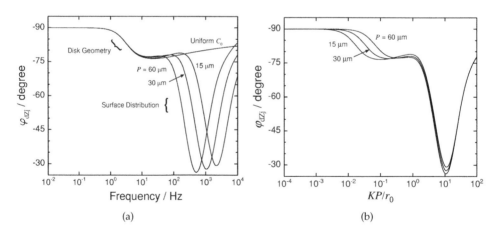

Figure 13.36: Imaginary-impedance-derived phase angle for a disk electrode within an insulating plane: a) phase angle as a function of frequency and b) phase angle as a function of dimensionless frequency based on the averaged capacitance. Taken from Alexander et al.[264]

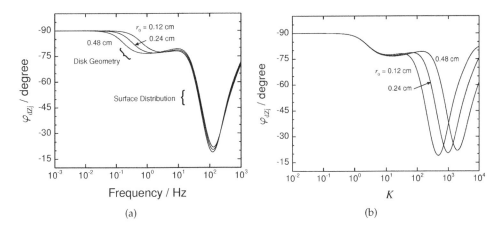

Figure 13.37: Imaginary-impedance-derived phase angles values calculated from the impedance data with the disk radius as a parameter: a) as a function of frequency and b) as a function of dimensionless frequency $K = \omega \langle C_0 \rangle \pi r_0 / 4\kappa$. Taken from Alexander et al.[264]

in the period of the distribution caused the deviation to shift to lower frequencies. The same results are presented in Figure 13.36(b) as a function of dimensionless frequency in which the period of the distribution was used as the characteristic length. The frequency dispersion due to the surface heterogeneity was superposed, indicating that the period of the capacitance distribution is the appropriate characteristic length. The results obtained for a recessed electrode provided in Figure 13.33 show only the frequency dispersion due to the surface distribution of capacitance.

The imaginary-impedance-derived phase angle is presented in Figure 13.37(a) as a function of frequency with the radius of the disk as a parameter. The period of the distribution was fixed at 60 μm. Changes in the radius of the electrode influenced only the frequency dispersion associated with the geometry of the disk. The imaginary-impedance-derived phase angle is shown in Figure 13.37(b) as a function of dimensionless frequency, equation (13.102). The frequency dispersion associated with disk geometry was superposed, and the frequency dispersion associated with the capacitance distribution was not.

The characteristic frequency at which frequency dispersion occurs for a planar disk electrode exhibiting a distribution of capacitance may be expressed as

$$f_{c,r_0} = \frac{\kappa}{2\pi \langle C_0 \rangle r_0} \tag{13.103}$$

where $\langle C_0 \rangle$ represents the surface-averaged capacitance. The characteristic frequency at which frequency dispersion begins due to the capacitance distribution may be expressed as

$$f_{c,P} = \frac{\kappa}{2\pi \langle C_0 \rangle P} \tag{13.104}$$

in which the period of the distribution is the characteristic length associated with the

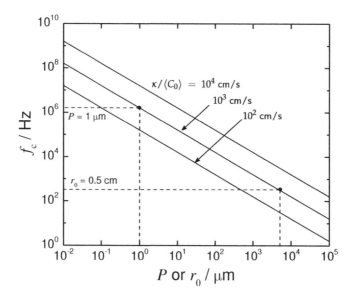

Figure 13.38: The frequency $K4P/\pi r_0 = 1$ at which the surface heterogeneity influences the impedance as a function of distribution period and disk radius with $\kappa/\langle C_0 \rangle$ as a parameter. Taken from Alexander et al.[264]

distribution. The frequency at which dispersion begins is presented in Figure 13.38 as a function of both the disk radius as well as the period of the capacitance distribution with the ratio, $\kappa/\langle C_0 \rangle$, as a parameter. The period of the distribution may be associated with the average width of grain sizes within a noncrystalline surface. For a 0.5 cm radius electrode in a system with $\kappa/\langle C_0 \rangle = 1$ cm/s (corresponding, for example, to $\langle C_0 \rangle = 20$ μF/cm^2 and a 0.16 mM sodium chloride concentration) and an average grain size of 1 μm, the frequency at which dispersion occurs due to the disk geometry is 0.3 Hz; whereas, the frequency at which dispersion occurs due to the nonuniform capacitance is 1.6 kHz. The frequency dispersion due to the capacitance distribution will always occur at greater frequencies than the frequency dispersion due to the disk geometry.

Example 13.6 **Distributions of Capacitance:** *Show that the characteristic frequency associated with a distribution of capacitance on a disk electrode is larger than that associated with the geometry of the electrode.*

Solution: *The characteristic frequency at which frequency dispersion occurs for a planar disk electrode exhibiting a distribution of capacitance may be expressed as*

$$f_{c,r_0} = \frac{\kappa}{2\pi\langle C_0 \rangle r_0} \tag{13.105}$$

where $\langle C_0 \rangle$ represents the surface-averaged capacitance. The characteristic frequency at which *frequency dispersion begins due to the capacitance distribution may be expressed as*

$$f_{c,P} = \frac{\kappa}{2\pi \langle C_0 \rangle P} \tag{13.106}$$

in which the period of the distribution is the characteristic length associated with the distribution. The ratio is given by

$$\frac{f_{c,P}}{f_{c,r_0}} = \frac{r_0}{P} \tag{13.107}$$

The period of cpacitance variation must be smaller than the radius of the disk; thus, $f_{c,P} > f_{c,r_0}$.

13.3.3 Reactivity

Under some conditions, distributions of rate constants for electrochemical reactions give rise to frequency dispersion at frequencies below that associated with the disk geometry. Alexander et al.[265] showed that a nonuniform charge-transfer resistance associated with a single electrochemical reaction is not associated with frequency dispersion. A distribution of rate constants for reactions coupled by an adsorbed intermediate, however, does lead to frequency dispersion at low frequencies associated with the relaxation of the adsorbed intermediate. This result is in agreement with the observation of Wu et al.[249,250] who found that the nonuniform current and potential distribution associated with a disk geometry may give rise to a low-frequency dispersion (i.e., CPE-like behavior) for reactions coupled by an adsorbed intermediate.

For each form of surface heterogeneity leading to roughness and capacitance distributions, there exists a characteristic frequency at which dispersion is induced. The characteristic frequency is inversely proportional to the characteristic length associated with the type of surface heterogeneity. As the characteristic length decreases, the frequency dispersion shifts to higher frequencies. For a disk electrode within an insulating plane, the characteristic length is the radius of the disk. An approximate characteristic length of a ring electrode can be expressed as a function of the width of the ring. For surface roughness, the characteristic length is the product of the square of the roughness factor and the period of the surface irregularities. The characteristic length which describes a distribution of capacitance is the period of the distribution. In all of these cases, the characteristic length is small, leading to frequency dispersions that occur at the upper limit or outside the measurable frequency range.

13.4 Characteristic Dimension for Frequency Dispersion

The material presented in Section 13.2 on geometry-induced current and potential distributions and Section 13.3 on electrode surface property distributions lead to a common conclusion: that frequency dispersion associated with distributions of ohmic resistance or capacitance occurs above a characteristic frequency. This characteristic frequency is related, in turn, to the inverse of a characteristic dimension. For design of

Table 13.1: Characteristic dimensions and frequencies associated with frequency dispersion caused by nonuniform ohmic resistance or capacitance.

Experimental System	ℓ_c / cm	f_c / Hz
Rectangular electrode ($d_1 \times d_2$)[266]	$\dfrac{d_1 d_2}{d_1 + d_2}$	$\dfrac{1}{2\pi} \dfrac{\kappa(d_1 + d_2)}{C_0 d_1 d_2}$
Disk electrode[125, 147]	r_0	$\dfrac{1}{2\pi} \dfrac{\kappa}{C_0 r_0}$
Disk electrode (local CPE)[248]	r_0	$\dfrac{1}{2\pi} \left(\dfrac{\kappa}{Q r_0} \right)^{1/\alpha}$
Rough disk electrode[262, 265]	$f_r r_0$	$\dfrac{1}{2\pi} \dfrac{\kappa}{C_0 f_r r_0}$
Ring electrode[267]	$\dfrac{r_2 - r_1}{1 + (r_1/r_2)^2}$	$\dfrac{\kappa(1 + (r_1/r_2)^2)}{2\pi C_0 (r_2 - r_1)}$
Roughness[262, 265]	$f_r^2 P$	$\dfrac{\kappa}{2\pi C_0 f_r^2 P}$
Intermittent roughness[265]	$f_r f_p P$	$\dfrac{\kappa}{2\pi C_0 f_r f_p P}$
Distribution of capacity[264]	P	$\dfrac{\kappa}{2\pi \langle C_0 \rangle P}$

an experiment, it is useful to understand the relationship between system scale and the appearance of frequency dispersion that can confound meaningful interpretation of impedance measurements.

The characteristic dimensions and frequencies associated with frequency dispersion caused by nonuniform ohmic resistance or capacitance are summarized in Table 13.4 for a number of electrode geometries. For long rectangular electrodes, e.g., $d_2 \gg d_1$, the characteristic dimension approaches the shorter dimension, e.g., $\ell_{c,rect} \rightarrow d_1$. For disk electrodes, the frequency associated with geometry-induced time-constant dispersion decreases as $\ell_{c,disk}$ increases. The characteristic dimension for ring electrodes, taken from Cleveland et al.,[267] is approximate. Cleveland et al.[267] provide a multiplicative correction factor that varies from unity at $\ell_{c,ring} \rightarrow 0$ to 1.2 for $\ell_{c,ring} \rightarrow 1$.

13.5   Convective Diffusion Impedance at Small Electrodes

Small electrodes are currently used to study fast electrochemical kinetics or as flow measurement devices in chemical engineering systems. In the latter case, the first experimental and theoretical studies appeared in the early fifties. The goal of these studies was to achieve probes sensitive to the local wall velocity gradient

$$\beta_y = \frac{\partial v_x}{\partial y} \tag{13.108}$$

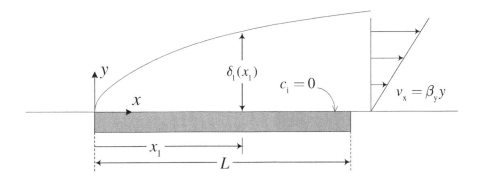

Figure 13.39: Schematic representation of flow past a small electrode where the coordinate y is in the direction perpendicular to the plane of the electrode (x, y).

The well-known property of those probes is that the limiting diffusion current is proportional to $\beta_y^{1/3}$ under steady-state conditions.[268] For use in electrochemical engineering, an increasing interest is now focused on the nonsteady behavior of those small electrodes under conditions of fluctuating velocity gradient $\beta_y(t)$.

It is possible to deduce hydrodynamic information from the limiting current measurement, either in quasi-steady state where $I(t) \propto \beta_y^{1/3}(t)$ or, at higher frequency, in terms of spectral analysis. In the latter case, it is possible to obtain the velocity spectra from the mass-transfer spectra, where the transfer function between the mass-transfer rate and the velocity perturbation is known. However, in most cases, charge transfer is not infinitely fast, and the analysis also requires knowledge of the convective-diffusion impedance, i.e., the transfer function between a concentration modulation at the interface and the resulting flux of mass under steady-state convection.

13.5.1 Analysis

A schematic representation of a small electrode embedded in an insulating wall is given in Figure 13.39 on which a fast electrochemical reaction occurs under mass-transport limitation, i.e., $c_i(0) = 0$. The length of this electrode in the mean flow direction is small enough that the diffusion layer thickness δ_i is very small, thus minimizing the effect of the normal velocity component. The normal velocity is proportional to y^2, whereas the longitudinal velocity component is proportional to y, where y is the coordinate normal to the wall. Under these conditions, the boundary-layer approximations apply. Using a local frame of reference (x, y) attached to the electrode, the mass conservation equation governing the concentration distribution c_i of a species transported by convective diffusion can be expressed as

$$\frac{\partial c_i}{\partial t} + \beta_y y \frac{\partial c_i}{\partial x} = D_i \frac{\partial^2 c_i}{\partial y^2} \tag{13.109}$$

where $v_x = \beta_y y$. For simplification, the electrode can be considered to be sufficiently small that the flow is uniform in the diffusion layer and β_y is independent of the space coordinates. The time-average solution for the concentration distribution has been given by Lévêque.[268] As shown in Example 2.2, the concentration in the diffusion layer can be expressed as

$$\bar{c}_i = \frac{c_i(\infty) - c_i(0)}{3^{2/3}\Gamma(4/3)} \int_0^{y/\delta_i} \exp\left(-\frac{\eta^3}{9}\right) d\eta + c_i(0) \tag{13.110}$$

where $3^{2/3}\Gamma(4/3)\delta_i$ represents the local value of the diffusion layer thickness with $\delta_i = (D_i x/\beta_y)^{1/3}$, $0 \le x \le L$, and $c_i(\infty)$ is the concentration in the bulk. On a small rectangular electrode of width W, the steady flux is

$$\overline{N}_i = W \int_0^L D_i \frac{\partial \bar{c}_i}{\partial y} dx = \frac{3^{1/3} \left(c_i(\infty) - c_i(0)\right) D_i^{2/3}\beta_y^{1/3}L^{2/3}W}{2\Gamma(4/3)} \tag{13.111}$$

or

$$\overline{N}_i = 0.80755(c_i(\infty) - c_i(0)) D_i^{2/3}\beta_y^{1/3}L^{2/3}W \tag{13.112}$$

Equation (13.112) shows that the steady mass-transfer-controlled flux of a reacting species is proportional to the cube root of the velocity gradient, i.e., $\overline{N}_i \propto \beta_y^{1/3}$.

Example 13.7 Flux on a Small Circular Electrode: *Derive an expression for the steady flux on a circular small electrode of radius R.*

Solution: *Equation (13.110) can be used to calculate the flux on a circular small electrode by summing along z, the effect of elementary rectangular strips. In this case, x contained in the definition of δ_i must be replaced by $(x - x_1)$, which actually corresponds in the local Cartesian frame of reference to the distance from the leading edge of any elementary strip. The position of the leading edge $x_1(z)$ is a function of R and z as*

$$x_1(z) = R - \sqrt{R^2 - z^2} \tag{13.113}$$

Thus, the expression of the flux is

$$\overline{N}_{\text{circ}} = \int_{-R}^R dz \int_{R-\sqrt{R^2-z^2}}^{R+\sqrt{R^2-z^2}} D_i \frac{\partial \bar{c}_i}{\partial y} dx \tag{13.114}$$

or

$$\overline{N}_{\text{circ}} = 0.84 \frac{3^{1/3} \left(c_i(\infty) - c_i(0)\right) D_i^{2/3}\beta_y^{1/3}(2R)^{5/3}}{2\Gamma(4/3)} \tag{13.115}$$

Comparison with equation (13.111) reveals that the flux at a circular electrode is 84 percent smaller than that at a square electrode with L = W = 2R.

13.5.2 Local Convective Diffusion Impedance

The nonsteady part of the mass balance equation (13.109) may be written as:

$$j\omega\widetilde{c}_i - \frac{\partial^2\widetilde{c}_i}{\partial y^2} + y\beta_y\frac{\partial\widetilde{c}_i}{\partial x} = 0 \tag{13.116}$$

The boundary conditions for the nonsteady equations are

$$\begin{aligned} \widetilde{c}_i &= \widetilde{c}_i(0) \quad \text{for} \quad y=0 \text{ and } x \geq x_1 \\ \frac{\partial\widetilde{c}_i}{\partial y} &= 0 \quad \text{for} \quad y=0 \text{ and } x \geq x_1 \\ \widetilde{c}_i &= 0 \quad \text{for} \quad y \to \infty \text{ and all } x \end{aligned} \tag{13.117}$$

A dimensionless concentration θ_i can be defined such that

$$\theta_i = \frac{\widetilde{c}_i}{\widetilde{c}_i(0)} \tag{13.118}$$

The dimensionless normal distance to the wall can be defined to be

$$\eta = y/\delta_i = y\left(\frac{\beta_y}{D_i(x-x_1)}\right)^{1/3} \tag{13.119}$$

where x_1 is the coordinate of the leading edge of the electrode as shown in Figure 13.39. Equation (13.119) represents a similarity variable (see Section 2.5). Introduction of η into equation (13.116) results in definition of a dimensionless position-dependent frequency given by

$$K_{x,i} = \omega\left(\frac{(x-x_1)^2}{\beta_y^2 D_i}\right)^{1/3} \tag{13.120}$$

In terms of equations (13.118) to (13.120), equation (13.116) becomes

$$jK_{x,i}\theta_i + \frac{2}{3}K_{x,i}\eta\frac{\partial\theta_i}{\partial K_{x,i}} - \frac{\eta^2}{3}\frac{\partial\theta_i}{\partial\eta} - \frac{\partial^2\theta_i}{\partial\eta^2} = 0 \tag{13.121}$$

The spatial dependence of the sinusoidal perturbation is evident in the definition of $K_{x,i}$.

As $K_{x,i}$ contains a dependence on the space coordinates, it is necessary to derive first the local solution. A solution in the form of a series can be obtained as

$$\theta_i = \sum_{m=0}^{\infty}(jK_{x,i})^m\theta_{i,m}(\eta) \tag{13.122}$$

For this solution the number of terms that play a role in the series increases with the frequency. Generally the solution given by equation (13.122) is used for the low-frequency solution, and the high-frequency solution is derived by another method.

Low-Frequency Solution

The elementary functions $\theta_{i,m}(\eta)$ are real and obey

$$\frac{\partial^2 \theta_{i,0}}{\partial \eta^2} + \frac{\eta^2}{3}\frac{\partial \theta_{i,0}}{\partial \eta} = 0 \tag{13.123}$$

and

$$\frac{\partial^2 \theta_{i,m}}{\partial \eta^2} + \frac{\eta^2}{3}\frac{\partial \theta_{i,m}}{\partial \eta} - \frac{2m\eta}{3}\theta_{i,m} = \theta_{i,m-1} \tag{13.124}$$

The boundary conditions at $\eta = 0$ are $\theta_i(0) = 1$, $\theta_{i,0}(0) = 1$, and $\theta_{i,m}(0) = 0$ for all $m > 0$. The boundary condition at $\eta \to \infty$ is $h(\infty) = 0$. In fact, since the only observable quantity is the interfacial flux, only its expression is needed, i.e.,

$$\frac{d\theta_i}{d\eta}\bigg|_0 = \sum_{m=0}^{\infty} (jK_{x,i})^m \frac{d\theta_{i,m}}{d\eta}\bigg|_0 \tag{13.125}$$

The terms $d\theta_{i,m}/d\eta|_0$ are tabulated by Deslouis et al.[269] for $0 \leq \updownarrow \leq 79$.

High-Frequency Solution

Since the concentration modulation is rapidly damped close to the wall at high frequencies, the convective term can be disregarded and equation (13.121) becomes

$$jK_{x,i}\theta_i - \frac{\partial^2 \theta_i}{\partial \eta^2} = 0 \tag{13.126}$$

The solution to equation (13.126) can be found using the methods described in Section 2.2.

Due to the boundary conditions ($\theta_i = 0$ when $\eta \to \infty$ and $\theta_i = 1$ when $\eta = 0$), the analytic solution is, as given for equation (2.38),

$$\theta_i = \exp\left[-(jK_{x,i})^{1/2}\eta\right] \tag{13.127}$$

The local dimensionless diffusion impedance is obtained as

$$\frac{z_D}{z_D(0)} = \left[\frac{d\theta_i}{d\eta}\bigg|_0\right]^{-1} = -(jK_{x,i})^{-1/2} \tag{13.128}$$

which is a normalized Warburg impedance as described in Example 11.1 on page 249. As seen in Figure 13.40, the high-frequency solution (13.125) and the low-frequency solution (13.128) present a satisfactory overlap for $6 \leq K_{x,i} \leq 13$.

13.5.3 Global Convective Diffusion Impedance

The dimensionless impedance of a small electrode can be defined by summing the effects of the local convective-diffusion impedance, i.e.,

$$\frac{Z_D(0)}{Z_D} = \int\!\!\int z_D^{-1} dxdz = -\int\!\!\int \frac{1}{\tilde{c}_i(0)}\frac{\partial \tilde{c}_i}{\partial y}\bigg|_0 dxdz = -\int\!\!\int \frac{dh}{d\eta}\bigg|_0 \frac{dxdz}{\delta_i(x)} \tag{13.129}$$

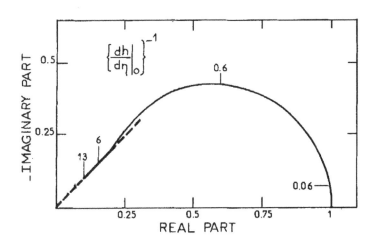

Figure 13.40: Local normalized diffusion impedance for the small electrode given in Figure 13.39. The solid line represents the low-frequency solution (equation (13.125)), and the dashed line represents the high-frequency solution (equation (13.128)). Overlap is obtained for $6 \leq K_{x,i} \leq 13$, with the dimensionless frequency $K_{x,i}$ given by equation (13.120). (Taken from Delouis et al.[269])

For a rectangular electrode of length L and of width W, the expression of the impedance is given by

$$\frac{Z_D(0)}{Z_D} = -W \int_0^L \left(\sum_{m=0}^\infty (jK_{x,i})^m \left. \frac{d\theta_{i,m}}{d\eta} \right|_0 \right) \frac{dx}{\delta_i(x)} \qquad (13.130)$$

By using the dimensionless frequency

$$K_i = \omega \left(L^2 / D_i \beta_y^2 \right)^{1/3} \qquad (13.131)$$

equation (13.130) can be expressed as

$$\frac{Z_D(0)}{Z_D} = \left(\frac{\beta_y L^2}{D_i} \right)^{1/3} WD_i H(K_i) \qquad (13.132)$$

with

$$H(K_i) = \frac{3}{2K_i} \int_0^{K_i} \left. \frac{d\theta_i}{d\eta} \right|_0 dK_{x,i} \qquad (13.133)$$

In the low-frequency range, the expression of $H(K_i)$ is obtained from the series expansion (13.125)

$$H(K_i) = \frac{3}{2} \sum_{m=0}^\infty \frac{(jK_i)^m}{m+1} \left. \frac{d\theta_{i,m}}{d\eta} \right|_0 \qquad (13.134)$$

In the high-frequency range, the integration must be split in two parts since the leading edge of an electrode will be always under a low-frequency regime. Indeed, the local thickness of the diffusion layer, equal to $3^{2/3}\Gamma(4/3)\delta_i(x)$ and thus proportional to $x^{1/3}$,

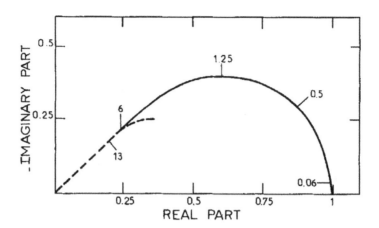

Figure 13.41: Normalized global convective-diffusion impedance for a small rectangular electrode. The solid line represents the low-frequency solution (equation (13.134)), and the dashed line represents the high-frequency solution (equation (13.137)). Overlap is obtained for $6 \leq K_i \leq 13$, with the dimensionless frequency K_i given by equation (13.131). (Taken from Delouis et al.[269])

is very small at the leading edge, and $K_{x,i}$ remains there always small even for high values of $\omega/2\pi$. Thus,

$$H(K_i) = \frac{3}{2K_i} \left(\int_0^{\sigma_1} \left. \frac{d\theta_i}{d\eta} \right|_0 dK_{x,i} + \int_{\sigma_1}^{K_i} \left. \frac{d\theta_i}{d\eta} \right|_0 dK_{x,i} \right) \qquad (13.135)$$

The first integral corresponds to the low-frequency regime and the second one to the high-frequency regime where equations (13.125) and (13.128) must be used, respectively. Equation (13.135) becomes

$$H(K_i) = -\frac{B(\sigma_1)}{K_i} + (jK_i)^{1/2} \qquad (13.136)$$

The term $B(\sigma_1)$ has been calculated for $\sigma_1 \leq 13$, and $B(\sigma_1)$ was found to be constant and equal to 0.25j in the frequency range $6 \leq \sigma_1 \leq 13$. This result means that equation (13.136) is valid for $K_i \geq 6$ and therefore can be written as

$$H(K_i) = -\frac{0.25j}{K_i} + (jK_i)^{1/2} \qquad (13.137)$$

As a consequence, a fair overlap between equation (13.134) and equation (13.137) is obtained for $6 \leq K_i \leq 13$ as shown in Figure 13.41.

Remember! **13.2** *Not all time-constant distributions give rise to a CPE.*

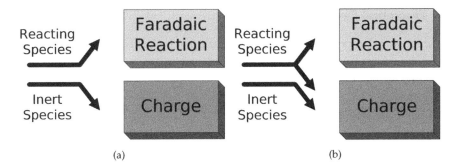

Figure 13.42: Schematic representation illustrating the contribution of the reacting species to the charging of the electrode-electrolyte interface corresponding to: a) decoupled charging and faradaic currents and b) coupled charging and faradaic currents.

13.6 Coupled Charging and Faradaic Currents

A controversy from the late 1960s over the correct method for developing deterministic models for impedance response was largely neglected by the electrochemical community until it was raised in 2012 by Nisancioglu and Newman.[270] In electrochemical systems, the passage of current through an electrode can be attributed to both faradaic reactions and to double-layer charging. As proposed by Sluyters,[271] the two processes are usually considered separately for simulating impedance response. The total current is subsequently obtained by adding the double-layer charging current to the faradaic current. This approach was criticized by Delahay and co-workers[272–274] because part of the flux of reacting species should, in principle, contribute to the charging of the interface as well as to the faradaic reaction. Nevertheless, the accepted procedure for model development has been to assume that the faradaic and charging currents are independent.[91]

The opposing viewpoints are presented schematically in Figure 13.42. With decoupled charging and faradaic currents, the flux of reacting species contributes only to the faradaic reaction, as shown in Figure 13.42(a), and the charging current has contribution only from inert species. This corresponds to the currently accepted procedure for model development. The schematic representation of the case that allows coupling of charging and faradaic currents is presented in Figure 13.42(b). The reacting species are seen to contribute to both the faradaic reaction and, along with the inert species, to the charging current. The representation shown in Figure 13.42(b) is consistent with the arguments presented by Delahay.[272–274]

Relaxation of the assumption that faradaic and charging currents are independent requires coupling an explicit model of the double layer to the convective diffusion equations for each ionic species. Wu et al.[275] employed the framework provided by Nisancioglu and Newman[270] to show the influence of coupled charging and faradaic currents on the impedance response of a disk electrode.

13.6.1 Theoretical Development

A two-dimensional impedance model was developed by Wu et al.[275] to study the effect of nonuniform mass transfer for a rotating disk electrode geometry and the influence of coupled charging and faradaic currents on the impedance response.

Mass Transport in Dilute Solutions

Conservation of species accounting for mass transport in dilute solution is expressed as

$$\frac{\partial c_i}{\partial t} = -\nabla \cdot \mathbf{N}_i + R_i \tag{13.138}$$

where the flux is given by the Nernst-Planck equation

$$\mathbf{N}_i = -D_i \left(z_i c_i \frac{F}{RT} \nabla \Phi + \nabla c_i \right) + c_i \mathbf{v} \tag{13.139}$$

c_i is the concentration of species i, z_i is the charge number, D_i is the diffusivity, $\mathbf{v}$ is the mass-averaged velocity associated with the rotating disk, and R_i represents the production of species i by homogeneous reactions. In the absence of homogeneous reactions, $R_i = 0$, and, under assumption that the diffusion coefficients are uniform,

$$\frac{\partial c_i}{\partial t} + \mathbf{v} \cdot \nabla c_i = z_i D_i \frac{F}{RT} \nabla \cdot (c_i \nabla \Phi) + D_i \nabla^2 c_i \tag{13.140}$$

For a system with n species, n expressions in the form of equation (13.140) are required; whereas the conservation of charge implies that

$$\nabla \cdot \mathbf{i} = -\nabla \cdot \left(F \sum_i D_i z_i \left(z_i c_i \frac{F}{RT} \nabla \Phi + \nabla c_i \right) \right) = 0 \tag{13.141}$$

Equations (13.140) and (13.141) constitute a set of nonlinear differential equations.

In the frequency domain, each variable, including potential and the concentrations of each species, is described as shown in equation (13.58). The mass and charge conservation equations become

$$j\omega \tilde{c}_i + \mathbf{v} \cdot \nabla \tilde{c}_i = D_i \nabla \cdot \left(\nabla \tilde{c}_i + \frac{z_i F}{RT} \left(\bar{c}_i \nabla \tilde{\Phi} + \tilde{c}_i \nabla \bar{\Phi} \right) \right) \tag{13.142}$$

and

$$\nabla \cdot \left(\sum_i z_i D_i \left(\nabla \tilde{c}_i + \frac{z_i F}{RT} \left(\bar{c}_i \nabla \tilde{\Phi} + \tilde{c}_i \nabla \bar{\Phi} \right) \right) \right) = 0 \tag{13.143}$$

respectively, where the higher-order terms such as $\tilde{c}_i \nabla \tilde{\Phi}$ were neglected. At the electrode surface, the flux of each species may be expressed as

$$\tilde{N}_{i,y}(0) = -D_i \left(\left. \frac{d\tilde{c}_i}{dy} \right|_{y=0} + \frac{z_i F}{RT} \left(\bar{c}_i \left. \frac{d\tilde{\Phi}}{dy} \right|_{y=0} + \tilde{c}_i \left. \frac{d\bar{\Phi}}{dy} \right|_{y=0} \right) \right) \tag{13.144}$$

The correlation between the flux and the current oscillations at electrode boundary is discussed in two cases where the faradaic current and the double-layer charging current are considered with and without a priori separation of faradaic and charging currents.

Coupled Faradaic and Charging Currents

Under the assumption that the charge density on the metal surface q_m is dependent on the interfacial potential V and on the concentration $c_i(0)$ of each species i, located outside the diffuse region of charge, the variation of the surface charge density is given by

$$\mathrm{d}q_m = \left(\frac{\partial q_m}{\partial V}\right)_{c_i(0)} \mathrm{d}V + \sum_i \left(\frac{\partial q_m}{\partial c_i(0)}\right)_{V, c_{j \neq (i)(0)}} \mathrm{d}c_i(0) \tag{13.145}$$

Equation (13.145) is written in terms of $n+1$ parameters which are treated as properties of the interface. These can be expressed as

$$C_0 = \left(\frac{\partial q_m}{\partial V}\right)_{c_i(0)} \tag{13.146}$$

where C_0 is the usual differential capacitance, and

$$C_i = \left(\frac{\partial q_m}{\partial c_i(0)}\right)_{V, c_{j \neq (i)(0)}} \tag{13.147}$$

which can be expressed for species $i = 1, \ldots, n$. Both terms may be be obtained from detailed models of the diffuse double layer.

The current at the electrode surface may be expressed as

$$\tilde{\imath} = \mathrm{j}\omega \tilde{q}_m + \tilde{\imath}_F \tag{13.148}$$

where the oscillations of the surface charge density and the faradaic current density are approximated by Taylor series expansions about their steady values as

$$\tilde{q}_m = \left(\frac{\partial \bar{q}_m}{\partial V}\right)_{c_i(0)} \tilde{V} + \sum_i \left(\frac{\partial \bar{q}_m}{\partial c_i(0)}\right)_{V, c_{j \neq (i)(0)}} \tilde{c}_i(0) \tag{13.149}$$

and

$$\tilde{\imath}_F = \left(\frac{\partial \bar{\imath}_F}{\partial V}\right)_{c_i(0)} \tilde{V} + \sum_i \left(\frac{\partial \bar{\imath}_F}{\partial c_i(0)}\right)_{V, c_{j \neq (i)(0)}} \tilde{c}_i(0) \tag{13.150}$$

respectively. The surface flux can be expressed as

$$\tilde{N}_{i,y}(0) = -\frac{\partial \Gamma_i}{\partial c_i(0)} \frac{\partial c_i(0)}{\partial q_m} \mathrm{j}\omega \tilde{q}_m - \frac{s_i}{n\mathrm{F}} \tilde{\imath}_F \tag{13.151}$$

where s_i is the stoichiometric coefficient for the reaction and Γ_i is the surface concentration of species i. Equations (13.148) and (13.151) were applied as the boundary conditions to evaluate the impedance response without a priori assumption of the separation of faradaic and charging currents.

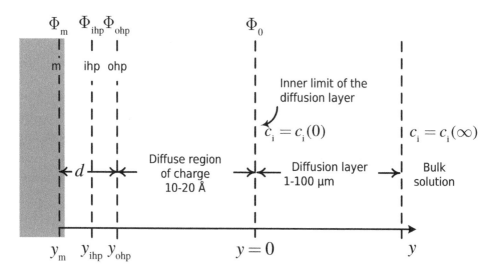

Figure 13.43: Schematic representation of the electrical double layer structure following a Stern-Gouy-Chapman model. The sketch is not drawn to scale. Taken from Wu et al.[275]

Double-Layer Model

The Stern-Gouy-Chapman model was used by Wu et al.[275] to describe the electrical behavior in the diffuse double layer. Under the assumption that ion-specific adsorption could be neglected, values for C_0 and C_i, equations (13.146) and (13.147), respectively, were obtained as functions of potential and the concentration of individual ionic species at the outer limit of the diffuse part of the double layer. The structure of the electrical double layer is illustrated in Figure 13.43 where the metal layer (M) is associated with the excess charge at the metal surface, the outer Helmholtz plane (OHP) is the plane of closest approach for solvated ions, and the inner Helmholtz plane (IHP) is the plane associated with ions adsorbed to the electrode.

As the thickness of the double layer is very thin, it is often considered to be a part of the electrode-electrolyte interface. The interfacial region as a whole obeys electrical neutrality such that the excess surface charge density at electrode is balanced by the surface charges at the IHP and in the diffuse part of the double layer, i.e.,

$$q_\mathrm{m} + q_\mathrm{ihp} + q_\mathrm{d} = 0 \tag{13.152}$$

The surface charge density is related to the surface excess concentration of charged species by

$$q_\mathrm{m} = - \left(q_\mathrm{ihp} + q_\mathrm{d} \right) = -\mathrm{F} \sum_i z_i \Gamma_i \tag{13.153}$$

The mean electrostatic potentials at the metal surface, the IHP, and the OHP are denoted by Φ_m, Φ_ihp, and Φ_ohp, respectively. The surface concentrations $c_i(0)$ and potential Φ_0 used in the rate expressions are usually evaluated at the outer limit of the diffuse layer, or the inner limit of the diffusion layer.

The ionic concentrations in the diffuse part of the double layer are assumed to have a Boltzmann distribution, i.e.,

$$c_i = c_i(\infty) \exp\left(-\frac{z_i F \Phi}{RT}\right) \tag{13.154}$$

Poisson's equation provides the relationship between the concentrations and potential as

$$\frac{d^2\Phi}{dy^2} = -\frac{F}{\varepsilon_d \varepsilon_0} \sum_i z_i c_i = -\frac{F}{\varepsilon_d \varepsilon_0} \sum_i z_i c_i(\infty) \exp\left(-\frac{z_i F \Phi}{RT}\right) \tag{13.155}$$

where y is the distance from the electrode, ε_d is the dielectric constant in the diffuse layer, and ε_0 is the permittivity of vacuum ($\varepsilon_0 = 8.8542 \times 10^{-14}$ F/cm). Integration of the Poisson equation gives the relation of potential gradient to the surface charge density in the diffuse layer as

$$\left.\frac{d\Phi}{dy}\right|_{y_{ohp}} = \frac{q_d}{\varepsilon_d \varepsilon_0} \tag{13.156}$$

where y_{ohp} is located at the inner limit of the diffuse part of the double layer, as indicated in Figure 13.43. At equilibrium, the potential approaches zero far from the electrode surface, and Poisson's equation can then be solved by applying the boundary conditions to yield the charge density in the diffuse layer, i.e.,

$$q_d = \mp \left\{ 2RT\varepsilon_d\varepsilon_0 \sum_i c_i(\infty) \left[\exp\left(\frac{-z_i F \Phi_{ohp}}{RT}\right) - 1\right] \right\}^{1/2} \tag{13.157}$$

The upper sign is used if the potential is positive and, conversely, the lower sign is used if the potential is negative.

In the absence of ion-specific adsorption on the IHP, i.e., $q_{ihp} = 0$, the surface concentration is related to the individual charge density within the diffuse part of the double layer by

$$\Gamma_i = \frac{q_{d,i}}{z_i F} \tag{13.158}$$

Following the convention adopted by Gibbs (see, for example, Chapter 7 in reference[116]) the surface charge associated with a given species i within the diffuse part of the double layer can be expressed as

$$q_{d,i} = z_i F \int_{y_{ohp}}^{\infty} (c_i - c_i(\infty)) \, dy \tag{13.159}$$

where c_i is given by equation (13.154). Integration over y is accomplished by use of the definition of electric field, i.e.,

$$E = -\frac{d\Phi}{dy} \tag{13.160}$$

where E is obtained as a function of Φ from equations (13.156) and (13.157). The charge density associated with individual species is therefore obtained by an integral over the potential drop in the diffuse region as

$$q_{d,i} = \mp \int_0^{\Phi_{ohp}} \frac{z_i F c_i(\infty) \left[\exp\left(\frac{-z_i F \Phi}{RT} \right) - 1 \right]}{\left\{ \frac{2RT}{\varepsilon_d \varepsilon_0} \sum_k c_k(\infty) \left[\exp\left(\frac{-z_k F \Phi}{RT} \right) - 1 \right] \right\}^{1/2}} d\Phi \tag{13.161}$$

The above expressions for charge densities were derived by assuming a true equilibrium of the system in which the ionic concentrations at the outer limit of the diffuse layer are the same as the bulk values, i.e., $c_i(0) = c_i(\infty)$, and the potential at the outer limit of the diffuse layer is equal to that of a reference electrode placed at infinity; thus, for a reference electrode of the same kind as the working electrode, $\Phi_0 = 0$.

When the net current flowing to the electrode is not equal to zero, the system is not at equilibrium, i.e., $\Phi_0 \neq 0$. Equations (13.157) and (13.161) become

$$q_d = \mp \left\{ 2RT \varepsilon_d \varepsilon_0 \sum_i c_i(0) \left[\exp\left(\frac{-z_i F (\Phi_{ohp} - \Phi_0)}{RT} \right) - 1 \right] \right\}^{1/2} \tag{13.162}$$

and

$$q_{d,i} = \mp \int_0^{\Phi_{ohp}} \frac{z_i F c_i(0) \left[\exp\left(\frac{-z_i F \Phi}{RT} \right) - 1 \right]}{\left\{ \frac{2RT}{\varepsilon_d \varepsilon_0} \sum_k c_k(0) \left[\exp\left(\frac{-z_k F \Phi}{RT} \right) - 1 \right] \right\}^{1/2}} d\Phi \tag{13.163}$$

respectively. The charge densities in the diffuse layer are now related to the concentrations and potential at the outer limit of the diffuse layer, which can be obtained by solving the mass and charge conservation equations outside the diffuse region of charge.

The evaluation of the surface charge density in equation (13.157) or (13.162) requires additional information in the electrical double layer. Gauss's law relates the surface charge density to the electric field within the OHP by

$$q_m = -q_d = \frac{\varepsilon \varepsilon_0}{\delta} (\Phi_m - \Phi_{ohp}) \tag{13.164}$$

where ε is the dielectric constant between the metal surface and the OHP and δ is the distance between the metal surface and the OHP, as shown in Figure 13.43. Equation (13.164) can be used as a second equation to solve for q_m and Φ_{ohp} in the electrical double layer.

Decoupled Faradaic and Charging Currents

The conventional approach is to consider the double-layer charging and the faradaic currents to be separable quantities, as is done in most of published works (see, e.g., reference[13]). The total current density is expressed by

$$i = i_\mathrm{C} + i_\mathrm{F} = C_0 \frac{\mathrm{d}V}{\mathrm{d}t} + i_\mathrm{F} \tag{13.165}$$

or, in frequency domain,

$$\widetilde{i} = \mathrm{j}\omega C_0 \widetilde{V} + \widetilde{i}_\mathrm{F} = \mathrm{j}\omega \left(\frac{\partial q_\mathrm{m}}{\partial V} \right)_{c_i(0)} \widetilde{V} + \widetilde{i}_\mathrm{F} \tag{13.166}$$

The contribution of mass flux in charging the double layer is neglected when the faradaic current and the charging current are considered separately. Thus,

$$\widetilde{N}_{i,y}(0) = -\frac{s_i}{n\mathrm{F}} \widetilde{i}_\mathrm{F} \tag{13.167}$$

Comparison of equation (13.167) with equation (13.151) shows that the first term in equation (13.151) is neglected. Under the assumption that there is a priori separation of faradaic and charging currents, equations (13.166) and (13.167) are used as the boundary conditions to evaluate the transient responses of potential and concentrations.

13.6.2 Numerical Method

Numerical methods were used to solve the sets of equations corresponding to the steady-state distributions of concentration and the model for the diffuse double layer. These provided the parameters presented in equations (13.146) and (13.147) that were needed for the numerical solution of the coupled convective diffusion equations in frequency domain.

Steady-State Calculations

Equations (13.140) and (13.141) were solved under the assumption of a steady state using the finite-element package COMSOL® Multiphysics® with the Nernst-Planck module in a 2-D axial symmetric coordinate system. The domain size was identical to that used for the impedance calculations presented in section 13.6.2. The values for the concentrations at the inner limit of the diffusion layer were used as input for the double-layer model described in section 13.6.2, and, as the problem is nonlinear due to the explicit treatment of the migration term, the steady-state values of potential and concentration were used for the impedance calculations described in section 13.6.2.

Double-Layer Properties

The nonlinear equations relating to the charge and potential distributions in the electrical double layer were solved by using the Newton-Raphson method. For nonequilibrium systems, a local equilibrium was assumed in which the concentrations and

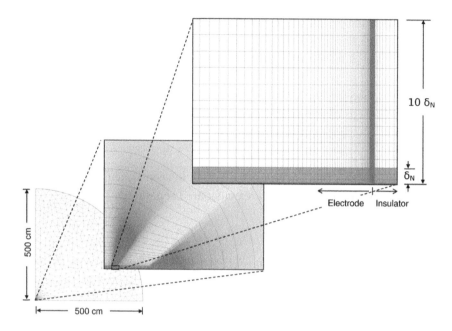

Figure 13.44: Representation of the different mesh scales used for the finite-element calculations. Taken from Wu et al.[275]

potential at the outer limit of the diffuse layer were given by the steady-state solution of the convective diffusion equations. The radially dependent surface concentrations and potential gave a radial distribution of charge over the electrode surface. The surface-averaged value of charge was obtained from the surface-averaged values of concentration and potential. The charge in the diffuse layer associated with individual species was obtained by subsequent numerical integration of equation (13.161) or (13.163), using the MATLAB® integration function, the adaptive Gauss-Kronrod quadrature method. The thickness of the compact layer between the metal surface and the OHP was assumed to be 3 Å. The dielectric constant in the compact region within the OHP was approximately 6, according to Bockris,[276] for a fully oriented water layer next to the electrode surface. The dielectric constant in the diffuse layer was assumed to be 78, which is the value of water at room temperature.

Impedance Calculations

The equations were solved for a rotating disk electrode by using the finite-element package COMSOL® Multiphysics® with the Nernst-Planck module in a 2-D axial symmetric coordinate system. The domain size was 2,000 times larger than the disk radius in order to meet the assumption that the counterelectrode was located infinitely far from the electrode surface. The meshed domain used to calculate the coupled solution for potential and concentrations is shown in Figure 13.44. A coarse mesh was used for the domain distant from the electrode to reduce physical memory usage and calculation time. A finer mesh was applied in the region that is 20 times larger than the disk radius to capture the variation of potential in the vicinity of the electrode. Since the concentration of ionic species varies only in a small distance above the electrode surface, a much finer mesh was constructed in the region that is 10 times larger than the characteristic thickness of the diffusion layer

$$\delta_N = \left(\frac{3D_i}{av}\right)^{1/3} \left(\frac{\nu}{\Omega}\right)^{1/2} \tag{13.168}$$

where $a = 0.51023$ is the coefficient for the first term in the velocity expansion for the rotating disk, ν is the kinematic viscosity, and Ω is the disk rotation speed.[116] This approach captured the concentration variation at the electrode surface.

Steady-state solutions were used in the impedance model in the form of lookup tables from which appropriate values could be obtained by interpolation. The use of the same meshed domain in the steady-state and the impedance models reduced the error from interpolation between nodal points. The distributions of the double-layer capacitance and other thermodynamic parameters at electrode boundary were applied in the impedance model to evaluate the impedance response for the case for coupled faradaic and charging currents. For the case with decoupled faradaic and charging currents, a surface-averaged double-layer capacitance

$$\langle C_0 \rangle = \frac{1}{\pi r_0^2} \int_0^{r_0} C_0(r) r dr \tag{13.169}$$

was used to determine the charging current that did not include the contribution of mass transfer in the diffuse region of charge.

The velocity used in the expression for convective diffusion consisted of an interpolation formula that followed the three-term velocity expansion for radial and axial velocity components appropriate for the velocity boundary layer near the disk electrode surface, equations (11.53) and (11.51), respectively, and the two-term expansion appropriate for the region far from the disk surface (see problem 11.5 on page 300).[181] The interpolation formula provided excellent accuracy near the electrode surface and showed good agreement with numerical solutions of the Navier-Stokes equations throughout the domain.

13.6.3 Consequence of Coupled Charging and Faradaic Currents

The normalized global impedances calculated under assumption of decoupled and coupled charging and faradaic currents are presented in Figures 13.45(a) and 13.45(b), respectively, for two electrolyte compositions: 0.1 M AgNO$_3$ and 1 M KNO$_3$, and 0.01 M AgNO$_3$ and 1 M KNO$_3$. The dimensionless frequency K is given by equation (13.62). The differences between results obtained under the assumption that charging and faradaic currents are decoupled or coupled may be observed at intermediate to high fre-

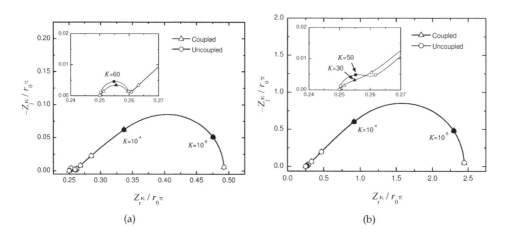

Figure 13.45: Normalized global impedances calculated under assumption of uncoupled and coupled charging and faradaic currents: a) 0.1 M $AgNO_3$ in a 1 M KNO_3 electrolyte and b) 0.01 M $AgNO_3$ in a 1 M KNO_3 electrolyte. Taken from Wu et al.[275]

quencies where both the faradaic and charging processes are important. At high frequencies, a depressed semicircle is seen for both cases. The degree of depression, however, is greater when faradaic and charging currents are assumed to be coupled.

The distinction between models obtained under the assumption that charging and faradaic currents are decoupled or coupled can be seen more clearly in Figure 13.46, where the logarithm of the absolute value of the imaginary part of the global impedance is presented as a function of the logarithm of frequency. As discussed by Orazem et al.[252] and in Section 18.3 on page 505, the slope of the curves can be related to the exponent of a constant-phase element. The slopes of the lines presented in Figure 13.46 are presented in Figure 13.47 with fraction of the mass-transfer-limited current density as a parameter.

The depressed semicircle obtained under the assumption that charging and faradaic currents are decoupled is very similar to that obtained by Huang et al.[125] for a secondary current distribution in which mass transfer effects were ignored. The slope at high frequency has a mild frequency dependence and approaches a value of $\alpha = 0.9$ for the decoupled results presented in both Figures 13.47(a) and 13.47(b). Huang et al.[125] reported similar values that were described as a pseudo-CPE behavior attributed to the influence of nonuniform current and potential distributions. The depressed semicircle obtained under the assumption that charging and faradaic currents are coupled corresponds to a high-frequency pseudo-CPE behavior with an exponent α between 0.5 and 0.6, which is much smaller than the values attributed to the influence of nonuniform current and potential distributions.[125] These figures suggest that an additional factor must be contributing to the frequency dispersion when charging and faradaic currents are assumed to be coupled.

The frequency dispersion associated with coupled charging and faradaic currents is

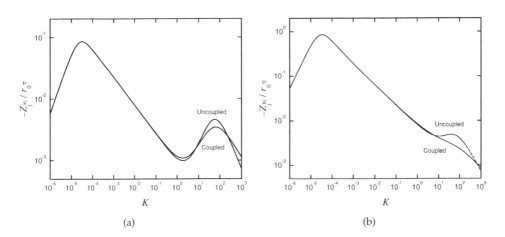

Figure 13.46: Normalized imaginary part of the global impedances calculated under assumption of uncoupled and coupled charging and faradaic currents: a) 0.1 M $AgNO_3$ in a 1 M KNO_3 electrolyte and b) 0.01 M $AgNO_3$ in a 1 M KNO_3 electrolyte. Taken from Wu et al.[275]

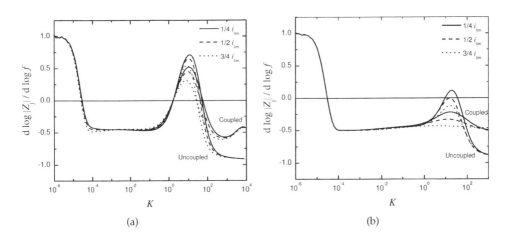

Figure 13.47: Derivative of the logarithm of the imaginary part of the global impedance with respect to the logarithm of frequency calculated under assumption of uncoupled and coupled charging and faradaic currents: a) 0.1 M $AgNO_3$ in a 1 M KNO_3 electrolyte and b) 0.01 M $AgNO_3$ in a 1 M KNO_3 electrolyte. Taken from Wu et al.[275]

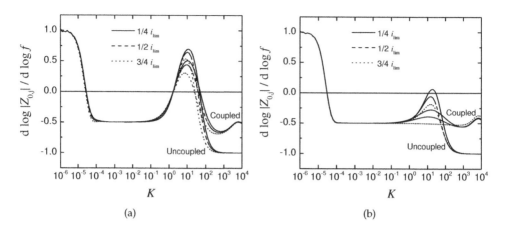

Figure 13.48: Derivative of the logarithm of the imaginary part of the global interfacial impedance with respect to the logarithm of frequency calculated under assumption of uncoupled and coupled charging and faradaic currents: a) 0.1 M $AgNO_3$ in a 1 M KNO_3 electrolyte and b) 0.01 M $AgNO_3$ in a 1 M KNO_3 electrolyte.

seen by examination of the global interfacial impedance for the uncoupled and coupled cases, shown in Figure 13.48. Under the assumption that charging and faradaic current may be treated independently, the derivative of the logarithm of the imaginary part of the global interfacial impedance with respect to the logarithm of frequency approaches -1 at high frequencies, as would be expected for an ideal faradaic reaction. In sharp contrast, the corresponding value obtained when charging and faradaic currents are coupled is between -0.7 and -0.5.

The simulations of the coupled charging and faradaic currents requires thermody-namic properties of the interface that are not readily available. Wu et al.[275] obtained them under assumption of a Stern-Gouy-Chapman model of a diffuse double layer in the absence of ion-specific adsorption. Use of a refined model will certainly alter the results. Their work serves to show, however, that the frequency dispersion associated with the coupling of charging and faradaic currents does not disappear in the presence of a large concentration of supporting electrolyte. This result contradicts the conclusions presented by Nisancioglu and Newman.[270] Even though a low concentration of silver ions must be associated with a small contribution to the charging current, the contribution to the charging current influences the contribution to the faradaic current and, therefore, influences the impedance response. It is evidently appropriate to decouple faradaic and charging currents for systems that are not influenced by mass transfer. The present work suggests, however, that the coupling of faradaic and charging currents should be considered when modeling the impedance response for systems influenced by mass transfer, even for large concentrations of supporting electrolyte. The work of Wu et al.[275] shows that the coupling of faradaic and charging currents results in a frequency dispersion that may appear as a pseudo-CPE behavior.

13.7 Exponential Resistivity Distributions

The electrochemical impedance of an oxide layer reveals an apparent CPE behavior in the high-frequency range, and the origin of the CPE behavior is generally attributed to a distribution of time constants. The influence of a distribution of time constants along the electrode surface (i.e., a 2-D distribution) was discussed in Section 13.2. A time-constant dispersion can also be attributed to a distribution along the dimension normal to the electrode surface (i.e., a 3-D distribution). The LEIS technique described in Section 7.5.2 on page 158 can be used to distinguish between a 2-D and a 3-D distribution.[277] With the LEIS technique, a 2-D distribution is characterized at high frequencies by pure capacitance behavior, and a 3-D distribution is characterized by an apparent CPE behavior. Of course, a local CPE behavior characteristic of a 3-D distribution can be also be involved in a 2-D current and potential distribution as discussed in Section 13.2.[248] For an oxide layer, the distribution of time constants in the direction normal to the electrode surface can be due to varying oxide composition.

The impedance of a layer can be considered to correspond to a large number of Voigt elements as represented in Figure 12.5 (page 310). The capacity $C(y)$ is the capacity of a dielectric with a thickness dy, and the resistance $R(y)$ corresponds to a layer of same thickness dy with a resistivity $\rho(y)$. The local impedance is obtained by integration along the y-direction from 0 to the coating thickness δ, i.e.,

$$Z_Y = \int_0^\delta \frac{\rho(y)}{1 + j\omega\varepsilon\varepsilon_0\rho(y)}\,dy \tag{13.170}$$

As shown in Table 5.6 on page 126, the dielectric constant ε varies in a narrow range for a metal oxide. Thus, to a first approximation, ε can be considered to be independent of y, and the resistivity ρ may be assumed to be a function of y.

For an oxide layer, Young[278] assumed that the nonstoichiometry of the oxide layer resulted in an exponential variation of the conductivity with respect to the normal distance to the electrode. This assumption may be expressed in terms of resistivity as

$$\rho(y) = \rho_0 \exp\left(-y/\lambda\right) \tag{13.171}$$

A normalized resistivity $\rho(y)/\rho_0$ is presented in Figure 13.49 for a film with a thickness of 8 nm and with δ/λ as a parameter.

The Young impedance for the gradient presented in equation (13.171) is[279,280,135]

$$Z_Y = \frac{\lambda}{j\omega\varepsilon\varepsilon_0} \ln\left(\frac{1 + j\omega\rho_0\varepsilon\varepsilon_0}{1 + j\omega\rho_0\varepsilon\varepsilon_0 \exp(-\delta/\lambda)}\right) \tag{13.172}$$

Remember! 13.3 *Not all depressed semicircles correspond to a CPE behavior.*

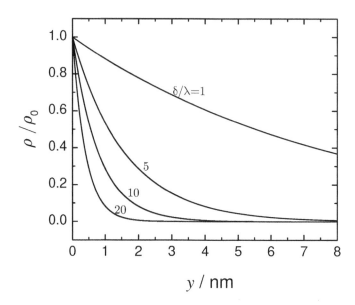

Figure 13.49: Normalized resistivity from equation (13.171) as a function of position with δ/λ as a parameter. The film thickness was assumed to be 8 nm.

In the low-frequency limit, application of L'Hôpital's rule yields

$$\lim_{\omega \to 0} Z_Y = \lambda \rho_0 \left(1 - \exp\left(-\delta/\lambda\right)\right) \tag{13.173}$$

which corresponds to a resistor. This result is in agreement with the direct integration of resistivity. In the high-frequency limit,

$$\lim_{\omega \to \infty} Z_Y = -j\frac{\delta}{\omega \varepsilon \varepsilon_0} \tag{13.174}$$

which corresponds to the impedance of the capacitor

$$C_Y = \frac{\varepsilon \varepsilon_0}{\delta} \tag{13.175}$$

given as equation (12.11) on page 310.

    An example of the Young impedance is plotted in Figure 13.50 in Nyquist format with δ/λ as a parameter. The model parameters were $\varepsilon = 12$, and $\rho_0 = 1 \times 10^9$ Ωcm. The dashed line represents the impedance response of an RC circuit. As compared to the dashed line, the Young impedance appears as a depressed semicircle, similar to what is obtained for a resistance in parallel with a CPE (see Chapter 14 on page 395). The Nyquist plot is not symmetric, especially for $\delta/\lambda > 5$. A clearer representation of the impedance response is given by the phase angle, presented in Figure 13.51 as a function of normalized frequency. For an RC circuit (dashed lines), the normalized frequency is given by $f/f_c = 2\pi RC$. For the Young model (solid lines), the normalized frequency is given by $f/f_c = 2\pi \varepsilon \varepsilon_0 \rho_0$. The phase angle is equal to zero at low

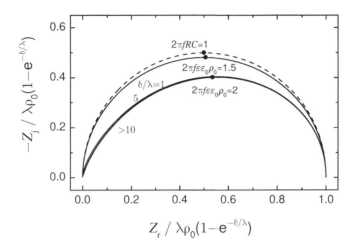

Figure 13.50: Young impedance given by equation (13.172) in Nyquist format with δ/λ as a parameter. The model parameters were $\varepsilon = 12$, and $\rho_0 = 1 \times 10^9$ Ωcm. The dashed line represents the impedance response of an RC circuit.

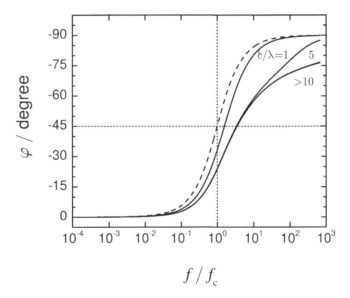

Figure 13.51: Phase angle associated with the Young impedance given by equation (13.172) as a function of normalized frequency with δ/λ as a parameter. The model parameters were $\varepsilon = 12$, and $\rho_0 = 1 \times 10^9$ Ωcm. The dashed represents the phase angle associated with an RC circuit.

frequency and approaches $-90°$ at high frequency. At no frequency is the phase angle independent of frequency. Clearly no constant phase angle is found, but, in spite of the depressed semicircle of Figure 13.50, the phase angle approaches $-90°$ at high frequencies.

In this case, a CPE could be used to describe the experimental data obtained in a limited frequency range, but, when a physical model is assumed, as for the Young impedance, true CPE behavior is not found. It should be noted that the CPE model corresponds to a specific distribution of time constants that may or may not correspond to a given physical situation. Local impedance measurements can give information about the nature of this distribution, whether 2-D, 3-D, or both. This example shows that not all depressed semicircles correspond to a CPE behavior.

Problems

13.1 Provide an analytic solution to equations (13.123) and (13.124), and write explicitly the equation corresponding to $m = 1$.

13.2 Consider a 0.25 cm radius Pt disk in a 0.1 M NaCl solution at room temperature. Estimate the frequency above which geometry-induced time-constant dispersion will influence the impedance response.

13.3 Consider a 0.25 cm radius steel disk covered with a native oxide layer. The electrolyte is a 0.1 M NaCl solution at room temperature. Estimate the frequency above which geometry-induced time-constant dispersion will influence the impedance response.

13.4 Consider a 0.25 cm radius steel disk covered with a polymer coating that has a thickness of 100 μm. The electrolyte is a 0.1 M NaCl solution at room temperature. Estimate the frequency above which geometry-induced time-constant dispersion will influence the impedance response.

13.5 The dimensionless form of equation (13.61) was used to justify the use of a dimensionless frequency, defined in equation (13.62), in which the characteristic dimension was the radius of the disk r_0. Beginning with

$$K = \omega C_0 R_e \tag{13.176}$$

and equation (13.73), identify an alternative characteristic dimension.

13.6 Find the equations that are necessary to solve the response of a small electrode to a rotation speed modulation.

13.7 A thin layer cell is comprised of an isolated plane at a very short distance ε from a working electrode. By considering the system with a cylindrical symmetry, calculate the impedance of a disk electrode in this configuration.

Table 13.2: Calculated characteristic frequencies for rectangular electrodes with dimension $d_1 \times d_2$.

f_c, kHz	d_1, mm	d_2, mm
21.59	0.04055	0.00405
9.81	0.04055	0.01014
5.89	0.04055	0.02028
3.92	0.04055	0.04055
3.11	0.02819	0.28190
2.94	0.04055	0.08110
2.35	0.04055	0.20275
2.22	0.04055	0.30413
2.16	0.04055	0.40550
2.11	0.04055	0.52715
1.41	0.07048	0.28190
0.85	0.14095	0.28190
0.56	0.28190	0.28190
0.42	0.56380	0.28190
0.34	1.40950	0.28190
0.32	2.11425	0.28190
0.31	2.81900	0.28190
0.30	4.22850	0.28190

13.8 Explore the limiting behavior of equation (13.19) when $\ell\sqrt{2\rho/rZ_{eq}}$ is very small. What independent parameters or combinations of parameters can be obtained by regression of this model to experimental data? To what geometry does this limit conform?

13.9 Explore the behavior of equation (13.19) when $\ell\sqrt{2\rho/rZ_{eq}}$ is neither very small nor very large. What independent parameters or combinations of parameters can be obtained by regression of this model to experimental data?

13.10 Numerical methods following Section 13.2.1 were used to generate the characteristic frequencies shown in Table 13.2 for a rectangular electrode of different dimensions with a counterelectrode located at infinity. The capacitance was assigned a value of 20 $\mu F/cm^2$, and the resistivity was 1,000 Ωcm. Show that the characteristic dimension associated with this characteristic frequency can be expressed as $d = d_1 d_2 /(d_1 + d_2)$.

13.11 By inspection of Table 13.4, identify an expression for the characteristic frequency associated with time-constant dispersion associated with a CPE on a rectangular electrode.

13.12 Find the smallest frequency for which geometry-induced time-constant dispersion will appear for an electrode for which three electrodes of different dimensions are connected in parallel:

(a)   a disk of radius $r_0 = 0.25$ cm, a rectangle with dimension 0.2 cm$\times$0.6 cm, and a hemispherical electrode with radius $r_0 = 0.25$ cm.

(b)   a disk of radius $r_0 = 0.25$ cm, a disk of radius $r_0 = 0.5$ cm, and a disk of radius $r_0 = 0.75$ cm.

(c)   three rectangles with dimensions 0.2×0.2 cm, 0.2×0.6 cm, and 0.2×1.2 cm.

13.13 Find the smallest frequency for which geometry-induced time-constant dispersion will appear for the following:

(a)   an electrode consisting of 20 disks of radius $r_0 = 10$ μm.

(b)   a single disk electrode of area equal to 20 disks of radius $r_0 = 10$ μm.

13.14 Consider the impedance response of a film in which the resistivity varies with the position perpendicular to the plane of the film according to

$$\rho(y) = \rho_0 \exp(-y/\lambda) \tag{13.177}$$

where λ is a characteristic dimension over which resistivity varies.

(a)   Plot the resulting resistivity as a function of position for a film of thickness $\delta = 8$ nm. You may assume that $\lambda = 4$ nm and $\rho_0 = 1 \times 10^9$ Ωcm.

(b)   Develop an expression for the impedance response of a film of thickness You may be guided by the discussion of the Young impedance within the textbook.

(c)   Plot the resulting impedance as a Nyquist plot for $\varepsilon = 12$, $\lambda = 4$ nm, $\delta = 8$ nm, and $\rho_0 = 1 \times 10^9$ Ωcm. Does this impedance represent a depressed semicircle? Does this impedance represent a CPE? You may use a plot of the absolute value of the imaginary part of the impedance as a function of frequency on a log-log scale to justify your answers.

13.15 The impedance for a system has been modeled by

$$Z = R_e + \frac{1}{(j\omega)^\alpha Q} \tag{13.178}$$

where $\alpha = 0.5$. Equation (13.178) is normally associated with a distribution of time constants through a film or along a surface, but other explanations may apply when $\alpha = 0.5$. Suggest two other possible physical explanations that could account for this result.

Chapter 14

Constant-Phase Elements

Electrical circuits invoking constant-phase elements (CPE) are often used to fit impedance data arising from a broad range of experimental systems. The CPE is often used simply as a way to improve the fit of a model to impedance data, justified by a vague assertion that a distribution of time constants is present in the system under investigation. The development in the present chapter shows that the CPE parameters have a physical meaning. The CPE may be seen as a special case of the time-constant dispersion presented in Chapter 13 on page 319 in which the time constants follow a particular distribution. The interpretation of CPE parameters in terms of physically meaningful properties such as capacitance requires an understanding of the nature of the time-constant distribution.

14.1 Mathematical Formulation for a CPE

The impedance for a film-covered electrode showing CPE behavior may be expressed in terms of ohmic resistance R_e, a parallel resistance $R_{||}$, and CPE parameters α and Q as

$$Z = R_e + \frac{R_{||}}{1 + (j\omega)^\alpha R_{||} Q} \tag{14.1}$$

When $\alpha = 1$, the system is described by a single time constant, and the parameter Q has units of capacitance. When $\alpha \neq 1$, Q has units of $s^\alpha/\Omega cm^2$ or $F/s^{(1-\alpha)} cm^2$, and the system shows behavior that has been attributed to surface heterogeneity[281,282] or to continuously distributed time constants.[283–287]

Under conditions that $(\omega)^\alpha R_{||} Q \gg 1$,

$$Z = R_e + \frac{1}{(j\omega)^\alpha Q} \tag{14.2}$$

which has the appearance of a blocking electrode. The term $R_{||}$ in equation (14.1) accounts for a resistance that may be attributed to different current pathways that exist in parallel to the dielectric response of a film. These may include interconnecting conductive phases within a solid matrix or pores at the bottom of which a reaction may take place at the electrolyte/metal interface.

14.2   When Is a Time-Constant Distribution a CPE?

Often the analysis of an electrochemical system is based upon a model that uses ordinary electrical elements such as resistors, capacitors, inductors, diffusion impedances, deLevie impedances, or Young impedances. When the regression of the data is not sufficient, one or more time-constant distributions are introduced to obtained a more accurate fit to the data. A priori, it is very difficult to know if the time-constant distribution corresponds to a CPE or to another form. A graphical analysis of experimental data can provide an answer to this question.

   If the hypothetic time-constant distribution is considered to play a role in the high frequency range, then a modified expression for the impedance can be developed. If the system is simple, the total impedance can be written under the form of equation (14.1).

$$Z_T(\omega) = R_e + \frac{R_{||}}{1 + (j\omega)^\alpha R_{||} Q} \tag{14.3}$$

Under conditions that $\omega^\alpha R_{||} Q \gg 1$, equation (14.3) corresponds to equation (14.2). In the general case, equation (14.3) may be expressed as

$$Z_T(\omega) = R_e + \frac{Z(\omega)}{1 + (j\omega)^\alpha Z(\omega) Q} \tag{14.4}$$

where, at high frequency, the impedance corresponds to equation (14.2).

   The ohmic-resistance-corrected phase angle of the impedance can be obtained using the graphical methods described in Section 18.2.1 on page 504. If, at high frequency, this phase angle is equal to $-90°$ and is independent of frequency, a CPE is not needed to describe the system. If the phase angle approaches a constant value $-90\alpha°$ at high frequency, the hypothesis of a CPE is valid.

Example 14.1 **Graphical Identification of CPE Behavior:** *Plot in Bode representation the impedance of the passive layer for a stainless steel electrode taken from Frateur et*

Remember! 14.1 *We introduce in this chapter the Constant-Phase Elephant. Remember, just as not all time-constant distributions are constant-phase distributions, not all elephants are Constant-Phase Elephants.*

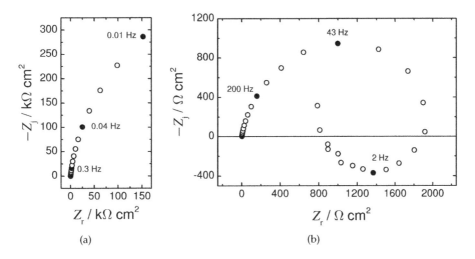

Figure 14.1: The impedance in Nyquist format: a) of the passive layer for a stainless steel electrode taken from Frateur et al.[288] and b) of the niobium-oxide covered niobium electrode taken from Cattarin et al.[289]

al.[288] *(shown in Figure 14.1(a)) and for a niobium oxide covered niobium electrode taken from Cattarin et al.[289] and shown in Figure 14.1(b). Plot the same data corrected for ohmic resistance. Develop conclusions regarding the use of a CPE to describe the data and the usefulness of different plots for extracting information concerning the data.*

Solution: *The Bode plot for a passive layer for a stainless steel is presented in Figure 14.2 and the Bode plot for niobium is presented in Figure 14.3. It is impossible to use Figures 14.2 and 14.3 to conclude whether the frequency dispersion can be attributed to a constant-phase element or to another type of distribution. The behavior in the high frequency range shown in Figures 14.2 and 14.3 is masked by the effect of the electrolyte resistance.*

Ohmic-resistance-corrected Bode plots are presented in Figure 14.4 for the passivated stainless steel. The ohmic-resistance-corrected modulus shown in Figure 14.4(a) may be superposed by a straight line that has a slope of −0.89. A deviation from the straight line is observed at a frequency of around 2,400 Hz, indicated by the vertical dashed line. This deviation corresponds to the frequency dispersion associated with the disk electrode geometry, as discussed in Section 13.2 on page 337. The ohmic-resistance-corrected phase angle shown in Figure 14.4(b) reveals a constant phase of about 80° over a frequency ranging from 1 Hz to 1 kHz. The phase angle is related to the exponent α in equations (14.1) and (14.2) by

$$\alpha = \frac{-\varphi_{adj}}{90°} \tag{14.5}$$

For this system, α = 0.89. For frequencies lower than 1 Hz, the ohmic-resistance-corrected phase is influenced by the real value of the oxide layer impedance at zero frequency and by the resistivity value at the metal-oxide interface (see equation (14.30)). For frequencies higher than 2,400 Hz, indicated by the vertical dashed line, the phase angle is influenced by the current and

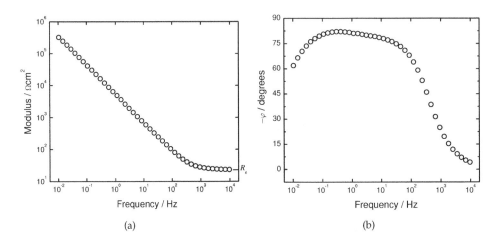

Figure 14.2: The impedance in Bode format of the high-frequency portion of the impedance data shown in Figure 14.1(a) for a stainless steel electrode:[288] a) modulus and b) phase.

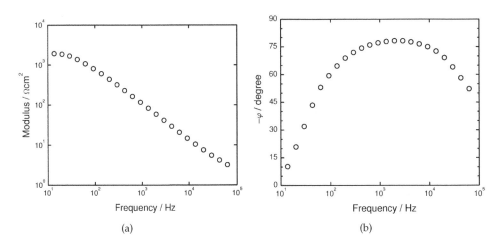

Figure 14.3: The impedance in Bode format of the high-frequency portion of the impedance data shown in Figure 14.1(b) for a niobium-oxide covered niobium electrode:[289] a) modulus and b) phase.

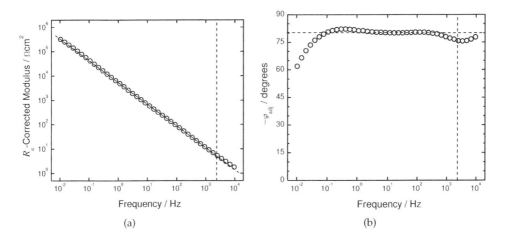

Figure 14.4: Ohmic-resistance-corrected Bode plots for the high-frequency portion of the impedance data shown in Figure 14.1(a) for a stainless steel electrode:[288] a) modulus and b) phase.

potential distributions associated with the disk geometry. The use of ohmic-resistance-corrected Bode plots allows a refined interpretation of the impedance response.

Ohmic-resistance corrected Bode plots are presented in Figure 14.5 for the niobium oxide. Unlike the data presented for steel in Figure 14.4(a), the data presented in Figure 14.5(a) cannot be superposed by a straight line. Thus, the ohmic-resistance-corrected modulus cannot be represented by a constant-phase element. The ohmic-resistance-corrected phase angle, presented in Figure 14.5(b), increases with frequency. Figures 14.4 and 14.5 suggest that a power-law model for distribution of resistivity, described in Section 14.4.4 and which corresponds to a CPE behavior, applies for the steel and a Young model providing an exponential distribution of resistivity should be considered for the niobium oxide (see Section 13.7 on page 389).

14.3 Origin of Distributions Resulting in a CPE

Several systems are presented in Chapter 13 (page 319) for which time-constant dispersions are observed, but these do not give rise to a constant-phase element. As constant-phase elements represent a special case of time-constant or frequency dispersions, the discussion of the origin of CPE behavior can be considered a restricted case of the dispersions described in Chapter 13.

Remember! 14.2 *Interpretation of a CPE in terms of physical parameters requires an understanding of the type of time-constant distribution causing the frequency dispersion.*

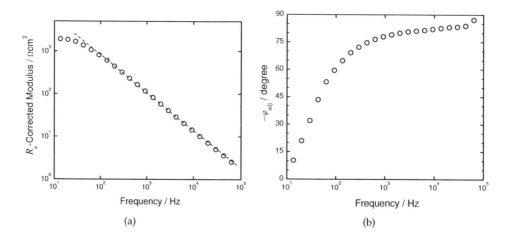

Figure 14.5: Ohmic-resistance-corrected Bode plots for the high-frequency portion of the impedance data shown in Figure 14.1(b) for a niobium oxide covered niobium electrode:[289] a) modulus and b) phase.

Jorcin et al.[277] used Local Electrochemical Impedance Spectroscopy (LEIS) to attribute CPE behavior, seen in the global measurements, to time constants that were distributed either along the electrode surface or in the direction perpendicular (or normal) to the electrode surface. Normal distributions of time constants can be expected in systems such as oxide films, organic coatings, and human skin and can be associated with the dielectric and resistive properties of a material. Surface distributions of film properties may also be expected. In the absence of additional information, such as may be obtained from local impedance measurements, one cannot conclude a priori whether CPE behavior associated with film-covered electrodes or membranes should be attributed to normal or surface distributions.

The extraction of effective capacitance from CPE model parameters presented by Brug et al.[149] was predicated on the assumption that capacitance was distributed along an electrode surface. As mentioned in Section 13.2.3 (page 342), Huang et al.[147] showed that a surface distribution of local ohmic impedance gave results in agreement with Brug et al.,[149] but this effect was seen only for frequencies larger than a specific value. It is more difficult to find an explanation for frequency dispersion covering a broad range of frequency that could be associated with surface distributions of time constants.

Electrode roughness is often assumed to give rise to CPE behavior, but, as discussed in Section 13.3.1 on page 349, the frequency at which roughness causes frequency dispersion is higher than that associated with the disk geometry.[262] Even for a recessed electrode, frequency dispersion was estimated to be observed at frequencies higher than 100 kHz for the degree of roughness associated with No. 1000 grit paper. Similarly, as discussed in Section 13.3 on page 349, a surface distribution of capacitance gives rise to frequency dispersion at frequencies higher than that associated with the disk geometry.[264]

Local surface distributions of reactivity are often assumed to cause CPE behavior. Alexander et al.[265] showed that a surface variation of charge transfer resistance gives rise to only an averaged surface reactivity. This conclusion is limited to electrochemical reactions that may be described in terms of a single charge-transfer resistance. Wu et al.[249,250] have discussed frequency dispersion associated with reactions coupled by an adsorbed intermediate. They found that the nonuniform current and potential distribution associated with a disk geometry may give rise to a low-frequency dispersion (i.e., CPE-like behavior) for reactions coupled by an adsorbed intermediate. Similarly, for a recessed electrode, surface distributions of rate constants for coupled reactions give rise to low-frequency dispersion surface distributions of rate constants for coupled reactions.[290]

Hirschorn et al.[291] have shown that a power-law distribution of resistivity yields CPE behavior. Comparison to independent determinations of film properties showed good agreement with values derived from the power-law model for oxides and human skin.[292,134] The power-law model has been applied as well to investigate water uptake in coatings.[293,294,137]

In addition to surface and normal distributions of time constants, other phenomena may also give rise to frequency dispersion. Orazem[295] suggested that the coupled behavior of charging and faradaic currents could give rise to a frequency dispersion that resembles CPE behavior.

14.4 Approaches for Extracting Physical Properties

The CPE parameters α and Q are representative of a physical system but do not have a direct connection to physical properties such as the thickness and dielectric constant of a film. As discussed in Section 5.7.2 on page 126, the dielectric constant or the film thickness may be obtained from

$$C_{\text{eff}} = \frac{\epsilon \epsilon_0}{\delta} \tag{14.6}$$

In equation (14.6), δ is the film thickness, ϵ is the dielectric constant, and ϵ_0 is the permittivity of vacuum with a value of $\epsilon_0 = 8.8542 \times 10^{-14}$ F/cm.

The challenge has been to find a correct way to extract an effective capacitance from CPE parameters. Four approaches are discussed in this section. These include:

1. simply equating C_{eff} to Q while ignoring the difference in units,

2. a formula derived from the work of Hsu and Mansfeld[296] based on the characteristic frequency of the impedance,

3. a formula developed by Brug et al.[149] for an assumed surface distribution of time constants, and

4. a model presented by Hirshorn et al.[291,292] based on a power-law distribution of resistivity.

The presentation in this section follows that of Orazem et al.[134]

Other methods have been used to extract capacitance values directly from impedance data without initial evaluation in terms of a CPE. Oh et al.[297] described a graphical method that applies only for capacitive systems. Oh and Guy[298,299] used

$$C_{\text{eff,OG}} = \frac{\tan \theta_{\text{c,norm}}}{2\pi R_{||} f_{\text{c,norm}}}$$

(14.7)

where $\theta_{\text{c,norm}}$ is the phase angle at the characteristic frequency for the impedance, $f_{\text{c,norm}}$. Equation (14.7) can be expressed as

$$C_{\text{eff,OG}} = \frac{\tan (\alpha \pi / 4)}{2\pi R_{||} f_{\text{c,norm}}}$$

(14.8)

and is mathematically equivalent to equation (14.14) multiplied by $\tan (\alpha \pi / 4)$. As the methods reported in references 298 and 299 do not make use of CPE parameters, they are not discussed further in this section.

14.4.1 Simple Substitution

In this approach C_{eff}, expressed in units of F/cm^2, is assigned the numerical value of Q, expressed in units of $F/s^{(1-\alpha)}cm^2$, while ignoring the difference in units. Thus, the approach consists of positing that

$$C_{\text{eff,Q}} = Q$$

(14.9)

This approach, often used when $1 > \alpha > 0.9$, was shown by Orazem et al.[134] to be inaccurate and should thus be avoided. Criteria that may be used to estimate the error in equation (14.9) are developed in Examples 14.4 and 14.5.

14.4.2 Characteristic Frequency: Normal Distribution

A normal distribution of time constants is represented schematically in Figure 14.6. The proper distribution of time constants is envisioned to lead to a CPE in parallel with a resistance. This resistance is presumed to be the resistance of a film.

As shown by Hirshorn et al.,[135] the relationship between CPE parameters and effective capacitance for a normal distribution of time constants requires an assessment of the characteristic time constant corresponding to the impedance of the film. The impedance of a film can be expressed in terms of a CPE by equation (14.1) where $R_{||}$ represents the parallel resistance. Alternatively,

$$Z = R_{\text{e}} + \frac{R_{||}}{1 + (jK_{\text{norm}})^{\alpha}}$$

(14.10)

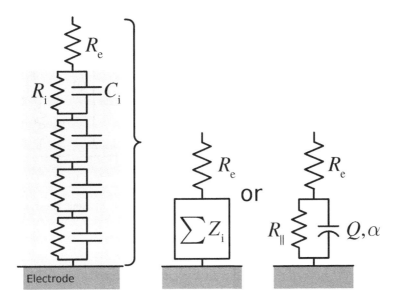

Figure 14.6: Schematic representation of a normal distribution of time constants leading to constant-phase behavior. The proper distribution of time constants is envisioned to lead to a CPE in parallel with a resistance. A correct representation of the CPE associated with a normal distribution of time constants is given in Figure 14.9.

where K_{norm} is a dimensionless frequency which can be expressed for an impedance as

$$K_{\text{norm}} = \left(R_{||}Q\right)^{1/\alpha}\omega \tag{14.11}$$

A characteristic frequency can be found for $K_{\text{norm}} = 1$ such that

$$f_{\text{c,norm}} = \frac{1}{2\pi\left(R_{||}Q\right)^{1/\alpha}} \tag{14.12}$$

The essential hypothesis of the approach presented in this section is that the characteristic frequency can also be expressed in terms of an effective capacitance as

$$f_{\text{c,norm}} = \frac{1}{2\pi R_{||}C_{\text{eff,norm}}} \tag{14.13}$$

By solving equations (14.12) and (14.13) for $C_{\text{eff,norm}}$, an expression for the effective capacitance associated with the CPE is found to be

$$C_{\text{eff,norm}} = Q^{1/\alpha}R_{||}^{(1-\alpha)/\alpha} \tag{14.14}$$

Equation (14.14) is equivalent to equation (3) in the work of Hsu and Mansfeld,[296] presented in terms of the characteristic angular frequency ω_{max}.

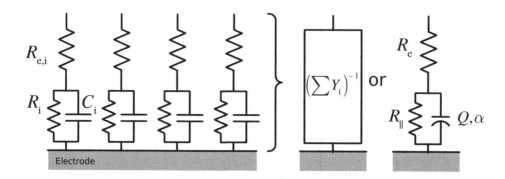

Figure 14.7: Schematic representation of a normal distribution of time constants leading to constant-phase behavior. The proper distribution of time constants is envisioned to lead to a CPE in parallel with a resistance.

Orazem et al.[134] showed that the effective capacitance derived from the characteristic frequency of the impedance does not provide good estimates for physical properties because the resistance associated with the characteristic frequency is independent of the dielectric properties of the film. Indeed, the parallel resistance cannot be the resistance of the film as the film resistance is given by the summation of resistance terms in Figure 14.6. A correct representation of the CPE associated with a normal distribution of time constants is given in Figure 14.9.

14.4.3   Characteristic Frequency: Surface Distribution

A surface distribution of time constants is represented schematically in Figure 14.7. As discussed in Section 13.3 on page 349, a distribution of time constants, either $R_{e,i}C_i$ or R_iC_i, may lead to frequency dispersion. The proper distribution of time constants is envisioned to lead to a CPE in parallel with a resistance. A similar representation may be made for a blocking electrode in which R_i and the parallel resistance $R_{||}$ are not observed.

In the case of a surface time-constant distribution, the global admittance response of the electrode includes additive contributions from each part of the electrode surface. Hirschorn et al.[135] showed the development of a relationship between capacitance and CPE parameters for a surface distribution of time constants. In agreement with Chassaing et al.[300] and Bidóia et al.,[301] they observed that the appearance of a CPE behavior associated with a surface distribution of time constants requires the contribution of an ohmic resistance. The influence of ohmic resistance on the CPE behavior is evident in Figure 14.8, where a distribution of time constants in the absence of an ohmic resistance yields only an effective resistance and an effective capacitance.

The development of Brug et al.[149] was applied toward surface distributions of the charge transfer resistance R_t associated with kinetic parameters. In the present work, the same approach is applied for a surface distribution for the properties of a film such

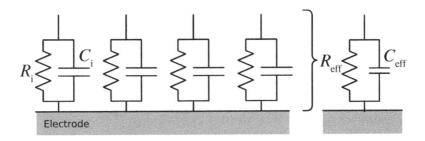

Figure 14.8: Schematic representation of a normal distribution of time constants in the absence of an ohmic resistance. This system yields neither frequency dispersion nor CPE behavior.

as its parallel resistance $R_{||}$ or capacitance. The relationship between CPE parameters and effective capacitance is obtained from the characteristic time constant associated with the admittance of the electrode. The admittance of the electrode can be expressed in terms of the CPE represented by equation (14.1) as

$$Y = \frac{1}{R_e} \left[1 - \frac{R_{||}}{R_e + R_{||}} \left(1 + \frac{R_e R_{||}}{R_e + R_{||}} Q(j\omega)^\alpha \right)^{-1} \right] \qquad (14.15)$$

Equation (14.15) can be expressed as

$$Y = \frac{1}{R_e} \left[1 - \frac{R_{||}}{R_e + R_{||}} \left(1 + (jK_{surf})^\alpha \right)^{-1} \right] \qquad (14.16)$$

where K_{surf} is a dimensionless frequency, expressed for an admittance as

$$K_{surf} = \left(\frac{R_e R_{||}}{R_e + R_{||}} Q \right)^{1/\alpha} \omega \qquad (14.17)$$

A characteristic frequency can be found for $K_{surf} = 1$ such that

$$f_{c,surf} = \frac{1}{2\pi \left(Q R_e R_{||} / (R_e + R_{||}) \right)^{1/\alpha}} \qquad (14.18)$$

This frequency depends on the ohmic and parallel resistances as well as the CPE parameters. When $R_e = 0$, the development yields the response of an ideal capacitance parallel to a resistance, and $f_{c,surf} \to \infty$.

The characteristic frequency can also be expressed in terms of an effective capacitance as

$$f_{c,surf} = \frac{1}{2\pi \left(\dfrac{R_e R_{||}}{R_e + R_{||}} C_{eff,surf} \right)} \qquad (14.19)$$

Equations (14.18) and (14.19) yield an expression for the effective capacitance as

$$C_{eff,surf} = Q^{1/\alpha} \left(\frac{R_e R_{||}}{R_e + R_{||}} \right)^{(1-\alpha)/\alpha} \qquad (14.20)$$

Equation (14.20) is equivalent to equation (20) derived by Brug et al.[149] for a surface distribution with a different definition of CPE parameters. In the limit that $R_{||}$ becomes much larger than R_e, equation (14.20) becomes

$$C_{\text{eff,surf}} = Q^{1/\alpha} R_e^{(1-\alpha)/\alpha} \qquad (14.21)$$

which is equivalent to equation (5) presented by Brug et al.[149] for a blocking electrode.

Equations (14.20), (14.21), and (14.14) have all the same form, but the resistance used in the calculations of C_{eff} is different in the three cases, being, respectively, the parallel combination of $R_{||}$ and R_e for equation (14.20), R_e for equation (14.21), and $R_{||}$ for equation (14.14).

14.4.4 Power-Law Distribution

A normal distribution of time constants is represented schematically in Figure 14.9. The proper distribution of time constants is envisioned to lead to a CPE. No parallel resistance is required as the resistance of the film is associated with the low-frequency asymptote for the summation of Voigt elements.

Hirschorn et al.[291,292] identified a relationship between CPE parameters and physical properties by regressing a measurement model[69,100] to synthetic CPE data. Following the procedure described by Agarwal et al.,[69,100] sequential Voigt elements were added to the model until the addition of an element did not improve the fit and one or more model parameters included zero within their 95.45 percent (2σ) confidence interval.

The concept was to identify the distribution of resistivity that, under the assumption that the dielectric constant is independent of position, would result in CPE behavior. The development is presented in detail in reference 291. The assumption of a uniform dielectric constant is not critical to the development summarized below. Musiani et al.[302] have shown that the results presented by Hirschorn et al.[291,292] apply, even when the assumption of a uniform dielectric constant is relaxed by allowing variation of ε in the region of low resistivity.

Example 14.2 Determination of Resistivity Distribution: *Show how regression of Voigt elements may be used to extract the resistivity distribution under the assumption that the resistivity varies smoothly from the metal-film interface to the film-electrolyte interface.*

Remember! 14.3 *For a surface distribution of time constants, the Brug formula given as equations (14.20) or (14.21) should be used to extract the double-layer capacitance.*

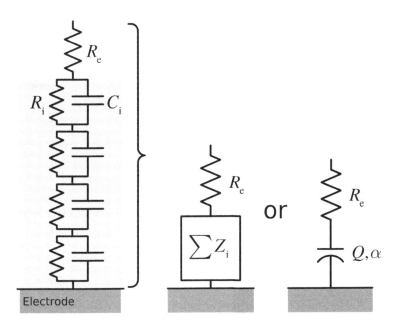

Figure 14.9: Schematic representation of a normal distribution of time constants leading to constant-phase behavior. The proper distribution of time constants is envisioned to lead to a CPE.

Solution: *The local capacitance shown in Figure 14.9 is related to local dielectric constant ϵ_i by*

$$C_i = \frac{\epsilon_i \epsilon_0}{d_i} \tag{14.22}$$

where ϵ_0 is the permittivity of vacuum and d_i is the thickness associated with element i. The local resistance can be expressed in terms of a local resistivity ρ_i as

$$R_i = \rho_i d_i \tag{14.23}$$

The time constant

$$\tau_i = \rho_i \epsilon_i \epsilon_0 \tag{14.24}$$

is independent of the element thickness. The variability of capacitance could be interpreted, for a uniform dielectric constant, as being a consequence of a changing element thickness. This interpretation has consequences for local resistivity. Following equations (14.23) and (14.24), the resistivity is given by

$$\rho_i = \frac{R_i}{d_i} = \frac{\tau_i}{\epsilon \epsilon_0} \tag{14.25}$$

Under the assumption that the resistivity has a smooth progression from one end of the film to the other, the resistivity values are sorted in descending order and the corresponding element thicknesses are added to yield the position. A resistivity distribution model can, therefore, be inferred from the regressed values for τ_i and C_i.

The resistivity associated with a CPE was found to follow a power-law profile, i.e.,

$$\frac{\rho}{\rho_\delta} = \zeta^{-\gamma} \tag{14.26}$$

where ζ is the dimensionless position $\zeta = y/\delta$, y represents the position through the depth of the film, and δ is the film thickness. The parameter ρ_δ is the resistivity at $\zeta = 1$, and γ is a constant indicating how sharply the resistivity varies. A distribution of resistivity that provides a bounded value for resistivity at $\zeta = 0$ was proposed to be

$$\frac{\rho}{\rho_\delta} = \left(\frac{\rho_\delta}{\rho_0} + \left(1 - \frac{\rho_\delta}{\rho_0} \right) \zeta^\gamma \right)^{-1} \tag{14.27}$$

where ρ_0 and ρ_δ are the boundary values of resistivity at the respective interfaces.

The impedance of the film can be written for an arbitrary resistivity distribution $\rho(y)$ as

$$Z_f(f) = \int_0^\delta \frac{\rho(y)}{1 + j\omega\varepsilon\varepsilon_0\rho(y)} dy \tag{14.28}$$

Equation (14.28) is similar to equation (12.9) on page 309 with the exception that ε is a constant in equation (14.28). A semi-analytic solution to equation (14.28) could be found for the resistivity profile given in equation (14.27) that applied under the conditions that $\rho_0 \gg \rho_\delta$ and $f < (2\pi\rho_\delta\varepsilon\varepsilon_0)^{-1}$. The solution was given as

$$Z_f(f) = g \frac{\delta\rho_\delta^{1/\gamma}}{(\rho_0^{-1} + j\omega\varepsilon\varepsilon_0)^{(\gamma-1)/\gamma}} \tag{14.29}$$

where g is a function of γ.

Equation (14.29) is in the form of the CPE for $f > (2\pi\rho_0\varepsilon\varepsilon_0)^{-1}$, i.e.,

$$Z_f(f) = g \frac{\delta\rho_\delta^{1/\gamma}}{(j\omega\varepsilon\varepsilon_0)^{(\gamma-1)/\gamma}} = \frac{1}{(j\omega)^\alpha Q} \tag{14.30}$$

Thus, for $(\rho_0\varepsilon\varepsilon_0)^{-1} < \omega < (\rho_\delta\varepsilon\varepsilon_0)^{-1}$, equation (14.29) yields the impedance given by the ohmic-resistance-compensated form of equation (14.2). Inspection of equation (14.30) suggests that

$$\alpha = \frac{\gamma - 1}{\gamma} \tag{14.31}$$

or $1/\gamma = 1 - \alpha$ where $\gamma \geq 2$ for $0.5 \leq \alpha \leq 1$. Numerical integration was used to develop the interpolation formula

$$g = 1 + 2.88(1 - \alpha)^{2.375} \tag{14.32}$$

A relationship among the CPE parameters Q and α and the dielectric constant ε, resistivity ρ_δ, and film thickness δ was found to be

$$Q = \frac{(\varepsilon\varepsilon_0)^\alpha}{g\delta\rho_\delta^{1-\alpha}} \tag{14.33}$$

Equations (14.6) and (14.33) yield an expression for the effective capacitance as

$$C_{\text{eff,PL}} = gQ \left(\rho_\delta \varepsilon \varepsilon_0\right)^{1-\alpha} \tag{14.34}$$

In addition to the CPE parameters Q and α, $C_{\text{eff,PL}}$ depends on the dielectric constant ε and the smaller value of the resistivity ρ_δ. Unlike the development for $C_{\text{eff,surf}}$ and $C_{\text{eff,norm}}$, the characteristic frequency is not invoked, and, in contrast with $C_{\text{eff,norm}}$, the results depend only on the high-frequency data.

Equation (14.34) can also be expressed as

$$C_{\text{eff,PL}} = (gQ)^{1/\alpha} \left(\delta \rho_\delta\right)^{(1-\alpha)/\alpha} \tag{14.35}$$

which shows a construction similar to equations (14.14), (14.20), and (14.21). The difference between the approaches based on the power-law and the characteristic frequency of the impedance is that the resistance $\delta \rho_\delta$ is related to the high-frequency portion of the spectrum; whereas, the resistance $R_\|$ in equation (14.14) is associated with the low-frequency part of the spectrum.

The power-law model development presented here suggests that the characteristic frequency of the impedance cannot give a correct value for the capacitance. An explanation may be seen from the hierarchy, presented in Figure 14.10, of impedance models based on the power-law model.[291, 292] The circuit presented in Figure 14.10(a), showing a resistance in parallel to the dielectric response of the film given by equation (14.29), represents a general model for the response of a film. If the parallel resistance is larger than the zero-frequency limit of equation (14.29), the model will be given by Figure 14.10(b), and the impedance response shown in Figure 14.10(c) will be asymmetric in a Nyquist format. If the parallel resistance is smaller than the zero-frequency limit of equation (14.29), the model will be given by Figure 14.10(d), and the impedance response shown in Figure 14.10(e) will be symmetric in a Nyquist format.

The demonstration given in Figure 14.10 is based on the power-law model for the dielectric response of a material. The applicability of the power-law model is supported by experimental data obtained for human skin and oxides on steel,[292, 134, 303] and by recent results on water uptake of coatings.[293]

If the impedance is symmetric, the resistance $R_\|$ in Figure 14.10(d) accounts for processes that are parallel to the dielectric response of the film. The effective capacitance of the dielectric material, in this case, is independent of the parallel resistance. If the impedance is asymmetric, as is seen for Figure 14.10(c), $Z_f(0)$ may be more closely related

---

Remember! 14.4 *For a normal distribution of time constants, the power-law model, e.g., equation (14.34), should be used to extract the double-layer capacitance.*

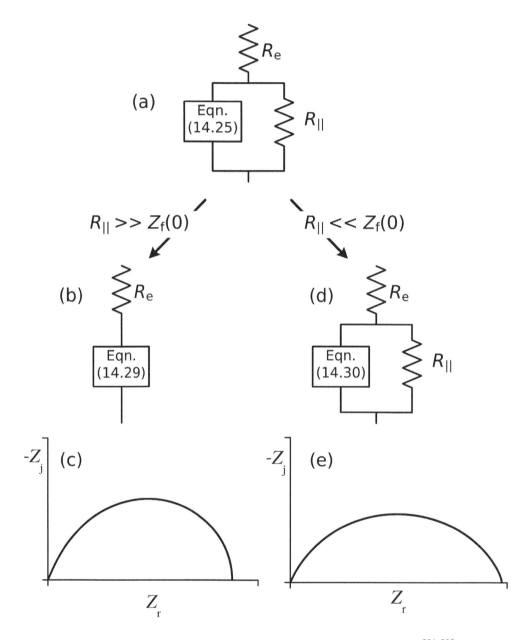

Figure 14.10: Hierarchy of impedance models based on the power-law model:[291, 292] a) circuit consisting of a resistance in parallel with the general power-law model expressed as equation (14.29); b) circuit resulting when $R_{||}$ is much larger than the zero-frequency limit of equation (14.29); c) Nyquist representation of the asymmetric impedance resulting from circuit b; d) circuit resulting when $R_{||}$ is much smaller than the zero-frequency limit of equation (14.29), thus allowing equation (14.29) to be replaced by equation (14.30); and e) Nyquist representation of the symmetric impedance resulting from circuit d. Taken from Orazem et al.[134]

Table 14.1: Approaches used to interpret CPE impedance response associated with normal or surface distributions of time constants.

Approach	Formula	Eqn.
Simple Substitution	$C_{\text{eff},Q} = Q$	(14.9)
Characteristic Frequency, Impedance	$C_{\text{eff,norm}} = Q^{1/\alpha} R_{\parallel}^{(1-\alpha)/\alpha}$	(14.14)
Characteristic Frequency, Admittance	$C_{\text{eff,surf}} = Q^{1/\alpha} \left(\dfrac{R_e R_{\parallel}}{R_e + R_{\parallel}} \right)^{(1-\alpha)/\alpha}$	(14.20)
Power-Law Normal Distribution	$C_{\text{eff,PL}} = g Q \left(\rho_\delta \epsilon \epsilon_0 \right)^{1-\alpha}$	(14.34)
	$C_{\text{eff,PL}} = (gQ)^{1/\alpha} \left(\delta \rho_\delta \right)^{(1-\alpha)/\alpha}$	(14.35)

to the dielectric properties, but a deeper analysis of the consequences of the power-law model shows that the capacitance is independent of this parameter. Introduction of equations (14.31) into equation (14.29) yields[292]

$$Z_f(f) = g \frac{\delta \rho_\delta^{(1-\alpha)}}{(\rho_0^{-1} + j\omega \varepsilon \varepsilon_0)^\alpha} \tag{14.36}$$

The asymptotic value of equation (14.36) as $f \to 0$ yields

$$Z_f(0) = g \delta \rho_\delta^{(1-\alpha)} \rho_0^\alpha \tag{14.37}$$

which may be used to obtain a value for ρ_0. The capacitance given by equation (14.34) is independent of ρ_0 and is, therefore, independent of $Z_f(0)$.

The formulae presented in the previous section are summarized in Table 14.1.

Example 14.3 Estimation of Film Thickness: *Orazem et al.[134] have reported (see Table 14.2) open-circuit impedance results for free-machining 18/8 stainless steel in an electrolyte consisting of 22 g/L boric acid with NaOH added to bring the pH to 7.2. Under the assumption that the dielectric constant is $\varepsilon = 12$, use the approaches described in Section 14.4 to estimate the film thickness.*

Solution:

a) Simple Substitution: *Entries are placed in Table 14.3 following equation (14.9). The thickness estimated by use of equation (14.6) is significantly smaller than that obtained by ex situ XPS analysis.*

b) Characteristic Frequency: Normal Distribution. *Equation (14.14) requires parameters Q, α, and $R_{\parallel}$ which are given in Table 14.2. The calculated capacitance is given in Table 14.3, and the resulting film thickness is significantly smaller than that obtained by XPS.*

Table 14.2: Regression results for free-machining 18/8 stainless steel in an electrolyte consisting of 22 g/L boric acid with NaOH added to bring the pH to 7.2.[134]

	As Received	After Proprietary Treatment
R_e, Ωcm^2	15.3	13.3
$R_{||}$, MΩcm^2	2.33	16.8
α	0.91	0.91
Q, μF/s$^{(1-\alpha)}$cm^2	11	30.5
δ by XPS, nm	6.3	2.5

Table 14.3: Estimations of oxide thickness for free-machining 18/8 stainless steel in an electrolyte consisting of 22 g/L boric acid with NaOH added to bring the pH to 7.2. The oxide thickness measured by XPS is given in Table 14.2.

	As Received		After Treatment	
Approach	C_{eff}, μF/cm^2	δ, nm (% error)	C_{eff}, μF/cm^2	δ, nm (% error)
Eqn. (14.9)	11	0.97 (-85%)	30.5	0.35 (-86%)
Eqn. (14.14)	15	0.72 (-89%)	59.0	0.18 (-93%)
Eqn. (14.20)	4.9	2.2 (-65%)	13.3	0.80 (-68%)
Eqn. (14.34)	1.8	5.9 (-6%)	4.0	2.7 (+8%)

c) Characteristic Frequency: Surface Distribution. *Equation (14.20) requires an ohmic resistance. In the case of the data presented in Table 14.2, $R_{\parallel} \gg R_e$, so the calculated capacitance depends only on the CPE parameters and the ohmic resistance. The resulting capacitance is given in Table 14.3. While the film thickness calculated from the capacitance is significantly smaller than that obtained by XPS, the value is more reasonable than that obtained by use of equations (14.9) or (14.14).*

d) Power-Law Distribution. *Equation (14.34) requires, in addition to CPE parameters Q and α, a value for the resistivity at the film-electrolyte interface, ρ_δ. The best approach is to calibrate for a given film. For example, Hirshorn et al.[292] reported a calibration of impedance measurements for a similar chromium-rich steel for which oxide-film thickness (obtained by XPS) and CPE parameters were available.[288] From this work, $\rho_\delta \approx 500 \ \Omega cm$. The corresponding values of capacitance are provided in Table 14.3. The film thickness calculated from the capacitance is in excellent agreement with the values obtained by XPS.*

While the power-law model provides an excellent means of interpreting CPE parameters in terms of the properties of films showing normal distribution of time constants, equation (14.34) requires a value for ρ_δ which is not generally known. Orazem et al.[303] discuss ways for bounding the value of the parameter, for calibration, and for comparative analysis in which the unknown parameter may be eliminated.

Bounds for Resistivity

The impedance response of the power law model exhibits capacitive behavior above a characteristic frequency

$$f_\delta = \frac{1}{2\pi\rho_\delta\varepsilon\varepsilon_0} \tag{14.38}$$

Under conditions that the value of ρ_δ is unknown for data showing high-frequency CPE behavior, an upper bound on its value can be defined because the characteristic frequency f_δ must be larger than the largest measured frequency f_{max}. Thus, a maximum value of ρ_δ can be obtained as

$$\rho_{\delta,max} = \frac{1}{2\pi\varepsilon\varepsilon_0 f_{max}} \tag{14.39}$$

A lower bound may be estimated on physical grounds.

Comparative Analysis

When a reliable value for ρ_δ is unavailable, useful information may still be obtained by taking ratios. For example, the ratio of film thickness for films 1 and 2 may be obtained from equation (14.33) to be

$$\frac{\delta_1}{\delta_2} = \frac{(\varepsilon_1\varepsilon_0)^{\alpha_1}}{(\varepsilon_2\varepsilon_0)^{\alpha_2}} \frac{g_2 Q_2 \rho_{\delta,2}^{(1-\alpha_2)}}{g_1 Q_1 \rho_{\delta,1}^{(1-\alpha_1)}} \tag{14.40}$$

If film properties ε and ρ_δ may, respectively, be assumed equal for the two samples,

$$\frac{\delta_1}{\delta_2} = \frac{g_2 Q_2}{g_1 Q_1} (\varepsilon \varepsilon_0 \rho_\delta)^{\alpha_1 - \alpha_2} \tag{14.41}$$

If $\alpha_1 = \alpha_2$, then $g_1 = g_2$, and values for ε and ρ_δ are not required. Thus,

$$\frac{\delta_1}{\delta_2} = \frac{Q_2}{Q_1} \tag{14.42}$$

For the results shown in Table 14.2, $\alpha_1 = \alpha_2$. In agreement with the analysis reported in Example 14.3, the ratio of film thicknesses obtained by XPS $\delta_1/\delta_2 = 2.52$ is approximately equal to the inverse ratio of CPE coefficient, i.e., $Q_2/Q_1 = 2.77$.

A similar approach is available when the dielectric constant is the desired quantity. Under the assumption that α, δ, and ρ_δ are the same for two samples 1 and 2,

$$\frac{\varepsilon_1}{\varepsilon_2} = \left(\frac{Q_1}{Q_2}\right)^{1/\alpha} \tag{14.43}$$

Equation (14.43) is very sensitive to differences in α because α and Q are highly correlated in the regression analysis. Thus, equation (14.43) should not be used when $\alpha_1 \neq \alpha_2$.

Example 14.4 Error by Assuming $C = Q$ for a Film: *Identify the error associated with assuming that $C = Q$ when the CPE behavior may be attributed to the properties of a dielectric film.*

Solution: *The results presented here show that, for normal distributions, the power-law model provides the most accurate assessment of CPE parameters in terms of physical properties. Thus, for a normal distribution, an appropriate figure of merit would be*

$$\frac{C_{\text{eff,PL}}}{C_{\text{eff,Q}}} = g (\rho_\delta \varepsilon \varepsilon_0)^{1-\alpha} \tag{14.44}$$

Equation (14.44) shows that the accuracy associated with setting $C = Q$ is not simply a function of α, but depends as well on other physical properties of the system under study.

The ratio $C_{\text{eff,PL}}/C_{\text{eff,Q}}$ is presented in Figure 14.11(a) as a function of α with typical values of $\rho_\delta \varepsilon \varepsilon_0$ as a parameter. The corresponding plot of $1 - C_{\text{eff,PL}}/C_{\text{eff,Q}}$ is presented in Figure 14.11(b) as a function of $1 - \alpha$. Figure 14.11(b) emphasizes the behavior near $\alpha = 1$. For the oxides films treated in the present work, $\rho_\delta \varepsilon \varepsilon_0$ has a value of 5×10^{-10} s. The error in using $C_{\text{eff,Q}}$ is 23 percent at $\alpha = 0.99$ and 100 percent at $\alpha = 0.97$. Thus, the assumption that $C = Q$ should not be used for normal distributions.

Example 14.5 Error by Assuming $C = Q$ for a Surface: *Identify the error associated with assuming that $C = Q$ when the CPE behavior may be attributed to distributed properties on the electrode surface.*

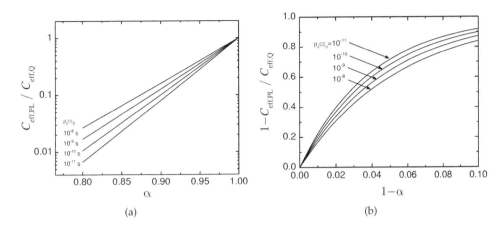

Figure 14.11: Correction factor, applicable for a normal distribution of time constants, for the assumption that $C = Q$: a) $C_{\text{eff,PL}}/C_{\text{eff,Q}}$ as a function of α with $\rho_\delta \epsilon \epsilon_0$ as a parameter and b) $1 - C_{\text{eff,PL}}/C_{\text{eff,Q}}$ as a function of $1 - \alpha$ with $\rho_\delta \epsilon \epsilon_0$ as a parameter.

Solution: *In previous work, Hirschorn et al.[135] showed that the model developed by Brug et al.[149] provided a good interpretation for cases where surface time constant distributions gave rise to CPE behavior. Thus, for a surface distribution, an appropriate figure of merit would be*

$$\frac{C_{\text{eff,surf}}}{C_{\text{eff,Q}}} = \left(Q \frac{R_e R_\parallel}{R_e + R_\parallel} \right)^{(1-\alpha)/\alpha} \tag{14.45}$$

Again, the accuracy associated with setting $C = Q$ is not simply a function of α, but depends as well on other physical properties of the system under study.

The factor $C_{\text{eff,surf}}/C_{\text{eff,Q}}$ is presented in Figure 14.12(a) as a function of α with the quantity $QR_e R_\parallel/(R_e + R_\parallel)$ as a parameter. To emphasize the behavior near $\alpha = 1$, the quantity $1 - C_{\text{eff,surf}}/C_{\text{eff,Q}}$ is presented in Figure 14.12(b) as a function of $1 - \alpha$. For $R_e = 10 \ \Omega cm^2$, $Q = 10^{-5} \ Fs^{\alpha-1}/cm^2$, and $R_\parallel \gg R_e$, the parameter $QR_e R_\parallel/(R_e + R_\parallel)$ has a value near $10^{-4} \ s^\alpha$. For $QR_e R_\parallel/(R_e + R_\parallel) = 10^{-4} \ s^\alpha$ and at a value $\alpha = 0.93$, $C_{\text{eff,Q}}$ will be a factor of 2 larger than $C_{\text{eff,surf}}$. The error will be near 10 percent at $\alpha = 0.99$. A comparison of Figures 14.11(b) and 14.12(b) shows that the error in assuming $C = Q$ for surface distributions is smaller at a given value of α than it is for normal distributions. However, the error is sufficiently large that the assumption that $C = Q$ is not recommended for surface distributions.

14.5 Limitations to the Use of the CPE

As compared to a parallel combination of a resistor and capacitor, the CPE is often able to provide a much better fit to most impedance data. The CPE can achieve this fit using only three parameters, which is only one parameter more than a typical *RC* couple. Some investigators allow α to take values from -1 to 1, thus treating the CPE as an

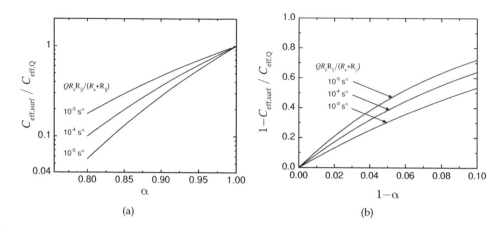

(a)   (b)

Figure 14.12: Correction factor, applicable for a surface distribution of time constants, for the assumption that $C = Q$: a) $C_{eff,surf}/C_{eff,Q}$ as a function of α with $QR_eR_{||}/(R_e + R_{||})$ as a parameter and b) $1 - C_{eff,surf}/C_{eff,Q}$ as a function of $1 - \alpha$ with $QR_eR_{||}/(R_e + R_{||})$ as a parameter.

extremely flexible fitting element. For $\alpha = 1$, the CPE behaves as a capacitor; for $\alpha = 0$, the CPE behaves as a resistor; and for $\alpha = -1$ the CPE behaves as an inductor (see Section 4.1.1 on page 79).

It must be emphasized that the mathematical simplicity of equations (14.1) and (14.2) is the consequence of a specific time-constant distribution. As shown in Chapter 13 (page 319), time-constant distributions can result from nonuniform mass transfer, geometry-induced nonuniform current and potential distributions, electrode porosity, and distributed properties of films. At first glance, the associated impedance responses may appear to have a CPE behavior, but, as shown in Example 14.1, the frequency dependence of the ohmic-resistance-corrected phase angle shows that the time-constant distribution differs from that of a CPE.

There are, therefore, two primary concerns with the use of the CPE for modeling impedance data:

1. While assuming that the time constant is distributed can be better than assuming that the time constant has a single value, the physical system may not follow the specific distribution implied in equations (14.1) and (14.2). The examples presented in Section 14.2 and Chapter 13 (page 319) illustrate systems for which a time-constant dispersion results that resembles that of a CPE, but with different distributions of time constants.

2. A satisfactory fit of a CPE to experimental data may not necessarily be correlated to the physical processes that govern the system. As shown in Section 4.4 on page 84, models for impedance are not unique; thus, an excellent fit to the data does not in itself guarantee that the model describes correctly the physics of a given system.

The graphical methods described in Chapter 18 (page 493) can be used to determine whether a system follows CPE behavior in a given frequency range.

Once a CPE behavior is identified, further interpretation in terms of system properties requires an assessment of the nature of the associated time-constant distribution. As discussed in Section 14.4.4, CPE behavior may arise from a variation of properties in the direction that is normal to the electrode surface. Such variability may be attributed, for example, to changes in the conductivity of oxide layers. For such systems, the power-law model described in Section 14.4.4 provides a useful means of extracting film properties from impedance data. The Young model,[278,304,305] which applies for an exponential distribution of resistivity and is described in Section 13.7, does not yield CPE behavior.

A CPE may arise as well from a distribution of properties along the surface of an electrode (see Section 14.4.3). As discussed in Section 13.3 (page 349), both roughness on a scale associated with a poorly polished electrode and distributions of capacitance will yield frequency dispersion, but at frequencies higher than that associated with the disk electrode geometry. Distributions of charge-transfer resistance associated with a single electrochemical reaction do not lead to frequency dispersion, but distributions of kinetic properties for reactions coupled by adsorbed intermediates will give rise to low-frequency dispersions. Thus, a surface distribution of kinetic properties for coupled reactions may account for CPE behavior. Jorcin et al.[277] have shown that local impedance measurements may be used to differentiate between CPE behavior associated with normal and surface distributions of properties.

Remember! 14.5 *While use of a CPE may lead to improved regressions, the meaning can be ambiguous, and the physical system may not follow the specific distribution implied in the CPE model.*

Problems

14.1   Why does the following represent a Constant-Phase Elephant?

14.2   With the aid of Table 4.1 on page 77 and under the assumption that Q has units of $F/s^{(1-\alpha)}cm^2$ and resistances have units of Ωcm^2, identify the units of capacitance given in the following:

  (a)   Equation (14.9)
  (b)   Equation (14.14)
  (c)   Equation (14.20)
  (d)   Equation (14.34)
  (e)   Equation (14.35)

14.3   Time-constant distributions leading to CPE behavior were described in Section 14.3 as being in the direction normal to the electrode surface or along the electrode surface.

  (a)   Under what conditions could a system show CPE behavior resulting from a distribution in only the direction normal to the electrode surface?
  (b)   Under what conditions could a system show CPE behavior resulting from a distribution along the electrode surface?

14.4   Starting from equation (14.10), verify equation (14.14).

14.5   Starting from equation (14.15), verify equation (14.20).

14.6   Starting from equation (14.29), verify equations (14.34) and (14.35).

14.7   Following the development of the power-law model in Section 14.4.4, calculate the film thickness for the data presented in Table 14.2.

  (a)   Assume that $\varepsilon = 12$ and that $\rho_\delta = 500\ \Omega cm$.
  (b)   Assume that $\varepsilon = 20$ and that $\rho_\delta = 500\ \Omega cm$.
  (c)   Assume that $\varepsilon = 12$ and that $\rho_\delta = 10,000\ \Omega cm$.

14.8   For the data presented in Table 14.2, use the Brug formula developed in Section 14.4.3 to estimate the film thickness.

14.9   For the data presented in Table 14.2, estimate the error in capacitance obtained if the film capacitance is assigned the numerical value for Q as discussed in Section 14.4.1.

14.10 Open-circuit electrochemical impedance measurements were performed on 0.25 mm diameter gold and platinum disk electrodes in deionized water. The frequency range was 54.5 Hz to 46.5 mHz. The resulting regression parameters are

Table 14.4: Regression results for the data collected on gold and platinum electrodes in deionized water.[306]

Parameter	Au	Pt
R_e, kΩcm^2	5.85±0.07	4.93±0.05
Q, μF/s$^{(1-\alpha)}$cm^2	27.6±0.3	27.6±0.3
α	0.66±0.01	0.69±0.01
$R_{\|\|}$, kΩcm^2	790±210	750±170

provided in Table 14.4.

(a)   Use the Brug formula developed in Section 14.4.3 to estimate the capacitance.

(b)   Following the development in Section 13.2 (page 337), determine the characteristic frequency at which time-constant dispersion may be expected. Comment on the relevance of the value of characteristic frequency.

Chapter 15

Generalized Transfer Functions

Electrochemical measurements are generally designed either to analyze an interfacial mechanism by kinetic characterization and chemical identification of the reaction intermediates or to estimate a parameter characteristic of some process (i.e., corrosion rate, deposition rate, and state of charge of a battery) from the measurement of a well-defined quantity.

Electrical techniques are extremely efficient for disentangling the coupling between mass-transport and chemical and electrochemical reactions or for performing a test, because they allow in situ study of the electrochemical system. The techniques placed at the electrochemist's disposal are founded on an application of signal processing to electrochemistry. By using a small-amplitude sine-wave perturbation, electrochemical systems can be considered to be linear, and they can be investigated on the basis of a frequency analysis of a transfer function involving at least one electrical quantity (current or potential). So far, most significant results have been obtained by measurements of the electrochemical impedance, which leads to kinetic characterization of the phenomena in terms of process rates (mass transport, electrochemical, or chemical reaction). More recently, use of nonelectrical quantities has been introduced in impedance spectroscopy, which complements those obtained by measuring the electrochemical impedance.

The object of this chapter is to provide a framework for the variety of electrical and nonelectrical impedance techniques that have emerged and to provide a consistent set of notations that can be used to describe the results of these measurements. The presentation follows the treatment of Gabrielli and Tribollet.[65]

15.1 Multi-input/Multi-output Systems

During the last 40 years, the measurement of the impedance of an electrode has become a technique widely used for investigating numerous interfacial processes. The interpretation of this quantity is based on models obtained from the equations governing the

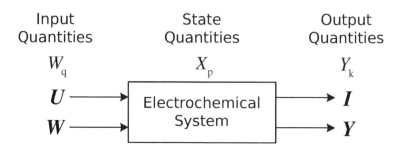

Figure 15.1: Schematic representation of an electrochemical system with input, output, and state variables. (Taken from Gabrielli and Tribollet.[65])

coupled transport and kinetic processes, which may include heterogeneous and/or homogeneous reaction steps. Although these models are able to explain many complex behaviors, they can be questioned so long as they are not supported by additional experimental evidence.

Unfortunately, in situ analytical identification of reaction intermediates or of surface layers remains extremely difficult. Therefore, it is convenient to complement the impedance measurement, which relates only electrical quantities, with the measurement of complex transfer functions of different natures, which involve other types of quantities. In fact, the modeling of these new quantities is carried out on the same basis as those that govern the electrochemical impedance. As an example, the ac impedance measured on the diffusion plateau of the anodic dissolution of metal is often very difficult to analyze because of the close coupling between kinetics and mass transport. By perturbing the rotation speed of the electrode and measuring the current, the electrohydrodynamic (EHD) impedance is obtained (see Chapter 16 on page 443). EHD impedance reveals directly the influence of mass transport on surface phenomena and thereby gives additional information on the process. The model that expresses the coupling between surface kinetics and mass transport can be tested against both the EHD impedance and the electrochemical impedance.

For any electrochemical system, the state of the electrochemical interface is defined by three classes of quantities that are shown in Figure 15.1:

1. The input quantities or constraints W_q impose the experimental conditions. These may include the electrode potential U (or overall current I), the temperature T, the pressure p, the rotation velocity of the electrode Ω, and the magnetic field B.

Remember! 15.1 *Measurement of different transfer functions provides a convenient in situ method to probe electrochemical systems.*

---

2. The values of state quantities X_p fix the state of the system. These may include the concentration of the reacting species c_i, the coverage of the electrode by the intermediate species γ_i, and the local interfacial potential $V(x, y)$.

3. The output quantities Y_k allow observation of the state of the system. These may include the current I (or potential U), the reflective power of the electrode surface, the mass added or removed from the electrode M, and the current flowing through a secondary electrode (often a concentric ring to a disk electrode) I_{ring}.

Each output quantity Y_k can be considered to be a function, which is often very complicated, of the constraints or input quantities.

$$Y_k = G(W_q) \qquad \text{(for q and k} = 1, 2 \ldots) \tag{15.1}$$

Equation (15.1) may be called the input–output relationship. In order to write equation (15.1), different steps involving the evolution equation of the state quantities generally have to be considered, i.e.,

$$\frac{dX_p}{dt} = H_p(X_\ell, W_q) \qquad \text{(for p, } \ell, \text{ and q} = 1, 2 \ldots) \tag{15.2}$$

$$Y_k = F_k(X_p, W_q) \qquad \text{(for k, p, and q} = 1, 2 \ldots) \tag{15.3}$$

The set of equations (15.2) governs the time change of the state quantities. The set of equations (15.3) allows the output quantities to be obtained. Equations (15.3) can be called the observation equations.

   In general, the functions H_p and F_k are nonlinear. These nonlinearities are usually due to the exponential activation of the electrochemical rate constants by the potential (see Section 5.4). In addition, even for time-invariant electrochemical systems, equations (15.2) can comprise either differential equations, when only kinetic equations are considered to be involved at the interface, or partial differential equations, when distributed processes occur in the bulk of the solution (such as may result from transport of the reacting species or a temperature gradient in the solution).

   A general time-dependent solution of the set of differential equations given by equations (15.2) and (15.3) is usually impossible or very difficult to obtain. This general solution corresponds, for example, to cyclic voltammetry or step responses. The steady-state solution of this set of equations obtained for $dX_p/dt = 0$ can be derived more easily, and then, if a small perturbation of the input quantities $dW_q(t)$ about the steady-state value $\overline{W}_q$ is imposed, a small change of the output quantities $dY_k(t)$ about the steady value $\overline{Y}_k$ is observed. A linearization of equations (15.2) and (15.3), for example, by neglecting the expansions above the first order, gives an approximate description of the real system in the time domain

$$\frac{dX}{dt} = \mathbf{A}dX + \mathbf{B}dW \tag{15.4}$$

$$dY = \mathbf{C}dX + \mathbf{D}dW \tag{15.5}$$

where $\mathbf{A}$, $\mathbf{B}$, $\mathbf{C}$, and $\mathbf{D}$ are matrixes whose components are equal to

$$A_{pq} = \left.\frac{\partial H_p}{\partial X_q}\right|_{X_{k,k\neq q},W_m} \tag{15.6}$$

$$B_{pq} = \left.\frac{\partial H_p}{\partial W_q}\right|_{X_k,W_{m,m\neq q}} \tag{15.7}$$

$$C_{kq} = \left.\frac{\partial F_k}{\partial X_q}\right|_{X_{k,k\neq q},W_m} \tag{15.8}$$

and

$$D_{kq} = \left.\frac{\partial F_k}{\partial W_q}\right|_{X_k,W_{m,m\neq q}} \tag{15.9}$$

respectively, and dX, dY, and dW are column vectors whose components are dX_p, dY_k, and dW_m, respectively.

When a small-amplitude sine wave perturbation is added to one input quantity, under linear conditions, each state quantity and each output quantity can be written formally for the quantity of interest χ as

$$\chi = \bar{\chi} + \mathrm{Re}\left\{\tilde{\chi}(\omega)\exp(j\omega t)\right\} \tag{15.10}$$

where $\mathrm{Re}\{\chi\}$ represents the real part of χ, and $\tilde{\chi}(\omega)$ is generally a complex quantity independent of the time but depending on the frequency, i.e.,

$$\tilde{\chi}(\omega) = |\tilde{\chi}(\omega)|\exp(j\varphi) \tag{15.11}$$

where φ is the phase shift (see Example 1.9 on page 20). With the previous notation, the perturbation of the quantity χ in the time domain, $d\chi(t)$, is related to the perturbation of the same quantity in the frequency domain, $\tilde{\chi}(\omega)$, by

$$d\chi(t) = \mathrm{Re}\{\tilde{\chi}(\omega)\exp(j\omega t)\} \tag{15.12}$$

Equations (15.6), (15.7), (15.8), and (15.9), which describe the system in the time domain, become, in the frequency domain,

$$j\omega\tilde{X} = \mathbf{A}\tilde{X} + \mathbf{B}\tilde{W} \tag{15.13}$$

and

$$\tilde{Y} = \mathbf{C}\tilde{X} + \mathbf{D}\tilde{W} \tag{15.14}$$

where $\tilde{X}$, $\tilde{Y}$, and $\tilde{W}$ are column vectors, with components $\tilde{X}_p$, $\tilde{Y}_k$, and $\tilde{W}_m$.

By eliminating $\tilde{X}$, the output quantity $\tilde{Y}$ is obtained as a function of the input quantity $\tilde{W}$

$$\tilde{Y} = (\mathbf{C}(j\omega\mathbf{J} - \mathbf{A})^{-1}\mathbf{B} + \mathbf{D})\tilde{W} \tag{15.15}$$

The matrix $(\mathbf{C}(j\omega\mathbf{J} - \mathbf{A})^{-1}\mathbf{B} + \mathbf{D})$ is the generalized transfer function of the electrochemical interface considered as a multi-input W_q/multi-output Y_k system. Each term of the matrix is an elementary transfer function and $\mathbf{J}$ is the identity matrix. The transfer function may be analyzed as a function of the static property space, which represents a linearized characterization of the system. The same information is obtained as would be obtained by analyzing the entire nonlinear electrochemical system, which is much more complex. As an example, for the electrical quantities the measurement of the impedance about each polarization point of the steady-state current-voltage curve leads to an exhaustive analysis of the electrochemical interface. No information is lost as compared to a large-amplitude perturbation technique such as cyclic voltammetry. It is equivalent to carrying out several impedance measurements all along the current-voltage curve or cyclic voltammetry at different sweep rates, but the mathematical analysis of the impedance measurements is much simpler due to the linearity of the equations.

The kth row of the matrix that gives the transfer function presented as equation (15.15) is

$$\tilde{Y}_k = \sum_P Z_{q,k} \, \tilde{W}_q \tag{15.16}$$

where $Z_{q,k}$ is the elementary transfer function between Y_k and W_q. Each elementary transfer function is generally a complex quantity, but, when the frequency tends toward zero, $Z_{q,k}$ tends toward $\partial Y_k / \partial W_q|_{W_p}$, which is the partial derivative of the steady-state solution $\overline{Y}_k$ with respect to $\overline{W}_q$ (e.g., polarization resistance $\partial U / \partial I|_{W_p}$ for the electrical quantities).

The proliferation of different and conflicting sets of notation for different impedance techniques makes necessary a unified approach for describing the transfer function resulting from all impedance measurements. A unified notation is presented in the following section for cases where the electrical properties (current or potential) are the measured output quantity and where the input forcing function is nonelectrical. A subsequent section addresses cases for which the input forcing function is electrical.

15.1.1 Current or Potential Are the Output Quantity

As I and U play a particular role in an electrochemical system, and both can be considered to be either input or output quantities, according to the type of regulation of

---

Remember! 15.2 *The small-signal transfer function may be analyzed at different values of the static property space, which represents a linearized characterization of the system. The same information is obtained as would be obtained by analyzing the entire nonlinear electrochemical system using a large-amplitude perturbation technique such as cyclic voltammetry, but the analysis is simpler.*

---

the polarization (potentiostat or galvanostat), $\widetilde{I}$ and $\widetilde{U}$ appear in either the input or the output vector columns. According to equation (15.16)

$$\widetilde{I} = Z^{-1}\widetilde{U} + \sum_m Z_{I,m}\widetilde{W}_m \tag{15.17}$$

where $Z^{-1}\widetilde{U}$ is an element of the summation shown in equation (15.16). Similarly,

$$\widetilde{U} = Z\widetilde{I} + \sum_m Z_{U,m}\widetilde{W}_m \tag{15.18}$$

In equations (15.17) and (15.18), Z is the usual electrochemical impedance, $Z_{I,m}$ is the elementary transfer function between the current response to the perturbation of the mth input quantity, and $Z_{U,m}$ is the elementary transfer function between the overvoltage response to the perturbation of the mth input quantity.

Equations (15.17) and (15.18) are simultaneously satisfied; therefore, by replacing $\widetilde{U}$ in equation (15.17) by its value given by equation (15.18),

$$\widetilde{I} = Z^{-1}\left(Z\widetilde{I} + \sum_m Z_{U,m}\widetilde{W}_m\right) + \sum_m Z_{I,m}\widetilde{W}_m \tag{15.19}$$

i.e.,

$$\sum_m (Z_{U,m}Z^{-1} + Z_{I,m})\widetilde{W}_m = 0 \tag{15.20}$$

Thus, for any input quantity m,

$$Z_{U,m}\,Z^{-1} + Z_{I,m} = 0 \tag{15.21}$$

Therefore, the elementary transfer functions that give the response of current and potential to the perturbation of the mth input quantity are related by the electrochemical impedance. For a given experiment, only one input quantity is sinusoidally modulated around a mean value, and all others are maintained constant by different regulations.

The notation $\widetilde{Y}_k/\widetilde{W}_m$ means that all input quantities W_p (with $p \neq m$) are fixed. Thus, equation (15.21) can be written as

$$\frac{\widetilde{U}}{\widetilde{W}_m} = -\frac{\widetilde{U}}{\widetilde{I}}\frac{\widetilde{I}}{\widetilde{W}_m} \tag{15.22}$$

According to this notation, $(\widetilde{U}/\widetilde{W}_m)$ is obtained under galvanostatic regulation (i.e., $\widetilde{I} = 0$), $\widetilde{U}/\widetilde{I}$ is the usual electrochemical impedance Z obtained for $\widetilde{W}_m = 0$, and $(\widetilde{I}/\widetilde{W}_m)$ is obtained in potentiostatic regulation ($\widetilde{U} = 0$). Equation (15.22) has been given previously in the particular cases of the speed modulation of a rotating disk electrode,[307] of a magnetic field modulation,[166] and for a temperature modulation.[308]

The experimental arrangement devised for the transfer-function measurements involving electrical input and output quantities is illustrated in Figure 15.2. The transfer-function analyzer generates the perturbing signal $dE(t)$ and the control device of the

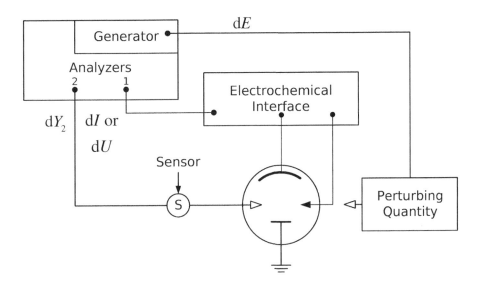

Figure 15.2: The experimental arrangement devised for the transfer-function measurements involving nonelectrical input quantity and electrical output quantity.

electrical quantity delivers the responses $dI(t)$ or $dU(t)$ of the current or potential to channel 1 of the analyzer. As shown in Figure 15.2, a sensor adapted to the observed output quantity Y_2 allows measurement of the response $dY_2(t)$ of this output quantity. The analyzer measures the transfer function between Y_2 and U ($\widetilde{Y}_2/\widetilde{U}$) or between Y_2 and I, i.e., $(\widetilde{Y}_2/\widetilde{I})$.

15.1.2 Current or Potential Are the Input Quantity

For any output k, if a galvanostatic regulation is considered, equation (15.16) can be written as

$$\widetilde{Y}_k = Z_{k,p}\widetilde{I} + \sum_m Z_{k,m}\widetilde{W}_m \tag{15.23}$$

where $Z_{k,p}$ is the elementary transfer function between the response of the kth output quantity and the perturbation of the current and $Z_{k,m}$ is the elementary transfer function between the response of the kth output and the perturbation of the mth input quantity.

If only the input I is modulated, the output Y_k and U can be written

$$\widetilde{Y}_k = Z_{k,I}\widetilde{I} \tag{15.24}$$

$$\widetilde{U} = Z\widetilde{I} \tag{15.25}$$

Then

$$\frac{\widetilde{Y}_k}{\widetilde{U}} = \frac{Z_{k,I}}{Z} = \frac{\widetilde{Y}_k}{\widetilde{I}}\frac{\widetilde{I}}{\widetilde{U}} \tag{15.26}$$

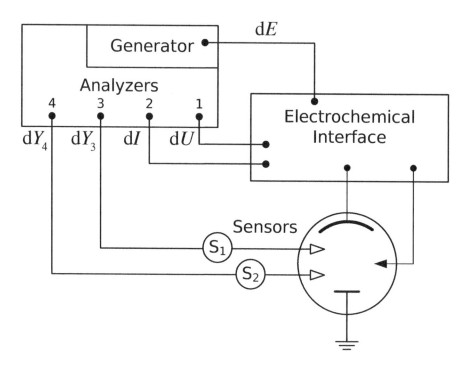

Figure 15.3: Principles of the experimental arrangement used for transfer-function measurements where current or potential is the input quantity.

where $\tilde{U}/\tilde{I}$ is the usual electrochemical impedance Z, $\tilde{Y}_k/\tilde{I}$ is the transfer function between Y_k and I, and $\tilde{Y}_k/\tilde{U}$ is the transfer function between Y_k and U. Note that, in the derivation of equation (15.22), current and potential are both considered to be the output; whereas, in the derivation of equation (15.26), potential was treated as an input, resulting in the change of sign. Thus, the only difference between equation (15.22) and equation (15.26) is a minus sign coming from the implicit function derivation.

The experimental arrangement devised for the transfer-function measurements involving a nonelectrical output quantity, the current or the potential being the input quantity is illustrated in Figure 15.3. The transfer-function analyzer generates the perturbing signal $dE(t)$ and the control device of the electrical quantity delivers the responses $dI(t)$ and $dU(t)$ of the current and potential to channels 1 and 2 of the transfer-

---

🐘 **Remember! 15.3** *Equation (15.22) provides the relationship between the usual electrochemical impedance response and the transfer function for cases where current or potential is the output. Equation (15.26) provides the corresponding relationship for cases where current or potential is the input. Equation (15.22) and equation (15.26) are distinguished by a minus sign resulting from the implicit function derivation.*

---

function analyzer. A sensor adapted to the observed output quantity Y_3 allows the response $dY_3(t)$ of this output quantity (see Figure 15.3). The transfer-function analyzer simultaneously measures the electrochemical impedance $\widetilde{U}/\widetilde{I}$ and the transfer function between Y_3 and U given as $\widetilde{Y}_3/\widetilde{U}$ and between Y_3 and I given as $\widetilde{Y}_3/\widetilde{I}$. If required, a second sensor, also shown in Figure 15.3, can measure the response $dY_4(t)$ of a second output quantity. This last feature makes this multi-transfer-function technique very flexible. The simultaneous measurement of the electrochemical impedance $\widetilde{U}/\widetilde{I}$ and of the transfer function $\widetilde{Y}_3/\widetilde{I}$ requires the use of a four-channel transfer-function analyzer.

15.1.3   Experimental Quantities

It is necessary to notice that the interfacial potential V and the faradaic current I_F are state quantities generally related to the observable experimental quantities U and I by

$$U = V + R_e I \tag{15.27}$$

and

$$I = I_F + C_{dl}\frac{dV}{dt} \tag{15.28}$$

where R_e is the electrolyte resistance and C_{dl} is the double-layer capacitance.

From a modeling point of view, the kinetic equations and the evolution equations generally involve I_F and V; thus, in a first step, it is more convenient to write the different equations by considering I_F and V to be the observable quantities (input or output) and, at the end, to use equations (15.27) and (15.28) to write the final relation with I and U.

15.2   Transfer Functions Involving Exclusively Electrical Quantities

The examples of transfer functions presented here involve exclusively electrical quantities but are distinct from the usual impedance measurements described in other parts of this book.

15.2.1   Ring–Disk Impedance Measurements

In the late sixties, Albery et al.[309] investigated extensively the rotating ring–disk electrode (rrde). The extension to the dynamic regime can, by studying the species collected on the ring, yield information on changes with time and potential of the number of species trapped on the disk surface during the electrode process. This approach was extended to the study of the collection efficiency response to a sine wave perturbation to the disk potential.[310]

The principle of an rrde is the following:

1. The species A transforms electrochemically into B at the disk, e.g.,

$$A \rightarrow B + n_{disk}e^- \tag{15.29}$$

where the charges carried by A and B are such that reaction (15.29) is electrically balanced.

2. The species B leaves the disk and is transported by convective diffusion to the ring.

3. The species B reacts at the ring and is transformed into P, e.g.,

$$B \rightarrow P + n_{\text{ring}} e^- \tag{15.30}$$

The product P may be identical to A, the initial species, as in the case of a redox system, or may be different.

The number of electrons concerned in reaction (15.29) or (15.30), i.e., n_{disk} or n_{ring}, is positive for an anodic and negative for a cathodic process. It can be seen that the rrde operates with three distinct processes: electrochemical reaction at the disk, coupling between the disk and the ring through the convective diffusion, and collection at the ring.

At the steady state, the collection efficiency is defined to be

$$\overline{\mathcal{N}} = \frac{\overline{i}_{\text{ring}}}{\overline{i}_{\text{disk}}} \tag{15.31}$$

In a similar way, the collection efficiency under an ac signal at the angular frequency ω is defined by

$$\mathcal{N}(\omega) = \frac{\widetilde{i}_{\text{ring}}}{\widetilde{i}_{\text{disk}}} \tag{15.32}$$

As the disk and the ring current are linked through three distinct steps, $\mathcal{N}(\omega)$ can hence be split as the product of three transfer functions, i.e.,

$$\mathcal{N}(\omega) = \frac{\widetilde{i}_{\text{ring}}}{\widetilde{i}_{\text{disk}}} = \frac{\widetilde{N}_{\text{B,disk}}}{\widetilde{i}_{\text{disk}}} \frac{\widetilde{N}_{\text{B,ring}}}{\widetilde{N}_{\text{B,disk}}} \frac{\widetilde{i}_{\text{ring}}}{\widetilde{N}_{\text{B,ring}}} = \mathcal{N}_{\text{disk}}(\omega) \mathcal{N}_{\text{t}}(\omega) / \mathcal{N}_{\text{ring}}(\omega) \tag{15.33}$$

where $N_{\text{B,disk}}$ is the flux of species B at the disk interface and $N_{\text{B,ring}}$ is the flux of species B at the ring interface. In equation (15.33), $\mathcal{N}_{\text{ring}}(\omega) = \widetilde{N}_{\text{B,ring}} / \widetilde{i}_{\text{ring}}$ is the kinetic efficiency of the ring process. If this process is sufficiently fast that it is controlled entirely by convective diffusion, then $\mathcal{N}_{\text{ring}}(\omega) = 1/n_{\text{ring}} F$, where F is Faraday's constant. The applicability of the rrde is in fact closely dependent on the possibility of finding a suitable ring material and electrochemical system for which $\mathcal{N}_{\text{ring}}(\omega)$ can be considered constant.

The term $\mathcal{N}_{\text{t}}(\omega) = \widetilde{N}_{\text{B,ring}} / \widetilde{N}_{\text{B,disk}}$ is determined completely by mass transport, and, therefore, $\mathcal{N}_{\text{t}}(\omega)$ can be reduced by the dimensionless frequency $p\text{Sc}_i^{1/3}$ (see Chapter 11 on page 243). No analytic expression is available, and a numerical solution must be derived for each geometry.

The term $\mathcal{N}_{\text{disk}}(\omega) = \widetilde{N}_{\text{B,disk}}/\widetilde{i}_{\text{disk}}$ is the kinetic emission efficiency, which depends only on the kinetics at the disk electrode. At the steady state,

$$\lim_{\omega \to 0} \mathcal{N}_{\text{disk}}(\omega) = \frac{1}{n_{\text{disk}}\text{F}} \tag{15.34}$$

In the transient regime, a fraction of the disk current $\widetilde{i}_{\text{disk}}$ may be stored at the disk surface as the charge $\widetilde{q}$, implied in the faradaic process through the formation of adsorbed intermediate species (2-D) or of films (3-D). From the conservation of electrical charge,

$$\widetilde{i}_{\text{disk}} = n_{\text{disk}}\text{F}\widetilde{N}_{\text{B,disk}} + j\omega\widetilde{q} \tag{15.35}$$

Equation (15.35) becomes

$$\frac{\widetilde{N}_{\text{B,disk}}}{\widetilde{i}_{\text{disk}}} = \frac{1}{n_{\text{disk}}\text{F}} - \frac{j\omega}{n_{\text{disk}}\text{F}}\frac{\widetilde{q}}{\widetilde{i}_{\text{disk}}} \tag{15.36}$$

By introducing the faradaic impedance of the disk electrode Z_{disk}, the transfer function can be written as

$$\mathcal{N}(\omega) = \mathcal{N}_0(\omega) - j\omega\frac{\widetilde{q}}{\widetilde{V}}Z_{\text{disk}}\mathcal{N}_{\text{t}}(\omega)n_{\text{ring}}\text{F} \tag{15.37}$$

where

$$\mathcal{N}_0(\omega) = \frac{n_{\text{ring}}}{n_{\text{disk}}}\mathcal{N}_{\text{t}}(\omega) \tag{15.38}$$

The transfer function $\widetilde{q}/\widetilde{V}$ can be determined from the faradaic impedance Z_{disk} and the transfer function $\mathcal{N}(\omega)$. It is an important kinetic parameter that allows evaluation of the frequency dependence of the amount of charge stored at the electrode surface.

15.2.2 Multifrequency Measurements for Double-Layer Studies

The faradaic impedance is linked in parallel to the double-layer capacitance C_{dl} and in series to the solution resistance R_{e} as illustrated in Figure 9.1. The double-layer capacitance is in general considered to be constant. It is frequently observed, however, that C_{dl} changes with the dc current at which the impedance measurements were carried out. The capacitance C_{dl} has been observed, for example, to increase with increasing current density.

The double-layer capacitance C_{dl} may be assumed to be linked to the surface coverage γ_i, where C_{dl} is a function of frequency due to the frequency dependence of γ_i. The kinetic description of electrochemical impedance involving the surface coverage of intermediates was especially promoted by Epelboin et al.[311] They considered that, for a reaction mechanism involving few reactions, some reaction intermediates adsorb following a Langmuir isotherm and are characterized by a surface coverage γ_i. In this framework, all the loops except the diffusion ones are semicircles centered on the real axis. These semicircles could be capacitive or inductive loops. To explore the role of surface coverage in the modulation of double-layer capacitance, a new technique was

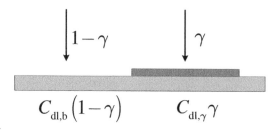

Figure 15.4: Schematic diagram showing the different values of the double-layer capacitance in the presence of a surface coverage.

invented to measure this frequency dependence, and in this way to verify directly the theory developed a long time ago.[92]

As suggested by Figure 15.4, the double-layer capacitance at the bare surface and at a surface covered by the reaction intermediate can be given as $C_{dl,b}$ and $C_{dl,\gamma}$, respectively. A related development is presented in Section 9.2.2 on page 196. Under the assumption that these double-layer capacitances are different, the effective capacitance can be given by

$$C_{dl} = C_{dl,b}(1 - \gamma) + C_{dl,\gamma}\gamma \qquad (15.39)$$

A capacitance transfer function based on the total double-layer capacitance can be shown to be a linear function of the surface coverage γ as

$$\frac{\tilde{C}_{dl}}{\tilde{V}} = (C_{dl,\gamma} - C_{dl,b})\frac{\tilde{\gamma}}{\tilde{V}} \qquad (15.40)$$

This conceptual approach can be illustrated for a simple system involving adsorbed intermediates such as is described in Section 10.4.1.

For a system without mass-transport dependence, $i_F = f(V, \gamma)$, and, according to equation (10.5), the admittance is

$$\frac{\tilde{i}_F}{\tilde{V}} = \frac{1}{Z_F} = \frac{1}{R_t} + \left(\frac{\partial f}{\partial \gamma}\right)_V \frac{\tilde{\gamma}}{\tilde{V}} \qquad (15.41)$$

The overall impedance is given by

$$Z(\omega) = R_e + \frac{1}{j\omega C_{dl} + 1/Z_F(\omega)} \qquad (15.42)$$

where C_{dl} is a function of γ_i according to equation (15.39). At high enough frequencies ω_{HF}, the contribution of the faradaic impedance to the overall impedance becomes negligible, and the current response $\tilde{i}(\omega_{HF})$ to a potential perturbing signal $\tilde{U}(\omega_{HF})$ is expressed as

$$\tilde{i}(\omega_{HF}) = \frac{\tilde{U}(\omega_{HF})}{R_e + 1/(j\omega_{HF}C_{dl})} \qquad (15.43)$$

The transfer function $\widetilde{\gamma}_i/\widetilde{V}$ and the transfer function $\widetilde{C}_{dl}/\widetilde{V}$ can be neglected at this frequency. Thus the capacitance C_{dl} may be considered to have its steady-state value. If R_e can be neglected with respect to $1/(j\omega_{HF}C_{dl})$, the capacitance is obtained from the expression

$$C_{dl} = \frac{1}{j\omega_{HF}} \frac{\widetilde{i}(\omega_{HF})}{\widetilde{U}(\omega_{HF})} \tag{15.44}$$

in which $\widetilde{U}(\omega_{HF})$ is the perturbing quantity, and, therefore, $\widetilde{U}(\omega_{HF})$ has a real value. As C_{dl} is also a real number, $\widetilde{i}(\omega_{HF})$ must have an imaginary value. For a high-frequency ω_{HF}, C_{dl} is directly proportional to $|\widetilde{i}(\omega_{HF})|$. If two additive perturbing signals $\widetilde{U}(\omega_{HF})$ and $\widetilde{U}(\omega)$ are superimposed, one gets the linear (first-order) response of current $\widetilde{i}$ as the sum of the elementary currents at frequencies ω_{HF} and ω

$$\widetilde{i} = \widetilde{i}(\omega_{HF}) + \widetilde{i}(\omega) = \widetilde{U}(\omega_{HF})j\omega_{HF}C_{dl} + \frac{\widetilde{U}(\omega)}{Z(\omega)} \tag{15.45}$$

If the capacitance measured at high-frequency ω_{HF} using equation (15.44) is modulated at a lower-frequency ω, $\widetilde{i}(\omega_{HF})$ will also be modulated at the frequency ω. This approach can yield simultaneous measurement of $Z(\omega)$ and $\widetilde{C}_{dl}(\omega)/\widetilde{U}(\omega)$.

An original experimental setup was devised by Antaño-Lopez et al.[92,312,313] to determine simultaneously the transfer functions $Z(\omega)$ and $\widetilde{C}_{dl}(\omega)/\widetilde{U}(\omega)$. The experimental setup simultaneously measured $\widetilde{U}(\omega)$, $\widetilde{i}(\omega)$, and $\widetilde{C}_{dl}(\omega)$. A lock-in amplifier determined the latter signal with the reference frequency ω_{HF} as illustrated in Figure 15.5. The sine-wave generator of the lock-in amplifier delivered a perturbing signal $\Delta E(\omega_{HF})$ to the voltage adder of the potentiostat. The generator of the frequency response analyzer delivered a perturbing signal $\Delta E(\omega)$ to the voltage adder of the potentiostat. Electronic summation of the two signals by the potentiostat synthesized the composite perturbation $(\Delta E(\omega) + \Delta E(\omega_{HF}))$. The electrochemical cell was connected as usual to the potentiostat. The potentiostat outputs two ac signals, ΔU and ΔI, both containing a composite ac signal produced by superposition of the frequencies ω_{HF} and ω. A low-pass filter eliminated the high-frequency signal at frequency ω_{HF} and both signals were sent back to the frequency response analyzer; $\Delta E(\omega_{HF})$ to input channel 2 and $\Delta I(\omega)$ to input channel 1. The current output signal was also sent to the input of the lock-in amplifier. The out-of-phase component of the current response at frequency ω_{HF} was linked through the low-pass filter to input channel 3 of the frequency response analyzer. By correlation at frequency ω, the frequency response analyzer calculated two transfer functions, $Z(\omega)$ from channels 1 and 2 and $\widetilde{C}_{dl}(\omega)/\widetilde{U}(\omega)$ from channels 3 and 2. In fact, the transfer function $\widetilde{C}_{dl}(\omega)/\widetilde{U}(\omega)$ must be corrected by the transfer function $F(\omega)$ corresponding to the instrumental frequency response of the built-in filters of the lock-in amplifier. This method was applied to the anodic dissolution of iron.[92] The authors verified experimentally the close correlation between the relaxations of the interface capacitance and that of faradaic current.

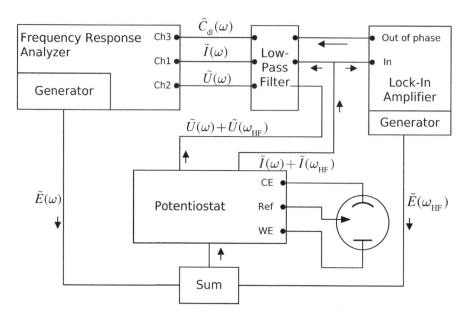

Figure 15.5: Experimental setup used to determine the variation of interface capacitance with a low-frequency perturbing signal.

15.3   Transfer Functions Involving Nonelectrical Quantities

The principle of the experimental arrangement devised for the transfer-function measurements involving a nonelectrical input quantity was given previously in Figures 15.2 and 15.3. In this section, a few examples are presented. The transfer function corresponding to the response of the electrochemical system to a perturbation of the rotation speed of a rotating disk electrode is given in Chapter 16.

15.3.1   Thermoelectrochemical (TEC) Transfer Function

The technique introduced by Citti et al.[314] and then by Rotenberg[315] used a vertical electrode heated by a laser beam or by an infrared diode. In this case, the same order of magnitude is obtained for the free convection and the thermal convection, and experiments can be performed with a sufficient accuracy. A fast redox reaction at a heated vertical electrode is considered. The motion of the solution is spontaneous and arises due to forces originating from heterogeneous reactions. Such forces result from density variation in the solution produced by both thermal and concentration gradients near

Remember! 15.4 *Transfer functions serve to isolate the influence of specific independent variables that contribute to the electrochemical impedance response of a system.*

the electrode.

A perturbation of the electrode temperature generates a perturbation of the velocity field and then a perturbation of the concentration field near the electrode. According to Fick's law

$$i_F = -nFD_i \left. \frac{dc_i}{dy} \right|_{y=0} \tag{15.46}$$

the TEC transfer function corresponding to the faradaic current can be divided in two terms

$$\frac{\tilde{i}_F}{\tilde{T}} = \bar{i}_F \frac{1}{D_i} \frac{\tilde{D}_i}{\tilde{T}} - nFD_i \frac{\left. \frac{d\tilde{c}_i}{dy} \right|_{y=0}}{\tilde{T}} \tag{15.47}$$

in agreement with the theory developed by Aaboubi et al.[308] and by Rotenberg.[315] Taking into account the temperature dependence of the diffusion coefficient in which $D_i \propto \exp\left(-A_i/RT\right)$, the first term becomes $A_i i_F/RT^2$, where A_i is the activation energy for diffusion. The second term is frequency dependent and is limited by mass transport. Rotenberg[315] introduced a term corresponding to the response of the charging current to a temperature perturbation: $\tilde{i}_C = j\omega(\tilde{q}/\tilde{T})$. The transfer function corresponding to the overall current is the sum of three terms, i.e.,

$$\frac{\tilde{i}}{\tilde{T}} = \frac{\tilde{i}_F}{\tilde{T}} + \frac{\tilde{i}_C}{\tilde{T}} = \frac{A_i \bar{i}_F}{RT^2} - nFD_i \frac{\left. \frac{d\tilde{c}_i}{dy} \right|_{y=0}}{\tilde{T}} + j\omega \frac{\tilde{q}}{\tilde{T}} \tag{15.48}$$

Motion of the solution in thermal laminar free convection is spontaneous and arises due to forces originating from heterogeneous reactions and from the release of heat from the electrode. Such forces follow the modification of the solution density caused by two phenomena. The concentration in the proximity of the reaction surface changes in the course of heterogeneous reaction and leads to changes in the density of solution. In addition, the release of heat induces variations of solution density from point to point as a result of nonuniform changes in the temperature of the solution. The density of the solution is a function of concentration and temperature and can be expressed by

$$\rho = \rho(\infty) + \left(\frac{\partial \rho}{\partial c_i} \right) (c_i - c_i(\infty)) + \left(\frac{\partial \rho}{\partial T} \right) (T - T(\infty)) \tag{15.49}$$

where $c_i(\infty)$, $T(\infty)$, and $\rho(\infty)$ are the concentration of reacting species, the temperature, and the density, respectively, measured in the bulk solution.

The body force acting on a unit of fluid volume is equal to ρg and changes from point to point in the solution. It is natural to consider that most of the change in concentration occurs in a very thin layer and most of the change in temperature occurs in a larger, but still thin layer. The changes in temperature and in concentration are the causes of fluid motion. Therefore, one can safely assume that fluid motion also occurs in this layer. Thus, the theories for the hydrodynamic boundary layer can also be applied to fluid motion in thermal free convection. In this case, the hydrodynamic boundary layer coincides with the thermal diffusion layer.

The concentration distribution of the electroactive species is determined from the solution of the convective-diffusion equation

$$\frac{\partial c_i}{\partial t} = -\nabla \cdot (D_i(y)\nabla c_i + \mathbf{v}c_i) \tag{15.50}$$

or

$$\frac{\partial c_i}{\partial t} + v_x \frac{\partial c_i}{\partial x} + v_y \frac{\partial c_i}{\partial y} = D_i(y)\frac{\partial^2 c_i}{\partial y^2} + \frac{\partial D_i}{\partial y}\frac{\partial c_i}{\partial y} \tag{15.51}$$

with the boundary conditions: for $y = 0$, $c_i = 0$; and for $y \to \infty$, $c_i = c_i(\infty)$.

The diffusion coefficient $D_i(y)$ is a function of temperature, and it varies with position near the electrode according to the local temperature variation. However, as the thermal layer thickness is about five times larger than the diffusion layer thickness, the diffusion coefficient has in fact a variation that can be assumed to be negligible within the mass-transfer diffusion layer corresponding to the integration domain of equation (15.51). Thus, in the following development, $D_i(y) = D_i$, and $\partial D_i/\partial y = 0$.

The tangential and normal velocity components are, respectively, v_x and v_y. According to Marchiano and Arvia,[316] the velocity components can be expressed as a function of dimensionless coordinate

$$v_x = 4\nu(\frac{g}{4\nu^2})^{1/2}\, x^{1/2}f'(\mu) \tag{15.52}$$

and

$$v_y = \nu(\frac{g}{4\nu^2})^{1/4}\frac{\mu f'(\mu) - 3f(\mu)}{x^{1/4}} \tag{15.53}$$

where $\mu = (g/4\nu^2)^{1/4}$, ν is the kinematic viscosity, g is the acceleration of gravity, and $f(\mu)$ can be written as a series development in μ, which can be limited near the electrode to the first term: $f(\mu) = \mathcal{A}(T)\mu^2$.

The expression of the steady-state local flux at a vertical electrode under simultaneous effects of concentration and thermal gradients is given by

$$N_i = -\,D_i\frac{d\overline{c}_i}{dy}\bigg|_{y=0} = D_i c_i(\infty)\frac{(\mathcal{A}Sc_i)^{1/3}}{\Gamma(4/3)}\left(\frac{g}{4\nu^2}\right)^{1/4}x^{-1/4} \tag{15.54}$$

As the electrode temperature is larger than the bulk temperature, a modulation of the temperature of the electrochemical interface induces a modulation of the thermal gradient in the solution while the bulk temperature is kept constant. Then the temperature-dependent parameters, like buoyancy forces, are modulated inside the thermal diffusion layer adjacent to the surface and, consequently, a modulation of the velocity is induced near the electrode. Therefore, the transient material balance equation may be written as

$$j\omega\widetilde{c}_i + \overline{v}_x\frac{\partial \widetilde{c}_i}{\partial x} + \overline{v}_y\frac{\partial \widetilde{c}_i}{\partial y} - D_i\frac{\partial^2 \widetilde{c}_i}{\partial y^2} = -\widetilde{v}_x\frac{d\overline{c}_i}{dx} - \widetilde{v}_y\frac{d\overline{c}_i}{dy} \tag{15.55}$$

with the boundary conditions that, for $y = 0$, $\widetilde{c}_i(0) = 0$; and, for $y \to \infty$, $\widetilde{c}_i = 0$.

Equation (15.55) is a partial differential equation in two-dimensions that can be solved by following the method described in Section 13.5.2. The first step is to use a set of dimensionless variables that can be defined as a dimensionless normal distance from the electrode $\xi = y/\delta(x)$ and an x-dependent dimensionless frequency $K_x = \omega\delta^2(x)/D_i$. Equation (15.55) can be written as

$$jK_x\widetilde{c}_i + \frac{4\xi K_x}{9}\frac{\partial\widetilde{c}_i}{\partial K_x} - \frac{\xi^2}{3}\frac{\partial\widetilde{c}_i}{\partial\xi} - \frac{\partial^2\widetilde{c}_i}{\partial\xi^2} = \frac{\xi^2}{3}\frac{d\bar{c}_i}{d\xi}\frac{\widetilde{A}}{\bar{A}} \tag{15.56}$$

The material balance equation is written as a partial differential equation of two variables: ξ and K_x. In the low-frequency range, a solution exists in the form of a series

$$\widetilde{c}_i = \sum_{m=0}^{\infty}(jK_x)^m h_m(\xi) \tag{15.57}$$

The solution for $h_m(\xi)$ is obtained by a method similar to that presented in Section 13.5.2.

In the high-frequency range, the temperature perturbation is rapidly damped close to the wall; thus, the convective term can be disregarded, and equation (15.55) reduces to

$$jK_x\widetilde{c}_i - \frac{\partial^2\widetilde{c}_i}{\partial\xi^2} = \frac{\xi^2}{3}\frac{d\bar{c}_i}{d\xi}\frac{\widetilde{A}}{\bar{A}} \tag{15.58}$$

Since the frequency is large, the distance over which a temperature perturbation prevails is small, and some simplifications can be made to facilitate analytic solution of equation (15.58).

The series corresponding to the low-frequency solution needs a number of terms that increases with frequency; thus, the number of terms was chosen to provide a sufficient overlap between the low- and high-frequency solutions. In the present case, 80 terms were used to provide an overlap for $8 \leq K_x \leq 11$. These 80 terms are given by Aaboubi et al.[308]

The previous equations allow determination of the local flux. To obtain the response of the electrode itself, it is necessary to integrate the local flux over the electrode surface. For a rectangular electrode,

$$\widetilde{i}_{rect} = L\int_0^\ell nFD_i\left.\frac{\partial\widetilde{c}_i}{\partial y}\right|_0 dx \tag{15.59}$$

and, for a circular electrode,

$$\widetilde{i}_{circ} = \int_{-R}^{+R} dz \int_{R-\sqrt{R^2-z^2}}^{R+\sqrt{R^2-z^2}} nFD_i\left.\frac{\partial\widetilde{c}_i}{\partial y}\right|_0 dx \tag{15.60}$$

The experimental TEC transfer function is compared with the model in Figure 15.6. The three terms of equation (15.48) appear clearly in this figure.

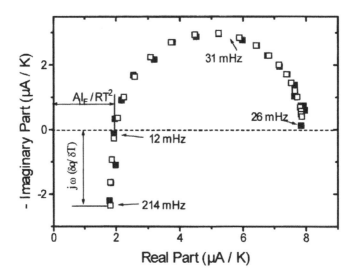

Figure 15.6: Experimental TEC transfer function compared with the model: □ represents the theoretical points and ■ the experimental points. (Taken from Aaboubi et al.[308])

15.3.2 Photoelectrochemical Impedance Measurements

Modulation of light intensity provides an attractive method for probing the response of photoactive materials and systems. The required sinusoidal modulation of the intensity of the incident light is achieved by using the apparatus shown in Figure 15.7, and the technique is referred to as intensity-modulated photocurrent spectroscopy (IMPS).[317] The intensity of the incident laser beam is modulated by an acousto-optic modulator driven by the dc-biased output of a frequency response analyzer. Up to now, measurements have been made only under potentiostatic control. The complex ratio of the ac component of the photocurrent to the incident modulated light flux is obtained by deriving a reference signal from a fast photodiode that samples the laser beam. The time-dependent flux of minority carriers into the surface follows the excitation profile with a delay less than 1 ns, so the transfer function between the flux of minority carriers and the illumination can be considered to be a real number. The net photocurrent response, which is made up of the instantaneous minority carrier flux and the coupled majority carrier flux, can be derived. Finally, the output current response is determined by taking into account the cell transfer function, which is determined by the combination of the space-charge capacitance C_{sc} and the solution resistance R_e.

The general transfer function appears as the product of three transfer functions, i.e.,

$$\frac{\widetilde{I}}{h\widetilde{\nu}} = \frac{\widetilde{F}_{minority}}{h\widetilde{\nu}} \times \frac{\widetilde{F}_{majority}}{\widetilde{F}_{minority}} \times \frac{\widetilde{I}}{\widetilde{F}_{majority}} \tag{15.61}$$

where the first transfer function is a real number, $h\widetilde{\nu}$ is the input quantity, $\widetilde{F}_{minority}$ and

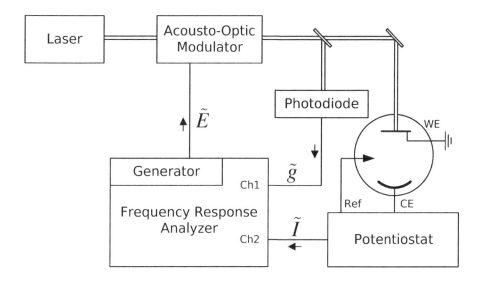

Figure 15.7: Experimental device for the measurement of impedance obtained by IMPS.

$\widetilde{F}_{majority}$ are state quantities, and $\widetilde{I}$ is the output quantity.

This frequency-response analysis offers unique insights into complex photoelectrode processes. The analysis of surface recombination and photocurrent multiplication has shown that it is possible to deconvolute the contributions to the photocurrent of minority, majority, and injected carriers. The dependence of the rates of surface processes on potential, solution composition, and surface orientation and preparation can now be studied in detail. The systems investigated include the reduction of oxygen at p-Gas, the photooxidation of Si in NH_4F,[317] and the anodic dissolution of InP.[318]

15.3.3 Electrogravimetry Impedance Measurements

The use of a quartz microbalance to measure the mass loading on one face of a quartz crystal through the change of its resonance frequency (often of the order of 6 MHz) in electrolytic medium was introduced at the beginning of the 1980s. If the electrode is polarized in a potentiostatic circuit, the microbalance can be used to measure the change of the mass of the reacting species involved in a reaction occurring on the electrode. The sensitivity reached ($\approx 10^{-9}$ g/cm^2) is sufficient to allow determination of mass changes associated with adsorbed reaction intermediates or ion insertion in films coating the electrode.

Under steady-state or quasi-steady-state operation, the mass change as a function of time is followed by measuring directly the frequency f_w of the quartz oscillator by means of a frequency counter. The use of this microbalance in a sinusoidal regime is carried out by measuring the difference between the frequencies f_w of the working oscillator and f_0 of a reference oscillator in air. This difference $df = f_w - f_0$, which is

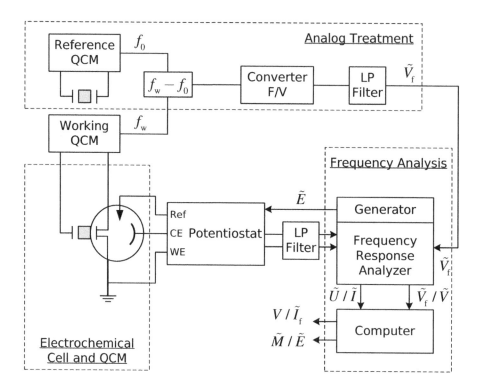

Figure 15.8: Experimental device for the measurement of the electrogravimetric transfer function.

proportional to the mass change and sinusoidal in linear regime, is converted to a voltage by means of a converter. The resulting signal can be simultaneously analyzed by the transfer-function analyzer with the current response $\tilde{I}$ to a potential perturbation $\tilde{E}$ generated by the analyzer. The electrochemical response between $\tilde{U}$ and $\tilde{I}$ is simultaneously measured, which allows measurement of the mass relaxation at the interface between the electrolyte and the electrode coating the quartz crystal.[11,319,12]

A schematic representation of the system is given in Figure 15.8. In this case, the output quantities are the current $Y_1 = I$ and the mass $Y_2 = M$. In addition to the charge and mass balances the equation governing the change of mass can be established

$$\frac{\mathrm{d}M}{\mathrm{d}t} = H(\theta_i, c_k, U) \tag{15.62}$$

The usual derivation based on the linearization procedure allows calculation of the electrochemical impedance and the $\tilde{M}/\tilde{U}$ transfer function. Two types of behavior occur, depending on whether the average mass of the electrode changes continuously with time.

1. When the mass increases or decreases (e.g., for deposition or dissolution of a metal, respectively) at a constant current, the low-frequency limit of $\tilde{M}/\tilde{U}$ tends to infinity.

2. When the mass does not change (e.g., a polymer film at zero current), the low-frequency limit of $\widetilde{M}/\widetilde{U}$ tends to a finite value.

The simultaneous measurement of the impedance and mass/potential transfer function leads to new information on the kinetics of the processes involved. It may lead to chemical identification of the species involved in the intermediate reaction steps by allowing the atomic masses of the adsorbed intermediates of the multistep reaction mechanisms to be estimated.

Problems

15.1 Derive the steady-state concentration gradient corresponding to equation (15.55).

15.2 Derive the high-frequency solution corresponding to equation (15.59) and, in particular, show that the phase shift is constant and equal to $-135°$.

15.3 Derive the relationship among the thermoelectrochemical transfer function measured under potentiostatic regulation, the thermoelectrochemical transfer function measured under galvanostatic regulation, and the electrochemical impedance.

15.4 Derive the equation relating the electrogravimetric transfer function measured under potentiostatic regulation, the electrogravimetric transfer function measured under galvanostatic regulation, and the electrochemical impedance.

Chapter 16

Electrohydrodynamic Impedance

Electrohydrodynamic (EHD) impedance provides a practical example of a generalized transfer function involving nonelectrical quantities. In this chapter, a rotating disk electrode is considered. The current is a function of the rotation speed, which means that the current is totally or partially limited by mass transport. This measurement technique, based on the analysis of the current response to a rotation speed perturbation, was proposed at the beginning of the seventies by Bruckenstein et al.[320] Very early, these authors suggested application of sinusoidal hydrodynamic modulations. Bruckenstein et al.[320] derived the first theoretical analysis of the problem by considering the response of the mass-transfer rate to a modulation of the angular velocity of a rotating disk electrode such that[321]

$$\Omega^{1/2} = \overline{\Omega}^{1/2} \left(1 + \left(\frac{\Delta\Omega}{\overline{\Omega}} \right)^{1/2} \cos\omega t \right) \tag{16.1}$$

where ω is the modulation frequency.

The concept that the response to a modulation of rotation speed Ω should be seen as a modulation of the square root of Ω was naturally supported by the results of the Levich theory in steady-state conditions.[176] However, due to the fact that $i = f(\Omega) = k\Omega^{1/2}$,

$$\frac{\partial f}{\partial \Omega} = \left(\frac{1}{2}\Omega^{-1/2} \right) \frac{\partial f}{\partial \Omega^{1/2}} \tag{16.2}$$

Equation (16.2) shows that the transfer function corresponding to a modulation of the angular velocity is directly proportional to the transfer function corresponding to a modulation of the square root of the angular velocity, and the coefficient of proportionality is $\frac{1}{2}\Omega^{-1/2}$. Therefore, after the pioneer works of Bruckenstein et al.,[320,321] a direct modulation of the angular velocity was considered. With the increasing development of impedance techniques, aided by development of increasingly sophisticated instrumentation,[63] Deslouis and Tribollet[322] promoted the use of impedance concept for this type of perturbation and introduced the electrohydrodynamic (EHD) impedance.

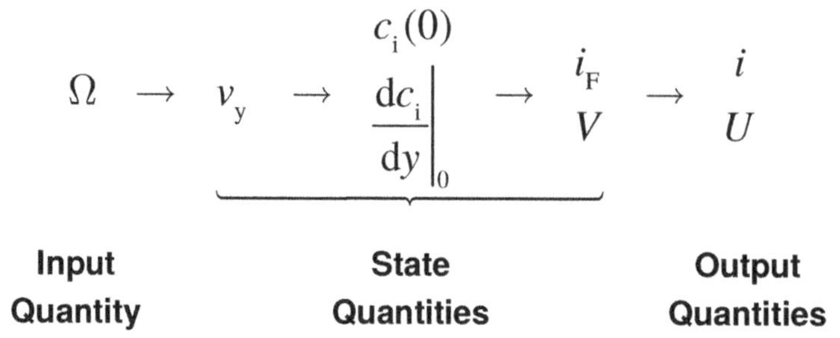

$$\Omega \;\rightarrow\; v_y \;\rightarrow\; \begin{array}{c} c_i(0) \\ \left.\dfrac{dc_i}{dy}\right|_0 \end{array} \;\rightarrow\; \begin{array}{c} i_F \\ V \end{array} \;\rightarrow\; \begin{array}{c} i \\ U \end{array}$$

Input	**State**	**Output**
Quantity	**Quantities**	**Quantities**

Figure 16.1: Schematic representation of the different transfer functions existing between the input quantity and the output quantities.

The EHD impedance is useful for analysis of electrochemical systems that are either partially or completely limited by mass transport. For a rotating disk electrode, the input quantities are, at least, one electrical quantity, e.g., overall current or electrode potential, and one nonelectrical quantity, i.e., the rotation speed of the rotating disk electrode Ω. For EHD impedance, the input quantity is the rotation speed. Under galvanostatic regulation, the output quantity is the electrode potential; under potentiostatic regulation, the output quantity is the overall current. To analyze this problem, the mass conservation equation must be considered with the normal velocity v_y near the electrode and the concentration of the involved species $c_i(0)$ as state quantities.

A perturbation of the rotation speed Ω induces a perturbation of the normal velocity v_y, which, in turn, induces a perturbation of the concentration field of the electroactive species and in particular of $c_i(0)$ and $\left.\frac{dc_i}{dy}\right|_0$. These last interfacial state quantities are linked to the surface-averaged faradaic current density i_F and to the interfacial potential V by the kinetic equations. Finally, by taking into account the double-layer capacitance and the electrolyte resistance, these last electrical quantities are linked to the output quantities of the average current density i and the electrode potential U. The relationship among input, state, and output quantities is shown schematically in Figure 16.1. The transfer functions associated with several applications of electrohydrodynamic impedance are developed in the subsequent sections.

Remember! 16.1 *Electrohydrodynamic impedance provides a means to isolate the influence of mass transfer the electrochemical impedance response of a system.*

16.1 Hydrodynamic Transfer Function

Through the von Karman transformation, the steady-state Navier-Stokes equations for a rotating disk can be expressed in terms of three coupled, nonlinear, ordinary differential equations as[116]

$$2F + H' = 0 \tag{16.3}$$

$$F^2 - G^2 + HF' - F'' = 0 \tag{16.4}$$

and

$$2FG + HG' - G'' = 0 \tag{16.5}$$

where F, H, and G represent the dimensionless radial, angular, and axial velocity components, respectively. Equations (16.3), (16.4), and (16.5) are functions only of the axial dimensionless distance $\zeta = y\sqrt{\overline{\Omega}/\nu}$ and can be solved subject to boundary conditions

$$F(0) = H(0) = 0 \tag{16.6}$$

$$G(0) = 1 \tag{16.7}$$

and

$$F(\infty) = G(\infty) = 0 \tag{16.8}$$

The steady flow field created by an infinite disk rotating at a constant angular velocity in a fluid with constant physical properties was presented in Chapter 11.

For the unsteady situation, the instantaneous value of rotation speed Ω can be defined by

$$\Omega = \overline{\Omega} + \text{Re}\{\widetilde{\Omega} \exp j\omega t\} \tag{16.9}$$

where $\omega/2\pi$ is the modulation frequency and $\widetilde{\Omega} = \Delta\Omega$ is a real number. Large-amplitude modulations induce a nonlinear flow response.[323,324] This nonlinear problem is outside the scope of an impedance study, which is defined here to be based on a linearized system response. A periodic flow generated by small oscillations of a body in a fluid at rest involves nonlinearities in the mass-transport problem or in the secondary flow and is therefore also outside the scope of this presentation. The electrohydrodynamic impedance concept is developed here following the work of Tribollet and Newman[184] for small-amplitude modulation such that $(\Delta\Omega \ll \overline{\Omega})$. Under these conditions, the system response is linear.

Following the development in Chapters 10 (page 207) and 11 (page 243) for a current or potential response to a perturbation, the radial, angular, and axial velocity components can be expressed as

$$v_r = r\overline{\Omega}[F(\zeta) + \frac{\Delta\Omega}{\overline{\Omega}}\text{Re}\{\widetilde{f} \exp j\omega t\}] \tag{16.10}$$

$$v_\theta = r\overline{\Omega}[G(\zeta) + \frac{\Delta\Omega}{\overline{\Omega}}\text{Re}\{\widetilde{g} \exp j\omega t\}] \tag{16.11}$$

and

$$v_y = r\overline{\Omega}[H(\zeta) + \frac{\Delta\Omega}{\overline{\Omega}}\mathrm{Re}\{\tilde{h}\exp j\omega t\}] \tag{16.12}$$

where $\tilde{f}$, $\tilde{g}$, and $\tilde{h}$ are complex functions. The equation of continuity and the unsteady Navier-Stokes equations are linearized, i.e., the quadratic terms, proportional to $(\Delta\Omega/\overline{\Omega})^2$, are neglected.

The time-dependent forms of equations (16.3), (16.4), and (16.5) may be written as

$$2\tilde{f} + \tilde{h}' = 0 \tag{16.13}$$

$$j\tilde{f}p + 2F\tilde{f} - 2G\tilde{g} + H\tilde{f}' + F'\tilde{h} = \tilde{f}'' \tag{16.14}$$

and

$$j\tilde{g}p + 2G\tilde{f} + 2F\tilde{g} + \tilde{h}G' + H\tilde{g}' = \tilde{g}'' \tag{16.15}$$

where $p = \omega/\overline{\Omega}$ is the dimensionless modulation frequency. Equations (16.13), (16.14), and (16.15) may be solved subject to the boundary conditions

$$\tilde{f}(0) = \tilde{h}(0) = 0 \tag{16.16}$$

$$\tilde{g}(0) = 1 \tag{16.17}$$

and

$$\tilde{f}(\infty) = 0 \tag{16.18}$$

As discussed in Section 1.2.2 on page 4, each complex function may be written as the sum of a real function and an imaginary function. The set of the three coupled equations (16.13), (16.14), and (16.15) then becomes a set of six coupled linear ordinary differential equations. By using Newman's method,[208] a numerical solution for the six equations can be obtained for each dimensionless frequency.

In a manner similar to that developed for the steady-state solution in Chapter 11 on page 243, the complex functions $\tilde{f}$, $\tilde{g}$, and $\tilde{h}$ can be written in terms of series expansions for small values of ζ. Of particular importance is the derivative $\tilde{f}'(0)$ obtained from equations (16.13), (16.14), and (16.15) and given in Table 16.1 for different values of dimensionless frequency p. The real and imaginary parts of the derivative $\tilde{f}'(0)$ are also presented in Figure 16.2 as a function of $p = \omega/\Omega$. These derivatives are essential to determination of the first coefficients of the series expansions. The other coefficients

---

Remember! 16.2 *The methods described in this section allow the numerical calculation of generalized functions that can be used, via a look-up table, to model any electrohydrodynamic impedance response.*

Table 16.1: Calculated values for real and imaginary parts of $\widetilde{f}'(0)$ as a function of dimensionless frequency p. (Taken from Deslouis and Tribollet.[325])

$p = \omega/\Omega$	$\mathrm{Re}\{\widetilde{f}'(0)\}$	$\mathrm{Im}\{\widetilde{f}'(0)\}$
0.0631	0.7652	−0.0130
0.1000	0.7650	−0.0206
0.1585	0.7645	−0.0329
0.2512	0.7630	−0.0527
0.3981	0.7579	−0.0849
0.6310	0.7410	−0.1356
1.0000	0.6943	−0.2035
1.5849	0.6020	−0.2642
2.5119	0.4832	−0.2842
3.9811	0.3748	−0.2652
6.3095	0.2906	−0.2297
10.000	0.2272	−0.1922

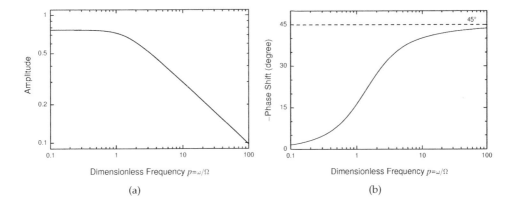

(a)   (b)

Figure 16.2: Dimensionless function $f'(0)$ in Bode representation as functions of dimensionless frequency $p = \omega/\Omega$: a) modulus and b) phase shift.

are deduced from the first one by using the set of equations (16.13), (16.14), and (16.15). In particular,

$$\tilde{h} = -\tilde{f}'(p)\zeta^2 + \frac{2}{3}\zeta^3 \tag{16.19}$$

where $\tilde{h}$ represents the complex function associated with the axial velocity.

By using the results of Sparrow and Gregg,[326] who solved the nonsteady flow problem, the low-frequency expression of $\tilde{f}'$ can be obtained following the present notation as

$$\tilde{f}' = (0.765345 - 0.023112p^2) - 0.204835p\mathrm{j} \tag{16.20}$$

As the frequency modulation tends toward zero, $r\tilde{f}_1$ tends toward the derivative of $\bar{v}_r$ with respect to $\overline{\Omega}$ (see equation 16.10) and $\tilde{f}_2$ tends toward zero. Therefore, $\tilde{f}'(0) = \frac{3}{2}\frac{dF}{d\zeta}\big|_0 = 0.765345$. Sharma[327] obtained an asymptotic solution for $\tilde{f}'$ appropriate for high frequencies as

$$\tilde{f}' = \frac{1}{\sqrt{2p}} - \mathrm{j}\left(\frac{1}{\sqrt{2p}} - \frac{0.313}{p}\right) \tag{16.21}$$

From this expression and as shown in Figure 16.2(b), the phase shift of f' tends toward $-45°$ as p tends toward infinity. Equation (16.20) may be used with an accuracy better than 1 percent for $p < 0.1$. Equation (16.21) may be used with an accuracy better than 1 percent for $p > 7$. Between these two p values, Table 16.1 must be used.

The hydrodynamic transfer function is given by

$$\frac{\tilde{v}_y}{\Delta\Omega} = r\tilde{h}(\zeta) = r\left(\tilde{f}'(p)\,\zeta^2 + \frac{2}{3}\zeta^3\right) \tag{16.22}$$

Equation (16.22) is implicit in the subsequent development of transfer functions involving concentrations of reacting species.

16.2   Mass-Transport Transfer Function

The mass-transfer problem for a rotating disk electrode at a constant rotation speed is presented in Section 11.3.2. Under modulation of the rotation speed of the electrode, equation (11.65) becomes

$$\mathrm{j}\omega\tilde{c}_i\mathrm{e}^{\mathrm{j}\omega t} + \bar{v}_y\frac{d\tilde{c}_i}{dy} + \bar{v}_y\frac{d\tilde{c}_i}{dy}\mathrm{e}^{\mathrm{j}\omega t} + \tilde{v}_y\frac{d\bar{c}_i}{dy}\mathrm{e}^{\mathrm{j}\omega t} - D_i\frac{d^2\bar{c}_i}{dy^2} - D_i\frac{d^2\tilde{c}_i}{dy^2}\mathrm{e}^{\mathrm{j}\omega t} = 0 \tag{16.23}$$

where the second-order term $\tilde{v}_y\frac{d\tilde{c}_i}{dy}\mathrm{e}^{2\mathrm{j}\omega t}$ is neglected in agreement with the hypothesis of linearity. The solution of the steady-state equation (11.55) (page 259) was given

---

Remember! 16.3 *The phase shift of the hydrodynamic transfer function f' tends toward $-45°$ as the dimensionless frequency $p = \omega/\Omega$ tends toward infinity.*

by equation (11.58). Upon cancelation of the steady-state terms and division by $e^{j\omega t}$, equation (16.23) becomes

$$j\omega \tilde{c}_i + \bar{v}_y \frac{d\tilde{c}_i}{dy} - D_i \frac{d^2\tilde{c}_i}{dy^2} = -\tilde{v}_y \frac{d\bar{c}_i}{dy} \tag{16.24}$$

By using the same dimensionless position ξ and the same dimensionless frequency K_i as was used to solve equation (11.66), equation (16.24) can be written in the form

$$\frac{d^2\tilde{c}_i}{d\xi^2} + \left(3\xi^2 - \left(\frac{3}{a^4}\right)^{1/3} \frac{\xi^3}{Sc_i^{1/3}} \right) \frac{d\tilde{c}_i}{d\xi} - jK_i\tilde{c}_i = \tag{16.25}$$

$$-\frac{\Delta\Omega}{\bar{\Omega}} \left(\frac{3\tilde{f}'(p)\xi^2}{a} - 2\left(\frac{3}{a^4}\right)^{1/3} \frac{\xi^3}{Sc_i^{1/3}} \right) \frac{d\bar{c}_i}{d\xi}$$

where only the first two terms of the velocity expansion were considered.

The solution of equation (16.25) can be obtained by the technique of reduction of order by setting $\tilde{c}_i = \lambda(\xi)\theta_i(\xi)$, where $\theta_i(\xi)$ is a solution of the homogeneous equation satisfying the boundary conditions (11.69) (see Section 11.3 on page 256). The term λ satisfies

$$\frac{d^2\lambda}{d\xi^2} + \left(3\xi^2 - \left(\frac{3}{a^4}\right)^{1/3} \frac{\xi^3}{Sc_i^{1/3}} + \frac{2\theta_i'}{\theta_i} \right) \frac{d\lambda}{d\xi} = \tag{16.26}$$

$$-\frac{\Delta\Omega}{\bar{\Omega}} \left(\frac{3\tilde{f}'(p)\xi^2}{a} - 2\left(\frac{3}{a^4}\right)^{1/3} \frac{\xi^3}{Sc_i^{1/3}} \right) \frac{1}{\theta_i} \frac{d\bar{c}_i}{d\xi}$$

Integration gives

$$\tilde{c}_i = K_2\theta_i + K_1\theta_i \int_0^\xi \frac{\exp\left(-\chi^3 + \left(\frac{3}{a^4}\right)^{1/3} \frac{\chi^4}{4Sc_i^{1/3}} \right)}{\theta_i^2(\chi)} d\chi \tag{16.27}$$

$$-\frac{\Delta\Omega}{\bar{\Omega}} \frac{d\bar{c}_i}{d\xi}\bigg|_0 \int_0^\xi \frac{\exp\left(-\chi^3 + \left(\frac{3}{a^4}\right)^{1/3} \frac{\chi^4}{4Sc_i^{1/3}} \right)}{\theta_i^2(\chi)}$$

$$\times \int_0^\chi \left(3\frac{\tilde{f}'(p)}{a}\chi_1^2 - \left(\frac{3}{a^4}\right)^{1/3} \frac{2\chi_1^3}{Sc_i^{1/3}} \right) \theta_i(\chi_1) d\chi_1 d\chi$$

where K_1 and K_2 are integration constants. The boundary condition $\theta_i(0) = 1$ at $\xi = 0$ yields the value of $K_2 = \tilde{c}_i(0)$. The value of K_1 can be obtained from the boundary condition that $\tilde{c}_i$ approaches zero as ξ approaches infinity, with the results

$$K_1 = \frac{\Delta\Omega}{\bar{\Omega}} \frac{d\bar{c}_i}{d\xi}\bigg|_0 W_i \tag{16.28}$$

Table 16.2: Coefficients for the calculation of W_i with a Schmidt number correction.[325]

$p\mathrm{Sc_i}^{1/3}$	t_1	t_2	t_3	t_4	t_5	t_6
0	0.6533	0	0.7788	0	−0.5961	0
0.1000	0.6513	−0.0397	0.7729	−0.0830	−0.5939	0.0408
0.1585	0.6484	−0.0626	0.7639	−0.1307	−0.5907	0.0644
0.2512	0.6410	−0.0983	0.7418	−0.2037	−0.5828	0.1010
0.3981	0.6230	−0.1525	0.6888	−0.3098	−0.5634	0.1564
0.6310	0.5807	−0.2291	0.5696	−0.4437	−0.5181	0.2341
1.0000	0.4905	−0.3204	0.3404	−0.5541	−0.4218	0.3245
1.5849	0.3325	−0.3873	0.0251	−0.5181	−0.2556	0.3834
2.5119	0.1380	−0.3680	−0.1905	−0.2833	−0.0586	0.3442
3.9811	−0.0036	−0.2576	−0.1686	−0.0415	0.0664	0.2101
6.3095	−0.0483	−0.1352	−0.0562	0.0334	0.0787	0.0807
10.0000	−0.0385	−0.0600	−0.0054	0.0169	0.0435	0.0195
15.8488	−0.0218	−0.0263	0.0008	0.0035	0.0178	0.0035
25.1187	−0.0112	−0.0122	0.0003	0.0006	0.0068	0.0006
39.8104	−0.0056	−0.0059	0.0001	0.0001	0.0026	0.0001
63.0952	−0.0028	−0.0029	0	0	0.0010	0
100.000	−0.0014	−0.0014	0	0	0.0004	0

where

$$W_i = \int_0^\infty \left(3\frac{\widetilde{f}'(p)}{a}\zeta^2 - \left(\frac{3}{a^4}\right)^{1/3} \frac{2\zeta_1^3}{\mathrm{Sc_i}^{1/3}} \right) \theta_i \, \mathrm{d}\zeta \tag{16.29}$$

is a dimensionless quantity whose value is worth recording.

The expansion for θ_i is given by equation (11.75) in terms of powers of $p\mathrm{Sc_i}^{1/3}$. The expansion for W_i is more complicated because, while the expansion of θ_i depends on $p\mathrm{Sc_i}^{1/3}$, $\widetilde{f}'(p)$ depends on p without $\mathrm{Sc_i}^{1/3}$. The expansion for W_i takes the form

$$W_i = \widetilde{f}'(p)(t_1 + jt_2) + \frac{1}{\mathrm{Sc_i}^{1/3}} \left[\widetilde{f}'(p)(t_3 + jt_4) + t_5 + jt_6 \right] \tag{16.30}$$

where the functions t_k, given in Table 16.2 as function of $p\mathrm{Sc_i}^{1/3}$, are calculated from the definitions

$$t_1 = \frac{3}{a} \int_0^\infty \zeta^2 \mathrm{Re}\{\theta_{i,0}\} \, \mathrm{d}\zeta \tag{16.31}$$

$$t_2 = \frac{3}{a} \int_0^\infty \zeta^2 \mathrm{Im}\{\theta_{i,0}\} \, \mathrm{d}\zeta \tag{16.32}$$

$$t_3 = \frac{3}{a} \int_0^\infty \xi^2 \mathrm{Re}\,\{\theta_{i,1}\}\, d\xi \tag{16.33}$$

$$t_4 = \frac{3}{a} \int_0^\infty \xi^2 \mathrm{Im}\,\{\theta_{i,1}\}\, d\xi \tag{16.34}$$

$$t_5 = -2 \left(\frac{3}{a^4}\right)^{1/3} \int_0^\infty \xi^3 \mathrm{Re}\,\{\theta_{i,0}\}\, d\xi \tag{16.35}$$

and

$$t_6 = -2 \left(\frac{3}{a^4}\right)^{1/3} \int_0^\infty \xi^3 \mathrm{Im}\,\{\theta_{i,0}\}\, d\xi \tag{16.36}$$

The concentration gradient at the wall is obtained from equation (16.28) where $K_2 = \widetilde{c}_i(0)$, $\theta_i'(0) = 1$, and K_1 is given by equation (16.28). The general result of this section is therefore a relationship between the concentration and the concentration derivative, both evaluated at the electrode surface. In terms of the dimensional distance y, this can be expressed as

$$\left.\frac{d\widetilde{c}_i}{dy}\right|_{y=0} = \frac{\widetilde{c}_i(0)}{\delta_i}\theta_i'(0) + \frac{\Delta\Omega}{\overline{\Omega}} \left.\frac{d\overline{c}_i}{dy}\right|_{y=0} W_i \tag{16.37}$$

where $-1/\theta_i'(0)$ is the dimensionless convective diffusion impedance as given in equation (11.86).

16.2.1 Asymptotic Solution for Large Schmidt Numbers

When the Schmidt number is infinitely large, W_i is reduced to $\widetilde{f}'(p)(t_1 + \mathrm{j}t_2)$ and appears as the product of a hydrodynamic transfer function $\widetilde{f}'(p)$ and a mass-transport transfer function $Z_c = t_1 + \mathrm{j}t_2$. The mass-transport transfer function Z_c is presented in Figure 16.3. It is easily verified that W_i approaches 0.5 when the frequency tends toward zero, in agreement with the exponent of the rotation speed in the Levich equation. This value of 0.5 is also verified if the complete expression of W_i is used. The complex function $2W_i$ is presented in Figure 16.4 in Bode format as a function of dimensionless frequency $p\mathrm{Sc}_i^{1/3}$ with Schmidt number Sc_i as a parameter.

16.2.2 Asymptotic Solution for High Frequencies

When the perturbation frequency is large, the distance over which a concentration wave proceeds is small. Thus, $\exp(-\xi^3)$ can be considered to be equal to one, and, the velocity modulation being rapidly damped close to the wall, the convective term can be disregarded in the homogeneous part of equation (16.25), which becomes

$$\frac{d^2\widetilde{c}_i}{d\xi^2} - \mathrm{j}K_i\theta_i = -\frac{\Delta\Omega}{\overline{\Omega}}\frac{3\widetilde{f}'(p)\xi^2}{a}\left.\frac{d\overline{c}_i}{d\xi}\right|_0 \tag{16.38}$$

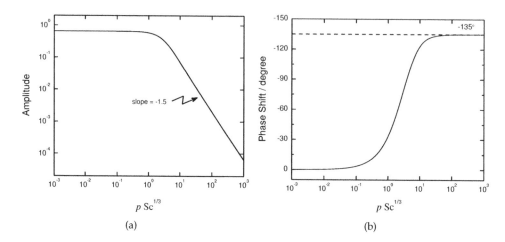

Figure 16.3: Dimensionless mass-transport-transfer function Z_c in Bode representation: a) modulus versus the dimensionless frequency $p\mathrm{Sc_i}^{1/3}$ and b) phase shift versus the dimensionless frequency $p\mathrm{Sc_i}^{1/3}$.

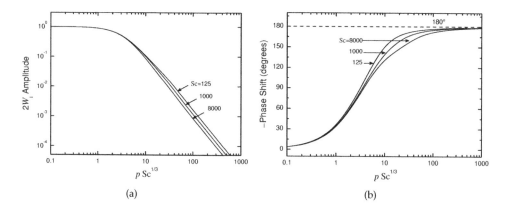

Figure 16.4: Dimensionless EHD impedance W_i in Bode representation with Schmidt number as a parameter: a) modulus versus the dimensionless frequency $p\mathrm{Sc_i}^{1/3}$ and b) phase shift versus the dimensionless frequency $p\mathrm{Sc_i}^{1/3}$.

The solution of the homogeneous equation is $\theta = \exp\left(-(jK_i)^{1/2}\zeta\right)$ and the solution of equation (16.38) may be written as

$$
\tilde{c}_i = \theta(\zeta) \left[\int_0^\zeta \theta^{-2}(\zeta') \right.
$$

$$
\left. \left(-\int_0^{\zeta'} \frac{\Delta\Omega}{\overline{\Omega}} \frac{3\tilde{f}'(p)}{a} \zeta''^2 \theta(\zeta'') \left. \frac{d\bar{c}_i}{d\zeta} \right|_0 d\zeta'' + K_1 \right) d\zeta' + \tilde{c}_i(0) \right] \tag{16.39}
$$

where

$$
K_1 = \frac{\Delta\Omega}{\overline{\Omega}} \frac{3\tilde{f}'(p)}{a} \left. \frac{d\bar{c}_i}{d\zeta} \right|_0 \int_0^\infty \zeta^2 \theta(\zeta) d\zeta \tag{16.40}
$$

and

$$
\left. \frac{d\tilde{c}_i}{dy} \right|_0 = \frac{\Delta\Omega}{\overline{\Omega}} \frac{\tilde{f}'(p)}{a} \left. \frac{d\bar{c}}{dy} \right|_0 \frac{6}{(jK_i)^{3/2}} \tag{16.41}
$$

At high frequencies, the derivative $\left. \frac{d\tilde{c}_i}{dy} \right|_0$ is proportional to the two complex quantities $\tilde{f}'(p)$ and $(jK_i)^{-1.5}$, and the phase shift is the sum of the phase shift of each complex quantity. The phase shift of $\tilde{f}'(p)$ tends toward $-45°$ when the dimensionless frequency p tends toward infinity (see equation (16.21)) and the phase shift of $(jK_i)^{-1.5}$ has a value of $-135°$. Thus, the phase shift of $\left. \frac{d\tilde{c}_i}{dy} \right|_0$ tends toward $-180°$ when the dimensionless frequency tends toward infinity. In the same way, the modulus of $\tilde{f}'(p)$ decreases with frequency with a slope of -0.5 in logarithmic coordinates (see Figure 16.2(a)), and the modulus of $(jK_i)^{-1.5}$ decreases with a slope of -1.5. The modulus of $\left. \frac{d\tilde{c}_i}{dy} \right|_0$ decreases with a slope of -2 as shown in Figure 16.4(a).

16.3 Kinetic Transfer Function for Simple Electrochemical Reactions

Under the assumption that the interface is uniformly reactive, $c_i(0)$ is independent of the radial coordinate. A single electrode reaction can be written in symbolic form as equation (10.1) (page 207). Following the treatment presented in Section 10.2.2, the faradaic current can be written in the form

$$
i_F = f(V, c_i(0)) \tag{16.42}
$$

For minor species, in the presence of supporting electrolyte and with neglect of double-layer adsorption of these minor species, the concentration gradient is related to the faradaic current density by

$$
D_i \left. \frac{\partial c_i}{\partial y} \right|_0 = \frac{s_i}{nF} i_F \tag{16.43}
$$

Following equation (10.5), a Taylor series expansion about the steady value can be written as

$$\tilde{i}_F = \left(\frac{\partial f}{\partial V}\right)_{c_i(0)} \tilde{V} + \sum_i \left(\frac{\partial f}{\partial c_i(0)}\right)_{V, c_j(0), j \neq i} \tilde{c}_i(0) \tag{16.44}$$

Following the discussion presented in Chapter 10 (page 207), the usual charge-transfer resistance R_t can be identified as the reciprocal of $(\partial f / \partial V)_{c_i(0)}$. Combination of equations (16.37), (16.43), and (16.44) yields

$$\tilde{V} = R_t \tilde{i}_F - \sum_i R_t \left(\frac{\partial f}{\partial c_i(0)}\right)_{V, c_j(0), j \neq i} \left[\frac{\delta_i}{\theta_i'(0)} \frac{s_i}{nFD_i} \tilde{i}_F - \frac{\Delta\Omega}{\overline{\Omega}} \frac{W_i \delta_i}{\theta_i'(0)} \frac{s_i \overline{i}_F}{nFD_i}\right] \tag{16.45}$$

or

$$\tilde{V} = R_t \tilde{i}_F + Z_D \tilde{i}_F + \frac{\Delta\Omega}{\overline{\Omega}} \frac{\overline{i}_F R_t}{nF} \sum_i \left(\frac{\partial f}{\partial c_i(0)}\right)_{V, c_j(0), i \neq j} \frac{W_i \delta_i}{\theta_i'(0)} \frac{s_i}{D_i} \tag{16.46}$$

where

$$Z_D = -R_t \sum_i \left(\frac{\partial f}{\partial c_i(0)}\right)_{V, c_j(0), j \neq i} \frac{\delta_i s_i}{nFD_i \theta_i'(0)} \tag{16.47}$$

The term Z_D represents the convective-diffusion impedance with its dimensionless form $-1/\theta_i'(0)$ (see Chapter 11 on page 243). The relationship among observable quantities corresponding to the general form of equation (15.22) can be found from equations (15.27), (15.28), and (16.46) to be

$$\tilde{U} = Z\tilde{I} + \frac{\Delta\Omega}{\overline{\Omega}} \frac{1}{1 + j\omega C_D(R_t + Z_D)} \frac{\overline{I} R_t}{nF} \sum_i \frac{\partial f}{\partial c_i(0)}\bigg|_{E, c_j(0), j \neq i} \frac{W_i \delta_i}{\theta_i'(0)} \frac{s_i}{D_i} \tag{16.48}$$

where

$$Z = R_e + \frac{R_t + Z_D}{1 + j\omega C_D(R_t + Z_D)} \tag{16.49}$$

The term Z represents the impedance of the usual Randles equivalent circuit (see Figure 10.5 on page 220).

16.4 Interface with a 2-D or 3-D Insulating Phase

The main hypotheses for developing the EHD impedance theory are that the electrode interface is uniformly accessible and the electrode surface has uniform reactivity. However, in many cases, real interfaces deviate from this ideal picture due, for example, either to incomplete monolayer adsorption leading to the concept of partial blocking (2-D adsorption) or to the formation of layers of finite thickness (3-D phenomena). These effects do not involve the interfacial kinetics on bare portions of the metal, which, for simplification, will be assumed to be inherently fast. The changes will affect only the local mass transport toward the reaction sites. Before presenting an application of practical interest, the theoretical EHD impedance for partially blocked electrodes and for electrodes coated by a porous layer will be analyzed.

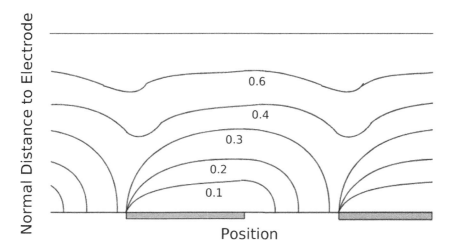

Figure 16.5: Concentration field over a partially blocked electrode. The active parts of the electrode are displayed in gray and the concentration is dimensionless $\theta = c/c(\infty)$. (Taken from Caprani et al.[328])

16.4.1 Partially Blocked Electrode

Integration of the nonstationary material balance equation requires that the steady-state concentration distribution be known. This is achieved by solving the steady-state part of

$$\frac{\partial c_i}{\partial t} + v_x \frac{\partial c_i}{\partial x} + v_y \frac{\partial c_i}{\partial y} = D_i \left(\frac{\partial^2 c_i}{\partial x^2} + \frac{\partial^2 c_i}{\partial y^2} \right) \tag{16.50}$$

where $v_x = \sqrt{v_r^2 + v_\theta^2}$ is the longitudinal velocity and $v_y = \beta_y(t)y^2$ is the normal velocity. The steady-state iso-concentration lines, deduced from a numerical solution,[328] are plotted in Figure 16.5. The boundary conditions are that $c_i = 0$ for $y = 0$ on the active surfaces (shown in gray in Figure 16.5) and $\partial c_i/\partial y = 0$ on the insulating surfaces. The concentration profiles shown in Figure 16.5 correspond to the effect of two neighboring microelectrodes. Such an arrangement describes locally a periodical distribution of active and passive sites. It appears that the concentration profiles of the two leading edges coincide reasonably well and one may assume that the memory effect on the concentration distribution is lost over one or a few active sites. The iso-concentration lines are parallel to the electrode surface at large values of y (far from the electrode). This condition is the same as seen for a uniformly accessible disk electrode. The iso-concentration lines close to the electrode are similar to the iso-concentration lines of an

Remember! 16.4 *Electrohydrodynamic impedance provides a means to reveal and quantify the influence of partially blocked electrodes and electrodes coated by porous layers.*

isolated microelectrode (see Section 13.5).

For a partially blocked electrode, the diagrams show two characteristic frequencies and, whatever the mean rotation speed Ω, fall onto a single curve when plotted versus a dimensionless frequency $p = \omega/\Omega$, where $\omega/2\pi$ is the frequency modulation. The high-frequency domain is characteristic of the response of the sum of the active sites, as if they were not coupled by their diffusion layers.

The decoupled behavior of the active sites at high frequency is phenomenologically based on the theoretical analysis given by Deslouis et al.[329] for the frequency response of a small electrode to a flow modulation. The response of a partially blocked electrode surface was obtained from numerical calculations.[328] The main result of this approach is that, at variance with the case of a uniform active disk electrode, such a surface shared between active and passive sites displays two characteristic frequencies, one in the low-frequency range corresponding to the uniform active disk response and one in the high-frequency range corresponding to the response of individual active sites, which are then assumed to behave without interactions.

The average active site dimension d_{act} was deduced from the two characteristic frequencies corresponding, respectively, to the high-frequency behavior and to the low-frequency behavior, p_{HF} and p_{LF}, respectively, i.e.,

$$d_{act} = 2.1^{3/2} R \left(p_{HF}/p_{LF} \right)^{-3/2} \tag{16.51}$$

where R is the electrode radius, and p_{HF} and p_{LF} are obtained from the intercept on the EHD amplitude Bode diagrams of the low-frequency horizontal plateau and the two lines characterizing the high-frequency behaviors of the disk electrode and of the microelectrodes, respectively.

Experimental verification of the calculation was provided by Silva et al.[330] with arrays of microelectrodes prepared by photolithography. A very good agreement was found between the common microelectrode dimension measured by SEM microscopy and calculated from applications of equation (16.51). The phase-shift response of the microelectrode array in Figure 16.6 clearly shows that two time constants are resolved: one in the low-frequency range approaches the active disk response and the other at higher frequency fairly coincides with the microelectrode response. This approach was further extended to a fully active rough surface by phenomenological analogy where the dimensions of a single protrusion in the electrode plane played the same role as the characteristic dimension of a plane active site in the 2-D frame.[332]

Remember! 16.5 *For a partially blocked electrode, all diagrams obtained at different rotation speeds merge on one diagram by using the normalized amplitude and the dimensionless frequency p. This is not observed for a coated electrode.*

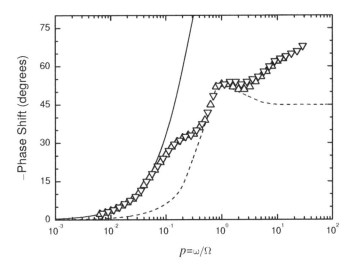

$$p=\omega/\Omega$$

Figure 16.6: Phase shift of the EHD impedance corresponding to the ferricyanide reduction on a platinum electrode (area 0.38 cm²) coated with a photoresist on which an array of circular sites of diameter d_{act} was patterned. $d_{act} = 649\ \mu m$, $\Omega = 96$ rpm (▼) and 375 rpm (▲). Theoretical curve for a single microelectrode (dashed line) or for the active disk electrode (solid line). (Taken from Deslouis and Tribollet.[331])

16.4.2 Rotating Disk Electrode Coated by a Porous Film

Porous nonreacting layers covering reacting metallic interfaces may slow down the mass transfer of diffusing species. This decrease includes the effect of the diffusivity $D_{f,i}$, as well as that of the layer thickness δ_f. This problem was discussed in a more general way in Section 11.6 on page 271.

A schematic representation of the system under investigation is presented in Figure 11.18 on 271. The concentration gradient is distributed between the fluid and the porous layer. In addition, the metal-layer interface is assumed to be uniformly reactive.

Two material balance equations can be written as follows:

1. In the porous layer, the concentration distribution $c_i^{(1)}$ is determined only by molecular diffusion following

$$\frac{\partial c_i^{(1)}}{\partial t} = D_{i,f}\frac{\partial^2 c_i^{(1)}}{\partial y^2} \tag{16.52}$$

2. In the fluid, the concentration distribution $c_i^{(2)}$ is governed by convective diffusion, i.e.,

$$\frac{\partial c_i^{(2)}}{\partial t} = D_i\frac{\partial^2 c_i^{(2)}}{\partial y_e^2} - v_y\frac{\partial c_i^{(2)}}{\partial y_e} \tag{16.53}$$

For simplicity, the origin of the coordinate y is taken to be at the metal-layer interface and that of y_e is at the layer-fluid interface ($y_e = y - \delta_f$).

Associated with equations (16.52) and (16.53) are boundary conditions that express the continuity of the concentration fields and of the fluxes for the steady state as well as for the time-dependent quantities. For $y = \delta_f$ or $y_e = 0$,

$$c_i^{(1)}(\delta_f) = c_i^{(2)}(0) \tag{16.54}$$

and

$$D_{f,i}\frac{\partial c_i^{(1)}}{\partial y} = D_i\frac{\partial c_i^{(2)}}{\partial y_e} \tag{16.55}$$

At $y_e = 0$,

$$v_y = 0 \tag{16.56}$$

At $y_e \to \infty$, $c_i^{(2)} \to c_i(\infty)$ (then $\bar{c}_i^{(2)} \to c_i(\infty)$ and $\hat{c}_i^{(2)} \to 0$).

Steady-State Solutions

Equation (16.52) is reduced to the simple form

$$\frac{d^2\bar{c}_i^{(1)}}{dy^2} = 0 \tag{16.57}$$

which leads to

$$J_i = D_{f,i}\frac{\bar{c}_i^{(1)}(\delta_f) - \bar{c}_i^{(1)}(0)}{\delta_f} \tag{16.58}$$

If a reaction of first order is assumed

$$J_i = K\bar{c}_i^{(1)}(0) \tag{16.59}$$

The flux in the electrolyte is given by

$$J_i = D_i\frac{c_i(\infty) - \bar{c}_i^{(2)}(0)}{\delta_{N,i}} \tag{16.60}$$

By eliminating $\bar{c}_i^{(1)}(0)$ and $\bar{c}_i^{(2)}(0)$ between these last three equations, one obtains

$$J_i = \frac{c_i(\infty)}{\dfrac{1}{k} + \dfrac{\delta_{N,i}}{D_i} + \dfrac{\delta_f}{D_{f,i}}} \tag{16.61}$$

Remember! 16.6 *In contrast to Reminder 16.5, for an electrode coated by a porous film, all diagrams obtained at different rotation speeds do not collapse to one diagram by using the normalized amplitude and the dimensionless frequency p.*

---

which can be written in the form

$$J_i^{-1} = J_{i,k}^{-1} + J_{i,\lim}^{-1} + J_{i,\Omega\to\infty}^{-1} \tag{16.62}$$

where

$$J_{i,k} = Kc_i(\infty) \tag{16.63}$$

is the kinetic flux,

$$J_{i,\lim} = D_i \frac{c_i(\infty)}{\delta_{N,i}} \tag{16.64}$$

is the limiting diffusion flux on a metallic surface, free from porous layer, and

$$J_{i,\Omega\to\infty} = D_{f,i} \frac{c_i(\infty)}{\delta_f} \tag{16.65}$$

is the limiting flux when the entire concentration gradient is located within the porous layer, i.e., when $\Omega \to \infty$.

The interest of using reciprocal values is that, as shown in Example 5.6 on page 106, the experimental plot of $1/J_i$ as a function of $1/\sqrt{\Omega}$ must be a straight line parallel to the line corresponding to Levich's result for the mass-transfer-limited case, which passes through the origin. The ordinate value of the intercept of this straight line at $1/\sqrt{\Omega} = 0$ is $1/J_{i,k} + 1/J_{i,\Omega\to\infty}$. In the particular case of a very fast reaction, $1/J_{i,k} \to 0$ and the value of the intercept is $D_{f,i}/\delta_f$.

AC and EHD Impedances

In the porous layer, the fluctuating part of equation (16.52) may be written as

$$\frac{d^2\widetilde{c}_i^{(1)}}{dy^2} - \frac{j\omega}{D_{f,i}}\widetilde{c}_i^{(1)} = 0 \tag{16.66}$$

The solution is

$$\widetilde{c}_i^{(1)} = M \exp\sqrt{\frac{j\omega}{D_{f,i}}y^2} + N \exp\sqrt{-\frac{j\omega}{D_{f,i}}y^2} \tag{16.67}$$

where M and N are integration constants obtained from the boundary conditions.

At the fluid-layer interface ($y_e = 0$), equation (16.37) can be applied since there is no additional process in the fluid, i.e.,

$$\left.\frac{d\widetilde{c}_i^{(2)}}{dy_e}\right|_{y_e=0} = \frac{\widetilde{c}_i^{(2)}(0)}{\delta_{N,i}}\theta_i'(0) + \frac{\Delta\Omega}{\overline{\Omega}}\left.\frac{d\overline{c}_i^{(2)}}{dy_e}\right|_{y_e=0} W_i \tag{16.68}$$

From the boundary conditions at $y = \delta_e$, the constants M and N may be eliminated and the general expression is obtained as

$$
\left.\frac{d\widehat{c}_i^{(1)}}{dy}\right|_0 = \frac{-\widehat{c}_i^{(1)}}{\delta_f} \frac{\left(\dfrac{j\omega\delta_f^2}{D_{f,i}}\right) Z_D Z_{D,f} + \dfrac{D_i}{D_{f,i}}\dfrac{\delta_f}{\delta_{N,i}}}{Z_D + \dfrac{D_i}{D_{f,i}}\dfrac{\delta_f}{\delta_{N,i}} Z_{D,f}}
$$

$$
+ \Delta\Omega \frac{\dfrac{1}{\cosh(j\omega\delta_f^2/D_{f,i})} \left.\dfrac{d\widehat{c}_i^{(1)}}{dy}\right|_0 \dfrac{W_i}{\overline{\Omega}}}{1 + \dfrac{D_i}{D_{f,i}}\dfrac{\delta_f}{\delta_{N,i}}\dfrac{Z_{D,f}}{Z_D}}
$$

(16.69)

where $Z_D = -1/\theta_i'(0)$ is the dimensionless convective diffusion in the solution and

$$
Z_{D,f} = \frac{\tanh\sqrt{j\omega\delta_f^2/D_{f,i}}}{\sqrt{j\omega\delta_f^2/D_{f,i}}}
$$

(16.70)

is the dimensionless diffusion impedance for a finite stagnant diffusion layer (see equation (11.20) on page 247). For $\Delta\Omega = 0$, equation (16.69) becomes similar to the result for a rotating disk presented as equation (11.128) on page 278.

It may be easily verified that when the layer effect is gradually decreased (i.e., $\delta_f \to 0$ and $D_{f,i} \to D_i$) one finds again the relation (16.37). At the opposite extreme, when $\Omega \to \infty$, then $\delta_{N,i} \to 0$, the relation becomes

$$
\left.\frac{d\widehat{c}_i^{(1)}}{dy}\right|_0 = -\frac{\widehat{c}_i^{(1)}(0)}{\delta_f}\frac{1}{Z_{D,f}}
$$

(16.71)

In Figures 16.7(a) and (b), the amplitude and the phase shift corresponding to equation (16.69) for different angular velocities show that, in contrast with the simple behavior of a bare electrode, the data are no longer reducible by the dimensionless frequency p. In fact $\left.\frac{d\widehat{c}_i^{(1)}}{dy}\right|_0 / \Delta\Omega$ contains both W_i and $(-1/\theta_i'(0))$, which depend on p, for a given Schmidt number, and also a function of $j\omega\delta_f^2/D_{f,i}$. Thus, an increase of Ω produces a shift of the Bode diagrams toward smaller p values, other parameters being kept constant. Frequency analysis provides both the diffusion time constant $\delta_f^2/D_{f,i}$ and the speed of diffusion $D_{f,i}/\delta_f$. Thus, independent estimates can be obtained for δ_f and $D_{f,i}$.

Example 16.1 2-D and 3-D Blocking: *Apply the concepts of electrohydrodynamic impedance to scale deposit in seawater. This system shows behavior associated with both partially blocked surfaces and diffusion through porous layers.*

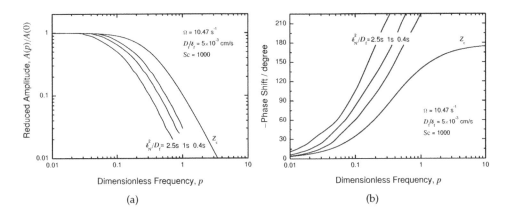

Figure 16.7: Concentrostatic EHD impedance as a function of the dimensionless frequency p with rotation speed as a parameter. (Taken from Deslouis et al.[202]) The line marked Z_c represents the EHD impedance on a bare electrode: a) dimensionless modulus and b) phase shift.

Solution: *Cathodic protection is widely used to protect immersed metallic structures from corrosion. When this technique is applied, in the range of -0.8 to -1.2 V (SCE), dissolved oxygen from seawater reduces onto metallic surface according to*

$$O_2 + 2H_2O + 2e^- \rightarrow H_2O_2 + 2OH^- \tag{16.72}$$

and

$$H_2O_2 + 2e^- \rightarrow 2OH^- \tag{16.73}$$

In addition, hydrogen evolution takes place following

$$H_2O + 2e^- \rightarrow H_2 + 2OH^- \tag{16.74}$$

The production of hydroxyl ions OH^- by reactions (16.72), (16.73), and (16.74) allows precipitation of magnesium hydroxide

$$Mg^{2+} + 2OH^- \leftrightarrows Mg(OH)_2 \downarrow \tag{16.75}$$

Moreover, these reactions lead to changes in inorganic carbonic equilibrium at the metallic interface

$$2OH^- + HCO_3^- \leftrightarrows H_2O + CO_3^{2-} \tag{16.76}$$

and allow precipitation of $CaCO_3$, i.e.,

$$Ca^{2+} + CO_3^{2-} \rightarrow CaCO_3 \downarrow \tag{16.77}$$

Calcareous deposition ($CaCO_3$ and $Mg(OH)_2$) on the metallic surface creates a diffusional barrier toward oxygen and thus decreases the energy needed to maintain efficient protection. Knowledge of the formation time and characteristics of such layers are then essential to improve cathodic protection monitoring.

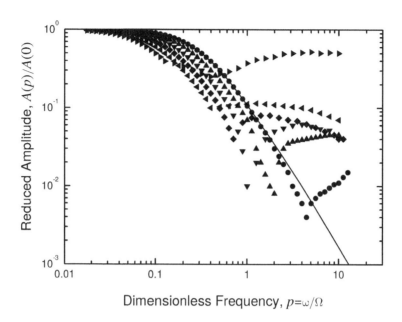

Figure 16.8: Modulus of the EHD impedance during the calcareous formation at -1.2 V(SCE) and 360 rpm. The solid line corresponds to $I = I_0$; $I/I_0 = 0.95$ (•), 0.87 (▲), 0.72 (▼), 0.55 (♦), 0.31 (◄), and 0.13 (►). (Taken from Deslouis and Tribollet.[331])

Studies in seawater have been carried out with a gold rotating disk electrode. The EHD impedances were recorded during the calcareous deposition for different potentials and different values of mean velocity. As an example, diagrams corresponding to -1.2 V(SCE) and 360 rpm are presented in Figure 16.8 as the logarithm of normalized modulus as a function of the logarithm of dimensionless frequency p. The corresponding dc current is indicated as a fraction of I_0, the dc current for a bare surface.

The time constant with calcareous deposits was found to be the same as for a bare surface in the low-frequency range with a slight shift of diagrams to lower reduced frequencies when the deposition time increases. At high frequency (above $p = 1$), a second time constant appeared. Both these features characterized the presence of a partially blocked surface with a moderate value of the active fraction of the electrode. The slight separation of the EHD diagrams observed in the low-frequency region, when the deposition time is increasing, accounted for the existence of diffusion through a porous layer. This correspond to the area covered by the $Mg(OH)_2$ layer.

The average crystal size d was calculated from EHD diagrams obtained for calcareous deposits formed at different applied potentials and at 360 rpm (Figure 16.9(a)) and for calcareous deposits formed at -1.2 V(SCE) and different electrode rotation speeds (Figure 16.9(b)). When analyzing these results, it seems that the applied potential does not affect the value of d. The crytal size is around 15 μm at the beginning of the scale formation and increases to around 30-40 μm for $I/I_0 \approx 0.5$. The values obtained at the end of deposition increase to 200 μm. The influence of stirring is clearly defined (Figure 16.9(b)) and is consistent with previous work on

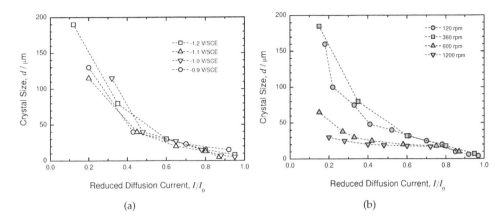

Figure 16.9: Variation of average crystal size d during the calcareous formation: a) at 360 rpm and different potentials and b) at -1.2 V(SCE) and different electrode rotation speeds. (Taken from Deslouis and Tribollet.[331])

CaCO$_3$ alone.[333]

In a general way, the values of the crystal dimension d are in good agreement with the crystal size shown in SEM pictures (Figure 16.10). The crystal sizes are close to 15 μm at the beginning of deposition and stabilize at around 30 μm at the end of their formation. In the present case, the dimension d would therefore allow the determination of the size of crystals at the beginning of calcareous formation, when they are randomly distributed on the surface and do not coalesce. The growth of crystals can then be observed until approximately I = 0.5I$_0$.

For smaller I / I$_0$ values, i.e., when rate of calcareous formation is larger, the d values increase very rapidly. Intuitively, one may suppose that this corresponds to overlapping of calcium carbonate crystals. The dimension d would then give the average distance between areas that remain active, and, therefore, the size of CaCO$_3$ aggregates. However, the corresponding theory is not yet available, and the above interpretation is speculative for the variations of d in the range 0 < I / I$_0$ < 0.5.

Electrohydrodynamic impedance characterization of calcareous deposits showed mainly partially blocked electrode behavior and allowed the estimation of the average size of characteristic sites of the interface. These results have been confirmed by ex situ SEM images.

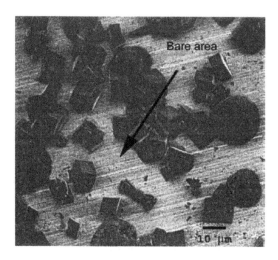

Figure 16.10: Calcareous deposit formed at -1.2 V(SCE) and 1200 rpm, $I/I_0 = 0.7$. (Taken from Deslouis and Tribollet.[331])

Problems

16.1   From equation (16.37), deduce the expressions (15.21) and (15.22).

16.2   Derive the EHD impedance for the anodic dissolution of copper in a chloride medium where the reactions proceed according to

$$Cu + Cl^- \rightleftarrows CuCl_{ads} + e^- \tag{16.78}$$

where $CuCl_{ads}$ is an adsorbed intermediate that reacts to form $CuCl_2^-$ by

$$CuCl_{ads} + Cl^- \rightleftarrows CuCl_2^- \tag{16.79}$$

The mass-transfer limitation is due only to $CuCl_2^-$.

16.3   Beginning with equation (16.48), write an expression for the electrohydrodynamic impedance under potentiostatic regulation $Z_{EHD,p}$, the electrohydrodynamic impedance under galvanostatic regulation $Z_{EHD,g}$, and the relationship between $Z_{EHD,p}$, $Z_{EHD,g}$, and the electrochemical impedance.

Part IV

Interpretation Strategies

Chapter 17

Methods for Representing Impedance

Impedance data are presented in different formats to emphasize specific classes of behavior. The impedance format emphasizes the values at low frequency, which typically are of greatest importance for electrochemical systems that are influenced by mass transfer and reaction kinetics. The admittance format, which emphasizes the capacitive behavior at high frequencies, is often employed for solid-state systems. The complex capacity format is used for dielectric systems in which the capacity is often the feature of greatest interest. The method of representing impedance data has great impact on the use of graphical methods to visualize and interpret data.

Chapters 1 and 4 provide a useful foundation for the material presented in this chapter. Summaries of identities for the imaginary number, relationships among complex variables, and relationships between polar and rectangular coordinates are presented in Tables 1.1, 1.2, and 1.3, respectively.

The methods for representing impedance data are illustrated here for the two simple *RC* electrical circuits shown in Table 17.1. The material presented in this chapter supports Chapter 18 (page 493), which covers graphical methods that may be used to extract quantitative information from more typical electrochemical impedance data.

17.1 Impedance Format

The impedance can be expressed as the complex ratio of potential and current contributions, i.e.,

$$Z = \frac{\widetilde{U}}{\widetilde{I}} \tag{17.1}$$

As discussed in Section 4.1.2, the impedance for a series arrangement of passive elements is additive. The impedance for the parallel arrangement of impedances Z_1 and Z_2 is given by equation (4.24).

Table 17.1: Summary of complex impedance, admittance, and capacitance characteristics for simple blocking and reactive circuits.

Circuit	(a)	(b)
Classification		
	blocking	reactive
Complex impedance		
Z_r	R_e	$R_e + \dfrac{R}{1 + (\omega RC)^2}$
Z_j	$-\dfrac{1}{\omega C}$	$-\dfrac{\omega CR^2}{1 + (\omega RC)^2}$
Time constant	none	RC
Complex admittance		
Y_r	$\dfrac{R_e\,(\omega C)^2}{1 + (\omega R_e C)^2}$	$\dfrac{R_e R\,(\omega C)^2 + 1 + R_e/R}{R\left[(1 + R_e/R)^2 + (\omega R_e C)^2\right]}$
Y_j	$\dfrac{\omega C}{1 + (\omega R_e C)^2}$	$\dfrac{\omega C}{(1 + R_e/R)^2 + (\omega R_e C)^2}$
Time constant	$R_e C$	$\dfrac{R_e C}{(1 + R_e/R)}$
Complex capacitance‡		
C_r	C	C
C_j	0	$-\dfrac{1}{R\omega}$
Time constant	none	none
Effective Capacitance $C_{eff} = -\dfrac{1}{\omega Z_j}$	C	$C + \dfrac{1}{\omega^2 R^2 C}$
Time constant	none	RC

‡Complex capacitance is determined for $Z - R_e$.

For a resistor R_e and capacitor C in series, shown in Table 17.1(a), the impedance is given by

$$Z = R_e - j\frac{1}{\omega C} \tag{17.2}$$

The real part of the impedance is independent of frequency, and the imaginary part of the impedance tends to $-\infty$ as frequency tends toward zero according to $1/\omega$. In fact,

$$-\omega Z_j = \frac{1}{C} \tag{17.3}$$

for all frequencies ω.

The system comprising the resistor R_e and capacitor C in series provides an example of a class of systems for which, at the zero-frequency or dc limit, current cannot pass. Such systems are considered to have a blocking or ideally polarizable electrode. Depending on the specific conditions, batteries, liquid mercury electrodes, semiconductor devices, passive electrodes, and electroactive polymers provide examples of systems that exhibit such blocking behavior.

For a resistor R_e in series with the parallel combination of R and capacitor C, shown in Table 17.1(b), the impedance is given by

$$Z = R_e + \frac{R}{1 + j\omega RC} \tag{17.4}$$

or

$$Z = R_e + \frac{R}{1 + (\omega RC)^2} - j\frac{\omega C R^2}{1 + (\omega RC)^2} \tag{17.5}$$

The real part of the impedance tends toward $R_e + R$ as frequency tends toward zero. The imaginary part of the impedance tends toward zero as frequency tends toward zero such that

$$-\lim_{\omega \to 0} \frac{Z_j}{\omega} = C R^2 \tag{17.6}$$

A characteristic angular frequency $\omega_{RC} = 1/(RC)$ can be identified for which the imaginary part of the impedance has a maximum value

$$-Z_j(\omega_{RC}) = \frac{R}{2} \tag{17.7}$$

The real part of the impedance tends toward R_e as frequency tends toward ∞, and the imaginary part of the impedance tends toward zero such that

$$-\lim_{\omega \to \infty} \omega Z_j = \frac{1}{C} \tag{17.8}$$

Remember! 17.1 *The impedance representation emphasizes values at low frequency and is often used for electrochemical systems for which information is sought regarding mass transfer and reaction kinetics.*

The high-frequency limit for the imaginary part of the impedance given in equation (17.8) is identical to that given for the series arrangement as equation (17.3) for all frequencies.

The reactive system shown in Table 17.1(b) may be considered to be an example of a class of systems for which, at the zero-frequency or dc limit, the resistance to passage of current is finite, and dc current can pass. Many electrochemical and electronic systems exhibit such nonblocking or reactive behavior. Even though the impedance response of the systems represented in this chapter is extremely simple as compared to that of typical electrochemical and electronic systems, the blocking and nonblocking systems comprise a broad cross-section of electrochemical and electronic systems. The concepts described can therefore be easily adapted to experimental data.

The resistor R and capacitor C shown in Table 17.1 can take on different meanings for different electrochemical systems. The resistance may, for example, be associated with the charge-transfer resistance of an electrochemical reaction, with the resistance of an oxide or porous layer, or with the electronic resistance of a semiconductor. The capacitor C may be associated with the double layer for an electrode in electrolyte, with surface capacitance of a film, or with the space-charge region of a semiconductor. The resistor R_e may be associated with the ohmic resistance of the electrolyte or with the frequency-independent resistance of a solid.

17.1.1 Complex-Impedance-Plane Representation

Impedance data are often represented in complex-impedance-plane or Nyquist format, shown in Figure 17.1. The data are presented as a locus of points, where each data point corresponds to a different measurement frequency. One disadvantage of the complex-impedance-plane format is that the frequency dependence is obscured. This disadvantage can be mitigated somewhat by labeling some characteristic frequencies. In fact, characteristic frequencies should always be labeled to allow a better understanding of the time constants of the underlying phenomena. In addition, the real and imaginary axes must have the same scale such that the plot of a semicircle has the appearance of a semicircle. Adherence to the orthonormal axes convention facilitates interpretation of spectra.

Due to the use of orthonormal axes in the Nyquist representation shown in Figure 17.1, the impedance for the reactive circuit of Table 17.1(b) has the appearance of a semicircle. The asymptotic limits of the real part of the impedance are R_e at high frequencies and $R_e + R$ at low frequencies. These limits are indicated in the complex-impedance-

---

Remember! 17.2 *Complex-impedance-plane or Nyquist plots should have orthonormal axes, and some characteristic frequencies should be labeled.*

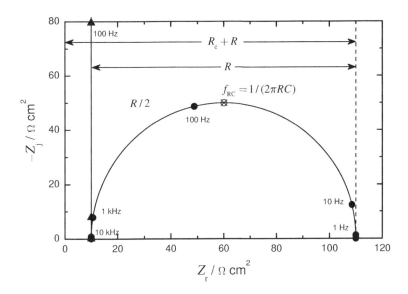

Figure 17.1: Impedance-plane or Nyquist representation of impedance data for $R_e = 10\ \Omega\text{cm}^2$, $R = 100\ \Omega\text{cm}^2$, and $C = 20\ \mu\text{F}/\text{cm}^2$. The blocking system of Table 17.1(a) is represented by ▲ and dashed lines, and the reactive system of Table 17.1(b) is represented by ● and solid lines.

plane plot, and the characteristic frequency for the system is shown at the peak where the negative imaginary impedance reaches a maximum value. For the system chosen here, $f_{RC} = 79.6$ Hz. The maximum value of the imaginary part of the impedance is equal to $R/2$ for a single RC circuit.

The real part of the impedance for the blocking circuit shown in Table 17.1(a) is equal to R_e for all frequencies. The tendency of the imaginary part to approach $-\infty$ as frequency tends toward zero appears as a vertical line in the complex-impedance-plane plot.

Complex-impedance-plane plots are very popular because the shape of the locus of points yields insight into possible mechanisms or governing phenomena. If the locus of points traces a perfect semicircle, for example, the impedance response corresponds to a single activation-energy-controlled process. A depressed semicircle indicates that a more detailed model is required, and multiple peaks provide clear indication that more than one time constant is required to describe the process. The significant disadvantages are that the frequency dependence is obscured, that low impedance values are obscured, and that apparent agreement between model and experimental data in impedance-plane format may obscure large differences in frequency and at low impedance values.

Example 17.1 Analysis of Two Nyquist Plots: *Consider the Nyquist plots for two systems. Each Nyquist plot, shown in Figure 17.2, comprises two capacitive loops with*

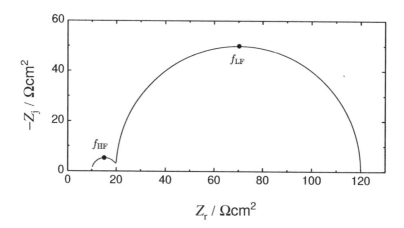

Figure 17.2: Nyquist plot for Example 17.1 showing an ohmic resistance of 10 Ωcm^2 and two capacitive loops with diameters of 10 Ωcm^2 for the high-frequency loop and 200 Ωcm^2 for the low-frequency loop.

diameters of 10 Ωcm^2 for the high-frequency loop and 200 Ωcm^2 for the low-frequency loop. The characteristic frequencies are $f_{HF} = 16$ kHz and $f_{LF} = 20$ Hz for the first system and $f_{HF} = 400$ Hz and $f_{LF} = 0.8$ Hz for the second system. Propose a physical model for each impedance diagram.

Solution: *The characteristic frequency f_c is equal to $1/2\pi RC$; thus, the value of the capacitance is $C = 1/2\pi R f_c$.*

a) System One: *For the first loop, the capacitance value is*

$$C_{HF} = \frac{1}{2\pi \times 10 \times 16,000} = 0.99 \ \mu F/cm^2 \tag{17.9}$$

For the second loop, the capacitance is

$$C_{LF} = \frac{1}{2\pi \times 200 \times 20} = 39.8 \ \mu F/cm^2 \tag{17.10}$$

Due to the values of the two capacitances, the high-frequency loop could be attributed to a layer and the low-frequency loop to a double layer in parallel with a charge-transfer resistance.

b) System Two: *For the second Nyquist plot, the capacitance of the first loop is*

$$C_{HF} = \frac{1}{2\pi \times 10 \times 400} = 39.8 \ \mu F/cm^2 \tag{17.11}$$

For the second loop the capacitance value is

$$C_{LF} = \frac{1}{2\pi \times 200 \times 0.8} = 0.995 \ mF/cm^2 \tag{17.12}$$

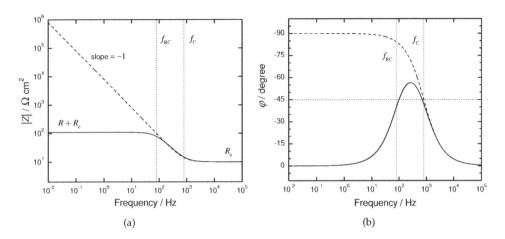

Figure 17.3: Bode representation of impedance data for $R_e = 10 \ \Omega cm^2$, $R = 100 \ \Omega cm^2$, and $C = 20 \ \mu F/cm^2$. The blocking system of Table 17.1(a) is represented by dashed lines, and the reactive system of Table 17.1(b) is represented by solid lines. Characteristic frequencies are noted as $f_{RC} = 1/(2\pi RC)$ and $f_c = 1/(2\pi R_e C)$; a) magnitude and b) phase angle.

The high-frequency capacitance may correspond to a double layer in parallel with a charge-transfer resistance. The value of the low-frequency capacitance is too large to be a real capacitance; this loop corresponds to a capacitance effect and could be attributed to mass transport or adsorption of electroactive species.

17.1.2   Bode Representation

The functionality with respect to frequency is seen more clearly in the Bode representation shown in Figures 17.3(a) and (b) for the magnitude and phase angle, respectively. Frequency is generally presented on a logarithmic scale to reveal the important behavior seen at lower frequencies. Note that, following customary practice, the frequency f given in Figure 17.3 has units of Hz (cycles/s); whereas the angular frequency ω used in the mathematical development has units of s^{-1} (radian/s). The conversion is given by $\omega = 2\pi f$.

The magnitude of the impedance of the blocking circuit of Table 17.1a can be expressed as

$$|Z| = \sqrt{R_e^2 + \left(\frac{1}{\omega C}\right)^2} \qquad (17.13)$$

The magnitude tends toward R_e as frequency tends toward ∞ and toward ∞ as $1/\omega$ as frequency tends toward zero. The magnitude is usually presented on a logarithmic scale as a function of frequency on a logarithmic scale, as shown in Figure 17.3(a). The slope of the line at low frequencies, therefore, has a value of -1 for the blocking electrode considered here. A value with magnitude smaller than unity could provide an indication of a blocking electrode with a distribution of characteristic time constants. Such systems are described in Chapter 18 on page 493.

The magnitude of the impedance of the reactive system of Table 17.1b can be expressed as

$$|Z| = \sqrt{\left(R_e + \frac{R}{1 + (\omega RC)^2}\right)^2 + \left(\frac{\omega CR^2}{1 + (\omega RC)^2}\right)^2} \tag{17.14}$$

The magnitude tends toward R_e as frequency tends toward ∞ and toward $R_e + R$ as frequency tends toward zero. The transition between low frequency and high frequency asymptotes has a slope of -1 on a log-log scale.

Following equation (4.32), the phase angle for the blocking configuration can be expressed as

$$\varphi = \tan^{-1}\left(\frac{-1}{\omega R_e C}\right) \tag{17.15}$$

The phase angle tends toward $-90°$ at low frequencies and toward zero at high frequencies. The phase angle at the characteristic angular frequency $\omega_c = 1/(R_e C)$ is equal to $-45°$ (see equation (17.15)).

The phase angle for the reactive configuration can be expressed as

$$\varphi = \tan^{-1}\left(\frac{-\omega R^2 C}{R + R_e \left(1 + (\omega RC)^2\right)}\right) \tag{17.16}$$

The phase angle tends toward zero at low frequencies, indicating that the current and potential are in phase. The phase angle tends toward zero at high frequencies as well, due to the influence of the leading resistor R_e in equation (17.16). Note that, while the characteristic angular frequency for this circuit is $\omega_{RC} = 1/(RC)$, the phase angle at the characteristic angular frequency ω_{RC} is given by

$$\varphi = \tan^{-1}\left(\frac{-1}{1 + 2R_e/R}\right) \tag{17.17}$$

which is equal to $-45°$ only if $R_e/R = 0$. The characteristic angular frequency for which the phase angle is equal to $-45°$ can be expressed as

$$\omega_c = \frac{1}{2R_e C}\left(1 \pm \sqrt{1 - 4\frac{R_e}{R}\left(1 + \frac{R_e}{R}\right)}\right) \tag{17.18}$$

but this expression is valid only if

$$\frac{R_e}{R} \leq \frac{\sqrt{2} - 1}{2} \tag{17.19}$$

For the case presented here, the phase angle reaches $-45°$ at frequencies of 100 and 696 Hz, values that have no direct correspondence to the characteristic frequency based on R of $f_{RC} = 79.6$ Hz or based on R_e of $f_c = 796$ Hz. The peak in the phase angle is seen at a characteristic frequency

$$f_c = \frac{1}{4\pi RC}\sqrt{1 + \frac{R}{R_e}} \tag{17.20}$$

which has a value of 264 Hz.

The popularity of the Bode representation stems from its utility in circuits analysis. The phase angle plots are sensitive to system parameters and, therefore, provide a good means of comparing model to experiment. The modulus is much less sensitive to system parameters, but the asymptotic values at low and high frequencies provide values for the dc and electrolyte resistance, respectively.

The Bode representation has drawbacks for electrochemical systems. The influence of electrolyte resistance confounds the use of phase angle plots such as shown in Figure 17.3(b) to estimate characteristic frequencies. In addition, Figure 17.3(b) shows that the current and potential are in phase at high frequencies; whereas, at high frequencies, the current and surface potential are exactly out of phase. This result is seen because, at high frequencies, the impedance of the surface tends toward zero, and the ohmic resistance dominates the impedance response. The electrolyte resistance, then, obscures the behavior of the electrode surface in the phase angle plots.

17.1.3 Ohmic-Resistance-Corrected Bode Representation

If an accurate estimate for electrolyte resistance $R_{e,est}$ is available, a modified Bode representation is possible as

$$|Z|_{adj} = \sqrt{(R_e - R_{e,est})^2 + \left(\frac{1}{\omega C}\right)^2} \tag{17.21}$$

and

$$\varphi_{adj} = \tan^{-1}\left(\frac{1}{(R_e - R_{e,est})\,\omega C}\right) \tag{17.22}$$

for the blocking configuration, and

$$|Z|_{adj} = \sqrt{\left((R_e - R_{e,est}) + \frac{R}{1 + (\omega RC)^2}\right)^2 + \left(\frac{\omega C R^2}{1 + (\omega RC)^2}\right)^2} \tag{17.23}$$

and

$$\varphi_{adj} = \tan^{-1}\left(\frac{\omega R^2 C}{R + (R_e - R_{e,est})\,(1 + (\omega RC)^2)}\right) \tag{17.24}$$

for the reactive configuration. The results are presented in Figures 17.4(a) and (b) for magnitude and phase angle, respectively. The current and potential for the blocking electrode are shown correctly in Figure 17.4(b) to be out of phase at all frequencies.

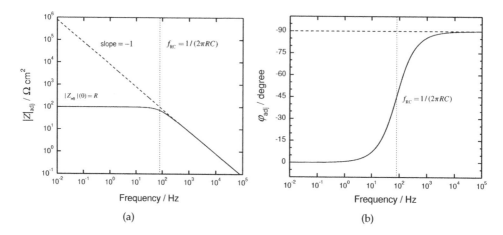

Figure 17.4: Electrolyte-resistance-corrected Bode representation of impedance data for $R_e = 10 \ \Omega cm^2$, $R = 100 \ \Omega cm^2$, and $C = 20 \ \mu F/cm^2$. The blocking system of Table 17.1(a) is represented by dashed lines, and the reactive system of Table 17.1(b) is represented by solid lines. The characteristic frequency is given as $f_{RC} = 1/(2\pi RC)$: a) magnitude and b) phase angle.

The current and potential for the reactive configuration is shown to be in phase at low frequencies and out of phase at high frequencies, and the phase angle has a value of $-45°$ at the characteristic angular frequency ω_{RC}. This approach is applied for more complicated systems in Section 18.2.1 and Example 18.2.

Interpretation of ohmic-resistance-corrected Bode plots requires caution. As seen in equation (17.24), nonzero values for $(R_e - R_{e,est})$ can give the appearance of an additional high-frequency relaxation process. When possible, an assessment of $R_{e,est}$ should be made independently of the regression.

17.1.4 Impedance Representation

Plots of the real and imaginary components of the impedance as functions of frequency are presented in Figures 17.5(a) and (b), respectively. The impedance representation has the significant advantage that the characteristic frequencies can be readily identified. Following equation (17.5), the real part of the impedance of the reactive configuration has a value of $R_e + R/2$ at ω_{RC}. The magnitude of the imaginary part has a maximum value at ω_{RC}, and this value is equal to $R/2$. The imaginary part of the impedance for the blocking circuit shows no characteristic time constant in this representation, and the real part of the impedance is independent of frequency.

The behavior at lower impedance values is emphasized when the impedance components are plotted on a logarithmic scale, as shown in Figures 17.6(a) and (b) for real and imaginary parts of the impedance, respectively. Figure 17.6(b) in particular provides a rich source of insight into the experimental system. As in Figure 17.5(b), the maximum value is seen at the characteristic frequency. The slopes at low and high fre-

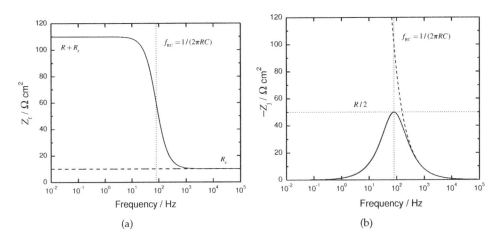

Figure 17.5: Real and imaginary parts of the impedance as a function of frequency for $R_e = 10\ \Omega\text{cm}^2$, $R = 100\ \Omega\text{cm}^2$, and $C = 20\ \mu\text{F}/\text{cm}^2$. The blocking system of Table 17.1(a) is represented by dashed lines, and the reactive system of Table 17.1(b) is represented by solid lines. The characteristic frequency is given as $f_{RC} = 1/(2\pi RC)$: a) real part of impedance and b) imaginary part of impedance.

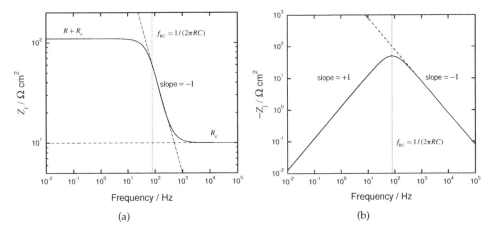

Figure 17.6: Real and imaginary parts of the impedance on a logarithmic scale as a function of frequency for $R_e = 10\ \Omega\text{cm}^2$, $R = 100\ \Omega\text{cm}^2$, and $C = 20\ \mu\text{F}/\text{cm}^2$. The blocking system of Table 17.1(a) is represented by dashed lines, and the reactive system of Table 17.1(b) is represented by solid lines. The characteristic frequency is given as $f_{RC} = 1/(2\pi RC)$: a) real part of impedance and b) imaginary part of impedance.

quency are $+1$ and -1, respectively, for the simple reactive system presented in Table 17.1(b). Departure from slopes of ± 1 provides an indication of distributed processes. Observation of multiple maxima shows that the data must be interpreted in terms of more than one process. Interpretation of Figures 17.5(b) and 17.6(b) in terms of characteristic frequencies is not confounded by the electrolyte resistance, as was seen for the Bode plots of phase angle.

As discussed in Chapter 21, the variances of stochastic errors are equal for real and imaginary parts of the impedance. Thus, another advantage of presenting real and imaginary parts of the impedance as a function of frequency is that comparison between data and levels of stochastic noise can be easily represented.

17.2   Admittance Format

The admittance can be expressed as the complex ratio of current and potential contributions, i.e.,

$$Y = \frac{1}{Z} = \frac{\tilde{I}}{\tilde{U}} \tag{17.25}$$

As discussed in Section 4.1.2, the admittance for a parallel arrangement of passive elements is additive.

Following equation (1.24), the admittance can be expressed in terms of real and imaginary components of the impedance as

$$Y = \frac{1}{Z} = \frac{Z_r}{Z_r^2 + Z_j^2} - j\frac{Z_j}{Z_r^2 + Z_j^2} \tag{17.26}$$

For the blocking system shown in Table 17.1(a), the impedance follows equation (4.26), and

$$Y = \frac{R_e (\omega C)^2}{1 + (\omega R_e C)^2} + j\frac{\omega C}{1 + (\omega R_e C)^2} \tag{17.27}$$

As angular frequency ω tends toward zero, the real admittance tends toward zero according to ω^2 such that

$$\lim_{\omega \to 0} \frac{Y_r}{\omega^2} = R_e C^2 \tag{17.28}$$

The imaginary admittance tends toward zero according to ω such that

$$\lim_{\omega \to 0} \frac{Y_j}{\omega} = C \tag{17.29}$$

Remember! 17.3 *The admittance representation emphasizes values at high frequency and is often used for solid-state systems for which information is sought regarding system capacitance. The admittance format has the advantage that it has a finite value for all frequencies, even for blocking electrodes.*

As angular frequency ω tends toward ∞, the real admittance tends toward $1/R_e$, and the imaginary admittance tends toward zero according to $1/\omega$ such that

$$\lim_{\omega \to \infty} \omega Y_j = \frac{1}{R_e^2 C} \tag{17.30}$$

The maximum value for the imaginary part of the admittance is found at a characteristic angular frequency $\omega_c = 1/(R_e C)$ to be $Y_j(\omega_c) = 1/(2R_e)$.

The corresponding development for the reactive system shown in Table 17.1(b) is somewhat more complicated. The admittance can be expressed as

$$Y = \frac{R_e R (\omega C)^2 + 1 + R_e/R}{R \left[(1 + R_e/R)^2 + (\omega R_e C)^2 \right]} + j \frac{\omega C}{(1 + R_e/R)^2 + (\omega R_e C)^2} \tag{17.31}$$

As angular frequency ω tends toward zero, the real part of the admittance tends toward $1/(R_e + R)$, and the imaginary part of the admittance tends toward zero according to ω such that

$$\lim_{\omega \to 0} \frac{Y_j}{\omega} = \frac{R^2 C}{(R_e + R)^2} \tag{17.32}$$

As angular frequency ω tends toward ∞, the real admittance tends toward $1/R_e$, and the imaginary admittance tends toward zero according to $1/\omega$ such that

$$\lim_{\omega \to \infty} \omega Y_j = \frac{1}{R_e^2 C} \tag{17.33}$$

The high-frequency limit is the same as is found for the series combination of the resistance R_e and capacitance C shown as equation (17.30). The parallel resistance R does, however, influence the value obtained for the imaginary part of the admittance at the characteristic frequency found for the impedance format, i.e.,

$$Y_j\left(\frac{1}{R_e C}\right) = \frac{1}{R_e \left[(1 + R_e/R)^2 + 1 \right]} \tag{17.34}$$

In addition, the value of the characteristic angular frequency is shifted slightly to larger values, i.e.,

$$\omega_c = \frac{1}{R_e C} \left(1 + \frac{R_e}{R} \right) \tag{17.35}$$

The maximum value for the imaginary admittance is given by

$$Y_j(\omega_c) = \frac{R}{2R_e(R + R_e)} \tag{17.36}$$

Thus, in the admittance plane, the locus of points for both the blocking and the single RC reactive system describes a semicircle.

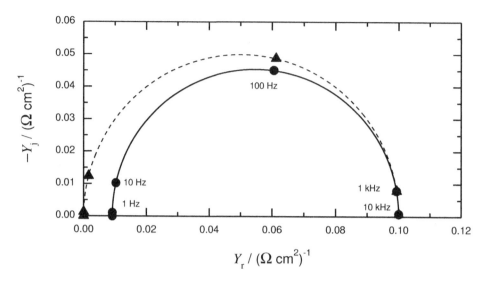

Figure 17.7: Admittance-plane representation for $R_e = 10$ Ωcm^2, $R = 100$ Ωcm^2, and $C = 20$ μF/cm^2. The blocking system of Table 17.1(a) is represented by ▲ and dashed lines, and the reactive system of Table 17.1(b) is represented by ● and solid lines.

17.2.1 Admittance-Plane Representation

Admittance-plane plots are presented in Figure 17.7 for the series and parallel circuit arrangements shown in Figure 4.3(a). The data are presented as a locus of points, where each data point corresponds to a different measurement frequency. As discussed for the impedance-plane or Nyquist representation (Figure 17.1), the admittance-plane format obscures the frequency dependence. This disadvantage can be mitigated somewhat by labeling some characteristic frequencies.

The high-frequency asymptote for the real part of the admittance is equal to $1/R_e$ for both blocking and reactive systems. The zero-frequency asymptote for the real part of the admittance is equal to zero for the blocking system and to $1/(R_e + R)$ for the reactive system. It is interesting to note that both the blocking and the reactive systems show a finite value at low frequencies. In contrast, the imaginary part of the impedance for the blocking circuit tends toward $-\infty$ as frequency tends toward zero in the impedance-plane plot shown in Figure 17.1. The presence of blocking behavior is revealed in Figure 17.7 by a high-frequency asymptote equal to zero for both real and imaginary parts of the impedance.

The maximum value for the imaginary part of the admittance is equal to $1/2R_e$ for the blocking system, and the characteristic angular frequency at the maximum is equal to $\omega_c = 1/R_e C$.

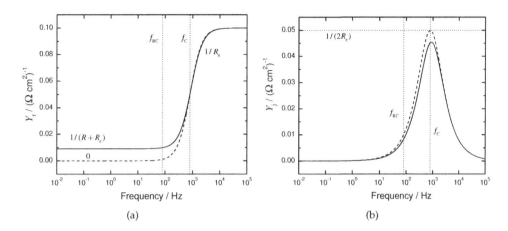

Figure 17.8: Real and imaginary parts of the admittance as a function of frequency for $R_e = 10 \,\Omega\text{cm}^2$, $R = 100 \,\Omega\text{cm}^2$, and $C = 20 \,\mu\text{F}/\text{cm}^2$. The blocking system of Table 17.1(a) is represented by dashed lines, and the reactive system of Table 17.1(b) is represented by solid lines. Characteristic frequencies are noted as $f_{RC} = 1/(2\pi RC)$ and $f_C = 1/(2\pi R_e C)$: a) real part of admittance and b) imaginary part of admittance.

17.2.2 Admittance Representation

The real and imaginary parts of the admittance are presented as functions of frequency in Figures 17.8(a) and (b), respectively. The low- and high-frequency limits can be interpreted in terms of the relationship between impedance and admittance for the respective circuit. The corresponding plots in logarithmic format are presented as Figures 17.9(a) and (b), respectively. Presentation of admittance on linear or logarithmic scales as functions of frequency shows the dependence on frequency, and as compared to impedance, has the advantage that the imaginary part of the admittance for a blocking circuit has a finite value at low frequencies. On a logarithmic scale, deviations from the expected slopes of ± 1 in Figure 17.9(b) or $+2$ for the blocking circuit in Figure 17.9(a) provide indications of processes characterized by distributed or multiple time constants.

The characteristic frequency evident as a peak for the imaginary part of the admittance in Figures 17.8(b) and 17.9(b) has a value of $f_c = 1/(2\pi R_e C)$ for the blocking system. As shown by equation (17.35), the presence of a faradaic process confounds use of graphical techniques to assess this characteristic frequency.

The admittance format is not particularly well suited for analysis of electrochemical and other systems for which identification of faradaic processes parallel to the capacitance represents the aim of the impedance experiments. When plotted in impedance format, the characteristic time constant is that corresponding to the faradaic reaction. When plotted in admittance format, the characteristic time constant is that corresponding to the electrolyte resistance, and that is obtained only approximately when faradaic reactions are present.

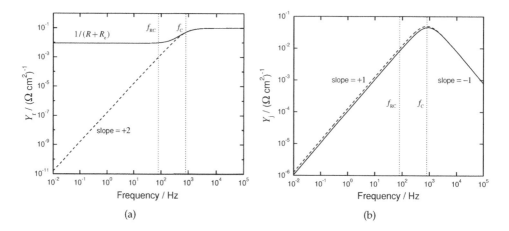

Figure 17.9: Real and imaginary parts of the admittance on a logarithmic scale as a function of frequency for $R_e = 10\ \Omega cm^2$, $R = 100\ \Omega cm^2$, and $C = 20\ \mu F/cm^2$. The blocking system of Table 17.1(a) is represented by dashed lines, and the reactive system of Table 17.1(b) is represented by solid lines. Characteristic frequencies are noted as $f_{RC} = 1/\left(2\pi RC\right)$ and $f_C = 1/\left(2\pi R_e C\right)$: a) real part of admittance and b) imaginary part of admittance.

As shown in Example 17.2, the admittance format is ideally suited for analysis of dielectric systems for which the leading resistance can be neglected entirely.

Example 17.2 Admittance of Dielectrics: *Find an expression for the admittance of the electrical circuit shown in Figure 17.10. Identify the characteristic frequencies.*

Solution: *The circuit corresponds to the dielectric response of a semiconductor device. The term C_{sc} represents the space-charge capacitance, and the terms $R_{t,1}$, $C_{t,1}$, $R_{t,2}$, and $C_{t,2}$ account for the potential-dependent occupancy of deep-level electronic states, which typically have a small concentration. The term R_L accounts for the leakage current, which would be equal to zero for an ideal dielectric.*

The admittance can be expressed as

$$Y(\omega) = \quad \frac{1}{R_L} + \frac{\omega^2 R_{t,1} C_{t,1}^2}{1 + \left(\omega R_{t,1} C_{t,1}\right)^2} + \frac{\omega^2 R_{t,2} C_{t,2}^2}{1 + \left(\omega R_{t,2} C_{t,2}\right)^2} \tag{17.37}$$

$$+ j\omega \left[C_{sc} + \frac{\omega R_{t,1} C_{t,1}^2}{1 + \left(\omega R_{t,1} C_{t,1}\right)^2} + \frac{\omega R_{t,2} C_{t,2}^2}{1 + \left(\omega R_{t,2} C_{t,2}\right)^2} \right]$$

As $\omega \to 0$, $Y_r \to 1/R_L$ and $Y_j \to \infty$. As $\omega \to \infty$, $Y_r \to 1/R_L + 1/R_{t,1} + 1/R_{t,2}$ and $Y_j \to 0$. The characteristic angular frequencies are $\omega_{t,1} = 1/\left(R_{t,1} C_{t,1}\right)$ and $\omega_{t,2} = 1/\left(R_{t,2} C_{t,2}\right)$. The characteristic angular frequencies could be easily identified from plots of real and imaginary parts of the admittance as functions of frequency.

The complex capacitance representation for this type of system is particularly interesting. See the discussion in Section 17.3 and Example 17.5.

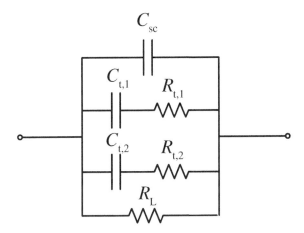

Figure 17.10: Electrical circuit analogue developed to account for the influence of two Shockley-Read-Hall electronic transitions through deep-level states.

17.2.3 Ohmic-Resistance-Corrected Representation

If the electrolyte resistance R_e is removed from the expression for the admittance, the admittance is simplified to

$$Y = 0 + j\omega C \tag{17.38}$$

for the series configuration resulting in only C and

$$Y = \frac{1}{R} + j\omega C \tag{17.39}$$

for the parallel (reactive) configuration of R and C. The resulting real and imaginary parts of admittance are presented in Figures 17.11(a) and (b), respectively. The imaginary components of admittance for the series and parallel configurations, equations (17.38) and (17.39), respectively, are identical and can be used to recover the capacitance at any given frequency. The use of the admittance format to obtain the capacitance for dielectric systems motivates the development of analysis in terms of complex capacitance, as presented in the subsequent section.

Remember! 17.4 *Like the admittance representation, the complex-capacitance representation emphasizes values at high frequency and is often used for solid-state and dielectric systems for which information is sought regarding system capacitance.*

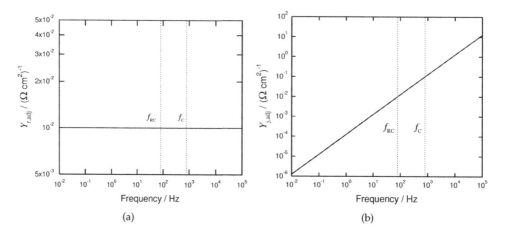

Figure 17.11: Real and imaginary parts of the ohmic-resistance-corrected admittance on a logarith-
mic scale as a function of frequency for $R_e = 10\ \Omega\mathrm{cm}^2$, $R = 100\ \Omega\mathrm{cm}^2$, and $C = 20\ \mu\mathrm{F/cm}^2$. The
blocking system of Table 17.1(a) is represented by dashed lines, and the reactive system of Table
17.1(b) is represented by solid lines. Characteristic frequencies are noted as $f_{RC} = 1/(2\pi RC)$ and
$f_c = (2\pi R_e C)$: a) real part of admittance and b) imaginary part of admittance.

17.3   Complex-Capacitance Format

The complex capacitance is defined for an impedance corrected by the ohmic resistance,
i.e.,

$$\mathcal{C}(\omega) = \mathcal{C}_r + j\mathcal{C}_j = \frac{1}{j\omega(Z - R_e)} \tag{17.40}$$

For the blocking system of Table 17.1(a), the complex capacitance yields directly the
capacitance. The corresponding development for the reactive system of Table 17.1(b)
yields

$$\mathcal{C}(\omega) = C - j\frac{1}{\omega R} \tag{17.41}$$

In equation (17.41), the real part of the complex capacitance is independent of frequency
and is equal to the capacitance.

The complex-capacitance given by equation (17.41) is presented in Figure 17.12(a).
The data are presented as a locus of points, where each data point corresponds to a
different measurement frequency. As discussed for the impedance- and admittance-
plane representations (Figures 17.1 and 17.7, respectively), the complex-capacitance-
plane format obscures the frequency dependence. This disadvantage can be mitigated
by labeling some characteristic frequencies. The high-frequency asymptote for the real
part of the complex capacitance yields the input value of $C = 20\ \mu\mathrm{F/cm}^2$. The behavior
of the complex capacitance is shown more clearly in Figure 17.12(b) where the real part
of the complex capacitance has a value corresponding to the input capacitance and the
slope of the imaginary part is equal to -1.

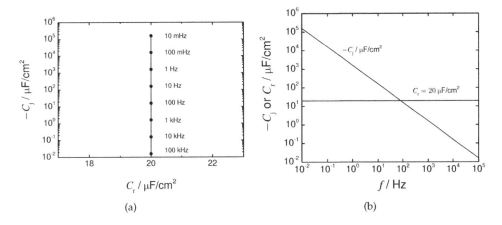

Figure 17.12: Complex capacitance for $R_e = 10 \ \Omega cm^2$, $R = 100 \ \Omega cm^2$, and $C = 20 \ \mu F/cm^2$ for the reactive system of Table 17.1(b): a) complex-capacitance plane plot and b) real and imaginary parts of the complex capacitance as functions of frequency.

As shown in equation (17.40), application of the complex capacitance requires correction for the ohmic resistance. In some cases, it is attractive to correct as well for the parallel resistance that is not associated with the dielectric response of the material, i.e.,

$$Z_\epsilon = \frac{R_p \left(Z - R_e \right)}{R_p - \left(Z - R_e \right)} \tag{17.42}$$

where

$$C_\epsilon(\omega) = \frac{1}{j\omega Z_\epsilon} \tag{17.43}$$

The approach represented by equations (17.42) and (17.43) has been used to analyze the impedance response of human skin.[334,16]

The high-frequency limit of the real part of the complex capacitance can provide the capacitance of an oxide film, from which the dielectric constant or film thickness may be determined. The following examples illustrate the approach for oxide films characterized by a distribution of resistivity.

Example 17.3 Complex Capacitance of the Young Model: *Show that, for the Young model presented in equation (13.172) on page 389, the real part of the complex capacitance approaches the film capacitance in the high-frequency limit. Assume that $\varepsilon = 12$, $\rho_0 = 10^{10} \ \Omega cm$, $\lambda = 2 \ nm$, and $\delta = 4 \ nm$.*

Solution: *The capacitance associated with the film is given as*

$$C = \frac{\varepsilon \varepsilon_0}{\delta} \tag{17.44}$$

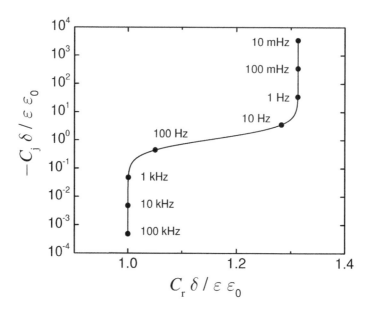

Figure 17.13: Scaled complex-capacitance-plane plot for the Young model presented as equation (13.172) on page 389. Model parameters were $\varepsilon = 12$, $\rho_0 = 10^{10}$ Ωcm, $\lambda = 2$ nm, and $\delta = 4$ nm. The complex capacitance is scaled by the value of the capacitance given as equation (17.44).

Application of equation (17.40) to equation (13.172) yields the results presented in Figure 17.13. The real part of the complex capacitance approaches the value of the film capacitance given as equation (17.44).

Example 17.4 Complex Capacitance of the Power-Law Model: *Show that, for the power-law model described in Section 14.4.4 on page 406, the real part of the complex capacitance approaches the film capacitance in the high-frequency limit. Assume that $\rho_0 = 10^{20}$ Ωcm, $\rho_\delta = 5 \times 10^8$ Ωcm, $\gamma = 6.67$ (corresponding to $\alpha = 0.85$), $\varepsilon = 10$, and $\delta = 10$ nm.*

Solution: *The impedance associated with the power-law model is given by equation (14.28) (page 408), where the resistivity is of the form of equation (14.27). Application of equation (17.40) to equation (14.28) yields the results presented in Figure 17.14. The transition from the CPE behavior to a pure capacitance occurs at a frequency*

$$f_\delta = \frac{1}{2\pi\rho_\delta\varepsilon\varepsilon_0} \qquad (17.45)$$

which has a value for the given parameters of $f_\delta = 360$ Hz. As shown in Figure 17.14, the departure from the CPE behavior given by the dashed line becomes apparent at frequencies smaller than f_δ.

The capacitance of the film, given as equation (17.44), has a value of $C = 0.885$ μF/cm². A closer view of the high-frequency portion of the complex capacitance, shown in Figure 17.15,

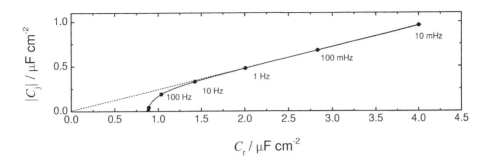

Figure 17.14: Complex-capacitance-plane plot for the power-law model presented as equation (14.28) on page 408. Model parameters were $\rho_0 = 10^{20}$ Ωcm, $\rho_\delta = 5 \times 10^8$ Ωcm, $\gamma = 6.67$ (corresponding to $\alpha = 0.85$), $\varepsilon = 10$, and $\delta = 10$ nm. The dashed line is the extrapolation of the region in which the impedance follows CPE behavior.

shows that the real part of the complex capacitance approaches the value of the film capacitance.

It should be noted that the extrapolation of the frequency domain in which the impedance behaves as a CPE crosses the origin, shown as a dashed line in Figure 17.14. Thus, the high-frequency asymptotic limit for the complex capacitance for a CPE does not yield a capacitance. The parameters chosen for Example 17.4 provided a frequency f_δ that was well within the experimental frequency range. A value of $\rho_\delta = 450$ Ωcm, as reported by Hirschorn et al.[292] for the oxide on a Fe17Cr stainless steel electrode, yields a frequency $f_\delta = 400$ MHz, well beyond the usual experimental frequency range. For such systems, the impedance may be represented by equation (14.30) on page 408. Extrapolation of the complex capacitance, as shown by the dashed lines in Figure 17.14, will yield a nonphysical value of zero for film capacitance.

Example 17.5 Complex Capacitance of Dielectrics: *Find an expression for the complex capacitance for the electrical circuit shown in Figure 17.10 and discussed in Example 17.2. Identify the limits and characteristic frequencies.*

Solution: *Insertion of equation (17.38) into the definition for complex capacitance*

$$C(\omega) = \frac{Y}{j\omega} \tag{17.46}$$

yields

$$C(\omega) = C_{sc} + \frac{C_{t,1}}{1 + (\omega R_{t,1} C_{t,1})^2} + \frac{C_{t,2}}{1 + (\omega R_{t,2} C_{t,2})^2} \tag{17.47}$$
$$- j\omega \left[\frac{1}{\omega R_L} + \frac{\omega R_{t,1} C_{t,1}^2}{1 + (\omega R_{t,1} C_{t,1})^2} + \frac{\omega R_{t,2} C_{t,2}^2}{1 + (\omega R_{t,2} C_{t,2})^2} \right]$$

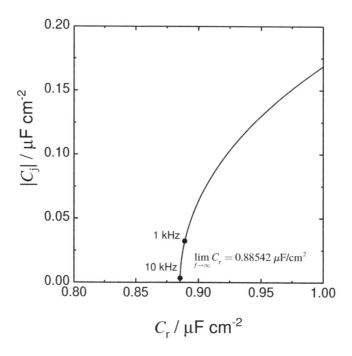

Figure 17.15: The high-frequency portion of the complex-capacitance-plane plot for the results presented in Figure 17.14.

As $\omega \to 0$, $C_r \to C_{sc} + C_{t_1} + C_{t,2}$. As $\omega \to \infty$, $C_r \to C_{sc}$. The characteristic angular frequencies are $\omega_{t,1} = 1/(R_{t,1}C_{t,1})$ and $\omega_{t,2} = 1/(R_{t,2}C_{t,2})$. The characteristic frequencies could be identified easily from plots of real and imaginary parts of the complex capacitance as functions of frequency.

17.4   Effective Capacitance

A representation of effective capacitance for electrochemical systems may be obtained directly from the imaginary part of the impedance as

$$C_{eff} = -\frac{1}{\omega Z_j} \tag{17.48}$$

In contrast to the complex capacitance described in Section 17.3, the effective capacitance described in equation (17.48) is defined to be a real number. As shown in Section

---

Remember! 17.5 *The effective capacitance provides a means of quantitatively determining the interfacial capacitance of a system.*

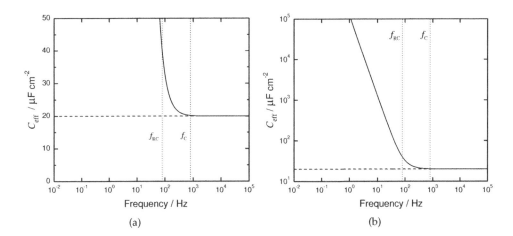

Figure 17.16: Effective capacitance as a function of frequency for $R_e = 10 \ \Omega\text{cm}^2$, $R = 100 \ \Omega\text{cm}^2$, and $C = 20 \ \mu\text{F/cm}^2$. The blocking system of Table 17.1(a) is represented by dashed lines, and the reactive system of Table 17.1(b) is represented by solid lines. The characteristic frequencies noted are $f_{RC} = 1/(2\pi RC)$ and $f_c = (2\pi R_e C)$: a) linear scale and b) logarithmic scale.

17.3 the high-frequency limit for the real part of the complex capacitance may be used in some cases to extract the capacitance. As shown in this section, the high-frequency limit of equation (17.48) also yields the capacitance. The advantages of equation (17.48) are that equation (17.48) does not require correction for the ohmic resistance, and that the high-frequency limit for a related expression,

$$Q_{\text{eff}} = \sin\left(\frac{\alpha \pi}{2}\right) \frac{-1}{Z_j(f) (\omega)^\alpha} \tag{17.49}$$

may be used to extract the value of Q for systems described by a CPE. As shown in Example 17.4, the extrapolation of a complex capacitance cannot yield a capacitance for a system following CPE behavior.

For the series circuit,

$$C_{\text{eff}} = C \tag{17.50}$$

and for the reactive circuit,

$$C_{\text{eff}} = C + \frac{1}{\omega^2 R^2 C} \tag{17.51}$$

Linear and logarithmic plots of effective capacitance are presented as functions of frequency in Figures 17.16(a) and (b), respectively. The high-frequency asymptote provides correct values for the double-layer capacitance for both blocking and reactive circuits. The characteristic angular frequency for the reactive system is seen in equation (17.51) to be $\omega_c = 1/(R_e C)$, and at this value, a factor of 2 (or 100 percent) error is seen for identification of double-layer capacitance. As seen in Figure 17.17, the error is diminished at frequencies larger than the characteristic frequency $f_{RC} = 1/(2\pi RC)$. At frequencies only one order of magnitude larger than f_{RC}, the error in assessment of

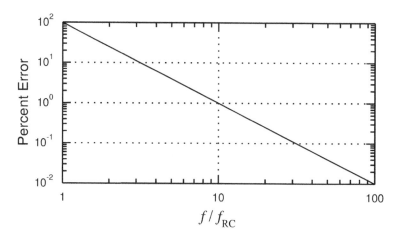

Figure 17.17: Error in assessment of the double-layer capacitance for the reactive system of Table 17.1(b) as a function of frequency scaled by the characteristic frequency $f_{RC} = 1/(2\pi RC)$.

double-layer capacitance is only 1 percent. Measurement at several different frequencies should be used to ensure that the capacitance is obtained at a frequency sufficiently larger than the largest characteristic relaxation frequency for the system.

Example 17.6 **Evaluation of Double-Layer Capacitance:** *Find the meaning of the effective capacitance obtained using equation (17.48) for the convective-diffusion impedance expressed as equation (10.77) on page 220, i.e.,*

$$Z(\omega) = R_e + \frac{R_t + Z_D(\omega)}{1 + j\omega C_{dl}(R_t + Z_D(\omega))} \tag{17.52}$$

where $Z_D(\omega)$ is a complex function of frequency that tends toward zero as frequency tends toward ∞.

Solution: *At high frequencies, all models for convective diffusion to a rotating disk approach the Warburg impedance, given as equation (11.28). Thus, the convective diffusion impedance can be expressed as $Z_D(\omega) = Z_D(0)/\sqrt{j\omega\tau}$. Following Example 1.5 (page 11), which provides a demonstration of methods used to determine the square root of complex numbers,*

$$Z_D(\omega) = Z_D(0)\frac{0.7071}{\sqrt{\omega\tau_D}}(1 - j) = A(\omega)(1 - j) \tag{17.53}$$

The impedance can be expressed as

$$Z = R_e + \frac{R_t + A(\omega)(1 - j)}{1 + j\omega C_{dl}(R_t + A(\omega)(1 - j))} \tag{17.54}$$

or

$$Z = R_e + \frac{R_t + A(\omega) - jA(\omega)}{1 + \omega C_{dl}A(\omega) + j\omega C_{dl}(R_t + A(\omega))} \tag{17.55}$$

After multiplying by the complex conjugate, the real and imaginary contributions are obtained as

$$Z_r = R_e + \frac{R_t + A(\omega)}{(1 + \omega C_{dl} A(\omega))^2 + \omega^2 C_{dl}^2 (R_t + A(\omega))^2} \tag{17.56}$$

and

$$Z_j = -\frac{A(\omega) + \omega C_{dl} \left(A^2(\omega) + (R_t + A(\omega))^2\right)}{(1 + \omega C_{dl} A(\omega))^2 + \omega^2 C_{dl}^2 (R_t + A(\omega))^2} \tag{17.57}$$

respectively. Following equation (17.48), the effective capacitance that can be obtained at the high-frequency limit is given as

$$C_{eff} = \lim_{\omega \to \infty} -\frac{1}{\omega Z_j} = C_{dl} \tag{17.58}$$

Thus, the high-frequency limit of the effective capacitance can be used to obtain the double-layer capacitance for even quite complicated systems. The reason this approach works is that, at high frequencies, the faradaic current is blocked, and all current passes through the double-layer capacitor.

Problems

17.1 Consider the electrical circuit given as Figure 4.6(a). Use a spreadsheet program to plot the results for $R_0 = 10\ \Omega$, $R_1 = 50\ \Omega$, $C_1 = 20\ \mu F$, $R_2 = 500\Omega$, and $C_2 = 10\ \mu F$ for the following representations:

(a) impedance in Nyquist, Bode, and ohmic-resistance-corrected Bode plots.

(b) admittance in Nyquist, real, and imaginary plots.

(c) complex capacitance in a Cole–Cole plot.

(d) effective capacitance as a function of frequency.

17.2 Use a spreadsheet program to plot the results of Problem 10.7 for the following representations:

(a) impedance in Nyquist, Bode, and ohmic-resistance-corrected Bode plots.

(b) admittance in Nyquist, real, and imaginary plots.

(c) complex capacitance in a Cole–Cole plot.

17.3 Consider the equation used for a so-called constant phase element, e.g.,

$$Z(\omega) = R_e + \frac{R_t}{1 + (j\omega)^\alpha R_t Q} \tag{17.59}$$

Plot the model results using the formats presented in this chapter, letting $R_e = 10\ \Omega cm^2$, $R_t = 100\ \Omega cm^2$, and $Q = 20\ \mu Fs^{\alpha-1}/cm^2$. Allow α to be a parameter such that $1 > \alpha > 0.5$.

17.4 Verify equation (17.58) in Example 17.6.

17.5 Electrochemical impedance spectroscopy has become a powerful tool to assess the response of human skin to external stimuli. Two approaches have been taken to analyze the skin impedance data. The skin properties have been assumed to be independent of spatial position but dependent on frequency, and, alternatively, skin properties have been assumed to be independent of frequency but dependent on spatial position.

(a) Under the assumption that the impedance response of human skin can be described by

$$Z_\epsilon = \frac{R}{1 + j\omega RC} \tag{17.60}$$

derive expressions for the frequency-dependent complex relative permittivity

$$\varepsilon(\omega) = \delta \frac{C(\omega)}{\epsilon_0} \tag{17.61}$$

(b) Express your results in terms of real frequency-dependent quantities: the real part of the relative permittivity and the resistivity.

Chapter 18

Graphical Methods

Graphical methods provide a first step toward interpretation and evaluation of impedance data. An outline of the graphical method is presented in Chapter 17 (page 467) for simple reactive and blocking circuits. The same concepts are applied here for systems that are more typical of practical applications. The graphical techniques presented in this chapter do not depend on any specific model. The approaches, therefore, can provide a qualitative interpretation. Surprisingly, even in the absence of specific models, values of such meaningful parameters as the double-layer capacitance can be obtained from high- and low-frequency asymptotes.

The methods for graphical representation and interpretation of impedance data are presented here for a circuit corresponding to an electrode coated by a porous layer, as presented in Figure 9.8 (page 201). This circuit is given in Figure 18.1 with capacitances replaced by constant-phase elements and Z_F replaced by R_t.

The parameters α and Q are associated with a constant-phase element (CPE) as discussed in Chapter 14 (page 395). When $\alpha = 1$, Q has units of a capacitance, e.g., $\mu F/cm^2$, and represents a capacity. When $\alpha \neq 1$, the system shows a behavior that has been attributed to surface heterogeneity or to distribution of time constants within the dielectric layer. The phase angle associated with a CPE is independent of frequency.

The parameter values used for the simulations presented here are given in Table 18.1. The parameters were chosen such that the high-frequency element has a characteristic frequency of 18 kHz and the low-frequency characteristic frequency is 40 Hz.

Table 18.1: Parameters values used for the simulations pertaining to the electrical circuit presented in Figure 18.1 for an electrode coated by a porous layer.

R_e, Ωcm^2	$R_{\ell,\gamma}$, Ωcm^2	$R_{t,\gamma}$, Ωcm^2	$Q_{dl,\gamma}$, $\mu F/s^{(1-\alpha)}cm^2$	α_{dl}	$Q_{\ell,\gamma}$, $\mu F/s^{(1-\alpha)}cm^2$	α_ℓ
100	100	10^5	4	1	0.0885	1
100	100	10^5	15.8	0.8	0.656	0.8

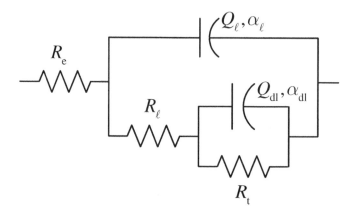

Figure 18.1: Electrical circuit corresponding to an electrode coated by a porous layer, as presented in Figure 9.8, with capacitances replaced by constant-phase elements and Z_F replaced by R_t.

18.1   Based on Nyquist Plots

As discussed in Section 17.1.1 (page 470), impedance data are often represented in complex-impedance-plane or Nyquist formats. This traditional representation of impedance data is given in Figure 18.2 for the circuit presented as Figure 18.1. The data are presented as a locus of points, where each data point corresponds to a different measurement frequency. The asymptotic limits of the real part of the impedance for the circuit shown in Figure 18.2 are R_e at high frequency and $R_e + R_{\ell,\gamma} + R_{t,\gamma}$ at low frequency.

Complex-impedance-plane plots are very popular because the shape of the locus of points yields insight into possible mechanisms or governing phenomena. If the locus of points traces a perfect semicircle, for example the impedance response corresponds to a single activation-energy-controlled process. A depressed semicircle indicates that a more detailed model is required, and the multiple loops shown in Figure 18.2 provide a clear indication that more than one time constant is required to describe the process. The significant disadvantages are that the frequency dependence is obscured, that low impedance values are obscured and that apparent agreement between model and experimental data in complex-impedance-plane format may obscure large differences in frequency and at low impedance values.

18.1.1   Characteristic Frequency

When the two loops are well separated, as is the case in Figure 18.2, an estimate of the capacity value can be obtained from the characteristic frequency defined in Chapter 17. The characteristic frequency for the high-frequency loop is 18 kHz, and the corresponding value is 0.4 Hz for the low-frequency loop. Then the corresponding capacitance

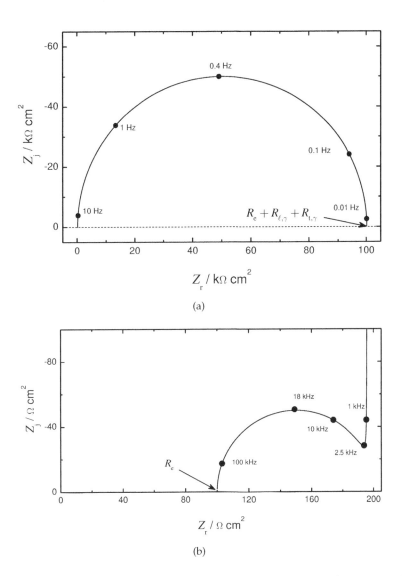

(a)

(b)

Figure 18.2: Nyquist plots corresponding to the circuit given in Figure 18.1 for an electrode coated by a porous layer. Parameter values taken from the first row of Table 18.1: a) low-frequency part and b) zoom of the high-frequency region.

values are

$$C_{\ell,\gamma} = \frac{1}{2\pi f_c R_{\ell,\gamma}} = 0.0885 \ \mu\text{F}/\text{cm}^2 \tag{18.1}$$

and

$$C_{\text{dl},\gamma} = \frac{1}{2\pi f_c R_{\text{t},\gamma}} = 4 \ \mu\text{F}/\text{cm}^2 \tag{18.2}$$

respectively. The value of the capacitance $C_{\ell,\gamma}$ corresponds to a dielectric layer. Under the assumption that the layer coats 100% of the electrode, the thickness of the layer could be obtained as

$$\delta = \frac{\varepsilon \varepsilon_0}{C_{\ell,\gamma}} = 10^{-5} \ \text{cm} \tag{18.3}$$

if the dielectric constant is assigned a value of $\varepsilon = 10$.

To estimate the fraction of the surface that is active, a value for capacitance of the bare electrode must be assumed. Under the assumption that $C_{\text{dl}} = 40 \ \mu\text{F}/\text{cm}^2$, the value of the active part of the electrode is given as

$$1 - \gamma = \frac{C_{\text{dl},\gamma}}{C_{\text{dl}}} = 0.1 \tag{18.4}$$

For an electrode with an area of 1 cm^2, the value of the pore resistance corresponds to a pore with a net surface area of 0.1 cm^2. The electrolyte resistivity inside the pore is given by

$$\rho_\ell = \frac{R_{\ell,\gamma}(1 - \gamma)}{\gamma} = 10^4 \ \Omega\text{cm} \tag{18.5}$$

For a disk electrode, the electrolyte resistance is given by equation (5.112) on page 115, which may be expressed as

$$\rho = \frac{4R_e}{\pi r_0} \tag{18.6}$$

where ρ is the resistivity of the electrolyte. For an electrode area of 1 cm^2, the radius $r_0 = 0.56$ cm, and the resistivity can be estimated to be $\rho = 226 \ \Omega\text{cm}$.

A relationship between electrolyte conductivity and ionic composition is given by equation (5.109) on page 114. As the resistivity within the pore is much bigger than the resistivity of the electrolyte, the composition of the electrolyte inside the pore has a much lower ionic strength than the composition in the bulk.

When the two loops are well separated but may be described as depressed semicircles, as in Figure 18.3, the characteristic frequency of each loop can also be used in order to get an approximation of the corresponding capacitance. For the low-frequency loop

---

Remember! 18.1 *The characteristic frequency of the impedance provides useful clues for model identification.*

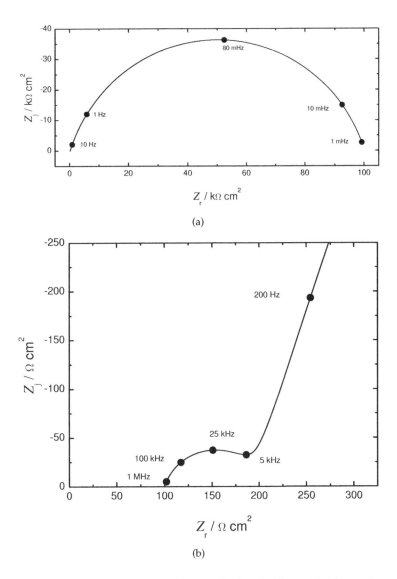

Figure 18.3: Nyquist plots corresponding to the circuit given in Figure 18.1 for an electrode coated by a porous layer. Parameter values taken from the second row of Table 18.1: a) low-frequency part and b) zoom of the high-frequency region.

Table 18.2: Values characteristic of capacitive loops presented by Frateur et al.[288] for an iron-chromium alloy in the presence of bovine serum albumin (BSA) in sulphuric acid aqueous solution after 30 minutes of exposure at the open-circuit potential. The values of R_e and Q presented below were not reported by Frateur et al. and were extracted from presented results.

c_{BSA}, mg/L	f_c, Hz	R_e, Ωcm^2	R_t, Ωcm^2	Q, $\mu F/cm^2 s^{1-\alpha}$	α
0	6.2	11.8	168	147	.88
10	6.2	3.1	367	219	.82
20	6.2	3.1	475	207	.82

shown in Figure 18.3(a), the characteristic frequency is 80 mHz and the corresponding resistance has a value $R_{\ell,\gamma} = 10^5$ Ωcm^2. The corresponding capacitance may be approximated to be

$$C_{\ell,\gamma} = \frac{1}{2\pi f_c R_{\ell,\gamma}} = 0.06 \ \mu F/cm^2 \tag{18.7}$$

The value of the capacitance $C_{\ell,\gamma}$ may be attributed to a dielectric layer.

For the high-frequency loop shown in Figure 18.3(b), the characteristic frequency is 25 kHz and the corresponding resistance has a value $R_{t,\gamma} = 100$ Ωcm^2. The corresponding capacitance may be approximated to be

$$C_{dl,\gamma} = \frac{1}{2\pi f_c R_{t,\gamma}} = 20 \ \mu F/cm^2 \tag{18.8}$$

respectively. The capacitance $C_{dl,\gamma}$ has a value that may be attributed to a double layer. Note that the values of capacitance extracted from a depressed semicircle by equations (18.7) and (18.8) are estimates that may be used to attribute features to physical phenomena. A formal determination of capacitance from data showing CPE behavior requires assumption of the physical origin of the time-constant distribution, as is discussed in Section 14.3 on page 399.

Example 18.1 Interpretation of Characteristic Frequency: *Frateur et al.[288] described the influence of bovine serum albumin (BSA) on corrosion and passivation of an iron-chromium alloy in aqueous sulphuric acid solutions. Some results are presented in Table 18.2. Find the fractional coverage of the electrode surface by BSA.*

Solution: *A preliminary estimation of capacitance may be obtained from the characteristic frequency following*

$$C_{f_c} = \frac{1}{2\pi f_c R_t} \tag{18.9}$$

The resulting values, shown in Table 18.3, range from 80 to 140 $\mu F/cm^2$ and are slightly larger than expected for a double-layer capacitance (Grahame[42] reported double-layer capacitance values from 10 to 90 $\mu F/cm^2$). As the capacitance value expected for a 3 nm-thick film, typical for

Table 18.3: Calculated values of capacitance and surface coverage extracted from the data presented in Table 18.2.

c_{BSA}, mg/L	C_{f_c}, $\mu F/cm^2$	$C_{dl,B}$, $\mu F/cm^2$	γ
0	80.3	56.5	—
10	140	51.3	0.092
20	108	47.9	0.15

oxides on a chromium steel, is around 3.5 $\mu F/cm^2$ (see Section 5.7.2 on page 126), the capacitance estimated from equation (18.9) may be assumed to correspond to that of a double layer on a metal electrode. Further refinement is needed because the data presented in Table 18.2 indicate that the capacitive loop could be modeled as a constant-phase element.

As discussed in Section 14.4, identification of the capacitance associated with a CPE requires assumption of the nature of the corresponding time-constant distribution. The estimated capacitance suggests that the time-constant distribution cannot be attributed to a film and must, therefore, be characterized by a surface distribution. Application of the Brug formula (equation (14.20) on page 405), corresponding to a surface distribution, yields the capacitance values given as $C_{dl,B}$ in Table 18.3. The capacitance values are well within the range expected for a double layer. The ratio of the capacitance in the presence of the BSA to that in the absence of BSA yields, through application of equation (18.4), values for the factional coverage of the surface by BSA. For a BSA concentration of 10 mg/L, the fractional coverage of the surface by BSA is estimated to be 0.092 and, for a BSA concentration of 20 mg/L, the fractional coverage of the surface by BSA is estimated to be 0.15.

18.1.2 Superposition

Superposition of Nyquist plots can be used to extract information concerning the evolution of the impedance according to a single parameter. Examples of parameters that may influence the impedance response include potential, elapsed time, and the rotation speed of a disk electrode for a system under mass-transport control.

Mass Transfer

Graphical methods can be used to extract information concerning mass transfer if the data are collected under well-controlled hydrodynamic conditions. The systems described in Chapter 11 that are uniformly accessible with respect to convective diffusion would be appropriate. The analysis would apply to data collected on a rotating disk electrode as a function of disk rotation speed, or an impinging jet as a function of jet velocity.

The experimental data presented in Figure 18.4(a) correspond to reduction of ferricyanide on a Pt rotating disk electrode.[99,335] The electrolyte consisted of 0.01 M $K_3Fe(CN)_6$ and 0.01 M $K_4Fe(CN)_6$ in 1 M KCl. The temperature was controlled at

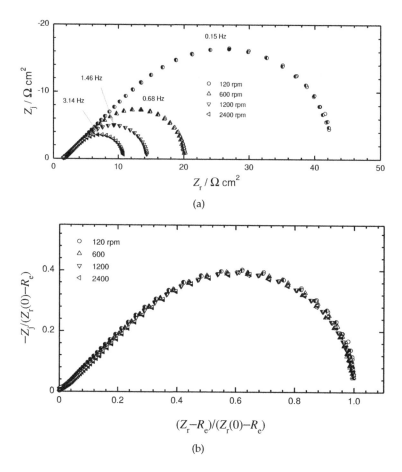

(a)

(b)

Figure 18.4: Impedance data as a function of frequency for reduction of ferricyanide on a rotating Pt disk electrode with rotation speed as a parameter: a) raw data and b) scaled data. (Taken from Durbha et al.[99, 335])

$25 \pm 0.1°C$. The electrode diameter was 5 mm, yielding a surface area of 0.1963 cm^2. A polishing technique was selected for the Pt disk electrode that provided the maximum steady mass-transfer-limited current. The electrode was polished with a 1,200-grit emery cloth, washed in deionized water, polished with alumina paste, and subjected to ultrasound cleaning in a 1:1 solution of water and ethyl alcohol. Experiments were conducted at $1/2$ of the mass-transfer-limited current.

The part of the impedance influenced by mass transfer can be isolated by subtracting the electrolyte resistance. The data presented in Figure 18.4(a) were then scaled by the zero frequency asymptotic value for the real part of the impedance, as seen in Figure 18.4(b). The superposition of impedance spectra is striking, but, in itself, does not constitute a proof that the impedance data were controlled by mass transfer. Scaled plots of impedance such as shown in Figure 18.4(b) provide, however, a useful means of identifying systems where the impedance has changed but the underlying physical phenomena are unchanged.

Evolution of Active Area

The experimental data presented in Figure 18.5(a) were obtained for an as-cast magnesium alloy AZ91 at the corrosion potential after different immersion times in 0.1 M NaCl.[158,336] The size of the loops shown in Figure 18.5(a) increases with elapsed time. Superposition of Nyquist plots may be used to determine whether the data support a hypothesis that growth of a thin dielectric MgO film decreased the active area, thereby causing the impedance to increase. The superposition of the scaled data, shown in Figure 18.5(b), shows that, while the impedance has changed with passage of time, the underlying physical phenomena are unchanged. Thus, the hypothesis that the active area has decreased is supported by the impedance data. Superposition of Nyquist plots was used by Baril et al.[158] to support a model invoking reduction of active area over time for pure magnesium in sodium sulfate solutions. The foundation for this use of superposition of Nyquist plots is developed in Section 9.2.2 on page 196.

18.2   Based on Bode Plots

Impedance data are often presented in the Bode representation, in which the modulus and phase angle are presented as function of frequency. The traditional representation, without correction for the ohmic impedance, is given in Figure 18.6 for the circuit

---

Remember! 18.2 *Scaled plots of impedance provide a useful means of identifying systems where the impedance has changed but the underlying physical phenomena are unchanged.*

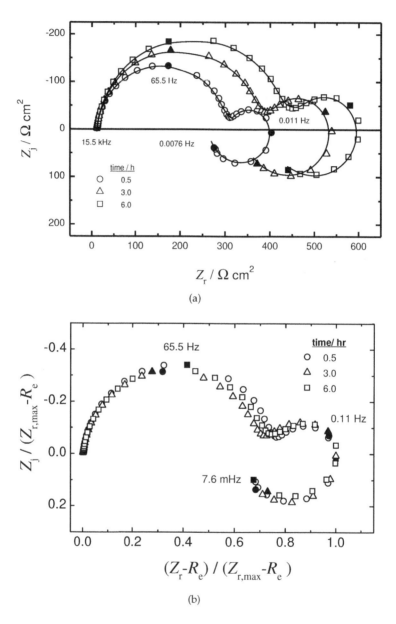

Figure 18.5: Nyquist plots for impedance data obtained for the AZ91 alloy at the corrosion potential after different immersion times in 0.1 M NaCl: a) Nyquist representation of the original data (the lines represent the measurement model fit to the complex data sets) and b) scaled Nyquist plots.

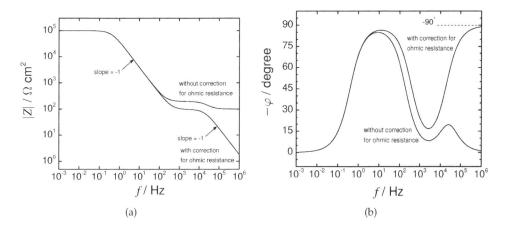

Figure 18.6: Bode representation of the impedance with and without correction for the ohmic impedance for the circuit given in Figure 18.1 for an electrode coated by a porous layer. Parameter values taken from the first row of Table 18.1: a) phase angle and b) modulus.

presented as Figure 18.1 with parameter values taken from the first row of Table 18.1. Similar plots are presented in Figure 18.7 for parameter values taken from the second row of Table 18.1. The Bode representation reveals the dependence on frequency of the magnitude (Figures 18.6(a) and 18.7(a)) and phase angle (Figures 18.6(b) and 18.7(b)). Frequency is generally presented on a logarithmic scale to reveal important behavior seen at lower frequencies. The magnitude of the impedance tends toward R_e as frequency tends toward ∞ and toward $R_e + R_t + R_\ell$ as frequency tends toward zero.

The phase angle, expressed as

$$\varphi = \tan^{-1}\left(\frac{Z_j}{Z_r}\right) \tag{18.10}$$

tends toward zero at low frequencies, indicating that the current and potential are in phase. The phase angle tends toward zero at high frequencies as well, due to the influence of the leading resistor R_e in the value of Z_r used in equation (18.10). No clear correspondence is seen between the Bode plots and the time constants of the impedance data.

The popularity of the Bode representation stems from its utility in circuits analysis. The phase angle plots are sensitive to system parameters and, therefore, provide a good means of comparing model to experiment. The modulus is much less sensitive to system parameters, but the asymptotic values at low and high frequencies provide values for the dc and electrolyte resistance, respectively.

For electrochemical systems exhibiting an ohmic or electrolyte resistance, however, the Bode representation has serous drawbacks. The influence of electrolyte resistance confounds the use of phase-angle plots, such as shown in Figure 18.6(b), to estimate characteristic frequencies. In addition, Figure 18.6(b) shows that the current and over-

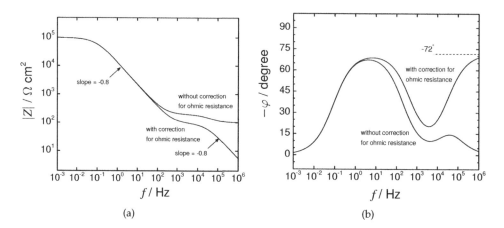

Figure 18.7: Bode representation of the impedance with and without correction for the ohmic impedance for the circuit given in Figure 18.1 for an electrode coated by a porous layer. Parameter values taken from the second row of Table 18.1: a) phase angle and b) modulus.

potential are in phase at high frequencies; whereas, at high frequencies, for $\alpha = 1$, the current and surface potential are exactly out of phase. Even when $\alpha \neq 1$, the current and surface potential are not in phase at high frequencies. The electrolyte resistance, therefore, obscures the behavior of the electrode surface in the phase angle plots.

18.2.1 Ohmic-Resistance-Corrected Phase

If an accurate estimate for electrolyte resistance $R_{e,est}$ is available, a modified phase angle is given by

$$\varphi_{adj} = \tan^{-1}\left(\frac{Z_j}{Z_r - R_{e,est}}\right) \tag{18.11}$$

The influence of the correction is seen in Figures 18.6(b) and 18.7(b) only for the high-frequency range where Z_r and R_e are of the same order of magnitude. In Figure 18.6(b), corresponding to $\alpha = 1$, the phase angle tends toward $-90°$; whereas, in Figure 18.7(b), corresponding to $\alpha \neq 1$, the phase angle tends toward $-90\alpha°$. The observation that, in the high-frequency range, the phase is independent of frequency reveals the high-frequency constant phase or CPE character of the impedance. Example 14.1 on page 396 shows how the ohmic-resistance-corrected phase angle may be used to ascertain the presence of CPE or another time-constant dispersion. This property of the impedance

Remember! *18.3 Interpretation of the phase angle in terms of interfacial properties is confounded by the contribution of the ohmic resistance. The adjusted phase angle, given in equation (18.11), reveals the behavior of the interface.*

is obscured in the traditional Bode representation.

18.2.2 Ohmic-Resistance-Corrected Magnitude

The ohmic-resistance-corrected modulus of the impedance, given by

$$|Z|_{adj} = \sqrt{(Z_r - R_{e,est})^2 + (Z_j)^2} \tag{18.12}$$

is presented in Figures 18.6(a) and 18.7(a). The slope of the corrected modulus yields valuable information concerning the existence of a CPE behavior. In Figure 18.6(a), where $\alpha = 1$, the slope is -1 and, in Figure 18.7(a), the slope is $-\alpha$. For the traditional Bode representation, these slopes in the high-frequency range are completely masked by the effect of the electrolyte resistance. While the slope of the ohmic-resistance-corrected modulus yields the value for α, the ohmic-resistance-corrected phase angle provides a more accurate value.

18.3 Based on Imaginary Part of the Impedance

The difficulty with using the Ohmic-resistance-corrected Bode plots presented in the previous section is that an accurate estimate is needed for the electrolyte resistance and that, at high frequencies, the difference $Z_r - R_{e,est}$ is determined by stochastic noise. The imaginary part of the impedance, presented in Figure 18.8, has the significant advantage that the characteristic frequencies can be readily identified at the peak values. The imaginary part of the impedance is independent of electrolyte resistance, so correction for ohmic resistance is not needed.

18.3.1 Evaluation of Slopes

Figure 18.8 provides a rich source of insight into the experimental system. The slope at low and high frequencies are $+1$ and -1, respectively, for Figure 18.8(a), and $+\alpha$ and $-\alpha$ for Figure 18.8(b). Observation of multiple maxima shows that the data must be interpreted in terms of more than one process. Interpretation of Figure 18.8 in terms of characteristic frequencies is not confounded by the electrolyte resistance, as was seen for the Bode plots of phase angle. When the different time constants are not well separated, the slope of the imaginary part of the impedance as a function of frequency in a logarithmic scale has physical meaning in terms of a CPE or capacitor only at high frequency.

Remember! 18.4 *The slope of the magnitude of the imaginary part of the impedance as a function of frequency in logarithmic coordinates yields the exponent α of a CPE.*

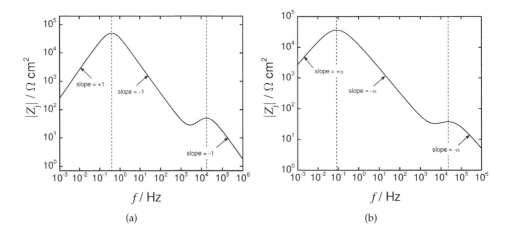

Figure 18.8: The magnitude of the imaginary part of the impedance as a function of frequency in logarithmic coordinates for the circuit given in Figure 18.1 for an electrode coated by a porous layer: a) parameter values taken from the first row of Table 18.1 and b) parameter values taken from the second row of Table 18.1.

18.3.2 Calculation of Derivatives

The high-frequency slope is frequently not sufficient to determine whether the curve is a straight line or if curvature exists. A more accurate way to obtain the α value is to plot the derivative of $\log(|Z_j|)$ with respect to $\log(f)$, such as is shown in Figure 18.9. The asymptotic limits of of $+1$ and -1, shown as low- and high-frequency limits, respectively, in Figure 18.9(a), represent pure capacitances. The corresponding values seen in Figure 18.9(b) are ± 0.8, representative of CPE behavior. These values are in agreement with the parameters presented in Table 18.1.

18.4 Based on Dimensionless Frequency

It is often useful to represent data in terms of scaled or dimensionless frequency. The dimensionless frequency corresponding to mass transfer can be expressed as $K = \omega \delta^2 / D_i$, presented as equation (11.16) on page 247. For a rotating disk, the diffusion layer thickness is replaced by the Nernst diffusion layer thickness, given by equation (11.72) on page 262, which is inversely proportional to the square root of the rotation speed. Thus, the dimensionless frequency for a rotating disk may be expressed as $p = \omega / \Omega$. The influence of the disk electrode geometry is represented by a dimensionless frequency $K = \omega C_0 r_0 / \kappa$ or $K = Q \omega^\alpha r_0 / \kappa$, given, respectively, in equations (13.62) (page 339) and (13.67) (page 339). Consideration of the appropriate dimensionless frequency provides useful information.

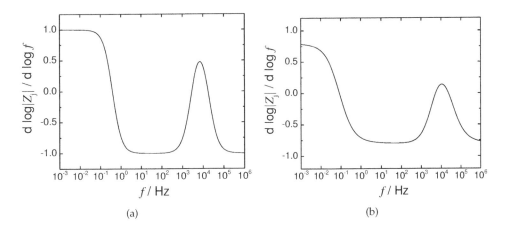

Figure 18.9: The derivative of the logarithm of the imaginary impedance magnitude with respect to the logarithm of frequency for the circuit given in Figure 18.1 for an electrode coated by a porous layer: a) parameter values taken from the first row of Table 18.1 and b) parameter values taken from the second row of Table 18.1.

18.4.1 Mass Transport

An evaluation of the role of mass transfer is obtained by plotting the scaled impedance values as a function of dimensionless frequency $p = \omega/\Omega$, in which the electrical perturbation frequency is scaled by the rotation speed. The real and imaginary parts of the scaled impedance, shown in Figures 18.10(a) and (b), respectively, are superposed at low frequencies. Thus, the impedance values are, at low frequencies, controlled by convective mass transfer to the rotating disk. Differences are seen at higher frequencies that can be attributed to electrode kinetics.

18.4.2 Geometric Contribution

Data presented by Huang et al.[248] for the impedance response of a glassy carbon electrode in KCl electrolyte are used here to demonstrate the application of dimensionless frequency to assess the role of geometry on impedance response. Impedance measurements were made at three different concentrations of KCl. The results obtained in 0.5 M, 0.06 M and 0.0065 M KCl are presented in Figure 18.11.

Following strategies presented in this chapter, a sequential graphical method was used to extract parameters from the experimental data. The ohmic resistance could be obtained from the high-frequency limit of the real part of the impedance, presented in Figure 18.12 as a function of frequency. A value for the electrolyte conductivity can be extracted from the ohmic resistance following

$$\kappa = \frac{1}{4r_0 R_e} \tag{18.13}$$

The real part of the impedance corresponding to the values obtained from equation

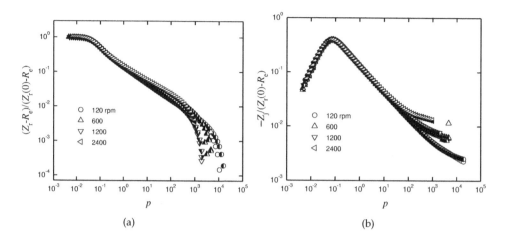

Figure 18.10: Impedance data from Figure 18.4 as a function of scaled frequency $p = \omega/\Omega$ for reduction of ferricyanide on a rotating Pt disk electrode: a) the real part of the impedance and b) the imaginary part of the impedance.

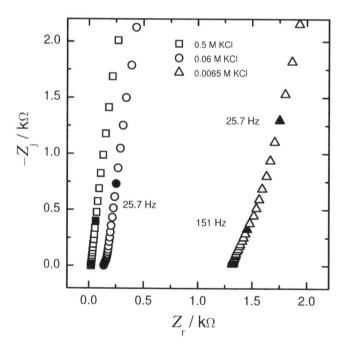

Figure 18.11: Impedance response of a glassy carbon disk in KCl electrolytes with concentration as a parameter. Only high frequencies are shown to emphasize the differences between measurements. Data taken from Huang et al.[248]

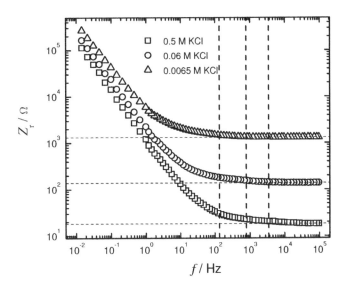

Figure 18.12: The real part of the impedance for a glassy carbon disk as a function of frequency with concentration as a parameter. The horizontal dashed lines correspond to the conductivity obtained from equation (18.13). The vertical dashed lines represent the characteristic frequencies at which the disk geometry influences the impedance. Data taken from Huang et al.[248]

Table 18.4: Parameter values for experimental results extracted from Figures 18.12, 18.13, and 18.14. The frequency at which $K = 1$ was obtained from equation (13.81).[248]

KCl Concentration, M	α	Q, $\mu F/cm^2 s^{1-\alpha}$	κ, S/cm	f, Hz at $K = 1$
0.5	0.93	22.5	0.0546	3470
0.06	0.92	13.2	0.00717	782
0.0065	0.88	9.2	0.000757	134

(18.13) is presented in Figure 18.12 as dashed lines. The resulting values for conductivity are presented in Table 18.4.

The ohmic-resistance-corrected phase angle (see Section 18.2.1) for the glassy carbon disk is presented in Figure 18.13 as a function of frequency with concentration as a parameter. The value of α was extracted from Figure 18.13 following

$$\alpha = -\frac{\max(\varphi_{\text{adj}})}{90} \qquad (18.14)$$

The values, reported in Table 18.4, are close to what may be obtained from the slope of the imaginary part of the impedance plotted in logarithmic scales as a function of frequency. The values for α obtained from Figure 18.2.1 are presented in Table 18.4.

Given a value for α, the effective CPE coefficient may be obtained from equation (18.15), described in a subsequent section. The Q_{eff} values for the glassy carbon data are presented in Figure 18.14 as a function of frequency. Dashed lines correspond to the

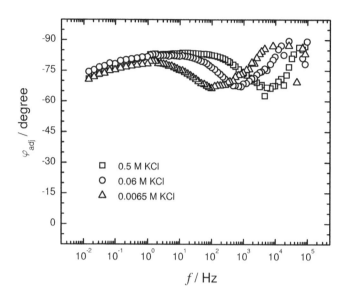

Figure 18.13: The ohmic-resistance-corrected phase angle for a glassy carbon disk as a function of frequency with concentration as a parameter. Data taken from Huang et al.[248]

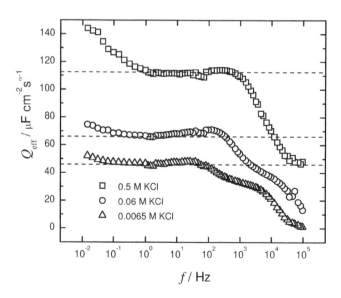

Figure 18.14: The effective CPE coefficient Q_{eff} for a glassy carbon disk as a function of frequency with concentration as a parameter. Dashed lines correspond to the values extracted from the figure. Data taken from Huang et al.[248]

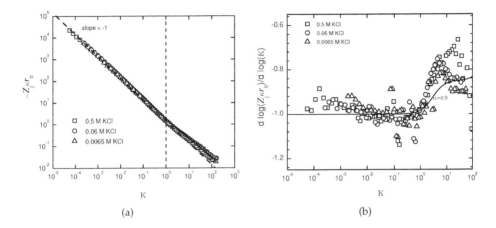

Figure 18.15: Dimensionless analysis for the impedance response of a glassy carbon disk in KCl electrolytes with concentration as a parameter: a) dimensionless imaginary part of the impedance as a function of dimensionless frequency and b) derivative of the logarithm of the dimensionless imaginary part of the impedance with respect to the logarithm of dimensionless frequency. The line corresponds to calculations for a disk electrode with CPE behavior corresponding to $\alpha = 0.9$, reported by Huang et al.[248]

values extracted from the figure. The resulting values for Q are presented in Table 18.4.

The values of α, κ, and Q presented in Table 18.4 were used to calculate, using equation (13.67), the frequency at which $K = 1$. These values, reported in Table 18.4, show that the disk geometry influences the impedance at frequencies as low as 135 Hz for the electrolyte containing 0.0065 M KCl. As shown in Figure 13.11 on page 343, for a given desired frequency range, the geometry effect may be avoided for electrolytes with poor conductivity by using a smaller electrode.

The values of α, κ, and Q presented in Table 18.4 were also used to calculate dimensionless values for impedance and frequency. The dimensionless imaginary part of the impedance is presented in Figure 18.15(a) as a function of dimensionless frequency. At low frequencies, the slope of the imaginary part of the dimensionless impedance plotted in logarithmic scales as a function of dimensionless frequency has a value of -1. When dimensional impedance is plotted as a function of frequency, f, the low-frequency slope has a value of $-\alpha$. The superposition of data for the three values of conductivity in Figure 18.15(a) is excellent, and the change in slope from a value of -1 appears at frequencies higher than $K = 1$. Thus, the features in the impedance data observed a frequencies higher than the critical values reported in Table 18.4 can be attributed to the geometry of the disk electrode.

The derivative of the logarithm of the dimensionless imaginary part of the impedance with respect to the logarithm of dimensionless frequency is presented in Figure 18.15(b). The dispersion of the data apparent in Figure 18.15(b) can be attributed to the fact that the derivative calculations were performed on experimental data (12 points per decade). The superposition of data for the three values of conductivity is in excellent agreement with the solid line, reported by Huang et al.[248] for a disk electrode with CPE behavior corresponding to $\alpha = 0.9$. The transitional frequency between low- and high-frequency response is in good agreement with the theoretical value of $K = 1$.

The analysis presented in this section identifies the frequency range suitable for further interpretation. As the impedance data obtained at frequencies above the critical values reported in Table 18.4 can be attributed to the geometry of the disk electrode, only the impedance obtained for frequencies below this value should be used for further data analysis.

18.5 System-Specific Applications

Applications of graphical methods are presented in this section for some specific systems. The discussion includes evaluation of capacitance or CPE coefficient for systems with well-separated time constants, use of asymptotic expressions to evaluate mass transfer to rotating disk electrodes, use of dimensionless impedance and frequency for systems governed by activation-energy-controlled reactions, application of Mott–Schottky plots for semiconductors, and application of Cole–Cole plots.

18.5.1 Effective CPE Coefficient

An effective capacitance, or, when $\alpha \neq 1$, an effective CPE coefficient may be obtained directly from the imaginary part of the impedance as

$$Q_{eff} = \sin\left(\frac{\alpha \pi}{2}\right) \frac{-1}{Z_j(f)(\omega)^\alpha} \tag{18.15}$$

When $\alpha = 1$, the CPE coefficient Q becomes a capacitance, and equation (18.15) can be written

$$C_{eff} = \frac{-1}{Z_j(f)(\omega)} \tag{18.16}$$

In contrast to the complex capacitance described in Section 17.3, the effective CPE coefficient and effective capacitance, described in equations (18.15) and (18.16), respectively, are defined to be real numbers.

The effective capacitance obtained from equation (18.16) with parameter values taken from the first row of Table 18.1 is presented in Figure 18.16(a). The impedance corresponding to the equivalent circuit of Figure 18.1 shows two well-separated characteristic frequencies. The effective capacitance values obtained from Figure 18.16(a) are in excellent agreement with the input values, as indicated as dashed lines in the figure. The effective CPE coefficient obtained from equation (18.15), shown in Figure 18.16(b), yields similarly good agreement with input values for Q.

For systems with poorly-separated characteristic frequencies, plots such as those presented in Figure 18.16 may be used to extract the CPE coefficient associated with the

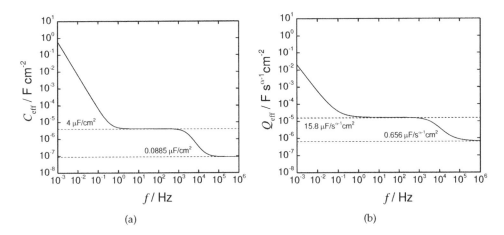

Figure 18.16: Effective capacitance (equation (18.16)) or CPE coefficient (equation (18.15)) for the circuit given in Figure 18.1 for an electrode coated by a porous layer: a) parameter values taken from the first row of Table 18.1 and b) parameter values taken from the second row of Table 18.1.

highest characteristic frequency. As shown in Figure 17.16 (page 489), the assessment should be made at frequencies significantly larger than the largest characteristic relaxation frequency for the system. At a frequency only one order of magnitude larger than f_{RC}, the error in assessment of CPE coefficient is only 1 percent. Measurement at several different frequencies should be used to ensure that the CPE coefficient is obtained at a frequency sufficiently larger than the largest characteristic relaxation frequency for the system.

Example 18.2 Plots for Mg Alloys: *Consider the global impedance response for an as-cast magnesium alloy AZ91 presented in Figure 18.17 for the AZ91 alloy at the corrosion potential after different immersion times in 0.1 M NaCl.[158,336] What quantitative information can be obtained without considering a detailed process model such as discussed in Example 10.6 on page 231?*

Solution: *The imaginary part of the impedance is plotted on a logarithmic scale in Figure 18.18. A line with slope -0.856 ± 0.007 is shown, which was fitted to the high-frequency data for $t = 0.5$ h of immersion. This slope has the value of $-\alpha$, and departure from -1 provides an indication of distributed processes. The low-frequency portion of the high-frequency capacitive*

Remember! 18.5 *Calculation of an effective capacitance or CPE coefficient according to equation (18.15) yields, in the high-frequency limit, properties associated with the electrode under study.*

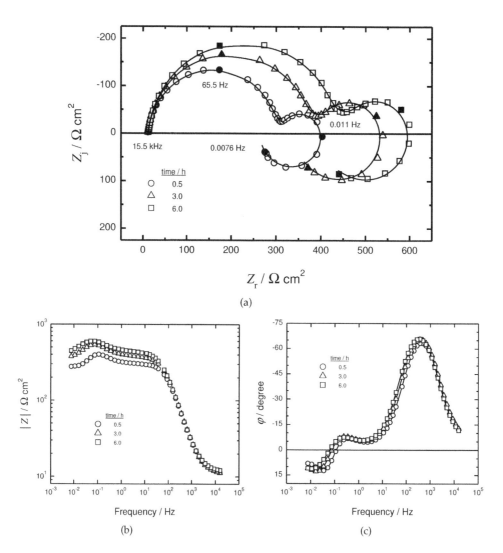

Figure 18.17: Traditional representation of impedance data obtained for the AZ91 alloy at the corrosion potential after different immersion times in 0.1 M NaCl: a) complex-impedance-plane or Nyquist representation (the lines represent the measurement model fit to the complex data sets); b) Bode representation of the magnitude of the impedance as a function of frequency; and c) Bode representation of the phase angle as a function of frequency. (Taken from Orazem et al.[252])

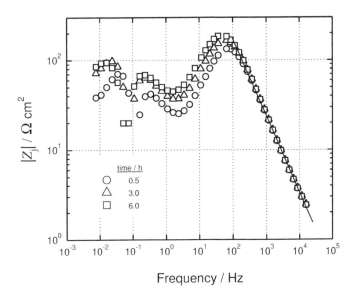

Figure 18.18: Imaginary part of the impedance as a function of frequency for the AZ91 alloy at the corrosion potential after different immersion times in 0.1 M NaCl. The line with slope -0.857 ± 0.007 was fitted to the high-frequency data for $t = 0.5$ h of immersion time. (Taken from Orazem et al.[252])

Table 18.5: Values for high-frequency component obtained from asymptotic values for the corrosion of an AZ91 Mg alloy.[252]

Immersion Time / h	0.5	3.0	6.0
α / dimensionless	0.856	0.872	0.877
$Q_{\mathrm{eff}}/\mu\mathrm{Fs}^{\alpha-1}\mathrm{cm}^{-2}$	22.7	19.1	18.5
$R_{\mathrm{e}}/\Omega\mathrm{cm}^2$	10.65	10.45	11.4

loop has a slope of 0.661 ± 0.008. The lack of symmetry suggests that the high-frequency capacitance is in parallel with other reactive processes. Observation of multiple maxima also shows that the data must be interpreted in terms of more than one process.

The slopes at the high-frequency asymptotes appear to be in good agreement for the three data sets, but a more detailed analysis reveals some trending. Values for the CPE exponent are provided in Table 18.5. A small increase in the value of α is evident as the immersion time increases.

The value of α can be used in equation (18.15) to obtain an apparent CPE coefficient Q_{eff}. The resulting values of Q_{eff} are presented in Figure 18.19. The absence of a clearly identifiable asymptote may be attributed to high-frequency instrumental artifacts. The values for the CPE coefficient provided in Table 18.5 represent the average over the values for the 10 highest frequencies. A small reduction in the value of Q_{eff} is evident as the immersion time increases.

The value of α obtained from Figure 18.18 can also be used to find the solution resistance R_{e}

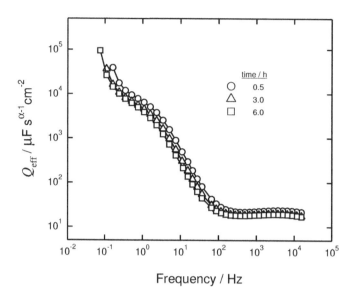

Figure 18.19: Effective CPE coefficient defined by equation (18.15) for the AZ91 alloy at the corrosion potential after different immersion times in 0.1 M NaCl. (Taken from Orazem et al.[252])

used in equation (18.11) to yield the expected asymptotic value for adjusted phase angle given by equation (18.11). The resulting ohmic-resistance-corrected phase angle and magnitude are given in Figure 18.20. The slope of the high-frequency asymptote for magnitude corrected for electrolyte resistance has a value of $-\alpha$. Values for the electrolyte resistance are provided in Table 18.5.

18.5.2   Asymptotic Behavior for Low-Frequency Mass Transport

A graphical method was reported by Tribollet et al.[175] that can be used to extract Schmidt numbers from experimental data in which the convective-diffusion impedance dominates. The technique accounts for the finite value of the Schmidt number. The concept is based on the observation that

$$\lim_{p \to 0} \left(\frac{d \, \text{Re} \{Z\}}{dp \, \text{Im} \{Z\}} \right) = \lim_{p \to 0} \left(\frac{d \, \text{Re} \left\{ -\frac{1}{\overline{\theta}_i(0)} \right\}}{dp \, \text{Im} \left\{ -\frac{1}{\overline{\theta}_i(0)} \right\}} \right) = \lambda \, \text{Sc}_i^{1/3} = s \qquad (18.17)$$

Thus, the quantity $\lambda \text{Sc}_i^{1/3}$ may be obtained by plotting $\text{Re}\{Z\}$ as a function of $p \, \text{Im}\{Z\}$, and the straight line of slope $s = \lambda \text{Sc}_i^{1/3}$ is fitted to the low-frequency data. The constant λ was obtained by taking into account the above development.

$$\lambda = \lim_{p \to 0} \left(\frac{d \, \text{Re} \left\{ -\frac{1}{\overline{\theta}_i(0)} \right\}}{dp \, \text{Sc}_i^{1/3} \text{Im} \left\{ -\frac{1}{\overline{\theta}_i(0)} \right\}} \right) = 1.2261 + 0.84 \text{Sc}_i^{-1/3} + 0.63 \text{Sc}_i^{-2/3} \qquad (18.18)$$

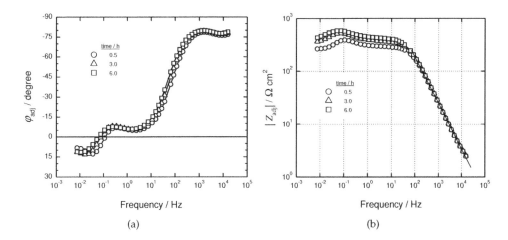

Figure 18.20: Ohmic-resistance-corrected Bode plots for the AZ91 alloy at the corrosion potential after different immersion times in 0.1 M NaCl, with values adjusted for the electrolyte resistance following equation (18.11): a) phase angle; and b) modulus. (Taken from Orazem et al.[252])

The value of the Schmidt number is then obtained by solving the equation

$$1.2261 \text{Sc}_i^{2/3} + (0.84 - s)\, \text{Sc}_i^{1/3} + 0.63 = 0 \qquad (18.19)$$

This approach is attractive because $d\,\text{Re}\,\{Z\}/dp\,\text{Im}\,\{Z\}$ is constant over a substantial frequency range.

An illustration of the asymptotic technique is presented in Figure 18.21 for the data presented in Figure 18.4(a). The scale presented in Figure 18.21(a) shows the approach to linearity as frequency tends toward zero. The data collected at different rotation speeds are superposed in Figure 18.21(a). Regression of the linear portion of the plot, shown in Figure 18.21(b), yields a slope of -13.52. Solution of equation (18.19) yields a Schmidt number equal to 1,091, in good agreement with the expected value of 1,100 for this system.

Regression of a process model for this system, reported by Orazem et al.,[99] yielded values for the Schmidt number that appeared to be a function of rotation speed. While the value obtained at 120 rpm of 1,114 was in good agreement with the value obtained by use of the low-frequency asymptotic relation, much larger values were obtained at higher rotation speeds. A regressed value of 1,222, for example, was reported for a rotation speed of 2,400 rpm. The superposition, at low frequencies, of data collected at all rotation speeds in Figure 18.21(a) suggests that the process model used by Orazem et al. did not account properly for a phenomenon evident at higher frequencies and that errors in fitting data at higher frequencies were propagated to lower frequencies in the form of an incorrect assessment of Schmidt number. The use of the asymptotic method can complement development of complete process models.

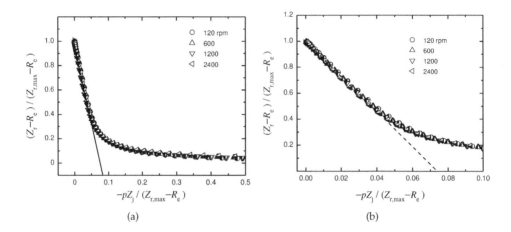

Figure 18.21: Illustration of the use of the method described by Tribollet et al.[175] to extract the Schmidt number from the low-frequency portion of the scaled measurement. Impedance data were taken from Figure 18.4: a) scale showing the approach to linearity as frequency tends toward zero and b) regression of the linear portion of the plot shown in part a. The slope of the line was −13.52, resulting in a Schmidt number equal to 1,091.

18.5.3 Arrhenius Superposition

Systems that are governed by reaction kinetics show well-defined behaviors as a function of temperature and potential. If the system can be described as being controlled by a single activation-energy-controlled process, graphical methods can be used to cause the data to superpose.

The excitation of electrons in semiconducting systems follows an Arrhenius temperature dependence, e.g.,

$$k = k_0 \exp\left(\frac{-\Delta E}{RT}\right) \tag{18.20}$$

where k_0 is a constant. As $R_t \propto 1/k_0$, for a system showing a single potential-dependent reaction, the charge-transfer resistance can be expected to follow

$$R_t = R_t^\circ \exp\left(\frac{\Delta E}{RT}\right) \tag{18.21}$$

Under the assumption that the capacitance is independent of temperature, the characteristic time constant for the system should follow the same Arrhenius temperature dependence, i.e.,

$$\tau_{RC} = \tau_{RC}^\circ \exp\left(\frac{\Delta E}{RT}\right) \tag{18.22}$$

Impedance data taken at different temperatures for systems governed by single-step activation-energy-controlled processes may be expected to superpose when properly normalized.

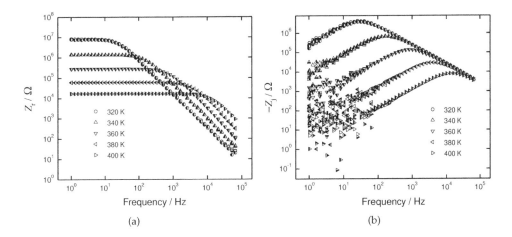

Figure 18.22: Impedance data as a function of frequency for an n-GaAs/Ti Schottky diode with temperature as a parameter: a) the real part of impedance and b) the imaginary part of impedance. (Taken from Jansen et al.[338])

Graphical techniques based on application of an Arrhenius relationship are illustrated here for an n-GaAs single crystal diode with a Ti Schottky contact and a Au, Ge, Ni Schottky contact at the eutectic composition. This material has been well characterized in the literature and, in particular, has a well-known EL2 deep-level state that lies 0.83 to 0.85 eV below the conduction band edge.[337] Experimental details are provided by Jansen et al.[338,339]

The experimental data are presented in Figures 18.22(a) and (b) for the real and imaginary parts of the impedance, respectively, for data collected at temperatures ranging from 320 to 400 K. The logarithmic scale used emphasizes the scatter seen in the imaginary impedance at low frequencies. The impedance response is seen to be a strong function of temperature. The impedance plane plots shown in Figure 18.23, for data collected at 320 and 340 K, show the classic semicircle associated with a single relaxation process.

The reduced presentation of the data given in Figure 18.24 shows that the impedance response is dominated by a single relaxation process with an activation energy of 0.827 eV. The impedance data collected at different temperatures were normalized by the maximum mean value of the real part of the impedance and plotted against a normalized frequency defined by

$$f^* = \frac{f \exp\left(\frac{E}{kT}\right)}{f^\circ} \tag{18.23}$$

where E was given a value of 0.827 eV, and the characteristic frequency f° was given a value of 2.964×10^{14} Hz such that the imaginary part of the normalized impedance values reached a peak value near $f^* = 1$. The data collected at different temperatures

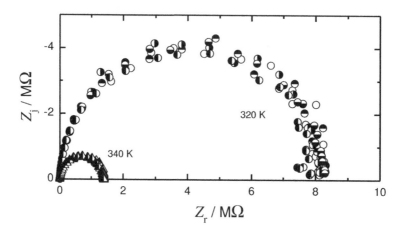

Figure 18.23: Impedance data in impedance-plane format for an n-GaAs/Ti Schottky diode with temperature as a parameter. (Taken from Jansen et al.[338])

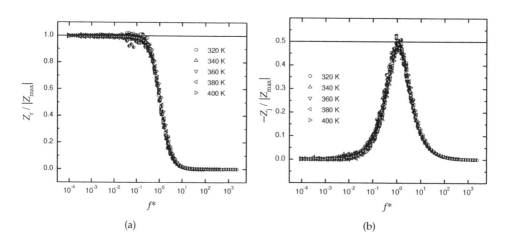

(a)   (b)

Figure 18.24: Impedance data from Figure 18.22 collected for an n-GaAs/Ti Schottky diode as a function of frequency $f^* = \frac{f}{f^\circ} \exp(E/kT)$: a) real part of impedance and b) imaginary part of impedance. (Taken from Orazem and Tribollet.[340])

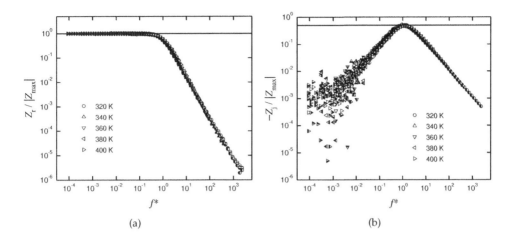

Figure 18.25: Impedance data from Figure 18.22 collected for an n-GaAs/Ti Schottky diode as a function of frequency $f^* = \frac{f}{f_\circ} \exp\left(E/kT\right)$: a) real part of impedance and b) imaginary part of impedance. (Taken from Orazem and Tribollet.[340])

are reduced to a single line. The extent to which the data are superposed is seen more clearly on the logarithmic scale shown in Figure 18.25.

It is worth noting that, while the data do superpose nicely in Figures 18.24 and 18.25, the impedance data do in fact contain information on minor activation-energy-controlled electronic transitions.[338,339] The information concerning these transitions can be extracted by regression of an appropriate process model using a weighting strategy based on the error structure of the measurement.

18.5.4 Mott–Schottky Plots

Graphical techniques can be applied for single-frequency measurements when the frequency selected excludes the contributions of confounding phenomena. For example, impedance measurements on a semiconductor diode at a sufficiently high frequency exclude the influence of leakage currents and of electronic transitions between deep-level and band-edge states. Thus, the capacitance can be extracted from the imaginary part of the impedance as

$$C = \frac{1}{\omega Z_j} \tag{18.24}$$

The problem is reduced to one of identifying the relationship between semiconductor properties and the capacitance as a function of applied potential. The mathematical development is presented in Section 12.6.

Mott-Shottky plots of $1/C^2$ as a function of potential are particularly useful at larger doping levels. As seen in Figure 12.7(a), $1/C^2$ is linear over a broad range of potential. The slope of $1/C^2$ with respect to potential is negative for an n-type semiconductor and

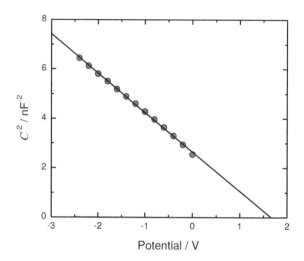

Figure 18.26: Mott–Schottky plot for an n-GaAs/Ti Schottky diode: $1/C^2$ as a function of applied potential.[341]

positive for a p-type semiconductor. The linear portion is given by equation (12.33) for an n-type semiconductor and by equation (12.34) for a p-type semiconductor.

An example of the use of Mott–Schottky plots is presented in Figure 18.26. The capacitance was measured at a frequency of 1 MHz applied across the semiconductor sample described in Figure 18.22 as a function of reverse bias with a DLTS spectrometer.[341] Deep-level states are not expected to influence the signal at such a high frequency; thus, the slope can be interpreted in terms of the doping level.

Deviations from straight lines in Mott–Schottky plots are frequently attributed to the influence of potential dependent charging of surface or bulk states. While deviations can also be attributed to nonuniform dopant concentrations, this interpretation is supported by analytic and numerical calculations of the contribution of defects to the space charge as a function of applied potential (see, e.g., Dean and Stimming[226] or Bonham and Orazem[342]).

18.5.5 High-Frequency Cole–Cole Plots

The complex capacitance plot is presented in Figure 18.27(a) for the for the circuit given in Figure 18.1 for an electrode coated by a porous layer and with parameter values taken from the first row of Table 18.1. The high-frequency limit for the real complex ca-

Remember! 18.6 *Deviations from ideal behavior can give important clues to the physical phenomena governing an experimental system.*

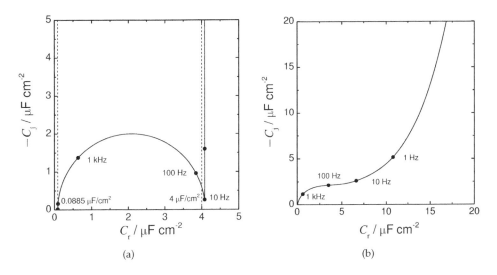

Figure 18.27: Complex capacitance plots for the for the circuit given in Figure 18.1 for an electrode coated by a porous layer: a) parameter values taken from the first row of Table 18.1 and b) parameter values taken from the second row of Table 18.1.

pacitance yields the input value of $0.0885 \ \mu F/cm^2$ that was used for the high-frequency loop. The low-frequency limit of the semicircle yields a value that approximates the input value of $4 \ \mu F/cm^2$ for the low-frequency loop. The ability to extract the low-frequency capacitance depends on the degree of separation of the time constants associated with the circuit given in Figure 18.1.

Values for capacitance could not be obtained from the complex capacitance plot presented in Figure 18.27(b) for parameter values taken from the second row of Table 18.1. As discussed in Section 17.3 (page 484), the CPE is not amenable to extraction of capacitance values from complex-capacitance plots. The complex capacitance plot is useful for systems displaying other time-constant distributions, such as the Young model shown in Example 17.3 on page 485.

18.6   Overview

The graphical representations presented here are intended to enhance analysis and to provide guidance for the development of appropriate physical models. While visual inspection of data alone cannot provide all the nuance and detail that can, in principle, be extracted from impedance data, the graphical methods described in this chapter can provide both qualitative and quantitative evaluation of impedance data.

The impedance-plane or Nyquist plots presented as Figures 18.2, 18.3, 18.4 and 18.5 provide a sense of the type of processes that govern the low-frequency behavior of the system. The shape of the low-frequency loop in Figure 18.4 is typically associated with mass-transfer effects. Three time constants are clearly evident in Figure 18.17(a).

Aside from values for asymptotic limits for the real part of the impedance, it is difficult to extract meaningful information from the traditional Bode plots when the ohmic resistance is not negligible. In contrast, the Bode plots of magnitude and phase angle corrected for ohmic resistance can be used to identify CPE behavior at high frequencies. Following equation (14.5), the high-frequency limit for the phase angle (Figures 18.6(b), 18.7(b), and 18.20(a)) can be used to extract values for the CPE coefficient α. The slope of the corrected modulus (Figures 18.6(a), 18.7(a), and 18.20(b)) can also be used to extract values for the CPE coefficient α.

The plots of the imaginary part of the impedance on a log-log scale shown in Figures 18.8 and 18.18 are particularly helpful. The slopes of the lines at low and high frequency in Figure 18.8 indicate clearly that two time constants can be discerned. In the case of the results shown in Figure 18.8(a), the high-frequency feature has capacitive characteristics that extend to the low-frequency feature. In the case of the results shown in Figure 18.8(b), the high-frequency feature has CPE characteristics that extend to the low-frequency feature. These results indicate that the two time constants are coupled through a double-layer capacitive or constant-phase element, respectively. Log-log plots of imaginary impedance can be used to distinguish between a depressed impedance-plane semicircle caused by a continuous distribution of time constants associated with a CPE and that caused by contributions of discrete processes with closely overlapping but discrete time constants.

For the AZ91 data, the log of the imaginary part of the impedance shown in Figure 18.18 indicated a CPE behavior, meaning that the high-frequency feature had the characteristic of a distributed time constant rather than several discrete time constants. The absence of symmetry for the high-frequency feature suggested that the capacitive behavior was in parallel to other reactive processes. The decrease in α with immersion time suggests that the surface became more homogeneous with the growth of corrosion product layers.

The effective CPE coefficient representation in Figures 18.16 and 18.19 yields, for $\alpha = 1$, information concerning the high-frequency capacitance of the system. In the case that $\alpha < 1$, Figures 18.16 and 18.19 yield an effective CPE coefficient Q_{eff} that can be related to the film capacitance through models of the distributed time constants (see Section 14.3 on page 399).

The plots presented here have general application. They have been useful for evaluating the high-frequency behavior associated with local impedance measurements, where low-frequency measurements were not reliable and the frequency range was therefore not sufficient to allow regression analysis with a more detailed mathematical model.[277] The graphical analysis showed that, while high-frequency CPE behavior was evident in global impedance measurements for a Mg AZ91 alloy disk electrode, the local impedance measured near the center of the disk exhibited ideal RC behavior. The CPE behavior was thereby attributed to a 2-D radial distribution of the charge-transfer resistance. The capacitance extracted from high-frequency asymptotic behavior was

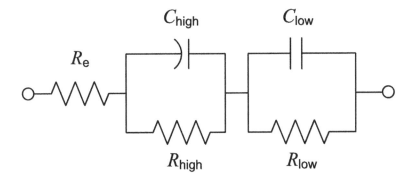

Figure 18.28: Equivalent circuit for Problem 18.1.

used to estimate the area sampled by the local impedance technique.

As shown in Section 13.2, logarithmic plots of the imaginary part of the impedance can be used to show the high-frequency pseudo-CPE behavior caused by nonuniform current distribution. A graphical analysis similar to that demonstrated in Example 18.2 was employed by Huang et al.[248] to obtain values for parameters R_e, α, and Q associated with the impedance of a glassy carbon disk in KCl solutions and an oxide-covered stainless steel electrode in 0.05 M NaCl + 0.005 M Na_2SO_4 electrolyte.

The graphical methods developed in this chapter provide guidance for model development and preliminary estimates for parameter values. As shown in Section 18.4.2, the graphical methods can be used to identify the range of frequencies that are unaffected by geometry-induced current and potential distributions. The next step in model development is to employ complex nonlinear regression, shown in Chapters 19 (page 527) and 20 (page 545).

Problems

18.1 Choose appropriate parameters the circuit presented in Figure 18.28. that will yield time constants in a measurable frequency range and develop plots similar to Figures 18.2, 18.6, 18.8, 18.9, and 18.16.

18.2 Consider the circuit presented in Figure 9.8 for an electrode covered by a film. Upon choosing appropriate parameters that will yield time constants in a measurable frequency range, develop plots similar to Figures 18.2, 18.6, 18.8, 18.9, and 18.16.

18.3 The value of the Schmidt number may be used to determine which species participate in an electrochemical reaction. Estimate the value for the Schmidt number that could be expected to be determined by impedance measurements at 25 °C for the system 0.01 M $K_3Fe(CN)_6$, 0.01 M $K_4Fe(CN)_6$, 1 M KCl:

(a) At the open circuit potential

(b) At the cathodic mass-transfer-limited current

Table 18.6: Experimental diffusion impedance data obtained at a rotation speed of 600 rpm.

f, Hz	Z_r, Ωcm^2	Z_j, Ωcm^2	f, Hz	Z_r, Ωcm^2	Z_j, Ωcm^2
0.07942	997.9	−65.1	3.16	309	−273
0.1	992.2	−85.2	3.98	270.6	−253.7
0.1258	991	−97.8	5.02	239	−259.2
0.1584	973	−118	6.3	208.6	−197.9
0.1996	967	−149	7.94	188.8	−179.1
0.2512	940	−190	10	165.2	−159.1
0.316	920	−231	12.58	147.3	−141.6
0.398	882	−263	15.84	131.9	−126.4
0.502	831	−320	19.94	116.3	−112.8
0.63	764	−324.6	25.12	103.7	−100.4
0.794	692	−347	31.624	91.9	−89.04
1	613	−354	39.812	82.3	−79.43
1.258	544	−342	50.12	73.14	−70.85
1.584	470	−348	79.43	59.7	−57.6
1.994	409	−328	100	52.11	−50.11
2.52	356	−303			

   (c) At the anodic mass-transfer-limited current

18.4 The experimental diffusion impedance data presented in Table 18.6 were obtained at a rotation speed of 600 rpm. Find the numerical value of the Schmidt number by using equation (18.19).

18.5 For simple reactions, the charge-transfer resistance may be given as a function of potential by equation (10.19). Develop a means of superposing impedance data taken at different potentials for such a system where the impedance response is given by equation (10.21).

18.6 Show how a linear dependence of capacitance on potential would influence the superposition developed in Problem 18.5.

Chapter 19

Complex Nonlinear Regression

Complex nonlinear least-squares (CNLS) regression techniques, with application to impedance measurements, were developed in the late 1960s.[55–57] The CNLS approach provides a significant improvement over ordinary nonlinear least-squares (NLS) because a common set of parameters can be estimated by simultaneous regression of the model to both real and imaginary data.[343] Since the Kramers–Kronig relations constrain the real and imaginary parts of the complex quantity, the appropriate regression strategy should allow for the correlation of the real and imaginary parts of the complex model through the Kramers–Kronig relations. This chapter provides an overview of issues associated with regression. For a more detailed discussion of the mechanics of the regression techniques, the reader is referred to standard textbooks.[106,344–347] The discussion by Press et al.[348] is also helpful.

19.1 Concept

Macdonald provides a historical perspective for the application of regression techniques to impedance spectroscopy.[3,4] The regression of models to impedance data generally employs a complex nonlinear application of the method of least squares.[349–351] Complex nonlinear least-squares regression was developed in the late 1960s as an extension of nonlinear least squares (NLS) regression techniques.[55,56] The use of CNLS is consistent with the expectation that real and imaginary components of impedance data satisfy the constraints of the Kramers–Kronig relations.[66,67,352] The CNLS approach provides an improvement over NLS techniques because a combined model parameter set is estimated by simultaneous regression of the model to both real and imaginary parts of the measured spectrum. Weighted CNLS was first applied to impedance data by Macdonald et al.[57,58] The concept of weighting is critical for impedance spectroscopy because the impedance is a strong function of frequency.[353,339]

The regression of a complex function $\hat{Z}$ to complex data Z can be expressed in a least-squares sense as the minimization of the sum of squares

$$S = \left(Z - \widehat{Z}(\omega|\mathbf{P})\right)^{\mathrm{T}} \mathbf{V}^{-1} \left(Z - \widehat{Z}(\omega|\mathbf{P})\right) \tag{19.1}$$

where $\mathbf{V}$ is a symmetric positive-definite variance-covariance matrix of the experimental stochastic errors, Z represents the complex impedance data measured at frequencies ω, and $\widehat{Z}(\omega|\mathbf{P})$ represents the complex model calculated for frequency ω as a function of a parameter vector $\mathbf{P}$.[345,347] Under the assumption that the covariance terms in $\mathbf{V}$ can be neglected, i.e., that $\mathbf{V}$ is a diagonal matrix, and under the additional assumption that residual errors are uncorrelated, equation (19.1) can be replaced by

$$S = \sum_{k=1}^{N_{\mathrm{dat}}} \left[\frac{\left(Z_r(\omega_k) - \widehat{Z}_r(\omega_k|\mathbf{P})\right)^2}{V_{r,k}} + \frac{\left(Z_j(\omega_k) - \widehat{Z}_j(\omega_k|\mathbf{P})\right)^2}{V_{j,k}} \right] \tag{19.2}$$

where N_{dat} represents the number of measured values, $V_{r,k}$ and $V_{j,k}$ represent the real and imaginary components, respectively, of the variance of the stochastic errors, $Z_r(\omega_k)$ and $Z_j(\omega_k)$ represent the real and imaginary parts of the impedance data measured at frequency ω_k, and $\widehat{Z}_r(\omega_k|\mathbf{P})$ and $\widehat{Z}_j(\omega_k|\mathbf{P})$ represent the real and imaginary parts of the model value calculated for frequency ω_k as a function of a parameter vector $\mathbf{P}$.

The statement that the covariance terms in $\mathbf{V}$ can be neglected implies both that stochastic errors at one frequency are uncorrelated with errors at another frequency and that errors in real and imaginary parts of the impedance at a given frequency are not correlated. Use of equation (19.2) is therefore predicated on the assumption that errors in real and imaginary parts of the impedance are not correlated. If equation (19.2) is used under conditions where the error covariance terms cannot be neglected, the incorrect error structure will be reflected in the parameter estimates. Carson et al.[102] used numerical simulations to show that, when a Fourier analysis algorithm was used to obtain the impedance response from time-domain signals containing normally distributed errors, the errors in the real and imaginary parts of the impedance were uncorrelated. In contrast, use of a phase-sensitive-detection algorithm yielded correlation between the error in the real and imaginary components of the impedance.[354]

🐘**Remember!** 19.1 *The weighting used for nonlinear regression employs an estimate for the variance of the data. As the variance of impedance data is strongly dependent on frequency, an independent assessment of the error structure is needed.*

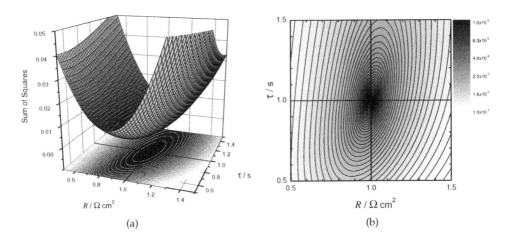

(a) (b)

Figure 19.1: The objective function, equation (19.3), for an RC circuit as a function of parallel resistor and capacitor values. The circuit parameters were $R_1 = 1\ \Omega$, and $\tau_1 = 1$ s. The synthetic data were calculated for frequencies ranging from 1 to 10^5 Hz at a spacing of 10 points per decade, and the noise was determined by machine precision. The objective function at the set parameter values was found to be equal to zero: a) 3-D perspective drawing of the contour surface and b) 2-D representation of the contour surface.

19.2 Objective Functions

The regression procedure involves minimization of equation (19.2), which can be expressed as

$$S(\mathbf{P}) = \sum_{k=1}^{N_{\text{dat}}} \frac{\left(Z_r(f_k) - \widehat{Z}_r(f_k|\mathbf{P})\right)^2}{\sigma_{r,k}^2} + \sum_{k=1}^{N_{\text{dat}}} \frac{\left(Z_j(f_k) - \widehat{Z}_j(f_k|\mathbf{P})\right)^2}{\sigma_{j,k}^2} \tag{19.3}$$

where the minimized value of $S(\mathbf{P})$ can be regarded to be the χ^2 statistic. An objective function of this type has the attractive feature of emphasizing data for which the confidence is high and deemphasizing data for which the confidence is low. Methods to assess the variance of the measurement are discussed in Chapter 21.

The objective function, equation (19.3), is presented in Figure 19.1(a) for an RC circuit as a function of parallel resistor and RC-time-constant values. The circuit parameters were $R = 1\ \Omega\text{cm}^2$, and $\tau_{\text{RC}} = 1$ s (see, e.g., Figure 4.3(b) and the corresponding Example 4.2). The synthetic data were calculated for frequencies ranging from 1 to 10^5 Hz at a spacing of 10 points per decade, and the noise was determined by machine precision.

The objective of the regression procedure is to identify the parameter values that minimize equation (19.3). The objective function, given as a response surface in Figure 19.1(a), was found to be equal to zero at the set parameter values. Solid lines have been drawn on the bottom contour map to indicate the values for which the function is minimized.

Several features of the optimization problem are apparent in Figure 19.1(a). The model is nonlinear with respect to parameters; nevertheless, the objective function is well behaved near the solution where it can be approximated by a quadratic function. The contours projected onto the base of the plot have an elliptical shape. The major axis of the ellipse does not lie along either axis.

Each of the function-minimization procedures involves some assessment of the objective function contour in parameter space. This contour is easily visualized in three dimensions, as shown in Figure 19.1(a), but the large number of adjustable parameters used in typical regressions makes graphical visualization cumbersome if not impossible. Numerical methods, such as those described in this chapter, are therefore required.

Example 19.1 Nonlinear Models: *Show that the equation for the impedance of a Voigt element is nonlinear with respect to parameters.*

Solution: *The impedance of a Voigt element can be given as*

$$Z(\omega) = R_0 + \sum_{k}^{K} \frac{R_k}{1 + j\omega\tau_k} \tag{19.4}$$

Calculate first and second derivatives with respect to parameters.

$$\frac{\partial Z}{\partial R_0} = 1; \qquad\qquad \frac{\partial^2 Z}{\partial R_0^2} = 0 \tag{19.5}$$

$$\frac{\partial Z}{\partial R_k} = \frac{1}{1 + j\omega\tau_k}; \qquad \frac{\partial^2 Z}{\partial R_k^2} = 0 \tag{19.6}$$

$$\frac{\partial Z}{\partial \tau_k} = -\frac{jR_k\omega}{(1 + j\omega\tau_k)^2}; \qquad \frac{\partial^2 Z}{\partial \tau_k^2} = -\frac{2R_k\omega^2}{(1 + j\omega\tau_k)^3} \tag{19.7}$$

The second derivatives of Z with respect to R_0 and R_k are equal to zero; thus, Z is linear with respect to the resistance parameters. The second derivatives of Z with respect to τ_k are not equal to zero; thus, Z is nonlinear with respect to the time constant τ_k.

19.3   Formalism of Regression Strategies

The nonlinear regression techniques discussed in Section 19.3.2 are extensions of the linear regression formalism described below. A more detailed description is provided by Press et al.[355]

19.3.1   Linear Regression

Consider a general model of the form

$$y(x) = \sum_{k=1}^{N_p} P_k X_k(x) \tag{19.8}$$

where $X_k(x)$ are arbitrary fixed functions of x called the basis functions and N_p represents the number of adjustable parameters P_k in the model. Equation (19.8) is linear with respect to parameters P_k even if $X_k(x)$ are nonlinear with respect to x.

A least-squares regression involves minimization of the objective function

$$\chi^2 = S(\mathbf{P}) = \sum_{i=1}^{N_{dat}} \frac{\left(y_i - \sum_{k=1}^{N_P} P_k X_k(x_i)\right)^2}{\sigma_i^2} \tag{19.9}$$

where y_i represents the measured values and σ_i represents the standard deviation of measurement i. At the minimum value, the derivative with respect to the parameters P_k vanishes. Thus,

$$\sum_{i=1}^{N_{dat}} \frac{\left(y_i - \sum_{j=1}^{N_P} P_j X_j(x_i)\right)}{\sigma_i^2} X_k(x_i) = 0 \tag{19.10}$$

Equation (19.10) represents a set of N_p equations of the form

$$\sum_{j}^{N_p} \alpha_{k,j} P_j = \beta_k \tag{19.11}$$

where

$$\beta_k = \sum_{i=1}^{N_{dat}} \frac{(y_i X_k(x_i))}{\sigma_i^2} \tag{19.12}$$

and

$$\alpha_{k,j} = \sum_{i=1}^{N_{dat}} \frac{(X_k(x_i) X_j(x_i))}{\sigma_i^2} \tag{19.13}$$

In vector form, equation (19.11) can be written

$$\boldsymbol{\alpha} \cdot \mathbf{P} = \boldsymbol{\beta} \tag{19.14}$$

or

$$\mathbf{P} = \boldsymbol{\alpha}^{-1} \cdot \boldsymbol{\beta}$$
$$= \mathbf{C} \cdot \boldsymbol{\beta} \tag{19.15}$$

The inverse matrix $\mathbf{C} = \boldsymbol{\alpha}^{-1}$ provides an estimate for the confidence intervals for the estimated parameters. The diagonal elements of $[\mathbf{C}]$ are the variances of the fitted parameters, i.e.,

$$\sigma_{P_j}^2 = C_{j,j} \tag{19.16}$$

The off-diagonal elements of $\mathbf{C}$, $C_{j,k}$, are the covariances between parameters P_j and P_k, which show the extent of correlation among parameters. This correlation is undesirable for a regression. It can appear when too many parameters are being sought in the regression, but correlation among parameters may sometimes be unavoidable due to the structure of the model.

19.3.2   Nonlinear Regression

Consider a general function $f(\mathbf{P}) = 0$ that is nonlinear with respect to parameters P_k. Under the assumption that $f(\mathbf{P})$ is twice continuously differentiable, a Taylor-series expansion about a parameter set $\mathbf{P}_0$ yields

$$f(\mathbf{P}) = f(\mathbf{P}_0) + \sum_{j}^{N_p} \frac{\partial f}{\partial P_j}\bigg|_{\mathbf{P}_0} \Delta P_j + \frac{1}{2}\sum_{j}^{N_p}\sum_{k}^{N_p} \frac{\partial^2 f}{\partial P_j \partial P_k}\bigg|_{\mathbf{P}_0} \Delta P_j \Delta P_k + \dots \tag{19.17}$$

Equation (19.17) is second order with respect to parameter increments ΔP_j and, therefore, describes a parabolic hypersurface. The assumption that, in the neighborhood of the minimum, the objective function can be described as a parabolic hypersurface, is supported by the results presented in Figure 19.1.

The optimal value for $\mathbf{P}$ is found when $f(\mathbf{P})$ has a minimum value. At the minimum, derivatives with respect to the parameter increments ΔP_i should be equal to zero; thus,

$$\frac{\partial f}{\partial \Delta P_j} = \frac{\partial f}{\partial P_j}\bigg|_{\mathbf{P}_0} + \sum_{k}^{N_p} \frac{\partial^2 f}{\partial P_j \partial P_k}\bigg|_{\mathbf{P}_0} \Delta P_k = 0 \tag{19.18}$$

Equation (19.18) represents a set of N_p equations of the form

$$\beta_j = \sum_{k}^{N_p} \alpha_{j,k} \Delta P_k \tag{19.19}$$

where

$$\beta_j = -\frac{1}{2}\frac{\partial f}{\partial P_j}\bigg|_{\mathbf{P}_0} \tag{19.20}$$

and

$$\alpha_{j,k} = \frac{1}{2}\frac{\partial^2 f}{\partial P_j \partial P_k}\bigg|_{\mathbf{P}_0} \tag{19.21}$$

where α and β are generally functions of parameters $\mathbf{P}$. In vector form, equation (19.19) can be written

$$\beta = \alpha \cdot \Delta P \tag{19.22}$$

or

$$\Delta P = \alpha^{-1} \cdot \beta$$
$$= \mathbf{C} \cdot \beta \tag{19.23}$$

The general formulation described above can now be applied to the nonlinear least-squares problem.

A least-squares regression involves minimization of the objective function

$$\chi^2 = \sum_{i=1}^{N_{dat}} \frac{\left(Z_i - \widehat{Z}(x_i|\mathbf{P})\right)^2}{\sigma_i^2} \tag{19.24}$$

where Z_i represents the measured values, $\widehat{Z}(x_i|\mathbf{P})$ represents the model, which now may be nonlinear with respect to the parameter vector $\mathbf{P}$, and σ_i represents the standard deviation of measurement i. Under the understanding that the function f to be minimized is given by equation (19.24), the gradient of the objective function with respect to parameters $\mathbf{P}$ is given by

$$\frac{\partial \chi^2}{\partial P_k} = -2 \sum_{i}^{N_{dat}} \frac{\left(Z_i - \widehat{Z}(x_i|\mathbf{P})\right)}{\sigma_i^2} \frac{\partial \widehat{Z}(x_i|\mathbf{P})}{\partial P_k} \tag{19.25}$$

or

$$\beta_k = \sum_{i}^{N_{dat}} \frac{\left(Z_i - \widehat{Z}(x_i|\mathbf{P})\right)}{\sigma_i^2} \frac{\partial \widehat{Z}(x_i|\mathbf{P})}{\partial P_k} \tag{19.26}$$

The components of equation (19.25) can be written in vector form as $\nabla \chi^2(\mathbf{P_0})$.

The second derivative of the objective function with respect to the parameters $\mathbf{P}$ is given by

$$\left. \frac{\partial^2 \chi^2}{\partial P_j \partial P_k} \right|_{\mathbf{P_0}} = \tag{19.27}$$

$$2 \sum_{i=1}^{N_{dat}} \frac{1}{\sigma_i^2} \left[\frac{\partial \widehat{Z}(x_i|\mathbf{P})}{\partial P_j} \frac{\partial \widehat{Z}(x_i|\mathbf{P})}{\partial P_k} - \left(Z_k - \widehat{Z}(x_i|\mathbf{P})\right) \frac{\partial^2 \widehat{Z}(x_i|\mathbf{P})}{\partial P_j \partial P_k} \right]$$

or $\nabla^2 \chi^2(\mathbf{P_0})$. Equation (19.27) is called the Hessian matrix, which involves both first and second derivatives of the model with respect to parameters. The matrix $\boldsymbol{\alpha}$ is equal to one-half of the Hessian matrix, i.e.,

$$\alpha_{j,k} = \sum_{i=1}^{N_{dat}} \frac{1}{\sigma_i^2} \left[\frac{\partial \widehat{Z}(x_i|\mathbf{P})}{\partial P_j} \frac{\partial \widehat{Z}(x_i|\mathbf{P})}{\partial P_k} - \left(Z_k - \widehat{Z}(x_i|\mathbf{P})\right) \frac{\partial^2 \widehat{Z}(x_i|\mathbf{P})}{\partial P_j \partial P_k} \right] \tag{19.28}$$

The second derivatives of the Hessian matrix are typically neglected in evaluation of equation (19.28), an action that is justified on two grounds. The first justification is that second derivatives are often small as compared to the first derivatives. For a linear problem, the second derivatives are identically equal to zero. The second justification is that the term is multiplied by $(y_k - y(x_i|\mathbf{P}))$, a term that, for a successful regression, should be uncorrelated with respect to x_i or to the model $y(x_i|\mathbf{P})$. Thus, the second derivative terms should tend to cancel when summed over all observations i. Accordingly,

$$\alpha_{j,k} \approx \sum_{i=1}^{N_{dat}} \frac{1}{\sigma_i^2} \left[\frac{\partial \widehat{Z}(x_i|\mathbf{P})}{\partial P_j} \frac{\partial \widehat{Z}(x_i|\mathbf{P})}{\partial P_k} \right] \tag{19.29}$$

is used in the nonlinear regression algorithms described in Section 19.4.

Equations (19.24) to (19.29) can be applied for complex nonlinear least-squares regression by concatenating the real and imaginary impedance data Z_i to form a data vector with length equal to twice the number of measured frequencies. A similar concatenation applies for the model values $\widehat{Z}(x_i|\mathbf{P})$. Press et al.[355] provide a very approachable discussion of the least-squares methods and their implementation.

19.4   Regression Strategies for Nonlinear Problems

Some common regression strategies are summarized in this section. For a more detailed discussion, the reader is referred to standard textbooks.[106,344–347]

19.4.1   Gauss–Newton Method

The Gauss–Newton method for solving the nonlinear set of equations (19.18) can be expressed as

$$\mathbf{P}_{j,\ell+1} = \mathbf{P}_{j,\ell} + \boldsymbol{\alpha}^{-1} \cdot \boldsymbol{\beta} \tag{19.30}$$

where ℓ is the interation counter, $\boldsymbol{\beta}$ is evaluated following equation (19.26), and $\boldsymbol{\alpha}$ is evaluated following equation (19.29). The Gauss–Newton method is characterized by quadratic convergence near the solution, but convergence can be very slow far from the solution. Quadratic convergence has the characteristic that the number of significant digits in the solution doubles at every iteration. In some cases, the Gauss–Newton method diverges and fails to yield a solution. Due to its extreme efficiency near the solution, the Gauss–Newton method provides the basis for most methods of nonlinear optimization.

19.4.2   Method of Steepest Descent

The method of steepest descent seeks a minimum value of the objective function by following the gradient of the objective function, i.e.,

$$\mathbf{P}_{j,\ell+1} = \mathbf{P}_{j,\ell} + \lambda \boldsymbol{\beta} \tag{19.31}$$

where $\boldsymbol{\beta}$ is evaluated following equation (19.26), and λ is a constant chosen to be sufficiently small as to avoid overrunning the minimum. A comparison of equations (19.30) and (19.31) shows that the matrix $\boldsymbol{\alpha}$ in the Gauss–Newton method modifies the method of steepest descent by accounting for the curvature of the objective function surface. For this reason, $\boldsymbol{\alpha}$ is called the curvature matrix.

   The method of steepest descent tends to be quite efficient far from the solution, but convergence can be painfully slow near the solution. Slow convergence is likely where the contours are attenuated and banana-shaped, i.e., where the method tends to change direction often with minimal changes in objective function value.

19.4.3   Levenberg–Marquardt Method

The Levenberg–Marquardt method described in this section represents a compromise between the Gauss–Newton method described in Section 19.4.1 and the method of steepest descent described in Section 19.4.2. The method of steepest descent is used far from the converged value, moving smoothly to the Gauss–Newton method as the solution is approached.

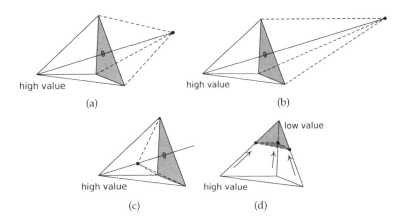

Figure 19.2: Schematic representation of the simplex algorithm. New points are denoted by closed circle •, and the vector mean of all except the highest value is denoted by an open circle ○: a) a reflection from the point with the highest value through the vector-mean of the remaining points; b) an expansion along the same line, taken if the resulting point yields a result that is lower than that seen at all other points; c) a contraction along the same line, taken if the reflection point yields a result that is worse than that seen at all other points; and d) a contraction among all dimensions toward the low point, taken if none of the actions taken yields a result that is better than than the highest value.

The critical concepts encompassed by the Levenberg–Marquardt method are the selection of the scaling factor for the method of steepest descent and an approach for making a smooth transition from one method to the other. The curvature matrix α is replaced by α' such that

$$\alpha'_{j,k} = \alpha_{j,k}(1+\lambda) \qquad \text{for} \quad j = k$$
$$\alpha'_{j,k} = \alpha_{j,k} \qquad \text{for} \quad j \neq k \qquad (19.32)$$

The equation solved is

$$\mathbf{P}_{j,\ell+1} = \mathbf{P}_{j,\ell} + (\boldsymbol{\alpha}')^{-1} \cdot \boldsymbol{\beta} \qquad (19.33)$$

When λ is large, α' is diagonally dominant, and the method approaches that described in equation (19.31). When λ is small, the method approaches that described in equation (19.30).

19.4.4 Downhill Simplex Strategies

The simplex method requires evaluations only of the objective function to be minimized. It does not require evaluation of the derivatives and does not require inversion of a matrix that may have a zero determinant. The simplex method can therefore be more robust than the strategies described in Sections 19.4.1, 19.4.2, and 19.4.3.

A schematic representation of the downhill simplex algorithm is presented in Figure 19.2.

- The objective function is evaluated at $N + 1$ points in parameter space, where N is the number of adjustable parameters. The geometrical figure that is formed is called a simplex.

- The function is evaluated at the reflection of the point with highest value through the vector-average of opposing surface, as shown in Figure 19.2(a) for an optimization in three parameters. If the new point gives a result better than the highest value, but not better than the lowest value, the highest point is replaced with the new point. The reflections are designed such that the volume of the simplex is conserved.

- If the new point gives a result better than all other points, a further extrapolation is made along this line, as shown in Figure 19.2(b). Again, if the new point gives a result better than the highest value, but not better than the lowest value, the highest point is replaced with the new point.

- If the new point gives a result that is worse than the highest point, a contraction is performed along this line, as shown in Figure 19.2(c). If the new point gives a result better than the highest value, but not better than the lowest value, the highest point is replaced with the new point.

- If none of the actions taken yields a result that is better than the highest value, a contraction is performed among all dimensions toward the low point, as shown in Figure 19.2(d). The new simplex is shown as a shaded volume.

Convergence can be slow, especially if the minimum exists in a long narrow valley in parameter space. Press et al.[348] describe several more efficient multidimension optimization strategies.

19.5   Influence of Data Quality on Regression

Regression procedures can be sensitive to presence of singular or near-singular matrixes. In such cases, the normal set of equations does not have a unique solution, and collinearity is said to exist. A regression problem is called ill conditioned if a small change in data causes large changes in estimates. Ill conditioning is undesirable in regression analysis because it leads to unreliable parameter estimates with large variances and covariances.[347]

Regression problems in impedance spectroscopy may become ill conditioned due to improper selection of measurement frequencies, excessive stochastic errors (noise) in the measured values, excessive bias errors in the measured values, and incomplete frequency ranges. The influences of stochastic errors and frequency range on regression are demonstrated by examples in this section. The issue of bias errors in impedance measurement is discussed in Chapter 22. The origin of stochastic errors in impedance measurements is presented in Chapter 21.

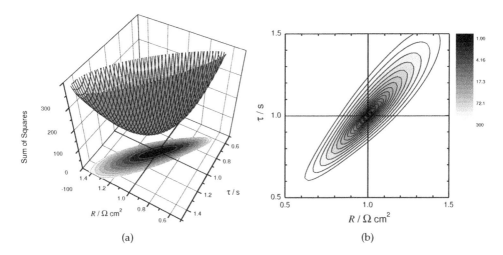

Figure 19.3: The objective function, equation (19.3), for an RC circuit as a function of parallel resistor and RC-time constant values. The circuit was the same as presented in Figure 19.1 with the exception that stochastic noise was added to the synthetic data with a standard deviation equal to 1 percent of the modulus. a) 3-D perspective drawing of the contour surface and b) 2-D representation of the contour surface.

19.5.1 Presence of Stochastic Errors in Data

The regression procedure is strongly influenced by stochastic errors or noise in the measurement. One effect is illustrated in comparison of Figure 19.1 to Figure 19.3, in which stochastic noise with a standard deviation equal to 1 percent of the modulus was added to the synthetic data. Solid lines have been drawn on the bottom contour map to indicate the values for which the function is minimized. The presence of stochastic errors in the data does not introduce roughness in the parabolic hypersurface, which is a function of parameter. The noise has the effect of increasing the value of the minimum that can be found from zero to a number typically larger than the degree of freedom of the problem. The hypersurface presented in Figure 19.3 has been scaled by the number of data points, and has a minimum value between 1 and 2. The parabolic shape of the surface is more apparent and the approach to the minimum has a steeper slope, as is evident in change in scale from 0 to 0.1 in Figure 19.1(b) to 1 to 300 in Figure 19.3(b).

19.5.2 Ill-Conditioned Regression Caused by Stochastic Noise

The presence of stochastic errors can impede sensitivity to minor parameters. The response surface for a Voigt circuit with $R_0 = 1$ Ωcm^2, $R_1 = 100$ Ωcm^2, $\tau_1 = 0.001$ s, $R_2 = 200$ Ωcm^2, $\tau_2 = 0.01$ s, $R_3 = 5$ Ωcm^2, and $\tau_3 = 0.05$ s is presented in Figure 19.4(a) as a function of $\log_{10}(R_3)$ and $\log_{10}(\tau_3)$. The synthetic impedance data were calculated for frequencies from 1 to 10^5 Hz at a spacing of 10 measurements per decade. The response surface is presented as a function of logarithm of parameters to extend

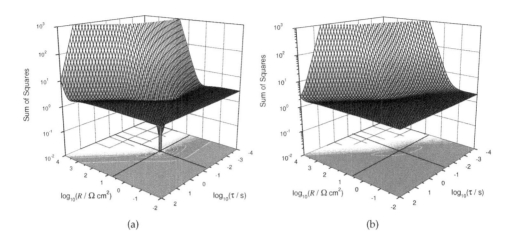

Figure 19.4: The objective function, equation (19.3), for a Voigt circuit with $R_0 = 1\ \Omega\text{cm}^2$, $R_1 = 100\ \Omega\text{cm}^2$, $\tau_1 = 0.001$ s, $R_2 = 200\ \Omega\text{cm}^2$, $\tau_1 = 0.01$ s, $R_3 = 5\ \Omega\text{cm}^2$, and $\tau_1 = 0.05$ s, presented as a function of R_3 and τ_3. The synthetic impedance data were calculated for frequencies from 1 to 10^5 Hz at a spacing of 10 measurements per decade: a) noise level determined by machine precision and b) stochastic noise added to the synthetic data with a standard deviation equal to 1 percent of the modulus.

the range of parameters seen.

Even though the line-shape corresponding to R_3 and τ_3 is not discernible from the impedance data shown in Figure 19.5, a minimum in the objective function for noise-free data is seen clearly in Figure 19.4(a). Thus, values for R_3 and τ_3 could be obtained by regression to noise-free data.

Addition of stochastic noise with standard deviation equal to 1 percent of the modulus (see symbols in Figure 19.5) is sufficient to obscure the effect of R_3 and τ_3 in the impedance data. A broad minimum is seen in Figure 19.4(b). Figure 19.4(a) yields a finite confidence interval for parameters R_3 and τ_3; whereas, Figure 19.4(b) yields confidence intervals that extend over several orders of magnitude and, therefore, includes zero. Statistically significant values for R_3 and τ_3, therefore, could not be obtained by regression to the noisy data. Thus, even though the parameters R_3 and τ_3 were used to generate the synthetic data, these parameters could not be recovered by regression to the noisy data.

Remember! 19.2 *The presence of noise in data can have a direct impact on model identification and on the confidence interval for the regressed parameters. The correctness of the model alone does not determine the number of parameters that can be obtained.*

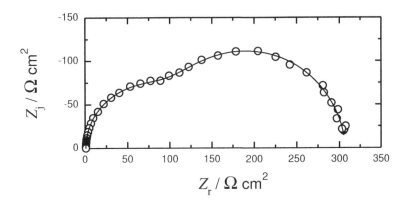

Figure 19.5: The impedance data used for Figure 19.4 in impedance-plane format. The solid line represents the noise-free data, and the symbols represent the data with added stochastic noise with a standard deviation equal to 1 percent of the modulus. Note that the third line-shape, with parameters $R_3 = 5 \ \Omega\text{cm}^2$ and $\tau_1 = 0.05$ s, is not readily seen, even for the noise-free solid line.

19.5.3   Ill-Conditioned Regression Caused by Insufficient Range

The frequency range of the measurement also has a direct impact on the number of parameters that can be identified by regression. Two impedance data sets are presented in Figure 19.6. The circles represent synthetic impedance data calculated for frequencies from 10^{-2} to 10^5 Hz at a spacing of 10 measurements per decade, and the triangles represent synthetic impedance data calculated for frequencies from 1 to 10^5 Hz at the same spacing of 10 measurements per decade. The impedance data were calculated for a Voigt circuit with $R_0 = 1 \ \Omega\text{cm}^2$, $R_1 = 100 \ \Omega\text{cm}^2$, $\tau_1 = 0.01$ s, $R_2 = 200 \ \Omega\text{cm}^2$, $\tau_2 = 0.1$ s, $R_3 = 100 \ \Omega\text{cm}^2$, and $\tau_3 = 10$ s. Noise with a standard deviation equal to 1 percent of the modulus was added.

The response surfaces corresponding to Figure 19.6 are presented as functions of parameters $\log_{10}(R_3)$ and $\log_{10}(\tau_3)$ in Figure 19.7(a) for the extended data set (10^{-2} to 10^5 Hz) and in Figure 19.7(b) for the truncated data set (1 to 10^5 Hz). The parameters R_3 and τ_3 can be readily identified from the response surface in Figure 19.7(a) but cannot be identified from Figure 19.7(b).

The response surfaces presented in this chapter have been given as a function of resistance and RC-time constant. Plots presented as functions of R and C have a similar appearance. The objective function, equation (19.3), is presented in Figure 19.8 for an RC circuit as a function of parallel resistor and capacitor values. The circuit parameters

---

Remember! **19.3** *The frequency range of the data can have a direct impact on model identification.*

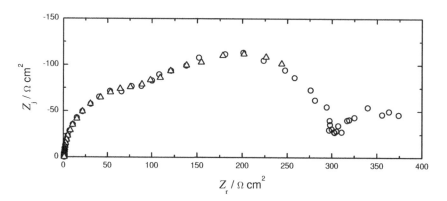

Figure 19.6: The impedance data for a Voigt circuit with $R_0 = 1$ Ωcm^2, $R_1 = 100$ Ωcm^2, $\tau_1 = 0.01$ s, $R_2 = 200$ Ωcm^2, $\tau_1 = 0.1$ s, $R_3 = 100$ Ωcm^2, and $\tau_1 = 10$ s. These data, with a standard deviation equal to 1 percent of the modulus, were used to generate Figure 19.7: $\circ$) synthetic impedance data calculated for frequencies from 10^{-2} to 10^5 Hz at a spacing of 10 measurements per decade and $\triangle$) synthetic impedance data calculated for frequencies from 1 to 10^5 Hz at a spacing of 10 measurements per decade.

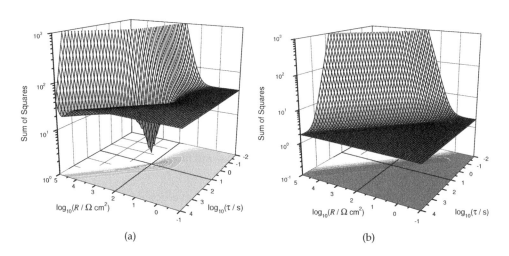

Figure 19.7: The objective function, equation (19.3), for the impedance data presented in Figure 19.6: a) synthetic impedance data collected from 10^{-2} to 10^5 Hz at a spacing of 10 measurements per decade and b) synthetic impedance data collected from 1 to 10^5 Hz at a spacing of 10 measurements per decade.

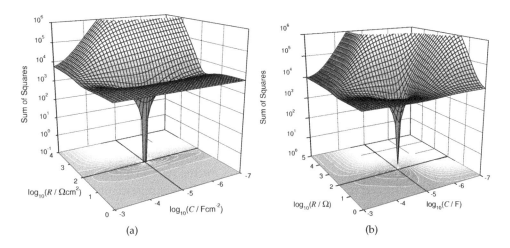

Figure 19.8: The objective function, equation (19.3), for an RC circuit as a function of parallel resistor and capacitor values. The circuit parameters were $R_0 = 1\ \Omega\mathrm{cm}^2$, $R_1 = 100\ \Omega\mathrm{cm}^2$, and $C_1 = 1 \times 10^{-5}$ F/cm^2: a) noise level determined by machine precision and b) noise added with a standard deviation equal to 1 percent of the modulus.

were $R_0 = 1\ \Omega\mathrm{cm}^2$, $R_1 = 100\ \Omega\mathrm{cm}^2$, and $C_1 = 1 \times 10^{-5}$ F/cm^2 (see Figure 4.3(b) and the corresponding Example 4.2). The objective of the regression procedure is to identify the parameter values that minimize the objective function. Solid lines have been drawn on the bottom contour map to indicate the values for which the function is minimized.

19.6   Initial Estimates for Regression

The complex nonlinear least-squares regression requires initial estimates for the parameter values. As shown in the contour plots given, for example, in Figures 19.4(a), 19.7(a), and 19.8, the objective function is insensitive to the parameters when the parameter value is far from the correct converged value. Good starting values will facilitate convergence, and poor starting values may result in convergence to a local minimum that does not represent the physics of the system.

Selection of an appropriate initial value for the time constant τ_k is critical for regression of the Voigt model (see equation (20.5)) to impedance data. Inductive loops can be modeled by the Voigt model by allowing the resistance values R_k to be negative while

---

Remember! 19.4 *The model identification problem is intricately linked to the error identification problem. Analysis of data requires analysis of the error structure of the measurement.*

---

forcing the time constant τ_k to be positive. It is important in this case that the time constant for the element with a negative resistance be consistent with the frequency range that shows inductive behavior. The time constant τ_k should be large, for example, to capture features at small frequencies.

Physical insight may provide good guidelines for selection of initial estimates for parameters used in process models such as diffusion coefficients and interfacial capacities. Another approach, described, for example, by Draper and Smith,[347] is to substitute into the postulated model a subset of the data equal to the number of parameters to be regressed and then solve the resulting equations for the parameters. The data points chosen should be spaced in frequency so as to capture the influence of each of the parameters. It can be helpful to explore the asymptotic limiting behavior where some of the parameters may have negligible influence on the impedance response. This approach ignores the scatter of the data, but the parameters so obtained can be used as initial guesses for the nonlinear regression.

19.7 Regression Statistics

The fitting procedure should provide three pieces of information:

- Parameter estimates

- Confidence intervals for the parameter estimates

- A statistical measure of the quality of the regression

The statistical measure of the quality of the regression is used to determine whether the model provides a meaningful representation of the data. The parameter estimates are reliable only if the model provides a statistically adequate representation of the data. The evaluation of the quality of the regression requires an independent assessment of the stochastic errors in the data, information that may not be available. In such cases, visual inspection of the fitting results may be useful. Issues associated with assessment of regression quality are discussed further in Section 19.7.2 and Chapter 20.

19.7.1 Confidence Intervals for Parameter Estimates

For linear regressions, the standard deviations for a parameter estimate P_j are given by

$$\sigma_j = \sqrt{C_{j,j}} \tag{19.34}$$

as described in Section 19.3.1. This result applies under the assumption that the fitting errors are normally distributed. As described in Chapter 21, the fitting errors comprise stochastic measurement errors, biased measurement errors, and errors associated with the inability of the model to describe the data.

Under the assumptions that the regression can be treated as being linear in the region of the χ^2 minimum and that fitting errors are normally distributed, equation (19.34) provides the standard deviation for parameter estimates in nonlinear regressions. Most commercial regression programs available for impedance spectroscopy provide both parameter estimates and the confidence interval based on the linear hypothesis. The size of the confidence interval relative to the parameter estimate can be used to determine the statistical significance of a parameter estimate. As described in Section 3.1.3, the probability is 67 percent that the parameter P_j lies within $P_j \pm \sigma_j$ and the probability is 95.45 percent that the parameter P_j lies within $P_j \pm 2\sigma_j$. Generally, the regression for a parameter is deemed not statistically meaningful if the confidence interval for that parameter includes zero.

A second and more generally applicable approach for determining the confidence interval for parameters is to create a surface of constant $\chi^2_{min} + \Delta\chi^2$. These surfaces resemble, for two parameters, the contours presented in Figure 19.1. The confidence level for a given $\Delta\chi^2$ can be obtained by Monte Carlo simulations, and the confidence interval for a specific parameter can be obtained by the projection of the $\Delta\chi^2$ onto the appropriate domain. Methods are described by Press et al.[355]

19.7.2 Statistical Measure of the Regression Quality

The minimum value of the χ^2 statistic is commonly used to provide a figure of merit for the quality of regression. As shown in equation (19.9), the χ^2 statistic accounts for the variance of the experimental data. In principle, for a model that well describes the data, the minimum value of the χ^2 statistic has a mean value of the degree of freedom ν with standard deviation $\sqrt{2\nu}$. Thus, for a good fit,

$$\frac{\chi^2_{min}}{\nu} = 1 \pm \sqrt{\frac{2}{\nu}} \tag{19.35}$$

The use of equation (19.35) to assess the quality of a regression is valid only with an accurate estimate of the variance of the stochastic errors in the impedance data.

An independent method to identify the stochastic errors of impedance data is described in Chapter 21. An alternative approach has been to use the method of maximum likelihood,[96] in which the regression procedure is used to obtain a joint estimate for the parameter vector **P** and the error structure of the data.[353,356] The maximum likelihood method is recommended under conditions where the error structure is unknown,[96] but the error structure obtained by simultaneous regression is severely constrained by the assumed form of the error-variance model. In addition, the assumption that the error variance model can be obtained by minimizing the objective function ignores the differences among the contributions to the residual errors shown in Chapter 21.

Problems

19.1   Create a synthetic data set for a system with a constant-phase element as given in equation (14.1) with parameters $R_e = 10$ Ωcm^2; $R_{||} = 10^3$ Ωcm^2; $Q = 100$ $s^\alpha/M\Omega cm^2$; and $\alpha = 0.7$ with frequencies ranging from 10^{-2} to 10^4 Hz spaced logarithmically with 10 measurements per decade. Use a commercial regression software or create your own to regress a Voigt model (see, e.g., equation (4.28)) to the data using modulus weighting. Find the maximum number of Voigt elements that can be added without including zero in the 2σ (95.45 percent) confidence interval.

19.2   Generate the synthetic data described in Problem 19.1, adding independent normally distributed random numbers $N(0,\sigma)$ to both real and imaginary parts of the impedance where $\sigma = a|Z|$ and the value of a is given below:

(a)   Generate the synthetic data set for $a = 0.01$. Perform the regression using modulus weighting for which $\sigma = 0.01|Z|$. Identify the value of the resulting χ^2 statistic and the number of parameters that can be obtained without including zero in the 2σ (95.45 percent) confidence interval.

(b)   Generate the synthetic data set for $a = 0.05$. Perform the regression using modulus weighting for which $\sigma = 0.01|Z|$. Identify the value of the resulting χ^2 statistic and the number of parameters that can be obtained without including zero in the 2σ (95.45 percent) confidence interval.

19.3   Generate the synthetic data described in Problem 19.1, adding independent normally distributed random numbers $N(0,\sigma)$ to both real and imaginary parts of the impedance where $\sigma = 0.01|Z|$.

(a)   Truncate the synthetic data set by removing the data below a frequency of 1 Hz. Perform the regression using modulus weighting for which $\sigma = 0.01|Z|$. Identify the value of the resulting χ^2 statistic and the number of parameters that can be obtained without including zero in the 2σ (95.45 percent) confidence interval.

(b)   Truncate the synthetic data set by removing the data below a frequency of 10 Hz. Perform the regression using modulus weighting for which $\sigma = 0.01|Z|$. Identify the value of the resulting χ^2 statistic and the number of parameters that can be obtained without including zero in the 2σ (95.45 percent) confidence interval.

19.4   Perform the regression for the data presented in Problem 19.2(a) using a weighting based on $\sigma = 0.05|Z|$. Compare the resulting value of the χ^2 statistic and the number of parameters that can be obtained without including zero in the 2σ (95.45 percent) confidence interval to that obtained for Problem 19.2(a).

Chapter 20

Assessing Regression Quality

The chapters presented in Part III describe development of mathematical models that can be used for interpretation of impedance measurements. These models may be regressed to data using the approaches presented in Chapter 19. A systematic approach is presented in this chapter to determine whether the model provides a statistically adequate description of the data.

20.1   Methods to Assess Regression Quality

Both quantitative and qualitative approaches may be used to assess the quality of a regression.

20.1.1   Quantitative Methods

As discussed in Section 19.7.2, the weighted χ^2 statistic defined in equation (19.24) provides a useful single numerical value for assessing the quality of a regression. As the differences between the observed and modeled values diminish, the χ^2 statistic becomes smaller. For a successful regression, the χ^2 statistic approaches the degree of freedom for the regression ν, given by equation (3.57). This assessment, however, requires an independent assessment of the stochastic errors in the impedance measurement. The influence of an inaccurate error analysis is illustrated in Section 20.2.1. The utility of an accurate error structure is illustrated in Sections 20.2.2 and 20.2.3.

In some cases, the models used for impedance are strictly defined. Others, such as the Voigt model, allow use of an arbitrary number of parameters. The fit of a Voigt model can be improved by sequentially adding RC elements, and the best model is achieved when the χ^2 statistic reaches a minimum value.

One potential concern is that there may be an inadequate sensitivity of the χ^2/ν statistic for identifying overfitting of data. Other criteria, such as the Akaike information criteria,[357–359] provide additional penalties for adding parameters to a model. The

Akaike performance index is given by

$$A_{PI} = \chi^2 \frac{1 + N_p/N_{dat}}{1 - N_p/N_{dat}} \tag{20.1}$$

where, as discussed below equation (19.8) on page 530, N_p is the number of parameters and N_{dat} is the number of data points. The related Akaike information criterion is given by

$$A_{IC} = \log\left(\chi^2 \left(1 + 2N_p/N_{dat}\right)\right) \tag{20.2}$$

As was discussed for the χ^2 statistic, the maximum number of parameters obtainable by regression is found where the Akaike information criteria reach minimum values.

Another quantitative assessment is obtained from the estimated confidence intervals for the regressed parameters, discussed in Section 19.7.1. The estimated confidence intervals, given by equation (19.34) under the assumptions that the regression can be treated as being linear in the region of the χ^2 minimum and that fitting errors are normally distributed, do not indicate the quality of the regression, but they do indicate whether the values obtained for a specific parameter are statistically significant. The probability is 95.45 percent that the parameter P_j lies within $P_j \pm 2\sigma_j$. Generally, the regression for a parameter is not statistically meaningful if the confidence interval for that parameter includes zero. Observation that the 95.45 percent confidence interval includes zero provides a third means of determining whether the model has too many parameters. Generally, all three approaches provide the same number of statistically significant parameters.

20.1.2 Qualitative Methods

The quality of a regression can also be assessed by visual inspection of plots. Of course, some plots are more sensitive than others to the level of agreement between model and experiment. As will be demonstrated in this chapter, the plot types can be categorized as given in Table 20.1. The comparison of plot types is presented in the subsequent sections for regression of models to a specific impedance data set.

20.2 Application of Regression Concepts

The data used for the present analysis were obtained by Durbha.[335] The electrolyte contained 0.01 M potassium ferricyanide, 0.01 M potassium ferrocyanide, and 1 M KCl in distilled water. A 5-mm-diameter Pt rotating disk was used as the working electrode, a Pt mesh was used as the counterelectrode, and a saturated calomel electrode was used as a reference electrode. The disk was rotated by a high-speed, low-inertia rotating disk apparatus developed at the CNRS.[360] The temperature of the electrolyte was controlled at $25.0 \pm 0.1°C$. The surface-polishing treatment followed wet polishing with 1,200-grit emery cloth, polishing with alumina paste, and ultrasonic cleaning. This was the

Table 20.1: Characterization of impedance plot types in terms of sensitivity to discrepancies between model and experimental values.

Poor Sensitivity	
Modulus	The Bode magnitude representation is singularly incapable of distinguishing between impedance models unless they provide extremely poor fits to impedance data.
Real	The real impedance representation is similarly insensitive to fit quality.
Modest Sensitivity	
Complex-impedance-plane	These plots are sensitive only for large impedance values. Impedance data at high-frequency values are typically obscured.
Imaginary	The imaginary impedance representation is modestly sensitive to fit quality.
Log-imaginary	These plots emphasize small values, and values of the slope that differ from ± 1 may suggest the need for new models.
Phase angle	The high-frequency behavior in these plots is counterintuitive due to the role of solution resistance.
Modified phase angle	Ohmic-resistance-corrected phase angle plots may confirm the existence of a CPE.
Excellent Sensitivity	
Residual error plots	Appearance of trending indicates a need to improve the model or to remove data that are inconsistent with the Kramers–Kronig relations.

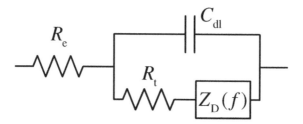

Figure 20.1: Electrical circuit corresponding to convective diffusion to a disk electrode.

procedure that gave the largest value for the mass-transfer-limited current and was therefore assumed to result in the smallest amount of surface blocking.

The potentials and currents were measured and controlled by a Solatron 1286 potentiostat, and a Solatron 1250 frequency response analyzer was used to apply the sinusoidal perturbation and to calculate the transfer function. The impedance data analyzed in this section were taken after 12 hours of immersion and were found by the methods described in Chapter 22 to be consistent with the Kramers–Kronig relations.

For the purpose of demonstrating the means of assessing regression quality, three models were applied for the analysis of the impedance data:

1. A Nernst stagnant-diffusion-layer model was used to account for the diffusion impedance. This model is often used to account for mass transfer in convective systems, even though it is well known that this model cannot account accurately for the convective diffusion associated with a rotating disk electrode.[211]

2. Following Agarwal et al.,[69, 100, 101] a measurement model based on the Voigt series was used to assess the error structure.

3. A refined process model was used that correctly accounts for convective diffusion to a rotating disk electrode under the assumption that the surface is uniformly accessible. This model also employs a constant-phase element to address complexities seen at high frequency.[99]

20.2.1 Finite-Diffusion-Length Model

The equivalent circuit presented in Figure 20.1 was regressed to the impedance data. The mathematical formulation for the model is given as

$$Z(f) = R_e + \frac{R_t + Z_D(f)}{1 + (j\omega C)(R_t + Z_D(f))} \tag{20.3}$$

where, under assumption of a Nernst stagnant diffusion layer (as shown in equation (11.20)),

$$Z_D(f) = Z_D(0) \frac{\tanh \sqrt{j\omega\tau}}{\sqrt{j\omega\tau}} \tag{20.4}$$

Table 20.2: Estimated χ^2/ν values for the regression of the model presented in Figure 20.1 to impedance data obtained for reduction of ferricyanide on a Pt rotating disk electrode.

Estimated Noise Level:	3 %	1 %	0.1 %	Measured
Model for σ	$0.03\lvert Z\rvert$	$0.01\lvert Z\rvert$	$0.001\lvert Z\rvert$	$9 \times 10^{-6}\lvert Z\rvert^2$
χ^2/ν	0.97	8.7	870	1,820

The regression was weighted by the error structure for the measurement determined using the measurement model approach described in Chapter 21.

Quantitative Assessment

Due to the appearance of the variance in equation (19.24), the χ^2 statistic provides a useful measure of the quality of a fit only if the variance of the measurement is known. The techniques described in Chapter 21 may be used to assess the standard deviation of an impedance measurement as a function of frequency. In the absence of such an assessment, researchers have used assumed error structures, but, in this case, the numerical value of the χ^2 statistic cannot be used to assess the quality of a regression. This situation is illustrated in Table 20.2, where the value of χ^2/ν is presented as a function of the assumed error structure for the regression of the model presented in equations (20.3) and (20.4) to the experimental data. Under the assumption that the standard deviation of the measurement was 3 percent of the modulus, $\chi^2/\nu = 0.97$. This value would indicate that the fitting errors were of the order of the noise in the measurement, indicating that the model provides an excellent fit to the data. On the other hand, assumption that the standard deviation of the measurement was 0.1 percent of the modulus yields $\chi^2/\nu = 870$. This value would indicate that the fitting errors were much larger than the noise in the measurement, indicating that the model provides an inadequate representation of the data. Thus, depending on the assumed error structure, the χ^2 statistic can be used to support or reject the model. In fact, for this regression and using the experimentally determined stochastic error structure, $\chi^2/\nu = 1,820$. The fitting errors were indeed much larger than the noise in the measurement.

Visual Inspection

A qualitative assessment of the fit quality can be obtained in the comparison of the calculated to experimental values shown in complex-impedance-plane or Nyquist format in Figure 20.2. The results at high frequency are not visible in Nyquist plots, but a clear

Remember! 20.1 *If the experimental error structure is not known, the numerical value of the χ^2 statistic cannot be used to assess the quality of a regression.*

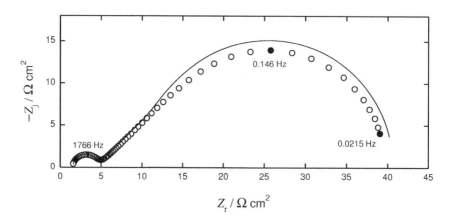

Figure 20.2: Comparison in complex-impedance-plane format of the model presented in Figure 20.1 to impedance data obtained for reduction of ferricyanide on a Pt rotating disk electrode.

discrepancy is evident at lower frequencies where the impedance values are larger and therefore more visible.

The discrepancy is not evident, however, in the Bode magnitude plot presented in Figure 20.3(a). In fact, the Bode magnitude representation is singularly incapable of distinguishing between impedance models unless they provide extremely poor fits to impedance data. The Bode phase-angle representation in Figure 20.3(b) is more sensitive and, for the present example, reveals discrepancies at both high and intermediate frequencies.

The sensitivity of the impedance representation, presented in Figures 20.4(a) and (b) for real and imaginary parts respectively, is somewhat comparable to that seen for the Bode representation. The real part of the impedance is as insensitive to model quality as is the Bode modulus, and the imaginary impedance plots show a discrepancy between model and experiment at intermediate frequencies.

The greatest sensitivity is observed for plots of residual errors. Residual errors normalized by the value of the impedance are presented in Figures 20.5(a) and (b), respectively, for the real and imaginary parts of the impedance. The experimentally measured standard deviation of the stochastic part of the measurement is presented as dashed lines in Figure 20.5. The interval between the dashed lines represents the 95.4 percent confidence interval for the data ($\pm 2\sigma$). Significant trending is observed as a function of frequency for residual errors of both real and imaginary parts of the impedance.

The plot of the absolute value of the imaginary part of the impedance with respect

Remember! 20.2 *The Bode magnitude and real impedance plots are relatively insensitive to the quality of the fit of a model to impedance data.*

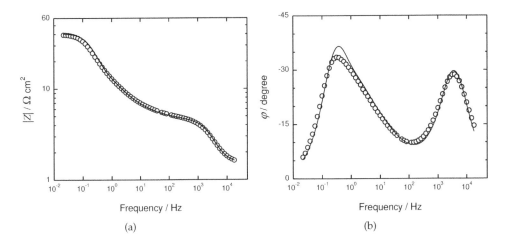

Figure 20.3: Comparison in Bode format of the model presented in Figure 20.1 to impedance data obtained for reduction of ferricyanide on a Pt rotating disk electrode: a) modulus and b) phase angle.

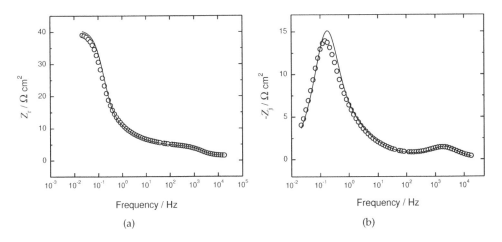

Figure 20.4: Comparison in impedance format of the model presented in Figure 20.1 to impedance data obtained for reduction of ferricyanide on a Pt rotating disk electrode: a) real and b) imaginary.

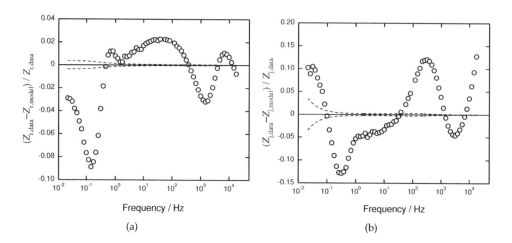

Figure 20.5: Normalized residual errors for the fit of the model presented in Figure 20.1 to impedance data obtained for reduction of ferricyanide on a Pt rotating disk electrode: a) real and b) imaginary.

to frequency on a logarithmic scale, shown in Figure 20.6, shows discrepancies at small impedance values, in particular at high frequency. As described in Section 18.3 (page 505), the imaginary part of the impedance at high frequency appears as a straight line in Figure 20.6, but the slope of this line differs from the value of -1 constrained by the model. This result suggests that the high-frequency capacitive loop may be modeled as a CPE.

20.2.2   Measurement Model

The measurement model method for distinguishing between bias and stochastic errors is based on using a generalized model as a filter for nonreplicacy of impedance data. The measurement model is composed of a superposition of line shapes that can be arbitrarily chosen subject to the constraint that the model satisfies the Kramers–Kronig relations. The model presented in Figure 21.8, composed of Voigt elements in series with a solution resistance, i.e.,

$$Z = R_0 + \sum_{k=1}^{K} \frac{R_k}{1 + j\omega\tau_k} \tag{20.5}$$

has been shown to be a useful measurement model. With a sufficient number of parameters, the Voigt model is able to provide a statistically significant fit to a broad variety of impedance spectra.[69] In the context of the present chapter, it provides a fit quality that can be the goal for model development.

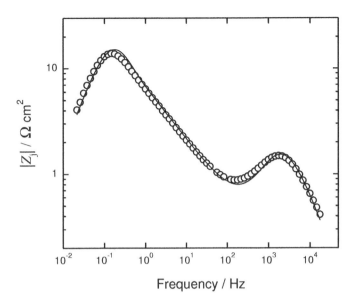

Figure 20.6: Comparison in log-imaginary-impedance format of the model presented in Figure 20.1 to impedance data obtained for reduction of ferricyanide on a Pt rotating disk electrode.

Quantitative Assessment

The measurement comprised real and imaginary parts of the impedance for 70 frequencies; thus, the vector of data for complex regression included 140 data. The measurement model with 11 Voigt elements (23 parameters) yielded the smallest value of χ^2, the smallest value of the Akaike information criteria, and the largest number of parameters with 95.45 percent confidence intervals that did not include zero. The degree of freedom for this problem was $\nu = 140 - 23 = 117$. The regression was weighted according to the error structure obtained from repeated impedance measurements and using the measurement model (Section 21.5) to filter imperfect replication. The χ^2 statistic for this regression yielded a value of $\chi^2/\nu = 1.22$ with standard deviation $\sqrt{2/\nu} = 0.13$, which suggests that the residual errors are nearly of the same order as the stochastic errors.

Visual Inspection

The fit of the measurement model with 11 Voigt elements is presented in complex-impedance-plane format in Figure 20.7. The discrepancies evident in Figure 20.2 for the model presented in Section 20.2.1 are not apparent in Figure 20.7.

An excellent agreement is seen as well for the Bode format plots in Figures 20.8(a) and (b). As discussed in Section 20.2.1, however, a good agreement between model and experiment in modulus plots does not provide a reliable assessment of fit quality. The agreement between model and experiment in Figure 20.8(b) is better than that seen in

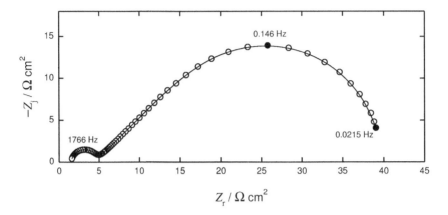

Figure 20.7: Comparison in complex-impedance-plane format of the measurement model to impedance data obtained for reduction of ferricyanide on a Pt rotating disk electrode.

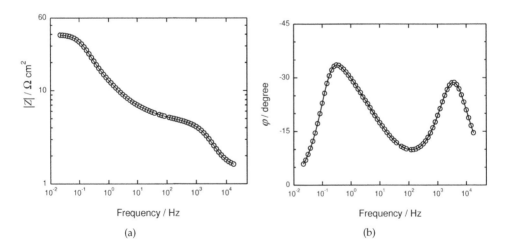

Figure 20.8: Comparison in Bode format of the measurement model to impedance data obtained for reduction of ferricyanide on a Pt rotating disk electrode: a) modulus; and b) phase angle.

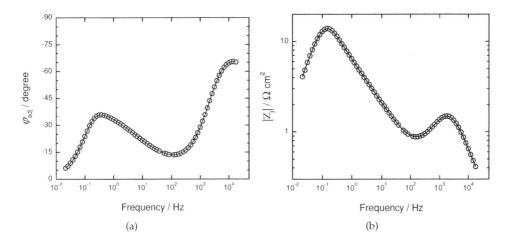

Figure 20.9: Comparison of the measurement model to impedance data obtained for reduction of ferricyanide on a Pt rotating disk electrode: a) ohmic-resistance-corrected phase angle and b) log-imaginary impedance.

Figure 20.3(b).

The ohmic-resistance-corrected phase-angle plot, see Section 18.2.1 on page 504, provides confirmation of the conclusion, drawn from Figure 20.6, that CPE behavior is evident at high frequencies. The constant phase-angle shown in Figure 20.9(a) reaches a value of $-67°$ at high frequency, corresponding to a CPE exponent value of $\alpha = 0.75$. The value of the slope in the high-frequency range of the data shown in Figure 20.6 is -0.75, also corresponding to $\alpha = 0.75$. The measurement model is seen to be fully capable of representing the CPE behavior evident in Figure 20.9(b).

The real and imaginary parts of the impedance are presented in Figures 20.10(a) and (b), respectively. An excellent agreement between model and experiment is observed, but, as discussed in Section 20.2.1, a good agreement between model and experiment in real impedance plots does not provide a reliable assessment of fit quality.

The residual errors are presented in Figures 20.11(a) and (b) for the real and imaginary parts of the impedance, respectively. The dashed lines represent the experimentally determined noise level of the measurement. The scales used to present the results in Figure 20.11 are in stark contrast to the scales used in Figure 20.5. The residual error plots show that the measurement model provides a substantially better fit to the data than does the finite-diffusion-length model.

20.2.3   Convective-Diffusion-Length Model

The quantitative and qualitative analysis presented in Section 20.2.1 demonstrates that the finite-diffusion-layer model provides an inadequate representation for the impedance response associated with a rotating disk electrode. The presentation in Section 20.2.2 demonstrates that a generic measurement model, while not providing a physical

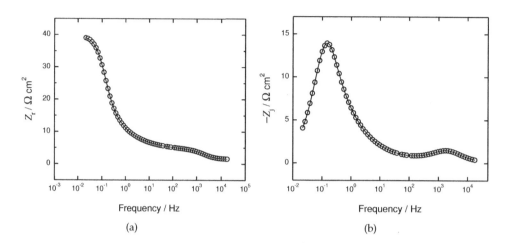

Figure 20.10: Comparison in impedance format of the measurement model to impedance data obtained for reduction of ferricyanide on a Pt rotating disk electrode: a) real and b) imaginary.

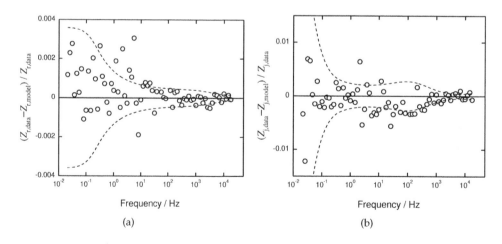

Figure 20.11: Normalized residual errors for the fit of the convective-diffusion models presented in Figure 20.12 to impedance data obtained for reduction of ferricyanide on a Pt rotating disk electrode: a) real and b) imaginary.

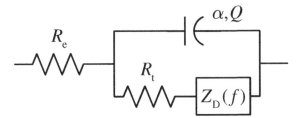

Figure 20.12: Electrical circuit corresponding to convective diffusion to a disk electrode with high-frequency CPE behavior.

interpretation of the disk system, can provide an adequate representation of the data. Thus, an improved mathematical model can be developed.

Refined models for mass transfer to a disk electrode are presented in Section 11.3. The equivalent circuit presented in Figure 20.12 was regressed to the impedance data. The mathematical formulation for the model is given as equation (10.77). Four models were considered for the convective-diffusion term $Z_D(f)$:

1. The finite-diffusion model given as equation (20.4)

2. The one-term convective-diffusion model consisting of the first term in equation (11.86)

3. The two-term convective-diffusion model consisting of the first two terms in equation (11.86)

4. The three-term convective-diffusion model consisting of all terms given in equation (11.86)

The three-term convective-diffusion model provides the most accurate solution to the one-dimensional convective-diffusion equation for a rotating disk electrode. The one-dimensional convective-diffusion equation applies strictly to the mass-transfer-limited plateau where the concentration of the mass-transfer-limiting species at the surface can be assumed to be both uniform and equal to zero. As described elsewhere,[123,361] the concentration of reacting species is not uniform along the disk surface for currents below the mass-transfer-limited current, and the resulting nonuniform convective transport to the disk influences the impedance response.[362,363]

Quantitative Assessment

The regression was weighted by the variance of the measurement, determined from experimental data using the measurement model analysis described in Section 21.5. Because the weighting was based on the experimental stochastic error structure, the χ^2/ν statistic presented in Table 20.3 provides a meaningful quantitative assessment of the quality of the regression. Ideally, as discussed in Section 19.7.2, χ^2/ν should have a

Table 20.3: χ^2/ν values for the regression of the models presented in Figure 20.1 to impedance data obtained for reduction of ferricyanide on a Pt rotating disk electrode.

Model:	No CPE	With CPE				Voigt
	Finite	Finite	1-Term	2-Term	3-Term	Model
ν	135	134	134	134	134	117
χ^2/ν	1,820	107	46.7	46.2	46.2	1.22

value of unity for an excellent regression. The χ^2/ν statistic had a value of 1,820 for the regression using the finite-diffusion model described in Section 20.2.1. Inclusion of a CPE to describe the high-frequency capacitive loop improved the value to $\chi^2/\nu = 107$. The large values of χ^2/ν are consistent with the understanding that these models are not accurate because they describe diffusion through a stagnant layer of fluid and do not account for the distribution of axial velocity within the diffusion layer.

The one-term numerical convective-diffusion model, which treats the Schmidt number as being infinitely large, shows significant improvement, yielding $\chi^2/\nu = 46.7$. Addition of a second term yields a slightly smaller number, i.e., $\chi^2/\nu = 46.2$, but addition of yet another term makes no improvement in the quality of the regression. Nevertheless, it was possible to obtain a χ^2/ν statistic with value close to unity. The regression of a Voigt measurement model, which has no particular physical interpretation, yielded $\chi^2/\nu = 1.22$. It is important to note that each of the four models employing a CPE has the same number of adjustable parameters. The difference between the models is due to the different accuracy of the solution to the one-dimensional convective-diffusion equation.

The result obtained with the Voigt measurement model shows that it is possible to obtain a regression using passive elements that describes the data to within the noise level of the measurement. The observation that the three-term model did not improve the regression shows that the regression cannot be improved by refining the solution to the one-dimensional convective-diffusion equation. Instead, the assumption of radial uniformity, implicit in the one-dimensional model, must be relaxed.

Visual Inspection

As described in Sections 20.2.1 and 20.2.2, the quality of the regressions can be assessed to varying degrees of success by inspection of plots. The Nyquist or complex-impedance-plane representation given in Figure 20.13 reveals the difference between

---

Remember! 20.3 *Careful examination of the regression statistics and residual errors can guide model development.*

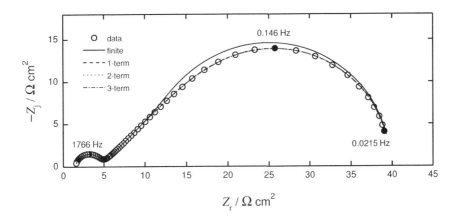

Figure 20.13: Comparison in complex-impedance-plane format of the convective-diffusion models presented in Figure 20.12 to impedance data obtained for reduction of ferricyanide on a Pt rotating disk electrode.

the finite-diffusion-length model and the models based on numerical solution of the convective-diffusion equation, but cannot be used to distinguish the models based on one-term, two-term, and three-term expansions.

The logarithmic plot presented in Figure 20.14 cannot be used to distinguish the model based on finite-stagnant-diffusion layer from the convective-diffusion models based on one-term, two-term, and three-term expansions. The logarithmic presentation in Figure 20.14 does show that each of the models considered used a CPE to account for high-frequency dispersion.

As shown in Figure 20.15(a), plots of the magnitude of the impedance cannot distinguish any of the models considered in this section, whereas the difference between the finite-diffusion-length model and the models based on numerical solution of the convective-diffusion equation are apparent in plots of the phase angle (Figure 20.15(b)). Figure 20.15(b) cannot be used, however, to distinguish the models based on one-term, two-term, and three-term expansions.

As shown in Figure 20.16(a), plots of the real part of the impedance cannot distinguish any of the models considered in this section, whereas the difference between the finite-diffusion-length model and the models based on numerical solution of the convective-diffusion equation are apparent in plots of the imaginary part of the impedance (Figure 20.16(b)). Figure 20.16(b) cannot be used, however, to distinguish the models based on one-term, two-term, and three-term expansions.

As described in Sections 20.2.1 and 20.2.2, plots of the residual errors provide the most sensitive assessment of fit quality. The normalized residual errors for the regressions treated in the section are presented in Figures 20.17(a) and (b) for the real and imaginary parts of the impedance, respectively. The distinction between the finite-diffusion-length model and the models based on numerical solution of the convective-

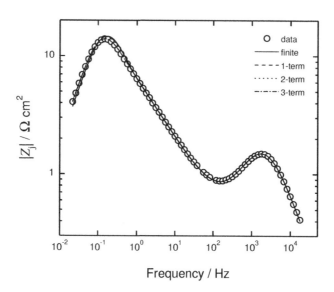

Figure 20.14: Comparison in log-imaginary-impedance format of the convective-diffusion models presented in Figure 20.12 to impedance data obtained for reduction of ferricyanide on a Pt rotating disk electrode.

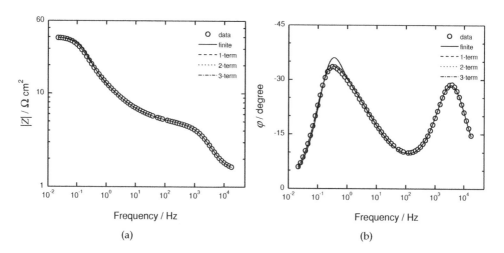

Figure 20.15: Comparison in Bode format of the convective-diffusion models presented in Figure 20.12 to impedance data obtained for reduction of ferricyanide on a Pt rotating disk electrode: a) modulus and b) phase angle.

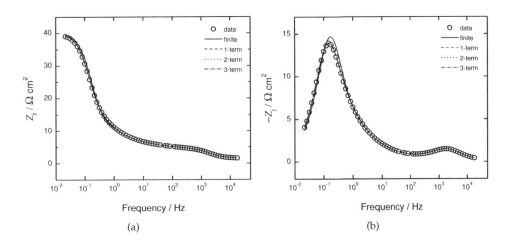

Figure 20.16: Comparison in impedance format of the convective-diffusion models presented in Figure 20.12 to impedance data obtained for reduction of ferricyanide on a Pt rotating disk electrode: a) real and b) imaginary.

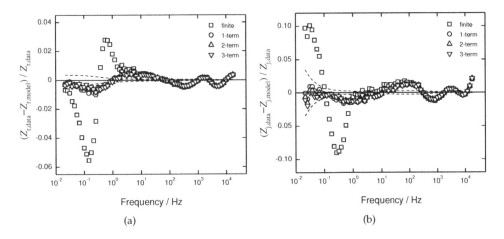

Figure 20.17: Normalized residual errors for the fit of the convective-diffusion models presented in Figure 20.12 to impedance data obtained for reduction of ferricyanide on a Pt rotating disk electrode: a) real and b) imaginary.

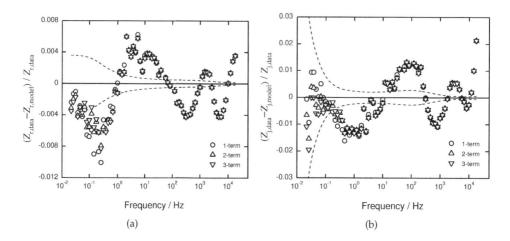

Figure 20.18: Normalized residual errors for the fit of the convective-diffusion models presented in Figure 20.12 to impedance data obtained for reduction of ferricyanide on a Pt rotating disk electrode: a) real and b) imaginary.

diffusion equation are readily apparent. The differences between the models based on numerical solution of the convective-diffusion equation are not so apparent on the scale presented in Figures 20.17(a) and (b).

The real and imaginary normalized residual errors presented in Figures 20.18(a) and (b), respectively, are similar for the convective diffusion models based on one-term, two-term, and three-term velocity expansions. The distinction between an excellent fit and a merely good fit can be seen by comparison between the residual errors presented in Figure 20.18, which are based on solution of the one-dimensional convective-diffusion equation, and those presented in Figure 20.11, which are based on regression by a Voigt measurement model. The normalized residual errors in Figure 20.18 show significant trending, whereas the normalized residual errors in Figure 20.11 do not. This assessment, which is based on visual inspection of the plots of normalized residual errors, is reflected as well in the quantitative assessment provided by the χ^2/ν statistic presented in Table 20.3.

Problems

20.1 Compare the plots described in Table 20.1 for the regression mentioned in Problem 19.2.

20.2 Plot χ^2/ν as a function of the number of Voigt elements for the regressions mentioned in Problem 19.2.

20.3 Statistics textbooks show different ways to visualize residual errors. A useful alternative representation involves scaling the residual errors by the standard deviation. Create such plots for the regression mentioned in Problem 19.2.

20.4 Monte Carlo simulation can be used to explore the influence of random noise on the parameter estimation. Consider the following protocol:

(1) Use a measurement model to identify error structure.

(2) Use a measurement model on Kramers–Kronig-consistent data set to get a model for data.

(3) Add to the model values obtained in the previous step a normally distributed random error with an experimentally determined standard deviation.

(4) Fit the resulting model to each synthetic data set, and obtain a distribution of parameter values.

Contrast this approach to that presented in Examples 3.3 and 3.4.

Part V

Statistical Analysis

Chapter 21

Error Structure of Impedance Measurements

The regression procedures described in Chapter 19 require, in addition to an adequate deterministic model, a quantitative assessment of measurement characteristics. The weighting strategy should generally account for the magnitude of stochastic errors. In addition, the regressed data set either should include only data that have not been corrupted by bias errors, or, as an alternative approach, could incorporate bias errors into the weighting strategy.

While the nature of the error structure of the measurements is often ignored or understated in electrochemical impedance spectroscopy, recent developments have made possible experimental identification of error structure. Quantitative assessment of stochastic and experimental bias errors has been used to filter data, to design experiments, and to assess the validity of regression assumptions.

21.1 Error Contributions

The error contributions to an impedance measurement can expressed in terms of the difference between the observed value $Z_{ob}(\omega)$ and a model value $Z_{mod}(\omega)$ as

$$Z_{ob}(\omega) - Z_{mod}(\omega) = \varepsilon_{res}(\omega) = \varepsilon_{fit}(\omega) + \varepsilon_{stoch}(\omega) + \varepsilon_{bias}(\omega) \qquad (21.1)$$

where $\varepsilon_{res}(\omega)$ represents the residual error, $\varepsilon_{fit}(\omega)$ is the systematic error that can be attributed to inadequacies of the model, $\varepsilon_{stoch}(\omega)$ is the stochastic error with expectation $E\{\varepsilon_{stoch}(\omega)\} = 0$, and $\varepsilon_{bias}(\omega)$ represents the systematic experimental bias error that cannot be attributed to model inadequacies. Typically, the impedance is a strong function of frequency and can vary over several orders of magnitude over the experimentally accessible frequency range. The stochastic errors of the impedance measurement are strongly heteroscedastic, which, in this case, means that the variance of the sto-

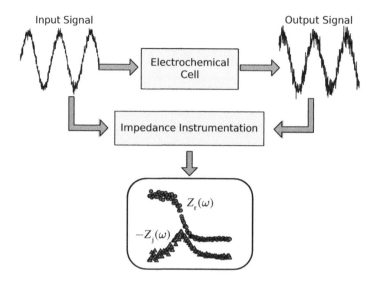

Figure 21.1: A schematic representation of the propagation of time-domain errors through an electrochemical cell and impedance instrumentation to the frequency domain.

chastic errors is a strong function of frequency. Selection of an appropriate weighting strategy is, therefore, critical for interpretation of data.

A distinction is drawn in equation (21.1) between stochastic errors that are randomly distributed about a mean value of zero, errors caused by the lack of fit of a model, and experimental bias errors that are propagated through the model. The problem of interpretation of impedance data is therefore defined to consist of two parts: one of identification of experimental errors, which includes assessment of consistency with the Kramers–Kronig relations (see Chapter 22 on page 595), and one of fitting (see Chapter 19 on page 527), which entails model identification, selection of weighting strategies, and examination of residual errors. The error analysis provides information that can be incorporated into regression of process models. The experimental bias errors, as referred to here, may be caused by nonstationary processes or by instrumental artifacts.

21.2   Stochastic Errors in Impedance Measurements

While the error terms given in equation (21.1) are functions of frequency, it is important to note that the signals used to generate impedance are functions of time, not frequency. The schematic representation shown as Figure 21.1 illustrates the process. The error $\varepsilon_{\text{stoch}}(\omega)$ arises from an integration of time-domain signals that contain noise originating from instrumental sources, thermal fluctuations of resistivity, thermal fluctuations of the concentration of species and the rates of electrochemical reactions, and macroscopic events such as pitting and bubble nucleation.

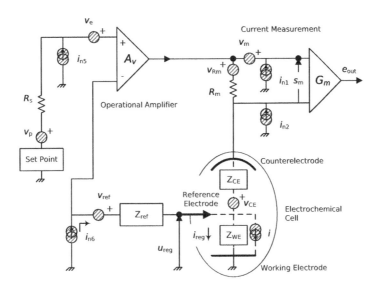

Figure 21.2: A schematic representation of an electrochemical cell under potentiostatic regulation, with sources of potential noise indicated as shaded circles and sources of current noise indicated as shaded double circles (see Gabrielli et al.[364]).

21.2.1 Stochastic Errors in Time-Domain Signals

Following Gabrielli et al.,[364] the potential imposed between the working electrode and the reference electrode

$$U_{\text{reg}} = \overline{U}_{\text{reg}} + u_{\text{reg}} \tag{21.2}$$

contains a stochastic noise u_{reg}. The noise in the regulation signal contains contributions from both potential and current sources that are identified in Figure 21.2. These noise contributions can be expressed conveniently in the frequency domain as

$$u_{\text{reg}} = \frac{Z_{\text{WE}}}{Z_{\text{WE}}A_v + Z_{\text{WE}} + R_m + Z_{\text{CE}}} \tag{21.3}$$
$$\left\{ A_v \left[v_e + v_p + R_s i_{n5} + v_{\text{ref}} \right] - v_{R_m} - v_{\text{CE}} - (R_m + Z_{\text{CE}}) (i - i_{n6}) + R_m i_{n2} \right\}$$

where Z_{WE} is the impedance of the working electrode, Z_{CE} is the impedance of the counterelectrode, Z_{ref} is the impedance of the reference electrode, A_v is the gain of the operational amplifier, R_m is the resistance of the current measurement circuit, R_s is the resistance of the potential control circuit, v_e, v_p, v_{ref}, v_{R_m}, and v_{CE} represent voltage

---

Remember! 21.1 *The stochastic errors in impedance measurements arise from an integration of time-domain signals that contain noise originating from the electrochemical cell and the instrumentation.*

---

noise contributions shown as shaded circles in Figure 21.2, and i, i_{n1}, i_{n2}, i_{n5}, and i_{n6} represent current noise contributions shown as shaded double circles in Figure 21.2. The electrochemical noise i arises from molecular-scale fluctuations.[365,366] The current noise contributions act through resistors R_m and R_s to generate voltage noise contributions.

Under the assumptions that the gain of the operational amplifier A_V is large and that the reference electrode impedance Z_{ref} is large, the dominant stochastic error in the regulation signal consists of additive contributions from the reference electrode and the operational amplifier as

$$u_{reg} = v_e + v_p + R_s i_{n5} + v_{ref} \tag{21.4}$$

As equation (21.4) is not a function of frequency, it applies to both time and frequency domains.

The regulation noise induces, through the electrochemical cell impedance, a parasitic current fluctuation $i_{reg}(t)$ that can be calculated from

$$i_{reg}(t) = \text{IFFT}\left\{\text{FFT}\{u_{reg}(t)\}/Z(\omega)\right\} \tag{21.5}$$

where the notation $\text{IFFT}\{x\}$ represents the inverse Fourier transform of the function x, and $\text{FFT}\{x\}$ represents the Fourier transform of the function x. The potential difference across the inputs to the current follower is given as

$$S_m = \overline{S}_m + s_m \tag{21.6}$$

where

$$s_m = v_m + v_{R_m} + R_m \left(i + i_{reg} - i_{n2} - i_{n6}\right) \tag{21.7}$$

The response of the current measurement channel is given by $E_{out} = G_m S_m$, and the noise in the current measurement channel is given by

$$e_{out} = G_m \left[v_m + v_{R_m} + R_m \left(i + i_{reg} - i_{n2} - i_{n6}\right)\right] \tag{21.8}$$

Thus, the noise in both current and potential measurement channels consists of sums of noise contributions. A similar development has been presented for zero-resistance ammeters.[367]

The investigation of noise sources suggests that instrumental and electrochemical noise can be represented by stochastic signals added to the time-averaged measured and controlled signals. The instrumental noise sources can be assumed to have a very high frequency; thus the evaluations of the instrumental noise at any two instances in time, t and $t + \tau$, are uncorrelated. The added signals should be statistically independent with the exception of the term $i_{reg}(t)$, which is correlated with both $e_{reg}(t)$ and $i_{reg}(t + \tau)$.

For most potentiostats, the assumption $A_V \gg 1$ becomes invalid at frequencies above 1 to 10 kHz. In this case, the noise terms are still additive, but the interaction between the gain of the operational amplifier and the cell impedance results in additional correlation between the input and output channels.

21.2.2 Transformation from Time Domain to Frequency Domain

The influence of noise on the impedance response can be illustrated by an extension of the analysis presented in Section 7.3 on page 143. The current density response to a 10-mV-amplitude ($b_a \Delta V = 0.19$) sinusoidal potential input is presented in Figure 21.3 for the system presented in Section 7.3 with parameters $C_{dl} = 31\ \mu F/cm^2$, $Fk_a = nFk_c = 0.14\ mA/cm^2$, $b_a = 19.5\ V^{-1}$, $b_c = 19.5\ V^{-1}$, and $\overline{V} = 0.1\ V$. These parameters yield a value of charge-transfer resistance $R_t = 51.08\ \Omega cm^2$ and a characteristic frequency of 100 Hz. The potential and current signals were scaled by ΔV and $\Delta I = \Delta V / |Z|$, respectively. The results presented in Figure 21.3 can be compared to the results presented in Figure 7.4 for a linear response to a 1 mV potential perturbation and to Figure 8.2 for a nonlinear response to a 40 mV potential perturbation. The slight phase shift evident in Figure 8.2(b) for a 10 Hz input signal is almost obscured by the noise in Figure 21.3(b). The larger phase shift associated with higher frequencies, e.g., Figures 8.2(c) and (d) for frequencies of 100 Hz and 10 kHz, respectively, can be discerned in Figures 21.3(c) and (d).

While the presence of additive errors obscures the phase differences in the signals presented in Figure 21.3, repeated sampling at a given frequency allows identification of the transfer-function response, as shown in the corresponding Lissajous plots presented in Figure 21.4. The linear response presented in Figure 21.4(a) for measurement at 1 mHz can be compared to the slightly broader Lissajous plot presented in Figure 21.4(b) for 10 Hz. An elliptical shape can be seen at the characteristic frequency of 100 Hz in Figure 21.4(c), and a perfectly circular response is evident for 10 kHz in Figure 21.4(d). The ideal Lissajous response for these frequencies is given by the 1 mV curves in Figure 8.3.

Figures 21.3 and 21.4 illustrate the transformation of time-domain stochastic errors to the frequency domain. The impedance can be calculated, for example, by the phase-sensitive-detection methods presented in Section 7.3.2 or the Fourier analysis presented in Section 7.3.3. The nature of the errors in the frequency domain will be influenced by the characteristics of the methods used to transform the time-domain signals to the frequency domain.

21.2.3 Stochastic Errors in Frequency Domain

The magnitude of the stochastic errors in impedance measurements depends on the selection of experimental parameters as detailed in Chapter 8. The simulation results described by Carson et al.[145, 102, 354] in particular provide insight into differences between commonly used impedance instrumentation, including methods based on Fourier analysis[102] and on phase-sensitive detection.[354]

Some general properties for stochastic errors have been established for impedance measurements through experimental observation and simulations. The results described here correspond to additive time-domain errors. The comparison between

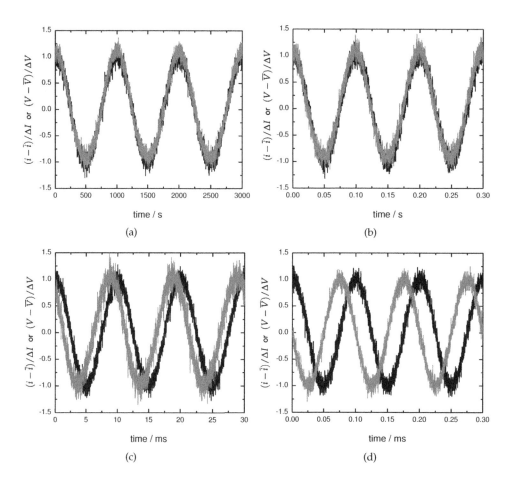

Figure 21.3: The current density response to a sinusoidal potential input for the system presented in Section 7.3 with parameters $C_{dl} = 31$ μF/cm^2, $nFk_a = nFk_c = 0.14$ mA/cm^2, $b_a = 19.5$ V^{-1}, $b_c = 19.5$ V^{-1}, and $\overline{V} = 0.1$ V: a) 1 mHz; b) 10 Hz; c) 100 Hz; and d) 10 kHz. The signals include normally distributed additive errors with a magnitude of 10 percent of the magnitude of the respective signal. The black line represents the potential input and the grey line represents the resulting current density, scaled by $\Delta I = \Delta V / |Z|$.

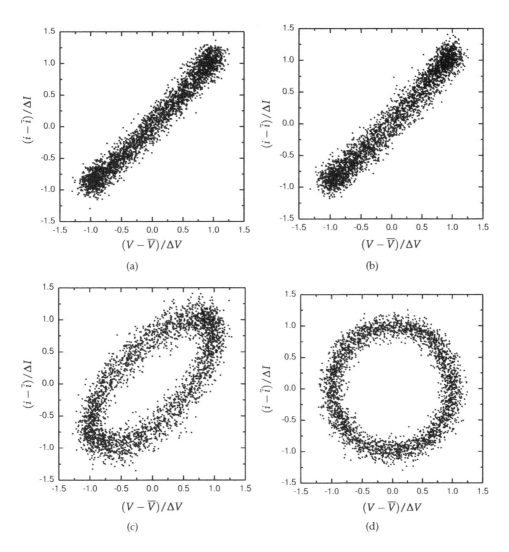

Figure 21.4: Lissajous plots for the system presented in Figure 21.3: a) 1 mHz; b) 10 Hz; c) 100 Hz; and d) 10 kHz. The potential and current signals were scaled by ΔV and $\Delta I = \Delta V/|Z|$, respectively.

simulations and experimental results obtained via Fourier analysis supports the suggestion[364] that the nature of experimental time-domain errors is likely to be additive rather than proportional:[102]

- Impedance measurements are, in general, heteroscedastic, which means that the variance of the stochastic errors is a strong function of frequency. It is important, therefore, to use a weighting strategy that accounts for the frequency dependence of the stochastic errors.

- The measurement technique may introduce undesired correlation in the impedance results. Carson et al.[354] showed that a phase-sensitive detection technique employing only one reference signal yielded significant correlation between the real and imaginary parts of the stochastic error structure of the impedance. In a companion paper, Carson et al.[145] showed that the different statistical properties obtained with phase-sensitive-detection (lock-in amplifier) simulations could be attributed in part to bias errors introduced when the square-wave reference signal was in phase with the measured signal. Modern phase-sensitive-detection instruments employ more than one reference signal and may thereby avoid this undesired correlation.

- When time-domain errors are additive, Fourier analysis techniques provide statistical properties that are intrinsic to transfer-function measurements.

- In the absence of instrument-induced correlations, stochastic errors in the frequency domain are normally distributed. The appearance of a normal distribution of frequency-domain stochastic errors can be regarded to be a consequence of the central limit theorem applied to the methodology used to measure the complex impedance.[97] This result validates an essential assumption routinely used during regression analysis of impedance (and other frequency-domain) data.

- In the absence of bias errors, the errors in the real and imaginary impedance are uncorrelated and the variances of the real and imaginary parts of the complex impedance are equal. Some specific identities are given in Table 21.1.

In a general sense, the frequency-domain error structure is determined by the nature of errors in the time-domain signals and by the method used to process the time-domain data into the frequency domain. The cell impedance influences the frequency-dependence of the variance of the measurements, but the cell impedance does not influence whether the variances of real and imaginary components are equal or whether errors in the real and imaginary components are uncorrelated.

The statistical properties described above for frequency-domain stochastic errors are based on the equations for the instruments actually used for the measurement

Table 21.1: Statistical properties of impedance values obtained by Fourier analysis techniques.[102]

$$\sigma_{Z_r}^2 = \sigma_{Z_j}^2 \qquad (21.9)$$

$$\sigma_{Z_r Z_j} = 0 \qquad (21.10)$$

$$\sigma_{|Z| \varphi} = 0 \qquad (21.11)$$

$$\sigma_{|Z|}^2 = |Z|^2 \sigma_\varphi^2 \qquad (21.12)$$

of complex quantities. While the statistical properties are developed here for electrochemical impedance spectroscopy, they are also valid for measurement of other complex quantities so long as the complex quantities are measured through similar physical principles.

21.3 Bias Errors

Bias errors are systematic errors that do not have a mean value of zero and that cannot be attributed to an inadequate descriptive model of the system. Bias errors can arise from instrument artifacts, parts of the measured system that are not part of the system under investigation, and nonstationary behavior of the system. Some types of bias errors lead the data to be inconsistent with the Kramers–Kronig relations. In those cases, bias errors can be identified by checking the impedance data for inconsistencies with the Kramers–Kronig relations. As some bias errors are themselves consistent with the Kramers–Kronig relations, the Kramers–Kronig relations cannot be viewed as providing a definitive tool for identification of bias errors.

21.3.1 Instrument Artifacts

Limitations of instruments such as potentiostats can influence the measured impedance response. Such influences can be expected at impedance extremes. For example, the impedance response of low-impedance systems such as fuel cells and batteries shows artifacts at high frequency. High-frequency artifacts have also been attributed to reference electrodes. In many cases, instrument artifacts lead to inconsistencies with the Kramers–Kronig relations; however, this is not always the case. As discussed in Sec-

Remember! 21.2 *Bias errors in impedance measurements can arise from instrument artifacts, parts of the measured system that are not part of the system under investigation, and nonstationary behavior of the system.*

tion 8.3.2, the experimentalist is encouraged to confirm the high-frequency behavior of electrochemical systems by measuring the impedance of an electrical circuit exhibiting the measured response and by comparing the high-frequency response to limiting values obtained by other experimental methods.

21.3.2 Ancillary Parts of the System under Study

The impedance response of low-impedance systems may include the finite impedance behavior of wires and connectors. These may be considered, from the perspective of model identification, as yielding artifacts in the measured response. Such artifacts may be simply resistive but may also exhibit a capacitive or even an inductive frequency dependence. Such artifacts will be generally consistent with the Kramers–Kronig relations.

21.3.3 Nonstationary Behavior

Most electrochemical systems show some nonstationary behavior due, for example, to growth of surface films, changes in concentrations of reactants or products in the electrolyte, or changes in surface reactivity. As discussed in Section 21.3.4, the issue is not whether a system is perfectly stationary, but, rather, whether the system has changed substantially during the course of the impedance measurement. The Kramers–Kronig relations are particularly useful for identification of artifacts introduced by nonstationary behavior. These artifacts are most visible at low frequencies but can be seen at all frequencies if the system change is sufficiently rapid.

21.3.4 Time Scales in Impedance Spectroscopy Measurements

Three measurement time scales apply to impedance measurements. The first time scale is that required for measuring a set of replicated measurements. Such a set of measurements is shown in Figure 21.5 for reduction of ferricyanide on a rotating Pt disk electrode. The time required for the set of three measurements was 3581 s (1 h).

The time required to measure a set of N_{set} replicated measurements can be given by

$$\tau_{set} = \sum_{k=1}^{N_{set}} \tau_{scan,k} \tag{21.13}$$

where $\tau_{scan,k}$ is the time required for each individual scan. The characteristic frequency for the series of impedance scans is given by

$$f_{set} = \frac{1}{\tau_{set}} \tag{21.14}$$

Systems containing stochastic errors with frequency much smaller than f_{set} may appear to be stationary on the time scale of the impedance measurements.

The time required for each frequency scan can be seen in Figure 21.6(a). The average

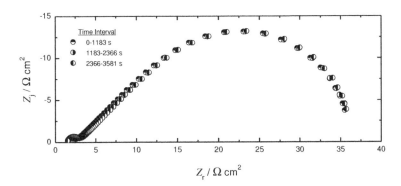

Figure 21.5: A series of three repeated impedance measurements. The data were collected for reduction of ferricyanide on a rotating Pt disk electrode.

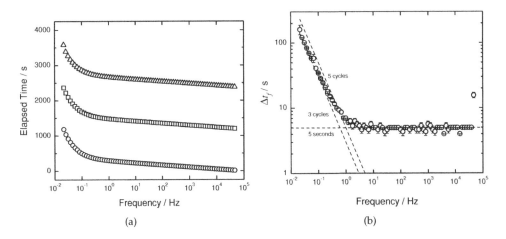

(a)   (b)

Figure 21.6: Elapsed time for the measurements presented in Figure 21.5: a) elapsed time and b) time required for measurement at each frequency. The standard deviation reflects variability among the three impedance measurements.

time required for each scan was 1194 s (0.33 h). The time required for a measurement at each individual frequency is illustrated in Figure 21.6(b). At low frequencies, the time required generally corresponds to three or four cycles, but at high frequencies, a much larger number of cycles is needed to account for the smaller signal-to-noise level.

The time required to measure a complete spectrum encompassing N_{scan} frequencies is given by

$$\tau_{\text{scan}} = \sum_{k=1}^{N_{\text{scan}}} \frac{N_k}{f_k} \tag{21.15}$$

where N_k represents the number of cycles needed to make the measurement at frequency f_k. The characteristic frequency corresponding to an impedance scan is given by

$$f_{\text{scan}} = \frac{1}{\tau_{\text{scan}}} \tag{21.16}$$

Stochastic errors with frequency larger than f_{scan} and smaller than f_k/N_k will generate a bias error in the measurement. The resulting spectrum will show inconsistencies with the Kramers–Kronig relations (see Chapter 22 on page 595). Stochastic errors with frequency much smaller than f_{scan} may also generate data that are inconsistent with the Kramers–Kronig relations.

The time required for measurement at each frequency depends on the type of measurement made and on the experimental parameters. For example, the time required for a measurement at each individual frequency is illustrated in Figure 21.7 for EHD measurements (see Chapter 16 on page 443). The extreme noise level observed at high frequencies increases greatly the time required to achieve a given closure error in the measurements.

If N_k cycles are needed for the measurement at frequency f_k, the characteristic frequency for an impedance measurement at a given frequency is given by

$$f_{N_k} = \frac{f_k}{N_k} \tag{21.17}$$

Stochastic errors with frequency much larger than f_{N_k} will appear as a stochastic error in the measurement; whereas, stochastic errors with frequency much smaller than f_{N_k} will appear as a bias error in the measurement. The significance of the bias error will depend on the comparison between the error frequency and the characteristic frequency of the entire measurement of the spectrum.

If the system is evolving very rapidly, changes can occur during the time in which one data point is collected. Impedance spectroscopy may not be a feasible experimental technique for such systems. For systems showing a slower rate of change, the impedance at each frequency may be measurable, but significant change can occur between the start and end of a complete frequency scan. These types of nonstationarities result in the data being inconsistent with the Kramers–Kronig relations. The issues arising

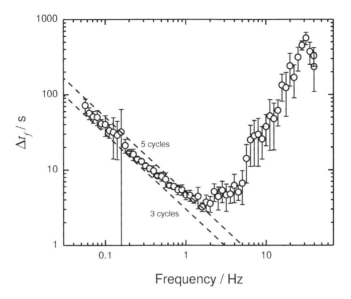

Figure 21.7: Time required for measurement at each frequency for EHD measurements. The standard deviation reflects variability between two measurements. The data were collected for reduction of ferricyanide on a rotating Pt disk electrode.

out of these inconsistencies are discussed in Chapter 22. At a still slower rate of evolution, the change in the system during one complete scan may be small enough to be ignored, but discernable differences can be seen between successive spectra. Such pseudo-stationary impedance scans are typically observed for even the most stationary electrochemical systems.

21.4 Incorporation of Error Structure

Three approaches have been documented in the literature for incorporating the error structure of impedance data into interpretation strategies. One approach has been to assume a standard form for the stochastic errors. Two models are commonly used. Zoltowski[368] and Boukamp[59,369] advocated use of modulus weighting. Use of a modulus weighting strategy invokes an assumption that the standard deviation is proportional to the frequency-dependent modulus $|Z(\omega)|$ of the impedance, i.e.,

$$\sigma_{Z_r}(\omega) = \sigma_{Z_j}(\omega) = \alpha_M |Z(\omega)| \tag{21.18}$$

where $\sigma_{Z_r}(\omega)$ and $\sigma_{Z_j}(\omega)$ represent the standard deviation of the real and imaginary parts of the impedance, respectively. The parameter α_M is assumed to be independent of frequency and is often arbitrarily assigned a value based on an assumed noise level of the measurement. Macdonald et al.[343,353,356] advocated use of a modified proportional weighting strategy, i.e.,

$$\sigma_{Z_r}^2(\omega) = \alpha^2 + \sigma^2 \left|F'(\omega,\theta)\right|^{2\zeta} \tag{21.19}$$

and

$$\sigma_{Z_j}^2(\omega) = \alpha^2 + \sigma^2 \left|F''(\omega,\theta)\right|^{2\zeta} \tag{21.20}$$

where α, σ, and ζ are error structure parameters, $F'(\omega,\theta)$ and $F''(\omega,\theta)$ are real and imaginary parts of the model immittance function, respectively, and θ is a vector of model parameters. The immittance is a general term that can represent either the admittance or impedance of an electrical circuit. There are fundamental differences between the two commonly used standard weighting strategies. Under equation (21.18), $\sigma_{Z_r} = \sigma_{Z_j}$, whereas equations (21.19) and (21.20) state that, in general, $\sigma_{Z_r} \neq \sigma_{Z_j}$ unless errors are assumed to be independent of frequency (i.e., $\sigma = 0$) or unless $F'(\omega,\theta) = F''(\omega,\theta)$.

A second approach has been to use the regression procedure to obtain an estimate for the error structure of the data.[353,356] A sequential regression is employed in which the parameters for an assumed error structure model, e.g., equations (21.19) and (21.20), are obtained directly from regression to the data.[353] In more recent work, the error variance model was replaced by

$$\sigma_{Z_r}^2(\omega) = \alpha^2 + \left|F'(\omega,\theta)\right|^{2\zeta} \tag{21.21}$$

and

$$\sigma_{Z_j}^2(\omega) = \alpha^2 + \left|F''(\omega,\theta)\right|^{2\zeta} \tag{21.22}$$

where parameters α and ζ are obtained by regression, and an extension of modulus weighting can be obtained by replacing the functions $F'(\omega,\theta)$ and $F''(\omega,\theta)$ with $\left|F(\omega,\theta)\right|$.[370] Independent of the assumed form of the error variance model, the assumption that the error variance model can be obtained by minimizing the objective function ignores the differences among the contributions to the residual errors shown in equation (21.1). The error structure obtained by simultaneous regression is also severely constrained by the assumed form of the error-variance model.

The third approach is to use experimental methods to assess the error structure. Independent identification of error structure is the preferred approach, but even minor nonstationarity between repeated measurements introduces a significant bias error in the estimation of the stochastic variance. Dygas and Breiter report on the use of intermediate results from a frequency-response analyzer to estimate the variance of real and imaginary components of the impedance.[371] Their approach allows assessment of the variance of the stochastic component without the need for replicate experiments. The drawback is that their approach cannot be used to assess bias errors and is specific to a particular commercial impedance instrumentation. Van Gheem et al.[372,373] have proposed a structured multi-sine signal that can be used to assess stochastic and bias errors without use of replicated measurements.

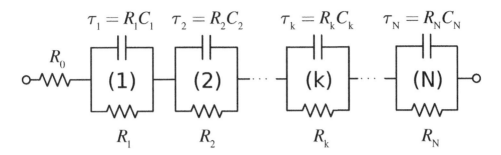

Figure 21.8: A schematic representation of a Voigt circuit used by Agarwal et al.[69, 100, 101] as a measurement model.

Agarwal et al.,[69,100,101] showed that measurement models are generally applicable and can be used to estimate both stochastic and bias errors of a measurement from imperfectly replicated impedance measurements. Orazem et al.[374] used a measurement model approach to show that a general model for the error structure could take the form

$$\sigma_{Z_r} = \sigma_{Z_j} = \alpha|Z_j| + \beta|Z_r - R_e| + \gamma\frac{|Z|^2}{R_m} + \delta \tag{21.23}$$

where α, β, γ, and δ are constants to be determined for a specific potentiostat, set of measurement parameters, and electrochemical system. Standard deviations for practical systems have been reported that are on the order of 0.2 to 0.04 percent of the modulus.[99] The drawback of the measurement model approach is that replicated measurements are required. The measurement model approach is presented in Section 21.5.

21.5   Measurement Models for Error Identification

The measurement model method for distinguishing between bias and stochastic errors is based on using a generalized model as a filter for nonreplicacy of impedance data. The measurement model is composed of a superposition of line-shapes that can be arbitrarily chosen. The model shown in Figure 21.8, composed of Voigt elements in series with a solution resistance, has been shown to be a useful and general measurement model.

While the line-shapes parameters may not be unequivocally associated with a set of deterministic or theoretical parameters for a given system, the measurement model approach has been shown to adequately represent the impedance spectra obtained for a large variety of electrochemical systems.[69] The line-shape models represent the low-frequency stationary components of the impedance spectra (in a Fourier sense). Regardless of their interpretation, the measurement model representation can be used to filter and thus identify the nonstationary (drift) and high-frequency (noise) components contained in the same impedance spectrum.

At first glance, it may not be obvious that such an approach should work. It is well known, for example, that the impedance spectrum associated with an electrochemical reaction limited by the rate of diffusion through a stagnant layer (either the Warburg or the finite-layer diffusion impedance) can be approximated by an infinite number of RC circuits in series (the Voigt model). In theory, then, a measurement model based on the Voigt circuit should require an infinite number of parameters to adequately describe the impedance response of any electrochemical system influenced by mass transfer.

In practice, stochastic errors (or noise) in the measurement limit the number of Voigt parameters that can be obtained from experimental data. An infinite number of Voigt parameters cannot be obtained even from synthetic data because round-off errors limit the information content of the calculation.[69] The residual errors associated with fitting a Voigt model to experimental impedance data that are influenced by mass transfer can, with appropriate weighting, be made to be of the order of the stochastic noise in the measurement. A Voigt circuit, or any equivalent circuit, can therefore yield an appropriate measurement model for electrochemical impedance spectra. It is evident that the measurement model composed of Voigt elements may not be the most parsimonious or efficient model for a given spectrum. In effect, by using the measurement model, one takes advantage of the noise present in any measurement, which limits the number of parameters that can be resolved, and the large number of measured frequencies as compared to the number of resolvable parameters.

Thus the Voigt circuit can provide an adequate description of impedance data influenced by mass transfer or by distributed-time-constant phenomena such as is described in Chapter 14. In addition, inductive loops can be fitted by a Voigt circuit by using a negative resistance and capacitance in an element. Such an element will have a positive RC time constant. The Voigt circuit serves as a convenient generalized measurement model.

The use of measurement models to check for the consistency of the experimental data with the Kramers–Kronig relations was first proposed by Agarwal et al.[69,375] Boukamp and Macdonald described the use of distributed relaxation time models as measurement models for assessing consistency of data with the Kramers–Kronig relations.[376] The approach taken was similar to that of Agarwal et al. with the exception that assumed rather than experimentally measured error structures were used to weight the regressions. A linearized application of measurement models was suggested by Boukamp that eliminated the need for sequential increase in the number of line-shape parameters at the expense of using one line-shape for every frequency measured.[377] The application of such a linearized model is also constrained by the need for an independent assessment of the level of noise in the measurement.

The use of measurement models to identify consistency with the Kramers–Kronig relations is equivalent to the use of Kramers–Kronig transformable circuit analogues. An important advantage of the measurement model approach is that it identifies a small set of model structures that are capable of representing a large variety of observed behaviors or responses. The problem of model discrimination is therefore significantly reduced. The inability to fit an impedance spectrum by a measurement model can be attributed to the failure of the data to conform to the assumptions of the Kramers–Kronig relations rather than the failure of the model. The measurement model approach, however, does not eliminate the problem of model multiplicity or model equivalence over a given frequency range. The reduced set of model structures identified for the measurement model makes it feasible to conduct studies aimed at identification of the error structure, the propagation of error through the model and through the Kramers–Kronig transformation, and issues concerning parameter sensitivity and correlation.

Another significant advantage of the measurement model approach is that the resulting models can be transformed analytically (in the Kramers–Kronig sense). This means that, in contrast to the other approaches for evaluation of consistency (e.g., fitting to polynomials), the real and imaginary parts of the impedance are related through a finite set of common parameters. The measurement models can therefore be used as statistical observers; that is, adequate identification and estimation of the model parameters over a given experimental region, e.g., a range of frequencies in the imaginary domain, will allow the description (or observation) of the behavior of the system over another region, i.e., the real domain. The selection of experimental region used for this evaluation will take advantage of the relative parameter sensitivity in the real and imaginary domains.

It should be noted that the error analysis methods using measurement models are sensitive to data outliers. Occasionally, outliers can be attributed to external influences. Most often, outliers appear near the line frequency and at the beginning of an impedance measurement. Data collected within ± 5 Hz of the line frequency and its first harmonic (e.g., 50 and 100 Hz in Europe or 60 and 120 Hz in the United States) should be deleted. Startup transients cause some systems to exhibit a detectable artifact at the first frequency measured. This point, too, should be deleted.

21.5.1 Stochastic Errors

If a single model were to be regressed to all spectra showing nonstationary behavior, the resulting residual error would include contributions from the drift between scans as well as the lack of fit of the model, instrumental bias errors, and stochastic errors, i.e.,

$$\varepsilon_{res}(\omega) = \varepsilon_{fit}(\omega) + \varepsilon_{inst}(\omega) + \varepsilon_{drift}(\omega) + \varepsilon_{stoch}(\omega) \qquad (21.24)$$

Remember! 21.3 *The error analysis methods using measurement models are sensitive to data outliers.*

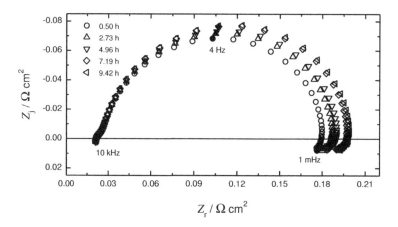

Figure 21.9: Five scans of impedance data collected at a current density of 0.2 A/cm^2 for a 5 cm^2 Polymer-Electrolyte-Membrane (PEM) fuel cell.[378]

Direct calculation of the variance of the resulting residual errors would lead to a value that includes the contribution of the changing baseline.

To eliminate the contribution of the changing baseline, a measurement model is regressed to each scan using the maximum number of parameters that can be resolved from the data. The parameters for the measurement model for each data set will be slightly different because the system changes from one experiment to the other. Hence, by regressing the measurement model to individual data sets separately, the effects of the change of the experimental conditions from one experiment to another are incorporated into the measurement model parameters. The variance of the real and imaginary residual errors can therefore be obtained as a function of frequency and provide a good estimate for the variance of the stochastic noise in the measurement. Thus,

$$\sigma_Z^2(\omega) = \frac{1}{N-1} \sum_{k=1}^{N} \left(\varepsilon_{\text{res},m,k}(\omega) - \bar{\varepsilon}_{\text{res},m}(\omega) \right)^2 \qquad (21.25)$$

where $\varepsilon_{\text{res},m,k}$ represents the residual error for scan k obtained from a model m. Shukla and Orazem[103] have shown that the estimate for the variance of the stochastic noise obtained in this way is independent of the measurement model used.

Example 21.1 **Identification of Stochastic Errors:** *Find the stochastic contribution to the error structure for the repeated impedance measurements shown in Figure 21.9. Measurements were obtained under galvanostatic modulation.*

Solution: *The measurement model analysis is described in detail by Roy and Orazem.[378] The measurement model represented in Figure 21.8 can be expressed mathematically as*

$$Z = R_e + \sum_k \frac{R_k}{1 + j\omega\tau_k} \qquad (21.26)$$

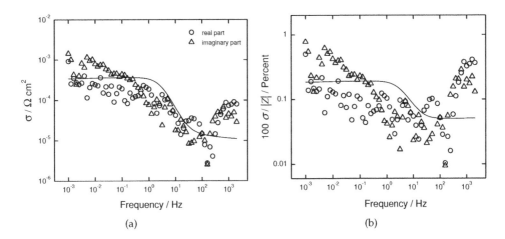

Figure 21.10: Estimates for the standard deviation of the real and imaginary parts of the stochastic measurement errors corresponding to Figure 21.9:[378] a) in units of Ωcm^2 and b) in units of percent, normalized to the modulus of the impedance.

where $\tau_k = R_k C_k$. *Following the procedure described by Agarwal et al.,[100] the measurement model given as (21.26) was fit individually to each of the spectra shown in Figure 21.26 using a frequency-independent weighting. The frequency-independent weighting employed is an appropriate initial weighting strategy for data collected under galvanoscopic modulation with a fixed perturbation amplitude. For data collected under potentiostatic modulation with a fixed perturbation amplitude, the initial weighting should be based on the modulus of the impedance.*

The number of parameters was constrained by the need to have the same number of parameters for each spectrum and the requirement that no parameter had a $\pm 2\sigma$ (95.45%) confidence interval that included zero. Typically, six Voigt elements could be regressed to a spectrum. In agreement with equation (21.25), the standard deviation of the residual errors was used as an estimate for the standard deviation of the stochastic measurement errors.

The resulting estimates for the standard deviation of the real and imaginary parts of the stochastic measurement errors are shown in Figure 21.10(a). The standard deviations for real and imaginary parts of the stochastic contribution to the error structure range between 10^{-5} Ωcm^2 at high frequency and 10^{-3} Ωcm^2 at low frequency. These contributions, normalized by the magnitude of the impedance, are presented in Figure 21.10(b). The standard deviation of the stochastic errors are on the order of 0.2 percent of the magnitude of the impedance. The larger values at high and low frequencies may be attributed to factors that caused lack of consistency with the Kramers–Kronig relations, as discussed in Example 21.2 Application of the error-structure model given as equation (21.23) yielded

$$\sigma_r = \sigma_j = 0.679 \frac{|Z|^2}{R_m} \tag{21.27}$$

where $R_m = 100$. This expression was used to weight the subsequent regression to assess the consistency with the Kramers–Kronig relations.

21.5.2 Bias Errors

There are several ways to assess consistency with the Kramers–Kronig relations. In principle, since the Voigt model is itself consistent with the Kramers–Kronig relations, the ability to fit this model to data within the noise level of the measurement should indicate that the data are consistent. An ambiguity exists when the data are not fully consistent, because the lack of fit of the model could be due to causes other than inconsistency with the Kramers–Kronig relations. Some other possible causes could be that the number of frequencies measured was insufficient to allow regression with a large enough number of Voigt parameters or that the initial guesses for the nonlinear regression could be poorly chosen.

While in principle a complex fit of the measurement model could be used to assess the consistency of impedance data, sequential regression to either the real or the imaginary provides greater sensitivity to lack of consistency. The optimal approach is to fit the model to the component that contains the greatest amount of information. The decision as to which component to fit is constrained by two conflicting considerations:

1. The standard deviations of the real and imaginary parts of the impedance are equal; therefore, the noise level represents a large percentage of the imaginary part of the impedance in the asymptotic limits where the imaginary impedance tends toward zero. In some cases, the value of the imaginary impedance can fall below the signal-to-noise threshold.

2. The imaginary part of the impedance is much more sensitive to contributions of minor line-shapes than is the real part of the impedance. Typically, more Voigt line-shapes can be resolved when fitting to the imaginary part of the impedance than can be resolved when fitting to the real part.

The solution resistance cannot be obtained by fitting the measurement model to the imaginary part of the impedance. The solution resistance is treated as an arbitrarily adjustable parameter when fitting to the imaginary part of the impedance.

Regression to one component with subsequent prediction of the other component provides a more sensitive method to assess consistency. A procedure for this analysis is described below:

1. Perform a fit to the imaginary part of the spectrum using error structure weighting. Increase the number of line-shapes used until the maximum number of statistically significant parameters is obtained. Ideally, the ratio of the sum of squares to the noise level should be within the F-test bounds given by the program.

2. Use a Monte Carlo simulation to identify the frequency-dependent confidence interval for the model prediction.

3. Examine the imaginary residual errors to determine whether they fall within the error structure. Should a few points lie outside the error structure at intermittent

frequency values, do not be concerned. Assess prediction of the real part of the impedance by examining real residual plots with confidence intervals displayed. Real residual data points that are outside the confidence interval are considered to be inconsistent with the Kramers–Kronig relations and should be removed from the data set.

4. Typically, the number of line-shapes that can be determined in a complex fit will increase when data inconsistent with the Kramers–Kronig relations are removed. Deletion of data that are strongly influenced by bias errors increases the amount of information that can be extracted from the data. In other words, the bias in the complete data set induces correlation in the model parameters, which reduces the number of parameters that can be identified. Removal of the biased data results in a better-conditioned data set that enables reliable identification of a larger set of parameters.

Experimental data can, therefore, be checked for consistency with the Kramers–Kronig relations without actually integrating the equations over frequency, avoiding the concomitant quadrature errors. The use of measurement models does require an implicit extrapolation of the experimental data set, but the implications of the extrapolation procedure are quite different from extrapolations reported in the literature. The extrapolations done with measurement models are based on a common set of parameters for the real and imaginary parts and on a model structure that has been shown to adequately represent the observations. The confidence in the extrapolation using measurement models is, therefore, higher. For the application to a preliminary screening of the data, the use of measurement models is superior to the use of more specific electrical circuit analogues, because one can determine whether the residual errors are due to an inadequate model, to failure of data to conform to the Kramers–Kronig assumptions, or to experimental noise. The algorithm proposed by Agarwal et al.,[69,100,101] in conjunction with error structure weighting, provides a robust way to check for consistency of impedance data.

Example 21.2 **Identification of High-Frequency Bias Errors:** *Use the measurement model to evaluate high-frequency bias errors for the repeated impedance measurements shown in Figure 21.9.*

Solution: *The measurement model approach developed by Agarwal et al.[101] was used to assess the consistency of high-frequency data with the Kramers–Kronig relations.[378] The Voigt model was fit to the real part of the measurement with a weighting based on the error structure identified in Example 21.1. The parameters so obtained were then used to predict the imaginary part of the measurement, and a confidence interval for the prediction was calculated based on the estimated confidence intervals for the regressed parameters. Data that fell outside of the confidence interval were deemed to be inconsistent with the Kramers–Kronig relations.*

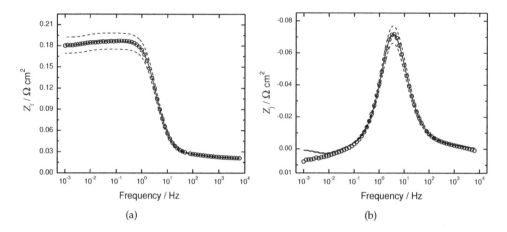

Figure 21.11: Regression of the Voigt model to the real part of the impedance corresponding to the second of five scans given in Figure 21.9: a) fit to the real part of the measurement and b) prediction of the imaginary part. The symbol represents the experimental data, the solid line represents the measurement model fit, and the thin dashed lines represent confidence intervals. Data taken from Roy and Orazem.[378]

This process is illustrated in Figure 21.11 for the second impedance scan shown in Figure 21.9. The fit to the real part of the impedance is given in Figure 21.11(a) where the thin dashed lines represent the confidence interval for the regression. The prediction of the imaginary part of the measurement, given in Figure 21.11(b), is excellent at intermediate frequencies, but a discrepancy is seen at both high and low frequencies. Regression to the real part of the impedance generally provides fewer parameters than does regression to the imaginary part. For this reason, the discrepancy seen at low frequencies was not considered to be significant.[101] The discrepancy at high frequency is seen where the real part of the impedance approaches asymptotically a finite value corresponding to a solution resistance.

The discrepancy is seen more clearly in the plots of normalized residual error given in Figure 21.12(a) for the fitting errors and in Figure 21.12(b) for the prediction errors. The normalization by the experimental value of the impedance causes the confidence-interval lines shown in Figure 21.12(b) to tend toward $\pm\infty$ at the point where the imaginary impedance changes sign. The analysis shows that the six highest frequencies fell outside the 95.45% confidence interval. These data were removed from the regression set. The conclusion that these points were inconsistent with the Kramers–Kronig relations is supported by the observation that the number of parameters that could be obtained from a complex regression increased when the high-frequency data were removed. In other words, deletion of data that were strongly influenced by bias errors increased the amount of information that could be extracted from the data. The bias in the complete data set induced correlation in the model parameters, which reduced the number of parameters which could be identified. Removal of the biased data resulted in a better conditioned data set that enables reliable identification of a larger set of parameters.[101]

For all measurements, data measured at frequencies above 1 kHz were found to be incon-

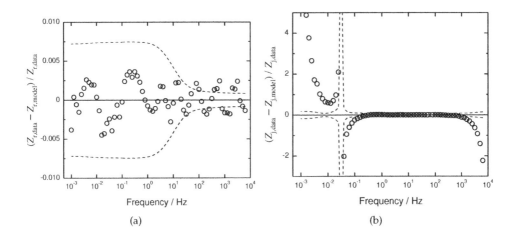

Figure 21.12: Normalized residual errors for the regression shown in Figure 21.11: a) fit to the real part of the measurement, where dashed lines show the $\pm 2\sigma$ bound for the stochastic error structure and b) prediction of the imaginary part where dashed lines indicate the 95.4 % confidence interval. Data taken from Roy and Orazem.[378]

sistent with the Kramers–Kronig relations. These data were removed from the data used in subsequent regressions. It is important to note that removal of data for which the imaginary impedance had a positive value was not sufficient to eliminate inconsistency with the Kramers–Kronig relations. As shown in Figure 21.13, the influence of the artifact extended well into the domain in which the imaginary impedance had a negative value. The filled symbols in Figure 21.13 correspond to data that were deemed inconsistent with the Kramers–Kronig relations.

The high-frequency bias errors identified in Example 21.2 can be attributed to instrumentation and/or wires and electrical connections. Typically, the frequency at which high-frequency bias errors are found depends on the magnitude of the impedance. The maximum impedance of the fuel cell analyzed in Example 21.2 was on the order of 0.2 Ωcm^2 or, for nominal area of 5 cm^2, 0.04 Ω. The frequency at which instrumentally associated bias errors are observed can be much larger for systems characterized by a larger impedance. As recommended on page 187, systematic instrument errors may be identified by measurement of the impedance response of an electrical circuit exhibiting the same impedance magnitude and characteristic frequencies as seen in the measured impedance response.

As discussed in Section 8.2.1 on page 170, the frequency range for an experiment is not determined solely by the instrument specification. The analysis presented in Example 21.2 shows that the high-frequency limit can be constrained by the impedance characteristics of the experimental system under study. The low-frequency limit may similarly be constrained by nonstationary behavior.

 Example 21.3 Identification of Low-Frequency Bias Errors: *Use the measure-*

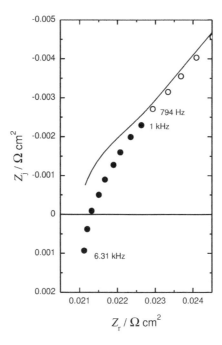

Figure 21.13: Nyquist representation of impedance data showing the inconsistency observed at high frequency. The filled symbols correspond to data that were deemed inconsistent with the Kramers–Kronig relations.[378]

ment model to evaluate low-frequency bias errors for the repeated impedance measurements shown in Figure 21.9.

Solution: *To test the consistency of the impedance data at low frequency, the imaginary part of the impedance data may be fit by the measurement model using the weighting strategy based on the empirical model for error structure developed in Example 21.1. The parameter set identified can then be used to predict the real part of the impedance. The Voigt measurement model was regressed to the imaginary part of the impedance data corresponding to the first scan of the impedance data presented in Figure 21.9. The results, given in Figure 21.14(a), show that the measurement model could provide an excellent fit to the imaginary part of the data, even at the low frequencies that revealed inductive loops, characterized by positive values of imaginary impedance. The parameter values obtained from regression to the imaginary part of the impedance were used to predict the real part, as shown in Figure 21.14(b). The dashed lines shown in Figure 21.14(b) represent the upper and lower bounds of the 95.45% ($\pm 2\sigma$) confidence interval obtained for the model prediction. The low-frequency data that are outside the confidence interval can therefore be considered inconsistent with the Kramers–Kronig relations.*

A more precise view of the regression quality and the level of agreement with the predicted values can be seen in plots of residual errors. The normalized residual error for the regression to the imaginary part of the impedance is shown in Figure 21.15(a), where the dashed lines

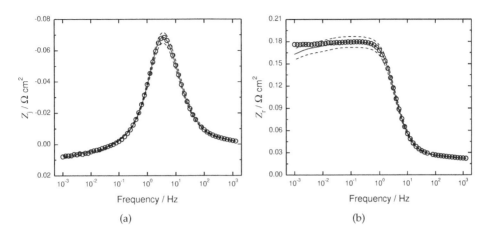

Figure 21.14: Regression of the Voigt model to the imaginary part of the impedance corresponding to the first of five scans given in Figure 21.9: a) fit to the imaginary part of the measurement and b) prediction of the real part. The symbols represent the experimental data, the solid line represents the measurement model fit, and the thin dashed lines represent confidence intervals. Data taken from Roy and Orazem.[378]

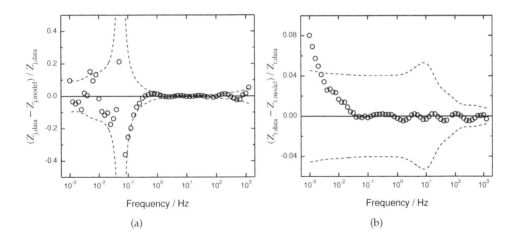

Figure 21.15: Normalized residual errors for the regression shown in Figure 21.14: a) fit to the imaginary part of the measurement, where dashed lines show the $\pm 2\sigma$ bound for the stochastic error structure and b) prediction of the real part where dashed lines indicate the 95.4 % confidence interval. Data taken from Roy and Orazem.[378]

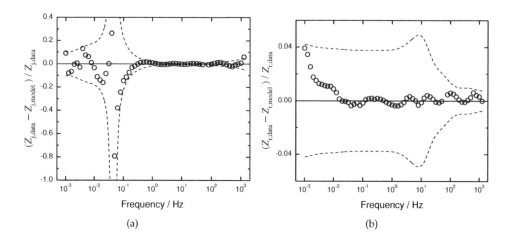

Figure 21.16: Normalized residual errors for the regression to the imaginary part of the impedance corresponding to the second of five scans given in Figure 21.9: a) fit to the imaginary part of the measurement, where dashed lines show the $\pm 2\sigma$ bound for the stochastic error structure and b) prediction of the real part where dashed lines indicate the 95.45% confidence interval. Data taken from Roy and Orazem.[378]

indicate upper and lower bounds for the stochastic noise level for the measurement. The dashed lines were calculated as $\pm 2\sigma$, where σ was obtained from equation (21.27). The normalization by the experimental value of the impedance causes the dashed lines to tend toward $\pm\infty$ at the point where the imaginary impedance changes sign. The quality of the regression is indicated by the observation that the residual errors for the regression fall within the noise level of the experiment. The normalized residual errors for the predicted real value are shown in Figure 21.15(b), where the solid line represents the upper and lower bounds of the 95.45% ($\pm 2\sigma$) confidence interval obtained for the model prediction. A lack of agreement between predicted and experimental values is seen for frequencies below 0.05 Hz. The data for the four lowest frequencies are seen to fall outside the confidence interval for the prediction. These points can be described as being inconsistent with the Kramers–Kronig relations.

Similar analyses of bias error were performed for subsequent impedance scans. The normalized residual error for the imaginary part of the second scan is shown in Figure 21.16(a), and the associated predicted error in the real part is shown in Figure 21.16(b). The agreement between predicted and experimental values is better for the second scan, shown in Figure 21.16(b), than for the first, shown in Figure 21.15(b). All data shown in Figure 21.16(b) fall inside the 95.45% confidence interval for the prediction. The second and subsequent scans were found to be consistent with the Kramers–Kronig relations.

The approach presented in this section is part of an overall assessment of measurement errors. The measurement model is used as a filter for lack of replicacy to obtain a quantitative value for the standard deviation of the measurement as a function of frequency. The mean error identified in this way is equal to zero; thus, the standard deviation of the measurement does not incorporate the bias errors. In contrast, the standard deviation of repeated impedance measurements typically includes a significant contribution from bias errors because perfectly replicate measurements can rarely be made for electrochemical systems. Since the line-shapes of the measurement model satisfy the Kramers–Kronig relations, the Kramers–Kronig relations then can be used as a statistical observer to assess the bias error in the measurement.

Problems

21.1 Consider a system with a single electrochemical reaction with $R_e = 10\,\Omega$, $R_t = 100\,\Omega$, and $R_t C_{dl} = 1$ s. The noise level of the measurement is often assumed to follow equation (21.18) with $\sigma = 0.03\,|Z|$.

(a) Plot the standard deviation of the real and imaginary parts of the impedance measurement as functions of frequency.

(b) Plot the normalized standard deviation of the real and imaginary parts of the impedance measurement, scaled to the respective impedance component, as functions of frequency.

(c) Plot the weighting factor $w = 1/\sigma^2$ for the real and imaginary parts of the impedance measurement as functions of frequency.

21.2 Use Figure 21.6 to estimate the time required for an impedance measurement from 100 kHz to 0.1 Hz with eight measurements per decade. How does this value change if the measurement requires 10 measurements per decade?

21.3 Use Figure 21.6 to estimate the time required for an impedance measurement from 100 kHz to 0.1 mHz with eight measurements per decade. How does this value change if the measurement requires three measurements per decade?

21.4 Use a spreadsheet program such as Microsoft Excel® or a computational programming environment such as MATLAB® to reproduce the results presented in Figures 21.3 and 21.4.

21.5 Add normally distributed stochastic errors to the time-domain potential and current signals for the system described in Example 7.2. Then apply the Fourier analysis to calculate the impedance response at the characteristic frequency. Repeat this process, refreshing the random numbers used, so as to calculate the standard deviation of the resulting impedance. How does this result depend on the number of cycles used for the integration?

Chapter 22

The Kramers–Kronig Relations

The Kramers–Kronig relations are integral equations that constrain the real and imaginary components of complex quantities for systems that satisfy conditions of linearity, causality, and stability. These relationships, derived independently by Kronig[66,352] and Kramers,[67,379] were initially developed from the constitutive relations associated with the Maxwell equations for description of an electromagnetic field at interior points in matter.[380]

The fundamental constraints are that the system be stable, in the sense that perturbations to the system do not grow, that the system responds linearly to a perturbation, and that the system be causal in the sense that a response to a perturbation cannot precede the perturbation. The Kramers–Kronig relationships were found to be entirely general with application to all frequency-domain measurements that could satisfy the above constraints. Bode extended the concept to electrical impedance and tabulated various extremely useful forms of the Kramers–Kronig relations.[110]

22.1   Methods for Application

The Kramers–Kronig relations are extremely general and have been applied to a wide variety of research areas. In the field of optics, for which the validity of the Kramers–Kronig is not in question, the relationship between real and imaginary components has been exploited to complete optical spectra. In other areas where data cannot be assumed to satisfy the requirements of the Kramers–Kronig relations, the equations presented in Table 22.1 have been used to check whether real and imaginary components

---

Remember! 22.1 *The Kramers–Kronig relations apply for systems that are linear, causal, and stable. The condition of stationarity is implicit in the requirement of causality.*

Table 22.1: Compilation of various forms of the Kramers–Kronig relations.

$$Z_j(\omega) = \frac{1}{\pi} \int_{-\infty}^{+\infty} \frac{Z_r(x)}{x - \omega} dx \tag{22.1}$$

$$Z_r(\omega) = Z_r(\infty) - \frac{1}{\pi} \int_{-\infty}^{+\infty} \frac{Z_j(x)}{x - \omega} dx \tag{22.2}$$

$$Z_j(\omega) = \frac{2\omega}{\pi} \int_{0}^{\infty} \frac{Z_r(x)}{x^2 - \omega^2} dx \tag{22.3}$$

$$Z_r(\omega) = Z_r(\infty) - \frac{2}{\pi} \int_{0}^{\infty} \frac{x Z_j(x)}{x^2 - \omega^2} dx \tag{22.4}$$

$$Z_j(\omega) = \frac{2\omega}{\pi} \int_{0}^{\infty} \frac{Z_r(x) - Z_r(\omega)}{x^2 - \omega^2} dx \tag{22.5}$$

$$Z_r(\omega) = Z_r(\infty) - \frac{2}{\pi} \int_{0}^{\infty} \frac{x Z_j(x) - \omega Z_j(\omega)}{x^2 - \omega^2} dx \tag{22.6}$$

$$\varphi(\omega) = -\frac{2\omega}{\pi} \int_{0}^{\infty} \frac{\ln\left(|Z(x)|\right)}{x^2 - \omega^2} dx \tag{22.7}$$

$$\varphi(\omega) = -\frac{2\omega}{\pi} \int_{0}^{\infty} \frac{\ln\left(|Z(x)|\right) - \ln\left(|Z(\omega)|\right)}{x^2 - \omega^2} dx \tag{22.8}$$

$$\varphi(\omega) = -\frac{1}{2\pi} \int_{0}^{\infty} \ln\left|\frac{x - \omega}{x + \omega}\right| \frac{d\ln|Z(x)|}{dx} dx \tag{22.9}$$

$$\ln\left(|Z(\omega)|\right) = \frac{2}{\pi} \int_{-\infty}^{+\infty} \frac{\varphi(x)}{x - \omega} dx \tag{22.10}$$

$$\mathrm{Re}\left\{\ln\left(Z(\omega)\right)\right\} = \mathrm{Re}\left\{\ln\left(Z(0)\right)\right\} + \frac{2\omega^2}{\pi} \int_{0}^{\infty} \frac{\mathrm{Im}\left\{\ln\left(Z(x)\right)\right\}}{x\left(x^2 - \omega^2\right)} dx \tag{22.11}$$

$$\mathrm{Im}\left\{\ln\left(Z(\omega)\right)\right\} = -\frac{2\omega}{\pi} \int_{0}^{\infty} \frac{\mathrm{Re}\left\{\ln\left(Z(x)\right)\right\}}{x^2 - \omega^2} dx \tag{22.12}$$

of complex variables are internally consistent. Failure of the Kramers–Kronig relations is assumed to correspond to a failure within the experiment to satisfy one or more of the constraints of linearity, stability, or causality.

In principle, the Kramers–Kronig relations can be used to determine whether the impedance spectrum of a given system has been influenced by bias errors caused, for example, by instrumental artifacts or time-dependent phenomena. Although this information is critical to the analysis of impedance data, the Kramers–Kronig relations have not found widespread use in the analysis and interpretation of electrochemical impedance spectroscopy data due to difficulties with their application. The integral relations require data for frequencies ranging from zero to infinity, but the experimental frequency range is necessarily constrained by instrumental limitations or by noise attributable to the instability of the electrode.

The Kramers–Kronig relations have been applied to electrochemical systems by direct integration of the equations, by experimental observation of stability and linearity, by regression of specific electrical circuit models, and by regression of generalized measurement models.

22.1.1 Direct Integration of the Kramers–Kronig Relations

Direct integration of the Kramers–Kronig relations involves calculating one component of the impedance from the other, e.g., the real component of impedance from the measured imaginary component. The result is compared to the experimental values obtained. The integral equations, for example, equation (22.6), require integration from 0 to ∞. One difficulty in applying this approach, as shown in Figure 22.2, is that the measured frequency range may not be sufficient to allow integration over the frequency limits of zero to infinity. Therefore, discrepancies between experimental data and the impedance component predicted through application of the Kramers–Kronig relations could be attributed to the use of a frequency domain that is too narrow, as well as to the failure to satisfy the constraints of the Kramers–Kronig equations. An interpolation function is required to allow extrapolation of the integrand into the experimentally inaccessible frequency regime.

A second issue is that the interpolation function must satisfy equations (22.76) or (22.79). It is clear that a suboptimal interpolation, such as the straight-line interpolation shown in Figure 22.3, will bias the estimate of the impedance through the Kramers–Kronig relations.

Two approaches for interpolation function have been used. In one, polynomials, e.g., in powers of ω^n, are fit to impedance data. Usually, a piecewise regression is required. While piecewise polynomials are excellent for smoothing, the best example being splines, they are not very reliable for extrapolation and result in a relatively large number of parameters. A second approach is to use interpolation formulas based on the expected asymptotic behavior of a typical electrochemical system.

22.1.2 Experimental Assessment of Consistency

Experimental methods can be applied to check whether impedance data conform to the Kramers–Kronig assumptions. A check for linear response can be made by observing whether spectra obtained with different magnitudes of the forcing function are replicates or by measuring higher-order harmonics of the impedance response (see Section 8.2.2). Stationary behavior can be also be identified experimentally by replication of the impedance spectrum. Spectra are replicates if the spectra agree to within the expected frequency-dependent measurement error. If the experimental frequency range is sufficient, the extrapolation of the impedance spectrum to zero frequency can be compared to the corresponding values obtained from separate steady-state experiments. The experimental approach to evaluating consistency with the Kramers–Kronig relations shares constraints with direct integration of the Kramers–Kronig equations. Because extrapolation is required, the comparison of the dc limit of impedance spectra to steady-state measurement is possible only for systems for which a reasonably complete spectrum can be obtained. Experimental approaches for verifying consistency with the Kramers–Kronig relations by replication are further limited in that, without an a priori estimate for the confidence limits of the experimental data, the comparison is more qualitative than quantitative. A method is therefore needed to evaluate the error structure, or frequency-dependent confidence interval, for the data that would be obtained in the absence of nonstationary behavior. Such methods are described in Chapter 21.

22.1.3 Regression of Process Models

Electrical circuits consisting of passive and distributed elements can be shown to satisfy the Kramers–Kronig relations. Therefore, successful regression of an electrical circuit analogue to experimental data implies that the data must also satisfy the Kramers–Kronig relations. This approach has the advantage that integration over an infinite frequency domain is not required; therefore, a portion of an incomplete spectrum can be identified as being consistent without use of extrapolation algorithms.

Perhaps the major problem with the use of electrical circuit models to determine consistency is that interpretation of a poor fit is ambiguous. A poor fit could be attributed to inconsistency of the data with the Kramers–Kronig relations or to use of an inadequate model or to regression to a local rather than global minimum (perhaps caused by a poor initial guess). A second unresolved issue deals with the regression

---

Remember! 22.2 *An insufficient experimental frequency range makes direct integration of the Kramers–Kronig relations problematic. Regression-based approaches, such as use of a measurement model, are preferred.*

---

itself, i.e., selection of the weighting to be used for the regression, and identification of a criterion for a good fit. A good fit could be defined by residual errors that are of the same size as the noise in the measurement, but, in the absence of a means of determining the error structure of the measurement, such a criterion is speculative at best.

22.1.4 Regression of Measurement Models

The concept of the measurement model as a tool for assessment of the error structure was applied to impedance spectroscopy initially by Agarwal et al.[69,100,101] The measurement model is described in greater detail in Chapter 21.

From the perspective of the approach proposed here, the use of measurement models to identify consistency with the Kramers–Kronig relations is equivalent to the use of Kramers–Kronig transformable circuit analogues, discussed in Section 22.1.3. An important advantage of the measurement model approach is that it identifies a small set of model structures that are capable of representing a large variety of observed behaviors or responses. The problem of model discrimination is therefore significantly reduced. The inability to fit an impedance spectrum by a measurement model can be attributed to the failure of the data to conform to the assumptions of the Kramers–Kronig relations rather than the failure of the model. The measurement model approach, however, does not eliminate the problem of model multiplicity or model equivalence over a given frequency range. The reduced set of model structures identified by the measurement model makes it feasible to conduct studies aimed at identification of the error structure, the propagation of error through the model and through the Kramers–Kronig transformation, and issues concerning parameter sensitivity and correlation.

The use of measurement models is superior to the use of polynomial fitting because fewer parameters are needed to model complex behavior and because the measurement model satisfies the Kramers–Kronig relations implicitly. Experimental data can, therefore, be checked for consistency with the Kramers–Kronig relations without actually integrating the equations over frequency, avoiding the concomitant quadrature errors. The use of measurement models does require an implicit extrapolation of the experimental data set, but the implications of the extrapolation procedure are quite different from extrapolations reported in Section 22.1.1. The extrapolation done with measurement models is based on a common set of parameters for the real and imaginary parts and on a model structure that has been shown to adequately represent the observations. The confidence in the extrapolation using measurement models is, therefore, higher.

The use of measurement models is superior to the use of more specific electrical circuit analogues because one can determine whether the residual errors are due to an inadequate model, to failure of data to conform to the Kramers–Kronig assumptions, or to experimental noise.

22.2   Mathematical Origin

The development presented here for the complex impedance, $Z = Z_r + jZ_j$, is general and can be applied, for example, to the complex refractive index, the complex viscosity, and the complex permittivity. The derivation for a general transfer function G follows that presented by Nussenzveig.[380] The development for the following analysis in terms of impedance follows the approach presented by Bode.[110]

22.2.1   Background

The fundamental constraints for the system transfer functions associated with the assumptions of linearity, casuality, and stability are described in this section.

Theorem 22.1 (Linearity) The output is a linear function of the input. *A general output function $x(t)$ can be expressed as a linear function of the input $f(t)$ as*

$$x(t) = \int_{-\infty}^{+\infty} g\left(t, t'\right) f(t') dt' \tag{22.13}$$

where $g(t, t')$ provides the relationship between input and output functions.

Theorem 22.2 (Time-Translation Independence) The output depends only on the input. *This means that, if the input signal is advanced or delayed by a time increment, the output will be advanced or delayed by the same time increment. Thus, $x(t + \tau)$ corresponds to $f(t + \tau)$ such that $g(t, t')$ can depend only on the difference between t and t'. Equation (22.13) can be written*

$$x(t) = \int_{-\infty}^{+\infty} g\left(t - t'\right) f(t') dt' \tag{22.14}$$

where $g(t - t')$ provides the relationship between input and output functions. The assumption of time-translation independence can be considered to be a consequence of the assumption of primitive causality, discussed in Theorem 22.3.

The function $g(t - t')$ can be expressed in terms of frequency through use of the Fourier integral transform,[93] i.e.,

$$G(\omega) = \int_{-\infty}^{+\infty} g(\tau) \exp\left(-j\omega\tau\right) d\tau \tag{22.15}$$

Similarly, the input can be expressed as

$$F(\omega) = \int_{-\infty}^{+\infty} f(\tau) \exp\left(-j\omega\tau\right) d\tau \tag{22.16}$$

and the output can be expressed as

$$X(\omega) = \int_{-\infty}^{+\infty} x(\tau) \exp(-j\omega\tau) \, d\tau \qquad (22.17)$$

The Fourier transform can be written with either negative or positive arguments in the exponential. The positive argument yields a positive imaginary impedance for the response of a capacitor to a potential perturbation, whereas the negative argument yields a negative imaginary impedance. The form used here yields results that are consistent with the electrical engineering conventions.

In frequency domain, the output can be expressed as a function of the input as

$$X(\omega) = G(\omega)F(\omega) \qquad (22.18)$$

where $G(\omega)$ acts as a transfer function. Equation (22.15) can be expressed as

$$G(\omega) = \int_{-\infty}^{+\infty} g(\tau) \left(\cos(\omega\tau) - j\sin(\omega\tau) \right) d\tau \qquad (22.19)$$

For a real-valued time-domain function $g(\tau)$, the function G has conjugate symmetry (see Section 1.3) such that

$$G(-\omega) = G_r + jG_j = \overline{G}(\omega) = G_r - jG_j \qquad (22.20)$$

Thus, $G_r(-\omega) = G_r(\omega)$ and $G_j(-\omega) = -G_j(\omega)$. The real part of G, G_r, is an even function of frequency, and the imaginary part G_j is an odd function of frequency.

The above development, expressed for general input, output, and transfer functions, can be presented in terms of impedance $Z(\omega)$. The results stemming from equation (22.19) indicate that the impedance $Z(\omega)$ has conjugate symmetry and can be expanded in powers of frequency ω according to

$$Z(\omega) = Z_{r,0} + jZ_{j,0}\omega + Z_{r,1}\omega^2 + jZ_{j,1}\omega^3 + \ldots \qquad (22.21)$$

for $\omega \to 0$ and

$$Z = Z_r(\infty) + \frac{jZ_{j,\infty}}{\omega} + \frac{Z_{r,1}^*}{\omega^2} + \frac{jZ_{j,1}^*}{\omega^3} + \ldots \qquad (22.22)$$

for $\omega \to \infty$ where $Z_{r,0}$, $Z_{j,0}$, $Z_r(\infty)$, and $Z_{j,\infty}, \ldots$ are coefficients in the corresponding power series expansion. Equations (22.21) and (22.22) apply to electrical circuits such as the Voigt circuit used by Agarwal et al.[69] as a measurement model.

Example 22.1 Verification of Equations (22.21) and (22.22): *Show that the assumption that the real part of the impedance is an even function of frequency and the imaginary* *is an odd function of frequency is consistent with the impedance response of a Voigt series (see Figure 21.8 on page 581).*

Solution: *The impedance response can be expressed in terms of resistance R_k and capacitance C_k by*

$$Z = R_0 + \sum_{k=1}^{n} \frac{R_k}{1 + j\omega C_k R_k} \tag{22.23}$$

The real and imaginary parts of the impedance can be written as

$$Z_r = R_0 + \sum_{k=1}^{n} \frac{R_k}{1 + \omega^2 C_k^2 R_k^2} \tag{22.24}$$

and

$$Z_j = -\omega \sum_{k=1}^{n} \frac{C_k R_k^2}{1 + \omega^2 C_k^2 R_k^2} \tag{22.25}$$

respectively. The real part of the impedance is an even function of ω, and the imaginary part is an odd function of ω. In the low-frequency limit ($\omega \to 0$),

$$Z_r = R_0 + \sum_{k=1}^{n} R_k - \omega^2 \sum_{k=1}^{n} R_k^3 C_k^2 \tag{22.26}$$

and

$$Z_j = -\omega \sum_{k=1}^{n} C_k R_k^2 + \omega^3 \sum_{k=1}^{n} C_k^3 R_k^4 \tag{22.27}$$

The sum of equations (22.26) and (22.27) yields the first four terms of equations (22.21).
    In the high-frequency limit ($\omega \to \infty$),

$$Z_r = R_0 + \frac{1}{\omega^2} \sum_{k=1}^{n} \frac{1}{C_k^2 R_k} \tag{22.28}$$

and

$$Z_j = -\frac{1}{\omega} \sum_{k=1}^{n} \frac{1}{C_k} \tag{22.29}$$

The sum of equations (22.28) and (22.29) yields the first three terms of equation (22.22).

Example 22.2 Application of Equations (22.21) and (22.22): *For a Voigt model, find values for the four terms of equation (22.21) and for the four terms of equation (22.22).*

Solution: *The sum of equations (22.26) and (22.27) yields the first four terms of equation (22.21):*

$$Z_{r,0} = R_0 + \sum_{k=1}^{n} R_k \tag{22.30}$$

$$Z_{j,0} = -\sum_{k=1}^{n} C_k R_k^2 \tag{22.31}$$

$$Z_{r,1} = -\sum_{k=1}^{n} R_k^3 C_k^2 \tag{22.32}$$

and

$$Z_{j,1} = \sum_{k=1}^{n} C_k^3 R_k^4 \tag{22.33}$$

The sum of equations (22.28) and (22.29) yields the first three terms of equation (22.22):

$$Z_{r,\infty} = R_0 \tag{22.34}$$

$$Z_{j,\infty} = -\sum_{k=1}^{n} \frac{1}{C_k} \tag{22.35}$$

and

$$Z_{r,1}^* = \sum_{k=1}^{n} \frac{1}{C_k^2 R_k} \tag{22.36}$$

The fourth term is not given by equations (22.28) and (22.29). It can be obtained from equation (22.25) for large value of ω, i.e.,

$$Z_j = -\omega \sum_{k=1}^{n} \frac{C_k R_k^2}{1 + \omega^2 C_k^2 R_k^2} \tag{22.37}$$

Upon division of numerator and denominator by $\omega^2 C_k^2 R_k^2$,

$$Z_j = -\sum_{k=1}^{n} \frac{1}{\omega C_k} \frac{1}{1 + \frac{1}{\omega^2 C_k^2 R_k^2}} \tag{22.38}$$

$$= -\sum_{k=1}^{n} \frac{1}{\omega C_k} \left(1 - \frac{1}{\omega^2 C_k^2 R_k^2} \right)$$

Thus,

$$Z_j = -\sum_{k=1}^{n} \frac{1}{\omega C_k} + \sum_{k=1}^{n} \frac{1}{\omega^3 C_k^3 R_k^2} \tag{22.39}$$

and

$$Z_{j,1}^* = \sum_{k=1}^{n} \frac{1}{C_k^3 R_k^2} \tag{22.40}$$

which is the fourth term in equation (22.22).

Theorem 22.3 (Primitive Causality) The output response to a perturbation cannot precede the input perturbation. *If $f(t) = 0$ for $t < 0$, then $x(t) = 0$ for $t < 0$. This implies that $g(\tau) = 0$ for $t < 0$. Thus, equation (22.15) can be written*

$$G(\omega) = \int_{0}^{+\infty} g(\tau) \exp(-j\omega\tau) \, d\tau \tag{22.41}$$

This important result means that G is analytic and continuous within the negative imaginary frequency plane.

The assumptions presented above, including equation (22.41), are not sufficient for derivation of dispersion relations. The value of the function $G(\omega)$ must be constrained as $\omega \to 0$. In order for G to represent a causal transform, G must satisfy[380]

$$\lim_{\omega \to \infty} G(\omega) \to 0 \qquad (22.42)$$

A function G that satisfies equation (22.42) can be shown, by use of Cauchy's Integral Formula (Theorem A.3), to be a causal transform. The properties of G implicit in Theorems 22.1-22.3 and equation (22.42) allow derivation of dispersion relations such as the Kramers–Kronig relations. Equation (22.42), however, does not apply for impedance or related electrochemical transfer functions. A weaker restriction can be applied, as discussed in Theorem 22.4.

Theorem 22.4 (Stability) The response to an input impulse cannot increase with time. *This statement can be expressed as requiring that the total output energy cannot exceed the total input energy.[380] A consequence of this statement is that $G(\omega)$ is bounded, i.e.,*

$$|G(\omega)|^2 \le A \qquad (22.43)$$

where A is a constant.

The function G that satisfies Theorem 22.4 is not a causal transform because the imaginary part of G does not contain all the information needed to obtain the real part. An additive constant cannot be determined by the transformation. By subtracting an additive constant, a modified function is obtained that is a causal transform.

22.2.2   Application of Cauchy's Theorem

As shown in equation (22.22), the real part of the impedance tends toward a finite value as frequency tends toward infinity. The transfer function $Z(x) - Z_r(\infty)$ tends toward zero with increasing frequency. As $Z(x)$ is analytic, Cauchy's integral theorem, given as Theorem (A.2) on page 635, can be written as

$$\oint (Z(x) - Z_r(\infty))\, dx = 0 \qquad (22.44)$$

where x is an independent and continuous variable that represents the complex frequency.

The first step in the analysis is to combine Z with some other function in such a way that the result vanishes as rapidly as $1/\omega^2$ as ω tends to infinity. In this way, the contribution of the large semicircular path to the contour integral (see, e.g., Figure 22.1) can be neglected.

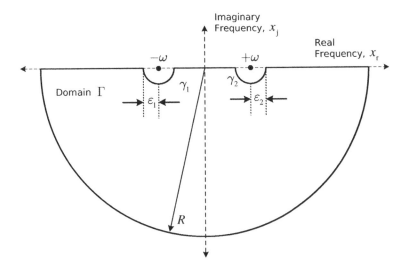

Figure 22.1: Domain of integration for application of Cauchy's integral formula. Poles are placed at frequencies $\pm\omega$ on the real frequency axis.

22.2.3 Transformation from Real to Imaginary

In order to evaluate the real or imaginary part of impedance at a particular frequency, a pole is created at ω, as shown in Figure 22.1. From considerations of symmetry, a complementary pole is created at $-\omega$. If $Z_r(\omega)$, the value assumed by Z_r at $x = \omega$, is subtracted from Z, the resulting contour integral can be written as:

$$\oint \left[\frac{Z(x) - Z_r(\omega)}{x - \omega} - \frac{Z(x) - Z_r(\omega)}{x + \omega} \right] dx = 0 \qquad (22.45)$$

Note that the constant used in equation (22.44), $Z_r(\infty)$, cancels in the integrand such that

$$\lim_{x \to \infty} \frac{(Z(x) - Z_r(\omega))}{x} \to 0 \qquad (22.46)$$

according to $1/x^3$ rather than $1/x$ (see equation (22.22)). Equation (22.45) is equal to zero because, following the condition of causality, the integrand is analytic in the domain considered and the path of integration (as shown in Figure 22.1) is a closed loop (see Theorem (A.2) on page 635). Under the assumptions that $\varepsilon_1 \to 0$ and $\varepsilon_2 \to 0$ at γ_1

and γ_2, respectively, the contributions to equation (22.45) are given as

$$
\oint \left[\frac{Z(x) - Z_r(\omega)}{x - \omega} - \frac{Z(x) - Z_r(\omega)}{x + \omega} \right] dx =
$$

$$
\int_{-\infty}^{+\infty} \left(\frac{Z(x) - Z_r(\omega)}{x - \omega} - \frac{Z(x) - Z_r(\omega)}{x + \omega} \right) dx
$$

$$
+ \int_{\gamma_1} \left(\frac{Z(x) - Z_r(\omega)}{x - \omega} - \frac{Z(x) - Z_r(\omega)}{x + \omega} \right) dx \qquad (22.47)
$$

$$
+ \int_{\gamma_2} \left(\frac{Z(x) - Z_r(\omega)}{x - \omega} - \frac{Z(x) - Z_r(\omega)}{x + \omega} \right) dx
$$

$$
+ \int_{\Gamma} \left(\frac{Z(x) - Z_r(\omega)}{x - \omega} - \frac{Z(x) - Z_r(\omega)}{x + \omega} \right) dx = 0
$$

where $\int_{-\infty}^{+\infty} f(x) dx$ represents the Cauchy principal value of the integral, defined in equation (A.27) on page 639. As shown in Figure 22.1, integration at γ_1 and γ_2 refers to semicircular paths centered at $-\omega$ and $+\omega$, respectively, and integration at Γ refers to integration along the large semicircle of radius R.

The concepts developed in Section A.3.2 (page 639) can be used to evaluate the contributions to equation (22.47). Under the assumption that $Z(x) - Z(\omega)$ approaches a finite value as $x \to \infty$,

$$
\lim_{R \to \infty} \int_{\Gamma} \left(\frac{Z(x) - Z_r(\omega)}{x - \omega} - \frac{Z(x) - Z_r(\omega)}{x + \omega} \right) dx = 0 \qquad (22.48)
$$

where R is the radius of the semicircular path shown in Figure 22.1. Therefore, the non-zero contributions to the integral in equation (22.45) result from the path of integration along the real frequency axis and at the two small semicircular indentations.

If the radii ε_1 and ε_2 of the semicircular paths γ_1 and γ_2 approach zero, the term $1/(x + \omega)$ dominates along path γ_1, and $1/(x - \omega)$ is the dominant term along path γ_2. From an application of Cauchy's Integral Formula, Theorem (A.3), to a half-circle,

$$
- \int_{\gamma_1} \frac{Z(x) - Z_r(\omega)}{x + \omega} dx = -j\pi \left(Z(-\omega) - Z_r(\omega) \right) \qquad (22.49)
$$

and

$$
\int_{\gamma_2} \frac{Z(x) - Z_r(\omega)}{x - \omega} dx = j\pi \left(Z(\omega) - Z_r(\omega) \right) \qquad (22.50)
$$

Since $Z_j(\omega)$ and $Z_r(\omega)$ are odd and even functions of frequency, respectively, the value of $Z(x)$ evaluated at ω is

$$
\lim_{x \to \omega} Z(x) = Z_r(\omega) + jZ_j(\omega) \qquad (22.51)
$$

and the value at $-\omega$ is given by

$$\lim_{x \to -\omega} Z(x) = Z_r(-\omega) + jZ_j(-\omega)$$

$$= Z_r(\omega) - jZ_j(\omega) \tag{22.52}$$

The first three terms on the right-hand side of equation (22.47) provide non-zero contributions.

Thus, equation (22.47) takes the form

$$\int_{-\infty}^{+\infty} \left(\frac{(Z_r(x) + jZ_j(x)) - Z_r(\omega)}{x - \omega} - \frac{(Z_r(x) + jZ_j(x)) - Z_r(\omega)}{x + \omega} \right) dx \tag{22.53}$$

$$= -2\pi Z_j(\omega)$$

Equation (22.53) can be expressed as

$$2\omega \int_{-\infty}^{+\infty} \left(\frac{(Z_r(x) + jZ_j(x)) - Z_r(\omega)}{x^2 - \omega^2} \right) dx = -2\pi Z_j(\omega) \tag{22.54}$$

The integral from $-\infty$ to $+\infty$ can be separated as

$$\int_{-\infty}^{0} \left(\frac{(Z_r(x) + jZ_j(x)) - Z_r(\omega)}{x^2 - \omega^2} \right) dx + \int_{0}^{+\infty} \left(\frac{(Z_r(x) + jZ_j(x)) - Z_r(\omega)}{x^2 - \omega^2} \right) dx \tag{22.55}$$

$$= -\frac{\pi}{\omega} Z_j(\omega)$$

The odd and even properties of the imaginary and real components of $Z(x)$ are again used to yield

$$Z_j(\omega) = -\frac{2\omega}{\pi} \int_{0}^{\infty} \frac{Z_r(x) - Z_r(\omega)}{x^2 - \omega^2} dx \tag{22.56}$$

Evaluation of the integral at $x \dashrightarrow \omega$ poses no particular problem because, by l'Hôpital's rule,

$$\lim_{x \to \omega} \frac{Z_r(x) - Z_r(\omega)}{x^2 - \omega^2} = \frac{1}{2x} \frac{dZ_r(x)}{dx} = \frac{1}{2} \frac{dZ_r(x)}{d\ln(x)} \tag{22.57}$$

A similar development can be used to obtain the real part of the impedance as a function of the imaginary part.

22.2.4 Transformation from Imaginary to Real

As shown in equation (22.22), the real part of the impedance tends toward a finite value as frequency tends toward infinity. To allow the integrand to tend toward zero with increasing frequency according to $1/\omega^2$, the integral is written in terms of

$$Z^*(x) = Z(x) - Z_r(\infty) \tag{22.58}$$

Thus, an equation corresponding to equation (22.45) can be written as

$$\oint \left[\frac{xZ^*(x) - j\omega Z_j(\omega)}{x - \omega} - \frac{xZ^*(x) - j\omega Z_j(\omega)}{x + \omega} \right] dx = 0 \tag{22.59}$$

where poles are created at $\pm\omega$, as shown in Figure 22.1. The contribution around the contour of radius R vanishes as $R \to \infty$.

The integrals along the semicircular paths γ_1 and γ_2 yield

$$-\int_{\gamma_1} \frac{xZ^*(x) - j\omega Z_j(\omega)}{x + \omega} dx = -j\pi \left((-\omega)Z^*(-\omega) - j\omega Z_j(\omega) \right) \tag{22.60}$$

and

$$\int_{\gamma_2} \frac{xZ^*(x) - j\omega Z_j(\omega)}{x - \omega} dx = j\pi \left(\omega Z^*(\omega) - j\omega Z_j(\omega) \right) \tag{22.61}$$

respectively.

The odd and even character of the imaginary and real parts are used to obtain the value of $xZ^*(x)$ at $\pm\omega$, i.e.,

$$\lim_{x \to \omega} xZ^*(x) = \omega Z_r(\omega) - \omega Z_r(\infty) + j\omega Z_j(\omega) \tag{22.62}$$

and the value at $-\omega$ is given by

$$\lim_{x \to -\omega} xZ^*(x) = -\omega Z_r(-\omega) + \omega Z_r(\infty) - j\omega Z_j(-\omega)$$
$$= -\omega Z_r(\omega) + \omega Z_r(\infty) + j\omega Z_j(\omega) \tag{22.63}$$

Thus, equation (22.59) takes the form

$$\int_{-\infty}^{+\infty} \left(\frac{xZ^*(x) - j\omega Z_j(\omega)}{x - \omega} - \frac{xZ^*(x) - j\omega Z_j(\omega)}{x + \omega} \right) dx \tag{22.64}$$
$$= -2j\pi\omega \left(Z_r(\omega) - Z_r(\infty) \right)$$

Equation (22.64) can be expressed as

$$2\omega \int_{-\infty}^{+\infty} \left(\frac{x \left(Z_r(x) - Z_r(\infty) + jZ_j(x) \right) - j\omega Z_j(\omega)}{x^2 - \omega^2} \right) dx \tag{22.65}$$
$$= -2j\pi \left(Z_r(\omega) - Z_r(\infty) \right)$$

The integral from $-\infty$ to $+\infty$ can be expressed as the sum of integrals from $-\infty$ to 0 and 0 to $+\infty$. The odd and even properties of the imaginary and real components of $Z(x)$ are again used to yield

$$Z_r(\omega) = Z_r(\infty) - \frac{2}{\pi} \int_0^\infty \frac{xZ_j(x) - \omega Z_j(\omega)}{x^2 - \omega^2} dx \tag{22.66}$$

Evaluation of the integral of equation (22.66) at $x \to \omega$ poses no particular problem because, by l'Hôpital's rule,

$$\lim_{x \to \omega} \frac{xZ_j(x) - \omega Z_j(\omega)}{x^2 - \omega^2} = \frac{1}{2x} \frac{x \, dZ_j(x)}{dx} = \frac{1}{2} \frac{d x Z_j(x)}{d \ln(x)} \qquad (22.67)$$

Through equation (22.66), the imaginary part of the impedance can be used to obtain the real part of the impedance if the asymptotic value at high frequency is known.

22.2.5 Application of the Kramers–Kronig Relations

Some forms of the Kramers–Kronig relations are presented in Table 22.1. Equations (22.1) and (22.2), called the Plemelj formulas, are obtained directly from consideration of causality in a linear system. Equations (22.3) and (22.4) are mathematically equivalent to equations (22.5) and (22.6), respectively, because

$$\int_0^\infty \frac{1}{x^2 - \omega^2} dx = 0 \qquad (22.68)$$

A series of equations have been developed to relate the phase angle to the modulus, represented by equations (22.7), (22.8), (22.9), and (22.10). Equations (22.11) and (22.12) were developed by Ehm et al.[381] in terms of the natural logarithm of the complex impedance. Some key relationships among the real and imaginary components of the natural logarithm of the complex impedance and phase angle are given as equations (1.128) and (1.129) in Table 1.7.

There are inherent difficulties in the direct application of the Kramers–Kronig relations to experimental data. Some of these difficulties are evident in Figure 22.2 where the integrand of equation (22.5) is given as a function of frequency for a single Voigt element. The major difficulty in applying this approach is that the measured frequency range may not be sufficient to allow integration over the frequency limits of zero to infinity. Therefore, discrepancies between experimental data and the impedance component predicted through application of the Kramers–Kronig relations could be attributed to use of a frequency domain that is too narrow as well as to failure to satisfy the constraints of the Kramers–Kronig equations.

A second issue is that the integral equations do not account explicitly for the stochastic character of experimental data. This requires solution of the Kramers–Kronig relations in an expectation sense, as discussed in the following section. The methods used to address the incomplete sampled frequency range are described in Section 22.1.

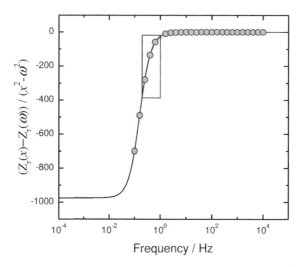

Figure 22.2: The integrand of equation (22.5) as a function of frequency for a single Voigt element ($R_0 = 10\ \Omega$, $R_1 = 1000\ \Omega$, $C_1 = 10^{-3}$ F). Symbols represent synthetic impedance data collected at 5 points per decade over a frequency range of 0.1-10,000 Hz. The equations are evaluated at a frequency of 1 Hz. The box identifies the region of the plot expanded as Figure 22.3.

22.3 The Kramers–Kronig Relations in an Expectation Sense

The contribution of stochastic error $\varepsilon(\omega)$ to the observed value of the impedance at any given frequency ω can be expressed as

$$Z_{ob}(\omega) = Z(\omega) + \varepsilon(\omega) = (Z_r(\omega) + \varepsilon_r(\omega)) + j(Z_j(\omega) + \varepsilon_j(\omega)) \qquad (22.69)$$

where $Z(\omega)$, $Z_r(\omega)$, and $Z_j(\omega)$ represent the error-free values of the impedance that conform exactly to the Kramers–Kronig relations. The measurement error $\varepsilon(\omega)$ is a complex stochastic variable such that

$$\varepsilon(\omega) = \varepsilon_r(\omega) + j\varepsilon_j(\omega) \qquad (22.70)$$

At any frequency ω the expectation of the observed impedance $E\left(Z_{ob}(\omega)\right)$, defined in equation (3.1) on page 45, is equal to the value consistent with the Kramers–Kronig relations, i.e.,

$$E\left\{Z_{ob}(\omega)\right\} = Z(\omega) \qquad (22.71)$$

Remember! *22.3 Impedance data that do not satisfy the Kramers–Kronig relations must violate at least one of the required conditions of stability, linearity, and causality. Satisfaction of the Kramers–Kronig relations is a necessary but not sufficient condition for meeting conditions of stability, linearity, and causality.*

---

if and only if both

$$E\{\varepsilon_r(\omega)\} = 0 \qquad (22.72)$$

and

$$E\{\varepsilon_j(\omega)\} = 0 \qquad (22.73)$$

Equations (22.72) and (22.73) are satisfied for errors that are stochastic and do not include the effects of bias.

22.3.1 Transformation from Real to Imaginary

Equation (22.56) can be applied to obtain the imaginary part from the real part of the impedance spectrum only in an expectation sense, i.e.,

$$E\{Z_j(\omega)\} = \frac{2\omega}{\pi} E\left\{ \int_0^\infty \frac{Z_r(x) - Z_r(\omega) + \varepsilon_r(x) - \varepsilon_r(\omega)}{x^2 - \omega^2} \right\} dx \qquad (22.74)$$

For a given evaluation of equation (22.74),

$$Z_j^{kk}(\omega) + \varepsilon_j^{kk}(\omega) = \frac{2\omega}{\pi} \left[E\left\{ \int_0^\infty \frac{Z_r(x) - Z_r(\omega) + \varepsilon_r(x) - \varepsilon_r(\omega)}{x^2 - \omega^2} dx \right\} \right.$$
$$\left. + \int_0^\infty \frac{\varepsilon_r(x)}{x^2 - \omega^2} dx \right] \qquad (22.75)$$

where $Z_j^{kk}(\omega)$ represents the value of the imaginary part of the impedance obtained by evaluation of the Kramers–Kronig integral equation and $\varepsilon_j^{kk}(\omega)$ represents the error in the evaluation of the Kramers–Kronig relations caused by the second integral on the right-hand side.

It is evident from equation (22.75) that, for the expected value of the observed imaginary component to approach its true value in the Kramers–Kronig sense, it is necessary that $E\left\{\varepsilon_j^{kk}(\omega)\right\} = 0$. Thus, the requirements are that equation (22.72) be satisfied and that

$$E\left\{ \frac{2\omega}{\pi} \int_0^\infty \frac{\varepsilon_r(x)}{x^2 - \omega^2} \right\} dx = 0 \qquad (22.76)$$

Requirements (22.72) and (22.76) place well-defined constraints on the evaluation of the Kramers–Kronig integral equations.

For the first condition to be met, it is necessary that the process be stationary in the sense of replication at every measurement frequency. The second condition can be satisfied in two ways. In the hypothetical case where all frequencies can be sampled, the expectation can be carried to the inside of the integral, and equation (22.76) results directly from equation (22.72).

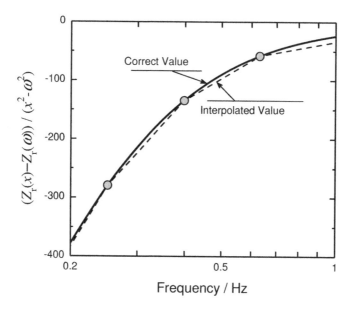

Figure 22.3: A region of Figure 22.2 is expanded to demonstrate the discrepancy between a straight-line interpolation between data points and the model that conforms to the interpolation of the data. System parameters are as given in Figure 22.2.

In the more practical case where the impedance is sampled at a finite number of frequencies, $\varepsilon_r(x)$ represents the error between an interpolated function and the "true" impedance value at frequency x. This error is seen in Figure 22.3, where a region of Figure 22.2 was expanded to demonstrate the discrepancy between a straight-line interpolation between data points and the model that conforms to the interpolation of the data. This error is composed of contributions from the quadrature and/or interpolation errors and from the stochastic noise at the measurement frequency ω. Effectively, equation (22.76) represents a constraint on the integration procedure. In the limit that quadrature and interpolation errors are negligible, the residual errors $\varepsilon_r(x)$ should be of the same magnitude as the stochastic noise $\varepsilon_r(\omega)$.

22.3.2 Transformation from Imaginary to Real

The Kramers–Kronig relations for obtaining the real part from the imaginary part of the spectrum can be expressed as equation (22.66), which, in terms of expectations, becomes

$$\mathrm{E}\left(Z_r(\omega) - Z_r(\infty)\right) = -\frac{2}{\pi}\mathrm{E}\left\{\int_0^\infty \frac{xZ_j(x) - \omega Z_j(\omega) + x\varepsilon_j(x) - \omega\varepsilon_j(\omega)}{x^2 - \omega^2}\mathrm{d}x\right\} \qquad (22.77)$$

For a given evaluation of equation (22.77),

$$
Z_r^{kk}(\omega) - Z_r(\infty) + \varepsilon_r^{kk}(\omega) = \varepsilon_{Z_r(\infty)}^{kk}
$$

$$
- \frac{2\omega}{\pi} \left[E \left\{ \int_0^\infty \frac{x Z_j(x) - \omega Z_j(\omega) + x\varepsilon_r(x) - \omega\varepsilon_r(\omega)}{x^2 - \omega^2} dx \right\} \right. \tag{22.78}
$$

$$
\left. + \int_0^\infty \frac{x\varepsilon_j(x)}{x^2 - \omega^2} dx \right]
$$

Following the discussion of equation (22.75), the necessary conditions for application of the Kramers–Kronig relations are that equation (22.73) be satisfied and that

$$
E \left\{ \frac{2}{\pi} \int_0^\infty \frac{x\varepsilon_j(x)}{x^2 - \omega^2} \right\} dx = 0 \tag{22.79}
$$

Under the assumption that the impedance is sampled at a finite number of frequencies, $\varepsilon_j(x)$ represents the error between an interpolated function and the "true" impedance value at frequency x. This term is composed of contributions from the quadrature and/or interpolation errors and from the stochastic noise at the measurement frequency ω. Similar to equation (22.76), equation (22.79) provides a well-defined constraint on the integration procedure.

Problems

22.1 Use a circuit with $R_e = 10 \, \Omega\text{cm}^2$, $R_t = 100 \, \Omega\text{cm}^2$, and $C_{dl} = 20 \, \mu\text{F/cm}^2$ to verify the equations given in Table 22.1. This problem requires use of a spreadsheet program such as Microsoft Excel® or a computational programming environment such as MATLAB®.

(a) Equation (22.1)

(b) Equation (22.2)

(c) Equation (22.3)

(d) Equation (22.4)

(e) Equation (22.5)

(f) Equation (22.6)

(g) Equation (22.7)

(h) Equation (22.8)

(i) Equation (22.9)

(j) Equation (22.10)

(k) Equation (22.11)

(l) Equation (22.12)

22.2 Verify that the impedance response that includes a CPE, expressed as

$$Z = R_e + \frac{R_{||}}{1 + (j\omega)^{\alpha} R_{||} Q} \tag{22.80}$$

satisfies the Kramers–Kronig relations.

22.3 Using the results of Problem 8.6, explore whether the Kramers–Kronig relations can be used to detect a nonlinearity in the impedance response. Discuss the implication of your result on experimental design.

22.4 Why are the assumptions of linearity, causality, and stability necessary for the derivation of the Kramers–Kronig relations?

22.5 What is a causal transform? Why is the impedance Z not a causal transform?

22.6 It is sometimes said that a finite impedance is needed in order for application of the Kramers–Kronig relations to an electrochemical system. Yet, blocking electrodes do not have a finite impedance. Do the Kramers–Kronig relations apply for blocking electrodes? If so, how can they be applied?

Part VI

Overview

Chapter 23

An Integrated Approach to Impedance Spectroscopy

The foundation of this textbook is a philosophy that integrates experimental observation, model development, and error analysis. This approach is differentiated from the usual sequential model development for given impedance spectra[3,4] by its emphasis on obtaining additional supporting observations to guide model selection, use of error analysis to guide regression strategies and experimental design, and use of models to guide selection of new experiments.

23.1 Flowcharts for Regression Analysis

While impedance spectroscopy can be a sensitive tool for analysis of electrochemical and electronic systems, an unambiguous interpretation of spectra cannot be obtained by examination of raw data. Instead, interpretation of spectra requires development of a model that accounts for the impedance response in terms of the desired physical properties. Model development should take into account both the impedance measurement and the physical and chemical characteristics of the system under study.

It is useful to envision a flow diagram for the measurement and interpretation of experimental measurements such as impedance spectroscopy. Barsoukov and Macdonald[3,4] proposed such a flow diagram for a general characterization procedure consisting of two blocks comprising the impedance measurement, three blocks comprising a physical (or process) model, one block for an equivalent electrical circuit, and blocks labeled curve fitting and system characterization. They suggested that impedance data may be analyzed for a given system by using either an exact mathematical model based on a plausible physical theory or a comparatively empirical equivalent circuit. The parameters for either model can be estimated by complex nonlinear least-squares regression. The authors observed that ideal electrical circuit elements represent ideal lumped constant properties, whereas the physical properties of electrolytic cells are often distributed. The distribution of cell properties motivates use of distributed impedance elements such as constant-phase elements (CPE). An additional problem with equivalent circuit analysis, which the authors suggest is not shared by the direct comparison to the theoretical model, is that circuit models are ambiguous and different models may provide equivalent fits to a given spectrum. The authors suggest that identification of the appropriate equivalent circuit can be achieved only by employing physical intuition and by carrying out several sets of measurements with different conditions.

A similar flow diagram was presented by Huang et al.[382] for solid-oxide fuel cells (SOFCs). The diagram accounts for the actions of measuring impedance data, modeling, fitting the model, interpreting the results, and optimizing the fuel cell for power generation. The authors emphasize that the interpreting action depends more on the electrochemical expertise of the researchers than on a direct mapping from model parameters to SOFC properties.

While helpful, the flow diagrams proposed by Barsoukov and Macdonald[4] and Huang et al.[382] are incomplete because they do not account for the role of independent assessment of experimental error structure and because they do not emphasize the critical role of supporting experimental measurements. Orazem et al.[383] proposed a flow diagram consisting of three elements: experiment, measurement model, and process model. The measurement model was intended to assess the stochastic and bias error structure of the data; thus, their diagram accounts for the independent assessment of experimental error structure. Their diagram does not, however, account for supporting nonimpedance measurements, and the use of regression analysis, while implied, is not shown explicitly. The objective of this chapter is to formulate a comprehensive approach that can be applied to measurement and interpretation of impedance spectra.

23.2 Integration of Measurements, Error Analysis, and Model

A refined philosophical approach toward the use of impedance spectroscopy is outlined in Figure 23.1, where the triangle evokes the concept of an operational amplifier for which the potential of input channels must be equal. Sequential steps are taken until the model provides a statistically adequate representation of the data to within the independently obtained stochastic error structure. The different aspects that comprise the philosophy are presented in this section.

Remember! 23.1 *The philosophical approach of this textbook integrates experimental observation, model development, and error analysis.*

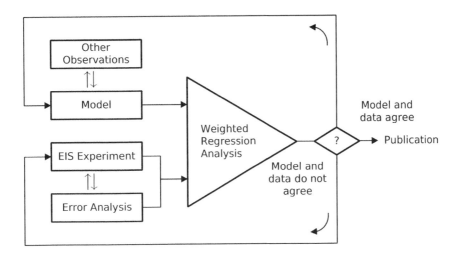

Figure 23.1: Schematic flowchart showing the relationship between impedance measurements, error analysis, supporting observations, model development, and weighted regression analysis. (Taken from Orazem and Tribollet.[340])

23.2.1   Impedance Measurements Integrated with Error Analysis

All impedance measurements should begin with measurement of a steady-state polarization curve. The steady-state polarization curve is used to guide selection of an appropriate perturbation amplitude and can provide initial hypotheses for model development. The impedance measurements can then be made at selected points on the polarization curve to explore the potential dependence of reaction rate constants. Impedance measurements can also be performed at different values of state variables such as temperature, rotation speed, and reactant concentration. Impedance scans measured at different points of time can be used to explore temporal changes in system parameters. Some examples include growth of oxide or corrosion-product films, poisoning of catalytic surfaces, and changes in reactant or product concentration.

The impedance measurements should also be conducted in concert with error analysis with emphasis on both stochastic and systematic bias errors. The bias errors can be defined to be those that result in data that are inconsistent with the Kramers–Kronig relations. An empirical error analysis can be undertaken using the measurement model approach suggested by Agarwal et al.[69,100,101,384] It should be noted, however, that this approach is not definitive, because Kramers–Kronig-consistent artifacts can be caused by electrical leads and the electronics. Use of dummy cells can help identify artifacts that are consistent with the Kramers–Kronig relations. As an alternative approach for identification of the stochastic part of the error structure, Dygas and Breiter have shown that impedance instrumentation could, in principle, provide standard deviations for the impedance measurements at each frequency.[385]

The feedback loop shown in Figure 23.1 between EIS experiment and error analysis indicates that the error structure is obtained from the measured data and that knowledge of the error structure can guide improvements to the experimental design. The magnitude of perturbation, for example, should be selected to minimize stochastic errors while avoiding inducing a nonlinear response. The frequency range should be selected to sample the system time constants, while avoiding bias errors associated with nonstationary phenomena. In short, the experimental parameters should be selected so as to minimize the stochastic error structure while, at the same time, allowing for the maximum frequency range that is free of bias errors.

23.2.2   Process Models Developed Using Other Observations

The model identified in Figure 23.1 represents a process model intended to account for the hypothesized physical and chemical character of the system under study. From the perspective embodied in Figure 23.1, the objective of the model is not to provide a good fit with the smallest number of parameters. The objective is rather to use the model to gain a physical understanding of the system. The model should be able to account for, or at least be consistent with, all experimental observations. The supporting measurements therefore provide a means for model identification. The feedback loop shown in Figure 23.1 between model and other observations is intended to illustrate that the supporting measurements guide model development and the proposed model can suggest experiments needed to validate model hypotheses. The supporting experiments can include both electrochemical and nonelectrochemical measurements.

   Numerous scanning electrochemical methods such as scanning reference electrodes, scanning tunneling microscopy, and scanning electrochemical microscopy can be used to explore surface heterogeneity. Scanning vibrating electrodes and probes can be used to measure local current distributions. Local electrochemical impedance spectroscopy provides a means of exploring the distribution of surface reactivity. Measurements can be performed across the electrode at a single frequency to create an image of the electrode or, alternatively, performed at a given location to create a complete spectrum. Other experiments may include in situ and ex situ surface analysis, chemical analysis of electrolytes, and both in situ and ex situ visualization and/or microscopy. Transfer-function methods such as electrohydrodynamic (EHD) impedance spectroscopy allow isolation of the phenomena that influence the electrochemical impedance response.[65]

Remember! 23.2 *Impedance spectroscopy is not a stand-alone technique. Other observations are required to validate a given interpretation of the impedance spectra.*

23.2.3 Regression Analysis in Context of Error Structure

The goal of the operation represented by the triangle in Figure 23.1 is to develop a model that provides a good representation of the impedance measurements to within the noise level of the measurement. The error structure for the measurement clearly plays a critical role in the regression analysis. The weighting strategy for the complex nonlinear least-squares regression should be based on the variance of the stochastic errors in the data, and the frequency range used for the regression should be that which has been determined to be free of bias errors. In addition, knowledge of the variance of the stochastic measurement errors is essential to quantify the quality of the regression.

Sequential steps are taken until the model provides a statistically adequate representation of the data to within the independently obtained stochastic error structure. The comparison between model and experiment can motivate modifications to the model or to the experimental parameters.

23.3 Application

Two systems are described in the following examples that illustrate the approach outlined in Figure 23.1.

Example 23.1 Model Identification: *Two models have been proposed for data collected for an n-GaAs Schottky diode at different temperatures. One model uses a CPE that accounts for implicit distributed relaxation processes.[287] This model fits the data very well with a minimal number of parameters. A second model accounts for discrete energy levels and provides a fit of equivalent regression quality at each temperature.[338,339] Which model extracts the more useful information from the measurements?*

Solution: *Orazem et al.[339] and Jansen et al.[338] described the impedance response for an n-type GaAs Schottky diode with temperature as a parameter. The system consisted of an n-GaAs single crystal with a Ti Schottky contact at one end and a Au, Ge, Ni Schottky contact at the eutectic composition at the other end. This material has been well characterized in the literature and, in particular, has a well-known EL2 deep-level state that lies 0.83 to 0.85 eV below the conduction band edge.[337] Experimental details are provided by Jansen et al.[338]*

The experimental data are presented in Figures 18.22(a) and (b) for the real and imaginary parts of the impedance, respectively, for temperatures ranging from 320 to 400 K. The loga-

---

Remember! 23.3 *The object of modeling is not to provide a good fit with the smallest number of parameters. The object is rather to use the model to learn about the physics and chemistry of the system under study.*

rithmic scale used in Figure 18.22 emphasizes the scatter seen in the imaginary impedance at low frequencies. The impedance response is seen to be a strong function of temperature. The impedance-plane plots shown in Figure 18.23, for data collected at 320 and 340 K, show the classic semicircle associated with a single relaxation process.

Jansen and Orazem showed that the impedance data could be superposed as given in Figure 18.24 (page 520).[338] The impedance data collected at different temperatures were normalized by the maximum mean value of the real part of the impedance and plotted against a normalized frequency defined by equation (18.23), where $E = 0.827$ eV, and the characteristic frequency $f°$ was assigned a value of 2.964×10^{14} Hz such that the imaginary part of the normalized impedance values reached a peak value near $f^* = 1$. The data collected at different temperatures are reduced to a single line. The extent to which the data are superposed is seen more clearly on the logarithmic scale shown in Figure 18.25. The superpositions shown in Figures 18.24 and 18.25 suggest that the system is controlled by a single-activation-energy-controlled process.

A closer examination of Figure 18.24(b) reveals that the maximum magnitude of the scaled imaginary impedance is slightly less than 0.5, whereas the corresponding value for a single-activation-energy-controlled process should be identically 0.5. Regression analysis using the traditional weighting strategies under which the standard deviation of the experimental values was assumed to be proportional to the modulus of the impedance, $\sigma_r = \sigma_j = \alpha|Z|$, to the magnitude of the respective components of the impedance, $\sigma_r = \alpha_r|Z_r|$ and $\sigma_j = \alpha_j|Z_j|$, or independent of frequency, $\sigma_r = \sigma_j = \alpha$, yielded one dominant RC time constant with only a hint that other parameters could be extracted. Jansen et al.[338,339] used the measurement model approach described by Agarwal et al.[69,100,101,384] to identify the stochastic error structure for the impedance data. When the data were regressed using this error structure for a weighting strategy, additional parameters could be resolved, revealing additional activation energies. Thus, while the data do superpose nicely in Figures 18.24 and 18.25, the impedance data do in fact contain information on minor activation-energy-controlled electronic transitions.[338,339] The information concerning these transitions could be extracted by regression of an appropriate process model using a weighting strategy based on the error structure of the measurement.

Two models have been proposed for the data presented above. Macdonald[287] proposed a distributed-time-constant model that accounts for distributed relaxation processes. This model fits the data very well and has the advantage that it requires a minimal number of parameters. A second model, presented in Figure 23.2, accounts for discrete energy levels and provides a fit of equivalent regression quality at each temperature.[338,339] In Figure 23.2, C_n is the space-charge capacitance, R_n is a resistance that accounts for a small but finite leakage current, and the parameters $R_1 \ldots R_k$ and $C_1 \ldots C_k$ are attributed to the response of discrete deep-level energy states. Parameters corresponding to deep-level states were added sequentially to the model, subject to the constraint that the 2σ (95.45 percent) confidence interval for each of the regressed parameters may not include zero. Including the space-charge capacitance and leakage resistance, four resistor-capacitor pairs could be obtained from the impedance data collected at 300, 320, and 340 K; three resistor-capacitor pairs could be obtained from the data collected at 360, 380, and 400 K; and two resistor-capacitor pairs could be obtained from the data collected at 420

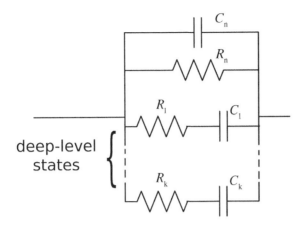

Figure 23.2: Electrical circuit corresponding to the model presented by Jansen et al.[338] in which C_n is the space-charge capacitance, R_n is a resistance that accounts for a small but finite leakage current, and the parameters $R_1 \ldots R_k$ and $C_1 \ldots C_k$ are attributed to the response of discrete deep-level energy states. (Taken from Orazem and Tribollet.[340])

K.[338] *This model has the disadvantage that up to eight parameters are required, depending on the temperature, as compared to the three parameters required for the distributed-time-constant model. The question to be posed, then, is "which is the better model for the measurements?"*

If the goal of the regression is to provide the most parsimonious model for the data, the model with the smallest number of parameters and a continuous distribution of activation energies is the best model. If the goal of the regression is to provide a quantitative physical description of the system for which the data were obtained, additional measurements are needed to determine whether the activation energies are discrete or continuously distributed. In this case, deep-level transient spectroscopy (DLTS) measurements indicated that the n-GaAs diode contained discrete deep-level states. In addition, the energy levels and state concentrations obtained by regression of the second model were consistent with the results obtained by DLTS.[338] Thus, the second model with a larger number of parameters provides the more useful description of the GaAs diode. The choice between the two models could not be made without the added experimental evidence.

Example 23.1 demonstrates the importance of coupling experimental observation, model development, and error analysis. The measurements conducted at different temperatures allowed identification of discrete activation energies for electronic transitions. Use of a weighting strategy based on the observed stochastic error structure increased the number of parameters that could be obtained from the regression analysis. Thus, four discrete activation-energy-controlled processes could be identified, but at the expense of a corresponding model that required eight parameters. A regression of almost the same quality could be obtained under the assumption of a continuous distribution of activation energies, and this model required only three parameters. Discrimination between the two models requires additional experimental observations,

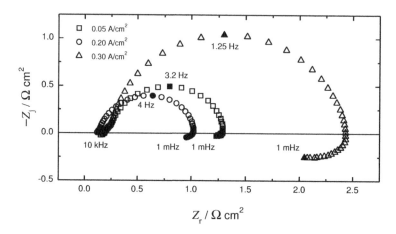

Figure 23.3: Electrochemical impedance results obtained for a single-cell PEM fuel cell with current density as a parameter. (Taken from Roy et al.[389])

such as the DLTS identification of electronic transitions involving discrete deep-level states.

Example 23.2 Models to Propose Experiments: *Electrochemical impedance data obtained for PEM fuel cells often reveal low-frequency inductive loops that have been attributed to parasitic reactions in which the* Pt *catalyst reacts to form* PtO *and subsequently forms* Pt$^+$ *ions.[378] Suggest experiments that could be used to support or reject this model.*

Solution: *Low-frequency inductive features[386–388] are commonly seen in impedance spectra for PEM fuel cells. Makharia et al.[386] suggested that side reactions and intermediates involved in the fuel cell operation can be possible causes of the inductive loop seen at low frequency. However, such low-frequency inductive loops could also be attributed to nonstationary behavior, or, due to the time required to make measurements at low frequencies, nonstationary behavior could influence the shapes of the low-frequency features.*

A typical result is presented in Figure 23.3 for the impedance response of a single 5 cm^2 PEMFC with hydrogen and air as reactants.[389] The measurements were conducted in gala-vanostatic mode for a frequency range of 1 kHz to 1 mHz with a 10 mA peak-to-peak sinusoidal perturbation. Roy and Orazem[378] used the measurement model approach developed by Agar-wal et al.[69,100,101,384] to demonstrate that, for the fuel cell under steady-state operation, the low-frequency inductive loops seen in Figure 23.3 were consistent with the Kramers–Kronig relations. Therefore, the low-frequency inductive loops could be attributed to process character-istics and not to nonstationary artifacts.

Roy et al.[389] proposed that the low-frequency inductive loops observed in PEM fuel cells could be caused by parasitic reactions in which the Pt *catalyst reacts to form* PtO *and sub-sequently forms* Pt$^+$ *ions. They also showed that a reaction involving formation of hydrogen*

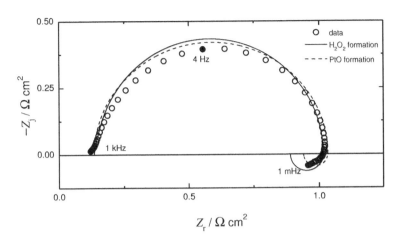

Figure 23.4: Comparison of the impedance response for a PEM fuel cell operated at 0.2 A/cm^2 to model predictions generated using a reaction sequence involving formation of hydrogen peroxide and a reaction sequence involving formation of PtO. (Taken from Roy et al.[389])

peroxide could yield the same inductive features. A comparison between the two models and the experimental results is shown in Figure 23.4, and the corresponding values for the real and imaginary parts of the impedance are presented as a function of frequency in Figures 23.5(a) and (b), respectively. The model calculations were not obtained by regression but rather by simulation using approximate parameter values. Regression was not used because the model was based on the assumption of a uniform membrane-electrode assembly (MEA), whereas the use of a serpentine flow channel caused the reactivity of the MEA to be very nonuniform. The parameters were first selected to reproduce the current-potential curve and then the same parameters were used to calculate the impedance response for each value of current density. The potential (or current) dependence of model parameters was that associated with the Tafel behavior assumed for the electrochemical reactions.

While reaction parameters were not identified by regression to impedance data, the simulation presented by Roy et al.[389] demonstrates that side reactions proposed in the literature can account for low-frequency inductive loops. Indeed, the results presented in Figures 23.4 and 23.5 show that both models can account for low-frequency inductive loops. Other models can also account for low-frequency inductive loops so long as they involve potential-dependent adsorbed intermediates.[311] It is generally understood that equivalent circuit models are not unique and have, therefore, an ambiguous relationship to physical properties of the electrochemical cell. As shown by Roy et al.,[389] even models based on physical and chemical processes are ambiguous. In the present case, the ambiguity arises from uncertainty as to which reactions are responsible for the low-frequency inductive features.

Resolution of this ambiguity requires additional experiments. The processes and reactions hypothesized for a given model can suggest experiments to support or reject the underlying hypothesis. For example, the proposed formation of PtO is consistent with a decrease in the

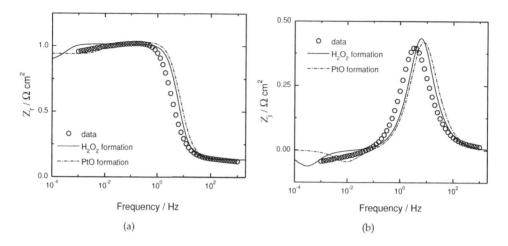

Figure 23.5: Comparison of the impedance response for a PEM fuel cell operated at 0.2 A/cm² to model predictions generated using a reaction sequence involving formation of hydrogen peroxide and a reaction sequence involving formation of PtO: a) Real part of impedance and b) imaginary part of impedance. (Taken from Roy et al.[389])

active catalytic surface area and a loss of Pt ions in the effluent. Cyclic voltammetry after different periods of fuel cell operation could be used to explore reduction in electrochemically active area. Inductively coupled plasma mass spectroscopy (ICPMS) could be used to detect residual Pt ions in the fuel cell effluent, and ex situ techniques could detect formation of PtO in the catalyst layer. A different set of experiments could be performed to explore the hypothesis that peroxide formation is responsible for the inductive loops. Platinum dissolution has been observed in PEM fuel cells,[390] and peroxide formation has been implicated in the degradation of PEM membranes.[391–393] Thus, it is likely that both reactions are taking place and contributing to the observed low-frequency inductive loops.

Example 23.2 demonstrates the utility of the error analysis for determining consistency with the Kramers–Kronig relations. In this case, the low-frequency inductive loops were found to be consistent with the Kramers–Kronig relations at frequencies as low as 0.001 Hz so long as the system had reached a steady-state operation. The mathematical models that were proposed to account for the low-frequency features were based on plausible physical and chemical hypotheses. Nevertheless, the models are ambiguous and require additional measurements and observations to identify the most appropriate for the system under study.

The philosophy described here and embodied in Figure 23.1 cannot always be followed to convergence. Often the hypothesized model is inadequate and cannot reproduce the experimental results. Even if the proposed reaction sequence is correct, surface heterogeneities may introduce complications that are difficult to model. The accessible frequency range may be limited at high frequency for systems with a very small impedance. The accessible frequency range may be limited at low frequency for systems subject to significant nonstationary behavior. The experimentalist may need to accept a large level of stochastic noise for a system with a large impedance. In cases where the models are unable to explain all features of the experiment, the graphical methods presented in Chapter 18 on page 493 can nevertheless yield quantitative information.

Problems

23.1 Suggest experiments that could be used to support or reject the model described in:
  (a) Example 10.4
  (b) Example 10.5
  (c) Problem 10.4
  (d) Problem 10.5
  (e) Problem 10.6
  (f) Example 23.1

23.2 Several prominent leaders in impedance research have proposed creating a catalog of impedance spectra with unique interpretations associated with different shapes of the impedance diagrams. Discuss the potential and limitations of such a catalog.

23.3 Explain the role of error analysis in:
  (a) Experimental design
  (b) Regression analysis
  (c) Model identification

23.4 Why is impedance spectroscopy not considered to be a standalone technique? Are other electrochemical techniques such as cyclic voltammetry considered to be standalone? Explain your answers.

Part VII

Reference Material

Appendix A

Complex Integrals

Chapter 1 provides a framework for the analysis of complex variables. This section provides a summary of important definitions before describing complex integration in greater detail. This work provides support for the discussion of the Kramers–Kronig relations in Chapter 22. For more detailed analysis, the reader is directed to specialized textbooks.[88–90,93]

A.1 Definition of Terms

Definition A.1 (Limit) *The limit of a function $f(z)$ at $z \to z_0$,*

$$\lim_{z \to z_0} f(z) = \ell \tag{A.1}$$

means that, given any $\varepsilon > 0$, there exists a δ such that $|f(z) - \ell| < \varepsilon$ when $0 < |z - z_0| < \delta$.

Note that z may approach z_0 from any direction in the complex plane, and the value of the limit ℓ is independent of the direction of approach.

Definition A.2 (Continuous Function) *A function $f(z)$ is continuous at z_0 if*

$$\lim_{z \to z_0} f(z) = f(z_0) \tag{A.2}$$

A function $f(z)$ is continuous if it is continuous at every point of its domain.

Definition A.3 (Derivative of a Function) *The derivative of a function $f(z)$ at z_0 is*

$$f'(z_0) = \lim_{z \to z_0} \frac{f(z) - f(z_0)}{z - z_0} \tag{A.3}$$

If $f(z)$ has a derivative at z_0, $f(z)$ is differentiable at z_0.

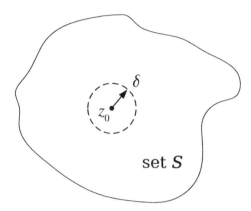

set S

Figure A.1: Representation of the δ-neighborhood of the point z_0 within set S.

Definition A.4 (Neighborhood of a Point) *The set of all points z that satisfy the inequality*

$$|z - z_0| < \delta \tag{A.4}$$

is called the δ-neighborhood of the point z_0.

A δ-neighborhood of the point z_0 is illustrated in Figure A.1.

Definition A.5 (Set) *A point z_0 is an interior point of a set S if there exists some neighborhood of z_0 that contains only points of S. A point z_0 is an exterior point of the set S if there exists some neighborhood of z_0 that contains only points that are not in set S. If z_0 is neither an exterior nor an interior point of S, it must be a boundary point of S.*

Definition A.6 (Open and Closed Sets) *An open set contains only interior points. If a set contains its boundary points, it is a closed set.*

Definition A.7 (Curve) *A curve C is described by a function*

$$z = z_r(t) + jz_j(t) \tag{A.5}$$

where $z_r(t)$ and $z_j(t)$ are continuous functions of parameter t defined for $a \le t \le b$. If $z(a) = z(b)$, curve C is closed. If curve C does not cross itself, curve C is simple.

Definition A.8 (Connected Sets) *An open set is connected if every pair of points z and w can be joined by a polygonal line that exists entirely within S.*

Two open sets are shown in Figure A.2. Set A is connected because a continuous line can be drawn between any two points within the set. Set B is not connected because it consists of two separated sections across which a continuous line cannot be drawn. Similarly, the set of all points that do not lie on the closed curve $|z| = 1$ is an open set which is not connected.

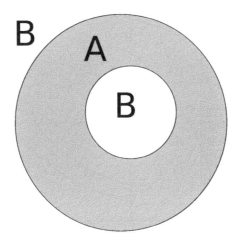

Figure A.2: Representation of two open sets. Set A is connected because a continuous line can be drawn between any two points within the set. Set B is not connected because it consists of two separated sections across which a continuous line cannot be drawn.

Definition A.9 (Domain) *An open set that is connected is a domain.*

The set A is a domain; whereas, set B is not a domain because it is not connected.

Definition A.10 (Region) *A domain, together with or without its boundary points, is a region. If the region contains all the domain and its boundary points, it is a closed region.*

The set A, with or without its boundary points, is a region; whereas B is not because set B is not a domain.

Definition A.11 (Analytic Function) *If function $f(z)$ is differentiable at every point within a neighborhood of z_0, $f(z)$ is analytic at z_0.*

Analytic functions are also called regular and holomorphic. If $f(z)$ is analytic for all finite values of z, it is entire. If $f(z)$ is not analytic at a point z_0, the point z_0 is a singular point. The singular point is often referred to as being a singularity or a pole.

Example A.1 Analytic Functions: *Provide examples of functions that are analytic and nonanalytic.*

Solution:

- *Polynomial functions are everywhere analytic.*

- *The exponential function is everywhere analytic.*

- *The absolute value function is not everywhere analytic because it is not differentiable when the argument is equal to zero.*

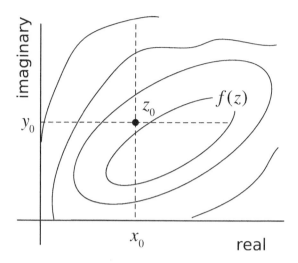

Figure A.3: Representation of two paths approaching z_0. The function $f(z)$ is presented in the form of a contour map as a function of real and imaginary parts of the complex independent variable z.

A.2   Cauchy–Riemann Conditions

The Cauchy–Riemann conditions describe the criteria under which a complex function is analytic.

Theorem A.1 (Cauchy–Riemann Conditions) *The complex function*

$$f(z) = u(x, y) + jv(x, y) \tag{A.6}$$

is analytic in a region G if and only if $\partial u/\partial x$, $\partial u/\partial y$, $\partial v/\partial x$, and $\partial v/\partial y$ are continuous and satisfy

$$\frac{\partial u}{\partial x} = \frac{\partial v}{\partial y} \quad \text{and} \quad \frac{\partial u}{\partial y} = -\frac{\partial v}{\partial x} \tag{A.7}$$

in the region G.

Following Definition A.11, the function $f(z)$ is analytic at z_0 if it is differentiable at every point in the neighborhood of z_0. Definitions A.1 and A.3 state that the value of the derivative should be independent of the path chosen. A general complex function $f(z)$ is presented in Figure A.3 in the form of a contour map as a function of real and imaginary parts of the complex independent variable z. The derivative along the real

Remember! **A.1** *The concept of an analytic function is critical to the derivation of the Kramers–Kronig relations as given in Section 22.2.*

axis df/dx at constant y is given by

$$f'(z_0) = \lim_{\Delta x \to 0} \frac{u(x_0 + \Delta x, y_0) - u(x_0, y_0)}{\Delta x}$$

$$+ \; j \lim_{\Delta x \to 0} \frac{v(x_0 + \Delta x, y_0) - v(x_0, y_0)}{\Delta x} \qquad (A.8)$$

where $\Delta z = \Delta x$. Equation (A.8) can be written

$$f'(z_0) = \frac{\partial u}{\partial x} + j\frac{\partial v}{\partial x} \qquad (A.9)$$

For the derivative along the imaginary axis at constant x, $\Delta z = j\Delta y$. Thus,

$$f'(z_0) = \lim_{\Delta y \to 0} \frac{u(x_0, y_0 + \Delta y) - u(x_0, y_0)}{j\Delta y}$$

$$+ \; j \lim_{\Delta y \to 0} \frac{v(x_0, y_0 + \Delta y) - v(x_0, y_0)}{j\Delta y} \qquad (A.10)$$

or

$$f'(z_0) = -j\frac{\partial u}{\partial y} + \frac{\partial v}{\partial y} \qquad (A.11)$$

As the values obtained for $f'(z_0)$ should be independent of direction, equation (A.7) must be satisfied. The additional requirement is that partial derivatives be continuous.

A.3 Complex Integration

An understanding of the use of the Kramers–Kronig relations requires an understanding of integration in a complex plane. While the subject leads to the very powerful tool of conformal mapping, the discussion is limited here to that needed to understand the derivation of the Kramers–Kronig relations.

A.3.1 Cauchy's Theorem

Definition A.12 (Simply Connected Domain) *A domain is simply connected if every simple closed curve C in the domain encloses only points of the domain.*

The meaning of Definition A.12 is more readily understood with the aid of Figure A.4. A simply connected domain is shown in Figure A.4(a). The simply connected domain of Figure A.4(a) was converted into a doubly connected domain by introducing a hole in the domain (see Figure A.4(b)). The doubly connected domain of Figure A.4(b) can be converted into a simply connected domain by introducing a pair of cuts, as shown in Figure A.4(c).

Theorem A.2 (Cauchy's Theorem) *If $f(z)$ is analytic in a simply connected domain D, and if C is a simple closed contour that lies in D, then*

$$\int_C f(z)dz = 0 \qquad (A.12)$$

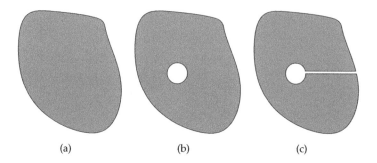

(a)   (b)   (c)

Figure A.4: The concept of connectedness: a) a simply connected domain; b) a doubly connected domain; and c) a doubly connected domain that has been converted into a simply connected domain by introducing a pair of cuts.

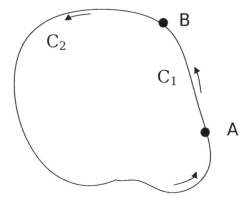

Figure A.5: Demonstration of the path independence for the value of the integral of an analytic function between two points in a simply connected domain.

The integral around the closed contour is also designated $\oint f(z)\mathrm{d}z$. A major consequence of Cauchy's theorem is that the value of the integral from one point to another is independent of the path. Two paths C_1 and C_2 between points A and B are shown in Figure A.5. The contour directions are the same; thus,

$$\int_C f(z)\mathrm{d}z = \int_{C_1} f(z)\mathrm{d}z + \int_{C_2} f(z)\mathrm{d}z \tag{A.13}$$

In order to have the direction of integration be from A to B for both contours, the direction of integration along contour C_2 is changed by integrating along $-C_2$. Thus,

Remember! **A.2** *The derivation of the Kramers–Kronig relations in Section 22.2 makes use of Cauchy's Theorem, given as Theorem A.2.*

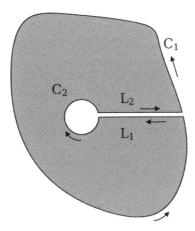

Figure A.6: Demonstration of the integration for a doubly connected domain that has been transformed into a simply connected domain.

from equation (A.12),

$$\int_{C_1} f(z)dz = \int_{-C_2} f(z)dz = -\int_{C_2} f(z)dz \qquad (A.14)$$

The value of the integral depends only on the end points of the integration.

As shown in Figure A.4, a multiply connected domain can be transformed into a singly connected domain by making suitable cuts. The integral around the closed contour of the new simply connected domain is expressed as the sum of contributions from the curves making up the boundary of the domain. An example is shown in Figure A.6. The integration path encompasses contours C_1, L_1, L_2, and C_2. Thus,

$$\int_C f(z)dz = \int_{C_1} f(z)dz + \int_{L_1} f(z)dz + \int_{C_2} f(z)dz + \int_{L_2} f(z)dz \qquad (A.15)$$

As the directions of integration along contours L_1 and L_2 are opposite, their contributions will cancel. Thus, following equation (A.12),

$$\int_{C_1} f(z)dz = -\int_{C_2} f(z)dz = \int_{-C_2} f(z)dz \qquad (A.16)$$

Example A.2 Application of Cauchy's Theorem: *Find the numerical value for the integral $\oint (z-a)^{-1}dz$ for the case where $z = a$ is outside the domain.*

Solution: *If the point $z = a$ lies outside the domain, the function $(z-a)^{-1}$ is analytic everywhere. Thus, via Cauchy's Theorem (Theorem A.2), $\oint (z-a)^{-1}dz = 0$.*

Example A.3 Special Case of Cauchy's Integral Formula: *Find the numerical value for the integral $\oint (z-a)^{-1} dz$ for the case where $z = a$ is inside the domain.*

Solution: *If the point $z = a$ lies inside the domain, the function $(z-a)^{-1}$ has a singularity at $z = a$. It can be enclosed by a circle of radius ε centered at $z = a$ such that, on the boundary of the domain, the function is analytic (see Figure A.6). From equation (A.16),*

$$\int_{C_1} \frac{1}{z-a} dz = \int_{-C_2} \frac{1}{z-a} dz \tag{A.17}$$

To evaluate the integral around the singularity, let

$$z = a + \varepsilon e^{j\theta}; \quad 0 \le \theta \le 2\pi \tag{A.18}$$

and

$$dz = j\varepsilon e^{j\theta} d\theta \tag{A.19}$$

The integral around C_2 becomes

$$\int_{-C_2} \frac{1}{z-a} dz = \int_0^{2\pi} \frac{j\varepsilon e^{j\theta}}{\varepsilon e^{j\theta}} d\theta = j \int_0^{2\pi} d\theta = 2j\pi \tag{A.20}$$

Thus,

$$\int_{C_1} \frac{1}{z-a} dz = 2j\pi \tag{A.21}$$

This result is a special case of Cauchy's integral formula.

Theorem A.3 (Cauchy's Integral Formula) *If $f(z)$ is analytic in a simply connected domain D, and if C is a simple positively oriented (counterclockwise) closed contour that lies in D, then, for any point z_0 that lies interior to C,*

$$\int_C \frac{f(z)}{z-z_0} dz = 2\pi j f(z_0) \tag{A.22}$$

Definition A.13 (Residue) *If z_0 is an isolated singular point of $f(z)$, then there exists a Laurent series*

$$f(z) = \sum_{n=-\infty}^{\infty} b_n (z-z_0)^n \tag{A.23}$$

Remember! A.3 *The derivation of the Kramers–Kronig relations in Section 22.2 makes use of Cauchy's integral formula for evaluation of a function at a singularity, given as Example A.3.*

---

valid for $0 < |z - z_0| < R$, for some positive value of R. The coefficient b_{-1} of $(z - z_0)^{-1}$ is the residue of $f(z)$ at z_0, designated as $\text{Res}\,[f(z), z_0]$. If C is a positively oriented simple closed curve in $0 < |z - z_0| < R$ that contains z_0, then

$$b_{-1} = \frac{1}{2\pi j} \int_C f(z)\mathrm{d}z \qquad (A.24)$$

Theorem A.4 (Cauchy's Residue Theorem) *If $f(z)$ is analytic in a simply connected domain D, except at a finite number of singular points $z_1, \ldots, z_k$ and if C is a simple positively oriented (counterclockwise) closed contour that lies in D, then*

$$\int_C f(z)\mathrm{d}z = 2\pi j \sum_{n=1}^{k} \text{Res}\,[f(z), z_n] \qquad (A.25)$$

A.3.2 Improper Integrals of Rational Functions

Definition A.14 (Cauchy Principal Value of Integral) *For real x, if $f(x)$ is a continuous function on $-\infty < x < +\infty$, the improper integral of $f(x)$ over $[-\infty, +\infty]$ is*

$$\int_{-\infty}^{\infty} f(x)\mathrm{d}x = \lim_{a \to -\infty} \int_a^0 f(x)\mathrm{d}x + \lim_{a \to \infty} \int_0^a f(x)\mathrm{d}x \qquad (A.26)$$

provided the limits exist. The Cauchy principal value of the integral over $[-\infty, +\infty]$ is given as

$$\fint_{-\infty}^{+\infty} f(x)\mathrm{d}x = \lim_{a \to \infty} \int_{-a}^{a} f(x)\mathrm{d}x \qquad (A.27)$$

Theorem A.5 (Evaluation of the Cauchy Principal Value) *If a function $f(z)$ is analytic in a simply connected domain D, except at a finite number of singular points $z_1, \ldots, z_k$, if $f(x) = P(x)/Q(x)$ where $P(x)$ and $Q(x)$ are polynomials, $Q(x)$ has no zeros, and the degree of $P(x)$ is at least two less than the degree of $Q(x)$, and if C is a simple positively oriented (counterclockwise) closed contour that lies in D, then*

$$\fint_{-\infty}^{+\infty} f(x)\mathrm{d}z = 2\pi j \sum_{n=1}^{k} \text{Res}\,[f(z), z_n] \qquad (A.28)$$

where $z_1, \ldots, z_k$ are poles of $f(z)$ that lie in the upper half-plane.

Example A.4 Poles on a Real Axis: *Evaluate the integral $\int_{-\infty}^{\infty} (z^2 - a^2)^{-1}\mathrm{d}x$ where a is a real number.*

Solution: *The function $\frac{1}{z^2 - a^2}$ has two poles on the real axis located at $x = a$ and $x = -a$, respectively. The integration will be performed on the contour shown in Figure A.7. The contour*

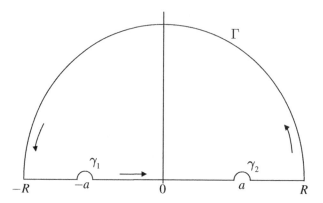

Figure A.7: Demonstration of integration for a domain with two poles on the real axis.

is chosen to avoid the singular points a and −a, and integration in the imaginary plane is at a value of R sufficiently large that it encircles all poles in the domain. A semicircle of radius ε is placed at the poles. The integral over the entire contour can be expressed as

$$\int_C \frac{1}{z^2-a^2}dz = \int_{-R}^{-a-\varepsilon} \frac{1}{x^2-a^2}dx + \int_{\gamma_1} \frac{1}{z^2-a^2}dz + \int_{-a+\varepsilon}^{a-\varepsilon} \frac{1}{x^2-a^2}dx+$$

$$\int_{\gamma_2} \frac{1}{z^2-a^2}dz + \int_{-a-\varepsilon}^{R} \frac{1}{x^2-a^2}dx + \int_{\Gamma} \frac{1}{z^2-a^2}dz \qquad (A.29)$$

$$= 2\pi j \sum_{n=1}^{2} \text{Res}\left[\frac{1}{z^2-a^2}, z_n\right]$$

$$= 0$$

For sufficiently large R,

$$\int_{\Gamma} \frac{1}{z^2-a^2}dz \to 0 \qquad (A.30)$$

Thus, in the limit that R → ∞ and ε → 0, equation (A.29) becomes

$$\int_{-\infty}^{\infty} \frac{1}{x^2-a^2}dx = -\int_{\gamma_1} \frac{1}{z^2-a^2}dz + -\int_{\gamma_2} \frac{1}{z^2-a^2}dz \qquad (A.31)$$

Thus, the integral along the real axis can be evaluated by evaluating the integrals along the contours that skirt the poles.

To evaluate the integral at γ_1, let

$$z = -a + \varepsilon e^{j\theta} \qquad (A.32)$$

Then,

$$\int_{\gamma_1} \frac{1}{z^2-a^2}dz = \int_{\pi}^{0} \frac{1}{\varepsilon e^{j\theta}\left(\varepsilon e^{j\theta}-2a\right)} j\varepsilon e^{j\theta} d\theta \qquad (A.33)$$

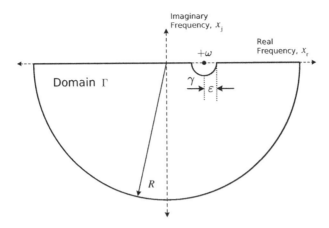

Figure A.8: Domain of integration for development application of Cauchy's integral formula. A pole is placed at frequency $+\omega$ on the real frequency axis.

As $\varepsilon \to 0$,

$$j \int_\pi^0 \frac{1}{\left(\varepsilon e^{j\theta} - 2a\right)} d\theta \to \frac{j\pi}{2a} \tag{A.34}$$

To evaluate the integral at γ_2, let

$$z = a + \varepsilon e^{j\theta} \tag{A.35}$$

Thus,

$$\int_{\gamma_2} \frac{1}{z^2 - a^2} dz = j \int_\pi^0 \frac{1}{\left(\varepsilon e^{j\theta} + 2a\right)} d\theta \to -\frac{j\pi}{2a} \tag{A.36}$$

Equation (A.31) can be expressed as

$$\int_{-\infty}^\infty \frac{1}{x^2 - a^2} dx = -\frac{j\pi}{2a} + \frac{j\pi}{2a}$$
$$= 0 \tag{A.37}$$

Problems

A.1   Show that e^z is an analytic function in any domain.

A.2   Show that the exponential of the complex conjugate $e^{\bar{z}}$ is not an analytic function in any domain.

A.3   Beginning with

$$\oint_\Gamma \frac{G(x)}{x - \omega} dx = 0 \tag{A.38}$$

derive equations (22.1) and (22.2). The domain Γ is shown in terms of complex frequency x in Figure A.8.

Appendix B

Tables of Reference Material

The following tables provide equations, constants, physical properties, and statistical values that may be useful for the student.

Complex Variables

Table 1.1   Identities for the imaginary number j.   5

Table 1.2   Relationships for complex variables $z = z_r + jz_j$ and $w = w_r + jw_j$.   7

Table 1.3   Relationships between polar and rectangular coordinates for the complex variable $z = z_r + jz_j$.   10

Table 1.4   Properties for the complex conjugates of $z = z_r + jz_j$ and $w = w_r + jw_j$.   15

Table 1.5   Properties for the absolute value of $z = z_r + jz_j$ and $w = w_r + jw_j$.   16

Table 1.6   Trigonometric and hyperbolic relationships for the complex variable $z = z_r + jz_j$.   18

Table 1.7   Functional relationships of complex variables commonly encountered in impedance spectroscopy, where x and y are real numbers, and $z = x + jy$.   20

Statistics

Table 3.1   Properties of the expectation where c is a constant and x and y are random variates.   46

Table 3.2   Properties of the variance where c is a constant and x and y are random variates.   47

Table 3.5   Student's t-test values.   63

Table 3.6 F-test values for comparison of variance of samples with equal degrees of freedom, i.e., $\nu_1 = \nu_2 = \nu$.   66

Table 3.10 χ^2-test values for degree of freedom ν and confidence level p.   72

Physical Properties

Table 5.4   Typical values of diffusion coefficients for ions at infinite dilution in 113
              water at 25 °C.

Table 5.6   Typical ranges of values for capacitance.   126

Table 12.1   Values for resistivity and dielectric constant at 293 K for solids.   304

Table 12.2   Physical properties for GaAs at 300 K.   314

Selected Electrochemical Notation

Table 4.1   Symbols, units, and relationships for quantities used in electrical cir- 77
              cuits.

Table 5.3   Definitions and notation for potentials used in electrochemical sys-   109
              tems.

Table 7.2   Definitions and notation for local impedance variables.   159

Relationships for Impedance

Table 10.1   Some useful relationships for the development of the impedance re-   212
              sponse associated with faradaic reactions.

Table 13.4   Characteristic dimensions and frequencies associated with frequency   370
              dispersion caused by nonuniform ohmic resistance or capacitance.

Table 14.1   Approaches used to interpret CPE impedance response associated   411
              with normal or surface distributions of time constants.

Table 17.1   Summary of complex impedance, admittance, and capacitance char-   468
              acteristics for simple blocking and reactive circuits.

Table 21.1   Statistical properties of impedance values obtained by Fourier analy- 575
              sis techniques.

Table 22.1   Compilation of various forms of the Kramers–Kronig relations.   596

Appendix C

List of Examples

The following examples are presented throughout the text to illustrate concepts, calculations, and derivations.

A Brief Introduction to Impedance Spectroscopy

Example 0.1	Characteristic Frequencies	xxxiv
Example 0.2	Impedance and Ohm's Law	xxxvii

Complex Variables

Example 1.1	Multiplication of Complex Numbers	6
Example 1.2	Division of Complex Numbers	7
Example 1.3	Rectangular Coordinates	9
Example 1.4	Polar Coordinates	10
Example 1.5	Square Roots of Complex Variables	11
Example 1.6	Square Roots of Complex Variables 2	13
Example 1.7	Blocking Constant-Phase Element	13
Example 1.8	Reactive Constant-Phase Element	14
Example 1.9	Exponential Form	20
Example 1.10	Verification of Expression for Impedance	21

Differential Equations

Example 2.1	Linear First-Order Differential Equation	26
Example 2.2	Convective Diffusion Equation	27
Example 2.3	Complex Roots for an ODE	30
Example 2.4	Diffusion in a Finite Domain	30

Example 2.5   Transmission Line   31
Example 2.6   Accurate Solution for the Convective Diffusion Equation   33
Example 2.7   Diffusion with Changing Film Thickness   37
Example 2.8   Convective Diffusion with Changing Film Thickness   37
Example 2.9   Diffusion in an Infinite Domain   39
Example 2.10   Foundation for Warburg Impedance   42

Statistics

Example 3.1   Mean and Variance for Replicated Measurements   47
Example 3.2   Error Propagation   56
Example 3.3   Continuation of Example 3.2   57
Example 3.4   Continuation of Example 3.3   57
Example 3.5   Evaluation of Impedance Data   65
Example 3.6   Evaluation of Chi-Squared Statistics   71

Electrical Circuits

Example 4.1   Impedance in Series   80
Example 4.2   Impedance in Parallel   81
Example 4.3   Bode Representation of Elemental Circuits   82
Example 4.4   Impedance Expression for a Nested Circuit   83

Electrochemistry

Example 5.1   Relationship between Flux and Current Density   93
Example 5.2   Butler Volmer and Tafel Equations   96
Example 5.3   Corrosion   100
Example 5.4   Water Oxidation and Reduction   102
Example 5.5   Mass-Transfer-Limited Current Density   105
Example 5.6   Application of the Koutecky–Levich Equation   106
Example 5.7   Rate Expression for Copper Dissolution   111
Example 5.8   Rate Expression for the Hydrogen Reaction   111
Example 5.9   Primary and Secondary Current Distributions   117
Example 5.10   Cell Potential Contributions   120

Electrochemical Instrumentation

Example 6.1   Negative Feedback   131
Example 6.2   Current Follower   132
Example 6.3   Voltage Adder   133

Experimental Methods

Example 7.1   Derivation of Lissajous Ellipse   146
Example 7.2   Lissajous Analysis   148
Example 7.3   Fourier Analysis   155

Experimental Design

Example 8.1   Electrode Connections   167
Example 8.2   Guideline for Linearity   170
Example 8.3   Influence of Ohmic Resistance on Linearity   174
Example 8.4   Influence of Capacitance on Linearity   180

Equivalent Circuit Analogs

Example 9.1   Method to Identify Corrosion Model   196
Example 9.2   Coated Electrode   197
Example 9.3   Time-Dependent Ohmic Resistance   204

Kinetic Impedance

Example 10.1   Impedance Derivation without Electrical Circuits   211
Example 10.2   Evaluation of Kinetic Current   218
Example 10.3   Diffusion with First-Order Reaction   220
Example 10.4   Iron in Anaerobic Solutions   222
Example 10.5   Iron in Aerobic Solutions   224
Example 10.6   Corrosion of Magnesium   231
Example 10.7   Diffusion of Two Species   234
Example 10.8   Corrosion of Copper in Chloride Solutions   236

Diffusion Impedance

Example 11.1   Warburg Impedance for Stagnant Diffusion Layer   249
Example 11.2   Insensitivity of Warburg Impedance to Film Thickness   250
Example 11.3   Propagation of Concentration Fluctuations into Film   250
Example 11.4   Diffusion or Capacitance   254
Example 11.5   Electrode Covered by Porous Film   273
Example 11.6   Diffusion through a Film with $\epsilon = 1.0$ 275
Example 11.7   Continuation of Example 11.6   276
Example 11.8   Diffusion Impedances in Series   278
Example 11.9   Impedance with Homogeneous Reactions   282

Example 11.10   Gerischer Impedance   287
Example 11.11   Dissolution of Copper through a CuCl Salt Layer   291

Impedance of Materials

Example 12.1   Geometric Capacitance   308
Example 12.2   Mott–Schottky Plots   316

Time-Constant Dispersion

Example 13.1   Submarine Telegraph Line Model   322
Example 13.2   Corrosion of Cast Iron in Drinking Water   325
Example 13.3   Impedance of Duplex Coating   330
Example 13.4   Characteristic Frequency for Disk   344
Example 13.5   Characteristic Frequency for Roughness   358
Example 13.6   Distributions of Capacitance   368
Example 13.7   Flux on a Small Circular Electrode   372

Constant-Phase Elements

Example 14.1   Graphical Identification of CPE Behavior   396
Example 14.2   Determination of Resistivity Distribution   406
Example 14.3   Estimation of Film Thickness   411
Example 14.4   Error by Assuming $C = Q$ for a Film   414
Example 14.5   Error by Assuming $C = Q$ for a Surface   414

Electrohydrodynamic Impedance

Example 16.1   2-D and 3-D Blocking   460

Methods for Representing Impedance

Example 17.1   Analysis of Two Nyquist Plots   471
Example 17.2   Admittance of Dielectrics   482
Example 17.3   Complex Capacitance of the Young Model   485
Example 17.4   Complex Capacitance of the Power-Law Model   486
Example 17.5   Complex Capacitance of Dielectrics   487
Example 17.6   Evaluation of Double-Layer Capacitance   490

Graphical Methods

Example 18.1   Interpretation of Characteristic Frequency   498
Example 18.2   Plots for Mg Alloys   513

Complex Nonlinear Regression

Example 19.1   Nonlinear Models   530

Error Structure of Impedance Measurements

Example 21.1   Identification of Stochastic Errors   584
Example 21.2   Identification of High-Frequency Bias Errors   587
Example 21.3   Identification of Low-Frequency Bias Errors   589

Kramers–Kronig Relations

Example 22.1   Verification of Equations (22.21) and (22.22)   601
Example 22.2   Application of Equations (22.21) and (22.22)   602

Integrated Approach

Example 23.1   Model Identification   621
Example 23.2   Models to Propose Experiments   624

Complex Integration

Example A.1   Analytic Functions   633
Example A.2   Application of Cauchy's Theorem   637
Example A.3   Special Case of Cauchy's Integral Formula   638
Example A.4   Poles on a Real Axis   639

List of Symbols

Roman

$\mathcal{A}$ temperature-dependent coefficient used in the development for thermoelectrochemical impedance, see Section 15.3.1

A   constant used for the velocity expansion for a rotating disk, A = 0.92486353, see Problem 11.5

a   constant used for the velocity expansion for a rotating disk (see Section 11.3.1), a = 0.510232618867

A lumped parameter in expression for impedance response of a reaction dependent on potential and surface coverage, equation (10.135)

A_i activation energy for diffusion, see Section 15.3.1

A_{IC} Akaike information criterion, see equation (20.2)

a_{IJ} constant used for the velocity expansion for an impinging jet with value determined by experiment, see Section 11.4.1

A_{op} open-loop gain for an operational amplifier, see Section 6.1

A_{PI} Akaike performance index, see equation (20.1)

A_v gain of the operational amplifier, see Figure 21.2 and equation (21.4)

B   constant used for the velocity expansion for a rotating disk, B = 1.20221175, see Problem 11.5

b   constant used for the velocity expansion for a rotating disk (see Section 11.3.1), b = −0.615922014399,

B lumped parameter in expression for impedance response of a reaction dependent on potential and surface coverage, equation (10.135)

B magnetic field, T, see Section 15.1

b kinetic parameter defined in equation (5.16), V^{-1}

C   covariance matrix for regression analysis, see equation (19.16)

$\mathcal{C}$ complex capacitance, see equation (17.40), F or F/cm^2

C	capacitance, F/cm^2 or F (1F = 1C/V)
C_0	interfacial capacitance, F/cm^2
C_{dl}	double-layer capacitance, F/cm^2 or F (1F = 1C/V)
CE	counterelectrode, see Figure 6.6
C_{eff}	effective capacitance, see equation (17.48), F or F/cm^2
c_i	volumetric concentration of species i, mol/cm^3
C_Y	oxide capacitance associated with the Young impedance, equation (13.172), F/cm^2
d	distance between potential sensing electrodes in the sensor used for local impedance spectroscopy measurements, see equation (7.53), cm
d_{act}	average diameter of the active portion of a partially blocked electrode, see equation (16.51), cm
D_i	diffusion coefficient for species i, cm^2/s
E	electronic energy, eV
E_c	conduction-band energy, eV
e_{out}	noise in the output from the current follower, V, see Figure 21.2 and equation (21.8)
E_{out}	potential output from the current follower, V
E_v	valence-band energy, eV
F	statistic used for testing equality of variances, see Section 3.3.3
F	Faraday's constant, $96,487$ C/equiv
$\tilde{f}$	complex oscillating variable used in expression for radial velocity for a rotating disk electrode, see equation (16.10)
F	dimensionless radial component of velocity for laminar flow to a disk electrode, see equation (11.48)
f	frequency, $f = \omega/2\pi$, Hz
$F(\omega)$	general frequency-dependent input function, see equation (22.16)
$f(t)$	general time-dependent input function, see equation (22.13)
f_i	activity coefficient for species i, dimensionless
f_{N_k}	characteristic frequency for measurement of impedance at a frequency f_k using N_k cycles, see equation (21.17), Hz
f_{scan}	characteristic frequency for measurement comprising a range of perturbation frequencies, see equation (21.16), Hz

f_{set}	characteristic frequency for measurement of a series of spectra, see equation (21.14), Hz
f^*	normalized frequency, see equation (18.23)
$\widetilde{g}$	complex oscillating variable used in expression for angular velocity for a rotating disk electrode, see equation (16.11)
G	dimensionless angular component of velocity for laminar flow to a disk electrode, see equation (11.49)
g	dimensionless parameter used for the power-law model, see equation (14.32)
g	gravitational acceleration, cm/s^2
$G(\omega)$	general transfer function, see equation (22.15)
$G(W)$	generic transfer function relating input and output quantities for an electrochemical system, see equation (15.1)
$g(t, t')$	function showing a linear relationship between input and output functions, see equation (22.13)
H	hypothesis, see Section 3.3
$\widetilde{h}$	complex oscillating variable used in expression for axial velocity for a rotating disk electrode, see equation (16.12)
H	dimensionless axial component of velocity for laminar flow to a disk electrode, see equation (11.50)
H_i	generic transfer function relating the time derivative of state quantities for an electrochemical system to the state and input quantities, see equation (15.2)
F_k	generic transfer function relating the output quantities for an electrochemical system to the state and input quantities, see equation (15.3)
I	current, A
i	current density, mA/cm^2
i_0	exchange current density, see equation (5.15), mA/cm^2
i_0	output current for an operational amplifier, see Figure 6.1(a), A
i_a	anodic current density, mA/cm^2
i_C	charging current density, mA/cm^2
i_c	cathodic current density, mA/cm^2
ΔI	amplitude of sinusoidal current signal, see, e.g., equation (4.7), A
i_F	faradaic current density, mA/cm^2

i_k kinetically controlled current density, $i_k = -k_c n F c_i(\infty) \exp(-b_c \eta_s)$, see equation (5.62), mA/cm^2

i_{lim} mass-transfer-limited current density, $i_{lim} = -n F D_i c_i(\infty)/\delta_i$, see equation (5.62), mA/cm^2

i_{n1} current noise contribution from differential amplifiers, A, see Figure 21.2 and equation (21.4)

i_{n2} current noise contribution from differential amplifiers, A, see Figure 21.2 and equation (21.4)

i_{n5} current noise contribution from differential amplifiers, A, see Figure 21.2 and equation (21.4)

i_{n6} current noise contribution from differential amplifiers, A, see Figure 21.2 and equation (21.4)

i_{probe} current density measured by a small local-impedance-spectroscopy sensor located near an electrode, see equation (7.53), mA/cm^2

iR_e ohmic potential drop between the solution adjacent to the working electrode and the location of a reference electrode, i.e., $iR_e = \Phi_0 - \Phi_{ref}$, see Table 5.3, V

i_{reg} parasitic current noise arising from noise in the voltage control, A, see Figure 21.2 and equation (21.5)

i_S power current for an operational amplifier, see Figure 6.1(a), A

$\mathbf{J}$ identity matrix

j imaginary number, $j = \sqrt{-1}$

J dimensionless exchange current density, see e.g., equation (5.114)

K dimensionless frequency associated with the geometry of a disk electrode, see equation (13.67)

K lumped-parameter rate constant for an electrochemical reaction that includes the equilibrium potential difference, see equation (10.16), mA/cm^2

K^* lumped-parameter rate constant for an electrochemical reaction, see equation (10.15), mA/cm^2

k rate constant for electrochemical reaction that excludes the exponential dependence on potential, see, e.g., equation (5.78)

K_i dimensionless frequency associated, for example, with convective diffusion of species i, see, e.g., equations (11.16) and (13.131)

k_M mass-transfer coefficient, cm/s

$K_{x,i}$ dimensionless position-dependent frequency associated, for example, with convective diffusion of species i, see equation (13.120)

ℓ	depth of a pore, cm
L	inductance, H ($1H = 1\text{Vs}^2/\text{C}$)
L	length of an electrode, cm
ℓ_c	characteristic dimension for geometry-induced frequency dispersion, see Section 13.4, cm
M	mass, g
M_i	symbol for the chemical formula of species i, see equation (5.5)
N	normal distribution function, see Section 3.1.3
N	collection efficiency for a ring–disk electrode, see equation (15.31)
n	electron concentration, cm^{-3}
n	number of electrons transferred in electrochemical reaction following the reaction stoichiometry given in equation (5.5), see equation (5.15), dimensionless
N_c	effective density of conduction-band states, cm^{-3}
N_{dat}	number of measured values, see equation (19.2)
N_i	flux of species i, $\text{mol}/\text{cm}^2\text{s}$
N_i	local mass-transfer flux, see Example 13.5, $\text{mol}/\text{cm}^2\text{s}$
N_k	number of cycles for an impedance measurement at frequency f_k, see equation (21.17)
N_p	number of parameters, see equation (19.8)
n_t	electron concentration in deep-level states, cm^{-3}
N_v	effective density of valence-band states, cm^{-3}
n_x	sample size used to calculate the standard error of sample distribution x, see equation (3.11)
P	parameter vector used in regression analysis, see equation (19.1)
P	probability, see Section 3.1.4
p	probability of observing the given sample result under the assumption that the null hypothesis is true, see Section 3.3.1
p	dimensionless frequency, $p = \omega/\Omega$
p	hole concentration, cm^{-3}
p	pressure, atm
$\mathbf{\Delta P}$	parameter vector for nonlinear regression analysis, see equation (19.19)
$p_{i,k}$	reaction order for anodic reactants i in reaction k, see equation (5.78)

Q	significance level at which the hypothesis that the variances are not equal can be rejected, see equation (3.49)
Q	CPE coefficient, see equation (14.1), $s^\alpha/\Omega cm^2$
q	charge, C/cm^2
$q_{i,k}$	reaction order for cathodic reactants i in reaction k, see equation (5.78)
R	universal gas constant, 8.3143 J/mol K
R	resistance, Ωcm^2 or Ω ($1\Omega = 1Vs/C$)
R	modulus, see equation (1.46)
r	radial coordinate, cm
r	rate of reaction, $mol/cm^2 s$
r_0	radius of a disk electrode, cm
Re	Reynolds number, $Re = \rho v 2R/\mu$, dimensionless
R_e	electrolyte or ohmic resistance, Ω or Ωcm^2
REF	reference electrode, see Figure 6.6
R_m	resistance of the current measurement circuit, Ω, see Figure 21.2 and equation (21.4)
R_s	resistance of the potential control circuit, Ω, see Figure 21.2 and equation (21.4)
R_t	charge-transfer resistance, Ωcm^2
S	weighted sum of squares, see equation (19.1)
s	sample standard deviation, see equation (3.9)
s	slope defined by equation (18.17)
s^2	sample variance, see equation (3.7)
Sc_i	Schmidt number, $Sc_i = \nu/D_i$, dimensionless
SE	standard error, see equation (3.11)
Sh	Sherwood number, $Sh = k_M d/D_i$, dimensionless
$s_{i,k}$	stoichiometric coefficient for species i in reaction k, see equation (5.5)
S_m	potential difference across the inputs to the current follower, V, see equation (21.6)
s_m	gain of the current follower, see Figure 21.2
s_m	noise in the potential difference across the inputs to the current follower, V, see Figure 21.2 and equation (21.6)

s_x standard error of sample distribution x, see equation (3.11)

t statistic used for testing equality of means, see Section 3.3.2

T temperature, K

T time corresponding to an integer number of cycles, s

t time, s

t_{cycle} period of a cycle at frequency f, s

t_i transfer function defined by equations (16.31) to (16.36)

U electrode potential with respect to a reference electrode, $U = \Phi_m - \Phi_{ref}$, see Table 5.3, V

u_i mobility of species i, related to diffusivity by equation (5.102)

V   variance-covariance matrix used in regression analysis, see equation (19.1)

V interfacial potential for the working electrode, see Table 5.3, V

V potential, V

v velocity, cm/s

V_+ potential at the positive input terminal for an operational amplifier, see Figure 6.1(a), V

V_- potential at the negative input terminal for an operational amplifier, see Figure 6.1(a), V

V_0 output potential for an operational amplifier, see Figure 6.1(a), V

$V_{0,k}$ interfacial potential under equilibrium conditions for reaction k, see Table 5.3, $V_{0,k} = (\Phi_m - \Phi_0)_{0,k}$, V

v_{CE} voltage noise contribution, V, see Figure 21.2 and equation (21.4)

ΔV amplitude of sinusoidal potential signal, see, e.g., equation (4.6), V

v_e voltage noise contribution, V, see Figure 21.2 and equation (21.4)

v_p voltage noise contribution, V, see Figure 21.2 and equation (21.4)

v_{ref} voltage noise contribution, V, see Figure 21.2 and equation (21.4)

v_{R_m} voltage noise contribution, V, see Figure 21.2 and equation (21.4)

V_S potential of the power leads for an operational amplifier, see Figure 6.1(a), V

W width of an electrode, cm

WE   working electrode, see Figure 6.6

W_i transfer function associated with electrohydrodynamic impedance of species i, see equation (16.29)

W_k generic input quantities for an electrochemical system, see Figure 15.1

$X(\omega)$ general frequency-dependent output function, see equation (22.17)

$x(t)$ general time-dependent output function, see equation (22.13)

X_k generic state quantities for an electrochemical system, see Figure 15.1

Y admittance, $Y = 1/Z, 1/\Omega cm^2$

y normal coordinate, cm

Y_k generic output quantities for an electrochemical system, see Figure 15.1

Z global impedance, see Table 7.2, Ωcm^2

z local impedance, see Table 7.2, Ωcm^2

Z_0 global interfacial impedance, see Table 7.2, Ωcm^2

z_0 local interfacial impedance, see Table 7.2, Ωcm^2

Z_c mass-transport transfer function defined in Section 16.2.1

Z_D diffusion impedance, Ωcm^2

Z_e global ohmic impedance, see Table 7.2, Ωcm^2

z_e local ohmic impedance, Ωcm^2

Z_{eq} interfacial impedance per unit pore area defined by equation (13.118)

Z_F faradaic impedance, Ωcm^2

z_i charge associated with species i

Z_k tabulated dimensionless values for diffusion impedance where k $=$ 1, 2, 3, see equation (11.86) for a rotating disk electrode and equation (11.98) for a submerged impinging jet

Z_Y Young impedance, equation (13.172), Ωcm^2

Greek

α CPE exponent, see equation (14.1), dimensionless

α constant used for the velocity expansion for a rotating disk, $\alpha = 0.88447411$, see Problem 11.5

α probability of incorrectly rejecting the null hypothesis when it is actually true, see Section 3.3.1

α symmetry factor used in electrode kinetics, see equation (5.15), dimensionless

$\boldsymbol{\alpha}$ tensor corresponding to the second derivative of the objective function with respect to parameter, see equation (19.21)

β Tafel slope, see, e.g., equation (5.19), V/decade of current

β	vector corresponding to the first derivative of the objective function with respect to parameter, see equation (19.20)
β_y	local wall velocity gradient, see equation (13.108)
Γ	maximum surface coverage, mol/cm^2
γ	exponent in power-law model, see equation (14.26)
γ	fractional surface coverage
δ	thickness, cm
ϵ	porosity
ε	dielectric constant
ε_{bias}	bias error, see equation (21.1), Ω
ε_{drift}	bias error associated with nonstationary behavior, see equation (21.24), Ω
ε_0	permittivity of vacuum, 8.8542×10^{-14} F/cm, see Section 5.7
ε_{fit}	error associated with model inadequacy, see equation (21.1), Ω
ε_{inst}	bias error associated with instrumental artifacts, see equation (21.24), Ω
ε_{stoch}	stochastic error, see equation (21.1), Ω
ζ	dimensionless position, $\zeta = y\sqrt{\Omega/\nu}$
η_c	concentration overpotential defined by equation (5.125), see Table 5.3, V
η	dimensionless axial position, $\eta = y\sqrt{a/\nu}$
η_s	surface overpotential for reaction k, $\eta_s = V - V_{0,k}$, see Table 5.3, V
Θ_i	dimensionless concentration, see equation (11.22)
θ_i	dimensionless oscillating part of the concentration, $\theta_i(y) = \tilde{c}_i/\tilde{c}_i(0)$
κ	conductivity, see, e.g., equation (5.109), S/cm
λ	Debye length, see equation (5.133), cm
λ	constant used for analysis of impedance at low frequency, see equation (18.18)
λ	constant used for regression analysis in the method of steepest descent, equation (19.31), and the Levenberg–Marquardt methods, equation (19.32)
μ	fluid viscosity, g/cm s
μ_i	electrochemical potential of species i, J/mol
μ_x	mean value of sample distribution x defined by equation (3.1)
ν	degree of freedom
ν	kinematic viscosity, $\nu = \mu/\rho$, cm^2/s

ν	wave number, see Section 15.3.2
ζ	dimensionless position
π	ratio of a circle's circumference to its diameter, $\pi = 3.14159265359$
ρ	fluid density, g/cm^3
ρ	resistivity, Ωcm
ρ_0	resistivity at position $y = 0$, Ωcm, see equation (14.27)
ρ_δ	resistivity at position $y = \delta$, Ωcm, see equation (14.27)
ρ_{sc}	charge density for a semiconductor, see equation (12.18), C/cm^3
σ_x	standard deviation of sample distribution x, see equation (3.9)
σ_{x_1,x_2}	covariance of sample distributions x_1 and x_2, see equation (3.15)
σ_x^2	variance of sample distribution x, see equation (3.7)
τ	time constant, e.g., $\tau = RC$, s
τ_{RC}	characteristic RC time constant, s
τ_{rz}	shear stress, N/cm^2
τ_{scan}	time required to measure a complete spectrum, see equation (21.15), s
τ_{set}	time required to measure a series of spectra, see equation (21.13), s
Φ	potential, V
ϕ	dimensionless stream function for flow to a submerged impinging jet, equations (11.87) and (11.88)
φ	phase angle for impedance, see equation (4.32)
Φ_0	potential of the electrolyte adjacent to the working electrode with respect to an unspecified but common reference potential, see Table 5.3, V
Φ_{m}	electrode potential with respect to an unspecified but common reference potential, see Table 5.3, V
Φ_{ref}	potential of a reference electrode with respect to an unspecified but common reference potential, see Table 5.3, V
φ	phase lag, see, e.g., equation (4.7)
χ^2	sum of squares used in regression analysis, see Sections 3.3.4 and 19.2
ψ	dimensionless parameter defined by equation (11.115)
Ω	rotation speed, s^{-1}
ω	angular frequency, $\omega = 2\pi f$, s^{-1}
ω_{c}	characteristic angular frequency, $\omega_{\text{c}} = 1/RC$, s^{-1}

General Notation

$\text{Im}\{X\}$ imaginary part of X

$\text{Re}\{X\}$ real part of X

$\overline{X}$ steady-state or time-averaged part of $X(t)$, see equation (10.2)

$\overline{z}$ complex conjugate of a complex number z, $\overline{z} = z_r - jz_j$

$\langle X \rangle$ spatially averaged value of X

$\widetilde{X}$ oscillating part of $X(t)$, see equation (10.2)

$\text{E}\{X\}$ expectation of $X(t)$, see equation (3.1) and Table 3.1

$\widehat{Z}$ model value for Z, see equation (19.1)

X' first derivative of $X(t)$ with respect to position

θ pertaining to the angular direction

Subscripts

0   located at the inner limit of the diffuse double layer

adj   pertaining to terms corrected for the ohmic resistance

a   pertaining to anodic reactions

c   pertaining to cathodic reactions

CE   pertaining to the counterelectrode, see Figure 6.6

cell   pertaining to the electrochemical cell, see Section 8.1.1

d   associated with the diffuse region of charge, see Figure 5.18 (page 125)

dl   double layer

HF   high frequency

i   pertaining to chemical species i

ihp   located at the inner Helmholtz plane, see Figure 5.18 (page 125)

IJ   impinging jet, see Section 11.4

j   imaginary

ℓ pertaining to a porous layer, see, e.g., Figure 9.8

LF   low frequency

m   located at the electrode surface, see Figure 5.18 (page 125)

mod   model value

ob   observed value

ohp   located at the outer Helmholtz plane, see Figure 5.18 (page 125) r   pertaining to the radial direction

r   real

ref   pertaining to a reference electrode, see Figure 6.6

WE   pertaining to the working electrode, see Figure 6.6

y   pertaining to the axial direction

Z_j pertaining to the imaginary part of the impedance

Z_r pertaining to the real part of the impedance

References

1. D. D. Macdonald, *Transient Techniques in Electrochemistry* (New York: Plenum Press, 1977).

2. C. Gabrielli, *Identification of Electrochemical Processes by Frequency Response Analysis*, Solartran Instrumentation Group Monograph, The Solartran Electronic Group Ltd., Farnborough, England (1980).

3. J. R. Macdonald, editor, *Impedance Spectroscopy: Emphasizing Solid Materials and Systems* (New York: John Wiley & Sons, 1987).

4. E. Barsoukov and J. R. Macdonald, editors, *Impedance Spectroscopy: Theory, Experiment, and Applications*, 2nd edition (Hoboken: John Wiley & Sons, 2005).

5. J. G. Saxe, *The Poetical Works of John Godfrey Saxe* (Boston: Houghton, Mifflin, 1892).

6. J. O. M. Bockris, D. Drazic, and A. R. Despic, "The Electrode Kinetics of the Deposition and Dissolution of Iron," *Electrochimica Acta*, **4** (1961) 325–361.

7. I. Epelboin and M. Keddam, "Faradaic Impedances: Diffusion Impedance and Reaction Impedance," *Journal of the Electrochemical Society*, **117** (1970) 1052–1056.

8. I. Frateur, C. Deslouis, M. E. Orazem, and B. Tribollet, "Modeling of the Cast Iron/Drinking Water System by Electrochemical Impedance Spectroscopy," *Electrochimica Acta*, **44** (1999) 4345–4356.

9. O. E. Barcia, O. R. Mattos, and B. Tribollet, "Anodic Dissolution of Iron in Acid Sulfate under Mass Transport Control," *Journal of the Electrochemical Society*, **139** (1992) 446–453.

10. A. B. Geraldo, O. E. Barcia, O. R. Mattos, F. Huet, and B. Tribollet, "New Results Concerning the Oscillations Observed for the System Iron-Sulphuric Acid," *Electrochimica Acta*, **44** (1998) 455–465.

11. C. Gabrielli, J. J. Garcia-Jareño, and H. Perrot, "Charge Compensation Process in Polypyrrole Studied by AC Electrogravimetry," *Electrochimica Acta*, **46** (2001) 4095–4103.

12. C. Gabrielli, J. J. Garcia-Jareño, M. Keddam, H. Perrot, and F. Vicente, "AC-Electrogravimetry Study of Electroactive Thin Films. II. Application to Polypyrrole," *Journal of Physical Chemistry B*, **106** (2002) 3192–3201.

13. M. E. Orazem and B. Tribollet, *Electrochemical Impedance Spectroscopy* (Hoboken: John Wiley & Sons, 2008).

14. V. F. Lvovich, *Impedance Spectroscopy: Applications to Electrochemical and Dielectric Phenomena* (Hoboken: John Wiley & Sons, 2012).

15. A. Lasia, *Electrochemical Impedance Spectroscopy and Its Applications* (New York: Springer, 2014).

16. S. Grimnes and Ø. G. Martinsen, *Bioimpedance and Bioelectricity Basics*, 3rd edition (Amsterdam: Academic Press, 2015).

17. M. Itagaki, *Electrochemical Impedance Method* (Tokyo: Maruzen Publishing Co., Ltd., 2008).

18. M. Itagaki, *Electrochemical Impedance Method*, 2nd edition (Tokyo: Maruzen Publishing Co., Ltd., 2011).

19. O. Heaviside, *Electrical Papers*, volume 1 (New York: MacMillan, 1894).

20. O. Heaviside, *Electrical Papers*, volume 2 (New York: MacMillan, 1894).

21. W. Nernst, "Methode zur Bestimmung von Dielektrizitätskonstanten," *Zeitschrift für Elektrochemie*, **14** (1894) 622–663.

22. C. Wheatstone, "An Account of Several New Instruments and Processes for Determining the Constants of a Voltaic Circuit," *Philosophical Transactions of the Royal Society of London*, **133** (1843) 303–327.

23. C. Wheatstone, *The Scientific Papers of Sir Charles Wheatstone* (London: Taylor and Francis, 1879).

24. J. Hopkinson and E. Wilson, "On the Capacity and Residual Charge of Dielectrics as Affected by Temperature and Time," *Philosophical Transactions of the Royal Society of London. Series A.*, **189** (1897) 109–135.

25. J. Dewar and J. A. Fleming, "A Note on Some Further Determinations of the Dielectric Constants of Organic Bodies and Electrolytes at Very Low Temperatures," *Proceedings of the Royal Society of London*, **62** (1898) 250–266.

26. C. H. Ayres, "Measurement of the Internal Resistance of Galvanic Cells," *Physical Review (Series I)*, **14** (1902) 17–37.

27. A. Finkelstein, "Über Passives Eisen," *Zeitschrift für Physikalische Chemie*, **39** (1902) 91–110.

28. E. Warburg, "Über das Verhalten sogenannter unpolarisirbarer Elektroden gegen Wechselstrom," *Annalen der Physik und Chemie*, **67** (1899) 493–499.

29. E. Warburg, "Über die Polarisationscapacität des Platins," *Annalen Der Physik*, **6** (1901) 125–135.

30. A. Fick, "Über Diffusion," *Annalen Der Physik*, **170** (1855) 59–86.

31. F. Krüger, "Über Polarisationskapazität," *Zeitschrift für Physikalische Chemie*, **39** (1902) 91–110.

32. R. E. Remington, "The High-frequency Wheatstone Bridge As a Tool in Cytological Studies: With Some Observations on the Resistance and Capacity of the Cells of the Beet Root," *Protoplasma*, **5** (1928) 338–399.

33. H. Fricke and S. Morse, "The Electric Resistance and Capacity of Blood for Frequencies between 800 and 4 1/2 Million Cycles," *Journal of General Physiology*, **9** (1925) 153–167.

34. H. Fricke, "The Electric Capacity of Suspensions with Special Reference to Blood," *Journal of General Physiology*, **9** (1925) 137–152.

35. J. F. McClendon, R. Rufe, J. Barton, and F. Fetter, "Colloidal Properties of the Surface of the Living Cell: 2. Electric Conductivity and Capacity of Blood to Alternating Currents of Long Duration and Varying in Frequency from 260 to 2,000,000 Cycles per Second," *Journal of Biological Chemistry*, **69** (1926) 733–754.

36. K. S. Cole, "Electric Phase Angle of Cell Membranes," *Journal of General Physiology*, **15** (1932) 641–649.

37. E. Bozler and K. S. Cole, "Electric Impedance and Phase Angle of Muscle in Rigor," *Journal of Cellular and Comparative Physiology*, **6** (1935) 229–241.

38. K. S. Cole, "Electric Impedance of Suspensions of Spheres," *Journal of General Physiology*, **12** (1928) 29–36.

39. H. Fricke, "The Theory of Electrolytic Polarization," *Philosophical Magazine*, **14** (1932) 310–318.

40. K. S. Cole and R. H. Cole, "Dispersion and Absorption in Dielectrics 1: Alternating Current Characteristics," *Journal of Chemical Physics*, **9** (1941) 341–351.

41. A. Frumkin, "The Study of the Double Layer at the Metal–Solution Interface by Electrokinetic and Electrochemical Methods," *Transactions of the Faraday Society*, **33** (1940) 117–127.

42. D. C. Grahame, "The Electrical Double Layer and the Theory of Electrocapillarity," *Chemical Reviews*, **41** (1947) 441–501.

43. D. C. Grahame, "Die Elektrische Doppelschicht," *Zeitschrift für Elektrochemie*, **59** (1955) 773–778.

44. P. I. Dolin and B. V. Ershler, "Kinetics of Processes on the Platinum Electrode: I. The Kinetics of the Ionization of Hydrogen Adsorbed on a Platinum Electrode," *Acta Physicochimica Urss*, **13** (1940) 747–778.

45. J. E. B. Randles, "Kinetics of Rapid Electrode Reactions," *Discussions of the Faraday Society*, **1** (1947) 11–19.

46. I. Epelboin, "Etude des Phénomènes Electrolytiques à l'aide de Courants Alternatifs Faibles et de Fréquence Variable," *Comptes Rendus Hebdomadaires des Séances de l'Académie des Sciences*, **234** (1952) 950–952.

47. H. Gerischer and W. Mehl, "Zum Mechanismus der Kathodischen Wasserstof-fabscheidung an Quecksilber, Silber, und Kupfer," *Zeitschrift für Elektrochemie*, **59** (1955) 1049–1059.

48. A. Frumkin, "Adsorptionserscheinungen und Elektrochemishe Kinetik," *Zeitschrift für Elektrochemie*, **59** (1955) 807–822.

49. I. Epelboin and G. Loric, "Sur un Phénomène de Résonance Observé en Basse Fréquence au Cours des Electrolyses Accompagnées d'une Forte Surtension Anodique," *Journal de Physique et le Radium*, **21** (1960) 74–76.

50. R. de Levie, "Electrochemical Responses of Porous and Rough Electrodes," in *Advances in Electrochemistry and Electrochemical Engineering*, P. Delahay, editor, volume 6 (New York: Interscience, 1967) 329–397.

51. J. S. Newman, "Frequency Dispersion in Capacity Measurements at a Disk Electrode," *Journal of the Electrochemical Society*, **117** (1970) 198–203.

52. E. Levart and D. Schuhmann, "Sur la Détermination Générale de L'impédance de Concentration (Diffusion Convective et Réaction Chimique) pour une Electrode à Disque Tournant," *Journal of Electroanalytical Chemistry*, **53** (1974) 77–94.

53. R. D. Armstrong, R. E. Firman, and H. R. Thirsk, "AC Impedance of Complex Electrochemical Reactions," *Faraday Discussions*, **56** (1973) 244–263.

54. I. Epelboin, M. Keddam, and J. C. Lestrade, "Faradaic Impedances and Intermediates in Electrochemical Reactions," *Faraday Discussions*, **56** (1973) 264–275.

55. R. J. Sheppard, B. P. Jordan, and E. H. Grant, "Least Squares Analysis of Complex Data with Applications to Permittivity Measurements," *Journal of Physics D—Applied Physics*, **3** (1970) 1759–1764.

56. R. J. Sheppard, "Least-Squares Analysis of Complex Weighted Data with Dielectric Applications," *Journal of Physics D—Applied Physics*, **6** (1973) 790–794.

57. J. R. Macdonald and J. A. Garber, "Analysis of Impedance and Admittance Data for Solids and Liquids," *Journal of the Electrochemical Society*, **124** (1977) 1022–1030.

58. J. R. Macdonald, J. Schoonman, and A. P. Lehnen, "The Applicability and Power of Complex Nonlinear Least Squares for the Analysis of Impedance and Immittance Data," *Journal of Electroanalytical Chemistry*, **131** (1982) 77–95.

59. B. A. Boukamp, "A Nonlinear Least Squares Fit Procedure for Analysis of Immittance Data of Electrochemical Systems," *Solid State Ionics, Diffusion & Reactions*, **20** (1986) 31–44.

60. A. A. Aksut, W. J. Lorenz, and F. Mansfeld, "The Determination of Corrosion Rates by Electrochemical DC and AC Methods: II. Systems with Discontinuous Steady State Polarization Behavior," *Corrosion Science*, **22** (1982) 611–619.

61. F. Mansfeld, M. W. Kendig, and S. Tsai, "Corrosion Kinetics in Low Conductivity Media: I. Iron in Natural Waters," *Corrosion Science*, **22** (1982) 455–471.

62. W. H. Smyrl and L. L. Stephenson, "Digital Impedance for Faradaic Analysis: 3. Copper Corrosion in Oxygenated 0.1N HCl," *Journal of the Electrochemical Society*, **132** (1985) 1563–1567.

63. C. Deslouis, I. Epelboin, C. Gabrielli, and B. Tribollet, "Impédance Electromécanique Obtenue au Courant Limite de Diffusion à Partir d'une Modulation Sinusoidale de la Vitesse de Rotation d'une Electrode à Disque," *Journal of Electroanalytical Chemistry*, **82** (1977) 251–269.

64. C. Deslouis, I. Epelboin, C. Gabrielli, P. S.-R. Fanchine, and B. Tribollet, "Relationship between the Electrochemical Impedance and the Electrohydrodynamical Impedances Measured Using a Rotating Disc Electrode," *Journal of Electroanalytical Chemistry*, **107** (1980) 193–195.

65. C. Gabrielli and B. Tribollet, "A Transfer Function Approach for a Generalized Electrochemical Impedance Spectroscopy," *Journal of the Electrochemical Society*, **141** (1994) 1147–1157.

66. R. de L. Kronig, "On the Theory of Dispersion of X-Rays," *Journal of the Optical Society of America and Review of Scientific Instruments*, **12** (1926) 547–557.

67. H. A. Kramers, "Die Dispersion und Absorption von Röntgenstrahlen," *Physikalishce Zeitschrift*, **30** (1929) 522–523.

68. D. D. Macdonald and M. Urquidi-Macdonald, "Application of Kramers-Kronig Transforms in the Analysis of Electrochemical Systems: 1. Polarization Resistance," *Journal of the Electrochemical Society*, **132** (1985) 2316–2319.

69. P. Agarwal, M. E. Orazem, and L. H. García-Rubio, "Measurement Models for Electrochemical Impedance Spectroscopy: 1. Demonstration of Applicability," *Journal of the Electrochemical Society*, **139** (1992) 1917–1927.

70. F. Dion and A. Lasia, "The Use of Regularization Methods in the Deconvulation of Underlying Distributions in Electrochemical Processes," *Journal of Electroanalytical Chemistry*, **475** (1999) 28–37.

71. M. E. Orazem, P. K. Shukla, and M. A. Membrino, "Extension of the Measurement Model Approach for Deconvolution of Underlying Distributions for Impedance Measurements," *Electrochimica Acta*, **47** (2002) 2027–2034.

72. Z. Stoynov, "Differential Impedance Analysis — An Insight into the Experimental Data," *Polish Journal of Chemistry*, **71** (1997) 1204–1210.

73. D. Vladikova, Z. Stoynov, and L. Ilkov, "Differential Impedance Analysis on Single Crystal and Polycrystalline Yttrium Iron Garnets," *Polish Journal of Chemistry*, **71** (1997) 1196–1203.

74. Z. Stoynov and B. S. Savova-Stoynov, "Impedance Study of Non-Stationary Systems: Four-Dimensional Analysis," *Journal of Electroanalytical Chemistry and Interfacial Electrochemistry*, **183** (1985) 133–144.

75. C. Gabrielli, editor, *Proceedings of the First International Symposium on Electrochemical Impedance Spectroscopy*, volume 35:10 of *Electrochimica Acta* (1990).

76. D. D. MacDonald, editor, *Proceedings of the Second International Symposium on Electrochemical Impedance Spectroscopy*, volume 38:14 of *Electrochimica Acta* (1993).

77. J. Vereecken, editor, *Proceedings of the Third International Symposium on Electrochemical Impedance Spectroscopy*, volume 41:7-8 of *Electrochimica Acta* (1996).

78. O. R. Mattos, editor, *Proceedings of the Fourth International Symposium on Electrochemical Impedance Spectroscopy*, volume 44:24 of *Electrochimica Acta* (1999).

79. F. Deflorian and P. L. Bonora, editors, *Proceedings of the Fifth International Symposium on Electrochemical Impedance Spectroscopy*, volume 47:13-14 of *Electrochimica Acta* (2002).

80. M. E. Orazem, editor, *EIS-2004: Proceedings of the Sixth International Symposium on Electrochemical Impedance Spectroscopy*, volume 51:8-9 of *Electrochimica Acta* (2006).

81. N. Pébère, editor, *EIS-2007: Proceedings of the Seventh International Symposium on Electrochemical Impedance Spectroscopy*, volume 53 of *Electrochimica Acta* (2008).

82. J. S. Fernandes and F. Montemor, editors, *8th International Symposium on Electrochemical Impedance Spectroscopy (EIS 2010)*, volume 56 of *Electrochimica Acta* (2011).

83. M. Itagaki, editor, *9th International Symposium on Electrochemical Impedance Spectroscopy (EIS 2013)*, volume 131 of *Electrochimica Acta* (2014).

84. R. S. Lillard, P. J. Moran, and H. S. Isaacs, "A Novel Method for Generating Quantitative Local Electrochemical Impedance Spectroscopy," *Journal of the Electrochemical Society*, **139** (1992) 1007–1012.

85. D. D. Macdonald, "Reflections on the History of Electrochemical Impedance Spectroscopy," *Electrochimica Acta*, **51** (2006) 1376–1388.

86. M. Sluyters-Rehbach and J. H. Sluyters, "Sine Wave Methods in the Study of Electrode Processes," in *Electroanalytical Chemistry*, A. J. Bard, editor, volume 4 (New York: Marcel Dekker, 1970) 1–128.

87. A. Lasia, "Electrochemical Impedance Spectroscopy and Its Applications," in *Modern Aspects of Electrochemistry*, R. E. White, B. E. Conway, and J. O. M. Bockris, editors, volume 32 (New York: Plenum Press, 1999) 143–248.

88. R. Churchill and J. Brown, *Complex Variables and Applications*, 6th edition (New York: McGraw Hill, 1990).

89. G. Cain, *Complex Analysis* (Atlanta: Georgia Institute of Technology, 1999).

90. C. F. C. M. Fong, D. D. Kee, and P. N. Kalomi, *Advanced Mathematics for Applied and Pure Sciences* (Amsterdam: Gordon and Breach Science Publishers, 1997).

91. M. Sluyters-Rehbach, "Impedances of Electrochemical Systems: Terminology, Nomenclature, and Representation: I. Cells with Metal Electrodes and Liquid Solutions," *Pure and Applied Chemistry*, **66** (1994) 1831–1891.

92. R. Antaño-López, M. Keddam, and H. Takenouti, "A New Experimental Approach to the Time-Constants of Electrochemical Impedance: Frequency Response of the Double Layer Capacitance," *Electrochimica Acta*, **46** (2001) 3611–3617.

93. C. R. Wylie, *Advanced Engineering Mathematics*, 3rd edition (New York: McGraw-Hill, 1966).

94. M. R. Spiegel, *Applied Differential Equations* (Englewood Cliffs: Prentice Hall, 1967).

95. J. Stewart, *Calculus*, 6th edition (Pacific Grove: Brooks Cole, 2007).

96. G. W. Snedecor and W. G. Cochran, *Statistical Methods*, 6th edition (Ames: the Iowa State University Press, 1961).

97. G. E. P. Box, J. S. Hunter, and W. G. Hunter, *Statistics for Experimenters: An Introduction to Design, Data Analysis and Model Building*, 2nd edition (New York: John Wiley & Sons, 2005).

98. D. C. Montgomery, *Design and Analysis of Experiments*, 7th edition (Hoboken: John Wiley & Sons, 2009).

99. M. E. Orazem, M. Durbha, C. Deslouis, H. Takenouti, and B. Tribollet, "Influence of Surface Phenomena on the Impedance Response of a Rotating Disk Electrode," *Electrochimica Acta*, **44** (1999) 4403–4412.

100. P. Agarwal, O. D. Crisalle, M. E. Orazem, and L. H. García-Rubio, "Measurement Models for Electrochemical Impedance Spectroscopy: 2. Determination of the Stochastic Contribution to the Error Structure," *Journal of the Electrochemical Society*, **142** (1995) 4149–4158.

101. P. Agarwal, M. E. Orazem, and L. H. García-Rubio, "Measurement Models for Electrochemical Impedance Spectroscopy: 3. Evaluation of Consistency with the Kramers-Kronig Relations," *Journal of the Electrochemical Society*, **142** (1995) 4159–4168.

102. S. L. Carson, M. E. Orazem, O. D. Crisalle, and L. H. García-Rubio, "On the Error Structure of Impedance Measurements: Simulation of Frequency Response Analysis (FRA) Instrumentation," *Journal of the Electrochemical Society*, **150** (2003) E477–E490.

103. P. K. Shukla, M. E. Orazem, and O. D. Crisalle, "Validation of the Measurement Model Concept for Error Structure Identification," *Electrochimica Acta*, **49** (2004) 2881–2889.

104. C. M. Grinstead and J. L. Snell, *Introduction to Probability* (Providence: American Mathematical Society, 1999).

105. M. Abramowitz and I. A. Stegun, *Handbook of Mathematical Functions* (New York: Dover Publications, 1972).

106. G. E. P. Box and N. R. Draper, *Empirical Model-Building and Response Surfaces* (New York: John Wiley & Sons, 1987).

107. C. Deslouis, O. Gil, B. Tribollet, G. Vlachos, and B. Robertson, "Oxygen as a Tracer for Measurements of Steady and Turbulent Flows," *Journal of Applied Electrochemistry*, **22** (1992) 835–842.

108. Student, "The Probable Error of a Mean," *Biometrika*, **6** (1908) 1–25.

109. S. Erol and M. E. Orazem, "The Influence of Anomalous Diffusion on the Impedance Response of $LiCoO_2$—C Batteries," *Journal of Power Sources*, **293** (2015) 57–64.

110. H. W. Bode, *Network Analysis and Feedback Amplifier Design* (New York: D. Van Nostrand, 1945).

111. E. Gileadi, *Electrode Kinetics for Chemists, Chemical Engineers, and Materials Scientists* (New York: VCH Publishers, 1993).

112. Š. K. Lovrić, "Working Electrodes," in *Electroanalytical Methods: Guide to Experiments and Applications*, F. Scholz, editor (New York: Springer, 2010) 273–290.

113. S. R. Waldvogel, A. Kirste, and S. Mentizi, "Use of Diamond Films in Organic Electrosynthesis," in *Synthetic Diamond Films: Preparation, Electrochemistry, Characterization and Applications*, E. Brillas and C. Huitle, editors, Wiley Series on Electrocatalysis and Electrochemistry (John Wiley & Sons, 2011) 483–510.

114. M. Panizza and G. Cerisola, "Application of Diamond Electrodes to Electrochemical Processes," *Electrochimica Acta*, **51** (2005) 191–199.

115. J. S. Newman, *Electrochemical Systems*, 2nd edition (Englewood Cliffs: Prentice Hall, 1991).

116. J. S. Newman and K. E. Thomas-Alyea, *Electrochemical Systems*, 3rd edition (Hoboken: John Wiley & Sons, 2004).

117. E. McCafferty, *Introduction to Corrosion Science* (New York: Springer-Verlag, 2010).

118. D. A. Jones, *Principles and Prevention of Corrosion* (Upper Saddle River: Prentice Hall, 1996).

119. B. E. Conway, *Electrochemical Data* (Amsterdam: Elsevier, 1952).

120. U. R. Evans, *The Corrosion of Metals* (London: Edward Arnold, 1924).

121. A. J. Bard and L. R. Faulkner, *Electrochemical Methods: Fundamentals and Applications* (New York: John Wiley & Sons, 1980).

122. V. Jovancicevic and J. O. M. Bockris, "The Mechanism of Oxygen Reduction on Iron in Neutral Solutions," *Journal of the Electrochemical Society*, **133** (1986) 1797–1807.

123. J. S. Newman, "Current Distribution on a Rotating Disk below the Limiting Current," *Journal of the Electrochemical Society*, **113** (1966) 1235–1241.

124. J. S. Newman, "Resistance for Flow of Current to a Disk," *Journal of the Electrochemical Society*, **113** (1966) 501–502.

125. V. M.-W. Huang, V. Vivier, M. E. Orazem, N. Pébère, and B. Tribollet, "The Apparent CPE Behavior of a Disk Electrode with Faradaic Reactions," *Journal of the Electrochemical Society*, **154** (2007) C99–C107.

126. C. Wagner, "Theoretical Analysis of the Current Density Distribution in Electrolytic Cells," *Journal of the Electrochemical Society*, **98** (1951) 116–128.

127. C. B. Diem, B. Newman, and M. E. Orazem, "The Influence of Small Machining Errors on the Primary Current Distribution at a Recessed Electrode," *Journal of the Electrochemical Society*, **135** (1988) 2524–2530.

128. I. Frateur, V. M.-W. Huang, M. E. Orazem, N. Pébère, B. Tribollet, and V. Vivier, "Local Electrochemical Impedance Spectroscopy: Considerations about the Cell Geometry," *Electrochimica Acta*, **53** (2008) 7386–7395.

129. D.-T. Chin, "Convective Diffusion on a Rotating Spherical Electrode," *Journal of the Electrochemical Society*, **118** (1971) 1434–1438.

130. K. Nisancioglu and J. S. Newman, "Current Distribution on a Rotating Sphere below the Limiting Current," *Journal of the Electrochemical Society*, **121** (1974) 241–246.

131. O. E. Barcia, J. S. Godinez, L. R. S. Lamego, O. R. Mattos, and B. Tribollet, "Rotating Hemispherical Electrode: Accurate Expressions for the Limiting Current and the Convective Warburg Impedance," *Journal of the Electrochemical Society*, **145** (1998) 4189–4195.

132. P. K. Shukla and M. E. Orazem, "Hydrodynamics and Mass-Transfer-Limited Current Distribution for a Submerged Stationary Hemispherical Electrode under Jet Impingement," *Electrochimica Acta*, **49** (2004) 2901–2908.

133. P. K. Shukla, M. E. Orazem, and G. Nelissen, "Impedance Analysis for Reduction of Ferricyanide on a Submerged Hemispherical Ni270 Electrode," *Electrochimica Acta*, **51** (2006) 1514–1523.

134. M. E. Orazem, B. Tribollet, V. Vivier, S. Marcelin, N. Pébère, A. L. Bunge, E. A. White, D. P. Riemer, I. Frateur, and M. Musiani, "Dielectric Properties of Materials Showing Constant-Phase-Element (CPE) Impedance Response," *Journal of the Electrochemical Society*, **160** (2013) C215–C225.

135. B. Hirschorn, M. E. Orazem, B. Tribollet, V. Vivier, I. Frateur, and M. Musiani, "Determination of Effective Capacitance and Film Thickness from CPE Parameters," *Electrochimica Acta*, **55** (2010) 6218–6227.

136. E. A. White, M. E. Orazem, and A. L. Bunge, "Characterization of Damaged Skin by Impedance Spectroscopy: Mechanical Damage," *Pharmaceutical Research*, **30** (2013) 2036–2049.

137. A. S. Nguyen, M. Musiani, M. E. Orazem, N. Pébère, B. Tribollet, and V. Vivier, "Impedance Analysis of the Distributed Resistivity of Coatings in Dry and Wet Conditions," *Electrochimica Acta*, **179** (2015) 452–459.

138. A. C. West, *Electrochemistry and Electrochemical Engineering. An Introduction* (Charleston: CreateSpace Independent Publishing Platform, 2012).

139. G. Prentice, *Electrochemical Engineering Principles* (Englewood Cliffs: Prentice Hall, 1991).

140. J. O. M. Bockris and A. K. N. Reddy, *Modern Electrochemistry: Ionics*, volume 1, 2nd edition (New York: Plenum Press, 1998).

141. J. O. M. Bockris and A. K. N. Reddy, *Modern Electrochemistry: Electrodics*, volume 2, 2nd edition (New York: Plenum Press, 2000).

142. K. B. Oldham, J. C. Myland, and A. M. Bond, *Electrochemical Science and Technology* (Hoboken: John Wiley & Sons, 2012).

143. C. M. A. Brett and A. M. O. Brett, *Electrochemistry: Principles, Methods, and Applications* (Cary: Oxford University Press, 1993).

144. T. F. Fuller and J. N. Harb, *Electrochemical Engineering* (Hoboken: John Wiley & Sons, 2017).

145. S. L. Carson, M. E. Orazem, O. D. Crisalle, and L. H. García-Rubio, "On the Error Structure of Impedance Measurements: Series Expansions," *Journal of the Electrochemical Society*, **150** (2003) E501–E511.

146. M. E. Van Valkenburg, *Network Analysis*, 3rd edition (Englewood Cliffs: Prentice-Hall, 1974).

147. V. M.-W. Huang, V. Vivier, M. E. Orazem, N. Pébère, and B. Tribollet, "The Apparent CPE Behavior of an Ideally Polarized Blocking Electrode: A Global and Local Impedance Analysis," *Journal of the Electrochemical Society*, **154** (2007) C81–C88.

148. F. Zou, D. Thierry, and H. S. Isaacs, "High-Resolution Probe for Localized Electrochemical Impedance Spectroscopy Measurements," *Journal of the Electrochemical Society*, **144** (1997) 1957–1965.

149. G. J. Brug, A. L. G. van den Eeden, M. Sluyters-Rehbach, and J. H. Sluyters, "The Analysis of Electrode Impedances Complicated by the Presence of a Constant Phase Element," *Journal of Electroanalytical Chemistry*, **176** (1984) 275–295.

150. M. Eisenberg, C. W. Tobias, and C. R. Wilke, "Ionic Mass Transfer and Concentration Polarization at Rotating Electrodes," *Journal of the Electrochemical Society*, **101** (1954) 306–319.

151. P. K. Shukla, *Stationary Hemispherical Electrode under Submerged Jet Impingement and Validation of the Measurement Model Concept for Impedance Spectroscopy*, Ph.D. dissertation, University of Florida, Gainesville, FL (2004).

152. J. Diard, B. LeGorrec, and C. Montella, "Deviation from the Polarization Resistance Due to Non-Linearity: 1. Theoretical Formulation," *Journal of Electroanalytical Chemistry*, **432** (1997) 27–39.

153. J. Diard, B. L. Gorrec, and C. Montella, "Deviation from the Polarization Resistance Due to Non-Linearity: 2. Application to Electrochemical Reactions," *Journal of Electroanalytical Chemistry*, **432** (1997) 41–52.

154. J. Diard, B. LeGorrec, and C. Montella, "Deviation from the Polarization Resistance Due to Non-Linearity: 3. Polarization Resistance Determination from Non-Linear Impedance Measurements," *Journal of Electroanalytical Chemistry*, **432** (1997) 53–62.

155. P. T. Wojcik, P. Agarwal, and M. E. Orazem, "A Method for Maintaining a Constant Potential Variation during Galvanostatic Regulation of Electrochemical Impedance Measurements," *Electrochimica Acta*, **41** (1996) 977–983.

156. P. T. Wojcik and M. E. Orazem, "Variable-Amplitude Galvanostatically Modulated Impedance Spectroscopy as a Tool for Assessing Reactivity at the Corrosion Potential without Distorting the Temporal Evolution of the System," *Corrosion*, **54** (1998) 289–298.

157. R. Pollard and T. Comte, "Determination of Transport Properties for Solid Electrolytes from the Impedance of Thin Layer Cells," *Journal of the Electrochemical Society*, **136** (1989) 3734–3748.

158. G. Baril, G. Galicia, C. Deslouis, N. Pèbére, B. Tribollet, and V. Vivier, "An Impedance Investigation of the Mechanism of Pure Magnesium Corrosion in Sodium Sulfate Solutions," *Journal of the Electrochemical Society*, **154** (2007) C108–C113.

159. O. Devos, C. Gabrielli, and B. Tribollet, "Nucleation-Growth Process of Scale Electrodeposition: Influence of the Mass Transport," *Electrochimica Acta*, **52** (2006) 285–291.

160. O. Devos, C. Gabrielli, and B. Tribollet, "Simultaneous EIS and In Situ Microscope Observation on a Partially Blocked Electrode: Application to Scale Electrodeposition," *Electrochimica Acta*, **51** (2006) 1413–1422.

161. L. Beaunier, I. Epelboin, J. C. Lestrade, and H. Takenouti, "Etude Electrochimique, et par Microscopie Electronique à Balayage, du Fer Recouvert de Peinture," *Surface Technology*, **4** (1976) 237–254.

162. L. Bousselmi, C. Fiaud, B. Tribollet, and E. Triki, "Impedance Spectroscopic Study of a Steel Electrode in Condition of Scaling and Corrosion: Interphase Model," *Electrochimica Acta*, **44** (1999) 4357–4363.

163. M. T. T. Tran, B. Tribollet, V. Vivier, and M. E. Orazem, "On the Impedance Response of Reactions Influenced by Mass Transfer," *Russian Journal of Electrochemistry*, (2017) in press.

164. M. Stern and A. L. Geary, "Electrochemical Polarization: I. A Theoretical Analysis of the Shape of Polarization Curves," *Journal of the Electrochemical Society*, **104** (1957) 56–63.

165. L. Péter, J. Arai, and H. Akahoshi, "Impedance of a Reaction Involving Two Adsorbed Intermediates: Aluminum Dissolution in Non-Aqueous Lithium Imide Solutions," *Journal of Electroanalytical Chemistry*, **482** (2000) 125 – 138.

166. O. Devos, O. Aaboubi, J. P. Chopart, E. Merienne, A. Olivier, C. Gabrielli, and B. Tribollet, "EIS Investigation of Zinc Electrodeposition in Basic Media at Low Mass Transfer Rates Induced by a Magnetic Field," *Journal of Physical Chemistry B*, **103** (1999) 496–501.

167. R. S. Cooper and J. H. Bartlett, "Convection and Film Instability: Copper Anodes in Hydrochloric Acid," *Journal of the Electrochemical Society*, **105** (1958) 109–116.

168. A. L. Bacarella and J. C. Griess, Jr., "The Anodic Dissolution of Copper in Flowing Sodium Chloride Solutions between 25° and 175°C," *Journal of the Electrochemical Society*, **120** (1973) 459–465.

169. A. Moreau, "Etude du Mecanisme d'Oxydo-Reduction du Cuivre dans les Solutions Chlorurees Acides: II. Systemes Cu; CuCl; $CuCl^{-2}$ et Cu; $Cu_2(OH)_3$; Cl^-; CuCl; Cu^{2+}," *Electrochimica Acta*, **26** (1981) 1609–1616.

170. H. P. Lee and K. Nobe, "Kinetics and Mechanisms of Cu Electrodissolution in Chloride Media," *Journal of the Electrochecmical Society*, **133** (1986) 2035–2043.

171. C. Deslouis, B. Tribollet, G. Mengoli, and M. M. Musiani, "Electrochemical Behaviour of Copper in Neutral Aerated Chloride Solution. I. Steady-State Investigation," *Journal of Applied Electrochemistry*, **18** (1988) 374–383.

172. C. Deslouis, B. Tribollet, G. Mengoli, and M. M. Musiani, "Electrochemical Behaviour of Copper in Neutral Aerated Chloride Solution: II. Impedance Investigation," *Journal of Applied Electrochemistry*, **18** (1988) 384–393.

173. A. K. Hauser and J. S. Newman, "Singular Perturbation Analysis of the Faradaic Impedance of Copper Dissolution Accounting for the Effects of Finite Rates of a Homogeneous Reaction," *Journal of the Electrochecmical Society*, **136** (1989) 2820–2831.

174. O. E. Barcia, O. R. Mattos, N. Pébère, and B. Tribollet, "Mass-Transport Study for the Electrodissolution of Copper in 1M Hydrochloric Acid Solution by Impedance," *Journal of the Electrochemical Society*, **140** (1993) 2825–2832.

175. B. Tribollet, J. S. Newman, and W. H. Smyrl, "Determination of the Diffusion Coefficient from Impedance Data in the Low Frequency Range," *Journal of the Electrochemical Society*, **135** (1988) 134–138.

176. V. G. Levich, *Physicochemical Hydrodynamics* (Englewood Cliffs: Prentice Hall, 1962).

177. R. B. Bird, W. E. Stewart, and E. N. Lightfoot, *Transport Phenomena* (New York: John Wiley & Sons, 1960).

178. C. Ho, I. D. Raistrick, and R. A. Huggins, "Application of A-C Techniques to the Study of Lithium Diffusion in Tungsten Trioxide Thin Films," *Journal of the Electrochemical Society*, **127** (1980) 343–350.

179. C. Gabrielli, O. Haas, and H. Takenouti, "Impedance Analysis of Electrodes Modified with a Reversible Redox Polymer Film," *Journal of Applied Electrochemistry*, **17** (1987) 82–90.

180. T. von Kármán, "Über Laminaire und Turbulente Reibung," *Zeitschrift für angewandte Mathematik und Mechanik*, **1** (1921) 233–252.

181. W. G. Cochran, "The Flow Due to a Rotating Disc," *Proceedings of the Cambridge Philosophical Society*, **30** (1934) 365–375.

182. E. Levart and D. Schuhmann, "Analyse du Transport Transitoire sur un Disque Tournant en Regime Hydrodynamique Laminaire et Permanent," *International Journal of Heat and Mass Transfer*, **17** (1974) 555–566.

183. J. S. Newman, "Schmidt Number Correction for the Rotating Disk," *Journal of Physical Chemistry*, **70** (1966) 1327–1328.

184. B. Tribollet and J. S. Newman, "The Modulated Flow at a Rotating Disk Electrode," *Journal of the Electrochemical Society*, **130** (1983) 2016–2026.

185. C. Deslouis, C. Gabrielli, and B. Tribollet, "An Analytical Solution of the Non-steady Convective Diffusion Equation for Rotating Electrodes," *Journal of the Electrochemical Society*, **130** (1983) 2044–2046.

186. B. Tribollet and J. S. Newman, "Analytic Expression for the Warburg Impedance for a Rotating Disk Electrode," *Journal of the Electrochemical Society*, **130** (1983) 822–824.

187. E. Levart and D. Schuhmann, "General Determination of Transition Behavior of a Rotating Disc Electrode Submitted to an Electrical Perturbation of Weak Amplitude," *Journal of Electroanalytical Chemistry*, **28** (1970) 45.

188. D.-T. Chin and C. Tsang, "Mass Transfer to an Impinging Jet Electrode," *Journal of the Electrochemical Society*, **125** (1978) 1461–1470.

189. J. M. Esteban, G. Hickey, and M. E. Orazem, "The Impinging Jet Electrode: Measurement of the Hydrodynamic Constant and Its Use for Evaluating Film Persistency," *Corrosion*, **46** (1990) 896–901.

190. C. B. Diem and M. E. Orazem, "The Influence of Velocity on the Corrosion of Copper in Alkaline Chloride Solutions," *Corrosion*, **50** (1994) 290–300.

191. M. E. Orazem, J. C. Cardoso, Filho, and B. Tribollet, "Application of a Submerged Impinging Jet for Corrosion Studies: Development of Models for the Impedance Response," *Electrochimica Acta*, **46** (2001) 3685–3698.

192. H. Cachet, O. Devos, G. Folcher, C. Gabrielli, H. Perrot, and B. Tribollet, "In Situ Investigation of Crystallization Processes by Coupling of Electrochemical and Optical Measurements: Application to $CaCO_3$ Deposit," *Electrochemical and Solid-State Letters*, **4** (2001) C23–C25.

193. O. Devos, C. Gabrielli, and B. Tribollet, "Nucleation-Growth Processes of Scale Crystallization under Electrochemical Reaction Investigated by In Situ Microscopy," *Electrochemical and Solid-State Letters*, **4** (2001) C73–C76.

194. M. T. Scholtz and O. Trass, "Mass Transfer in a Nonuniform Impinging Jet: I. Stagnation Flow-Velocity and Pressure Distribution," *AIChE Journal*, **16** (1970) 82–90.

195. M. T. Scholtz and O. Trass, "Mass Transfer in a Nonuniform Impinging Jet: II. Boundary Layer Flow-Mass Transfer," *AIChE Journal*, **16** (1970) 90–96.

196. C. Chia, F. Giralt, and O. Trass, "Mass Transfer in Axisymmetric Turbulent Impinging Jets," *Industrial Engineering and Chemistry*, **16** (1977) 28–35.

197. F. Giralt, C. Chia, and O. Trass, "Characterization of the Impingement Region in an Axisymmetric Turbulent Jet," *Industrial Engineering and Chemistry*, **16** (1977) 21–27.

198. F. Baleras, V. Bouet, C. Deslouis, G. Maurin, V. Sobolik, and B. Tribollet, "Flow Measurement in an Impinging Jet with Three-Segement Microelectrodes," *Experiments in Fluids*, **22** (1996) 87–93.

199. D. R. Gabe, G. D. Wilcox, J. Gonzalez-Garcia, and F. C. Walsh, "The Rotating Cylinder Electrode: Its Continued Development and Application," *Journal of Applied Electrochemistry*, **28** (1998) 759–780.

200. A. C. West, "Comparison of Modeling Approaches for a Porous Salt Film," *Journal of the Electrochemical Society*, **140** (1993) 403–408.

201. E. L'Hostis, C. Compère, D. Festy, B. Tribollet, and C. Deslouis, "Characterization of Biofilms Formed on Gold in Natural Seawater by Oxygen Diffusion Analysis," *Corrosion*, **53** (1997) 4–10.

202. C. Deslouis, B. Tribollet, M. Duprat, and F. Moran, "Transient Mass Transfer at a Coated Rotating Disk Electrode: Diffusion and Electrohydrodynamical Impedances," *Journal of the Electrochemical Society*, **134** (1987) 2496–2501.

203. E. Remita, B. Tribollet, E. Sutter, V. Vivier, F. Ropital, and J. Kittel, "Hydrogen Evolution in Aqueous Solutions Containing Dissolved CO_2: Quantitative Contribution of the Buffering Effect," *Corrosion Science*, **50** (2008) 1433–1440.

204. T. Tran, B. Brown, S. Nešić, and B. Tribollet, "Investigation of the Electrochemical Mechanisms for Acetic Acid Corrosion of Mild Steel," *Corrosion*, **70** (2014) 223–229.

205. S. J. Updike and G. P. Hicks, "The Enzyme Electrode," *Nature*, **214** (1967) 986–988.

206. A. Heller and B. Feldman, "Electrochemical Glucose Sensors and Their Applications in Diabetes Management," *Chemical Reviews*, **108** (2008) 2482–2505.

207. M. Harding, *Mathematical Models for Impedance Spectroscopy*, Ph.D. dissertation, University of Florida, Gainesville, FL (2017).

208. J. S. Newman, "Numerical Solution of Coupled, Ordinary Differential Equations," *Industrial and Engineering Chemistry Fundamentals*, **7** (1968) 514–517.

209. B. A. Boukamp and H. J. M. Bouwmeester, "Interpretation of the Gerischer Impedance in Solid State Ionics," *Solid State Ionics*, **157** (2003) 29–33.

210. H. Gerischer, "Wechselstrompolarisation Von Elektroden Mit Einem Potentialbestimmenden Schritt Beim Gleichgewichtspotential," *Zeitschrift fur Physikalische Chemie*, **198** (1951) 286–313.

211. C. Deslouis, I. Epelboin, M. Keddam, and J. C. Lestrade, "Impédance de Diffusion d'un Disque Tournant en Régime Hydrodynamique Laminaire. Etude Expérimentale et Comparaison avec le Modèle de Nernst," *Journal of Electroanalytical Chemistry*, **28** (1970) 57–63.

212. R. E. White, C. M. Mohr, Jr., and J. S. Newman, "The Fluid Motion Due to a Rotating Disk," *Journal of the Electrochemical Society*, **123** (1976) 383–385.

213. W. M. Haynes, editor, *CRC Handbook of Chemistry and Physics*, 97th edition (CRC Press, 2017).

214. J. F. Shackelford, *Introduction to Materials Science for Engineers*, 5th edition (Prentice–Hall, 2000).

215. L. H. Van Vlack, *Elements of Materials Science and Engineering* (Boston: Addison-Wesley, 1975).

216. U. Kaatze, "Complex Permittivity of Water as a Function of Frequency and Temperature," *Journal of Chemical & Engineering Data*, **34** (1989) 371–374.

217. V. V. Daniel, *Dielectric Relaxation* (London: Academic Press, 1967).

218. J. Q. Shang and J. A. Umana, "Dielectric Constant and Relaxation Time of Asphalt Pavement Materials," *Journal of Infrastructure Systems*, **5** (1999) 135–142.

219. K. S. Cole and R. H. Cole, "Dispersion and Absorption in Dielectrics 2: Direct Current Characteristics," *Journal of Chemical Physics*, **10** (1942) 98–105.

220. R. D. Armstrong and W. P. Race, "Double Layer Capacitance Dispersion at the Metal-Electrolyte Interphase in the Case of a Dilute Electrolyte," *Electroanalytical Chemistry and Inter:facial Electrochemistry*, **33** (1971) 285–290.

221. S. R. Morrison, *Electrochemistry at Semiconductor and Oxidized Metal Electrodes* (New York: Plenum Press, 1980).

222. R. Memming, "Processes at Semiconductor Electrodes," in *Comprehensive Treatise of Electrochemistry*, volume 7 (New York: Plenum Press, 1983) 529–592.

223. A. S. Grove, *Physics and Technology of Semiconductor Devices* (New York: John Wiley & Sons, 1967).

224. S. M. Sze, *Physics of Semiconductor Devices* (New York: John Wiley & Sons, 1969).

225. J. S. Blakemore, *Semiconductor Statistics* (New York: Dover Publications, 1987).

226. M. H. Dean and U. Stimming, "Capacity of Semiconductor Electrodes with Multiple Bulk Electronic States: I. Model and Calculations for Discrete States," *Journal of Electroanalytical Chemistry*, **228** (1987) 135–151.

227. A. J. Bard, R. Memming, and B. Miller, "Terminology in Semiconductor Electrochemistry and Photoelectrochemical Energy Conversion (Recommendations 1991)," *Pure and Applied Chemistry*, **63** (1991) 569–596.

228. A. J. Nozik and R. Memming, "Physical Chemistry of Semiconductor–Liquid Interfaces," *Journal of Physical Chemistry*, **100** (1996) 13061–13078.

229. P. J. Nahin, *Oliver Heaviside: Sage in Solitude* (New York: IEEE Press, 1988).

230. W. Thomson, "On the Theory of the Electric Telegraph," *Proceedings of the Royal Society of London*, **7** (1855) 382–399.

231. O. Heaviside, "On the Extra Current," *Philosophical Magazine*, **2** (1876) 135–145.

232. O. Heaviside, "On the Speed of Signalling through Heterogeneous Telegraph Circuits," *Philosophical Magazine*, **3** (1877) 211–221.

233. O. Heaviside, "On the Theory of Faults in Cables," *Philosophical Magazine*, **8** (1879) 60–177.

234. E. Weber and F. Nebecker, *The Evolution of Electrical Engineering* (New York: IEEE Press, 1994).

235. O. Heaviside, "On Resistance and Conductance Operators, and Their Derivatives, Inductance and Permittance, Especially in Connexion with Electric and Magnetic Energy," *Philosophical Magazine*, **24** (1887) 479–502.

236. R. Jurczakowski, C. Hitz, and A. Lasia, "Impedance of Porous Au-Based Electrodes," *Journal of Electroanalytical Chemistry*, **572** (2004) 355–366.

237. M. Keddam, C. Rakotomavo, and H. Takenouti, "Impedance of a Porous Electrode with an Axial Gradient of Concentration," *Journal of Applied Electrochemistry*, **14** (1984) 437–448.

238. A. Lasia, "Porous Eelectrodes in the Presence of a Concentration Gradient," *Journal of Electroanalytical Chemistry*, **428** (1997) 155–164.

239. C. Hitz and A. Lasia, "Experimental Study and Modeling of Impedance of the HER on Porous Ni Electrodes," *Journal of Electroanalytical Chemistry*, **500** (2001) 213–222.

240. Y. Gourbeyre, B. Tribollet, C. Dagbert, and L.Hyspecka, "A Physical Model for Anticorrosion Behavior of Duplex Coatings: Corrosion, Passivation, and Anodic Films," *Journal of the Electrochemical Society*, **153** (2006) B162–B168.

241. M. Itagaki, Y. Hatada, I. Shitanda, and K. Watanabe, "Complex impedance spectra of porous electrode with fractal structure," *Electrochimica Acta*, **55** (2010) 6255–6262.

242. E. Remita, E. Sutter, B. Tribollet, F. Ropital, X. Longaygue, C. Taravel-Condat, and N. Desamais, "A Thin Layer Cell Adapted for Corrosion Studies in Confined Aqueous Environments," *Electrochimica Acta*, **52** (2007) 7715–7723.

243. C. Gabrielli, M. Keddam, N. Portail, P. Rousseau, H. Takenouti, and V. Vivier, "Electrochemical Impedance Spectroscopy Investigations of a Microelectrode Behavior in a Thin-Layer Cell: Experimental and Theoretical Studies," *Journal of Physical Chemistry B*, **110** (2006) 20478–20485.

244. E. Remita, D. Boughrara, B. Tribollet, V. Vivier, E. Sutter, F. Ropital, and J. Kittel, "Diffusion Impedance in a Thin-Layer Cell: Experimental and Theoretical Study on a Large-Disk Electrode," *The Journal of Physical Chemistry C*, **112** (2008) 4626–4634.

245. K. Nisancioglu and J. S. Newman, "The Transient Response of a Disk Electrode," *Journal of the Electrochemical Society*, **120** (1973) 1339–1346.

246. K. Nisancioglu and J. S. Newman, "The Short-Time Response of a Disk Electrode," *Journal of the Electrochemical Society*, **121** (1974) 523–527.

247. P. Antohi and D. A. Scherson, "Current Distribution at a Disk Electrode during a Current Pulse," *Journal of the Electrochemical Society*, **153** (2006) E17–E24.

248. V. M.-W. Huang, V. Vivier, I. Frateur, M. E. Orazem, and B. Tribollet, "The Global and Local Impedance Response of a Blocking Disk Electrode with Local CPE Behavior," *Journal of the Electrochemical Society*, **154** (2007) C89–C98.

249. S.-L. Wu, M. E. Orazem, B. Tribollet, and V. Vivier, "Impedance of a Disk Electrode with Reactions Involving an Adsorbed Intermediate: Local and Global Analysis," *Journal of the Electrochemical Society*, **156** (2009) C28–C38.

250. S.-L. Wu, M. E. Orazem, B. Tribollet, and V. Vivier, "Impedance of a Disk Electrode with Reactions Involving an Adsorbed Intermediate: Experimental and Simulation Analysis," *Journal of the Electrochemical Society*, **156** (2009) C214–C221.

251. G. Sewell, *The Numerical Solution of Ordinary and Partial Differential Equations* (New York: John Wiley & Sons, 2005).

252. M. E. Orazem, N. Pébère, and B. Tribollet, "Enhanced Graphical Representation of Electrochemical Impedance Data," *Journal of the Electrochemical Society*, **153** (2006) B129–B136.

253. S.-L. Wu, M. E. Orazem, B. Tribollet, and V. Vivier, "The impedance response of rotating disk electrodes," *Journal of Electroanalytical Chemistry*, **737** (2015) 11–22.

254. T. Borisova and B. V. Ershler, "Determination of the Zero Voltage Points of Solid Metals from Measurements of the Capacity of the Double Layer," *Zhurnal Fizicheskoi Khimii*, **24** (1950) 337–344.

255. B. B. Mandelbrot, *The Fractal Geometry of Nature* (San Franscisco: Freeman, 1982).

256. A. L. Mehaute and G. Crepy, "Introduction to Transfer and Motion in Fractal Media: The Geometry of Kinetics," *Solid State Ionics*, **9** (1983) 17–30.

257. L. Nyikos and T. Pajkossy, "Fractal Dimension and Fractional Power Frequency-Dependent Impedance of Blocking Electrodes," *Electrochimica Acta*, **30** (1985) 1533–1540.

258. T. Pajkossy, "Impedance of Rough Capacitive Electrodes," *Journal of Electroanalytical Chemistry*, **364** (1994) 111–125.

259. Z. Kerner and T. Pajkossy, "On the Origin of Capacitance Dispersion of Rough Electrodes," *Electrochimica Acta*, **46** (2000) 207–211.

260. T. Pajkossy, "Impedance Spectroscopy at Interfaces of Metals and Aqueous Solutions — Surface Roughness, CPE and Related Issues," *Solid State Ionics*, **176** (2005) 1997–2003.

261. B. Emmanuel, "Computation of AC Responses of Arbitrary Electrode Geometries from the Corresponding Secondary Current Distributions: A Method Based on Analytic Continuation," *Journal of Electroanalytical Chemistry*, **605** (2007) 89–97.

262. C. L. Alexander, B. Tribollet, and M. E. Orazem, "Contribution of Surface Distributions to Constant-Phase-Element (CPE) Behavior: 1. Influence of Roughness," *Electrochimica Acta*, **173** (2015) 416–424.

263. S. Trassati and R. Parsons, "Interphases in Systems of Conducting Phases," *Pure and Applied Chemistry*, **58** (1986) 437–454.

264. C. L. Alexander, B. Tribollet, and M. E. Orazem, "Contribution of Surface Distributions to Constant-Phase-Element (CPE) Behavior: 2. Capacitance," *Electrochimica Acta*, **188** (2016) 566–573.

265. C. L. Alexander, B. Tribollet, and M. E. Orazem, "Influence of Micrometric-Scale Electrode Heterogeneity on Electrochemical Impedance Spectroscopy," *Electrochimica Acta*, **201** (2016) 374–379.

266. K. Davis, C. L. Alexander, and M. E. Orazem, "Influence of Geometry-Induced Frequency Dispersion on the Impedance of Rectangular Electrodes," (2017) in preparation.

267. Y.-M. Chen, C. L. Alexander, C. Cleveland, and M. E. Orazem, "Influence of Geometry-Induced Frequency Dispersion on the Impedance of Ring Electrodes," (2017) in preparation.

268. M. A. Lévêque, "Les Lois de la Transmission de Chaleur par Convection," *Annales Des Mines*, **13** (1928) 201–299.

269. C. Deslouis, B. Tribollet, and M. A. Vorotyntsev, "Diffusion-Convection Impedance at Small Electrodes," *Journal of the Electrochemical Society*, **138** (1991) 2651–2657.

270. K. Nisancioglu and J. S. Newman, "Separation of Double-Layer Charging and Faradaic Processes at Electrodes," *Journal of the Electrochemical Society*, **159** (2012) E59–E61.

271. J. H. Sluyters, "On the Impedance of Galvanic Cells I. Theory," *Recueil des Travaux Chimiques des Pays-Bas Journal of the Royal Netherlands Chemical Society*, **79** (1960) 1092–1100.

272. P. Delahay, "Electrode Processes without a Priori Separation of Double-Layer Charging," *Journal of Physical Chemistry*, **70** (1966) 2373–2379.

273. P. Delahay and G. G. Susbielle, "Double-Layer Impedance of Electrodes with Charge-Transfer Reaction," *Journal of Physical Chemistry*, **70** (1966) 3150–3157.

274. P. Delahay, K. Holub, G. G. Susbielle, and G. Tessari, "Double-Layer Perturbation without Equilibrium between Concentrations and Potential," *Journal of Physical Chemistry*, **71** (1967) 779–780.

275. S.-L. Wu, M. E. Orazem, B. Tribollet, and V. Vivier, "The Influence of Coupled Faradaic and Charging Currents on Impedance Spectroscopy," *Electrochimica Acta*, **131** (2014) 3–12.

276. J. O. M. Bockris and A. K. N. Reddy, *Modern Electrochemistry: An Introduction to an Interdisciplinary Area* (New York: Plenum Press, 1970).

277. J.-B. Jorcin, M. E. Orazem, N. Pébère, and B. Tribollet, "CPE Analysis by Local Electrochemical Impedance Spectroscopy," *Electrochimica Acta*, **51** (2006) 1473–1479.

278. L. Young, "Anodic Oxide Films 4: The Interpretation of Impedance Measurements on Oxide Coated Electrodes on Niobium," *Transactions of the Faraday Society*, **51** (1955) 1250–1260.

279. H. Göhr, "Contributions of Single Electrode Processes to the Impedance," *Berichte der Bunsengesellschaft für physikalische Chemie*, **85** (1981) 274–280.

280. H. Göhr, J. Schaller, and C. A. Schiller, "Impedance Studies of the Oxide Layer on Zircaloy after Previous Oxidation in Water Vapor At 400°C," *Electrochimica Acta*, **38** (1993) 1961–1964.

281. Z. Lukacs, "The Numerical Evaluation of the Distortion of EIS Data Due to the Distribution of Parameters," *Journal of Electroanalytical Chemistry*, **432** (1997) 79–83.

282. Z. Lukacs, "Evaluation of Model and Dispersion Parameters and Their Effects on the Formation of Constant-Phase Elements in Equivalent Circuits," *Journal of Electroanalytical Chemistry*, **464** (1999) 68–75.

283. J. R. Macdonald, "Generalizations of Universal Dielectric Response and a General Distribution-of-Activation-Energies Model for Dielectric and Conducting Systems," *Journal of Applied Physics*, **58** (1985) 1971–1978.

284. J. R. Macdonald, "Frequency Response of Unified Dielectric and Conductive Systems Involving an Exponential Distribution of Activation Energies," *Journal of Applied Physics*, **58** (1985) 1955–1970.

285. R. L. Hurt and J. R. Macdonald, "Distributed Circuit Elements in Impedance Spectroscopy: A Unified Treatment of Conductive and Dielectric Systems," *Solid State Ionics*, **20** (1986) 111–124.

286. J. R. Macdonald, "Linear Relaxation: Distributions, Thermal Activation, Structure, and Ambiguity," *Journal of Applied Physics*, **62** (1987) R51–R62.

287. J. R. Macdonald, "Power-Law Exponents and Hidden Bulk Relaxation in the Impedance Spectroscopy of Solids," *Journal of Electroanalytical Chemistry*, **378** (1994) 17–29.

288. I. Frateur, L. Lartundo-Rojas, C. Méthivier, A. Galtayries, and P. Marcus, "Influence of Bovine Serum Albumin in Sulphuric Acid Aqueous Solution on the Corrosion and the Passivation of an Iron-Chromium Alloy," *Electrochimica Acta*, **51** (2006) 1550–1557.

289. S. Cattarin, M. Musiani, and B. Tribollet, "Nb Electrodissolution in Acid Fluoride Medium," *Journal of the Electrochemical Society*, **149** (2002) B457–B464.

290. C. L. Alexander, B. Tribollet, and M. E. Orazem, "Contribution of Surface Distributions to Constant-Phase-Element (CPE) Behavior: 3. Reactions Coupled by an Adsorbed Intermediate," *Electrochimica Acta*, (2016) in preparation.

291. B. Hirschorn, M. E. Orazem, B. Tribollet, V. Vivier, I. Frateur, and M. Musiani, "Constant-Phase-Element Behavior Caused by Resistivity Distributions in Films: 1. Theory," *Journal of the Electrochemical Society*, **157** (2010) C452–C457.

292. B. Hirschorn, M. E. Orazem, B. Tribollet, V. Vivier, I. Frateur, and M. Musiani, "Constant-Phase-Element Behavior Caused by Resistivity Distributions in Films: 2. Applications," *Journal of the Electrochemical Society*, **157** (2010) C458–C463.

293. S. Amand, M. Musiani, M. E. Orazem, N. Pébère, B. Tribollet, and V. Vivier, "Constant-Phase-Element Behavior Caused by Inhomogeneous Water Uptake in Anti-Corrosion Coatings," *Electrochimica Acta*, **87** (2013) 693–700.

294. M. Musiani, M. E. Orazem, N. Pébère, B. Tribollet, and V. Vivier, "Determination of Resistivity Profiles in Anti-corrosion Coatings from Constant-Phase-Element Parameters," *Progress in Organic Coatings*, **77** (2014) 2076–2083.

295. M. E. Orazem, "The Influence of Coupled Faradaic and Charging Currents on Impedance Spectroscopy," in *24éme Forum sur les Impedances Electrochimiques*, H. Perrot, editor, volume 24 (Paris, France: CNRS, 2013) 5–15.

296. C. H. Hsu and F. Mansfeld, "Technical Note: Concerning the Conversion of the Constant Phase Element Parameter Y_0 into a Capacitance," *Corrosion*, **57** (2001) 747–748.

297. S. Y. Oh, L. Lyung, D. Bommannan, R. H. Guy, and R. O. Potts, "Effect of Current, Ionic Strength and Temperature on the Electrical Properties of Skin," *Journal of Controlled Release*, **27** (1993) 115–125.

298. S. Y. Oh and R. H. Guy, "Effect of Enhancers on the Electrical Properties of Skin: The Effect of Azone and Ethanol," *Journal of Korean Pharmaceutical Sciences*, **24** (1994) S41–S47.

299. S. Y. Oh and R. H. Guy, "The Effect of Oleic Acids and Polypropylene Glycol on the Electrical Properties of Skin," *Journal of Korean Pharmaceutical Sciences*, **24** (1994) 281–287.

300. E. Chassaing, B. Sapoval, G. Daccord, and R. Lenormand, "Experimental Study of the Impedance of Blocking Quasi-Fractal and Rough Electrodes," *Journal of Electroanalytical Chemistry*, **279** (1990) 67–78.

301. E. D. Bidóia, L. O. S. Bulhões, and R. C. Rocha-Filho, "Pt/HClO$_4$ Interface CPE: Influence of Surface Roughness and Electrolyte Concentration," *Electrochimica Acta*, **39** (1994) 763–769.

302. M. Musiani, M. E. Orazem, N. Pébère, B. Tribollet, and V. Vivier, "Constant-Phase-Element Behavior Caused by Coupled Resistivity and Permittivity Distributions in Films," *Journal of the Electrochemical Society*, **158** (2011) C424–C428.

303. M. E. Orazem, B. Tribollet, V. Vivier, D. P. Riemer, E. A. White, and A. L. Bunge, "On the Use of the Power-Law Model for Interpreting Constant-Phase-Element Parameters," *Journal of the Brazilian Chemical Society*, **25** (2014) 532–539.

304. L. Young, *Anodic Oxide Films* (New York: Academic Press, 1961).

305. C. A. Schiller and W. Strunz, "The Evaluation of Experimental Dielectric Data of Barrier Coatings by Means of Different Models," *Electrochimica Acta*, **46** (2001) 3619–3625.

306. C. Cleveland, S. Moghaddam, and M. E. Orazem, "Nanometer-Scale Corrosion of Copper in Deaerated Deionized Water," *Journal of the Electrochemical Society*, **161** (2014) C107–C114.

307. C. Deslouis and B. Tribollet, "Flow Modulation Technique and EHD Impedance: A Tool for Electrode Processes and Hydrodynamic Studies," *Electrochimica Acta*, **35** (1990) 1637–1648.

308. O. Aaboubi, I. Citti, J.-P. Chopart, C. Gabrielli, A. Olivier, and B. Tribollet, "Thermoelectrochemical Transfer Function under Thermal Laminar Free Convection at a Vertical Electrode," *Journal of the Electrochemical Society*, **147** (2000) 3808–3815.

309. W. J. Albery and M. L. Hitchman, *Ring-Disc Electrodes* (Oxford: Clarendon Press, 1971).

310. N. Benzekri, M. Keddam, and H. Takenouti, "AC Response of a Rotating Ring-Disk Electrode: Application to 2-D and 3-D Film Formation in Anodic Processes," *Electrochimica Acta*, **34** (1989) 1159–1166.

311. I. Epelboin, C. Gabrielli, M. Keddam, and H. Takenouti, "A Model of the Anodic Behavior of Iron in Sulphuric Acid Medium," *Electrochimica Acta*, **20** (1975) 913–916.

312. R. Antaño-López, M. Keddam, and H. Takenouti, "Interface Capacitance at Mercury and Iron Electrodes in the Presence of Organic Compound," *Corrosion Engineering, Science & Technology*, **39** (2004) 59–64.

313. E. Larios-Durán, R. Antaño-López, M. Keddam, Y. Meas, H. Takenouti, and V. Vivier, "Dynamics of Double-layer by AC Modulation of the Interfacial Capacitance and Associated Transfer Functions," *Electrochimica Acta*, **55** (2010) 6292–6298.

314. I. Citti, O. Aaboubi, J. P. Chopart, C. Gabrielli, A. Olivier, and B. Tribollet, "Impedance for Laminar Free Convection and Thermal Convection to a Vertical Electrode," *Journal of the Electrochemical Society*, **144** (1997) 2263–2271.

315. Z. A. Rotenberg, "Thermoelectrochemical Impedance," *Electrochimica Acta*, **42** (1997) 793–799.

316. S. L. Marchiano and A. J. Arvia, "Diffusional Flow under Non-Isothermal Laminar Free Convection at a Thermal Convective Electrode," *Electrochimica Acta*, **13** (1968) 1657–1669.

317. L. M. Peter, J. Li, R. Peat, H. J. Lewerenz, and J. Stumper, "Frequency Response Analysis of Intensity Modulated Photocurrents at Semiconductor Electrodes," *Electrochimica Acta*, **35** (1990) 1657–1664.

318. I. E. Vermeir, W. P. Gomes, B. H. Erné, and D. Vanmaekelbergh, "The Anodic Dissolution of InP Studied by the Optoelectrical Impedance Method: 2. Interaction between Anodic and Chemical Etching of InP in Iodic Acid Solutions," *Electrochimica Acta*, **38** (1993) 2569–2575.

319. C. Gabrielli, J. J. Garcia-Jareño, M. Keddam, H. Perrot, and F. Vicente, "AC-Electrogravimetry Study of Electroactive Thin Films. I. Application to Prussian Blue," *Journal of Physical Chemistry B*, **106** (2002) 3182–3191.

320. S. Bruckenstein, M. I. Bellavan, and B. Miller, "Electrochemical Response of a Disk Electrode to Angular Velocity Steps," *Journal of the Electrochemical Society*, **120** (1973) 1351–1356.

321. K. Tokuda, S. Bruckenstein, and B. Miller, "The Frequency Response of Limiting Currents to Sinusoidal Speed Modulation at a Rotating Disk Electrode," *Journal of the Electrochemical Society*, **122** (1975) 1316–1322.

322. C. Deslouis, C. Gabrielli, P. S.-R. Fanchine, and B. Tribollet, "Electrohydrodynamical Impedance on a Rotating Disk Electrode," *Journal of the Electrochemical Society*, **129** (1982) 107–118.

323. D. T. Schwartz, "Multilayered Alloys Induced by Fluctuating Flow," *Journal of the Electrochemical Society*, **136** (1989) 53C–56C.

324. D. T. Schwartz, T. J. Rehg, P. Stroeve, and B. G. Higgins, "Fluctuating Flow with Mass-Transfer Induced by a Rotating-Disk Electrode with a Superimposed Time-Periodic Modulation," *Physics of Fluids A–Fluid Dynamics*, **2** (1990) 167–177.

325. C. Deslouis and B. Tribollet, "Flow Modulation Techniques in Electrochemistry," in *Advances in Electrochemical Science and Engineering*, H. Gerischer and C. W. Tobias, editors (New York: VCH Publishers, 1991) 205–264.

326. E. M. Sparrow and J. L. Gregg, "Flow about an Unsteadily Rotating Disc," *Journal of the Aerospace Sciences*, **27** (1960) 252–256.

327. V. Sharma, "Flow and Heat-Transfer Due to Small Torsional Oscillations of a Disk about a Constant Mean," *Acta Mechanica*, **32** (1979) 19–34.

328. A. Caprani, C. Deslouis, S. Robin, and B. Tribollet, "Transient Mass Transfer at Partially Blocked Electrodes: A Way to Characterize Topography," *Journal of Electroanalytical Chemistry*, **238** (1987) 67–91.

329. C. Deslouis, O. Gil, and B. Tribollet, "Frequency Response of Electrochemical Sensors to Hydrodynamic Fluctuations," *Journal of Fluid Mechanics*, **215** (1990) 85–100.

330. C. R. S. Silva, O. E. Barcia, O. R. Mattos, and C. Deslouis, "Partially Blocked Surface Studied by the Electrohydrodynamic Impedance," *Journal of Electroanalytical Chemistry*, **365** (1994) 133–138.

331. C. Deslouis and B. Tribollet, "Recent Developments in the Electro-Hydrodynamic (EHD) Impedance Technique," *Journal of Electroanalytical Chemistry*, **572** (2004) 389–398.

332. C. Deslouis, N. Tabti, and B. Tribollet, "Characterization of Surface Roughness by EHD Impedance," *Journal of Applied Electrochemistry*, **27** (1997) 109–111.

333. C. Deslouis, D. Festy, O. Gil, G. Rius, S. Touzain, and B. Tribollet, "Characterization of Calcareous Deposits in Artificial Sea Water by Impedance Techniques: I. Deposit of $CaCO_3$ without $Mg(OH)_2$," *Electrochimica Acta*, **43** (1998) 1891–1901.

334. T. Yamamoto and Y. Yamamoto, "Dielectric Constant and Resistivity of Epidermal Stratum Corneum," *Medical and Biological Engineering*, **14** (1976) 494–500.

335. M. Durbha, *Influence of Current Distributions on the Interpretation of the Impedance Spectra Collected for Rotating Disk Electrode*, Ph.D. dissertation, University of Florida, Gainesville, Florida (1998).

336. G. Baril, C. Blanc, and N. Pébère, "AC Impedance Spectroscopy in Characterizing Time-Dependent Corrosion of AZ91 and AM50 Magnesium Alloys Characterization with Respect to Their Microstructures," *Journal of the Electrochemical Society*, **148** (2001) B489–B496.

337. G. M. Martin and S. Makram-Ebeid, "The Mid–Gap Donor Level EL2 in GaAs," in *Deep Centers in Semiconductors*, S. T. Pantelides, editor (New York: Gordon and Breach Science Publishers, 1986) 455–473.

338. A. N. Jansen, P. T. Wojcik, P. Agarwal, and M. E. Orazem, "Thermally Stimulated Deep-Level Impedance Spectroscopy: Application to an n-GaAs Schottky Diode," *Journal of the Electrochemical Society*, **143** (1996) 4066–4074.

339. M. E. Orazem, P. Agarwal, A. N. Jansen, P. T. Wojcik, and L. H. García-Rubio, "Development of Physico-Chemical Models for Electrochemical Impedance Spectroscopy," *Electrochimica Acta*, **38** (1993) 1903–1911.

340. M. E. Orazem and B. Tribollet, "An Integrated Approach to Electrochemical Impedance Spectroscopy," *Electrochimica Acta*, **53** (2008) 73607366.

341. A. N. Jansen, *Deep-Level Impedance Spectroscopy of Electronic Materials*, Ph.D. dissertation, University of Florida, Gainesville, Florida (1992).

342. D. B. Bonham and M. E. Orazem, "A Mathematical Model for the Influence of Deep-Level Electronic States on Photoelectrochemical Impedance Spectroscopy: 2. Assessment of Characterization Methods Based on Mott-Schottky Theory," *Journal of the Electrochemical Society*, **139** (1992) 126–131.

343. J. R. Macdonald and L. D. Potter, Jr., "A Flexible Procedure for Analyzing Impedance Spectroscopy Results: Description and Illustrations," *Solid State Ionics*, **23** (1987) 61–79.

344. P. R. Beington, *Data Reduction and Error Analysis for the Physical Sciences* (New York: McGraw-Hill, 1969).

345. G. A. F. Seber, *Linear Regression Analysis* (New York: John Wiley & Sons, 1977).

346. H. W. Sorenson, *Parameter Estimation: Principles and Problems* (New York: Marcel Dekker, 1980).

347. N. R. Draper and H. Smith, *Applied Regression Analysis*, 3rd edition (New York: Wiley Interscience, 1998).

348. W. H. Press, S. A. Teukolsky, W. T. Vetterling, and B. P. Flannery, *Numerical Recipes in C: The Art of Scientific Computing*, 2nd edition (New York: Cambridge University Press, 1992).

349. A. M. Legendre, *Nouvelles Méthodes pour la Détermination des Orbites des Comètes: Appendice sur la Méthode des Moindres Carrés* (Paris: Courcier, 1806).

350. C. F. Gauss, "Theoria Combinationis Observationum Erroribus Minimis Obnoxiae," *Werke*, **4** (1821) 3–26.

351. C. F. Gauss, "Supplementum Theoriae Combinationis Observationum Erroribus Minimis Obnoxiae," *Werke*, **4** (1826) 104–108.

352. R. de L. Kronig, "Dispersionstheorie im Röntgengebiet," *Physikalische Zeitschrift*, **30** (1929) 521–522.

353. J. R. Macdonald, "Impedance Spectroscopy: Old Problems and New Developments," *Electrochimica Acta*, **35** (1990) 1483–1492.

354. S. L. Carson, M. E. Orazem, O. D. Crisalle, and L. H. García-Rubio, "On the Error Structure of Impedance Measurements: Simulation of Phase Sensitive Detection (PSD) Instrumentation," *Journal of the Electrochemical Society*, **150** (2003) E491–E500.

355. W. H. Press, S. A. Teukolsky, W. T. Vetterling, and B. P. Flannery, *Numerical Recipes in C*, 2nd edition (Cambridge: Cambridge University Press, 1992).

356. J. R. Macdonald and W. J. Thompson, "Strongly Heteroscedastic Nonlinear Regression," *Communications in Statistics: Simulation and Computation*, **20** (1991) 843–885.

357. A. M. Awad, "Properties of the Akaike Information Criterion," *Microelectronics Reliability*, **36** (1996) 457–464.

358. T.-J. Wu and A. Sepulveda, "The Weighted Average Information Criterion for Order Selection in Time Series and Regression Models," *Statistics & Probability Letters*, **39** (1998) 1–10.

359. Y. Sakamoto, M. Ishiguso, and G. Kitigawa, *Akaike Information Criterion Statistics* (Boston: D. Reidel, 1986).

360. B. Robertson, B. Tribollet, and C. Deslouis, "Measurement of Diffusion Coefficients by DC and EHD Electrochemical Methods," *Journal of the Electrochemical Society*, **135** (1988) 2279–2283.

361. M. Durbha and M. E. Orazem, "Current Distribution on a Rotating Disk Electrode below the Mass-Transfer Limited Current: Correction for Finite Schmidt Number

and Determination of Surface Charge Distribution," *Journal of the Electrochemical Society*, **145** (1998) 1940–1949.

362. P. W. Appel, *Electrochemical Systems: Impedance of a Rotating Disk and Mass Transfer in Packed Beds*, Ph.D. dissertation, University of California, Berkeley, Berkeley, California (1976).

363. M. Durbha, M. E. Orazem, and B. Tribollet, "A Mathematical Model for the Radially Dependent Impedance of a Rotating Disk Electrode," *Journal of the Electrochemical Society*, **146** (1999) 2199–2208.

364. C. Gabrielli, F. Huet, and M. Keddam, "Investigation of Electrochemical Processes by an Electrochemical Noise-Analysis: Theoretical and Experimental Aspects in Potentiostatic Regime," *Electrochimica Acta*, **31** (1986) 1025–1039.

365. C. Gabrielli, F. Huet, and M. Keddam, "Fluctuations in Electrochemical Systems: 1. General-Theory on Diffusion-Limited Electrochemical Reactions," *Journal of Chemical Physics*, **99** (1993) 7232–7239.

366. C. Gabrielli, F. Huet, and M. Keddam, "Fluctuations in Electrochemical Systems: 2. Application to a Diffusion-Limited Redox Process," *Journal of Chemical Physics*, **99** (1993) 7240–7252.

367. U. Bertocci and F. Huet, "Noise Resistance Applied to Corrosion Measurements: III. Influence of Instrumental Noise on the Measurements," *Journal of the Electrochemical Society*, **144** (1997) 2786–2793.

368. P. Zoltowski, "The Error Function for Fitting of Models to Immitance Data," *Journal of Electroanalytical Chemistry*, **178** (1984) 11–19.

369. B. A. Boukamp, "A Package for Impedance/Admittance Data Analysis," *Solid State Ionics, Diffusion & Reactions*, **18-19** (1986) 136–140.

370. J. R. Macdonald, *CNLS (Complex Nonlinear Least Squares) Immitance Fitting Program LEVM Manual: Version 7.11*, Houston, Texas (1999).

371. J. R. Dygas and M. W. Breiter, "Variance of Errors and Elimination of Outliers in the Least Squares Analysis of Impedance Spectra," *Electrochimica Acta*, **44** (1999) 4163–4174.

372. E. V. Gheem, R. Pintelon, J. Vereecken, J. Schoukens, A. Hubin, P. Verboven, and O. Blajiev, "Electrochemical Impedance Spectroscopy in the Presence of Non-Linear Distortions and Non-Stationary Behavior: I. Theory and Validation," *Electrochimica Acta*, **49** (2006) 4753–4762.

373. E. V. Gheem, R. Pintelon, A. Hubin, J. Schoukens, P. Verboven, O. Blajiev, and J. Vereecken, "Electrochemical Impedance Spectroscopy in the Presence of Non-Linear Distortions and Non-Stationary Behavior: II. Application to Crystallographic Pitting Corrosion of Aluminum," *Electrochimica Acta*, **51** (2006) 1443–1452.

374. M. E. Orazem, T. E. Moustafid, C. Deslouis, and B. Tribollet, "The Error Structure of Impedance Spectroscopy Measurements for Systems with a Large Ohmic Re-

sistance with Respect to the Polarization Impedance," *Journal of the Electrochemical Society*, **143** (1996) 3880–3890.

375. P. Agarwal, O. C. Moghissi, M. E. Orazem, and L. H. García-Rubio, "Application of Measurement Models for Analysis of Impedance Spectra," *Corrosion*, **49** (1993) 278–289.

376. B. A. Boukamp and J. R. Macdonald, "Alternatives to Kronig-Kramers Transformation and Testing, and Estimation of Distributions," *Solid State Ionics*, **74** (1994) 85–101.

377. B. A. Boukamp, "A Linear Kronig-Kramers Transform Test for Immittance Data Validation," *Journal of the Electrochemical Society*, **142** (1995) 1885–1894.

378. S. K. Roy and M. E. Orazem, "Error Analysis for the Impedance Response of PEM Fuel Cells," *Journal of the Electrochemical Society*, **154** (2007) B883–B891.

379. H. A. Kramers, "La Diffusion de la Lumière pas les Atomes," in *Atti Congresso Internazionale Fisici, Como*, volume 2 (1927) 545–557.

380. H. M. Nussenzveig, *Causality and Dispersion Relations* (New York: Academic Press, 1972).

381. W. Ehm, H. Göhr, R. Kaus, B. Roseler, and C. A. Schiller, "The Evaluation of Electrochemical Impedance Spectra Using a Modified Logarithmic Hilbert Transform," *ACH-Models in Chemistry*, **137** (2000) 145–157.

382. Q.-A. Huang, R. Hui, B. Wang, and J. Zhang, "A Review of AC Impedance Modeling and Validation in SOFC Diagnosis," *Electrochimica Acta*, **52** (2007) 8144–8164.

383. M. E. Orazem, P. Agarwal, and L. H. García-Rubio, "Critical Issues Associated with Interpretation of Impedance Spectra," *Journal of Electroanalytical Chemistry*, **378** (1994) 51–62.

384. M. E. Orazem, "A Systematic Approach toward Error Structure Identification for Impedance Spectroscopy," *Journal of Electroanalytical Chemistry*, **572** (2004) 317–327.

385. J. R. Dygas and M. W. Breiter, "Measurements of Large Impedances in a Wide Temperature and Frequency Range," *Electrochimica Acta*, **41** (1996) 993–1001.

386. R. Makharia, M. F. Mathias, and D. R. Baker, "Measurement of Catalyst Layer Electrolyte Resistance in PEFCs Using Electrochemical Impedance Spectroscopy," *Journal of the Electrochemical Society*, **152** (2005) A970–A977.

387. O. Antoine, Y. Bultel, and R. Durand, "Oxygen Reduction Reaction Kinetics and Mechanism on Platinum Nanoparticles inside Nafion," *Journal of Electroanalytical Chemistry*, **499** (2001) 85–94.

388. Y. Bultel, L. Genies, O. Antoine, P. Ozil, and R. Durand, "Modeling Impedance Diagrams of Active Layers in Gas Diffusion Electrodes: Diffusion, Ohmic Drop Effects and Multistep Reactions," *Journal of Electroanalytical Chemistry*, **527** (2002) 143–155.

389. S. K. Roy, M. E. Orazem, and B. Tribollet, "Interpretation of Low-Frequency Inductive Loops in PEM Fuel Cells," *Journal of the Electrochemical Society*, **154** (2007) B1378–B1388.

390. R. M. Darling and J. P. Meyers, "Kinetic Model of Platinum Dissolution in PEM-FCs," *Journal of the Electrochemical Society*, **150** (2003) A1523–A1527.

391. C. F. Zinola, J. Rodriguez, and G. Obal, "Kinetic of Molecular Oxygen Electroreduction on Platinum Modified by Tin Underpotential Deposition," *Journal of Applied Electrochemistry*, **31** (2001) 1293–1300.

392. A. Damjanovic and V. Brusic, "Electrode Kinetic of Oxygen Reduction on Oxide Free Pt Electrode," *Electrochimica Acta*, **12** (1967) 615–628.

393. V. O. Mittal, H. R. Kunz, and J. M. Fenton, "Is H_2O_2 Involved in the Membrane Degradation Mechanism in PEMFC," *Electrochemical and Solid-State Letters*, **9** (2006) A229–A302.

Author Index

The references for this book are listed on page 663 in numerical order of appearance. The corresponding authors of the works are presented in this index. The numbers refer to the cited works given on page 663.

Aaboubi, O. 166, 308, 314
Abramowitz, M. 105
Agarwal, P. 69, 100, 101, 155, 338, 339, 375, 383
Akahoshi, H. 165
Aksut, A. A. 60
Albery, W. J. 309
Alexander, C. L. 262, 264–267, 290
Amand, S. 293
Antaño-López, R. 92, 312, 313
Antohi, P. 247
Antoine, O. 387, 388
Appel, P. W. 362
Arai, J. 165
Armstrong, R. D. 53, 220
Arvia, A. J. 316
Awad, A. M. 357
Ayres, C. H. 26

Bacarella, A. L. 168
Baker, D. R. 386
Baleras, F. 198
Barcia, O. E. 9, 10, 131, 174, 330
Bard, A. J. 86, 121, 227
Baril, G. 158, 336
Barsoukov, E. 4
Bartlett, J. H. 167
Barton, J. 35
Beaunier, L. 161
Beington, P. R. 344
Bellavan, M. I. 320
Benzekri, N. 310

Bertocci, U. 367
Bidóia, E. D. 301
Bird, R. B. 177
Blajiev, O. 372, 373
Blakemore, J. S. 225
Blanc, C. 336
Bockris, J. O. M. 6, 87, 122, 140, 141, 276
Bode, H. W. 110
Bommannan, D. 297
Bond, A. M. 142
Bonham, D. B. 342
Bonora, P. L. 79
Borisova, T. 254
Bouet, V. 198
Boughrara, D. 244
Boukamp, B. A. 59, 209, 369, 376, 377
Bousselmi, L. 162
Bouwmeester, H. J. M. 209
Box, G. E. P. 97, 106
Bozler, E. 37
Breiter, M. W. 371, 385
Brett, A. M. O. 143
Brett, C. M. A. 143
Brillas, E. 113
Brown, B. 204
Brown, J. 88
Bruckenstein, S. 320, 321
Brug, G. J. 149
Brusic, V. 392
Bulhões, L. O. S. 301
Bultel, Y. 387, 388
Bunge, A. L. 134, 136, 303

Cachet, H. 192
Cain, G. 89
Caprani, A. 328
Cardoso, J. C., Filho 191
Carson, S. L. 102, 145, 354
Cattarin, S. 289
Cerisola, G. 114
Chassaing, E. 300
Chen, Y.-M. 267
Chia, C. 196, 197
Chin, D.-T. 129, 188
Chopart, J. P. 166, 314
Churchill, R. 88
Citti, I. 308, 314
Cleveland, C. 267, 306
Cochran, W. G. 96, 181
Cole, K. S. 36–38, 40, 219
Cole, R. H. 40, 219
Compère, C. 201
Comte, T. 157
Conway, B. E. 87, 119
Cooper, R. S. 167
Crepy, G. 256
Crisalle, O. D. 100, 102, 103, 145, 354

Daccord, G. 300
Dagbert, C. 240
Damjanovic, A. 392
Daniel, V. V. 217
Darling, R. M. 390
Davis, K. 266
de L. Kronig, R. 66, 352
de Levie, R. 50
Dean, M. H. 226
Deflorian, F. 79
Delahay, P. 50, 272–274
Desamais, N. 242
Deslouis, C. 8, 63, 64, 99, 107, 158, 171,
    172, 185, 198, 201, 202, 211, 269, 307,
    322, 325, 328–333, 360, 374
Despic, A. R. 6
Devos, O. 159, 160, 166, 192, 193
Dewar, J. 25
Diard, J. 152–154
Diem, C. B. 127, 190
Dion, F. 70
Dolin, P. I. 44
Draper, N. R. 106, 347

Drazic, D. 6
Duprat, M. 202
Durand, R. 387, 388
Durbha, M. 99, 335, 361, 363
Dygas, J. R. 371, 385

Ehm, W. 381
Eisenberg, M. 150
Emmanuel, B. 261
Epelboin, I. 7, 46, 49, 54, 63, 64, 161, 211,
    311
Erné, B. H. 318
Erol, S. 109
Ershler, B. V. 44, 254
Esteban, J. M. 189
Evans, U. R. 120

Fanchine, P. S.-R. 64, 322
Faulkner, L. R. 121
Feldman, B. 206
Fenton, J. M. 393
Fernandes, J. S. 82
Festy, D. 201, 333
Fetter, F. 35
Fiaud, C. 162
Fick, A. 30
Finkelstein, A. 27
Firman, R. E. 53
Flannery, B. P. 348, 355
Fleming, J. A. 25
Folcher, G. 192
Fong, C. F. C. M. 90
Frateur, I. 8, 128, 134, 135, 248, 288, 291,
    292
Fricke, H. 33, 34, 39
Frumkin, A. 41, 48
Fuller, T. F. 144

Gabe, D. R. 199
Gabrielli, C. 2, 11, 12, 63–65, 75, 159, 160,
    166, 179, 185, 192, 193, 243, 308, 311,
    314, 319, 322, 364–366
Galicia, G. 158
Galtayries, A. 288
Garber, J. A. 57
Garcia-Jareño, J. J. 11, 12, 319
García-Rubio, L. H. 69, 100–102, 145, 354,
    375, 383

Gauss, C. F. 350, 351
Geary, A. L. 164
Genies, L. 388
Geraldo, A. B. 10
Gerischer, H. 47, 210, 325
Gheem, E. V. 372, 373
Gil, O. 107, 329, 333
Gileadi, E. 111
Giralt, F. 196, 197
Godinez, J. S. 131
Göhr, H. 279, 280, 381
Gomes, W. P. 318
Gonzalez-Garcia, J. 199
Gorrec, B. L. 153
Gourbeyre, Y. 240
Grahame, D. C. 42, 43
Grant, E. H. 55
Gregg, J. L. 326
Griess, J. C., Jr. 168
Grimnes, S. 16
Grinstead, C. M. 104
Grove, A. S. 223
Guy, R. H. 297–299

Haas, O. 179
Harb, J. N. 144
Harding, M. 207
Hatada, Y. 241
Hauser, A. K. 173
Haynes, W. M. 213
Heaviside, O. 19, 20, 231–233, 235
Heller, A. 206
Hickey, G. 189
Hicks, G. P. 205
Higgins, B. G. 324
Hirschorn, B. 135, 291, 292
Hitchman, M. L. 309
Hitz, C. 236, 239
Ho, C. 178
Holub, K. 274
Hopkinson, J. 24
Hsu, C. H. 296
Huang, Q.-A. 382
Huang, V. M.-W. 125, 128, 147, 248
Hubin, A. 372, 373
Huet, F. 10, 364–367
Huggins, R. A. 178
Hui, R. 382

Huitle, C. 113
Hunter, J. S. 97
Hunter, W. G. 97
Hurt, R. L. 285

Ilkov, L. 73
Isaacs, H. S. 84, 148
Ishiguso, M. 359
Itagaki, M. 17, 18, 83, 241

Jansen, A. N. 338, 339, 341
Jones, D. A. 118
Jorcin, J.-B. 277
Jordan, B. P. 55
Jovancicevic, V. 122
Jurczakowski, R. 236

Kaatze, U. 216
Kalomi, P. N. 90
Kaus, R. 381
Keddam, M. 7, 12, 54, 92, 211, 237, 243, 310–313, 319, 364–366
Kee, D. D. 90
Kendig, M. W. 61
Kerner, Z. 259
Kirste, A. 113
Kitigawa, G. 359
Kittel, J. 203, 244
Kramers, H. A. 67, 379
Krüger, F. 31
Kunz, H. R. 393

Lamego, L. R. S. 131
Larios-Durán, E. 313
Lartundo-Rojas, L. 288
Lasia, A. 15, 70, 87, 236, 238, 239
Lee, H. P. 170
Legendre, A. M. 349
LeGorrec, B. 152, 154
Lehnen, A. P. 58
Lenormand, R. 300
Lestrade, J. C. 54, 161, 211
Levart, E. 52, 182, 187
Lévêque, M. A. 268
Levich, V. G. 176
Lewerenz, H. J. 317
L'Hostis, E. 201
L.Hyspecka 240

Li, J. 317
Lightfoot, E. N. 177
Lillard, R. S. 84
Longaygue, X. 242
Lorenz, W. J. 60
Loric, G. 49
Lovrić, Š. K. 112
Lukacs, Z. 281, 282
Lvovich, V. F. 14
Lyung, L. 297

Macdonald, D. D. 1, 68, 85
Macdonald, J. R. 3, 4, 57, 58, 283–287, 343, 353, 356, 370, 376
Makharia, R. 386
Makram-Ebeid, S. 337
Mandelbrot, B. B. 255
Mansfeld, F. 60, 61, 296
Marcelin, S. 134
Marchiano, S. L. 316
Marcus, P. 288
Martin, G. M. 337
Martinsen, Ø. G. 16
Mathias, M. F. 386
Mattos, O. R. 9, 10, 78, 131, 174, 330
Maurin, G. 198
McCafferty, E. 117
McClendon, J. F. 35
Meas, Y. 313
Mehaute, A. L. 256
Mehl, W. 47
Membrino, M. A. 71
Memming, R. 222, 227, 228
Mengoli, G. 171, 172
Mentizi, S. 113
Merienne, E. 166
Méthivier, C. 288
Meyers, J. P. 390
Miller, B. 227, 320, 321
Mittal, V. O. 393
Moghaddam, S. 306
Moghissi, O. C. 375
Mohr, C. M., Jr. 212
Montella, C. 152–154
Montemor, F. 82
Montgomery, D. C. 98
Moran, F. 202
Moran, P. J. 84

Moreau, A. 169
Morrison, S. R. 221
Morse, S. 33
Moustafid, T. E. 374
Musiani, M. 134, 135, 137, 289, 291–294, 302
Musiani, M. M. 171, 172
Myland, J. C. 142

Nahin, P. J. 229
Nebecker, F. 234
Nelissen, G. 133
Nernst, W. 21
Nešić, S. 204
Newman, B. 127
Newman, J. S. 51, 115, 116, 123, 124, 130, 173, 175, 183, 184, 186, 208, 212, 245, 246, 270
Nguyen, A. S. 137
Nisancioglu, K. 130, 245, 246, 270
Nobe, K. 170
Nozik, A. J. 228
Nussenzveig, H. M. 380
Nyikos, L. 257

Obal, G. 391
Oh, S. Y. 297–299
Oldham, K. B. 142
Olivier, A. 166, 308, 314
Orazem, M. E. 8, 13, 69, 71, 80, 99–103, 109, 125, 127, 128, 132–137, 145, 147, 155, 156, 163, 189–191, 248–250, 252, 253, 262, 264–267, 275, 277, 290–295, 302, 303, 306, 338–340, 342, 354, 361, 363, 374, 375, 378, 383, 384, 389
Ozil, P. 388

Pajkossy, T. 257–260
Panizza, M. 114
Pantelides, S. T. 337
Parsons, R. 263
Peat, R. 317
Pébère, N. 81, 134, 137, 174, 252, 277, 294, 336
Perrot, H. 11, 12, 192, 295, 319
Péter, L. 165
Peter, L. M. 317
Pintelon, R. 372, 373

Pollard, R. 157
Portail, N. 243
Potter, L. D., Jr. 343
Potts, R. O. 297
Prentice, G. 139
Press, W. H. 348, 355

Race, W. P. 220
Raistrick, I. D. 178
Rakotomavo, C. 237
Randles, J. E. B. 45
Reddy, A. K. N. 140, 141, 276
Rehg, T. J. 324
Remington, R. E. 32
Remita, E. 203, 242, 244
Riemer, D. P. 134, 303
Rius, G. 333
Robertson, B. 107, 360
Robin, S. 328
Rocha-Filho, R. C. 301
Rodriguez, J. 391
Ropital, F. 203, 242, 244
Roseler, B. 381
Rotenberg, Z. A. 315
Rousseau, P. 243
Roy, S. K. 378, 389
Rufe, R. 35

Sakamoto, Y. 359
Sapoval, B. 300
Savova-Stoynov, B. S. 74
Saxe, J. G. 5
Schaller, J. 280
Scherson, D. A. 247
Schiller, C. A. 280, 305, 381
Scholtz, M. T. 194, 195
Scholz, F. 112
Schoonman, J. 58
Schoukens, J. 372, 373
Schuhmann, D. 52, 182, 187
Schwartz, D. T. 323, 324
Seber, G. A. F. 345
Sepulveda, A. 358
Sewell, G. 251
Shackelford, J. F. 214
Shang, J. Q. 218
Sharma, V. 327
Sheppard, R. J. 55, 56

Shitanda, I. 241
Shukla, P. K. 71, 103, 132, 133, 151
Silva, C. R. S. 330
Sluyters, J. H. 86, 149, 271
Sluyters-Rehbach, M. 86, 91, 149
Smith, H. 347
Smyrl, W. H. 62, 175
Snedecor, G. W. 96
Snell, J. L. 104
Sobolik, V. 198
Sorenson, H. W. 346
Sparrow, E. M. 326
Spiegel, M. R. 94
Stegun, I. A. 105
Stephenson, L. L. 62
Stern, M. 164
Stewart, J. 95
Stewart, W. E. 177
Stimming, U. 226
Stoynov, Z. 72–74
Stroeve, P. 324
Strunz, W. 305
Student 108
Stumper, J. 317
Susbielle, G. G. 273, 274
Sutter, E. 203, 242, 244
Sze, S. M. 224

Tabti, N. 332
Takenouti, H. 92, 99, 161, 179, 237, 243,
    310–313
Taravel-Condat, C. 242
Tessari, G. 274
Teukolsky, S. A. 348, 355
Thierry, D. 148
Thirsk, H. R. 53
Thomas-Alyea, K. E. 116
Thompson, W. J. 356
Thomson, W. 230
Tobias, C. W. 150, 325
Tokuda, K. 321
Touzain, S. 333
Tran, M. T. T. 163
Tran, T. 204
Trass, O. 194–197
Trassati, S. 263
Tribollet, B. 8–10, 13, 63–65, 99, 107, 125,
    128, 131, 134, 135, 137, 147, 158–160,

162, 163, 166, 171, 172, 174, 175,
184–186, 191–193, 198, 201–204, 240,
242, 244, 248–250, 252, 253, 262, 264,
265, 269, 275, 277, 289–294, 302, 303,
307, 308, 314, 322, 325, 328, 329,
331–333, 340, 360, 363, 374, 389
Triki, E. 162
Tsai, S. 61
Tsang, C. 188

Umana, J. A. 218
Updike, S. J. 205
Urquidi-Macdonald, M. 68

van den Eeden, A. L. G. 149
Van Valkenburg, M. E. 146
Van Vlack, L. H. 215
Vanmaekelbergh, D. 318
Verboven, P. 372, 373
Vereecken, J. 77, 372, 373
Vermeir, I. E. 318
Vetterling, W. T. 348, 355
Vicente, F. 12, 319
Vivier, V. 125, 128, 134, 135, 137, 147, 158,
163, 203, 243, 244, 248–250, 253, 275,
291–294, 302, 303, 313
Vlachos, G. 107
Vladikova, D. 73
von Kármán, T. 180

Vorotyntsev, M. A. 269

Wagner, C. 126
Waldvogel, S. R. 113
Walsh, F. C. 199
Wang, B. 382
Warburg, E. 28, 29
Watanabe, K. 241
Weber, E. 234
West, A. C. 138, 200
Wheatstone, C. 22, 23
White, E. A. 134, 136, 303
White, R. E. 87, 212
Wilcox, G. D. 199
Wilke, C. R. 150
Wilson, E. 24
Wojcik, P. T. 155, 156, 338, 339
Wu, S.-L. 249, 250, 253, 275
Wu, T.-J. 358
Wylie, C. R. 93

Yamamoto, T. 334
Yamamoto, Y. 334
Young, L. 278, 304

Zhang, J. 382
Zinola, C. F. 391
Zoltowski, P. 368
Zou, F. 148

Subject Index

Numbers written in bold font refer to the page where the corresponding entry is defined. Numbers followed by an "e" refer to the page where the corresponding entry can be found in an example, and numbers followed by a "p" refer to the page where the corresponding entry can be found in a problem.

activation energy
    controlled process, 471, 621e
    diffusion, 435
    electrochemical, 423
    graphical method, 518–521, 526p
admittance, 468, 478–483, 491p
    dielectrics, 482e
    ohmic-resistance corrected, 483
    plots, 481–483
adsoption
    reaction intermediates
      history, xlii–xliii
      kinetics, 226, 231e, 240, 241p,
        421–423, 431–433, 464p, 624e
    specific
      absence of, 313, 453
      history, xlii–xliii
      partial blocking, 454
adsorbed intermediates, 341, 346
alternative hypothesis, **60**
Argand diagram, **5**, 12
axisymmetric electrode, 244, 266

bias error, 170, 184–187, **567**, 575–579,
    587e, 589e, 618
black box, xxxiii
blind men, *see* elephant, parable of
Bode plots, 473–475, 491p, 501–505,
    525p, 550, 553, 559, 563p
    magnitude, 473

modulus, 473
ohmic-resistance corrected, 396e,
    475–476, 491p, 504, 505, 555,
    563p
phase angle, 474
Boltzmann distribution, 312
boundary condition, 26, 27, 29, 30, 33,
    39
    axisymmetric electrode, 244
    blocking electrode, 338
    blocking electrode with CPE, 339
    convective diffusion, 261
    coupled reactions, 341
    faradaic reactions, 339
    hydrodynamic, 445
    primary current distribution, 115
    reflective, 251–256
    secondary current distribution,
      116
    transmissive, 245–251
Brug formula, 406, 411, 418p
Butler–Volmer kinetics, **95**, 144

capacitance
    complex, 468
    constant-phase element
      Brug formula, 406, 411, 418p
      is $C = Q$?, 414e, 418p
      normal distribution, 403, 411,
        418p

power-law distribution, 409, 411, 418p

surface distribution, 406, 411, 418p

dielectric, 126, 127p

double layer, 122–126, 127p, 318p

effective, 468

film, 310

geometric, 307–309, 317p

reflective boundary condition, 254e

space-charge, 317, 318p

transfer function, 431–433

values, 126

capacitor, **76**

impedance response, 79

properties, 468

Cauchy's integral formula, 638e

Cauchy's theorem, 637e

application, 604–609, 614p

definition, 635–639

Cauchy–Riemann conditions, 634–635

causality, 600, 603, 614p

cell design

current distribution, 169

electrode connections, 167e

flow configurations, 167–168

reference electrode placement, 165

cell potential, 107, 119, 120e

central limit theorem, 50, 52–55

chain rule, 36–38

characteristic dimension

disk electrode, 339, 370

distributed capacitance, 364, 370

rectangular electrode, 370

ring electrode, 370

rough disk electrode, 353, 370

roughness, 355, 370

characteristic frequency, xxxive

admittance, 405

convective diffusion, 285

dielectric relaxation, 306

disk electrode, 370, 418p

distributed capacitance, 364, 370

double-layer, xxxve

faradaic reaction, xxxve, 343

geometric capacitance, 308e

graphical analysis, 506–512

homogeneous reaction, 285

impedance, 403

mass transfer, xxxve

normal distribution, 402–404

Nyquist plots, 494–499

ohmic impedance, 343, 344e, 418p

power-law distribution, 406–413

rectangular electrode, 370

ring electrode, 370

rough disk electrode, 353, 370

roughness, 354, 370

surface distribution, 404–406

charge

dielectric, 126

double layer, 108, 109, 122–126

electric, 76, 90

charge density, 312

charge transfer

reaction, 22p, 90, 207

relation to Tafel slope, 213

resistance, 22p, 127p, 144p, 173p, 194, 209, 210, 340

charge-transfer resistance, 216

chi-squared test, 70

circuit calculations

Bode representation, 82e

nested elements, 83e

parallel, 81e

series, 80e

closed loop

condition, 131

gain, 132

coating

capacitance, 126, 162, 163p

EHD response, 457, 461e
impedance, 201, 204e, 205, 206p,
    275, 276e, 325e, 439
Cole–Cole
    plots, 484–488, 522–523
    relaxation, 307
complex capacitance, 468, 484–488,
    491, 492p
    constant-phase element, 487
    dielectrics, 487e
    power-law model, 486e
    Young impedance, 485e
complex integrals, 631–641
    Cauchy–Riemann conditions, 634–
        635
    definitions, 631–633
complex integration
    Cauchy's integral formula, 638e
    Cauchy's theorem, 635–639
    poles, 639e
complex number, **5**
    division, 7e
    Euler equation, 13, 14e
    exponential, 20, 21e
    multiplication, 6e
    polar coordinates, 10e
    rectangular coordinates, 9e
    square roots, 11c
complex ohmic impedance, 342, 346
complex relative permittivity, 492p
complex variables, 4–21
concentration overpotential, 120
concentration-dependent reaction,
    213–222
conductivity, 114
conservation of species, 113
constant
    Faraday's, 209
    permittivity of vacuum, 122
    universal gas, 209
constant-phase element, 395–417, 491p

application, 396e, 411e, 418p,
    544p, 618
capacitance
    Brug formula, 406, 411, 418p
    is $C = Q$?, 414e, 418p
    normal distribution, 403, 411,
        418p
    power-law distribution, 409,
        411, 418p
    surface distribution, 406, 411,
        418p
    complex capacitance, 487
    concept, 83
    Euler equation, 13, 14e
    extracting physical properties,
        401–415
    film thickness, 411e, 418p
    formula, 22p
    history, xlii, xlvi
    identification, 396
    is $C = Q$?, 402, 414e, 418p
    limitations, 415–417
    Mathematical Formulation, 395
    normal distribution, 402–404, 418p
    origin, 399–401
    power-law distribution, 73p, 406–
        413
    regression, 555–562
    surface distribution, 404–406, 418p
    units, 418p
constant-phase elephant, 396, 418p
convective-diffusion impedance
    global, 374
    impinging jet, 268–269, 301p
    local, 373
    nonuniformly accessible electrode,
        370–376
    rotating cylinder, 270
    rotating disk, 261–266, 300p
    uniformly accessible electrode,
        244–298

coordinate transformation
  chain rule, 36–38
  modulated film thickness, 37e
  rotational elliptic, 337, 342
copper corrosion, 236e
corrosion
  cast iron, 325e
  copper, 94, 96e, 100e, 111e, 240p,
    291e, 464p
  dynamic surface films, 291e
  free-machining steel, 411e, 418p
  iron, 95
    aerobic, 224e
    anaerobic, 222e
  magnesium, 231e, 241p, 513e
  niobium, 396e
  silver, 240p
  stainless steel, 396e, 411e, 418p
  with salt films, 202, 204e, 205,
    206p
corrosion current
  Stern–Geary relation, 224, 240p
corrosion potential, 95, 97–99, 195,
    196e, 206p
coupled
  diffusion impedance, 277–278
  electrochemical reactions, 226–234
  electrochemical/chemical reac-
    tions, 234–240
  faradaic and charging current,
    377–388
  mass transfer, 271–278, 301p
    coating porosity of unity, 275,
      276e
    conservation equations, 271–272
    diffusion impedance, 277–278
    diffusion impedances in series,
      278e
    Koutecky–Levich equation,
      273e
    steady state, 272–276

covariance, 47
current
  addition, 195–201
  density, 113
    faradaic, 208
  distribution, 117e, 169
    mass-transfer-controlled, 117
    nonuniform, 167, 270, 337
    primary, 114–115
    secondary, 116
    tertiary, 116
  follower, 132e
  kinetic, 105, 215, 218e
  mass-transfer-limited, 105, 215,
    218e

Debye length, 125, 317p
deposition
  zinc, 234e
dielectric capacitance, 126
dielectric response
  homogeneous media, 304–309
  nonhomogeneous media, 309–310,
    317p
dielectrics
  admittance, 482e
  complex capacitance, 487e
differential equation
  complex roots, 30e
  diffusion, 27e, 30e, 39e
  diffusion impedance, 42e
  homogeneous, 25, 26, 27e, 29–31
  linear ODE, 26e
  nonhomogeneous, 25, 26e, 32–36
diffusion coefficients, 113
diffusion impedance, 209
  convective
    finite Schmidt number, 264–266,
      268–269, 300p
    infinite Schmidt number, 263
    Nernst hypothesis, 262

nonuniformly accessible electrode, 370–376

    rotating cylinder, 270

finite-length, 233e, 245–256, 582

in series, 278e

nonuniformly accessible electrode, 370–376

rotating cylinder, 270

Warburg, xlii, 11e, 13e, 39e, 43p, 249, 250e, 328, 374, 490e, 582

diffusion resistance, 217–219

dimensionless frequency

    convective diffusion, 261

    disk geometry, 339

    distributed capacitance, 364

    film diffusion, 247

    global convective diffusion, 375

    graphical representation, 506–512

    local convective diffusion, 373

    normal time-constant distribution, 403

    rough disk geometry, 357

    roughness, 354

    surface time-constant distribution, 405

disk electrode

    blocking, 338

    blocking with CPE, 339

    characteristic dimension, 339, 370

    characteristic frequency, 343, 344e, 370

    complex ohmic impedance, 342, 346

    coupled reactions, 341, 346

    faradaic reactions, 339

    frequency dispersion

        geometry-induced, 337–347

        roughness-induced, 349–360

    nonuniform current distribution, 337–347

    numerical method, 342

distributed capacitance

    characteristic dimension, 364, 370

    characteristic frequency, 364, 370

distributed time constant

    geometry-induced, 344e, 358e

    mass transfer, 372e

    porous electrode, 325e

distribution

    central limit theorem, 50, 52–55

    current, *see* current distribution

    normal, **49**

    standard normal, **49**

double layer, 108, 109

    capacitance, 122–126

    model, 380–382

downhill simplex, 535–536

duplex coating, 330e

dynamic surface films, 291–298

effective capacitance, 468, 488–491, 491p, 512–516, 525p

effective CPE coefficient, 512–516, 525p

electric field, 312

electrical circuit

    corrosion potential, 195, 196e, 206p

    current addition, 195–201

    partially blocked electrode, 196, 197e, 206p

    porous layers, 201

    potential addition, 201–205

    superposition of Nyquist plots, 199

    two porous layers, 202, 205, 206p

electrical double layer, 108, 109, 122–126

electrical properties

    conductivity, 304

    dielectric constant, 304

    materials, 303–304

electrochemical

cell
    comparison to resistor, xxxviie,
        89–91
    potential, 120e
    interface, 133–136, 137p
    kinetics, 96e, 100e, 102e, 111e
    potential, 125
    reaction, 90, 207
electrode
    axisymmetric, 244, 266
    connections, 167e
    impinging jet, 266–269, 301p
        diffusion impedance, 268–269,
            301p
        fluid mechanics, 266–267
        mass transfer, 268
        properties, 168
    nonuniformly accessible, 370–376
    ring–disk, 429–431
    rotating cylinder
        diffusion impedance, 270
        fluid mechanics, 269
        mass transfer, 269–270
        properties, 168
    rotating disk
        diffusion impedance, 261–266,
            300p
        fluid mechanics, 256–258, 300p
        mass transfer, 259–260, 300p
        properties, 167
        state variables, 157
    rotating hemisphere
        properties, 168
    uniformly accessible, 244–245
electrogravimetry transfer function,
        439–441, 441p
electrohydrodynamic
    impedance, 443–463
        blocking, 460e
        partially blocked, 455–456
        porous film, 457–460

    transfer function, **451**
electrolyte conductivity, 114
electroneutrality, 114, 116, 122, 313
elementary transfer function, 425
elephant
    constant-phase, 396, 418p
    parable of, xxix–xxx
    remember!, xxv
equilibrium potential, 110
error propagation, 55–59, 71p, 73p
    linear, 56, 57e
    nonlinear, 57e
error structure, 567–593, 627p
    bias, 575–579
    contributions, 567
    integration with measurement,
        619
    regression, 621
    stochastic, 568–575, 593p
errors
    bias, 170, 184–187, **567**, 575–579,
        618
    fitting, **567**
    propagation, 55–59, 71p, 73p
    residual, **567**
    stochastic, 149, 170, 184–185, 266,
        478, 528, 537–538, 544p, **567**,
        568–575, 593p, 609–613, 618,
        621
Euler equation, 13, 14e
Evans diagram, 99
exchange current density, **95**, 99, 110,
        111e, 116, 141, 188p
expectation, **45**, 55, 567, 610–613
experimental design
    bias error, 185–187
    frequency range, 170
    information content, 187
    instrument bias, 186
    instrumentation parameters, 184–
        188

linearity, 188p
modulation technique, 182
nonstationary effects, 186
oscilloscope, 183
signal-to-noise, 184
experimental methods
comparison, 156
Fourier analysis, 152–156, 162p
frequency domain, 143–156
Lissajous analysis, 144–150, 156, 162, 163p
Lissajous ellipse, 146e
local electrochemical impedance spectroscopy, 158–162
lock-in amplifier, 151–152, 156
phase-sensitive detection, 151–152, 156
specialized techniques, 157–162
transfer-function analysis, 157
transient response, 142–143
experimental requirements
linearity, 170–182
Lissajous analysis, 182
modulation technique, 182
steady state, 142
experimental systems
calcareous deposits, 460e
cathodic protection, 460e
constant-phase element, 396e, 411e, 418p
corrosion, 97
cast iron, 325e
copper, 94, 96e, 100e, 111e, 236e, 240p, 291e, 464p
iron, 95, 222e, 224e
magnesium, 231e, 241p, 513e
silver, 240p
with salt films, 202, 204e, 205, 206p
CuCl, 236e
current distribution, 117e

deposition
zinc, 234e
dielectrics, 482e
duplex coating, 330e
dynamic surface films, 291–298
ferricyanide reduction
diffusion with first-order reaction, 220e
electrohydrodynamic impedance, 456, 578
error structure, 47e, 65e
rotating disk, 499–517, 525p, 546–562, 576
ferrocyanide oxidation
rotating disk, 525p
free-machining steel, 411e, 418p
GaAs, 317p
deep-level states, 621e
Schottky diode, 518–522, 621e
homogeneous chemical reactions, 278–290
human skin, 492p
hydrogen evolution, 90, 98, 102e, 111e, 120e
lithium battery, 73p
mass-transfer-controlled reactions, 499
niobium, 396e
oxygen evolution, 90, 95, 102e, 120e
partially blocked electrode, 501
PEM fuel cell, 73, 74p, 624e
peroxide oxidation, 93e
porous layer, 493
porous layers
circuit model, 201
stainless steel, 396e, 411e, 418p
thin-layer cell, 333–337
transdermal drug delivery, 188p
Young impedance, 396e

f-test, 64–70, 73, 74p
faradaic
    current density, 208
    impedance, 193, 209
    reaction
        characteristic frequency, 343
        expression, 92, 207
        rate, 109
Faraday
    cage, 185, **187**
    constant, 209, 430
ferricyanide reduction
    diffusion with first-order reaction,
        220e
    electrohydrodynamic impedance,
        456
    error structure, 47e, 65e
    rotating disk, 499–517, 525p, 546–
        562
        electrohydrodynamic impe-
            dance, 578
        impedance, 576
ferrocyanide oxidation
    rotating disk, 525p
film
    capacitance, 310
    diffusion impedance
        reflective boundary condition,
            251–256, 300p
        transmissive boundary condi-
            tion, 245–251, 300p
        with exchange of electroactive
            species, 245–251, 300p
        without exchange of electroac-
            tive species, 251–256, 300p
    resistance, 310
    thickness, 411e, 418p
finite-length diffusion impedance,
    233e, 245–256, 582
fitting error, **567**
flat-band potential, 312

flow configurations, 167–168
fluid mechanics
    impinging jet, 266–267
    rotating cylinder, 269
    rotating disk, 256–258, 300p
flux
    dilute solutions, 113
    natural convection, 436
    rotating disk, 300p
    small circular electrode, 372e
Fourier analysis, 152–156, 162p
    multiple frequency, 155, 156
    single frequency, 152–156
frequency dispersion, 206p
    coupled faradaic and charging
        currents, 385–388
    electrode geometry, 337–347
    exponential resistivity distribu-
        tion, 389–392
    roughness, 349–360
    surface property distributions,
        349–369
    Young impedance, 389–392
frequency domain, 143–156
frequency range, 170

GaAs, 317p
galvanostat, 135, 137p
galvanostatic regulation, 182, 188p
Gauss–Newton method, 534
generalized heterogeneous reaction,
    92, 207
generalized transfer function, 421–425
geometric capacitance, 307–309, 317p
geometry-induced frequency disper-
    sion
    characteristic frequency, 343, 344e
    disk, 337–347
Gerischer impedance, 287e, 301p
global
    convective diffusion impedance,

374
    impedance, **160**, 342, 347, 385
    interfacial impedance, **161**, 162, 163p, 347, 388
    ohmic impedance, **161**, 344
graphical analysis, 493–525
    admittance, 482e
    Arrhenius, 518–521, 526p
    capacitance, 471e
    characteristic frequency, 506–512
    Cole–Cole, 522–523
    complex capacitance, 485–487e
    corrosion of magnesium, 513e
    dimensionless frequency, 506–512
    effective capacitance, 490e, 512–516, 525p
    effective CPE coefficient, 512–516, 525p
    low-frequency mass transfer, 516–517, 526p
    Mott–Schottky, 521–522
    Nyquist plots, 471e
    scaled-frequency superposition, 518–521, 526p
    Schmidt number, 516–517, 526p
graphical representation
    Bode plots, 473–475, 491p, 501–505, 525p, 550, 553, 559, 563p
    complex-admittance plane, 480
    complex-impedance plane, 470–473, 491p, 494–501, 525p, 549, 553, 558, 563p
    imaginary part of admittance, 481–483
    imaginary part of impedance, 476–478, 505, 525p, 550, 552, 555, 559
    Nyquist plots, 470–473, 480, 491p, 494–501, 525p, 549, 553, 558, 563p
    ohmic-resistance-corrected Bode plots, 475–476, 491p, 504, 505, 555, 563p
    real part of admittance, 481–483
    real part of impedance, 476–478, 550, 555, 559
    residual error, 550, 555, 559, 563p

Helmholtz
    double layer, 123
    plane, 108, 123
history
    coupled faradaic and charging current, 377
    electrode roughness, 349
    impedance, xli–xlviii
    regression techniques, 527
    submarine telegraph cable, 319
homogeneous
    differential equation, 4, 25, 26, 27e, 29–31
    reaction, 278–290, 325
        Gericher impedance, 287e, 301p
        linear kinetics, 287e, 301p
        nonlinear kinetics, 282e
hydrodynamic transfer function, 445–448
hydrogen evolution, 90, 111e
hypothesis
    alternative, **60**
    null, **60**
hypothesis tests, 59
    t-test, 65e
    chi-squared, 71e
    f-test, 64–70, 73, 74p
    Student's t-test, 61–62, 73, 74p
    table
        t-test values, 63
        chi-squared test values, 72
        f-test values, 66

identity matrix, 425
imaginary number, **4**

imaginary-impedance-derived phase
    angle, 352
immittance, **580**
impedance, 468
    adsorbed intermediates, xxxix
    applications, xxxix–xl
    comparison to resistance, xxxviie
    corrosion, xl
    diffusion, 209
    electroactive polymers, xl
    electrohydrodynamic, 443–463
    faradaic, 209
    faradaic , 193
    history, xli–xlviii
    interfacial , 193
    magnitude, 473
    modulus, 473
    phase angle, 474
    philosophical approach, 617–627
    plots, 476–478, 505, 525p, 550, 552,
        555, 559
    representation, 467–478
    viscosity gradients, xl
impedance interpretation
    circuits, 204e
    kinetics, 222e, 224e, 231e
    mass transfer, 220e, 234e, 236e,
        249, 250e, 254e, 275, 276e,
        278e
    Mott–Schottky plots, 316e
impedance measurement
    error structure, 619
    Fourier analysis, 155e
    linearity, 170e, 174e, 180e
    Lissajous analysis, 148e
    not stand-alone, 620
    not standalone, 627p
    other observations, 620
impedance models
    constant-phase element, 13, 14e,
        22p

convective diffusion
    finite Schmidt number, 264–266,
        268–269, 300p
    impinging jet, 268–269, 301p
    infinite Schmidt number, 263
    Nernst hypothesis, 262
    nonuniformly accessible elec-
        trode, 370–376
    rotating cylinder, 270
finite-length diffusion, 233e, 245–
    256, 582
lack of uniqueness, 84, 85, 416,
    627p
pore-in-pore, 23p, 328–333
pore-in-pore-in-pore, 330
porous electrodes, 322–328
thin-layer cell, 333–337
transmission line, 319–337
Warburg, xlii, 11e, 13e, 39e, 43p,
    249, 250e, 328, 374, 490e, 582
impinging jet electrode, 266–269, 301p
    convective diffusion impedance,
        268–269, 301p
    fluid mechanics, 266–267
    mass transfer, 268
    properties, 168
inductance, 226, 231e
inductor, **76**, 79
information content, 187
input quantities, **422**, 444
input-output relationship, **423**
instrument bias, 186
instrumentation parameters, 184–188
interfacial impedance, 193
interfacial potential, 108
intermediates
    history, xlii–xliii
    kinetics, 226, 231e, 240, 241p, 421–
        423, 431–433, 464p, 624e
irrational number, **3**
IUPAC convention, 5

kinetic transfer function, 453–454
kinetics
    Butler–Volmer, **95**, 144
    reaction intermediates, 226, 231e,
        240, 241p, 421–423, 431–433,
        464p, 624e
    Tafel, 209, 214, 223, 227, 232e,
        235e, 625
    Tafel kinetics, **96**, 171e
    Tafel slope, **96**, 141, 188p
Koutecky–Levich equation
    coupled mass transfer, 273e, 300p
    impedance, 215
    steady state, **105**, 106e, 273
Kramers–Kronig relations, 595–599
    application, 156, 187, 527, 547, 548,
        568, 575–576, 595–599, 609,
        614p
    Cauchy's theorem, 604–609, 614p
    causality, 600, 603, 614p
    direct integration, 597, 614p
    expectation sense, 610–613
    experimental assessment, 598
    history, xliii, xlvi
    integral forms, 596, 614p
    linearity, 600, 614p
    measurement model, 552, 563,
        586–593, 599, 619–620, 624
    process model, 598
    stability, 604, 614p
    time-translation independence,
        600

Laplace's equation, 114
Levenberg–Marquardt method, 534–
    535
Levich equation, 27e, 106, 218, 292,
    300p, 443, 451
    plot, 107
    Schmidt number correction, 33e
linear regression, 530–531

linearity, 170–182
    guideline, 170e
    influence of capacitance, 180e
    influence of ohmic resistance, 174e
    Kramers–Kronig relations, 600,
        614p
    Tafel slope, 188p
Lissajous
    analysis, 144–150, 156, 162, 163p
    ellipse, 146e
    to detect nonlinearity, 182
local convective diffusion impedance,
    373
local electrochemical impedance spec-
    troscopy, 158–162
    global impedance, **160**
    global interfacial impedance, **161**,
        162, 163p
    global ohmic impedance, **161**
    local impedance, **160**
    local interfacial impedance, **160**
    local ohmic impedance, **161**
local impedance, **160**
local interfacial impedance, **160**
local ohmic impedance, **161**
lock-in amplifier, 151–152, 156
Lord Kelvin, 319

magnitude
    impedance, 473
    ohmic-resistance corrected, 396e,
        475, 491p, 505, 563p
mass transfer, 105, 106e
    impinging jet, 268
    relation to faradaic current, 93e
    rotating cylinder, 269–270
    rotating disk, 259–260, 300p
mass transport, 113–118
mass-transfer transfer function, 448–
    453

mass-transfer-controlled current distribution, 117

mass-transfer-limited current density, 105, 106e

materials
    Cole–Cole relaxation, 307
    dielectric relaxation, 304–307
    electrical properties, 303–304
    geometric capacitance, 307–309

mathematical equivalence, 84, 87p

mean, **45**

measurement model, 552–555
    error identification, 581–593
    even and odd properties, 601, 602e
    high-frequency bias errors, 587e
    integration with measurement, 619–620
    Kramers–Kronig relations, 586–593, 599
    low-frequency bias errors, 589e
    Monte Carlo methods, 563p
    resistivity distribution, 406e
    statistics, 563p
    stochastic errors, 584e

method of steepest descent, 534

mixed potential, **95**, 97–99, 102e

model identification, 621e

modulation technique, 182

modulus
    impedance, 473
    ohmic-resistance corrected, 396e, 475, 491p, 505, 563p

Monte Carlo methods, 563p

Mott–Schottky
    analysis, 311–317
    plots, 316e, 317, 318p, 521–522

multiple independent reactions, 222–225

negative feedback, 131e

Nernst hypothesis
    diffusion layer thickness, 27e
    Gerischer impedance, 287e
    rotating disk, 262, 300p

Nernst–Einstein equation, 113

nonhomogeneous differential equation, 25, 26e, 32–36

noninverting amplifier, 137p

nonlinear regression, 532–536, 544p, 563p

nonstationary effects, 186

nonuniform current distribution, 167, 270, 337

nonuniformly accessible electrode, 370–376

normal distribution, **49**

null hypothesis, **60**

number
    complex, **5**
    imaginary, **4**
    irrational, **3**
    rational, **3**
    real, **5**

numerical method, 342, 351, 355, 360, 383

Nyquist plots, 470–473, 480, 491p, 494–501, 525p, 549, 553, 558, 563p
    characteristic frequency, 494–499
    superposition, 199, 499–501

objective function, 529–530

observation equations, **423**

Ohm's law, xxxviie, 114

ohmic impedance
    characteristic frequency, 343, 344e
    high and low frequency, 346
    high frequency, 342

ohmic potential drop, 108, 119

ohmic resistance, xxxv, 308, 317p, 326, 340

ohmic-resistance corrected

admittance, 483
Bode plots, 396e
magnitude, 396e, 475, 491p, 505, 563p
modulus, 396e, 475, 491p, 505, 563p
phase angle, 396e, 475, 504
open-loop gain, 132
operational amplifier
closed-loop, 131
closed-loop gain, 132
current follower, 132e
electrochemical interface, 133–136, 137p
galvanostat, 135, 137p
ideal, 129–131
negative feedback, 131e
noninverting amplifier, 137p
open-loop gain, 132
potentiostat, 134–136
voltage adder, 133e
orthonormal axes, 470
oscilloscope, 183
output quantities, **423**, 444
overpotential
concentration, 120
surface, **95**, 101, 102e, 108, 119
oxygen evolution, 90, 95

partial differential equation, 38–41
partially blocked electrode, 196, 197e, 206p
PEM fuel cell, 624e
permittivity of vacuum, 122
phase angle, 503
imaginary-impedance-derived, 352
impedance, 474
ohmic-resistance corrected, 396e, 475, 504
phase-sensitive detection, 151–152, 156

photoelectrochemical transfer function, 438–439
plots
Bode, 473–475, 491p, 501–505, 525p, 550, 553, 559, 563p
Cole–Cole, 484–488, 522–523
impedance, 476–478, 505, 525p, 550, 552, 555, 559
Mott–Schottky , 316e, 317, 318p, 521–522
Nyquist, 470–473, 480, 491p, 494–501, 525p, 549, 553, 558, 563p
ohmic-resistance-corrected Bode, 475–476, 491p, 504, 505, 555, 563p
residual error, 550, 555, 559, 563p
Poisson's equation, 311, 317p
poles on a real axis, 639e
pore-in-pore model, 328–333
pore-in-pore-in-pore model, 330
porous electrode, 322–328
corrosion, 325e
duplex coating, 330e
porous film, 245–256
circuit model, 201
coupled with convection, 271–278, 301p
potential
addition, 201–205
cell, 107, 119, 120e
corrosion, **95**, 97–99
drop, ohmic, 108, 119
equilibrium, 110
interfacial, 108
mixed, **95**, 97–99
reference electrode, 107
window, 90, 103e, 120e, 127p
working electrode, 107
potential-dependent reaction, 209–213
potentiostat, 134–136
potentiostatic regulation, 182

power-law model
  complex capacitance, 486e
  development, 406–413
  error propagation, 73p
primary current distribution, 114–115,
    117e
primary resistance
  disk electrode, 115
proposed experiments, 624e

rational number, **3**
reaction
  homogeneous, 278–290, 301p, 325
  intermediates
    history, xlii–xliii
    kinetics, 226, 231e, 240, 241p,
      421–423, 431–433, 464p, 624e
  mechanism
    concentration dependent, 213–
      222
    coupled, 226–240
    multiple independent, 222–225
    potential dependent, 209–213
real number, **5**
rectangular electrode
  characteristic dimension, 370
  characteristic frequency, 370
reference electrode
  placement, 165
  potential, 107
reflective boundary condition, 251–
    256, 300p
regression
  application, 546–562
  downhill simplex, 535–536
  error structure, 621
  Gauss–Newton, 534
  influence of bias errors, 627p
  influence of data quality, 536–541,
    544p, 627p
  influence of insufficient range,

    539–541, 544p
  influence of stochastic errors, 537–
    538, 544p, 627p
  initial estimates, 541–542
  Levenberg–Marquardt, 534–535
  linear, 530–531
  measurement model, 552–555,
    563p
  nonlinear, 532–536, 544p, 563p
  objective function, 529–530
  quality, 545–546
  statistics, 542–543
  steepest descent, 534
  strategies, 530–536
  weighting, 527, 528, 579–581, 593p
relative permittivity, 492p
residual error, **567**
  data, 47e
  plots, 550, 555, 559, 563p
  uncorrelated, 49
resistance
  charge-transfer, 194, 209, 216, 340
  diffusion, 217–219
  film, 310
  ohmic, xxxv, 308, 317p, 326, 340
resistivity, 336, 492p
  distribution, 73p, 317p, 389–392,
    406–413
  film, 204
  materials, 303–304
  pore, 202
resistor, **75**
  comparison to electrochemical
    cell, 89–91
  comparison to impedance, xxxviie
  impedance response, 79
  properties, 468
ring electrode
  characteristic dimension, 370
  characteristic frequency, 370
ring–disk transfer function, 429–431

rotating cylinder
    diffusion impedance, 270
    fluid mechanics, 269
    mass transfer, 269–270
    properties, 168
rotating disk, 256–266
    convective diffusion impedance,
        261–266, 300p
    coupled faradaic and charging
        current, 377–388
    fluid mechanics, 256–258, 300p
    mass transfer, 259–260, 300p
    mass-transfer-limited current, 27e,
        33e
    nonuniform current distribution,
        337–347
    primary resistance, 115, 127p
    properties, 167
    state variables, 157
    velocity expansion, 257, 300p
rotating hemisphere
    properties, 168
rotational elliptic coordinates, 337
rough disk electrode
    characteristic dimension, 353, 370
    characteristic frequency, 353, 370
roughness
    characteristic dimension, 355, 370
    characteristic frequency, 354, 370
roughness-induced frequency disper-
        sion, 349–360

sample
    covariance, **47**
    mean, **45**
    standard deviation, **46**
    variance, **45**
Schmidt number, 73p
Schottky diode
    GaAs, 621e
        Arrhenius-based plots, 518–521

Mott–Schottky Plots, 521–522
scroll, *find* "EIS", xxxive
secondary current distribution, 116,
        117e
signal-to-noise, 184
significance level, **60**
similarity transformation, 38–41
skin, human, 492p
small circular electrode, 372e
space-charge region, 311
stability, 604, 614p
standard deviation, **46**
standard error, **46**
standard normal distribution, **49**
state quantities, **423**, 444
steady state, 142
Stern–Geary relation, 224, 240p
stochastic error, 149, 170, 184–185, 266,
        478, 528, 537–538, 544p, **567**,
        568–575, 584e, 593p, 609–613,
        618, 621
stoichiometric coefficient, 207
Student's *t*-test, 61–62, 73, 74p
superposition
    Nyquist plots, 199, 499–501
    phase angle, 353, 354
surface overpotential, **95**, 101, 102e,
        108, 119

Tafel
    kinetics, **96**, 171e, 209, 214, 223,
        227, 232e, 235e, 625
    slope, **96**, 141
        relation to charge-transfer resis-
        tance, 213
        relation to linearity, 188p
        Stern–Geary relation, 224, 240p
telegraph cable
    above ground, 319
    impedance response, 322
    submarine, 319

telegrapher's equations, 321–322

tertiary current distribution, 116

thermoelectrochemical transfer function, 434–437, 441p

thin-layer cell, 333–337

time constant, xxxive, 468

time-constant dispersion
    normal distribution, 402–404
    power-law distribution, 406–413
    surface distribution, 404–406

time-translation independence, 600

transdermal drug delivery, 188p

transfer function, **xxxiii**, xxxiv–xxxix
    admittance, 478–483, 491p
    capacitance, 431–433
    complex capacitance, 484–488, 491, 492p
    effective capacitance, 488–491, 491p, 512–516, 525p
    effective CPE coefficient, 512–516, 525p
    electrogravimetry, 439–441, 441p
    electrohydrodynamic, **451**
    elementary , 425
    general approach, 421–425
    hydrodynamic, 445–448
    impedance, 467–478
    input quantities, **422**, 444
    kinetic, 453–454
    mass transfer, 448–453
    output quantities, **423**, 444
    photoelectrochemical, 438–439
    ring–disk, 429–431
    state quantities, **423**, 444
    thermoelectrochemical, 434–437, 441p

transfer-function analysis, 157

transformation
    Kramers–Kronig , 595–599
    coordinate, 36–38
    rotational elliptic coordinates, 337

similarity, 38–41
    time domain to frequency domain, 143–156, 571–575, 593p
    von Karman, 256, 445

transient response, 142–143

transmission line
    impedance, 31e
    models, 319–337

transmissive boundary condition, 245–251, 300p

uniformly accessible electrode, 244–245

universal gas constant, 209

variance, **45**

variate, **45**

Voigt model
    even and odd properties, 601, 602e
    impedance, 80e, 537
    inductive loops, 542
    Kramers–Kronig analysis, 599
    nonlinearity, 530e
    regression, 541, 544p, 552–555, 563p, 581–593, 599
    resistivity distribution, 406e
    statistical analysis, 71e, 552–555, 581–593

voltage adder, 133e

von Karman transformation, 256, 445

Warburg impedance, xlii, 11e, 39e, 43p, 249, 250e, 328, 374, 490e, 582

weighting, 579–581, 593p

William Thomson, *see* Lord Kelvin

working electrode potential, 107

Young impedance, 389–392, 396e
    complex capacitance, 485e
    resistivity distribution, 317p

zinc deposition, 234e

THE ELECTROCHEMICAL SOCIETY SERIES

Corrosion Handbook
Edited by Herbert H. Uhlig

Electrochemistry in Biology and Medicine
Edited by Theodore Shedlovsky

Arcs in Inert Atmospheres and Vacuum
William Erik Kuhn

Modern Electroplating, Second Edition
Frederick Lowenheim

First International Conference on Electron and Ion Beam Science and Technology
Robert Bakish

Vapor Deposition
Carroll F. Powell, Joseph H. Oxley, and John Milton Blocher

The Electron Microprobe
Edited by T. D. McKinley, K. F. J. Heinrich, and D. B. Wittry

Chemical Physics of Ionic Solutions
Edited by B. E. Conway and R. G. Barradas

High-Temperature Materials and Technology
Edited by Ivor E. Campbell and Edwin M. Sherwood

Alkaline Storage Batteries
S. Uno Falk and Alvin J. Salkind

Free Radical Substitution Reactions
K. U. Ingold and B. P. Roberts

Zinc-Silver Oxide Batteries
Edited by Arthur Fleischer and J. J. Lander

The Primary Battery (in Two Volumes)
Volume I *Edited by* George W. Heise and N. Corey Cahoon
Volume II *Edited by* N. Corey Cahoon and George W. Heise

Modern Electroplating, Third Edition
Edited by Frederick A. Lowenheim

Lead-Acid Batteries
Hans Bode
Translated by R. J. Brodd and Karl V. Kordesch

Thin Films-Interdiffusion and Reactions
Edited by J. M. Poate, M. N. Tu, and J. W. Mayer

Lithium Battery Technology
Edited by H. V. Venkatasetty

Quality and Reliability Methods for Primary Batteries
P. Bro and S. C. Levy

Techniques for Characterization of Electrodes and Electrochemical Processes
Edited by Ravi Varma and J. R. Selman

Electrochemical Oxygen Technology
Kim Kinoshita

Synthetic Diamond: Emerging CVD Science and Technology
Edited by Karl E. Spear and John P. Dismukes

Corrosion of Stainless Steels, Second Edition
A. John Sedriks

CAD Method for Industrial Assembly: Concurrent Design of Products, Equipment and Control Systems
A. Delchambre

Fundamentals of Electrochemical Deposition
Milan Paunovic and Mordechay Schlesinger

Semiconductor Wafer Bonding: Science and Technology
Q.-Y. Tong and U. Gösele

Modern Electroplating, Fourth Edition
Mordechay Schlesinge and Milan Paunovic

Atmospheric Corrosion
Christofer Leygraf and Thomas Graedel

Electrochemical Systems, Third Edition
John Newman and Karen E. Thomas-Alyea

Fundamentals of Electrochemistry, Second Edition
Vladimir S. Bagotsky

Fundamentals of Electrochemical Deposition, Second Edition
Milan Paunovic and Mordechay Schlesinger

Fuel Cells: Problems and Solutions
Vladimir S. Bagotsky

Modern Electroplating, Fifth Edition
Mordechay Schlesinger and Milan Paunovic

Uhlig's Corrosion Handbook, Third Edition
Edited by R. Winston Revie

Fuel Cells: Problems and Solutions, Second Edition
Vladimir S. Bagotsky

Lithium Batteries: Advanced Technologies and Applications
Bruno Scrosati, K. M. Abraham, Walter A. van Schalkwijk, and Jusef Hassoun

Electrochemical Power Sources: Batteries, Fuel Cells, and Supercapacitors
Vladimir S. Bagotsky, Alexander M. Skundin, and Yurij M. Volfkovich

Molecular Modeling of Corrosion Processes: Scientific Development and Engineering Applications
Christopher D. Taylor and Philippe Marcus

Atmospheric Corrosion, Second Edition
Christofer Leygraf, Inger Odnevall Wallinder, and Johan Tidblad

Electrochemical Impedance Spectroscopy, Second Edition
Mark E. Orazem and Bernard Tribollet

Printed and bound by CPI Group (UK) Ltd, Croydon, CR0 4YY

16/04/2025

14658538-0002